Lecture Notes in Computer Science 9597

Commenced Publication in 1973
Founding and Former Series Editors:
Gerhard Goos, Juris Hartmanis, and Jan van Leeuwen

More information about this series at http://www.springer.com/series/7407

Giovanni Squillero
Paolo Burelli et al. (Eds.)

Applications of Evolutionary Computation

19th European Conference, EvoApplications 2016
Porto, Portugal, March 30 – April 1, 2016
Proceedings, Part I

Springer

Editors

see next page

ISSN 0302-9743　　　　　ISSN 1611-3349　(electronic)
Lecture Notes in Computer Science
ISBN 978-3-319-31203-3　　　ISBN 978-3-319-31204-0　(eBook)
DOI 10.1007/978-3-319-31204-0

Library of Congress Control Number: 2016933215

LNCS Sublibrary: SL1 – Theoretical Computer Science and General Issues

Printed on acid-free paper

This Springer imprint is published by Springer Nature
The registered company is Springer International Publishing AG Switzerland

Volume Editors

Giovanni Squillero
Politecnico di Torino, Italy
giovanni.squillero@polito.it

Paolo Burelli
Aalborg University, Copenhagen,
    Denmark
pabu@create.aau.dk

Jaume Bacardit
Newcastle University, UK
jaume.bacardit@newcastle.ac.uk

Anthony Brabazon
University College Dublin, Ireland
anthony.brabazon@ucd.ie

Stefano Cagnoni
University of Parma, Italy
cagnoni@ce.unipr.it

Carlos Cotta
Universidad de Málaga, Spain
ccottap@lcc.uma.es

Ivanoe De Falco
ICAR/CNR, Italy
ivanoe.defalco@na.icar.cnr.it

Antonio Della Cioppa
University of Salerno, Italy
adellacioppa@unisa.it

Federico Divina
Universidad Pablo de Olavide, Seville,
    Spain
fdivina@upo.es

A.E. Eiben
VU University Amsterdam,
    The Netherlands
a.e.eiben@vu.nl

Anna I. Esparcia-Alcàzar
Universitat Politècnica de València,
    Spain
aesparcia@pros.upv.es

Francisco Fernàndez de Vega
University of Extremadura, Spain
fcofdez@unex.es

Kyrre Glette
University of Oslo, Norway
kyrrehg@ifi.uio.no

Evert Haasdijk
VU University Amsterdam,
    The Netherlands
e.haasdijk@vu.nl

J. Ignacio Hidalgo
Universidad Complutense de Madrid, Spain
hidalgo@ucm.es

Ting Hu
Memorial University, St. John's, NL, Canada
ting.hu@mun.ca

Michael Kampouridis
University of Kent, UK
m.kampouridis@kent.ac.uk

Paul Kaufmann
University of Paderborn, Germany
paul.kaufmann@gmail.com

Michalis Mavrovouniotis
De Montfort University, UK
mmavrovouniotis@dmu.ac.uk

Antonio M. Mora García
Universidad de Granada, Spain
amorag@geneura.ugr.es

Trung Thanh Nguyen
Liverpool John Moores University, UK
T.T.Nguyen@ljmu.ac.uk

Robert Schaefer
AGH University of Science
  and Technology, Poland
schaefer@agh.edu.pl

Kevin Sim
Edinburgh Napier University, UK
k.sim@napier.ac.uk

Ernesto Tarantino
ICAR/CNR, Italy
ernesto.tarantino@na.icar.cnr.it

Neil Urquhart
Edinburgh Napier University, UK
n.urquhart@napier.ac.uk

Mengjie Zhang
Victoria University of Wellington,
  New Zealand
mengjie.zhang@ecs.vuw.ac.nz

Preface

This two-volume set contains the proceedings of EvoApplications 2016, the 19th European Conference on the Applications of Evolutionary Computation, that was held between March 30 and April 1, 2016, in Porto, Portugal, as a part of EVO*.

Over two decades, EVO* has become Europe's premier event in the field of evolutionary computing. Under the EVO* umbrella, EvoAPPS aims to show the modern applications of this research, ranging from proof of concepts to industrial case studies; EuroGP focuses on genetic programming; EvoCOP targets evolutionary computation in combinatorial optimization; and EvoMUSART is dedicated to evolved and bio-inspired music, sound, art, and design. The proceedings for all of these events are available in the LNCS series.

If EVO* coalesces four different conferences, EvoAPPS exhibits an even higher granularity: It started in 1998 as a collection of small independent workshops and it slowly brought together such vibrant and heterogeneous communities into a single organic event. At the same time, the scientific contributions started to show content more suited to a conference, and workshops evolved into tracks. The change is not over yet: As the world of evolutionary computation is in a constant turmoil, EvoAPPS is mutating to adapt. The scope broadened to include all nature-inspired and bio-inspired computational techniques and computational intelligence in general. New tracks appear every year, while others are merged or suspended.

The conference provides a unique opportunity for students, researchers, and professionals to meet and discuss the applicative and practical aspects of evolutionary computation, and to link academia to industry in a variety of domains.

The 2016 edition comprised 13 tracks focusing on different application domains: EvoBAFIN (business analytics and finance); EvoBIO (computational biology); EvoCOMNET (communication networks and other parallel and distributed systems); EvoCOMPLEX (complex systems); EvoENERGY (energy-related optimisation); EvoGAMES (games and multi-agent systems); EvoIASP (image analysis, signal processing, and pattern recognition); EvoINDUSTRY (real-world industrial and commercial environments); EvoNUM (continuous parameter optimization); EvoPAR (parallel architectures and distributed infrastructures); EvoRISK (risk management, security, and defence); EvoROBOT (evolutionary robotics); and EvoSTOC (stochastic and dynamic environments).

This year, we received 115 high-quality submissions, most of them well suited to fit in more than one track. We selected 58 papers for full oral presentation, while 17 works were given limited space and were shown as posters. All such contributions, regardless of the presentation format, appear as full papers in these two volumes (LNCS 9597 and 9598).

Many people contributed to this edition: We express our gratitude to the authors for submitting their works, and to the members of the Program Committees for devoting such huge effort to review papers pressed by our tight schedule.

The papers were submitted, reviewed, and selected through the MyReview conference management system, and we are grateful to Marc Schoenauer (Inria, Saclay-Île-de-France, France) for providing, hosting, and managing the platform.

We thank the local organizers, Penousal Machado and Ernesto Costa (University of Coimbra, Portugal), as well as the Câmara Municipal do Porto and Turismo do Porto for the local assistance.

We thank Pablo García Sánchez (Universidad de Granada, Spain) for maintaining the EVO* website and handling publicity.

We thank the invited speakers, Richard Forsyth and Kenneth Sörensen, for their inspiring presentations.

We thank the Institute for Informatics and Digital Innovation at Edinburgh Napier University, UK, for the coordination and financial administration.

And we express our gratitude to Jennifer Willies for her dedicated and continued involvement in EVO*. Since 1998, she has been essential for building our unique atmosphere.

February 2016

Giovanni Squillero	Paolo Burelli
Jaume Bacardit	Anthony Brabazon
Stefano Cagnoni	Carlos Cotta
Ivanoe De Falco	Antonio Della Cioppa
Federico Divina	A.E. Eiben
Anna I. Esparcia-Alcázar	Francisco Fernández de Vega
Kyrre Glette	Evert Haasdijk
J. Ignacio Hidalgo	Ting Hu
Michael Kampouridis	Paul Kaufmann
Michalis Mavrovouniotis	Antonio M. Mora Garcia
Trung Thanh Nguyen	Robert Schaefer
Kevin Sim	Ernesto Tarantino
Neil Urquhart	Mengjie Zhang

Organization

Organizing Committee

EvoApplications Coordinator

Giovanni Squillero   Politecnico di Torino, Italy

EvoApplications Publication Chair

Paolo Burelli   Aalborg University, Copenhagen, Denmark

Local Chairs

Penousal Machado   University of Coimbra, Portugal
Ernesto Costa   University of Coimbra, Portugal

Publicity Chair

Pablo García-Sánchez   University of Granada, Spain

Webmaster

Pablo García Sánchez   University of Granada, Spain

EvoBAFIN Chairs

Anthony Brabazon   University College Dublin, Ireland
Michael Kampouridis   University of Kent, UK

EvoBIO Chairs

Jaume Bacardit   Newcastle University, UK
Federico Divina   Universidad Pablo de Olavide, Seville, Spain
Ting Hu   Memorial University, St. John's, NL, Canada

EvoCOMNET Chairs

Ivanoe De Falco   ICAR/CNR, Italy
Antonio Della Cioppa   University of Salerno, Italy
Ernesto Tarantino   ICAR/CNR, Italy

EvoCOMPLEX Chairs

Carlos Cotta   Universidad de Málaga, Spain
Robert Schaefer   AGH University of Science and Technology, Poland

EvoENERGY Chairs

Paul Kaufmann   University of Paderborn, Germany
Kyrre Glette   University of Oslo, Norway

EvoGAMES Chairs

Paolo Burrelli   Aalborg University of Copenhagen, Denmark
Antonio M. Mora García Universidad de Granada, Spain

EvoIASP Chairs

Stefano Cagnoni   University of Parma, Italy
Mengjie Zhang   Victoria University of Wellington, New Zealand

EvoINDUSTRY Chairs

Kevin Sim   Edinburgh Napier University, UK
Neil Urquhart   Edinburgh Napier University, UK

EvoNUM Chair

Anna I. Esparcia-   Universitat Politècnica de València, Spain
  Alcázar

EvoPAR Chairs

Francisco Fernández   University of Extremadura, Spain
  de Vega
J. Ignacio Hidalgo   Universidad Complutense de Madrid, Spain

EvoRISK Chair

Anna I. Esparcia-   Universitat Politècnica de València, Spain
  Alcázar

EvoROBOT Chairs

Evert Haasdijk   VU University Amsterdam, The Netherlands
A.E. Eiben   VU University Amsterdam, The Netherlands

EvoSTOC Chairs

Michalis   De Montfort University, UK
  Mavrovouniotis
Trung Thanh Nguyen   Liverpool John Moores University, UK

Program Committees

Robert K. Abercrombie   Oak Ridge National Laboratory, USA [EvoRISK]
Rami Abielmona   University of Ottawa, Canada [EvoRISK]
Eva Alfaro   Instituto Tecnológico de Informàtica, Spain [EvoBAFIN]
Jhon E. Amaya   UNET, Venezuela [EvoCOMPLEX]
Michele Amoretti   University of Parma, Italy [EvoIASP]
Anca Andreica   Universitatea Babeş-Bolyai, Romania [EvoCOMPLEX]
Ignacio Arnaldo   MIT, USA [EvoPAR]
Maria Arsuaga Rios   CERN [EvoINDUSTRY]
Farrukh Aslam Khan   National University of Computer and Emerging Sciences,
                           Pakistan [EvoCOMNET]
Jason Atkin   University of Nottingham, UK [EvoINDUSTRY]
Joshua Auerbach   Ecole Polytechnique Fédérale de Lausanne, Switzerland
                           [EvoROBOT]
Lucia Ballerini   University of Dundee, UK [EvoIASP]
Tiago Baptista   Universidade de Coimbra, Portugal [EvoCOMPLEX]
Bahriye Basturk Akay   Erciyes University, Turkey [EvoINDUSTRY]
Hans-Georg Beyer   Vorarlberg University of Applied Sciences, Austria
                           [EvoNUM]
Leonardo Bocchi   University of Florence, Italy [EvoIASP]
Anthony Brabazon   University College Dublin, Ireland [EvoBAFIN]
Juergen Branke   University of Warwick, UK [EvoSTOC]
Nicolas Bredeche   Institut des Systémes Intelligents et de Robotique, France
                           [EvoROBOT]
Paolo Burelli   Aalborg University, Denmark [EvoGAMES]
David Camacho   Universidad Autónoma de Madrid, Spain [EvoGAMES]
Jose Carlos Ribeiro   Politechnique Institute of Leiria, Portugal [EvoPAR]
Nabendu Chaki   University of Calcutta, India [EvoRISK]
Ying-ping Chen   National Chiao Tung University, Taiwan [EvoNUM]
Kay Chen Tan   National University of Singapore [EvoRISK]
Hui Cheng   Liverpool John Moores University, UK [EvoSTOC]
Anders Christensen   University Institute of Lisbon, ISCTE-IUL, Portugal
                           [EvoROBOT]
Mario Cococcioni   NATO Undersea Research Centre, Italy [EvoRISK]
Jose Manuel Colmenar   URJC, Spain [EvoPAR]
Ernesto Costa   University of Coimbra, Portugal [EvoSTOC]
Antonio Córdoba   Universidad de Sevilla, Spain [EvoCOMPLEX]
Fabio D'Andreagiovanni   Zuse Institute Berlin, Germany [EvoCOMNET]
Sergio Damas   European Center for Soft Computing, Spain [EvoIASP]

Fabio Daolio	Shinshu University, Japan [EvoIASP]
Christian Darabos	University of Pennsylvania, USA [EvoBIO]
Ivanoe De Falco	ICAR - CNR, Italy [EvoIASP]
Antonio Della Cioppa	University of Salerno, Italy [EvoIASP]
Laura Dipietro	Cambridge, USA [EvoIASP]
Josep Domingo-Ferrer	Rovira i Virgili University, Spain [EvoRISK]
Stephane Doncieux	Institut des Systémes Intelligents et de Robotique, France [EvoROBOT]
Marco Dorigo	Université Libre de Bruxelles, Belgium [EvoROBOT]
Jitesh Dundas	Edencore Technologies, Indian Institute of Technology, India [EvoBIO]
Marc Ebner	Ernst Moritz Arndt Universität Greifswald, Germany [EvoIASP, EvoNUM]
Andries P. Engelbrecht	University of Pretoria, South Africa [EvoSTOC]
A. Sima Etaner-Uyar	Istanbul Technical University, Turkey [EvoSTOC]
Thomas Farrenkopf	University of Applied Sciences, Mittelhessen, Germany [EvoINDUSTRY]
Carlos Fernandes	ISR-Lisbon, Portugal [EvoCOMPLEX]
Stenio Fernandes	Federal University of Pernambuco, UFPE, Brazil [EvoRISK]
Florentino Fernández	Universidad de Vigo, Spain [EvoBIO]
Antonio J. Fernandez Leiva	University of Málaga, Spain [EvoGAMES]
Antonio Fernández-Ares	Universidad de Granada, Spain [EvoGAMES]
Gianluigi Folino	ICAR-CNR, Italy [EvoPAR]
Francesco Fontanella	University of Cassino, Italy [EvoIASP]
Alex Freitas	University of Kent, UK [EvoBIO]
José E. Gallardo	Universidad de Málaga, Spain [EvoCOMPLEX]
Pablo García Sànchez	University of Granada, Spain [EvoGAMES]
Antonios Gasteratos	Democritus University of Thrace, Greece [EvoCOMNET]
Carlos Gesherson	UNAM, Mexico [EvoCOMPLEX]
Mario Giacobini	Università di Torino, Italy [EvoBIO]
Raffaele Giancarlo	Università degli Studi di Palermo, Italy [EvoBIO]
Rosalba Giugno	University of Catania, Italy [EvoBIO]
Antonio Gonzalez Pardo	Basque Center for Applied Mathematics, Spain [EvoGAMES]
Casey Greene	Dartmouth College, USA [EvoBIO]
Michael Guckert	University of Applied Sciences, Mittelhessen, Germany [EvoINDUSTRY]
Johan Hagelbäck	Blekinge Tekniska Högskola, Sweden [EvoGAMES]
John Hallam	University of Southern Denmark, Denmark [EvoGAMES]
Heiko Hamann	University of Paderborn, Germany [EvoROBOT]
Jin-Kao Hao	University of Angers, France [EvoBIO, EvoCOMNET]
Jacqueline Heinerman	VU University Amsterdam, The Netherlands [EvoROBOT]
Malcom Heywood	Dalhousie University, Canada [EvoBAFIN]

Ronald Hochreiter	WU Vienna University of Economics and Business, Austria [EvoBAFIN]
Rolf Hoffmann	Technical University of Darmstadt, Germany [EvoCOMNET]
Joost Huizinga	University of Wyoming, USA [EvoROBOT]
Oscar Ibàñez	University of Granada, Spain [EvoIASP]
José Ignacio Hidalgo	Universidad Complutense de Madrid, Spain [EvoIASP]
Rodica Ioana Lung	Babes-Bolyai University, Germany [EvoGAMES]
Juan L. Jiménez Laredo	ILNAS/ANEC Normalisation, Luxembourg [EvoCOMPLEX, EvoPAR]
Michael Kampouridis	University of Kent, UK [EvoBAFIN]
Iwona Karcz-Dulęba	Politechnika Wrocławska, Poland [EvoCOMPLEX]
Ahmed Kattan	EvoSys.biz, Saudi Arabia [EvoBAFIN]
Shayan Kavakeb	Liverpool John Moores University, UK [EvoSTOC]
Edward Keedwell	University of Exeter, UK [EvoBIO]
Graham Kendall	University of Nottingham, UK [EvoCOMNET, EvoINDUSTRY]
Mario Koeppen	Kyushu Institute of Technology, Japan [EvoIASP]
Oliver Kramer	University of Oldenburg, Germany [EvoENERGY]
Wacław Kuś	Politechnika Śląska, Poland [EvoCOMPLEX]
William Langdon	University College London, UK [EvoNUM, EvoPAR]
Kenji Leibnitz	National Institute of Information and Communications Technology, Japan [EvoCOMNET]
Changhe Li	China University of Geosciences, China [EvoSTOC]
Antonios Liapis	University of Malta, Malta [EvoGAMES]
Federico Liberatore	Universidad Rey Juan Carlos, Spain [EvoGAMES]
Piotr Lipinski	University of Wroclaw, Poland [EvoBAFIN]
Francisco Luís Gutiérrez Vela	University of Granada, Spain [EvoGAMES]
Francisco Luna	Universidad de Málaga, Spain [EvoPAR]
Gabriel Luque	Universidad de Málaga, Spain [EvoCOMPLEX]
Evelyne Lutton	INRIA, France [EvoIASP]
Chenjie Ma	Fraunhofer Institute for Wind Energy and Energy System Technology, Germany [EvoENERGY]
Tobias Mahlmann	Lund University, Sweden [EvoGAMES]
Domenico Maisto	ICAR-CNR, Italy [EvoCOMNET]
Elena Marchiori	Radboud Universiteit van Nijmegen, The Netherlands [EvoBIO]
Davide Marocco	University of Naples, Italy [EvoCOMNET]
Ingo Mauser	FZI Karlsruhe, Germany [EvoENERGY]
Michalis Mavrovouniotis	De Montfort University, UK [EvoSTOC]
Michael Mayo	University of Waikato, New Zealand [EvoBAFIN]
Jorn Mehnen	Cranfield University, UK [EvoSTOC]
Juan J. Merelo	Universidad de Granada, Spain [EvoCOMPLEX, EvoNUM]

Pablo Mesejo Santiago	INRIA, France [EvoIASP]
Salma Mesmoudi	Institut des Systémes Complexes, France [EvoNUM]
Krzysztof Michalak	Wroclaw University of Economics, Poland [EvoBAFIN]
Martin Middendorf	University of Leipzig, Germany [EvoENERGY]
Jose Miguel Holguín	S2 Grupo, Spain [EvoRISK]
Maizura Mokhtar	Edinburgh Napier University, UK [EvoENERGY]
Jean-Marc Montanier	Barcelona Supercomputing Center, Spain [EvoROBOT]
Roberto Montemanni	IDSIA, Switzerland [EvoCOMNET]
Javier Montero	Universidad Complutense de Madrid, Spain [EvoRISK]
Frank W. Moore	University of Alaska Anchorage, USA [EvoRISK]
Antonio M. Mora García	University of Granada, Spain [EvoGAMES]
Maite Moreno	S2 Grupo, Spain [EvoRISK]
Vincent Moulton	University of East Anglia, UK [EvoBIO]
Jean-Baptiste Mouret	INRIA Larsen Team, France [EvoROBOT]
Nysret Musliu	Vienna University of Technology, Austria [EvoINDUSTRY]
Antonio Nebro	Universidad de Málaga, Spain [EvoCOMPLEX]
Ferrante Neri	De Montfort University, UK [EvoIASP, EvoNUM, EvoSTOC]
Frank Neumann	University of Adelaide, Australia [EvoENERGY]
Geoff Nitschke	University of Cape Town, South Africa [EvoROBOT]
Stefano Nolfi	Institute of Cognitive Sciences and Technologies, Italy [EvoROBOT]
Michael O'Neill	University College Dublin, Ireland [EvoBAFIN]
Una-May O'really	MIT, USA [EvoPAR]
Conall O'Sullivan	University College Dublin, Ireland [EvoBAFIN]
Kai Olav Ellefsen	University of Wyoming, USA [EvoROBOT]
Carlotta Orsenigo	Politecnico di Milano, Italy [EvoBIO]
Ender Ozcan	University of Nottingham, UK [EvoINDUSTRY]
Patricia Paderewski Rodríguez	University of Granada, Spain [EvoGAMES]
Peter Palensky	Technical University of Delft, The Netherlands [EvoENERGY]
Anna Paszyńska	Uniwersytet Jagielloński, Poland [EvoCOMPLEX]
David Pelta	University of Granada, Spain [EvoSTOC]
Sanja Petrovic	University of Nottingham, UK [EvoINDUSTRY]
Nelishia Pillay	University of KwaZulu-Natal, South Africa [EvoINDUSTRY]
Clara Pizzuti	ICAR CNR, Italy [EvoBIO]
Riccardo Poli	University of Essex, UK [EvoIASP]
Petr Pošík	Czech Technical University in Prague, Czech Republic [EvoNUM]
Mike Preuss	TU Dortmund, Germany [EvoGAMES, EvoNUM]
Abraham Prieto	University of La Coruña, Spain [EvoROBOT]
Jianlong Qi	Ancestry.com Inc., USA [EvoBio]
Michael Raymer	Wright State University, USA [EvoBIO]

Hendrik Richter	Leipzig University of Applied Sciences, Germany [EvoSTOC]
Diederik Roijers	University of Amsterdam, The Netherlands [EvoROBOT]
Simona Rombo	Università degli Studi di Palermo, Italy [EvoBIO]
Claudio Rossi	Universidad Politecnica De Madrid, Spain [EvoROBOT]
Guenter Rudolph	University of Dortmund, Germany [EvoNUM]
Jose Santos Reyes	Universidad de A Coruña, Spain [EvoBIO]
Sanem Sariel	Istanbul Technical University, Turkey [EvoINDUSTRY, EvoROBOT]
Ivo Fabian Sbalzarini	Max Planck Institute of Molecular Cell Biology and Genetics, Germany [EvoNUM]
Robert Schaefer	University of Science and Technology, Poland [EvoCOMNET]
Thomas Schmickl	University of Graz, Austria [EvoROBOT]
Marc Schoenauer	INRIA, France [EvoNUM]
Sevil Sen	Hacettepe University, Turkey [EvoCOMNET]
Noor Shaker	Aalborg University, Denmark [EvoGAMES]
Chien-Chung Shen	University of Delaware, USA [EvoCOMNET]
Bernhard Sick	University of Kassel, Germany [EvoENERGY]
Sara Silva	INESC-ID Lisbon, Portugal [EvoIASP]
Anabela Simões	Institute Polytechnic of Coimbra, Portugal [EvoSTOC]
Moshe Sipper	Ben-Gurion University, Israel [EvoGAMES]
Georgios Sirakoulis	Democritus University of Thrace, Greece [EvoCOMNET]
Stephen Smith	University of York, UK [EvoIASP]
Maciej Smołka	Akademia Górniczo-Hutnicza, Poland [EvoCOMPLEX]
Andy Song	RMIT, Australia [EvoIASP]
Stefano Squartini	Università Politecnica delle Marche, Italy [EvoENERGY]
Giovanni Squillero	Politecnico di Torino, Italy [EvoGAMES, EvoIASP]
Andreas Steyven	Edinburgh Napier University, UK [EvoINDUSTRY]
Kasper Stoy	IT University of Copenhagen, Denmark [EvoROBOT]
Guillermo Suárez-Tangil	Royal Holloway University of London, UK [EvoRISK]
Shamik Sural	Indian Institute of Technology, Kharagpur, India [EvoRISK]
Ernesto Tarantino	ICAR/CNR, Italy [EvoCOMNET]
Andrea Tettamanzi	University of Nice Sophia Antipolis/I3S, France [EvoBAFIN]
Olivier Teytaud	INRIA, France [EvoNUM]
Trung Thanh Nguyen	Liverpool John Moores University, UK [EvoSTOC]
Ruppa Thulasiram	University of Manitoba, Cananda [EvoBAFIN]
Jon Timmis	University of York, UK [EvoROBOT]
Renato Tinós	Universidade de São Paulo, Brazil [EvoSTOC]
Julian Togelius	New York University, USA [EvoGAMES]
Marco Tomassini	Lausanne University, Switzerland [EvoPAR, EvoCOMPLEX]

Alberto Tonda	Politecnico di Torino, Italy [EvoCOMPLEX, EvoGAMES]
Pawel Topa	AGH University of Science and Technology, Poland [EvoCOMNET]
Vicenç Torra	University of Skövde, Sweden [EvoRISK]
Krzysztof Trojanowski	Polish Academy of Sciences, Poland [EvoSTOC]
Andy Tyrrell	University of York, UK [EvoENERGY]
Roberto Ugolotti	Henesis srl, Italy [EvoIASP]
Ryan Urbanowicz	University of Pennsylvania, USA [EvoBIO]
Tommaso Urli	Csiro Data61, Australia [EvoGAMES]
Andrea Valsecchi	European Center of Soft Computing, Spain [EvoIASP]
Leonardo Vanneschi	Universidade Nova de Lisboa, Portugal [EvoBIO, EvoIASP]
Sebastien Varrette	Université du Luxemburg, Luxemburg [EvoPAR]
Nadarajen Veerapen	University of Stirling, UK [EvoINDUSTRY]
Roby Velez	University of Wyoming, USA [EvoROBOT]
Antonio Villalón	S2 Grupo, Spain [EvoRISK]
Marco Villani	University of Modena and Reggio Emilia, Italy [EvoCOMNET]
Markus Wagner	University of Adelaide, Australia [EvoENERGY]
Jaroslaw Was	AGH University of Science and Technology, Poland [EvoCOMNET]
Tony White	Carleton University, Canada [EvoCOMNET]
Alan Winfield	University of the West of England, UK [EvoROBOT]
Bing Xue	Victoria University of Wellington, New Zealand [EvoIASP, EvoBIO]
Shengxiang Yang	De Monfort University, UK [EvoINDUSTRY, EvoSTOC]
Georgios N. Yannakakis	University of Malta, Malta [EvoGAMES]
Xin Yao	University of Birmingham, UK [EvoSTOC]
Mengjie Zhang	Victoria University of Wellington, New Zealand [EvoBIO]
Nur Zincir-Heywood	Dalhousie University, Canada [EvoCOMNET, EvoRISK]

Sponsoring Organizations

Institute for Informatics and Digital Innovation at Edinburgh Napier University, UK
World Federation on Soft Computing (technical sponsor of the EvoCOMNET track)
Câmara Municipal do Porto, Portugal
Turismo do Porto, Portugal
University of Coimbra, Portugal

Contents – Part I

EvoCOMNET

EvoCOMPLEX

EvoENERGY

EvoGAMES

EvoIASP

EvoINDUSTRY

Contents – Part II

EvoBAFIN

Enhanced Multiobjective Population-Based Incremental Learning with Applications in Risk Treaty Optimization

Omar Andres Carmona Cortes[1]($\boxtimes$) and Andrew Rau-Chaplin[2]

[1] Informatics Department, Instituto Federal Do Maranhão, São Luis, MA, Brazil
omar@ifma.edu.br
[2] Risk Analytics Lab, Dalhousie University, Halifax, NS, Canada
arc@cs.dal.ca

Abstract. The purpose of this paper is to revisit the Multiobjective Population-Based Incremental Learning method and show how its performance can be improved in the context of a real-world financial optimization problem. The proposed enhancements lead to both better performance and improvements in the quality of solutions. Its performance was assessed in terms of runtime and speedup when parallelized. Also, metrics such as the average number of solutions, the average hypervolume, and coverage have been used in order to compare the Pareto frontiers obtained by both the original and enhanced methods. Results indicated that the proposed method is 22.1 % faster, present more solutions in the average (better defining the Pareto frontier) and often generates solutions having larger hypervolumes. The enhanced method achieves a speedup of 15.7 on 16 cores of a dual socket Intel multi-core machine when solving a Reinsurance Contract Optimization problem involving 15 Layers or sub-contracts.

Keywords: PBIL · Multiobjective optimization · Risk · Reinsurance

1 Introduction

Performance plays an important role in search algorithms because the more complex the problem, the harder discovering solutions in feasible time. This assertion is true especially in the industry where timely answers are required regardless the complexity of the search space. Thus, evolutionary algorithms, also known as search heuristics, can be a highly effective choice.

There are many heuristics search methods that can be applied to solve real-world problems, including Particle Swarm Optimization (PSO) [1], Differential Evolution (DE) [2], Genetic Algorithms (GA) [3], Evolution Strategies (ES) [4] and Population-Based Incremental Learning (PBIL) [5]. Whereas most heuristics search methods apply their genetic operations on a population of individual solutions, PBIL executes its operators in a special data structure called probability matrix that is responsible for creating the population in each iteration.

© Springer International Publishing Switzerland 2016
G. Squillero and P. Burelli (Eds.): EvoApplications 2016, Part I, LNCS 9597, pp. 3–18, 2016.
DOI: 10.1007/978-3-319-31204-0_1

Doing so, PBIL tends to be faster than other heuristic search methods, especially in discrete search spaces.

The first version of PBIL was introduced in 1994 in [5]. At that time, the probability matrix was only a vector where each position represented the probability of having a 0 or 1. The closer the probability to 1, the bigger the chance of creating this gene, while the opposite meant the chance of having a 0. Since, extensions have been proposed for continuous and base-n represented search spaces [6–8]. A version for a discrete space in the range [0, 1] was proposed in [9] and compared against DE and PSO in [10], showing that PBIL is often a very attractive heuristic search method. In [11], a multi-objective-based version of PBIL, called MOPBIL, was designed and applied to problems in reinsurance analytics.

An important problem in Reinsurance analytics is the Reinsurance Contract Optimization problem, where given the structure of a multi-layered reinsurance contract, we need to discover the best trade-offs between expected return and risk for the primary insurer. Such optimizations are key to developing the reinsurance risk hedging strategies that are so important in financial risk management [12].

In this paper we propose E-MOPBIL (Enhanced MOPBIL) which contains important enhancements to MOPBIL and apply it to the Reinsurance Contract Optimization problem in order to achieve faster and higher quality solutions. The Reinsurance Contract Optimization problem consists of, given a reinsurance contract formed by a fixed number of layers (subcontracts) and a simulated set of expected loss distributions (one per layer), plus a model of reinsurance costs, identifying optimal combinations of placements (i.e. percentage shares) in order to maximize the expected return while the associated risk value is minimized [11].

The remainder of this paper is organized as follows: Sect. 2 describes the risk optimization problem being studied in this work; Sect. 3 thoroughly explains the MOPBIL algorithm and the enhanced E-MOPBIL algorithm; Sect. 4 shows the comparative experiments; finally, Sect. 5 presents the conclusions and future work.

2   The RCO Problem

The reinsurance contract optimization problem is a particular kind of treaty optimization problem, which consists of a fixed number of contractual layers and a simulated set of expected loss distributions (one per layer), plus a model of reinsurance market costs [9]. Taking this into consideration, the task is to identify optimal combinations of shares (also called placements) in order to build a Pareto frontier that quantifies the best available trade-offs between expected return and risk [13]. In other words, insurance companies aim hedge their risk against potentially large claims, or losses [14]. Having these trade-offs the insurance companies are able to offer them to the reinsurance market.

The final purpose is to maximize the amount of expected return ($) received from the reinsurance company in case of massive claims and maximize the risk hedge to the reinsurance company at the same time. Doing so, the insurance company minimize the loss faced per year due to natural catastrophes.

In this context, (1) represents the problem in terms of optimization, where VaR is a risk metric, $\mathbf{R}$ is a function in term of placements (π) and E is the Expected Value[1]. For further details about the problems refer to [9] and [14].

$$maximize \ f_1(x) = VaR_\alpha(\mathbf{R}(\pi))$$
$$maximize \ f_2(x) = E[\mathbf{R}(\pi)] \tag{1}$$

3   MOPBIL

In this section we thoroughly describe the algorithm MOPBIL which is presented in Algorithm 1. The main inputs of the algorithm are the following parameters: number of iterations (N_G), number of best individuals and the slice size. The number of iterations is common in evolutionary algorithms, being one of the stop criteria can be adopted. The second one regards to the number of individuals who comprises the best population, which will be used to mutate the probability matrix. The last parameter is important to define the discretization of the search space. The smaller the slice size, the bigger the search space.

The algorithm starts estimating both the minimum and the maximum of the mean (expected return) in order to determine the interval of each slab. Actually, this is done by dividing the search space. One of the consequences of doing so is the possibility of parallelized the algorithm; however, in the original version the number of iterations is maintained; therefore, it is difficult to obtain speedup. On the other hand, the main idea behind this is to enhance the quality of solutions.

Then, when the *foreach* loop starts each slab produces a new population according to the probability matrix. The process is similar to the roulette wheel selection from genetic algorithms which is in essence a Monte Carlo simulation. For example, if the valid shares belong to the set {0, 0.1, 0.2, ..., 1.0}, a column will consist of 11 equal probabilities, thus if a contract is formed by 7 layers the probability matrix will be a matrix consisting of 11 rows and 7 columns with uniform probability. Each position in the matrix is called bucket and will be updated later in the algorithm in the mutation part of the algorithm. The number of buckets varies according to the slice size.

Afterward, new individuals are created according to the probability matrix and evaluated. The fitness is obtained firstly evaluating the expected return. If the expected return is outside from its boundaries [min, max] the risk is automatically set to zero. Given the evaluations, the new population is merged with the archive in order to identify all non-dominated solutions, rank and cluster them. All non-dominated individuals within the valid interval are considered valid and raked as 1. Then, valid dominated solutions are ranked as 2. Finally, individuals outside the limits are ranked ranging from 3 to *pop_size* according to the distance from the valid interval. The clustering is done using the current

---

[1] In probability theory, the expected value, usually denoted by E[X], refers to the value of a random variable X that we would "expect" to find out if we could repeat the random variable process an infinite number of times and take the average of the values obtained.

Input: N_G = number of generations; n_{best} = number of best individuals;
    slice_size=discretization;
Estimate the *min* and the *max* of the mean;
Divide the interval [min, max] into n slabs;
foreach *slab* **do**
| **while** *(N_G not reached)* **do**
| | Create the population using probability matrix;
| | Use the mean to determine the risk value;
| | **if** *(mean of one individual is outside the chunk)* **then**
| | | risk_value = 0;
| | **else**
| | | risk_value = compute(mean)
| | **end**
| | Merge archive and the new population; Determine the non-dominated
| | set;
| | Cluster the non-dominated set into k clusters;
| | Select k representative individuals;
| | Select worst individuals;
| | Insert the k individuals into the best population;
| | Identify the best and worst buckets;
| | Update and mutate the probability matrix;
| **end**
end
Combine the results of each slab;
Determine the Pareto frontier;

Algorithm 1. MOPBIL (Sketch)

Pareto frontier. If more than one individual belongs to the same cluster then the best one is selected to the $best_n$ sub-population based on the best risk, *i.e.*, that one which hedge more risk goes to the $best_n$ sub-population; however, if there are not enough non-dominated solutions then the $best_n$ set is filled up with valid dominated solutions $(rank = 2)$ or with the best ranked invalid solutions in the worst case scenario.

Having identified the best sub-population it is necessary to determine the best and worse buckets from each individual in order to mutate the probability matrix. The mutation is done in two steps. Firstly, a probability multiplier is computed based on both best and worst buckets according to the Algorithm 2, where roughly speaking if a best bucket coincides with a worst buckets the multiplier increases at a lower rate; otherwise, the multiplier might be larger than 1.

After the first step the probability matrix has to be normalized because the sum of a column can be either larger or lower than 1. Secondly, the probability matrix is mutated according to the Algorithm 3. Finally, the matrix has to be normalized again.

```
for (n=1 to rows in prob_matrix) do
    prob_multiplier ← 0 ;
    for (j = 1 to length(best_bucktes)) do
        if (worst_buckets == best_buckets_j) then
            prob_multiplier ← prob.multiplier + (1/length(best.buckets)) *
            ((1-n.learn.rate) + (4 * n.learn.rate)/n.buckets * abs( abs(n -
            best.buckets[j]) - n.buckets/2))
        else
            prob.multiplier ← prob.multiplier + (1/length(best.buckets)) *
            ((1-n.learn.rate2) + (4 * n.learn.rate2)/n.buckets * abs( abs(n -
            best.buckets[j]) - n.buckets/2))
        end
    end
    prob.matrix[n,j] ← prob.matrix[n,i]*prob.multiplier
end
```

Algorithm 2. Computing probability multiplier

```
for j = 1 to n.par do
    for i = 1 to I do
        mut.dir ← 0.0;
        if random(0,1) ≤ M_R then
            if random(0,1) ≤ 0.5 then
                mut.dir ← 1.0;
            end
            p_ij ← p_ij(1 − mut.shift) + mut.dir * mut.shift
        end
    end
end
```

Algorithm 3. Mutation

The mutation process showed in the Algorithms 2 and 3 can be represented by (2), where LF_{ijk} is the ith learning factor, as described in [6], for the kth best result for the jth variable.

$$p_{ij}^{NEW} = \sum_{k=1}^{q} p_{ij} \frac{LF_{ijk}}{q} \qquad (2)$$

3.1 Our Proposal: E-MOPBIL

Our first modification is to remove slabs. As a result, we do not have to compute boundaries which in fact is not time-consuming; however, now we do not have to deal with invalid points, *i.e.*, it is not necessary to rank them which demands to compute the distance between points and boundaries. Moreover, using boundaries as we increase the number of slabs we also increase the probability of creating invalid points affecting the quality of solutions in a parallel execution. The second modification regards to creating the best sub-population

($best_n$). Tests have demonstrated that the Pareto frontier formed by merging the new population and the archive on each iteration is enough to build the $best_n$ sub-population. Thirdly, we do not compute the worse buckets because it might reduce the increment on the probability of a promise bucket if they are the same; therefore, we use the Algorithm 4 for computing the probability multiplier. So, the final version is presented in the Algorithm 5.

---

for *(n=1 to rows in prob_matrix)* **do**
  prob_multiplier ← 0 ;
  for *(j = 1 to length(best_bucktes))* **do**
    prob_multiplier ← prob_multiplier + (1/length(best.buckets)) *
    ((1-n.learn.rate2) + (4 * n.learn.rate2)/n.buckets * abs(abs(n -
    best.buckets[j]) - n.buckets/2))
  end
  prob.matrix[n,j] ← prob.matrix[n,i]*prob_multiplier
end

Algorithm 4. Computing new probability multiplier

Input: N_G = number of generations; n_{best} = number of best individuals;
while *(N_G not reached)* **do**
  Create the population using probability matrix;
  Evaluate objective function on each member of the population;
  Merge archive and the new population;
  Determine the non-dominated set;
  Cluster the non-dominated set into k clusters;
  Select k representative individuals;
  Insert the k individuals into the best population;
  Update and mutate the probability matrix;
end
Determine the final Pareto frontier;

Algorithm 5. E-MOPBIL (Sketch based on our proposal)

As previously stated, MOPIBIL was designed for getting the best quality in terms of solutions dividing the search space into slabs. Doing so, the algorithm does not lead to an improvement regarding speedup when executed in more than one core. In our approach, we parallelized the iteration loop as illustrated in Algorithm 6, which means that the number of iterations is divided between processor units, thus the process is similar to that one presented in serial E-MOPBIL (Algorithm 5).

In order to make a fair comparison against our proposal of parallelizing the code, we also divided the number of iterations between slabs. Thus, the same number of calls to the evaluation function is done in both algorithms.

Input: N_G = number of generations; n_{best} = number of best individuals;
foreach *(N_G/#n_threads) **do** in parallel* **do**
  Create the population using probability matrix;
  Evaluate objective function on each member of the population;
  Merge archive and the new population;
  Determine the non-dominated set;
  Cluster the non-dominated set into k clusters;
  Select k representative individuals;
  Insert the k individuals into the best population;
  Identify the best buckets;
  Update and mutate the probability matrix;
end
Combine the results of each chunk of iterations;
Determine the Pareto frontier;

Algorithm 6. Parallel E-MOPBIL

4 Computational Experiments

4.1 Setup and Metrics

All tests were conducted using R version 3.2.1 on a Red Hat Linux 64-bit Operating in an Intel Xeon comprising of two Xeon processors E5-2650 running at 2.0 Ghz with 8 cores and hyper threading and 256 GB of memory. Considering 250 and 500 with a population size equals to 50. The following parameters were used:

- Population size = 50;
- Slice size = 0.05;
- Number of generations = 250, 500
- Best population = 3;
- learn.rate = 0.1,
- neg.learn.rate = 0.075,
- mut.prob = 0.02,
- mut.shift = 0.05,

In order to compare the algorithms we used the following metrics: number of solutions, hypervolume, coverage and generational distance. In the first one, the larger the number of solution the better the results tends to be; however, this metric is not enough to compare Pareto frontiers. All averages are calculated in 30 trials, allowing us to make inferences based on t-test.

The hypervolume depicts the volume of the dominated part of the curve, therefore, the bigger the hypervolume, the better the Pareto frontier might be. Mathematically, the hypervolume can be computed by (3).

$$hv = volume(\bigcup_{i=1}^{|Q|} v_i) \qquad (3)$$

The coverage represents the percentage of solutions which dominate at least one of the other solutions. Roughly speaking, $C(A, B)$ is the percentage of the solutions in B that are dominated by at least 1 solution in A [15], therefore, if $C(A, B) = 1$ then all solutions in A dominate B, and $C(A, B) = 0$ means the opposite. The coverage can be calculates as shown in (4).

$$C(A, B) = \frac{|\{b \in B | \exists a \in A : a \preceq b\}|}{|B|} \tag{4}$$

In terms of parallel computing the performance was evaluated using speedup, particularly we used the weak speedup depicted in (5) and suggested in [16], where T_1 is the time of the serial version and T_p is the time for running the code in p processor units. We used this kind of speedup because T_1 is the time for running the code in 1 thread, therefore, we do not need to guarantee that T_1 is obtained using the best possible implementation.

$$Speedup = \frac{T_1}{T_p} \tag{5}$$

4.2   Results

Figure 1 shows the final Pareto frontier obtained by E-MOPBIL and MOPBIL using 250 and 500 iterations in a 7-layered problem, in which the closer to zero the better the solutions. Visually, E-MOPBIL presents better results in both configurations which will be confirmed by metrics in Tables 1 and 2.

Table 1 shows metrics for 250 iterations and 7 layers in which we can see that the new algorithm is faster and presents better number of solutions, hypervolume and final number of solutions in the final Pareto frontier. In fact, the E-MOPBIL is 22.8 % faster than the original algorithm. A two-tailed t-test with an $alpha = 0.05$ of significance demonstrated that the differences are significant in the metrics number of solutions and time. Moreover, the coverage metric indicates that E-MOPBIL dominates 83 % of solutions from original MOPBIL against 3.5 % in the opposite direction.

Results regarding 500 iterations and 7 layers are shown in Table 2 where we can see that E-MOPBIL achieved better results in all metrics. Actually, the new algorithm is 21.5 % faster than the original MOPBIL, reaching better results in all metrics as demonstrated by the two-tailed t-test with an $alpha = 0.05$ of significance. The coverage indicates that E-MOPBIL dominates 83 % of solutions from the original algorithm instead of 1.4 % in the way around.

Figure 2 shows the final Pareto frontier for 250 and 500 iterations solving a 15-layered problem. Clearly, our proposal overcomes the original MOPBIL in both configurations. Tables 3 and 4 confirm this assertion.

A two-tailed t-test with an $alpha = 0.05$ of significance demonstrates that the differences between both algorithms are significant in the metrics number of solutions and time in both configurations. Actually, using 15 layers our proposal is 11.5 % faster and discover about twice the number of solutions per trial. In terms of coverage and 250 iterations, E-MOPBIL dominates 90.4 % of solutions

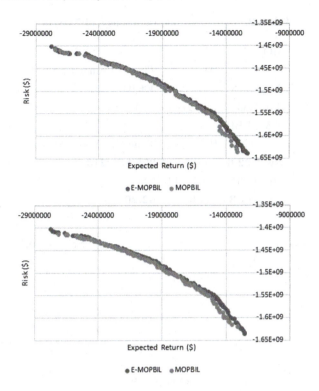

Fig. 1. Comparison between E-MOPBIL and MOPBIL for 250 and 500 iterations with 7 layers

Table 1. Metrics for 250 iterations and 7 layers

MOPBIL				
	#NS	HV	Time (s)	Final #NS
Avg	63.70	1.66E+15	201.45	114
Std	6.29	1.56E+14	2.43	-
E-MOPBIL				
	#NS	HV	Time (s)	Final #NS
Avg	**119.77**	**1.79E+15**	**162.53**	**218**
Std	20.24	2.66E+14	2.63	-
t	**-14.48**	-2.19	**59.53**	-

from MOPBIL, while in the opposite direction MOPBIL does not domain any solution from our proposal. Considering 500 iterations, E-MOPBIL dominates 87.2 % of solutions against 2.2 % from MOPBIL.

Table 2. Metrics for 500 iterations and 7 layers

MOPBIL				
	#NS	HV	Time (s)	Final #NS
Avg	75.63	1.74E+15	406.96	131
Std	5.33	1.13E+14	2.84	-
E-MOPBIL				
	#NS	HV	Time (s)	Final #NS
Avg	**122.6**	**1.90E+15**	**325.85**	**213**
Std	16.31	2.03E+14	6.55	-
t	**-14.35**	**-3.96**	**62.20**	-

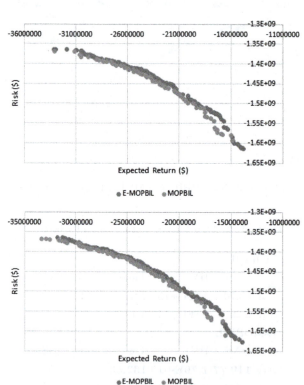

Fig. 2. Comparison between E-MOPBIL and MOPBIL for 250 and 500 iterations with 15 layers

4.3   Parallel Experiments

As previously stated, the original MOPBIL does not properly divided the tasks among workers, which means that all processor elements do the same amount of job. For instance, if we start an optimization using 250 iterations and 4 slabs

Table 3. Metrics for 250 iterations and 15 layers

MOPBIL

	#NS	HV	Time (s)	Final #NS
Avg	48.30	2.039E+15	408.96	74
Std	5.00	2.60E+14	2.39	-

E-MOPBIL

	#NS	HV	Time (s)	Final #NS
Avg	**100.67**	**2.12E+15**	**366.64**	178
Std	13.76	2.58E+14	3.15	-
t	**-19.59**	-1.29	**58.61**	-

Table 4. Metrics for 500 iterations and 15 layers

MOPBIL

	#NS	HV	Time (s)	Final #NS
Avg	52.87	2.23E+15	823.30	87
Std	5.53	2.50E+14	7.58	-

E-MOPBIL

	#NS	HV	Time (s)	Final #NS
Avg	**100.3**	2.07E+15	**736.47**	**198**
Std	19.47	3.67E+14	26.37	-
t	**-12.83**	1.93	**17.34**	-

(consequently 4 threads) the time will be in the average the same if we execute an optimization with the same number of iterations but using only 1 slab (1 thread). Experiments have demonstrated that time remains about 200 seconds on each trial regardless the number of slabs for 7 layers and 408 seconds for 15 layers, using 250 iterations in both cases.

The E-MOPBIL divides the number of iterations between threads. Thus, if we run an optimization with 500 iterations and 4 threads, each one will execute 125 iterations. In this context, Fig. 3 shows the speedup reached as we increase the thread count up to 16 threads in which we can see the reached speedup is similar up to 4 threads and from this point on the speedup is better using 500 iterations.

Figure 4 illustrates the speedup reached by the experiment running 250 and 500 iterations in a 15-layered problem, where we can observe that the performance is similar until about 8 threads, after that using 500 iterations tends to present better speedups.

Figure 5 presents the Pareto frontier for 250 and 500 iterations as we increase the thread count using 7 layers. In this case, we can observe a smooth difference in some parts of the Pareto frontier when 1, 2 and 16 threads are used.

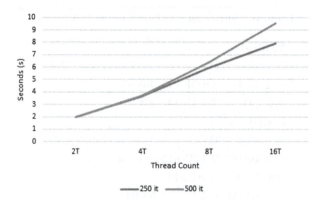

Fig. 3. Speedup achieved by E-MOPBIL for 250 and 500 iterations as we increase the thread count with 7 layers

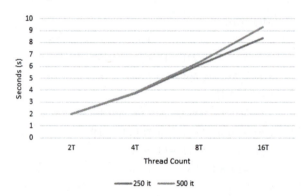

Fig. 4. Speedup achieved by E-MOPBIL for 250 and 500 iterations as we increase the thread count with 15 layers

This difference is expected because we divided the number of iterations between processor units. The coverage metric for 250 iterations and 7 layers indicates that using 1 thread it dominates 51.8 % from the Pareto frontier obtained by 16 threads, and 16.7 % on the contrary. On the other hand, the difference is not meaningful because the variation between these Pareto frontiers, computed by generational distance, is only 0.0006421748.

In order to make a fairer comparison between MOPBIL and E-MOPBIL, we divided the iterations that each slab can do based on the number of threads. Doing so, both algorithms MOPBIL and E-MOPBIL execute the same number of iterations; consequently, they perform the same number of calls to evaluation function. Figures 6 and 7 show the speedup reached by MOPBIL using 7 and 15 layers, respectively, where we can observe that dividing the number of iterations on each slab by the number of processor units almost leads to the ideal speedup.

Even though dividing the number of iteration from MOPBIL into slabs produces better speedups than E-MOPBIL, the quality of the solutions is affected

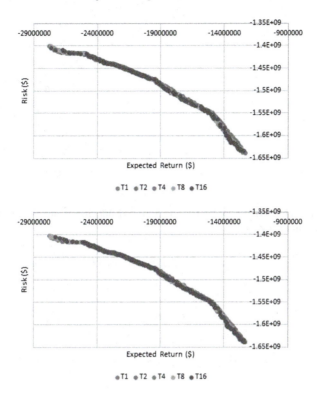

Fig. 5. Pareto frontier for 250 and 500 iterations as we the thread counting with 7 layers

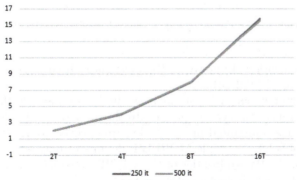

Fig. 6. Speedup achieved by MOPBIL for 250 and 500 iterations as we increase the thread count with 15 layers

as depicted in Fig. 8 in which we compare the Pareto frontier of 7 layers, 250 iterations, and 2 threads. In fact, the coverage indicates that E-MOPBIL domi-

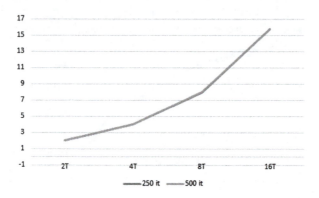

Fig. 7. Speedup achieved by MOPBIL for 250 and 500 iterations as we increase the thread count with 15 layers

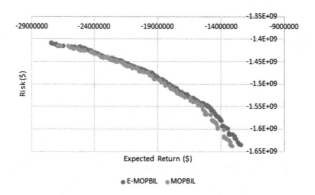

Fig. 8. Comparison against Pareto frontiers using 2 threads, 7 layers, and 250 iterations

nates 92.3 % of solutions from MOPBIL, and that MOPBIL does not dominate any solution from E-MOPBIL.

Although the number of iterations is smaller as we increase the number of threads, a difference still occurring using 16 thread and 500 iterations in a 15-layered problem as illustrated in Fig. 9. Regarding coverage, E-MOPBIL dominates 88.8 % of solutions, while MOBPBIL dominates only 0.6 % of solutions. In other words, the Pareto frontier of E-MOPBIL overcomes that one from the original algorithm in this configuration as well.

Table 5 shows metrics for the comparison between both algorithms using 16 threads, 500 iterations and 15 layers. The results validate that even though this slightly modified version of MOPBIL is faster in terms of execution time, E-MOPBIL produces more solutions, better hypervolume, and consequently a better Pareto frontier.

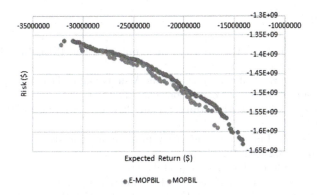

Fig. 9. Comparison against Pareto frontiers using 16 threads, 15 layers, and 500 iterations

Table 5. Metrics for 500 iterations, 15 layers, and 16 threads

MOPBIL				
	#NS	HV	Time (s)	Final #NS
Avg	29.43	1.61E+15	**52.32**	51
Std	3.90	2.85E+14	0.95	-
E-MOPBIL				
	#NS	HV	Time (s)	Final #NS
Avg	**93.93**	**2.62E+15**	78.97	220
Std	8.46	2.78E+14	3.83	-
t	**-37.90**	**13.86**	**-36.99**	-

5   Conclusions

This paper presented the Enhanced MOPBIL method, E-MOPBIL, and showed
how it achieves better results than the original method concerning both performance and quality of solution. E-MOPBIL is up to 21.5 % faster than the
original method when run sequentially using a single thread.

Furthermore, MOPBIL did not exhibit any speedup when run in a multicore
setting due to its structure and how it executed iterations over distinct segments
(slabs) of the parameter space. In E-MOPBIL the user has more flexibility in that
they can control how iterations are executed and trade-off small quality losses for
up to linear speed-up. This feature is attractive in many settings where timely
results are critical.

Acknowledgment. The authors would like to thank CNPq (Conselho Nacional de
Desenvolvimento Científico e Tecnológico) and IFMA (Instituto Federal de Educação,
Ciência e Tecnologia do Maranhão) for funding this research.

References

1. Kennedy, J., Eberhart, R.: Particle swarm optimization. IEEE Int. Conf. Neural Netw. **4**, 1942–1948 (1995)
2. Storn, R., Price, K.: Differential evolution: A simple and efficient adaptive scheme for global optimization over continuous spaces (1995)
3. Michalewicz, Z.: Genetic Algorithms + Data Structure = Evolution Programs. 3rd edn (1999)
4. Yao, X., Liu, Y., Lin, G.: Evolutionary programming made faster. IEEE Trans. Evol. Comput. **3**(2), 82–102 (1999)
5. Baluja, S.: Population based incremental learning (1994)
6. Servais, M., de Jager, G., Greene, J.R.: Function optimisation using multiple-base population based incremental learning. In: The Eighth Annual South African Workshop on Pattern Recognition, Rhodes University (1997)
7. Yuan, B., Gallagher, M.: Playing in continuous spaces: Some analysis and extension of population-based incremental learning. IEEE Congr. Evol. Comput. IEEE **17**, 443–450 (2003)
8. Bureerat, S.: Improved population-based incremental learning in continuous spaces. In: Gaspar-Cunha, A., Takahashi, R., Schaefer, G., Costa, L. (eds.) Soft Computing in Industrial Applications. AISC, vol. 96, pp. 77–86. Springer, Heidelberg (2011)
9. Cortes, O.A.C., Rau-Chaplin, A., Wilson, D., Cook, I., Gaiser-Porter, J.: Efficient optimization of reinsurance contracts using discretized PBIL. In: The Third International Conference on Data Analytics, pp. 18–24 (2013)
10. Cortes, O.A.C., Rau-Chaplin, A., Wilson, D., Gaiser-Porter, J.: On PBIL, DE and PSO for optimizationof reinsurance contracts. In: Esparcia-Alcázar, A.I., Mora, A.M. (eds.) EvoApplications 2014. LNCS, vol. 8602, pp. 227–238. Springer, Heidelberg (2014)
11. Brown, L., Beria, A.A., Cortes, O., Rau-Chaplin, A., Wilson, D., Burke, N., Gaiser-Porter, J.: Parallel MO-PBIL: Computing pareto optimal frontiers efficiently with applications in reinsurance analytics. In: 2014 International Conference on High Performance Computing Simulation (HPCS), pp. 766–775, July 2014
12. Wang, H., Cortes, O., Rau-Chaplin, A.: Dynamic optimization of multi-layered reinsurance treaties. In: The 30th ACM/SIGApp. Symposium On Applied Computing (2015)
13. Cortes, O., Rau-Chaplin, A., do Prado, P.F.: On VEPSO and VEDE for solving a treaty optimization problem. In: IEEE International Conference on Systems, Man and Cybernetics (SMC), pp. 2427–2432, October 2014
14. Cai, J., Tan, K.S., Weng, C., Zhang, Y.: Optimal reinsurance under VaR and CTE risk measures. Insurance: Mathematics and Economics, pp. 185–196 (2008)
15. Zhang, Q.H.: Moea/d: A multiobjective evolutionary algorithm based on decomposition. IEEE Trans. Evol. Comput. **11**(6), 712–731 (2007)
16. Alba, E.: Parallel evolutionary algorithms can achieve super-linear performance. Inf. Process. Lett. **82**(1), 7–13 (2002)

Genetic Programming with Memory
For Financial Trading

Alexandros Agapitos[1](✉), Anthony Brabazon[2], and Michael O'Neill[2]

[1] School of Computer Science and Informatics,
University College Dublin, Dublin, Ireland
alexandros.agapitos@ucd.ie
[2] School of Business, University College Dublin, Dublin, Ireland

Abstract. A memory-enabled program representation in strongly-typed Genetic Programming (GP) is compared against the standard representation in a number of financial time-series modelling tasks. The paper first presents a survey of GP systems that utilise memory. Thereafter, a number of simulations show that memory-enabled programs generalise better than their standard counterparts in most datasets of this problem domain.

1 Introduction

The problem of *sequence learning* is to discover the underlying function of a dynamic system, in order to be able either to produce the next step in a sequence produced by the system (sequence prediction), or correctly classify a sequence produced by the system (sequence classification). Sequence learning has numerous applications, e.g. stock market time-series prediction, protein structure prediction, speech and handwriting recognition. Genetic Programming (GP) [1] has been widely applied to sequence learning tasks. The vast majority of systems employ a sliding time-window of size W, where at each time-step t, values $\{s_{t-1}, \ldots, s_{t-W}\}$ from a sequence $\{s_i\}_{i=1}^{N}$ are used to populate an input vector of explanatory variables used to predict sequence value s_t. This input vector configuration biases towards the evolution of some sort of an *autoregressive* model.

Stateful Genetic Programming (GP) deals with the evolution of computer programs that utilise memory. Puzzlingly, very little work in sequence learning by means of stateful GP is found in the literature. The evolution of stateful sequence-processing programs is closely mirrored by the ongoing investigation of how to best evolve programs with memory or feedback loops in GP.

Any stateless program is essentially a function that maps input to output using the sliding time-window described earlier. In general, it is difficult to decide on the optimal window size. For tasks with long-term dependencies a large window is necessary. A solution might be to use a combination of several different-sized windows, but this is only applicable when the exact time-dependencies of the task are known a priori. Furthermore, fixed time-windows are inadequate when a task has variable time-dependencies. In one form or the other, stateless programs have an important drawback: only a fixed number of

© Springer International Publishing Switzerland 2016
G. Squillero and P. Burelli (Eds.): EvoApplications 2016, Part I, LNCS 9597, pp. 19–34, 2016.
DOI: 10.1007/978-3-319-31204-0_2

previous sequence values (the "context-length") can be taken into account in prediction/classification tasks. On the other hand, in principle, a stateful program can overcome the limited context-length by using memory that enables a dynamic program behaviour based on memory state.

We present an empirical study of time-series classification using stateful GP; the classification task is to discriminate between *go-long* or *go-short* signals in a financial trading system. The rest of the paper is structured as follows: Sect. 2 surveys previous work on the evolution of programs that use memory. Section 3 states the scope of current research. Section 4 introduces two stateful GP systems, and presents the experiment design. Section 5 analyses the results, whereas Sect. 6 concludes and discusses routes for future research.

2   Evolution of Programs with Memory

Previous work confirms that GP can automatically create programs that explicitly use memory. This section reviews previous work using a general taxonomy between *scalar memory* and *indexed memory*. In addition, a third subsection reviews previous work that investigated another form of memory via the concept of *recurrence* (i.e. implemented using time-delay feedback loops in a program).

In GP with scalar memory, the program is given access to a number of variables used as leaves in an expression-tree. In this way the same variable is accessed irrespective of program input. In GP with indexed memory, special primitives are available in the function-set which enable a parameterised access of memory elements based on an *index* value. Depending on the evaluated index, different variables can be accessed; thus different program input may trigger the access/manipulation of different variables. In summary, scalar memory allows for a "static" access to variables, whereas indexed memory allows for a "dynamic" access.

2.1   Scalar Memory

The use of memory in evolvable computer programs dates back to the work of Cramer [2] in 1985. Koza [3] (1992) used global registers that could be manipulated with storage operators. He presented an example where a single variable is used to maintain a running total during execution of a loop. Additionally, in [4], Koza used a small number of variables in a protein classification problem. Montana [5] (1994) presents two examples where GP is provided with local variables. Huelsbergen [6] (1996) used simple scalar memory in his machine-code GP system. Kirshenbaum (2000) presented work on the evolution of programs that use statically-scoped local variables [7]. Conrads et al. [8] (1998) applied linear GP with memory registers to speech time-series classification. Agapitos and Lucas [9] (2007) applied object-oriented GP to the task of evolving statistical operations on samples of data. Agapitos et al. [10] (2007) additionally evolved stateful control programs for a simulated car-racing task. They also experimented with the use of multi-objective optimisation to encourage the effective use of memory variables in car-racing controllers [11]. Poli et al. [12] (2008) introduced the concept *memory-with-memory* based on soft assignments of variables to values.

2.2   Indexed Memory

Teller (1994) introduced *indexed memory* [13]. Its implementation requires the use of an indexed array of memory cells, and the operations of `read` and `write` to enable memory access and manipulation. Using indexed memory Teller evolved programs that solve the problem of pushing blocks up against the boundaries of a world represented as a toroidal grid [14]. Jannink [15] (1994) studied the evolution of programs that use indexed memory to generate random numbers. Teller (1995) evolved programs for image classification [16] and face recognition [17]. In addition, Teller performed acoustic time-series classification [18].

Andre [19] (1994) tackled the problem of controlling an agent whose task is to collect all of the gold scattered in a five-by-five toroidal grid. Brave [20] (1996) studied a similar problem of an agent that explores the world and is required to produce a plan for reaching every arbitrary location in the world from every arbitrary starting point. Haynes and Wainwright [21] (1995) applied GP to evolve control programs for agents which have to survive in a simulated world containing mines. The agent's memory is a dynamically allocated linked-list.

The evolution of Abstract Data Types is investigated in the works of Langdon [22] (1996) and Bruce [23] (1996), who independently evolved stacks, queues, priority queues and linked-lists. Nordin and Banzhaf [24] (1996) used linear GP for sound compression. Spector and Luke [25] (1996) used indexed memory to implement *culture* in a GP run. Koza [26] (1999) generalised indexed memory into an *Automatically Defined Store*. O'Neill [27] (2002) investigated the use of indexed memory in Grammatical Evolution to evolve agents for the Tartarus world, and also evolve caching algorithms. Agapitos et al. [28] (2008) evolved classifiers that use indexed memory for EEG signal classification. Agapitos et al. [29] (2011) used indexed memory to represent models of racing tracks during the evolution of car-racing controllers. Finally, Agapitos et al. [30] (2011) used GP with indexed memory to evolve decision-trees for financial time-series classification.

2.3   Recurrent GP

Another form of memory in GP may be implemented via the notion of *recurrency* in terms of time-delay feedback loops. In sequence processing, time-delay feedback loops refer to a type of program input in which program outputs issued at time $t - 1$ are used as part of the input used in program execution at time t. In the case of an expression-tree, a feedback loop can be formed using an edge from a source tree-node to a destination tree-node. Most often, the source node is the root of the expression-tree (representing program output), which feeds back to the expression-tree using specially designed terminal or function nodes. Iba et al. [31] (1995) introduced special terminals which point at any non-terminal node within the tree. The value given by a terminal is the value at the indicated point in the tree during previous program execution. Sharman et al. [32] (1995) similarly used primitives to reference values returned by the root of an expression-tree or calculated at any of its constituent nodes at a previous program execution; explicit `push` functions within the expression-tree save the value at that point in the tree by pushing it onto a stack. A feedback loop that is formed using previously-issued program outputs is presented in the work of [8], which tackled the problem of sound classification out of raw time-series. Finally, the most recent study of feedback loops in GP for sequence learning is presented in the work of [33]. The GP system is allowed access to a vector composed of a combination of past sequence values and program outputs of previous executions. Input sequences are serially processed using a single-step rolling window.

3   Scope of Research

The literature review confirms that the task of sequence classification by means of GP with memory has received very limited attention [8, 18, 28, 30]. We hypothesise that the use of state variables in sequence learning problems is hindered when a fixed window of past sequence values is used as input. In such cases the search quickly converges to areas of the space containing autoregressive models, and ignores the use of memory. We speculate that autoregressive models form local optima in the fitness landscape and draw the attention of evolutionary search early in the process. On average, early-generation memory-enabled individuals are worse as compared against early-generation autoregressive programs, especially in cases where the number of previous sequence values form an adequate "context" for prediction/classification. Locating well-functioning stateful programs requires a more extensive search than in the case of search in spaces of stateless programs.

There is evidently a research gap between the evolution of sequence processing programs that are based on standard GP using a fixed number of past sequence value as input, and the evolution of programs with indexed memory. We believe that a method of serial processing of a sequence using a *single* sequence value as input will implicitly bias towards the use of memory. The computations that can be expressed using memory are likely to perform better than computations that merely represent variations of an autoregressive model of order one (given a single input value). The aim of this work is to test the efficiency of combining single-step sequence processing and memory-enabled GP. We will compare several stateful GP systems against standard *stateless* GP (no use of memory). The type of sequence learning problem that we address is that of financial time-series classification, which is also a novel application in the context of the proposed method.

4   Methods

4.1   GP Systems with Memory

In this study, program state is based on indexed-memory. Indexed-memory is implemented as an array of `double` variables of size M (i.e. `double[] Mem`). Operations of `read`, `write` and soft-assignment are included in the function-set to access memory.

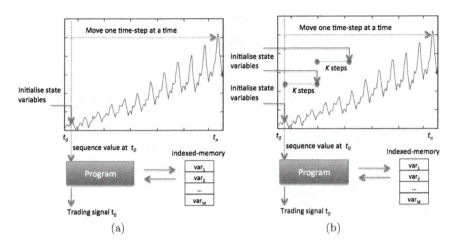

Fig. 1. Memory utilisation schemes. (a) Stateful-A. (b) Stateful-B.

Figure 1 (Stateful-A) presents the general case of a stateful sequence-processing program using a single-step rolling window. Every program in the population is allocated an individual array of `double` variables of size M. Variables are initialised to the value of zero prior to program execution. The sequence is linearly processed; at each time-step, a *single* value is fed as input to the program which outputs a value of type `double`, subsequently mapped to a trading signal. During each execution, a program is allowed to access memory. The method of initialising memory once at the beginning of sequence-processing can handle an arbitrary context-length for classification.

Figure 1 (Stateful-B) presents a different approach to the use of memory based on repeated initialisations of memory throughout sequence processing. Once a subset of sequence values of size K has been processed, memory is re-initialised to default values. The rest of the procedure is similar to that of Stateful-A. The aim is to experiment with a stateful system that can handle a context-length extending to localised parts of the entire sequence. This method requires the selection of the value for parameter K. For this purpose, one may adopt a principled approach like cross-validation. Another option is to use an adaptive value for K in different parts of the sequence. In this study we do not attempt any kind of optimisation for K, but report results using a range of values.

4.2   Feature Extraction

Technical Analysis (TA) indicators are used as filters to pre-process a *raw sequence* $\{s_i\}_{i=1}^N$ into a *feature sequence* $\{f_i\}_{i=1}^K$, with K < N. These indicators are [34,35]: (a) **simple moving average** (MA), (b) **trade break out** (TBO), (c) **filter** (FIL), (d) **volatility** (VOL), and **momentum** (MOM). TA indicators are parameterised with lag periods of certain time-steps. We used lag

Table 1. Strongly-typed language.

FUNCTION SET		
	Argument(s) type	**Return type**
Mathematical		
+, -, *	double, double	double
/ (protected)	double, double	double
log (protected)	double	double
sqrt (protected)	double	double
sin	double	double
Boolean logic		
and, or, nand, nor, xor	boolean, boolean	boolean
Memory access		
read	double	double
write, softAdd, softMul	double, double	double
Predicates		
$\geq$, $<$	double, double	boolean
Conditional		
IF-Then-Else	boolean, double, double	double
TERMINAL SET		
	Value	**Type**
Constants	$\{-10.0, -9.0, -8.0, \ldots, 8.0, 9.0, 10.0\}$	double
	MA(5), MA(10), ..., MA(195), MA(200)	double
	TBO(5), TBO(10), ..., TBO(195), TBO(200)	double
Input features	FIL(5), FIL(10), ..., FIL(195), FIL(200)	double
	MOM(5), MOM(10), ..., MOM(195), MOM(200)	double
	VOL(5), VOL(10), ..., VOL(195), VOL(200)	double

periods ranging from 5 to 200 time-steps, with a step of 5 time-steps. This generates 40 feature sequences using each indicator. Given 5 indicators, we generate 200 feature sequences in total. In order to synchronise the feature sequences, all technical indicators are invoked starting on sample s_{200} of the raw sequence. Features populate an input vector $X \in \mathbb{R}^{200}$. That is, at each time-step 200 feature-values are extracted, one value from each feature sequence respectively.

4.3   Program Language

Each program is given access to an array of variables Mem. Standard read(i) and write(i, value) primitives are included in the function-set. They operate as in [14]. read has one argument, which is evaluated to a memory index; it returns the value of memory at that index. write has two arguments; the first is evaluated to a memory index, while the second is evaluated to a value. write stores the value to the memory location and returns the value that has just overwritten (i.e. the value prior to the update).

In addition, we use soft-assignment operators; these are softAdd(i, value) and softMul(i, value). The former performs Mem[i]=Mem[i]*value, whereas the latter performs Mem[i]=Mem[i]*value. Both operators have two arguments; the first evaluates to a memory index, and the second evaluates to a soft update value. Both operators return the value stored in Mem[i] prior to the update. In order overcome the potential problem of IndexOufOfBounds exceptions in all four memory access operations above, we apply i modulo M prior to accessing memory. Here, i is the evaluated index in an operation, and M is the memory size. M is set to 10 in the experiments.

The GP system evolves programs with real-valued output. Table 1 presents the strongly-typed function and terminal sets.

4.4   Fitness Function

The fitness function (to be maximised) is the **Sharpe ratio**. For a benchmark trading strategy R_b with a constant risk-free return, the Sharpe ratio of trading strategy R_a is defined as $\mathbb{E}[R_a]/\sqrt{var[R_a]}$, where $\mathbb{E}[R_a]$ is the average daily return (ADR) and $\sqrt{var[R_a]}$ is the standard deviation of daily returns using R_a over a trading period T. In order to calculate $\mathbb{E}[R_a]$ and $var[R_a]$ we need to first generate the sequence of daily returns. A sequence of daily returns $\{r_t\}_{t=1}^{N-1}$ is produced out of a sequence of closing prices $\{s_t\}_{t=1}^{N}$ using $r_t = (s_t - s_{t-1})/s_{t-1}$, where s_t and s_{t-1} are the sequence values at time t and $t-1$ respectively. Positive program output is interpreted as *go-long* trading signal ($b = 1$), whereas negative program output is interpreted as *go-short* trading signal ($b = -1$). Let r_t be the daily return at time t, and let b_{t-1} be the trading signal generated by the program at time $t - 1$. Then $d_t = b_{t-1}r_t$ is the realised return at time t.

$\mathbb{E}[R_a]$ and $var[R_a]$ given a program R_a that produces a series of trading signals $\{b_t\}_{t=1}^{T}$ over a sequence of daily returns $\{r_t\}_{t=1}^{T}$ are calculated as:

$$\mathbb{E}[R_a] = \frac{1}{T} \sum_{t=2}^{T} b_{t-1}r_t \tag{1}$$

$$var[R_a] = \frac{1}{T-2} \sum_{t=2}^{T} (b_{t-1}r_t - \mathbb{E}[R_a])^2 \tag{2}$$

4.5   Financial Time-Series Datasets

The datasets used are the foreign exchange rate of EUR/USD, S&P500 and Nikkei225 daily indices for the period of 01/01/1990 to 31/03/2010. The first $2,500$ trading days are used for training, whereas the remaining $2,729$ are equally divided into two sets of $1,364$ days respectively. The *validation-set* is used for model selection, whereas the *test-set* is used to assess the generalisation performance. N best-of-generation programs are gathered during N generations of an evolutionary run for purposes of model selection. The output from a run is the best-of-generation program with the highest ADR on the validation-set.

4.6   Run Parameters

All GP systems are generational and elitist (1 % of population size). They use tournament selection with a tournament size of 4. The population size is set to $1,000$ individuals, and evolution proceeds for 50 generations. Ramped-half-and-half tree creation with a maximum depth of 5 is used to initialise a run. Throughout evolution, expression-trees are allowed to grow up to a maximum depth of 10. The evolutionary search is performed using subtree mutation, point mutation and subtree crossover. In point mutation an inner- or leaf-node is replaced random-uniformly from the set of available function and terminal nodes respectively, under type and arity constraints. The probability of a single node being mutated in point mutation is $1/size$, where $size$ is the number of nodes in an expression-tree. A probability governs the application between mutation and crossover, set to 0.7 in favour of mutation. When mutation is selected, subtree mutation is applied with a probability 0.6 relative to point mutation.

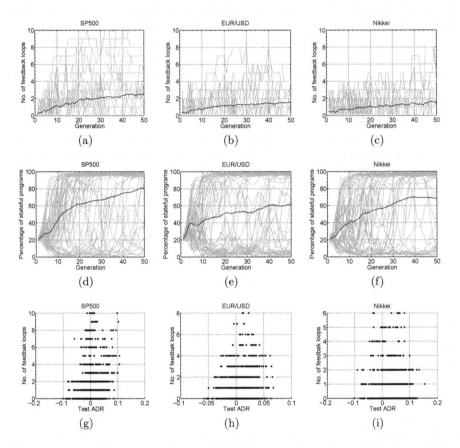

Fig. 2. First row shows the evolution of feedback loops in the best-of-generation individuals (50 runs, mean in bold). Second row shows the evolution of proportion of stateful programs in the population (50 runs, mean in bold). Third row presents scatter plots of test ADR versus number of feedback loops is best-of-generation individuals of 50 runs (2,500 points in total). All graphs are based on simulations using stateful-A.

4.7   Experiment Design

We will compare the generalisation performance of the stateful GP systems described in Sect. 4.1, and also that of a standard GP system that uses no memory (stateless). The stateless system does not include memory access primitives

(Table 1) in its function set; otherwise, function and terminal sets are similar to those of the stateful systems. In addition, we experiment with different time-steps K for memory re-initialisation in Stateful-B system; those of 20, 40, 60, 80, and 100 time-steps.

5   Results

5.1   Generalisation Performance

We performed 50 independent evolutionary runs for each of the three datasets using both stateful and standard GP (stateless) systems. Figure 3 shows the distribution of generalisation performance (measured as ADR over the test-set period) for the best-of-run selected programs in 50 runs. We performed Mann-Whitney U tests (included the Bonferroni correction to adjust for multiple comparisons) to test the significance on the difference of median test ADR between different systems. We set the level of significance at $p \leq 0.05$.

For the case of S&P500 (Fig. 3(a)), a statistically significant difference ($p = 0.046$) was found between stateless and stateful-B-60, with the latter outperforming the former. Overall, median generalisation performance of stateful systems is never worse than that of stateless GP, however not always statistically significant. Also, no statistically significant differences were found between the performance of stateful systems. Nonetheless, we observe that the median performance of stateful-B systems that utilise large time-gaps (K of 60, 80,100) in-between memory re-initialisations performed better than stateful-A.

For the case of EUR/USD (Fig. 3(b)), stateful-B-100 yielded a slightly higher median performance than stateless GP, however the difference is not statistically significant. The differences in the median performance of both stateless and stateful-B-100 against stateful-B-40 and stateful-B-60 are statistically significant with $p < 0.03$ and $p < 0.04$ respectively; stateless and stateful-B-100 perform better. Also, in all but the case against stateful-B-100, the stateless system yields higher median generalisation performance that the rest of stateful systems (although no statistical significance in difference).

For the case of Nikkei (Fig. 3(c)) results suggested that three stateful systems generalised significantly better than stateless GP. Statistically significant differences were found between stateless vs. stateful-B-40 ($p = 0.006$), stateless vs. stateful-B-60 ($p = 0.03$), and stateless vs. stateful-B-100 ($p = 0.009$). No statistically significant differences were found between stateful systems. However, stateful-B-100 that allows for a large time-gap in-between memory re-initialisations (K set to 100) yields a higher median test-set ADR than stateful-A.

5.2   Model Selection

The second column of Fig. 3 shows the distributions of generation numbers for model selection (generation number in which the best-of-run validation ADR occurs) in 50 runs for the three problems. We also plotted the evolution of best-of-generation test ADR in Fig. 4 in order to contrast it against the generations of

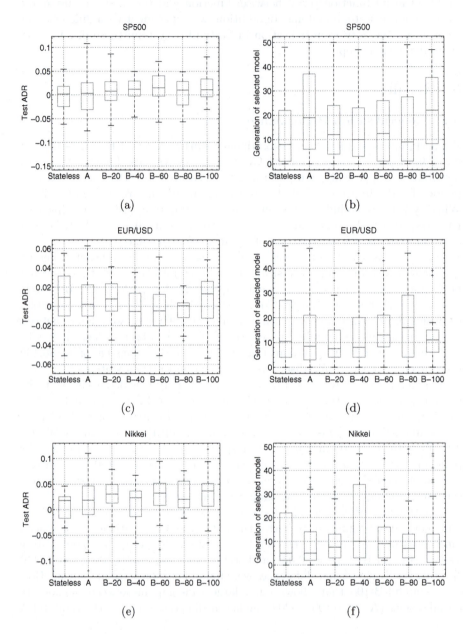

Fig. 3. First column shows the distributions of the test-set ADR (measure of generalisation) for different systems. Second column shows the distributions of generation numbers in which the validation-set ADR was the highest, and therefore model selection was performed.

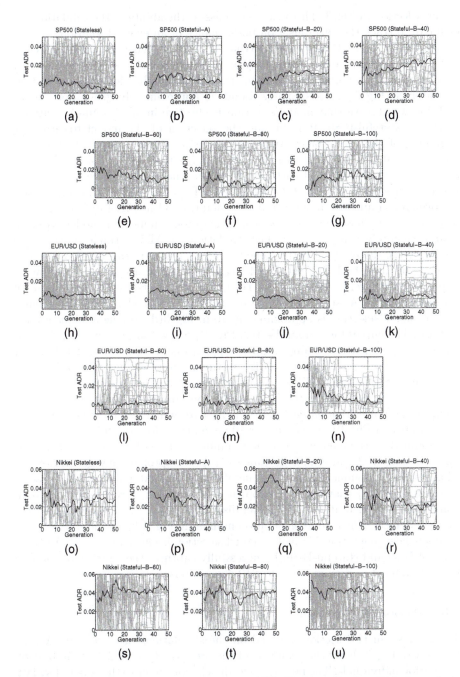

Fig. 4. Evolution of best-of-generation test-set ADR. Each graph plots 50 independent evolutionary runs, and the average is shown in bold. Figure 4(a) to Fig. 4(g) for S&P500. Figure 4(h) to Fig. 4(n) for EUR/USD. Figure 4(o) to Fig. 4(u) for Nikkei.

model selection in Fig. 3. This way we can assess the ability of the validation-set estimate to track the generalisation performance of an evolved solution.

For the case of S&P500 we observe that, on average, generalisation in stateless GP (Fig. 4(a)) improves approximately up to generation 10; loss of generalisation is seen after that point. The median generation number for model selection in stateless (Fig. 3(b)) is similarly found at approximately generation 10. In addition we observe that model selection is effective in all stateful-A, stateful-B-60 and stateful-B-80. It is biasing the selection of the model towards the beginning of the evolutionary runs (median generation of approximately 10). Indeed, in all three systems, the generalisation curves (Fig. 4) show that loss of generalisation is realised early in the evolutionary run. There are however two cases, those of stateful-B-20 and stateful-B-40 in which the generalisation performance improves on average past generation 30 in Fig. 4(c) and Fig. 4(d) respectively. In those runs, the median generation number for model selection is approximately generation 10. The validation-set ADR estimate is therefore hindering the discovery of better-generalising programs in this cases.

For the case of EUR/USD, Fig. 4 suggests that, on average, generalisation improves at the very early stages of runs, and then worsens for the systems of stateless, stateful-A, stateful-B-40, stateful-B-80. The median generation number for model selection is approximately 10 in Fig. 3(d) for the aforementioned systems. The curves for stateful-B-20 in Fig. 4(j) show that, on average, the generalisation performance improves up to approximately generation 20, however model selection is realised much earlier in the majority of runs (Fig. 3(d)).

Finally, for the case of Nikkei, in the systems of stateless, stateful-A, stateful-B-40, generalisation as a function of generation number attains a global maximum at the initial random generation (Fig. 4(p) and (r)). The median generation number in Fig. 3(f) is well under 10 generations for stateful-A. In the case of stateful-B-40 the median generation number is set to 10 generations and the upper-bound of the interquartile range is set to approximately generation 33. This results in significant loss of generalisation at that point. In the case of stateless, Fig. 4(o) shows that a drop in generalisation is realised at approximately generation 6 after which generalisation is shown to improve up to the final generation. The median generation number in Fig. 3(f) for the stateless system is bellow generation 5. Finally, model selection is effective for stateful-B-20, stateful-B-60 and stateful-B-80; it successfully captures the global maximum of test-set ADR at approximately after 10 generations.

5.3   Memory Utilisation

The analysis in this section is based on the simulations using stateful-A. The first row of Fig. 2 shows the evolution of the *number of feedback loops* in best-of-generation individuals. The number of feedback loops refer to the count of `write`, `softAdd` and `softMul` primitives that are found in an expression-tree. We use the term *feedback loop* instead of *memory access* to emphasise the fact that memory has been written to before it is accessed. We observe that on average the best-of-generation individuals contain at least one feedback loop, and that in the case of S&P500 the average feedback loops is set to 2.

The second row of Fig. 2 shows the evolution of the percentage of stateful programs in the population. A stateful program is one that contains at least one of the memory manipulation primitives write, softAdd or softMul. First we note that the initial proportion of stateful individuals is consistently set to approximately 20 % in all problems. Additionally we observe that, on average, in all three problems the percentage increases with increasing generation number, indicating that memory-based computations are fitter than stateless ones. Nonetheless, it is shown that in all problems there exist roughly two categories of runs. Those in which the population becomes quickly dominated by stateful programs, and those in which stateful programs are completely eliminated at the early stages of evolution. One explanation for this is due to the stochastic effects of selection in the early stages of evolution. It seems that in this problem domain both kinds of stateful and stateless programs have similar fitness in early stages of a run, thus selection is unable to discriminate between them. Subsequent sampling errors focus the search around either stateful of stateless programs, and evolution is unable to escape these local maxima.

Finally, the last row of Fig. 2 presents scatter plots of test ADR versus the number of feedback loops in all $2,500$ best-of-generation individuals accumulated during 50 evolutionary runs of 50 generations each. Our aim is to qualify the relationship between generalisation performance and memory utilisation. Figure 2(g) shows that in the case of S&P500 a number of feedback loops greater than 3 is mainly seen with positive test-set ADR. The Pearson correlation coefficient (PCC) of ADR vs. feedback loops is 0.25, suggesting a weak linear relationship between the two. In the case of EUR/USD (Fig. 2(h)) we see that a number of feedback loops greater than 4 is primarily associated with a positive test-set ADR. In this case PCC is 0.11 suggesting a very weak linear relationship between feedback loops and generalisation. Finally, Fig. 2(i) shows that positive test-set ADR is most often obtained from programs with a number of feedback loops greater than 3. The PCC is this case is -0.03, and suggests no relationship between feedback loops and generalisation.

6   Conclusion

Stateful GP for sequence processing using a single-step rolling-window approach was empirically confirmed to generalise similarly well with standard GP in one problem and outperform it in two other problems. Memory re-initialisation during sequence processing was shown to be advantageous and performed better than the case of no re-initialisation. In addition, it was shown that the frequency in which memory is re-initialised exerts an effect in generalisation performance; the less often the better. In practice, principled approaches like cross-validation should be employed for choosing the optimal frequency. Moreover, an interesting relationship was discovered between the number of memory accesses and generalisation performance; a positive out-of-sample ADR was obtained in the majority of programs that made use of more than 3 feedback loops in their execution for all three problems.

Validation-based model selection performs well in the majority of cases, however, there are cases where it cannot accurately estimate the generalisation performance. We believe that there is an interplay between the modelling technique used and the sampling bias that can be unintentionally introduced when choosing the hold-out dataset at random. The issue of sampling bias becomes particularly important in the case of sequence datasets in which the samples are not independently distributed but often exhibit time-dependencies. Furthermore, theory dictates that reducing the data sample-size during training can exert a negative impact on the generalisation performance of a learning system. Future work should focus on analytical methods for model selection. Analytical methods usually require the optimism of training error to be expressed as a function of the training sample-size and model complexity. Defining complexity measures for stateful programs is a promising topic for future work.

Finally, results suggested that there is an inherent difficulty in focusing the search in parts of the space populated by stateful individuals. Nearly-random stateless programs tend to perform equally well (if not better) as stateful ones in the early stages of evolution. Sampling errors in selection may eliminate stateful programs altogether from the population. Specifically, it was shown that there are many cases in which the search quickly concentrated solely around stateless programs. We strongly believe that if stateful programs are to be evolved, the exploration of "stateful" areas of the search space needs to be emphasised early in the run, giving more chances of sampling nearby areas of initially-unfit stateful programs. We are currently experimenting with a new type of dynamic multi-objective fitness function that introduces a bias towards stateful individuals (i.e. rewards the presence of memory-based primitives in an expression-tree) in the early stages of evolution, then gradually reduces this bias focussing solely on Sharpe ratio.

Acknowledgement. This publication has emanated from research conducted with the financial support of Science Foundation Ireland under Grant Number 08/SRC/FM1389.

References

1. Langdon, W.B., Poli, R.: Foundations of Genetic Programming. Springer-Verlag, Heidelberg (2002)
2. Cramer, N.L.: A representation for the adaptive generation of simple sequential programs. In: Grefenstette, J.J. (ed.), Proceedings of an International Conference on Genetic Algorithms and the Applications, Carnegie-Mellon University, pp. 183–187. Pittsburgh, PA, USA, 24–26 July 1985
3. Koza, J.: Genetic Programming: on the programming of computers by means of natural selection. MIT Press, Cambridge, MA (1992)
4. Koza, J.: Genetic Programming II: automatic discovery of reusable programs. MIT Press, Cambridge, MA (1994)

5. Montana, D.J.: Strongly typed genetic programming. BBN Technical Report #7866, Bolt Beranek and Newman Inc, 10 Moulton Street, Cambridge, MA 02138, USA, March 1994
6. Huelsbergen, L.: Toward simulated evolution of machine language iteration. In: Koza, J.R., Goldberg, D.E., Fogel, D.B., Riolo, R.L., eds.: Genetic Programming 1996: Proceedings of the First Annual Conference, Stanford University, CA, USA, pp. 315–320. MIT Press, 28–31 July 1996
7. Kirshenbaum, E.: Genetic programming with statically scoped local variables. Technical Report HPL-2000-106, Hewlett Packard Laboratories, Palo Alto, 11 August 2000
8. Conrads, M., Nordin, P., Banzhaf, W.: Speech sound discrimination with genetic programming. In: Banzhaf, W., Poli, R., Schoenauer, M., Fogarty, T.C. (eds.) EuroGP 1998. LNCS, vol. 1391, p. 113. Springer, Heidelberg (1998)
9. Agapitos, A., Lucas, S.: Evolving a statistics class using object oriented evolutionary programming. In: Ebner, M., O'Neill, M., Ekárt, A., Vanneschi, L., Esparcia-Alcázar, A.I. (eds.) EuroGP 2007. LNCS, vol. 4445, pp. 291–300. Springer, Heidelberg (2007)
10. Agapitos, A., Togelius, J., Lucas, S.M.: Evolving controllers for simulated car racing using object oriented genetic programming. In: Proceedings of the Genetic and Evolutionay Computation Conference (2007)
11. Agapitos, A., Togelius, J., Lucas, S.M.: Multiobjective techniques for the use of state in genetic programming applied to simulated car racing. In: Proceedings of IEEE CEC, pp. 1562–1569 (2007)
12. Poli, R., McPhee, N.F., Citi, L., Crane, E.: Memory with memory in genetic programming. J. Artif. Evol. Appl. **2009**, 429–433 (2009)
13. Teller, A.: Turing completeness in the language of genetic programming with indexed memory. In: Proceedings of the IEEE World Congress on Computational Intelligence. vol. 1, Orlando, Florida, USA, pp. 136–141. IEEE Press (27–29 June 1994) (1994)
14. Teller, A.: The evolution of mental models. In: Kinnear Jr., K.E. (ed.) Advances in Genetic Programming. MIT Press, Cambridge (1994)
15. Jannink, J.: Cracking and co-evolving randomizers. In: Kinnear Jr., K.E. (ed.) Advances in Genetic Programming, pp. 425–443. MIT Press, Cambridge (1994)
16. Teller, A., Veloso, M.: A controlled experiment: Evolution for learning difficult image classification. In: Pinto-Ferreira, C., Mamede, N.J. (eds.) Progress in Artificial Intelligence. LNCS, vol. 990, pp. 165–176. Springer, Heidelberg (1995)
17. Teller, A., Veloso, M.: Algorithm evolution for face recognition: What makes a picture difficult. In: International Conference on Evolutionary Computation, Perth, Australia, pp. 608–613. IEEE Press, 1–3 December 1995
18. Teller, A., Veloso, M.: Program evolution for data mining. Int. J. Expert Syst. **8**(3), 216–236 (1995)
19. Andre, D.: Evolution of mapmaking ability: Strategies for the evolution of learning, planning, and memory using genetic programming. In: Proceedings of the IEEE WCCI. vol. 1, pp. 250–255. Florida, USA (27–29 June 1994) (1994)
20. Brave, S.: The evolution of memory and mental models using genetic programming. In Koza, J.R., Goldberg, D.E., Fogel, D.B., Riolo, R.L., eds.: Genetic Programming 1996: Proceedings of the First Annual Conference, Stanford University, CA, USA, MIT Press (28–31 July 1996) 261–266

21. Haynes, T.D., Wainwright, R.L.: A simulation of adaptive agents in hostile environment. In: George, K.M., Carroll, J.H., Deaton, E., Oppenheim, D., Hightower, J. (eds.) Proceedings of the 1995 ACM Symposium on Applied Computing, pp. 318–323. USA, ACM Press, Nashville (1995)

22. Langdon, W.B.: Genetic Programming and Data Structures: Genetic Programming + Data Structures = Automatic Programming! vol. 1 of Genetic Programming. Kluwer, Boston, 24 April 1998

23. Bruce, W.S.: Automatic generation of object-oriented programs using genetic programming. In: GP 1996: Proceedings of the 1st Annual Conference

24. Nordin, P., Banzhaf, W.: Programmatic compression of images and sound. In: Koza, J.R., Goldberg, D.E., Fogel, D.B., Riolo, R.L. (eds.): Genetic Programming 1996: Proceedings of the First Annual Conference, Stanford University, CA, USA, pp. 345–350. MIT Press, 28–31 July 1996

25. Spector, L., Luke, S.: Cultural transmission of information in genetic programming. In: Koza, J.R., Goldberg, D.E., Fogel, D.B., Riolo, R.L. (eds.): Genetic Programming 1996: Proceedings of the First Annual Conference, Stanford University, CA, USA, pp. 209–214. MIT Press, 28–31 July 1996

26. Koza, J.R., Andre, D., Bennett III., F.H., Keane, M.: Genetic Programming 3: Darwinian Invention and Problem Solving. Morgan Kaufman

27. O'Neill, M., Ryan, C.: Investigations into memory in grammatical evolution. In: GECCO 2002, (ed.), pp. 141–144 (2002)

28. Agapitos, A., Dyson, M., Lucas, S.M., Sepulveda, F.: Learning to recognise mental activities: genetic programming of stateful classifiers for brain-computer interfacing. In: GECCO 2008: Proceedings of the 10th annual conference on Genetic and evolutionary computation (2008)

29. Agapitos, A., O'Neill, M., Brabazon, A., Theodoridis, T.: Learning environment models in car racing using stateful genetic programming. In: Proceedings of the IEEE Conference on Computational Intelligence and Games, Seoul, South Korea, pp. 219–226. IEEE (31 August - 3 September 2011) (2011)

30. Agapitos, A., O'Neill, M., Brabazon, A.: Stateful program representations for evolving technical trading rules. In: GECCO 2011: Proceedings of the 13th Annual Conference Companion on Genetic and Evolutionary Computation (2011)

31. Iba, H., de Garis, H., Sato, T.: Temporal data processing using genetic programming. In: Eshelman, L. (ed.): Genetic Algorithms: Proceedings of the Sixth International Conference (ICGA95), pp. 279–286, 15–19 July 1995

32. Sharman, K.C., Esparcia Alcazar, A.I., Li, Y.: Evolving signal processing algorithms by genetic programming. In: Zalzala, A.M.S. (ed.): First International Conference on Genetic Algorithms in Engineering Systems: Innovations and Applications, GALESIA. vol. 414, pp. 473–480. Sheffield, UK, IEE, 12–14 September 1995

33. Alfaro-Cid, E., Sharman, K., Esparcia-Alcazar, A.I.: Genetic programming and serial processing for time series classification. Evol. Comput. **22**(2), 265–285 (2014)

34. Brabazon, A., O'Neill, M.: Biologically Inspired Algorithms for Financial Modelling. Natural Computing Series. Springer, Heidelberg (2006)

35. Tsang, E.P.K., Li, J., Markose, S., Er, H., Salhi, A., Lori, G.: EDDIE in financial decision making. J. Manage. Econ. **20**, 101–112 (2000)

Improving Fitness Functions in Genetic Programming for Classification on Unbalanced Credit Card Data

Van Loi Cao[(✉)], Nhien-An Le-Khac, Michael O'Neill, Miguel Nicolau, and James McDermott

Natural Computing Research and Application Group,
University College Dublin, Dublin, Ireland
loi.cao@ucdconnect.ie,
{an.lekhac,m.oneill,miguel.nicolau,james.mcdermott2}@ucd.ie
http://ncra.ucd.ie

Abstract. Credit card classification based on machine learning has attracted considerable interest from the research community. One of the most important tasks in this area is the ability of classifiers to handle the imbalance in credit card data. In this scenario, classifiers tend to yield poor accuracy on the minority class despite realizing high overall accuracy. This is due to the influence of the majority class on traditional training criteria. In this paper, we aim to apply genetic programming to address this issue by adapting existing fitness functions. We examine two fitness functions from previous studies and develop two new fitness functions to evolve GP classifiers with superior accuracy on the minority class and overall. Two UCI credit card datasets are used to evaluate the effectiveness of the proposed fitness functions. The results demonstrate that the proposed fitness functions augment GP classifiers, encouraging fitter solutions on both the minority and the majority classes.

Keywords: Class imbalance · Credit card data · Fitness functions

1 Introduction

Credit cards have emerged as the preferred means of payment in response to continuous economic growth and rapid developments in information technology. The daily volume of credit card transactions reveals the shift from cash to card. Concurrently, losses due to credit card fraud and non-payment of bills have risen to billions of dollars globally each year. Therefore, there is great pressure on banks and financial organizations to improve the models that detect these problems in credit card data. Classification models are generally applied to these problems in credit card data, but the problem domain poses serious challenges for such models. Various benchmark machine learning techniques are explored in the literature, Decision Tree Induction (C45), Artificial Neural Networks, Class Weighted Support Vector Machines, Bayes Networks, Artificial Immune

© Springer International Publishing Switzerland 2016
G. Squillero and P. Burelli (Eds.): EvoApplications 2016, Part I, LNCS 9597, pp. 35–45, 2016.
DOI: 10.1007/978-3-319-31204-0_3

Systems, Genetic Algorithms and so on [1–3]. However, the credit card data exhibits unique characteristics which render the task extremely challenging for any machine learning technique. The most common characteristic is that the credit card datasets are highly unbalanced, that is they admit and uneven distribution of class transactions. The problem class (fraud or non-payment) is represented by only a small number of examples, so it is the minority class, while the non-problem (non-fraud, or full payment) class makes up the rest, so it is the majority class.

Using these datasets as training sets in the learning process can bias the learning algorithm resulting in poor accuracy on the minority class but high accuracy on the majority class [4]. This is because traditional training criteria such as overall success or error rate can be influenced by the larger number of instances from the majority class. Unfortunately, correctly identifying the minority (problem) class is the more important task. A small percentage of incorrectly classified examples from the minority class may lead to large monetary losses. Therefore, many methods for addressing unbalanced data have been developed to strengthen the classification algorithms for this issue in credit card data.

Generally speaking the approaches for handling unbalanced data can be categorized into two main groups, external and internal. Firstly, in the external approach an artificially balanced distribution of class examples for training is created by sampling from the original unbalanced data sets. For instance, over-sampling the minority class to boost representation, or editing of the majority class to decrease representation [5,6]. Although these approaches can perform effectively, they have drawbacks. For instance, the training process can be burdened with a computational overhead from not only sampling algorithms, but also adjusted training data sets. Secondly, in the internal approach, the focus is instead to apply learning algorithms to the original unbalanced datasets [7,8]. Interestingly, in Genetic Programming (GP), adapting the fitness function is known as a simple and efficient approach for addressing the imbalance in credit card approval data. Solutions with good classification accuracy on minority class and overall are rewarded with better fitness whereas those that are only biased toward majority class are penalized with poor fitness [9,10].

This paper aims to address the imbalance in credit card data by applying a natural computing technique, namely Genetic Programming (GP). Our focus is on adapting the fitness function to evolve GP classifiers with increased classification accuracy on the minority class (True Positive rate) while retaining high overall classification accuracy. The standard GP system is employed to investigate two fitness functions proposed by Bhowan [9], and based on the same two new fitness functions are presented. The experiments are conducted on two UCI standard datasets [11], the German Credit dataset, the Australian Credit Approval dataset, and the evaluation of these classifiers respects three basic measures, True Positive (TP), True Negative (TN) and overall accuracy. We also compare the results from GP classifiers evolved on these datasets to six benchmark machine learning methods including Support Vector Machines (c-SVM), two Bayesian Networks (BayesNet and NavieBayes), Decision Trees (J48) and two Artificial Neural Networks (Multilayer Perceptron and RBFNetwork).

The rest of this paper is organized as follows. In the next section, we give a short introduction to the genetic programming heuristic with emphasis on using GP for classification. In Sect. 3, we briefly review two fitness functions from [9] and propose two new fitness functions. This is followed by a section detailing the experimental settings. Experimental results are presented with discussion in Sect. 5. The paper concludes with highlights and future directions.

2 GP for Classification

2.1 Genetic Programming

Genetic Programming was popularized by Koza in the 1990s [12]. It is an evolutionary paradigm that is inspired by biological evolution to find good solutions to a diverse spectrum of problems in the form of computer programs. The solutions are represented as variable shape and size parse trees, which allows the solutions to be evolved with respect to their structure and behavior. The programs which together constitute a population represent different candidates to the problem and combined with other programs to create new hopefully better solutions. This process is repeated over a number of generations until a good solution is evolved.

In terms of classification, GP has the ability to represent solutions to a wide range of complex problems since the representation is highly flexible. GP can leverage this flexibility in its representation to tackle classification problems that confound standard approaches including Decision Tree Induction, Statistical Classifiers and Neural Networks. Therefore, GP methods are well adapted to some classification problems than alternative classification methods [13].

2.2 Classification Strategy

For classification of credit card data, the numeric (real) values returned by GP trees will be translated into class labels depending on whether the output of the GP classifier falls within the non-negative or negative real number. That is, if the output of the GP classifier is greater or equal to zero, the transaction is said to belong to the minority class (problem class), otherwise the transaction is predicted an instance of the majority class (non-problem class). Moreover, in order to achieve good accuracy on the minority class while retaining overall accuracy, the fitness function is adapted to evolve classifiers. Therefore, we examine two fitness functions from literature [9], and develop two new fitness functions. In this paper, we use the term GPout as representing the output from the GP tree.

3 Improving Fitness Function

In this section, we will investigate two fitness functions that were proposed by Bhowan [9] (f_{equal} and f_{errors}). f_{equal} is designed to treat the classification accuracy of both classes as equally important. f_{errors} is given by f_{equal} with the error on both classes. Thus f_{errors} is sensitive to not only the minority class accuracy but also the classification ability. With the aim of raising the minority class detection accuracy while retaining considerable overall classification accuracy, two new fitness functions are presented as the modification of f_{errors}.

3.1   Previous Fitness Function

Fitness Function 1. Equation 1 was presented in [9]. The classification accuracy of both classes is treated as equally important. Two aspects of this fitness function are studied in this paper. Firstly, we examine the effectiveness of balancing the TP and TN rates by treating as equally important on both classes. Secondly, the basic framework for evaluating the proposed fitness functions is described.

$$f_{equal} = \frac{TP}{TP + FN} + \frac{TN}{TN + FP} \tag{1}$$

Fitness Function 2. The fitness function f_{errors} proposed in [9] was formed by combining the fitness function in Eq. 1 with the error rate of each class, given by Eq. 3. The error function aims to differentiate between solutions which score the same classification accuracy on each class but with different internal classification models. Solutions with smaller levels of error for each class are closer to correctly labeling any incorrectly predicted examples. These solutions will have better classification models, and are favored over solutions with a larger levels of error [9].

$$f_{errors} = \frac{TP}{TP + FN} + \frac{TN}{TN + FP} + (1 - Err_{min}) + (1 - Err_{maj}) \tag{2}$$

where,

$$Err_c = \frac{(|P_c^{mx}| + |P_c^{mn}|)}{2}. \tag{3}$$

The error function for class c, Err_c, in Eq. 3, is estimated using the largest and smallest incorrect GPout values for a particular class, P_c^{mx} and P_c^{mn} respectively. The absolute value for incorrect GPout is taken because it is negative on the minority class. These values are scaled to between 1 and 0 where 1 indicates the highest level of error and 0 the lowest.

3.2   Proposed Fitness Function

Proposed Fitness Function 1. Equation 3 illustrates that Err_c is computed based on the largest and smallest incorrect GPout values on a particular class c. Therefore, if the values of $|P_c^{mx}|$ and $|P_c^{mn}|$ are extremely large, the level of error for class c will be very large even though there are few incorrect GPout values. This will give weight to the classification accuracy of the class with the extreme incorrect GPout values. Therefore, the fitness function f_{errors} is more sensitive to extreme incorrect GPout values.

In order to address this issue, Err_c in Eq. 3 is modified by using the mean of the all incorrect GPout values on particular class to estimate the level of error for the class, namely Err_c^{mean}. Where, P_{ci} and m are the $i-th$ incorrect GPout value on class c and the number of incorrect classification examples on class c respectively. By doing this, the fitness function $f_{errors\ mean}$ is less sensitive to the extreme incorrect GPout values than f_{errors}.

$$f_{errors\ mean} = \frac{TP}{TP + FN} + \frac{TN}{TN + FP} + (1 - Err_{min}^{mean}) + (1 - Err_{maj}^{mean}) \quad (4)$$

where,

$$Err_c^{mean} = \frac{\sum_{i=1}^{m}(|P_{ci}|)}{m}. \quad (5)$$

Proposed Fitness Function 2. When the distribution of the incorrect GPout values is skewed, the fitness function $f_{errors\ mean}$ may be still sensitive to the extreme values. In this case, the error function Err_c^{median} is proposed by using the median of the incorrect GPout values on particular class to estimate the error. Where, P_{ci} and m are the $i-th$ incorrect GPout value on class c and the number of incorrect classification examples on class c respectively. Equation 6 shows that the fitness function $f_{errors\ median}$ is partly contributed by Err_c^{median}, which is not sensitive to the extreme incorrect values and the skewed distribution of the incorrect values.

$$f_{errors\ median} = \frac{TP}{TP + FN} + \frac{TN}{TN + FP} + (1 - Err_{min}^{median}) + (1 - Err_{maj}^{median}). \quad (6)$$

where,

$$Err_c^{median} = \begin{cases} P_{c([\frac{m}{2}]+1)} : m = 2k + 1, k \in N, \\ \frac{P_{c([\frac{m}{2}])} + P_{c([\frac{m}{2}]+1)}}{2} : m = 2k, k \in N. \end{cases} \quad (7)$$

4   Experimental Settings

4.1   Credit Card Datasets

Two credit card datasets in UCI Machine Learning Repository [11] are employed that were previous used. The German credit dataset which contains categorical attributes was first procured by Prof. Hofmann. It was then converted into the numeric version by Strathclyde University. Each transaction is described by 24 attributes and one class attribute which classifies the transaction into the good or bad class. There are 700 instances for the good class and 300 instances for the bad class. For the German credit card dataset, we randomly select 50 percent on each class for the training set, other 50 percent on each class for the testing set in order to remain the imbalance between the good and bad class in the two sets.

The Australian Credit Approval dataset originated from Quinlan. All attribute names and values of the data had been changed to meaningless symbols for the confidentiality reasons. There are 690 instances each of which belongs to approval class or risk class (383 approval, 307 risk instances respectively). Each instance is described by 14 real-valued attributes and one class attribute. However, the data is slightly unbalanced, so we randomly sample only 30 percent for the training set, and 30 percent for the testing set on the risk class. Furthermore, we retain 50 percent for the training set and the testing set on the approval class. In order to weight all attributes equally, a Min-max normalization technique is employed to map the values of all attributes of the two datasets to the interval [0.0, 1.0].

4.2 Evolutionary Parameters

A set of parameter sweep experiments were carried out to search for the best evolutionary parameter settings for the credit card data. Five different settings for f_{equal} were tested on two datasets, with successively increasing population sizes, as illustrated in Figs. 1 and 2. The average fitness values are computed with respect to the different population sizes (100, 200, 300, 400 and 500) while keeping the number of generation at 1000 over 30 runs. As presented in Figs. 1 and 2, the parameters of Pop = 500 and Gen = 1000 produce the smallest fitness values on the both data sets over 1000 generations. Therefore, we use the settings for all experiments in this paper. Our experiments employed the standard GP system written on Java programming language with the standard parameter settings: crossover = 0.9, mutation = 0.1 and tournament size = 3. The evolutionary parameter settings are listed in Table 1.

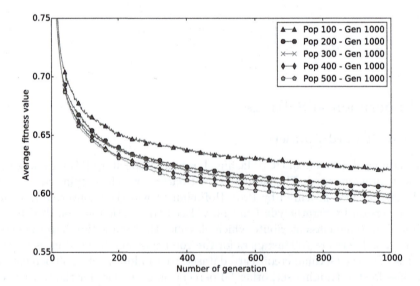

Fig. 1. Average fitness values across 30 runs for a range of parameter settings on the German Credit dataset.

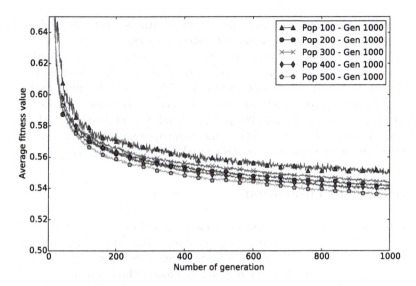

Fig. 2. Average fitness values across 30 runs for a range of parameter settings on the Australia Credit Approval dataset.

Table 1. Evolutionary parameter settings

Parameter	Value
Number of run	30
Population size	500
Number of generation	1000
Crossover Probability (P_{cross})	0.9
Mutation Probability (P_{mut})	0.1
Selection	Tournament
Tournament size	3
Function set	$\{+, -, \times, /\}$

4.3   Experiments

This section presents the settings for the experiments that were conducted in this study. Four sets of experiments are conducted in order to estimate the utility of the proposed fitness functions. The first aims to uncover how important balance between TP and TN rates are when the minority and majority classes are treated with equal importance. The result will be a basis on which to evaluate the classifiers in the following experiments. The second experiment is conducted to investigate how the error function, Eq. 3, of the fitness function f_{errors} effects the TP rate and classification ability. Finally, the third and fourth experiments will examine the extent to which the proposed fitness functions achieve high overall classification accuracy as the TP rate increases.

The effectiveness of the proposed fitness functions are also benchmarked against six well-known machine learning techniques. These techniques were trained and tested on the above datasets. The tested learning techniques include Support Vector Machines (c-SVM), Bayes Networks (BayesNet and NaivieBayes), Decision Trees (J48) and Artificial Neural Networks (Multilayer Perception and RBF networks). The results from applying the preceding learning methods to the above datasets are delineated in the following section.

5   Result and Discussion

This section presents the experimental results realized by applying GP to address the imbalance in credit card datasets. The experiments are implemented on the standard GP system with the evolutionary parameters reported in Table 1. The performance of each GP classifier is evaluated along three dimensions, TP, TN rates and overall accuracy. The results are shown in Tables 2 and 3.

Table 2. The true positive, true negative and accuracy rates of learning methods on the German Credit dataset.

Technique	TP Rate	TN Rate	Accuracy
c-SVM$_{(gamma=1,weights=0.45)}$	0.773	0.823	0.808
NaiveBayes	0.533	0.840	0.748
BayesNet	0.380	0.866	0.720
J48$_{(reducederrorprunning=T,numfolds=3)}$	0.507	0.866	0.758
RBFNetwork$_{(Seed=76,numcluster=9)}$	0.587	0.826	0.754
MultilayerPerceptron	0.760	0.886	0.848
GP$_{(equal)}$	0.757	0.758	0.757
GP$_{(errors)}$	0.823	0.616	0.678
GP$_{(errors\ mean)}$	0.770	0.745	0.752
GP$_{(errors\ median)}$	0.743	0.751	0.749

Tables 2 and 3 illustrate the results of applying GP algorithm with four different fitness functions, and six benchmark machine techniques on German Credit and Australian Credit Approval datasets. The implementations of the six benchmark techniques are performed in Weka [14]. The first subtable reports the results of various tested learning techniques, namely Support Vector Machine (c-SVM), two Bayesian Networks (NaiveBayes and BayesNet), Decision Tree (J48) and two Artificial Neural Networks (Multilayer Perceptron and RBFNetwork). The results of genetic programming classifiers with the four fitness functions

(f_{equal}, f_{errors}, $f_{error\ mean}$ and $f_{errors\ median}$) are presented in the last rows of these tables. The term $GP_{(i)}$ represents the GP classifier with its fitness function of f_i, where $i \in \{equal, errors, error\ mean, errors\ median\}$.

Tables 2 and 3 show that the classifier $GP_{(errors\ mean)}$ produces good results on both datasets. It not only achieves an increases in TP rate but also sustains good overall accuracy (0.770, 0.752 in Table 2, and 0.840, 0.864 Table 3) in comparison to the classifiers $GP_{(equal)}$ (0.757, 0.757 in Table 2, and 0.803, 0.877 in Table 3) and $GP_{(errors\ median)}$ (0.743, 0.749 in Table 2, and 0.803, 0.876 in Table 3). Although TP rates of the $GP_{(errors\ mean)}$ classifier are slightly lower than those of the $GP_{(errors)}$ classifier, the overall accuracy of $GP_{(errors\ mean)}$ is much higher than those of $GP_{(errors)}$.

Table 3. The true positive, true negative and accuracy rates of learning methods on the Australia Credit Approval dataset

Technique	TP Rate	TN Rate	Accuracy
c-SVM$_{(gamma=1, weights=0.45)}$	0.900	0.818	0.846
NaiveBayes	0.600	0.943	0.825
BayesNet	0.770	0.917	0.866
J48$_{(reducederrorprunning=T, numfolds=3)}$	0.840	0.885	0.870
RBFNetwork$_{(Seed=76, numcluster=9)}$	0.710	0.932	0.856
MultilayerPerceptron	0.780	0.932	0.880
GP$_{(equal)}$	0.803	0.915	0.877
GP$_{(errors)}$	0.842	0.853	0.849
GP$_{(errors\ mean)}$	0.840	0.877	0.864
GP$_{(errors\ median)}$	0.803	0.914	0.876

In comparison to the six benchmark machine learning techniques, the classifier $GP_{(errors\ mean)}$ performs better than some the machine learning techniques such as NaiveBayes, BayesNet and RBENetwork. The TP rates of the GP classifier $GP_{(errors\ mean)}$ (0.770 in Table 2 and 0.840 in Table 3) are much higher than those of some benchmark techniques (0.533 (NaiveBayes), 0.380 (BayesNet) or 0.587 (RBENetwork) in Table 2, and 0.600 (NaiveBayes), 0.770 (BayesNet) or 0.710 (RBENetwork) in Table 3). However, c-SVM produced the best results of TP rate and overall accuracy, 0.773 and 0.808 on the first dataset, and 0.900 and 0.846 on the second dataset respectively.

Overall, the results in this section demonstrates that GP can overcome the imbalance in credit card data, and judicious selection of fitness functions can make the search more efficient.

6   Conclusion

In this paper, we address the problem that practitioners often face when classifying credit card datasets. The problem is the imbalance between the problem (fraud or non-payment) class (minority class) and the non-problem (non-fraud, or full payment) class (majority class), which classifiers tend to yield poor accuracy on the minority class despite of high overall accuracy. In this paper, we propose two fitness functions to evolve GP classifiers with significantly increased classification accuracy on the minority class while retaining high overall classification accuracy. The GP classifiers are tested using the two UCI credit card datasets, and the classification abilities are also compared against six well-known learning techniques. The experimental results suggest that GP can handle the imbalance in credit card datasets, and judicious selection of fitness functions can make the search more efficient.

In terms of future work we plan to use the AUC and F-score measurements to evaluate the GP classifiers with proposed fitness functions.

Acknowledgements. This work is funded by Vietnam International Education Development (VIED) and by agreement with the Irish Universities Association.

References

1. Brabazon, A., Cahill, J., Keenan, P., Walsh, D.: Identifying online credit card fraud using artificial immune systems. In: 2010 IEEE Congress on Evolutionary Computation (CEC), pp. 1–7. IEEE (2010)
2. Duman, E., Ozcelik, M.H.: Detecting credit card fraud by genetic algorithm and scatter search. Expert Syst. Appl. **38**(10), 13057–13063 (2011)
3. Lu, Q., Ju, C.: Research on credit card fraud detection model based on class weighted support vector machine. J. Convergence Inf. Technol. **6**(1), 62–68 (2011)
4. Monard, M.C., Batista, G.E.: Learning with skewed class distributions. Adv. Log. Artif. Intell. Robot. LAPTEC 2002 **85**, 173 (2002)
5. Barandela, R., Sánchez, J.S., Garcia, V., Rangel, E.: Strategies for learning in class imbalance problems. Pattern Recogn. **36**(3), 849–851 (2003)
6. Kubat, M., Matwin, S., et al.: Addressing the curse of imbalanced training sets: one-sided selection. In: ICML, Nashville, USA, vol. 97, pp. 179–186 (1997)
7. Bradley, A.P.: The use of the area under the ROC curve in the evaluation of machine learning algorithms. Pattern Recogn. **30**(7), 1145–1159 (1997)
8. Caruana, R., Niculescu-Mizil, A.: Data mining in metric space: an empirical analysis of supervised learning performance criteria. In: Proceedings of the Tenth ACM SIGKDD International Conference on Knowledge Discovery and Data Mining, pp. 69–78. ACM (2004)
9. Bhowan, U., Zhang, M., Johnston, M.: Genetic programming for classification with unbalanced data. In: Esparcia-Alcázar, A.I., Ekárt, A., Silva, S., Dignum, S., Uyar, A.Ş. (eds.) EuroGP 2010. LNCS, vol. 6021, pp. 1–13. Springer, Heidelberg (2010)
10. Bhowan, U., Johnston, M., Zhang, M.: Developing new fitness functions in genetic programming for classification with unbalanced data. IEEE Trans. Syst. Man Cybern. Part B: Cybern. **42**(2), 406–421 (2012)

11. Lichman, M.: UCI Machine Learning Repository (2013)
12. Koza, J.R.: Genetic Programming: on the Programming of Computers by Means of Natural Selection, vol. 1. MIT press, Cambridge (1992)
13. Loveard, T., Ciesielski, V.: Representing classification problems in genetic programming. In: Proceedings of the 2001 Congress on Evolutionary Computation, vol. 2, pp. 1070–1077. IEEE (2001)
14. Hall, M., Frank, E., Holmes, G., Pfahringer, B., Reutemann, P., Witten, I.: The WEKA data mining software: An update. SIGKDD Explor. **11**(1), 10–18 (2009)

Evolving Classification Models for Prediction of Patient Recruitment in Multicentre Clinical Trials Using Grammatical Evolution

Gilyana Borlikova[1(✉)], Michael Phillips[2], Louis Smith[2], and Michael O'Neill[1]

[1] Natural Computing Research and Applications Group, School of Business,
University College Dublin, Dublin, Ireland
{gilyana.borlikova,m.oneill}@ucd.ie
[2] ICON Plc, Dublin, Ireland
{michael.phillips,louis.smith}@iconplc.com

Abstract. Successful and timely completion of prospective clinical trials depends on patient recruitment as patients are critical to delivery of the prospective trial data. There exists a pressing need to develop better tools/techniques to optimise patient recruitment in multicentre clinical trials. In this study Grammatical Evolution (GE) is used to evolve classification models to predict future patient enrolment performance of investigators/site to be selected for the conduct of the trial. Prediction accuracy of the evolved models is compared with results of a range of machine learning algorithms widely used for classification. The results suggest that GE is able to successfully induce classification models and analysis of these models can help in our understanding of the factors providing advanced indication of a trial sites' future performance.

Keywords: Clinical trials · Enrolment · Grammatical evolution · Grammar-based genetic programming

1 Introduction

Patient recruitment is a major bottleneck in conducting clinical trials [1,2]. As Chris Trizna writes in Chap. 19 of "Re-engineering clinical trials" [1] "No patients - No Data" [3]. The recent Tufts Center for the Study of Drug Development report [4] on patient enrolment shows, that though 89 % of trials eventually enrol the required number of patients, the timelines are usually pushed to nearly twice the original plan. Forty eight percent of sites in a given trial fail to enrol required number of patients resulting in the need to bring more sites into the study and extending overall enrolment timelines. Failure to enrol required number of subjects is one of the most frequent reasons for trials' discontinuation [5]. This situation makes optimisation of patient recruitment "a million dollar question" for pharma and CRO industries. In the work presented here, Grammatical Evolution [6,7], a grammar-based Genetic Programming system [8], was used to induce classification models for prediction of patient enrolment performance

© Springer International Publishing Switzerland 2016
G. Squillero and P. Burelli (Eds.): EvoApplications 2016, Part I, LNCS 9597, pp. 46–57, 2016.
DOI: 10.1007/978-3-319-31204-0_4

of investigators/sites in clinical trials. Though results of the best GE model selected based on the training fitness are below the results of the comparator models, the overall best test results obtained by GE models are comparable or even better than results of the comparator models. However, further work is needed to establish an approach to ensure a better way to select the most generalisable model out of the list produced by independent GE runs. The evolved GE models use only a subset of the predictor variables for classification and are human-interpretable, thus providing an insight into the factors behind the classification. These results illustrate applicability of GE and, more broadly, of evolutionary computation to real world business analytics problems. The rest of the paper is organised as follows. Section 2 provides a brief background to the problem and an overview of the related literature. Section 3 describes the dataset and design of the experiment. Section 4 describes and analyses the results, and Sect. 5 draws conclusions and future work directions.

2   Background

2.1   Subject Enrolment in Clinical Trials

Several components contribute to the success or failure of patient enrolment, such as investigator/site selection, complexity of the trial protocol and different strategies of patient engagement [1,2]. While a site/investigator may be good at recruiting one patient population, they may be not so good at another patient population. A non-performing site in one study may not be a non-performing site for all indications or therapeutic areas. Recently much attention has been given to the better ways of reaching target patient populations and maintaining patient engagement [1]. There is also a growing recognition that excessive complexity of trial protocols adversely affects subject enrolment and retention and attempts were made to better manage trial protocol complexity [1]. However, the process of improving patient enrolment remains an area of active business interest. Most trials are set up through a company's site selection function. Though a lot of empirical knowledge is usually accumulated by the professional site selection analysts, a lot of decisions are still made based on an individual analyst's experience and the use of online trial intelligence resources, rather than advanced analytics tools. At the same time the healthcare industry is gradually starting to adapt predictive business analytics techniques to improve processes and boost performance. There exists an urgent business need to capitalise on the advances in machine learning to develop methods able to address challenges of patient enrolment.

2.2   Different Approaches to Patient Enrolment Prediction

Most published research into patient enrolment concerns modelling enrolment rates and forecasting times that will take achieving certain number of enrolled patients. [9] developed detailed Gamma-Poisson mixture models of patient enrolment in multi-centre studies. There is also a considerable amount of research related to identifying patients and predicting potential enrolment eligibility from analysis of the existing patient databases [10] and electronic healthcare records [11]. A substantial number of various clinical trial recruitment support systems (CTRSS) was developed over the years utilising different technologies and algorithms and [12] provide a recent review of these systems. However, as [13] conclude after a systematic review of models to predict recruitment to multicentre clinical trials development of new better models is required.

2.3   Grammatical Evolution for Classification

Classification is one of the most used methods in machine learning and data mining [14]. The classification method consists in predicting the value of a categorical attribute (the class) based on the values of other attributes (predicting variables). An evolutionary learning technique of Genetic Programming (GP) [15, 16] has been successfully applied to a variety of classification problems [17]. Grammatical Evolution (GE) [6, 7] is a grammatical approach to GP [8]. In addition to the many features of GP that make it a very convenient technique for classification, use of grammar in GE allows for extra control of the syntax of evolved programs [18]. Previously GE was successfully applied to evolve classifiers for bond ratings from raw financial information, predict corporate failure and credit risk classification [19–21]. The evolved classifiers were competitive with the results produced by Neural Nets and the GE methodology was suggested to have general utility for rule-induction applications. This study extends GE methodology into the domain of prediction of patient enrolment in multicentre clinical trials.

2.4   Scope of Research

The goal of this study is to produce predictive models of future patient enrolment performance of investigators/sites to be selected to participate in a multicentre clinical trial. More specifically, we are interested in employing GE to evolve binary classification model capable of predicting future performance of investigators/sites. In this first study we use the raw unprocessed dataset with only a minor data preparation in order to establish the baseline of the possible model performance and leave a more sophisticated data pre-processing for the future experiments. We also conduct all experiments using only one cut of the data into the training and test set and will assess robustness of the produced models using multiple training/test splits in the future work.

3   Experimental Design

3.1   Model Data

The dataset used in all experiments was constructed based on a subset of the de-identified historical operational data provided by ICON Plc. on 21 Diabetes

Mellitus Type II Phase III trials. At the first stage of data preparation records with missing values were removed, as well as a few predictor variables with near-zero variance. The resultant dataset consisted of 1233 investigator/site related records and 42 predictor variables. The dataset contained 35 numerical variables and 7 categorical variables describing different aspects of investigator/site. The reference label divided all investigators/sites into two classes based on their patient enrolment performance. Prior to the beginning of experiments the balanced split of the data into a training and testing subsets (60/40 %) was performed using `createDataPartition` function of the CARET package in R [22]. In all experiments model training and tuning was performed using the training subset and then performance of the best models was tested on the test data subset to ascertain how well the evolved models generalise to unseen data.

3.2 Evolutionary Model Representation and Run Parameters

We decided to adopt a decision-tree type approach to constructing the GE classification model. The GE grammar used the function and terminal set detailed in Table 1. For this experiment we intentionally confined the function set to arithmetic operations (including protected division) to cover only linear transformations of the variables. Figure 1 shows the grammar used in the experiment.

Table 1. Function and Terminal sets of GE classifier

Function set	Terminal set
+, -, *, /, and, or, nor, xor, nand	35 numerical predictive variables: x1, ..., x35
equals, not_equals	3 categorical predictive variables: x36, x37, x38
less, greater, less_e, greater_e	4 Boolean predictive variables: x39, ..., x42
	21 random constants in -1.0, ..., 1.0 with 0.1 step

Table 2 details the evolutionary parameters used. The population was initialised using Sensible Initialisation [23] and the maximum derivation tree depth was set to 12. Invalid individuals were handled by reselection. For the Random Search (RS) experiment GE was run with 100 % crossover and 100 % mutation and elitisim (as below). Throughout the experiments classification error (number of misclassified records) was used to measure fitness.

3.3 Performance Measurement and Benchmarking

As the first step, results of the GE model were compared with the results returned by RS run under similar settings. Best and average fitness within each generation were used as a performance measure during the evolutionary run. Performance of the best models evolved over 50 generations was assessed by classification of the previously unseen test set. To benchmark the best evolved GE classification model its performance was compared to performance of a number of

```
<s> ::= if <pred> <out> <out>
<out> ::= <s> | <class>
<class> ::= 0 | 1

<pred> ::=  <bool_bool_comp> <pred> <pred>
          | <bool_num_comp> <expr_num> <expr_num>
          | <bool_bool_comp> <expr_bool> <expr_bool>

<bool_bool_comp> ::= and | or | nor | xor | nand

#Numerical
<bool_num_comp> ::= less | greater | less_e | greater_e
<expr_num> ::= <op_num> <expr_num> <expr_num> | <op_num> <var_num> <var_num> | <var_num>
<op_num> ::= + | - | * | /
<var_num> ::= <const_num> | <ft_num>
<ft_num> ::= x1|x2|x3|x4|x5|x6|x7|x8|x9|x10|x11|x12|x13|x14|x15|x16|x17|x18|x19|x20
            |x21|x22|x23|x24|x25|x26|x27|x28|x29|x30|x31|x32|x33|x34|x35
<const_num> ::= -1|-0.9|-0.8|-0.7|-0.6|-0.5|-0.4|-0.3|-0.2|-0.1
                |0.0|0.1|0.2|0.3|0.4 |0.5|0.6|0.7|0.8|0.9|1

#Categorical
<op_cat> ::= equals | not_equals
<var_cat> ::= <x36> | <x37> | <x38>
<x36> ::= 1|2|3|4|5|6|7|8|9|10|11|12|13|14|15|16|17|18|19|20|21
          |22|23|23|24|24|25|26|27|28|29|30|31|32|33|34|35|36|37
<x37> ::= 1|2|3
<x38> ::= 1|2|3|4|5

#Boolean
<expr_bool> ::= <op_bool> <expr_bool> <expr_bool>
              | <op_bool> <ft_bool> <ft_bool>
              | <ft_bool>
              | <op_cat> <var_cat>
<op_bool> ::= and | or | nor | xor | nand
<ft_bool> ::= x39 | x40 | x41 | x42
```

Fig. 1. Grammar used to construct a GE classifier

Table 2. Evolutionary parameter settings

Parameter	Value
Initialisation	sensible initialisation
Number of runs	30
Population size	1000
Number of generations	50
Selection	tournament
Tournament Size	5 (0.5 % of population size)
Replacement	generational
Elite size	1
Crossover	single point
Crossover Probability	0.9
Mutation	integer flip
Mutation Probability	1 event per individual
Max derivation tree depth	12

well-established machine learning algorithms often used in classification problems. The R CARET package [22] was used to train and tune the models and then to test their performance on the test set. Table 3 contains details of model settings. For illustration, performance of the best variant of each model during training is presented in Fig. 4A. The performance on the previously unseen test set (Fig. 4B and Table 4) was used to compare models using several statistics in addition to classification accuracy.

Table 3. Benchmark machine learning model settings

Model	CARET method	Parameter setting
Support vector machines with radial basis function kernel	svmRadial	sigma=0.01231675, cost=0.25
Classification and Regression Tree (CART)	rpart	Complexity parmeter=0
Multivariate Adaptive Regression Splines (MARS)	gcvEarth	product degree=1
Random Forest (RF)	rf	#randomly selected predictors=7
Nearest shrunken centroids (NSC)	pam	shrinkage threshold=2.231067
Simple Classification Rules (OneRule)	JRip (as per RWEKA)	-

4   Results and Analysis

We performed 30 independent evolutionary runs to evolve GE classification models. Results of this experiment are presented in Fig. 2. The best and average population fitness gradually improved over 50 GE generations. Figure 2B shows results of RS run for the same number of times. Best fitness of GE classifiers start to outperform results of random search from generation 8. This is further confirmed by testing evolved best individuals from generation 50 on the previously unseen test data (Fig. 2D). Though there is a substantial overlap in the test results between RS and GE individuals, the median fitness of GE individuals is substantially better than median fitness of individuals produced by RS. GE was also able to evolve the model which achieves the best classification overall.

When classification quality of the best individuals returned by 30 runs was assessed on the previously unseen test set it became apparent that different individuals returned best results on these two data subsets. The best individual as assessed by training fitness (0.273) performed rather poorly on the test set (test fitness 0.371); while another individual with a lesser training fitness (0.289) demonstrated the best test classification performance (test fitness 0.312). These results suggest a possibility of overfitting [21,24]. Figure 3 presents phenotypes

Grammatical Evolution

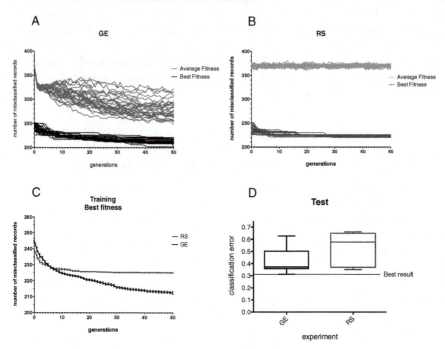

Fig. 2. Training and testing performance in Grammatical Evolution classification experiment. (A) Best and average fitness achieved by GE in each independent run over 50 generations during training. (B) - Best and average fitness achieved by Random Search (RS) in each independent run repeated for 50 generations during training. (C) Mean best fitness achieved by GE and RS during training, (mean sem). (D) Fitness (classification error) returned by evolved best-of-run evolutionary classifiers applied to the test data.

```
A:
[if    (or    x41    (equals    x37    3) )
[if    (and    (equals    x36    36 )    (not_equals    x36    24) )  1    0]
[if    (nor    (greater    (-    x13    -0.2    *    x25)    0.6)    ( greater    -1    0.0))    0
[if    (or    (nand    x39    (xor    x43    x41))    (nor    x39    x42) )  1    0]]]


B:
[if    (or    x41    (equals    x36    23))    0]
[if    (less_e    (+    -0.3    0.4)    x34)
[if    ( less    (+    (x14    *    0.4)    (x7    *    x6) )    x4 )  1    0]    0]
```

Fig. 3. Phenotypes of the best evolved GE classifiers. A: phenotype of the individual with the best training fitness, B: phenotype of the individual with the best fitness on the test dataset.

of these individuals. It can be observed that both classifiers use only 7 predictive variables out of the available 43 variables and both of them use the same 2 variables at the early splits. To provide a benchmark for the results obtained by GE, we compared them with the results obtained on the same dataset split (training and testing subsets), using 6 different machine learning methods. The models were trained and their parameters tuned using the train function from R CARET package. The trainControl function was used to specify 10-fold cross-validation as the re-sampling method. Table 3 provides details of the models and optimum parameters that were chosen during training. These settings were used to estimate model performance on the training and the test set. The levels of classification accuracy obtained by different models on the training and the previously unseen test set are presented in Fig. 4. All models, except NSC performed better than No Information Rate (NIR, 0.65) during training. As should be expected, performance of all models on the test set was lower than during training (NIR of test set 0.6308). All models struggled to achieve reliable classification of the test set. The best overall accuracy was achieved by SVM-RBF model (0.663), while the GE model selected on the basis of the best training fitness showed low generalisation to the test set (0.629). It is worth noting that GE was able to evolve another classifier that achieved much higher prediction accuracy on the unseen data (0.688), however performance of this classifier during training (as described above) prevents from the direct comparison of its results with the other models.

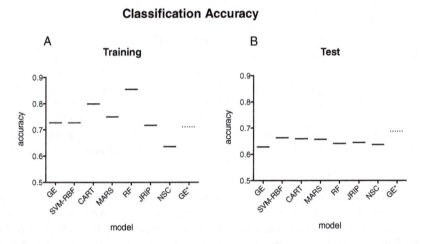

Fig. 4. Comparison of classification accuracy between different models. (A) Best performance during training. (B) Prediction on the test data. GE Grammatical Evolution, SVM Support Vector Machine with RBF (Gaussian) kernel, CART Classification And Regression Tree, MARS Multivariate Adaptive Regression Splines, JRIP One Rule, RF Random Forest, NSC Nearest Shrunken Centroids, GE* - Grammatical Evolution, best in test

There is no one universal best way to compare performance of different models, especially in the case of class imbalances [25]. Our dataset has some class imbalance (36/64 %) with our main class of interest being the minority class.

Classification Accuracy might not be a sufficient measure of model performance in this case. Table 4 gives detailed classification quality metrics obtained by models on the test set. Though SVM_RBF model delivers the best test accuracy and precision, MARS model achieves the best Kappa, while NSC model has the best Sensitivity, F-score and J-statistic, reflecting the superior ability of this model to correctly classify more instances of class 1, but this sensitivity comes at the cost of the ability to correctly classify instances of the second class and consequently, the overall accuracy of the model suffers. The results also show that the second examined GE classifier performs very well, in fact having the best accuracy, precision, Kappa and J-statistic, though does not reach the sensitivity and F-score of the NSC model. In conclusion, results captured in Table 4 highlight that ranking of the models - performance as classifiers greatly depend on the metric chosen, which in turn is dependent on the purpose of the classification (do we place an equal importance on the correct prediction of both classes, or are we specifically interested in correctly predicting instances of one of the classes even if it slightly degrades overall performance).

Table 4. Test Classification Quality Metrics (best result for each metric in bold, * - GE best in test results)

Model	Accuracy (training)	Accuracy (test)	Sensitivity (test)	Specificity (test)	Precision (test)	Kappa (test)	F-score (test)	J-stat (test)
GE	0.727	0.629	0.016	**0.987**	0.429	0.005	0.032	0.004
SVM_RBF	0.727	**0.663**	0.385	0.826	**0.565**	0.226	0.458	0.211
CART	0.800	0.659	0.412	0.804	0.551	0.228	0.472	0.216
MARS	0.750	0.657	0.445	0.781	0.544	**0.235**	0.489	0.227
JRIP	0.718	0.645	0.330	0.830	0.531	0.173	0.407	0.159
RF	**0.855**	0.641	0.456	0.749	0.516	0.210	0.484	0.205
NSC	0.637	0.637	**0.528**	0.701	0.508	0.227	**0.518**	**0.229**
GE*	0.711	**0.688***	0.379	0.868	**0.627***	**0.269***	0.473	**0.247***

We also briefly examined variable importance in different models. Both examined GE classifiers assigned investigators/sites to different classes based on 7 predictive variables. This is a considerably smaller number of variables compared to the lists of variables counted as important by the other top models, though the direct comparison is complicated by the way different models use categorical and numerical variables. Nevertheless, at least 4 of the variables highlighted by GE models appear in the list of variables with above 50 importance score of SVM-RBF, MARS and CART models. This fact further confirms soundness of the evolved GE classifiers.

5   Conclusions and Future Work

This paper adopted a classification approach to predict future performance of investigators/clinical sites in patient recruitment for clinical trials.

The developed GE-based method for classification was compared with a range of well-established machine learning algorithms of different complexity. Taken all together, results are very encouraging, suggesting that GE, in principle, is able to successfully evolve classification model in this scenario. The dataset, at least in its current raw form, proved to be challenging and neither studied model attained a high degree of classification accuracy. The GE classifier taken for the formal comparison with other machine learning algorithms (selected based on the best training performance) showed poor ability to generalise to the test set. However, results of the experiment assessing performance of all 30 evolved GE classifiers show that GE is able to evolve a classifier that is on par or even better than other models. In this first attempt at evolving GE classifier for this business problem we do not have any formal grounds to select the GE classifier that can show better generalisation, though we know that it exists in principle. Based on the observed results and the comparison of the phenotypes of the two examined GE classifiers we can speculate that the best classifier based on training is overfitted to the training set and does not generalise well as the result. This is not surprising, as we have chosen to use decision tree-like structures for our GE classifier and decision trees are known to tend to overfit [14]. Many successful classification algorithms based on decision trees use different techniques to try and avoid such overfitting, such as pruning or early stopping [14]. The issue of overfitting in GE and the generalisation of evolved GE solutions to unseen data were previously highlighted by [21], who examined early stopping as one of the possible approaches to these problems. [24] used monitoring of the generalisation performance of the best-of-generation individual to counteract model overtraining. In the future work we will explore avenues for incorporating early stopping or pruning into our GE set up to avoid overfitting. One of the advantages in the use of GE is a human-interpretable solution [18–20]. In this particular case if we allow ourselves to consider the phenotype of the GE classifier that returned the best results on the test set, we can see that it is not only human-interpretable, but has used only a fraction of predictor variables (7 out of 43) thus effectively performing feature selection simultaneously with classification. So, in depth investigations of these variables might prove to give some additional insights into the factors influencing investigator/site's success in patient enrolment. This study is the first attempt at evolving a GE classifier for this business problem. In this instance we have intentionally used the raw dataset without any data pre-processing in order to establish a baseline for GE model performance. In the future we plan to apply data pre-processing prior to analysis to check whether it will enhance model performance. We also believe that there might be a scope for further increase in the model accuracy through tailored grammar modifications. We also plan to test the use of different fitness functions in order to explore the avenues to tailor the classifier to the business need. The best of the evolved models can be used to help to screen out investigators/sites that have propensity to underperform and jeopardise the trial. Such screening at the early stages of the trial set up can facilitate clinical trial success and substantially reduce costs associated with the need to initiate and maintain low-performing sites and to bring in "rescue" sites later in the trial. The results clearly demonstrate that GE has the capacity to evolve classification models in this business domain.

Acknowledgments. The authors would like to thank Thomas O'Leary, Pamela Howard and Wilhelm Muehlhausen from ICON Plc. for critical reading of the manuscript and expert advice on patient recruitment and Dr. David Fagan, Dr. Alexandros Agapitos and Stefan Forstenlechner from the UCD Natural Computing Research and Applications Group for their insightful advice on GE methodology. This research is based upon work supported by ICON plc.

References

1. Schueler, P., Buckley, B. (eds.): Re-Engineering Clinical Trials. Best Practices for Streamlining the Development Process, 1st edn. Academic Press Elsevier, Amsterdam (2014)
2. Marks, L., Power, E.: Using technology to address recruitment issues in the clinical trial process. Trends Biotechnol. **20**(3), 105–109 (2002)
3. Trizna, C.: Chapter 9 - no patients, no data: patient recruitment in the 21st century. In: Re-Engineering Clinical Trials. Best Practices for Streamlining the Development Process, 1st edn, pp. 91–105. Academic Press Elsevier (2014)
4. Tufts: CSDD impact report - 89% of trials meet enrolment, but timelines slip, half of sites under-enrol. 15(1) (2013)
5. Kasenda, B., von Elm, E., You, J., Blumle, A., Tomonaga, Y., Saccilotto, R., Amstutz, A., Bengough, T., Meerpohl, J.J., Stegert, M., Tikkinen, K.A.O., Neumann, I., Carrasco-Labra, A., Faulhaber, M., Mulla, S.M., Mertz, D., Akl, E.A., Bassler, D., Busse, J.W., Ferreira-Gonzalez, I., Lamontagne, F., Nordmann, A., Gloy, V., Raatz, H., Moja, L., Rosenthal, R., Ebrahim, S., Schandelmaier, S., Xin, S., Vandvik, P.O., Johnston, B.C., Walter, M.A., Burnand, B., Schwenkglenks, M., Hemkens, L.G., Bucher, H.C., Guyatt, G.H., Briel, M.: Prevalence, characteristics, and publication of discontinued randomized trials. JAMA **311**, 1045–1052 (2014)
6. O'Neill, M., Ryan, C.: Grammatical Evolution: Evolutionary Automatic Programming in a Arbitrary Language, Genetic programming, vol. 4. Kluwer Academic Publishers, Dordrecht (2003)
7. Dempsey, I., O'Neill, M., Brabazon, A.: Foundations in Grammatical Evolution for Dynamic Environments. SCI, vol. 194. Springer, Heidelberg (2009)
8. McKay, R.I., Hoai, N.X., Whigham, P.A., Shan, Y., O'Neill, M.: Grammar-based genetic programming: a survey. Genet. Program. Evolvable Mach. **11**(3/4), 365–396 (2010). Tenth Anniversary Issue: Progress in Genetic Programming and Evolvable Machines
9. Anisimov, V.V., Fedorov, V.V.: Modelling, prediction and adaptive adjustment of recruitment in multicentre trials. Stat. Med. **26**(27), 4958–4975 (2007)
10. Aegerter, P., Bendersky, N., Tran, T.C., Ropers, J., Taright, N., Chatellier, G.: The use of drg for identifying clinical trials centers with high recruitment potential: a feasibility study. Stud. Health Technol. Inf. **205**, 783–787 (2014)
11. Kopcke, F., Lubgan, D., Fietkau, R., Scholler, A., Nau, C., Sturzl, M., Croner, R., Prokosch, H.U., Toddenroth, D.: Evaluating predictive modeling algorithms to assess patient eligibility for clinical trials from routine data. BMC Medical Informatics and Decision Making (2013)

12. Kopcke, F., Prokosch, H.U.: Employing computers for the recruitment into clinical trials: a comprehensive systematic review. J. Med. Internet Res. **16**(7), 161 (2014)
13. Barnard, K.D., Dent, L., Cook, A.: A systematic review of models to predict recruitment to multicentre clinical trials. BMC Medical Research Methodology 10(63) (2010)
14. Han, J., Kamber, M., Pei, J.: Data Mining: Concepts and Techniques, 3rd edn. MORGAN KAUFMANN, San Francisco (2011)
15. Koza, J.R.: Hierarchical genetic algorithms operating on populations of computer programs. In: Sridharan, N.S. (ed.) Proceedings of the Eleventh International Joint Conference on Artificial Intelligence IJCAI 1989, vol. 1, pp. 768–774. Detroit, MI, USA, Morgan Kaufmann, 20–25 August 1989
16. Koza, J.R., Keane, M.A., Streeter, M.J., Mydlowec, W., Yu, J., Lanza, G.: Genetic Programming IV: Routine Human-Competitive Machine Intelligence. Kluwer Academic Publishers, Dordrecht (2003)
17. Espejo, P.G., Ventura, S., Herrera, F.: A survey on the application of genetic programming to classification. IEEE Trans. Syst. Man Cybernetics, Part C: Appl. Rev. **40**(2), 121–144 (2010)
18. Nicolau, M., Saunders, M., O'Neill, M., Osborne, B., Brabazon, A.: Evolving interpolating models of net ecosystem CO_2 exchange using grammatical evolution. In: Moraglio, A., Silva, S., Krawiec, K., Machado, P., Cotta, C. (eds.) EuroGP 2012. LNCS, vol. 7244, pp. 134–145. Springer, Heidelberg (2012)
19. Brabazon, A., O'Neill, M.: Diagnosing corporate stability using grammatical evolution. Int. J. Appl. Math. Comput. Sci. **14**(3), 363–374 (2004)
20. Brabazon, A., O'Neill, M.: Credit classification using grammatical evolution. Informatica **30**(3), 325–335 (2006)
21. Tuite, C., Agapitos, A., O'Neill, M., Brabazon, A.: A preliminary investigation of overfitting in evolutionary driven model induction: implications for financial modelling. In: Di Chio, C., et al. (eds.) EvoApplications 2011, Part II. LNCS, vol. 6625, pp. 120–130. Springer, Heidelberg (2011)
22. Kuhn, M.: Building predictive models in r using the caret package. J. Stat. Softw. **28**(5), 1–26 (2008)
23. Ryan, C., Azad, R.M.A.: Sensible initialisation in grammatical evolution. In: Barry, A.M. (ed.) GECCO 2003: Proceedings of the Bird of a Feather Workshops, Genetic and Evolutionary Computation Conference, AAAI 142–145 (2003)
24. Agapitos, A., O'Neill, M., Brabazon, A.: Evolving seasonal forecasting models with genetic programming in the context of pricing weather-derivatives. In: Di Chio, C., et al. (eds.) EvoApplications 2012. LNCS, vol. 7248, pp. 135–144. Springer, Heidelberg (2012)
25. Kuhn, M., Johnson, K.: Applied Predictive Modeling. Springer, New York (2013)

Portfolio Optimization, a Decision-Support Methodology for Small Budgets

Igor Deplano[1], Giovanni Squillero[2], and Alberto Tonda[3(✉)]

[1] Politecnico di Torino, Corso Duca Degli Abruzzi 24, 10129 Torino, Italy
igor.deplano@gmail.com

[2] Politecnico di Torino, Corso Duca Degli Abruzzi 24, 10129 Torino, Italy
giovanni.squillero@polito.it

[3] INRA, UMR 782 GMPA,
1 Avenue Lucien Brétignières, 78850 Thiverval-grignon, France
alberto.tonda@grignon.inra.fr

Abstract. Several machine learning paradigms have been applied to financial forecasting, attempting to predict the market's behavior, with the final objective of profiting from trading shares. While anticipating the performance of such a complex system is far from trivial, this issue becomes even harder when the investors do not have large amounts of money available. In this paper, we present an evolutionary portfolio optimizer for the management of small budgets. The expected returns are modeled resorting to Multi-layer Perceptrons, trained on past market data, and the portfolio composition is chosen by approximating the solution to a multi-objective constrained problem. An investment simulator is then used to measure the portfolio performance. The proposed approach is tested on real-world data from Milan stock exchange, exploiting information from January 2000 to June 2010 to train the framework, and data from July 2010 to August 2011 to validate it. The presented tool is finally proven able to obtain a more than satisfying profit for the considered time frame.

Keywords: Portfolio optimization · Portfolio model · Financial forecasting · MLP · Multi-objective optimization · SOM · Artificial neural networks

1 Introduction

The recent diffusion of on-line trading platforms gave virtually anyone the possibility of investing in any stock exchange market, starting from any amount of money. The ever-growing number of small investors wishes to find the optimal strategies to manage their portfolio, that is, to select the best investment policy in term of minimum risk and maximum return, given their collection of investment tools, income, budget, and convenient time frame.

Portfolio management can be defined as the art of establishing the optimal composition of the investor's portfolio at each instant of time in an ever-changing

© Springer International Publishing Switzerland 2016
G. Squillero and P. Burelli (Eds.): EvoApplications 2016, Part I, LNCS 9597, pp. 58–72, 2016.
DOI: 10.1007/978-3-319-31204-0_5

scenario. Managing a portfolio can be expressed as a list of decisions: for each *tradable*, determine the amount to be bought or sold. The motivations behind each operation can be simplified as: buy/sell now the amount A of tradable T because in a limited time period its price will rise/fall. We call the expected return of such operation R^+, while the maximum loss caused by it is R^-.

The *risk profile* is a measure of how much an investor is afraid to lose money, or, conversely, is willing to take risks hoping for a higher return. The risk profile limits the set of financial instruments that can be used, since highly volatile instruments are considered riskier, and could be incompatible with the investor's preferences.

Generally speaking, the world of finance is extremely complex (for an introduction, see [1–3]). Over time, many types of tradable have been created, the most common of which are: *stocks, fixed income bonds, zero-coupon bonds, futures, options, export trading companies* (ETCs), *exchange-traded funds* (ETFs), *currency crosses, contract for difference* (CFD). For each type there are further differentiations, and while this proliferation can seem baffling, it is needed to meet different stakeholders' needs. For example, stockholders could secure the company's stock price at a fixed cost using *options*; or big farmers could exploit *grain futures* in *margination* as insurance for price variations. Such margination, a service offered by brokers, is the possibility to control capital C keeping only $\frac{C}{L}$ invested, where L is called the *leverage*. Indeed, leverage could be highly beneficial for the farmers in the example, because they would be required to immobilize a much smaller capital. Finally, investors trading such a wide assortment of instruments are necessary for the market liquidity: to be successful, a trade requires both the offer and the demand.

Commonly, investors wish to maximize the return on their portfolio, while minimizing the risks – however, a high return is frequently accompanied by a higher risk. In 1952, Markowitz started the so-called "modern portfolio theory", describing in a quantitative way how the diversification of assets could be used to minimize, but not eliminate, the overall risk without changing the portfolio expected return [4]. In Markowitz's original view, choosing an optimal portfolio is a *mean-variance optimization* problem, where the objective is to minimize the variance for a given mean. Despite its past popularity, the model is known to be based upon assumptions that do not hold in practice: investors are considered flawlessly risk-averse; and either the distribution of the rate of return is assumed multivariate normal, or the utility of the investor a quadratic function of the rate of return. It is now widely recognized that real-world portfolios do not follow a multivariate normal distribution.

More recently, researchers proposed different portfolio theories. The very same Markowitz improved his mean-variance idea in 1968 [5]; other improvements include *post-modern portfolio* [6]; and *behavioral portfolio* [7], that, differently from other approaches, takes into consideration investors who are not completely risk-adverse.

In this work, we study a decision-support tool for management strategies of small portfolios. The tool takes in input market data and, after a training phase, starts an investing simulation. During training, the performance expectation of each tradable over three days will be modeled by a separate neural network. Afterwards, for each day, the simulation uses the neural networks as oracles to optimize the portfolio management strategy, resorting to a multi-objective evolutionary algorithm.

The rest of the paper is organized as follows. Section 2 illustrates the financial real-world dataset used during the experience; the choice of architecture for the ANNs is discussed in Sect. 3; the proposed approach is presented in Sect. 4; while Sect. 5 shows the obtained results and concludes the paper.

2   Data

In this study, we consider end-of-day (EoD) data for about one hundred stocks from *Borsa Italiana* (Milan stock exchange), 40 % of them from medium-to-large companies and 60 % from small companies. Data are taken from *Yahoo! finance*[1], starting from January, 3rd 2000 to August, 2rd 2011. In the EoD data, *open* and *close* represent, respectively, the opening and closing price for a share, the prices traded in the opening and closing auction; *min* and *max* are, respectively, the minimum and the maximum price for that share traded during the day; while *vol* is the number of shares traded. To compare stocks performances from different companies, we choose to normalize EoD data resorting to the following formulas:

$$open'_i = \frac{open_i - close_{i-1}}{close_{i-1}} \tag{1}$$

$$close'_i = \frac{close_i - open_i}{open_i} \tag{2}$$

$$min'_i = \frac{min_i - open_i}{open_i} \tag{3}$$

$$max'_i = \frac{max_i - open_i}{open_i} \tag{4}$$

$$vol'_i = \frac{w_i - w_{i-1}}{w_{i-1}} \tag{5}$$

$$w_k = vol'_k \cdot \frac{max'_k - min'_k}{2} \tag{6}$$

The dataset is divided in two contiguous blocks: the beggining 90 % of the original data, 03/01/2000-01/06/2010; and the ending 10 %, 03/06/2010-02/08/2011. The first block is used for the neural networks, and it is further divided into the canonical: *training*, *validation* and *testing* sets; while the second block is used in the performances simulation, see Sect. 4.3.

---

[1] https://finance.yahoo.com/.

3   Forecasting: Architecture Selection

As common sense suggests, it is impossible to forecast rare, abrupt events which lead to sudden collapses of quotations, such as terrorist attacks, or sensational news on companies. This is a part of inherent market risk that cannot be avoided: however, that does not mean that profiting from trading is an unattainable goal.

In trading, there are different techniques that are widely used to forecast stock performances [8]. In summary, *fundamental analysis* studies company reports and macroeconomical data, and estimates future earnings, which also define the *fair stock price* [1,9,10]. It's a time-consuming activity that needs experience, better suited for long-term trading: widely used stocks comparison metrics for this approach are $\frac{price}{earnings}$, $\frac{price}{bookvalue}$, $\frac{price}{cashflow}$, $\frac{dividend}{price}$ (also known as *dividend yield*).

Technical analysis, also known as *chart trading* [11,12], studies prices charts, and gives buy and sell signals through combinations of indicators and oscillators. The most known methodologies of this kind are RSI (relative strength index), BB (Bollinger bands), moving averages (SMA, EMA, WMA), resistance, supports, and trends. Modeling financial time series is an intense research field for computer science as well, and applications are range from agent-based systems [13], to classifiers (EDDIE and $cAnt-miner_{PB}$ [14]), to genetic algorithms, and, more generally, evolutionary computation [15–23].

We are modeling how the investor's rate of return expectation (RRe) changes in the various market conditions, we test the multi-layer perceptron (MLP) prediction capability. MLP is a feed-forward artificial neural network (ANN) consisting of a layer of inputs, one or more hidden layers and one output layer. Every layer has a set of nodes, each one fully connected to the nodes on the next layer, forming a directed graph. MLPs are trained using the backpropagation algorithm, a supervised learning algorithm that map examples, each one being a pair of input and output features. MLP can be used for function approximation or as a classifier [24]. Briefly, for our purposes, the difference is in the meaning of the output: when MLP is used as function approximation, the output is the RRe codomain; when MLP is used as classifier, the output is a discretized RRe codomain, each output node represents a class, and the connected value is a belief measure. Higher values are linked to a higher confidence for the sample belonging to that class. Our choice of this machine learning paradigm is motivated by the good results obtained with MLP for financial forecasting in [25].

The Eni price is a common benchmark for financial forecasting in Italy, and here it is used to compare the performance of different MLP architectures and configurations. For each architecture we applied different input-output configurations, see Table 2 for a summary. The benchmark data comes from the dataset presented in Sect. 2, and it is composed by Eni and all companies in the oil and natural gas index. This dataset is divided in three parts: training, validation and testing (60 %, 20 %, 20 % of the original data, respectively).

Table 1. Training parameters, common to all tests.

epochs	100
goal	1e-3
learning rate	0.03
learning rate increment	1.005
learning rate decrement	0.095
momentum costant	0.75
max validation fails	20
minimum gradient	1e-10

Table 2. Input/output configurations (conf); the EoD input data is for Eni, oil and natural gas share index, output is for Eni only.

ref	input	output
1	10 days of *min,max,close,open,vol*	next 3 days of *close*
2	10 days of *max − min,close*	next 3 days of *close*
3	10 days of *close*	next 3 days of *close*

3.1 MLP as a Classifier

From start it is apparent that a MLP classifier has a more complex architecture than a MLP used for function approximation. The output classes are chosen clustering the *close performance* using k-means algorithm [27] over all the shares in the benchmark. The aim is to select the more realistic performance classes. After clustering trials from 20 to 60 clusters, we choose as reference 60 classes, because this configuration has a better resolution over the central and most frequently matched intervals. Table 2 reports the configurations and Table 3 the training functions, using 100-epoch and 500-epoch training. However, the correct class is rarely chosen first, mean errors are far from being acceptable and the solution scalability is minimal, due to the high training time needed. This architecture is thus abandoned, as the MLP function approximation yields better results in less time.

Table 3. Learning algorithms compared during the trials, all from the gradient descent family.

id	learning algorithm
traingd	back-propagation
traingdm	back-propagation with momentum
traingda	back-propagation with adaptive learning
traingdx	back-propagation with momentum and adaptive learning rate [26]

3.2 MLP for Function Approximation

The I/O configuration used is reported in Table 2, the neuron activation function is a hyperbolic tangent, and learning algorithms are reported in Table 3. The initialization algorithm that chooses initial weights and bias is *Nguyen-Widrow* (NW) [28]; for the first and third hidden layers we also use a random generator in the $(0, 1)$ interval. All training parameters are reported in Table 1.

Results are strongly dependent from initial weights and the network tends to fall into local minima, but, we noticed that a necessary condition for accepting the trained network is to test if the error distribution in the output layer belongs to a normal distribution family as in Fig. 1. Operatively, we can require that in all the three output nodes the error mean and variance should be lesser than e and $e*10$ respectively, where e starts from 0.1 and it is incremented by 30 % every 10 refusals.

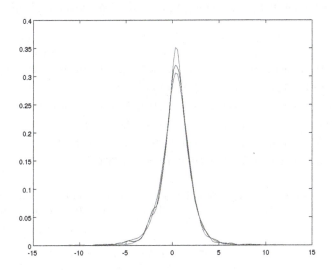

Fig. 1. An example of a fitting curve for error distribution of an acceptable NN.

Tables 4, 5, 6 show the best results among our tests: for our purposes, gradient descent back-propagation with adaptive learning performs generally better than the other techniques. Configurations 2 and 3 perform much better than configuration 1, both in final results and in training time, as they have less input features and hidden layers with less neurons. The result suggests that, for this ANN architecture, using the complete EoD data for each input day does not improve forecast capability.

Table 4. Best results for MLP as function approximation, configuration 1, mean(μ_i) and variance(σ_i) of error; randomly initialized NN have performed worst then NW. 1h, 2h, 3h are the number of hidden layer, number of neurons for each layer: ff1h(201), ff2h(200,100), ff3h(100,50,25). 100 input features, 3 output features.

	μ_1	σ_1	μ_2	σ_2	μ_3	σ_3
ff1h traingdx	0.7715	6.0441	1.1877	27.2118	1.2504	53.7264
ff1h traingda	0.1365	3.7334	0.2662	3.3989	0.1008	3.5990
ff1h traingdm	0.0373	3.6282	1.2129	39.7795	-6.7573	3.5380
ff2h traingda	0.1755	3.5330	0.1996	3.7211	0.2543	3.6210
ff3h traingda	0.3936	2.7554	0.4538	2.4185	0.2844	2.7811
ff3h traingda	0.4333	2.4198	0.4286	2.3372	0.4955	2.6574

Table 5. Best results for MLP as function approximation, configuration 2, number of neurons for each layer: ff1h(81), ff2h(80,40), ff3h(80,40,20). 40 input features, 3 output features.

	μ_1	σ_1	μ_2	σ_2	μ_3	σ_3
ff1h traingda	0.1308	4.4511	0.0265	4.7854	0.0907	4.9236
ff1h traingda	0.0057	4.8258	0.1355	4.2881	-0.0128	5.2964
ff3h traingda	0.4659	2.7963	0.3123	2.8014	0.3955	2.5120
ff3h traingda	0.4894	2.3433	0.4656	2.4899	0.3631	2.7598
ff3h traingda	0.3213	2.8939	0.3825	2.6646	0.4112	2.4843
ff3h traingda	0.3212	3.2860	0.4381	2.4763	0.3558	2.7143

Table 6. Best results for MLP as function approximation, configuration 3, number of neurons for each layer: ff1h(41), ff2h(40,20), ff3h(40,20,10). 20 input features, 3 output features.

	μ_1	σ_1	μ_2	σ_2	μ_3	σ_3
ff1hrnd traingda	0.0324	4.8329	0.2091	4.9226	-0.0220	7.5821
ff2h traingda	0.0116	5.3486	0.2755	3.8100	0.1281	4.3746
ff2h traingda	0.2132	3.7758	0.1946	3.8084	0.2032	4.0627
ff2h traingda	0.2362	4.1163	0.1349	4.0713	0.2003	3.2305
ff3h traingdx	0.1527	2.4093	0.0481	2.6952	-0.9912	2.5851
ff3h traingda	0.4323	2.9154	0.4333	2.5568	0.4406	2.6107
ff3hrnd traingda	0.2112	2.8737	0.3183	3.3552	0.4953	3.0535
ff3hrnd traingda	0.4120	2.7998	0.6899	2.0826	0.4799	2.8326

4   Proposed Approach

We propose a decision-support tool, a *portfolio optimizer* with a *simulator*, whose
optimization process is divided into two phases. The first step is devoted to
building, for each stock, a MLP that models the performance expectation for
the next three days. Being the core of the fitness function that will later be used,
this phase is crucial, as a good approximation will lead to good decisions. The
second step consists of searching for a performing stock combination, a Pareto
set of portfolios able to satisfy our model constraints. Finally, the *simulator* is
used to measure the strategy performance in simplified market conditions. It
must be noted that the simulator can also be used in real conditions as, for each
simulated day and for each stock, it provides the quantity and the book price to
sell and buy. The three phases are summarized in Fig. 2.

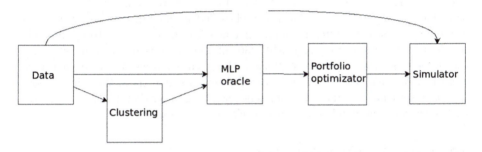

Fig. 2. High-level scheme of the proposed approach. Training data, properly clustered,
is used to train the MLP oracles that will try to predict the performance of company
shares. The optimizer uses a MOEA to obtain a trade-off between (predicted) risk
and reward. Finally, the simulator computes the effective reward obtained, exploiting
unseen validation data.

4.1   Phase I: Building the MLP Oracle

In the first phase, we build the MLP oracle, a procedure that is time consuming
and requires a considerable amount of data, for which we use the *base* dataset,
described in Sect. 2. From the considerations and the preliminary experiments
reported in Sect. 3, the better solution is MLP with data configuration 3, and 2
hidden layers with 35 and 20 nodes, respectively. This combination in empirical
tests provided the best results in terms of mean error and variance, with less
trials. The chosen training algorithm is the gradient descent backpropagation
with adaptive learning rate, while for the initialization, we use Nguyen-Widrow.
Data configuration 3 includes the last 10 days input of a target stock's *close
performance* with 10 days input of the related stocks. In the experiments reported
in Sect. 3, the related stocks were chosen by hand, but here we resort to an
automatic selection. To us, related stocks are stocks whose behavior patterns,

the sequence of closing performances in a 10-day range, are similar to the target stock. Fu, in [29], has shown how Self Organizing Maps (SOMs) are particularly apt at discovering shape patterns, and thus we exploit this methodology to define stock clusters.

Briefly, a SOM is an ANN with an input layer and a node matrix as output layer. The learning algorithm is unsupervised and defines in the output layer the concept of neighborhood: given a node, his neighborhood is composed by the near nodes, constrained by the layer topology and the distance measure. Learning means increasing the weight of a node and distance inverse-proportionally increase the weights of the nodes in the neighborhood, the neighborhood shrinks with time [24].

Our SOM configuration has 10 input nodes and a layer of 100 nodes in output, 100 training epochs, initial neighborhood distance 3, and a hexagonal layer topology. The *base* set defined in Sect. 2 is divided into blocks of 10 consecutive daily-close. Each block has every stock and it is used to train a SOM. It is natural that stocks in block x, which are classified in class y, can be assigned to a different class or to different neighbor classes, in block $x + 1$. To bypass this problem we use an a-priori algorithm [30–32] and mine frequent item-sets from the dataset of classes for every block without empty rows, where each row is a cluster. As each stock in a block can fall into a single class, the min-support used is lesser than 1 % and min-confidence used is greater than 95 %. Association rules discovered here define the related classes of stocks.

4.2   Phase II: MOEA Optimization

For the optimization step, we assume that the investor has a small budget, arbitrarily set to B €. Such an amount is small enough not to influence the overall market behavior, but still permits diversification with marginal transaction costs in the real world. No *margination* and no *short selling* are allowed. The investor does not want to lose more then $L\%$ budget daily.

The resulting model, an adaptation of Markowitz mean-variance model discussed in [33], is shown in Fig. 3. The model chooses portfolios composed by stocks whose most expected performance was uncorrelated in the last 30 days.

We optimize the following equations:

Equation 8 is the global correlation minimization between portfolio stocks. $\sigma_{i,j}$ is the covariance between performance of stock i and j in past 30 days — it does not make sense to consider a longer interval, given our short investing horizon: decisions can be changed daily, and we assume there are no liquidity problems. The equation must be minimized to penalize portfolios composed by highly correlated stocks.

Equation 7 is the expected return: q_i is an unknown model, and it represents the number of stocks i that will be bought tomorrow, where n is the stock number. Equation 11 is the third-day total expected return; Eq. 12 is the expected price for the buy, the average between the price for today, and tomorrow's expected price. It must be maximized.

Equation 10 expresses the constraint of maintaining the portfolio value to (at least) 95 % of the budget, preventing compositions that are expected to lose money. Equation 13 is the price that is expected after two days, which is the first possible day for eventually freeing the position; e_i is the mean error that the considered MLP has produced in past 30 days. Equation 9 is the budget constraint, where $pv_i(t + k)$, with $k = 1, 2, 3$ is the stock i performance, the output node k of our MLP oracle.

$$\max f_1 = \sum_{i=1}^{n} q_i \cdot (pf_i - pa_i)$$

$$\min f_2 = \sum_{i=1}^{n} \sum_{j=1}^{n} q_i \cdot q_j \cdot \sigma_{i,j}$$

$$\text{s.t.} \sum_{i=1}^{n} q_i \cdot p_i(t) \cdot (1 + pv_i(t + 1)) \le b(t)$$

$$b(t) \cdot l \le \sum_{i=1}^{n} q_i \cdot ps_i$$

$$pf_i = p_i(t) \cdot \prod_{k=0}^{3} (1 + pv_i(t + k))$$

$$pa_i = p_i(t) \cdot (1 + \frac{pv_i(t + 1)}{2})$$

$$ps_i = p_i(t) \cdot \prod_{k=0}^{2} (1 + pv_i(t + k) - e_i)$$

$$b > 0, b \in \Re, p_i(t) > 0, p_i(t) \in \Re, q_i \in \mathbb{N}$$

Fig. 3. Equations composing the model for portfolio optimization. $l = \frac{L}{100}$

This mixed-integer problem is non-linear and features several conflicting objectives. Multi-objective evolutionary algorithms (MOEA) [34,35] has been shown effective in finding a good approximation for the Pareto-front in such cases. We use NSGA-II [36], as it was already successfully exploited in [37], where it was demonstrated scalable and able to efficiently find solutions.

4.3 Phase III: Simulation

Our artificial investor builds its strategies for the next day after market closure. It is very disciplined and the following day, whatever happens, it will strictly follow the planned strategy. It desires a differentiated portfolio, it does not want to lose more then 5 %, and it wishes to maximize its returns. If the conditions aren't met, it can also stay out of the market, meaning that it will not hold any share. It is forbidden from taking short positions or borrowing money. Its budget for strategy optimization is defined as the remaining cash available for the current day, added to the money expected from the following day's liquidation of its actual portfolio. After the optimization, the artificial investor will have a set of Pareto optimal portfolios; it will chose the one that generate the least number of transactions, in order to minimize the transactions' cost, and execute the related strategy.

The market used in the simulations is a simplified version of a real market: there are no intra-day operation, no issues linked to partial order executions[2], no taxation, no issues related to sell-buy order synchronizations[3] as sell operations are executed before the buy operations, and no negotiation suspension[4].

When a sell operation is executed: if the open price is greater than the target price, the stock is sold at an open price; else, if the target price is between min e max, then the stock is sold at target price; otherwise, the sell operation is not executed. The quantity is the difference between the old portfolio and the new. When a buy operation is executed: if the open price is less than the target price, the operation is completed at open price; if the target price is between min e max, then the stuck is bought at target price; otherwise, the operation is not executed. The quantity depends on available cash, if the artificial investor does not possess enough resources, the quantity will be zero.

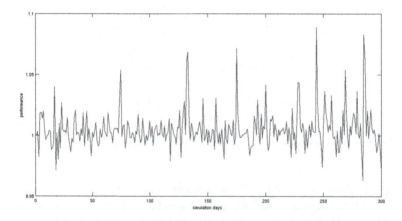

Fig. 4. The proposed approach portfolio's daily rate of return in a 300-day simulation. Despite considerable fluctuations, the approach is able to obtain a 0.445 % average daily rate of return.

---

[2] Obtaining the desired quantity at the wished price might not always be possible.

[3] In the real world, sometimes it is impossible to sell a stock in time, and as a result an investor might not have money available to buy another desired one.

[4] In a real market, a stock that has a bid-ask spread too wide could be suspended from the negotiation, and goes to auction, depending on market regulations.

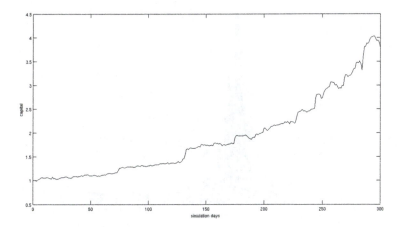

Fig. 5. Portfolio budget cumulative rate of return for each Euro invested.

5   Preliminary Results and Conclusions

In a first experimental evaluation, starting from a budget of 20,000 €, after 300 trading days simulated using the *simulation* data in Sect. 2, the results are quite interesting, see Figs. 4 and 5. The proposed approach is able to multiply its initial budget 3.788 times. Such a result might look astonishing, but we have to take into account the simplifications in the simulation (such as the absence of taxation), the market conditions[5], and the dataset composition. Still, it is important to notice that, even when the proposed portfolio choice is wrong, losses are contained (see Fig. 6).

It's not trivial to obtain a fair comparison of the proposed approach, given both the limited initial investment, that makes it impossible to weight it against large hedge funds, and the assumptions used in the simulation, that prevent our tool from performing several potentially rewarding intra-day operations, but also disregard taxation and costs for performing the transactions. Taking into account these real-world elements would cut the return rate of the approach approximately by 30 %, reducing it to a still considerable 194.6 % total performance.

In order to present at least an estimate for the goodness of the methodology, we use the Hall of Fame of ITCUP[6], a competition for Italian traders with starting budgets ranging from 2,000 to 50,000 €. The objective of the competition is to obtain the highest possible performance, expressed as a percentage of the initial investment, in a single month of trading. Data is available for years 2007-2015: for year 2011, our considered time frame, in one month (April) the

---

[5] FTSE-all share, 03/06/2010-02/08/2011. During this time frame, the index had considerable fluctuations, ranging from a maximum of 21600 reached before October 2010 to a fall to 19105 in November, up to a quote of 23167 in February 2011, and a final decrease to a minimum of 17270.

[6] http://www.itcup.it/, known as *Top Trader Cup* in 2011.

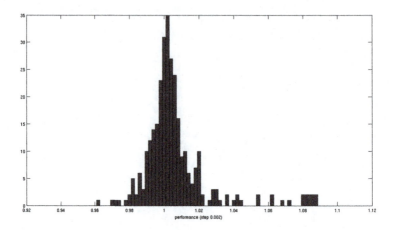

Fig. 6. Histogram of performances, bar steps are 0.002 wide.

champion obtained a performance of 123.54 % (investing 5,000 € and earning 11,176.96), second place a performance of 18.27 % and third place had 6.17 %. Our algorithm, even with the corrections to take into account taxation and costs, had a performance of 14.35 % for the same month, outperforming third place, and coming a little short of the second position. Thus, the final result obtained through the proposed approach seems realistic and consistent with this real-world data.

The proposed approach is easily scalable, because it does not require a real-time response: after the negotiation market closes, more then 15 hours are available for calculations, before the opening of the following day. The training part is computationally heavy, and on a normal notebook computer takes at least 2 calculation days, but a cluster would reduce the time to a fraction, and the training could be done once completely, and then upgraded monthly. Still, a complete run of the algorithm on the 300-day test set requires more than one day. We are currently running further experiments, to obtain a more reliable assessment of the algorithm's capabilities. Future works will focus on including all real-world conditions disregarded in the simulation, in order to obtain a better assessment of the proposed approach. Further studies must be performed on the stock performance forecasting, as this is the core of the problem, in particular the forecasting of inversion points as they are crucial to maximize returns.

References

1. Graham, B., Dodd, D.L.: Security analysis. McGraw-Hill New, York (2008)
2. Sheppard, K.: Financial econometrics notes. University of Oxford, New York (2010)
3. Bodie, Z., Kane, A., Marcus, A.J.: Investments (2014)
4. Markowitz, H.: Portfolio selection*. J. Finan. **7**(1), 77–91 (1952)
5. Markowitz, H.M.: Portfolio selection: efficient diversification of investments. Yale University Press, New Haven (1968)

6. Swisher, P., Kasten, G.W.: Post-modern portfolio theory. J. Finan. Plann. Denver **18**(9), 74 (2005)
7. Shefrin, H., Statman, M.: Behavioral portfolio theory. J. Finan. Quant. Anal. **35**(02), 127–151 (2000)
8. Oberlechner, T.: Importance of technical and fundamental analysis in the european foreign exchange market. Int. J. Finan. Econ. **6**(1), 81–93 (2001)
9. Abarbanell, J.S., Bushee, B.J.: Fundamental analysis, future earnings, and stock prices. J. Account. Res. **35**, 1–24 (1997)
10. Dechow, P.M., Hutton, A.P., Meulbroek, L., Sloan, R.G.: Short-sellers, fundamental analysis, and stock returns. J. Finan. Econ. **61**(1), 77–106 (2001)
11. Edwards, R.D., Magee, J., Bassetti, W.: Technical analysis of stock trends. CRC Press, Boca Raton (2007)
12. Bulkowski, T.N.: Encyclopedia of chart patterns. Wiley, New York (2011)
13. Neri, F.: Learning and predicting financial time series by combining natural computation and agent simulation. In: Chio, C., et al. (eds.) EvoApplications 2011, Part II. LNCS, vol. 6625, pp. 111–119. Springer, Heidelberg (2011)
14. Otero, F.E.B., Kampouridis, M.: A comparative study on the use of classification algorithms in financial forecasting. In: Esparcia-Alcázar, A.I., Mora, A.M. (eds.) EvoApplications 2014. LNCS, vol. 8602, pp. 276–287. Springer, Heidelberg (2014)
15. Lohpetch, D., Corne, D.: Outperforming buy-and-hold with evolved technical trading rules: Daily, weekly and monthly trading. In: Di Chio, C., Brabazon, A., Di Caro, G.A., Ebner, M., Farooq, M., Fink, A., Grahl, J., Greenfield, G., Machado, P., O'Neill, M., Tarantino, E., Urquhart, N. (eds.) EvoApplications 2010, Part II. LNCS, vol. 6025, pp. 171–181. Springer, Heidelberg (2010)
16. Hochreiter, R.: Evolutionary multi-stage financial scenario tree generation. In: Chio, C., et al. (eds.) EvoApplications 2010, Part II. LNCS, vol. 6025, pp. 182–191. Springer, Heidelberg (2010)
17. Gabrielsson, P., König, R., Johansson, U.: Evolving hierarchical temporal memory-based trading models. In: Esparcia-Alcázar, A.I. (ed.) EvoApplications 2013. LNCS, vol. 7835, pp. 213–222. Springer, Heidelberg (2013)
18. Loginov, A., Heywood, M.I.: On the utility of trading criteria based retraining in forex markets. In: Esparcia-Alcázar, A.I. (ed.) EvoApplications 2013. LNCS, vol. 7835, pp. 192–202. Springer, Heidelberg (2013)
19. Vassiliadis, V., Thomaidis, N., Dounias, G.: On the performance and convergence properties of hybrid intelligent schemes: Application on portfolio optimization domain. In: Chio, C., et al. (eds.) EvoApplications 2011, Part II. LNCS, vol. 6625, pp. 131–140. Springer, Heidelberg (2011)
20. Michalak, K.: Selecting best investment opportunities from stock portfolios optimized by a multiobjective evolutionary algorithm. In: Proceedings of the 2015 on Genetic and Evolutionary Computation Conference, pp. 1239–1246. ACM (2015)
21. Beasley, J.E., Meade, N., Chang, T.J.: An evolutionary heuristic for the index tracking problem. Eur. J. Oper. Res. **148**(3), 621–643 (2003)
22. Høyland, K., Wallace, S.W.: Generating scenario trees for multistage decision problems. Manage. Sci. **47**(2), 295–307 (2001)
23. Michalak, K., Filipiak, P., Lipinski, P.: Usage patterns of trading rules in stock market trading strategies optimized with evolutionary methods. In: Esparcia-Alcázar, A.I. (ed.) EvoApplications 2013. LNCS, vol. 7835, pp. 234–243. Springer, Heidelberg (2013)
24. Haykin, S., Lippmann, R.: Neural networks, a comprehensive foundation. Int. J. Neural Syst. **5**(4), 363–364 (1994)

25. Devadoss, A.V., Ligori, T.A.A.: Forecasting of stock prices using multi layer perceptron. Int. J. Comput. Algorithm **2**, 440–449 (2013)
26. Hagan, M.T., Demuth, H.B., Beale, M.H., et al.: Neural network design. Pws Pub, Boston (1996)
27. Arthur, D., Vassilvitskii, S.: k-means++: The advantages of careful seeding. In: Proceedings of the eighteenth annual ACM-SIAM symposium on Discrete algorithms, Society for Industrial and Applied Mathematics, pp. 1027–1035 (2007)
28. Nguyen, D., Widrow, B.: Improving the learning speed of 2-layer neural networks by choosing. In: Initial Values of the Adaptive Weights, International Joint Conference of Neural Networks, pp. 21–26 (1990)
29. Fu, T.C., Chung, F.l., Ng, V., Luk, R.: Pattern discovery from stock time series using self-organizing maps, Citeseer
30. Srikant, R., Vu, Q., Agrawal, R.: Mining association rules with item constraints. KDD **97**, 67–73 (1997)
31. Ye, Y., Chiang, C.C.: A parallel apriori algorithm for frequent itemsets mining. In: Fourth International Conference on Software Engineering Research, Management and Applications, 2006, IEEE, pp. 87–94 (2006)
32. Borgelt, C., Kruse, R.: Induction of association rules: Apriori implementation. In: Compstat, Physica-Verlag HD, pp. 395–400 (2002)
33. Chang, T.J., Meade, N., Beasley, J.E., Sharaiha, Y.M.: Heuristics for cardinality constrained portfolio optimisation. Comput. Oper. Res. **27**(13), 1271–1302 (2000)
34. Anagnostopoulos, K., Mamanis, G.: A portfolio optimization model with three objectives and discrete variables. Comput. Oper. Res. **37**(7), 1285–1297 (2010)
35. Branke, J., Scheckenbach, B., Stein, M., Deb, K., Schmeck, H.: Portfolio optimization with an envelope-based multi-objective evolutionary algorithm. Eur. J. Oper. Res. **199**(3), 684–693 (2009)
36. Deb, K., Pratap, A., Agarwal, S., Meyarivan, T.: A fast and elitist multiobjective genetic algorithm: Nsga-ii. IEEE Trans. Evol. Comput. **6**(2), 182–197 (2002)
37. Anagnostopoulos, K., Mamanis, G.: The mean-variance cardinality constrained portfolio optimization problem: An experimental evaluation of five multiobjective evolutionary algorithms. Expert Syst. Appl. **38**(11), 14208–14217 (2011)

Evolutionary Multiobjective Optimization for Portfolios in Emerging Markets: Contrasting Higher Moments and Median Models

Mai A. Ibrahim$^{(\boxtimes)}$, Mohammed El-Beltagy, and Motaz Khorshid

Operations Research and Decision Support Department,
Faculty of Computers and Information, Cairo University, Giza, Egypt
m.adel@fci-cu.edu.eg

Abstract. Multi-objective Evolutionary algorithms are well suited to Portfolio Optimization and hence have been applied in complex situations were traditional mathematical programming falls short. Often they were used in portfolios scenario of classical Mean-Variance which are not applicable to the Emerging Markets. Emerging Markets are characterized by return distributions that have shown to exhibit significance departure from normality and are characterized by skewness and fat tails. Therefore higher moments models and median models have been suggested in the literature for asset allocation in this case. Three higher moment models namely the Mean-Variance-Skewness, Mean-Variance-Skewness-Kurtosis, Mean-Variance-Skewness-Kurtosis for return and liquidity and three median models namely the Median-Value at Risk, Median-Conditional Value at Risk and Median-Mean Absolute Deviation are formulated as a multi-objective problem and solved using a multi-objective evolutionary algorithm namely the non-dominated sorting genetic algorithm II. The six models are compared and tested on real financial data of the Egyptian Index EGX. The median models were found in general to outperform the higher moments models. The performance of the median models was found to be better as the out-sample time increases.

Keywords: Median models · Higher moment models · Multi-objective evolutionary optimization · Non-dominated sorting genetic algorithm II · Egyptian stock exchange

1 Introduction

Evolutionary Algorithms have been used extensively for portfolio optimization problems. The majority of the researches done were using single-objective evolutionary algorithms but since the portfolio optimization problem is characterized by conflicting objectives in addition to being complex in nature the multi-objective evolutionary algorithms are more suitable for optimizing them.

© Springer International Publishing Switzerland 2016
G. Squillero and P. Burelli (Eds.): EvoApplications 2016, Part I, LNCS 9597, pp. 73–87, 2016.
DOI: 10.1007/978-3-319-31204-0_6

Since the Markowitz approach is not applicable under large departure from normality, a number of alternative models have been proposed. The emerging market returns distributions do not display normal patterns and have fatter tails relative to the normal distribution [1]. In general, emerging markets return distributions exhibit levels of skewness and kurtosis that constitute a significant departure from normality, therefore higher moments models have been introduced to account for the insufficiency of the description of a portfolio by only its first two moments [28,29]. The median model has been introduced as a robust statistic which is less affected by outliers than the mean [7]. Tail risk measures such as Value-at Risk (VaR) and Conditional Value-at-Risk (CVaR) have been introduced instead of variance to capture the effect of risk [26,27].

In this paper, six models that address the non-normality, asymmetry and fat-tails assumptions are formulated as multi-objective models with competing and conflicting objective functions and solved using the multi-objective evolutionary algorithm, Non-dominated Sorting GA (NSGA-II). The six models are the tri-objective Mean-Variance-Skewness (MVS), the quartic-objective Mean-Variance-Skewness-Kurtosis (MVSK), the eight-objective Mean-Variance-Skewness-Kurtosis (MVSK) for return and liquidity, the bi-objective Median-Value at Risk (MedVaR), Median-Conditional Value at Risk (MedCVaR) and Median Mean Absolute Deviation (MedMAD) models. The six models are compared and tested on real financial data from the Egyptian market. To the best of our knowledge, we are the first to apply the evolutionary algorithms in general and the multi-objective evolutionary algorithms in specific to both the median models and the MVSK models and also the first to solve multi-objective portfolio optimization problems with a high number of objectives than four objectives.

Therefore, the first objective of this paper is to conduct a performance comparison among the six models on emerging markets data, and to investigate whether the median models or the higher moments models perform better. The second objective is to represent a brief literature review on the MOEAs models used in portfolio optimization and on the techniques used to solve the higher moment models and median models and their shortcomings. The third objective is to explore if multi-objective genetic algorithms can efficiently and reliably solve multi-objective portfolio optimization problems with a high number of objectives than four objectives which have never been investigated.

The remainder of the paper is organized as follows. A brief literature review on the portfolio optimization models solved using MOEAs and on the existing approaches used to solve the higher moment models and median models and their shortcomings are provided in Sect. 2. Section 3 explains the six portfolio optimization models and their mathematical formulation. A description of the problem representation, constraint handling technique used and parameter setting are given in Sect. 4. The data, empirical results and discussions are presented in Sect. 5, while Sect. 6 states our conclusions.

2   Literature Review

2.1   Multi-objective Evolutionary Algorithms (MOEAs)

The motivation for using the EAs for portfolio optimization instead of the classical and exact techniques, is the large and non-continuous search space, when adding real world constraints to the problem as well as using new risk representation schemes [4]. The majority of the financial problems in general and the portfolio optimization in specific, that have been tackled with Evolutionary Algorithms (EAs), deal with a single-objective function, this is done by transforming the bi-objective problem to a single-objective.

In the last fifteen years, many authors have pointed out that many problems in finance involve multiple conflicting objectives, were the aim is no longer to identify one optimal solution, but a set of solutions representing the best trade-offs among the conflicting objectives. In addition to the multiple conflicting objectives, mathematical nature and the high complexity of the search spaces of the current financial models make MOEAs particularly suitable for dealing with these types of models [22]. According to Metaxiotis et al. [20] who conducted a comprehensive literature review on the application of Multi-objective Evolutionary algorithms for portfolio optimization, most of papers experimented with MOEA for solving bi-objective optimization problems were maximizing return and minimizing variance such as [3] others were using other risk measure other than variance as value-at-risk or expected shortfall [16]. Some other authors added a third objective to the mean and variance and solved the tri-objective using MOEAs example Fieldsend et al. [9] who added cardinality constraint as a third objective. Suksonghong et al. [33] have solved a portfolio optimization with four objectives including mean variance and skewness.

The main difference between the multi-objective evolutionary algorithms (MOEAs) and single-objective evolutionary algorithms (SOEAs) is that MOEAs simultaneously handle a set of solutions in a single run to identify a set of efficient solutions within limited computational time, while SOEAs needs several runs to do that. In addition SOEAs might be unable to deal with complex shapes of the Pareto front (such as non-convexity) while MOEAs don't require any rigid properties, such as linearity, convexity or continuity of neither the objective functions nor the constraints to tackle a certain problem [16]. MOEAs implement specific procedures that enforce diversity over Pareto front (such as niching) and avoid convergence to a single point on the front but SOEA might focus on a particular regions of the front and neglect others [22].

2.2   Median Models

Benati [7] has adopted the median instead of the mean in four models with risk measure as Value-at-Risk (VaR), Conditional Value-at-Risk (CVaR), Mean Absolute Deviation (MAD) and Maximum Loss (ML). Median is characterized by being a robust statistic which is less affected by biases than the mean. Skewness and fat-tails are some attributes that can affect the mean. As the median function is characterized by being piece-wise linear, non-differentiable and is a class of NP-hard problems, Benati [7] has reformulated the problem as a Mixed Integer Linear Programming(MILP) problem, solved by optimizing the median and adding the risk measure as a constraint. The results have shown greater diversification opportunities than obtained from the mean models.

2.3   Higher Moments Models

Markowitz mean-variance approach have assumed that asset returns follow multivariate normal distribution, and relied on only the first two moments i.e. mean and variance to describe the distribution of asset returns [19]. However most of the emerging markets stock returns are not normally distributed [6] and exhibit levels of skewness and kurtosis that constitute a significant departure from normality. Hence, incorporation of only the mean and variance, may underestimate the risk and this leads to inefficient portfolios, and accordingly higher moments (i.e. skewness and kurtosis) can't be neglected [14]. The typical investor is assumed to prefer high skewness and dislike high kurtosis.

Extending the mean-variance framework to higher moments for portfolio optimization comes with a couple of challenges. First, it is not easy to find a trade-off between the four objectives (i.e. mean, variance, skewness, kurtosis) because in the presence of the two high moments, the problem turns into a non-smooth and non-convex optimization problem [14]. In addition, the numerical solution becomes significantly more challenging, were mean-variance optimization may easily de solved, higher moment optimization requires more advanced algorithms [11].

Mean-Variance-Skewness Models (MVS). Canela et al. [8], Lai [18] and Prakash et al. [23] have constructed portfolio models that incorporated skewness and utilized Polynomial Goal Programming (PGP) to determine the optimal portfolio. They applied different formulations and used different data sets but they all concluded that the inclusion of skewness into an investor portfolio decision causes a major reallocation of wealth in the optimal portfolio. Harvey et al. [10] have used a traditional Bayesian decision theoretic framework to estimate risk and investigated the importance of incorporation of skewness into the portfolio selection problem. Yu et al. [35] proposed an efficient radial basis function (RBF) neural network-based methodology to solve the trade-off of the mean–variance–skewness model.

Mean-Variance-Skewness-Kurtosis Models (MVSK). The fourth moment, kurtosis didn't get much attention in literature as the skewness this may be due to the relatively slower development of techniques in dealing with the algebraic challenges associated with the kurtosis calculation. Kemalbay et al. [14], Lai et al. [17] and Mhiri [21] have augmented the dimension of portfolio selection in polynomial goal programming from mean-variance-skewness to mean-variance-skewness-kurtosis. All their results have indicated that the incorporation of the higher moments in the portfolio decision causes a major change in the construction of the optimal portfolios. Qi-fa et al. [24] proposed a dynamic portfolio selection model with higher moments risk and have shown that the dynamic portfolio is superior to the static portfolio.

Mean Variance Skewness Kurtosis Models for Return and Volatility.
Beardsley et al. [5] have extended Markowitz Portfolio Theory by incorporating
the four high moments of both return and liquidity into an investor's objective
function. They found it more adequate to add liquidity separately to the utility
function than internalizing it. They have shown that the optimal allocation of a
portfolio can change dramatically when higher moments of liquidity are added in
the investor's utility function. The problem was solved with the aid of Lagrange's
method and the consideration of the first-order and second-order optimality
conditions.

**Existing Approaches for Solving the Higher Moments Portfolio
Models.** There are several methods in the literature to attack multi-objective
optimization problems. One approach is to reduce the multi-objective problem to
a single-objective problem constrained by all the other objectives. This approach
has an advantage that the optimization problem is reduced to a scalar optimization problem and can be solved very efficiently if the optimization problem is
convex [2]. A shortcoming of this approach is the requirement of prior knowledge
about the given value of constraints which are subjective to the decision maker
and may sometimes guide the search algorithm to a wrong direction. In addition
it requires the changing of the values of preference constraints to generate a
number of solutions were one solution is obtained from one optimization run. To
solve the optimization problem, a large number of time-consuming computations
are required [13].

The second approach is called the weighed sum approach and it aggregates
all the objectives to a single objective with the use of relative importance weights
assigned to each objective. The weights can be used to express the relative significance of different criteria for the decision maker [25]. Similar to the first
approach, the weighed sum approach has the disadvantages of the requirement
of prior knowledge but this time about the preference for each objective which
are subjective to the decision maker. In addition, the values of these preference
coefficients have to be changed to generate a collection of solutions which is time
consuming [13].

The first and second approaches reduces the objectives to one single objective
either by using all the objectives but one, as constraints or by adding a weight
to each objective and summing. This defines the solution of the multi-objective
problem as a point on the Pareto set. But due to the contradicting nature of
the different objective functions, a set of non-dominated solutions should be
considered instead of single solution, were information about the whole Pareto
set is needed, to make a decision [14, 25].

The third approach is polynomial goal programming (PGP) which allows the
levels of preference toward an objective of an investor to be taken into account.
A new objective is formulated as the deviation from the goal value, then the
optimal value can be obtained. As for the polynomial goal programming, it is
characterized by its flexibility in incorporating the preferences of the investors
in addition to its computational simplicity [35]. Polynomial goal programming
approach shortcoming is the possibility of being trapped in a local instead of the
global minimum, if the choice of the initial estimates is not adequate [8].

3   Mathematical Formulation

N available assets are traded and their returns are observed for T time periods where $t = 1,......,T$. Let Z be a random variable representing portfolio returns, where r_{it} be the return of asset i in time period t and z_t represents portfolio returns in time t. The portfolio returns z_t are calculated as a weighted average of all asset returns r_{it} where the weights x_i represent the proportion of capital to be invested in asset i.

Mean-Variance Model (MV). The Markowitz MV model aims to maximize the return while minimizing the variance of the return [19]. Let z^{mean} be the mean of portfolio return were the portfolio return is calculated as $\sum_{i=1}^{N} x_i r_{it}$. Let σ_{ij} be the variance-covariance matrix which represents the covariance between returns of asset i and asset j. The problem is a bi-objective problem, maximizing mean 1 and minimizing variance 2 subject to the budget/unity 3 and non-negativity constraints 4.

$$\text{max mean} = z^{mean} \tag{1}$$

$$\text{min variance} = \sum_{i=1}^{N} x_i^2 \sigma_i^2 + \sum_{i=1}^{N}\sum_{j=1}^{N} x_i x_j \sigma_{ij}, \ (i \neq j) \tag{2}$$

$$s.t. \sum_{i=1}^{N} x_i = 1 \tag{3}$$

$$x_i \geq 0, \ i = 1, 2, ...N \tag{4}$$

3.1   Higher Moment Models

In general, investors will prefer high values for odd moments and low ones for even moments. Therefore the mean and skewness are being maximized while the variance and kurtosis are being minimized. The higher moments objectives are non-linear and non-convex [5] therefore dealing with the higher moments algebraically can be difficult to carry or even intractable.

Mean-Variance-Skewness Model (MVS). While the two dimensional matrix σ_{ij} represents the variance-covariance, the s_{iij}, s_{ijj} are the skewness-coskewness matrices of the joint distribution returns of asset i and j and can be visualized as a three dimensional space. The problem is a three objectives problem, maximizing mean 1, minimizing variance 2 and maximizing skewness 5 subject to the budget 3 and non-negativity constraints 4.

$$\text{max skewness} = \sum_{i=1}^{N} x_i^3 s_i^3 + 3 \sum_{i=1}^{N}(\sum_{j=1}^{N} x_i^2 x_j s_{iij} + \sum_{j=1}^{N} x_i x_j^2 s_{iij}, (i \neq j) \tag{5}$$

Mean-Variance-Skewness-Kurtosis Model (MVSK). The four dimensional kurtosis-cokurtosis matrices of the joint distribution returns of asset i and j are represented by k_{iiij}, k_{ijjj}, k_{iijj}. The problem is a four objectives problem, maximizing mean 1, minimizing variance 2, maximizing skewness 5 and minimizing kurtosis 6 subject to the budget 3 and non-negativity constraints 4.

$$min \ kurtosis = \sum_{i=1}^{N} x_i^4 k_i^4 + 4 \sum_{i=1}^{N} (\sum_{j=1}^{N} x_i^3 x_j k_{iiij} + \sum_{j=1}^{N} x_i x_j^3 k_{iijj})$$

$$+ 6 \sum_{i=1}^{N} \sum_{j=1}^{N} x_i^2 x_j^2 k_{iijj}, (i \neq j) \quad (6)$$

Mean-Variance-Skewness-Kurtosis Model for Return and Liquidity. In this model the liquidity and its higher-moments are incorporated in addition to the return and its higher-moments [5]. Let L_{it} denote the liquidity level of asset i in time period t and is measured by weekly trading volume (scaled down by dividing 10^9).The covariance between liquidity of asset i and asset j is represented by the variance-covariance matrix $(\sigma_{ij})_L$, while the $(s_{iij})_L$, $(s_{ijj})_L$ are the skewness-coskewness matrices of the joint distribution liquidity of asset i and j, $(k_{iiij})_L$, $(k_{ijjj})_L$, $(k_{iijj})_L$ are kurtosis-cokurtosis matrices of the joint distribution liquidity of asset i and j. The problem is an eight objectives problem, optimizing the four moments of return 1, 2, 5 and 6 in addition to optimizing the four moments of liquidity 7, 8, 9 and 10 subject to the budget 3 and non-negativity constraints 4.

$$max \ mean_L = \sum_{i=1}^{N} x_i L_{it} \quad (7)$$

$$min \ variance_L = \sum_{i=1}^{N} x_i^2 (\sigma_i^2)_L + \sum_{i=1}^{N} \sum_{j=1}^{N} x_i x_j (\sigma_{ij})_L, (i \neq j) \quad (8)$$

$$max \ skewness_L = \sum_{i=1}^{N} x_i^3 (s_i^3)_L + 3 \sum_{i=1}^{N} (\sum_{j=1}^{N} x_i^2 x_j (s_{iij})_L + \sum_{j=1}^{N} x_i x_j^2 (s_{iij})_L, (i \neq j)$$

$$(9)$$

$$min \ kurtosis = \sum_{i=1}^{N} x_i^4 (k_i^4)_L + 4 \sum_{i=1}^{N} (\sum_{j=1}^{N} x_i^3 x_j (kiiij)_L + \sum_{j=1}^{N} x_i x_j^3 (k_{iijj})_L)$$

$$+ 6 \sum_{i=1}^{N} \sum_{j=1}^{N} x_i^2 x_j^2 (kiijj)_L, (i \neq j) \quad (10)$$

3.2   Median Models

Median-Median Absolute Deviation Model (MedMAD). Both the Variance and the MAD are considered as symmetric risk measures but the MAD advantages over the Variance is that it does not make use of the covariance matrix of the asset returns, which is a very dense matrix in case of large-scale portfolios. Also, MAD results in portfolios with fewer assets which leads to lower transaction costs. The MAD model in [15] replaces the variance in the objective function of the MV model. Let z^{med} be the median of portfolio returns z_t as in [34]. The problem is a bi-objective problem, maximizing median 11 and minimizing MAD 12 subject to the budget 3 and non-negativity constraints 4.

$$\text{max median} = z^{med} \tag{11}$$

$$\text{min MAD} = \frac{1}{T}\sum_{t=1}^{T}|(z_t - z^{med})| \tag{12}$$

Median-Value at Risk (MedVaR). While the MAD and Variance are considered as symmetric risk measures, both the Value at Risk (VaR) and Conditional Value at Risk (CVaR) are considered as asymmetric risk measures (i.e. Quantile risk measures). The quantile or tail risk measures, are suitable to be applied when the portfolios return is not following the normal distribution as they are able to account for the fat-tails and asymmetry of the asset returns. VaR [12] is the maximum loss not exceeded with a given probability threshold (i.e. α) over a given period of time. The probability threshold lies between zero and one, and is set to $\alpha = 0.25$ [7]. The VaR function is a non-differentiable function and posses local minima and maxima therefore it is not suitable to apply standard optimization tools. The problem is a bi-objective problem, maximizing median 11 and minimizing VaR 13 subject to the budget 3 and non-negativity constraints 4.

$$\text{min VaR}_\alpha(Z) = -\min\{u\,|P(Z \le u) \ge \alpha\} \tag{13}$$

Median-Conditional Value at Risk (MedCVaR). Expected Shortfall or Conditional Value at Risk (CVaR) introduced by [26], represents expectation of the possible losses conditional on the loss being equal to or exceeding the Value at Risk (VaR). First the VaR of the random variable Z is calculated then the expectation of Z conditioned on taking values below VaR is calculated. In general the CVaR function is difficult to optimize. The probability threshold lies between zero and one, and is set to $\alpha = 0.1$ [7]. The problem is a bi-objective problem, maximizing median 11 and minimizing CVaR 14 subject to the budget 3 and non-negativity constraints 4.

$$\text{min CVaR}_\alpha(Z) = -E(Z|Z \le -VaR_\alpha(Z)) \tag{14}$$

4   Problem Representation and Constraints Handling

4.1   Problem Representation

Different types of representations have been proposed in the literature (i.e. representation with one real string and order-based representation), in our paper we follow the hybrid encoding scheme presented by [31,32]. A portfolio(one individual in the population) is represented by two strings, one binary and the other is real. The binary string shows which assets are included in the portfolio (assets with value 1) and which assets are not included in the portfolio (assets with value 0). While the real string represents the weights of each asset included in the portfolio. The length of both strings is equal to total number of assets in the problem. The operators such as crossover and mutation are performed independently for both strings. Before the evaluation process both strings have to be combined together so that the objective values can be calculated. This encoding scheme facilitates the removal and addition of different assets to a portfolio [30].

4.2   Constraint Handling

Initialization of the strings is done randomly then the repair algorithm is applied (normalization of the real string). The repair algorithm needs to be done before the evaluations for selections, and after crossover and mutation operations take place, since the reproduction operators cause deformation of the string. The unity/budget constraint (sum of all asset weight's equal one) in our problem is handled through the constraint handling procedure presented in [31,32].

4.3   Crossover and Mutation Operators

The crossover is a genetic operator that takes two parents (i.e. portfolios) and generates two offsprings by combining their parents' features. In our paper we use two types of crossover for the binary and real strings. As for the real string we use Simulated Binary Crossover SBX and for the binary one we use the Single Point Crossover.

Mutation is a genetic operator that produces random changes in chromosomes. Mutation operator helps in replacing genes that were lost during the selection process and some other times they provide the genes that were never in the initial population. Several mutation operators have been represented in the literature. In our paper for the real string we use Polynomial Mutation and for the binary one we use the Bit Flip Mutation with parametric chance Pmutation of being flipped.

4.4   Parameter Settings

For NSGA-II, The Population Size = 500, Number of Generation = 1000. Both, the binary crossover "Single Point Crossover" probability and the real crossover "Simulated Binary Crossover" are set to 0.8. Both the mutation probabilities of the binary mutation "Bit Flip Mutation" and the real mutation "Polynomial Mutation" are set to =0.01.

5 Data and Empirical Results

In this section the six models mentioned above which are the three median models MedMAD, MedVaR, MedCVaR and the three higher moments models which are MVS, MVSK, MVSK for return and volatility are used for asset allocation using real weekly prices for 21 main companies in EGX30- the main index of the Egyptian Stock Exchange-traded from July 2008 to November 2014 for a total of 335 trading weeks.

Using the time series of the 21 assets starting from July 2008 to November 2014, a number of data sets were constructed in order to compare median and higher moments models under different situations as different in-sample/out-sample splits and different sizes. The two splits used are the "two thirds to one third split" and the "half to half" split shown in Fig. 1. The size of the data sets varies between 150, 200, 225, 300 to 335. There are seven sets in the "two thirds to one third" split, were the first two-thirds of the in-sample data are used for optimization and one-third of out-sample data reserved for testing. Dataset0 spans the whole 335 trading weeks and DataSet1, Dataset2, Dataset3 and Dataset4 include a total of 150 trading weeks while DataSet5, Dataset6 include a total of 225 trading weeks. The "half to half split" contains 3 sets, DataSet7 that includes a total of 300 trading weeks, Dataset8, Dataset9 include a total of 200 trading weeks.

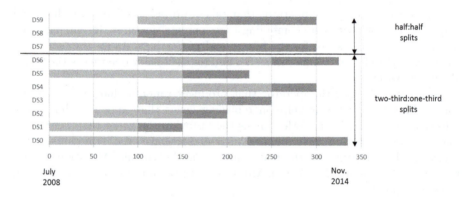

Fig. 1. Data sets splits

In general, median models have shown better performance than higher moments models. In specific, in the "half to half" split, the median models average mean returns of the out-sample were always higher in all the three sets. While in the "two thirds to one third" split the median models average mean return of out-sample were higher in 5 out of 7 datasets. The results are shown in Fig. 2.

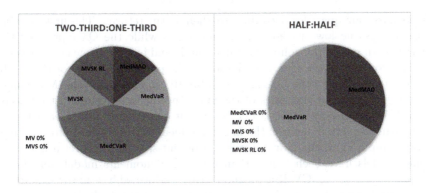

Fig. 2. The "two thirds to one third" and the "half to half" split

In Fig. 3 the P represents the best median model out-sample mean return percentage improvement over the best higher moments models out-sample mean return. The graph shows that median models in 8 out of 10 datasets have higher percentage P than the higher moments models.

We have also found that the median models have outperformed the mean models in all the datasets with long out-sample period which are Dataset5, Dataset6, Dataset7, Dataset8, Dataset9 and Dataset0 while the high moments models gave better returns in case of the datasets with smaller out-sample period.

The in-sample mean returns and out-sample mean returns of Dataset0 are compared, to do so the portfolios on the efficient frontier of each model are

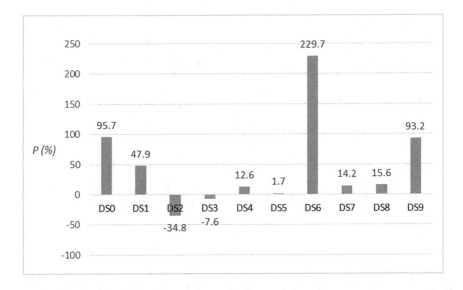

Fig. 3. Out-sample mean returns percentage improvement

divided into four quartiles according to their mean returns. The first quartile Q1 represents the low risk/low return portfolios while the Q4 represents the high risk/high return portfolios. The out-sample and in-sample mean returns are compared for each of the four quartiles for all the portfolio models for DataSet0. The results in Fig. 4 that belongs to Dataset0, confirm that the MV model does not suit the characteristics of the emerging markets return distributions. As the MV model does not consider fat tails it consistently predicted a higher return than the actual one (i.e. the in-sample mean return is much greater than the out-sample in all the four quartiles). The out-sample mean returns for the median models have been higher than the higher moments models for all the four quartiles. The MedCVaR is the model that performed best in three quartiles out of four. It provides actual mean returns (out-sample) higher than expected returns (in-sample) for the three low risk/low returns quartiles. Next to the MedCVaR comes the MedVaR in the out-sample mean returns.

Two concentration indexes HH index (Herfindahl-Hirschman) and Max index are considered to measure the diversification of the portfolio weights. The HH index is $\sum_{i=1}^{N} x_i^2$ and the Max index is $max \{x_i \,|i = 1, \ldots, N\}$. The maximum value for both indexes is 1 which indicates that the portfolio is composed of a single asset. In 5 out of 10 datasets the median models yielded a higher average HH index and Max index. The higher moments models however only yielded better

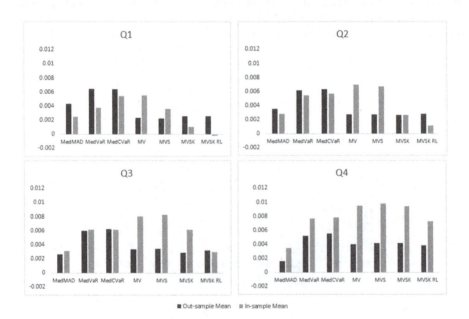

Fig. 4. Out-sample versus in-sample mean returs for four quartiles of different portfolio models for Dataset0

diversification in 2 out of ten sets, while the Mean-Variance model yielded in 3 out of 10 datasets better results. Hence no firm conclusion can be deduced with regards to the diversity of the median versus mean models.

Although the non-linear function CVaR is hard in general to optimize, it still has yielded good results over all the datasets in the "two thirds to one third" split. This probably reveals the suitability of the multi-objective evolutionary algorithms in dealing with complex models in portfolio optimization. In addition, multi-objective evolutionary optimization has helped to reveal structure and patterns of behavior across all the models. Also, it has created a unified framework for optimizing a number of portfolio models.

6 Conclusion

Multi-objective Evolutionary optimization were instrumental to solving the portfolio optimization problems with different characteristics, complexity and objectives. It aided in comparing the six different median and higher moments portfolio models regardless of the different characteristics as being non-convex, non-linear, non-continuous or including a large number of objectives up to eight. The simulation shows that the median models in general have better out-sample returns than higher moments models. With regards to return it has been observed that in datasets with large out-of sample duration the median models outperformed the higher moments models. Hence we conclude that for long term portfolio management, the median models might be preferable. In the future, some attempts should be done to choose the best value of α for both the MedVaR and MedCVaR models. The model comparisons should be done on different data sets from different emerging markets to be able to have a comprehensive assessment.

References

1. Aggarwal, R., Inclan, C., Leal, R.: Volatility in emerging stock markets. J. Finan. Quant. Anal. **34**(01), 33–55 (1999)
2. Anagnostopoulos, K., Mamanis, G.: A portfolio optimization model with three objectives and discrete variables. Comput. Oper. Res. **37**(7), 1285–1297 (2010)
3. Anagnostopoulos, K., Mamanis, G.: The mean variance cardinality constrained portfolio optimization problem: an experimental evaluation of five multiobjective evolutionary algorithms. Expert Syst. Appl. **38**(11), 14208–14217 (2011)
4. Aranha, C., Iba, H.: Application of a memetic algorithm to the portfolio optimization problem. In: Wobcke, W., Zhang, M. (eds.) AI 2008. LNCS (LNAI), vol. 5360, pp. 512–521. Springer, Heidelberg (2008)
5. Beardsley, X.W., Field, B., Xiao, M.: Mean-variance-skewness-kurtosis portfolio optimization with return and liquidity. Commun. Math. Finan. **1**(1), 13–49 (2012)
6. Bekaert, G., Erb, C.B., Harvey, C.R., Viskanta, T.E.: Distributional characteristics of emerging market returns and asset allocation. J. Portfolio Manag. **24**(2), 102–116 (1998)

7. Benati, S.: Using medians in portfolio optimization. J. Oper. Res. Soc. **66**(5), 1–12, 720–731 (2014)
8. Canela, M.A., Collazo, E.P.: Portfolio selection with skewness in emerging market industries. Emerg. Markets Rev. **8**(3), 230–250 (2007)
9. Fieldsend, J.E., Matatko, J., Peng, M.: Cardinality constrained portfolio optimisation. In: Yang, Z.R., Yin, H., Everson, R.M. (eds.) IDEAL 2004. LNCS, vol. 3177, pp. 788–793. Springer, Heidelberg (2004)
10. Harvey, C.R., Liechty, J.C., Liechty, M.W., Muller, P.: Portfolio selection with higher moments. Quant. Finan. **10**(5), 469–485 (2010)
11. Jondeau, E., Rockinger, M.: Optimal portfolio allocation under higher moments. Eur. Finan. Manag. **12**(1), 29–55 (2006)
12. Jorion, P.: Value at Risk: The new Benchmark for Controlling Market Risk. IRWIN, Chicago (1997)
13. Karoon, S.: Multi-objective portfolio selection with skewness preference: an application to the stock and electricity markets. Doctor of Philosophy, Faculty of Economics and Administration, University of Malaya, Kuala Lumpur (2014)
14. Kemalbay, G., Zkut, C.M., Franko, C.G.K., Ozkut, C., Franko, C.: Portfolio selection with higher moments: a polynomial goal programming approach to ISE30 index. Istanbul Univ. Econometrics Stat. e-J. **13**(1), 41–61 (2011)
15. Konno, H., Yamazaki, H.: Mean-absolute deviation portfolio optimization model and its applications to Tokyo stock market. Manag. Sci. **37**(5), 519–531 (1991)
16. Krink, T., Paterlini, S., et al.: Differential evolution for multiobjective portfolio optimization. Cent. Econ. Res. (RECent) **21**, 32 (2008)
17. Lai, K.K., Yu, L., Wang, S.: Mean variance skewness kurtosis based portfolio optimization. In: 1st International Multi-Symposiums on Computer and Computational Sciences, 2006, IMSCCS 2006, vol. 2, pp. 292–297. IEEE (2006)
18. Lai, T.Y.: Portfolio selection with skewness:a multiple-objective approach. Rev. Quant. Finan. Acc. **1**, 293–305 (1991)
19. Markowitz, H.: Portfolio selection. J. Finan. **7**(1), 77–91 (1952)
20. Metaxiotis, K., Liagkouras, K.: Multiobjective evolutionary algorithms for portfolio management: a comprehensive literature review. Expert Syst. Appl. **39**(4), 11685–11698 (2012)
21. Mhiri, M., Prigent, J.L.: International portfolio optimization with higher moments. Int. J. Econ. Finan. **2**(5), 157 (2010)
22. Ponsich, A., Jaimes, A.L., Coello, C., et al.: A survey on multiobjective evolutionary algorithms for the solution of the portfolio optimization problem and other finance and economics applications. IEEE Trans. Evol. Comput. **17**(3), 321–344 (2013)
23. Prakash, A.J., Chang, C.H., Pactwa, T.E.: Selecting a portfolio with skewness: recent evidence from US, European, and Latin American equity markets. J. Bank. Finan. **27**, 1375–1390 (2003)
24. Qi-fa, X., Cui-xia, J., Pu, K.: Dynamic portfolio selection with higher moments risk based on polynomial goal programming. In: International Conference on Management Science and Engineering, 2007, ICMSE 2007, vol. 14, pp. 2152–2157. IEEE (2007)
25. Radziukyniene, I., Zilinskas, A.: Approximation of pareto set in multi objective portfolio optimization. In: Ao, S.-L., Gerlan, L. (eds.) Advances in Electrical Engineering and Computational Science. LNEE, vol. 39, pp. 551–563. Springer, Heidelberg (2008)
26. Rockafellar, R.T., Uryasev, S.: Optimization of conditional value-at-risk. J. Risk **2**, 21–42 (2000)

27. Rockafellar, R.T., Uryasev, S.: Conditional value-at-risk for general loss distributions. J. Bank. Finan. **26**(7), 1443–1471 (2002)
28. Samuelson, P.A.: Efficient portfolio selection for pareto-levy investments. J. Finan. Quant. Anal. **2**(2), 107–122 (1967)
29. Scott, R.C., Horvath, P.A.: On the direction of preference for moments of higher-order than the variance. J. Finan. **35**(4), 915–919 (1980)
30. Skolpadungket, P.: Portfolio management using computational intelligence approaches: forecasting and optimising the stock returns and stock volatilities with fuzzy logic, neural network and evolutionary algorithms. Ph.D. thesis, Department of Computing, University of Bradford (2013)
31. Streichert, F., Ulmer, H., Zell, A.: Evaluating a hybrid encoding and three crossover operators on the constrained portfolio selection problem. In: Congress on Evolutionary Computation, 2004, CEC 2004, vol. 1, pp. 932–939. IEEE (2004)
32. Streichert, F., Ulmer, H., Zell, A.: Hybrid Representation for Compositional Optimization and Parallelizing MOEAs. Citeseer (2005)
33. Suksonghong, K., Boonlong, K., Goh, K.L.: Multi-objective genetic algorithms for solving portfolio optimization problems in the electricity market. Int. J. Electr. Power Energy Syst. **58**, 150–159 (2014)
34. Trzpiot, G., Majewska, J.: Investment decisions and portfolio classification based on robust methods of estimation. Oper. Res. Decisions **1**, 83–96 (2008)
35. Yu, L., Wang, S., Lai, K.K.: Neural network-based mean variance skewness model for portfolio selection. Comput. Oper. Res. **35**(1), 34–46 (2008)

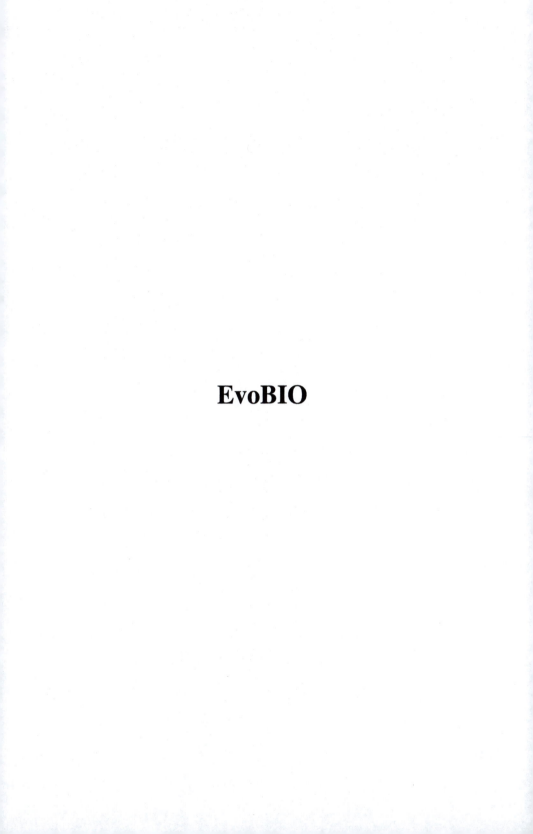

EvoBIO

On Combinatorial Optimisation in Analysis of Protein-Protein Interaction and Protein Folding Networks

David Chalupa[✉]

Department of Computer Science, University of Hull,
Cottingham Road, Hull HU6 7RX, UK
D.Chalupa@hull.ac.uk

Abstract. Protein-protein interaction networks and protein folding networks represent prominent research topics at the intersection of bioinformatics and network science. In this paper, we present a study of these networks from combinatorial optimisation point of view. Using a combination of classical heuristics and stochastic optimisation techniques, we were able to identify several interesting combinatorial properties of biological networks of the COSIN project. We obtained optimal or near-optimal solutions to maximum clique and chromatic number problems for these networks. We also explore patterns of both non-overlapping and overlapping cliques in these networks. Optimal or near-optimal solutions to partitioning of these networks into non-overlapping cliques and to maximum independent set problem were discovered. Maximal cliques are explored by enumerative techniques. Domination in these networks is briefly studied, too. Applications and extensions of our findings are discussed.

Keywords: Combinatorial optimisation · Protein-Protein interaction networks · Protein folding networks · Hybrid heuristics

1 Introduction

Bioinformatics has been a rapidly growing field in the last years. Certain biological problems can be modelled using networks, most notably gene regulatory networks [1] and protein-protein interaction (PPI) networks [2]. Solutions to network problems, which are relatively well studied in computer science, are often regarded as valuable for biologists [3].

In this paper, we present a unified study of combinatorial optimisation problems in analysis of PPI and protein folding (PF) networks. The aim of this paper is to explore the unique area at the intersection of two areas of applied evolutionary computation and computational intelligence in general. On one hand, it spans the computational intelligence in bioinformatics and on the other hand, we explore the biological networks using methodologies of evolutionary computation and heuristics in combinatorial optimisation.

G. Squillero and P. Burelli (Eds.): EvoApplications 2016, Part I, LNCS 9597, pp. 91–105, 2016.
DOI: 10.1007/978-3-319-31204-0_7

Contributions. Using a combination of classical and randomised search heuristics, we obtain high-quality solutions to some of the well-known combinatorial optimisation problems in PPI and PF networks, which are known to be NP-hard in general [4,5].

Experimental results are presented for networks of the European COSIN project [6]. For four different PPI networks, we obtain optimal solutions to maximum independent set and minimum vertex clique covering problem. We used a combination of greedy approximation algorithm for maximum independent set in sparse graphs [7] with a hybrid of iterated greedy (IG) clique covering and randomised local search (RLS) for maximum independent set [8]. For three of four PPI networks, we obtain optimal solutions to maximum clique and chromatic number problems using a hybrid of Brélaz's heuristic [9] with iterated greedy graph colouring algorithm [10]. To explore the minimum dominating set problem, we use a classical greedy approximation algorithm [11].

In addition, we apply the same techniques to a PF network, which is considerably larger than PPI networks. A reduced variant of this PF network is explored, too. We obtain that PF network has slightly different properties than PPI networks, which is probably related both to its size and structure. However, we obtained a very small gap between bounds for maximum clique size and chromatic number of this network, too.

The paper is organised as follows. In Sect. 2, we present an overview of the topic from several relevant perspectives. In Sect. 3, we present our approach to study of PPI and PF networks. In Sect. 4, we present the experimental results and their possible application. Finally, in Sect. 5, we formulate conclusions and summarise scientific problems, which remain open.

2   Combinatorial Optimisation Problems in Protein-Protein Interaction and Protein Folding Networks

There is a body of work concentrating on computer-scientific aspects of study of biological networks. In this section, we present an overview of relevant research and perspectives on our topic.

Protein-protein interaction (PPI) networks. Vertices of a PPI network represent proteins and edges represent interactions between them. These are constructed by molecular biologists usually as an outcome of two-hybrid screening experiments [3]. Analysis of PPI networks and their comparison represent common research topics [12], along with development of analytical software for biological networks [13]. In our experiments, we study public domain PPI network data of the European COSIN project [6]. These include PPI networks for bacterium *Escherichia Coli*, commonly found in gastrointestinal tract; nematode worm *Caenorhabditis elegans*; *Helicobacter pylori*, a bacteria associated with gastritis, usually found in upper gastrointestinal tract; and *Saccharomyces cerevisiae*, a commonly used species of yeast. PPI network data for yeast are a common subject of study [14,15].

Clustering of PPI networks. Probably the most well-known topic in computer-scientific research of PPI networks is represented by clustering of these networks, i.e. decomposition into relatively dense subgraphs. In PPI networks, this is motivated by the problems of complex and functional module detection, which aim to identify groups of mutually interacting proteins, which might often be involved in the same biological processes [16,17].

It is worth noting that biologists tend to distinguish between the term "complex" and "module". Complex in PPI network refers to a molecular machine of proteins, which bind to each other at the same time and space, while the term module refers to a group of mutually interacting proteins, which control certain cellular function, without taking the spatial and temporal aspect into account [18]. However, experimentally obtained PPI data often do not incorporate this information in the network. PPI network data are valuable in reconstruction of metabolic and signalling pathways [3], understanding of cell regulation, prediction of role of uncharacterised proteins and for possible therapy [18]. Multifunctional proteins have previously been revealed [19], i.e. discovery of overlapping modules is a relevant topic for PPI networks, too [20]. One way how they can be detected, is the use of clique merging [21].

Clustering of PPI networks has many similarities with detection of community structure in social networks [22]. Both areas suffer from existence of a large number of diverse clustering algorithms, using ideas ranging from information flow simulation [23], spectral properties of adjacency matrices [24,25], cost-based clustering [26], to stochastic optimisation techniques [18]. However, quality of such a clustering algorithm can be evaluated using a wide spectrum of metrics and multiple objective functions can be considered [27]. Both clustering quality and applicability of developed methods to large networks seem to be important [28]. It can be observed that different clustering algorithms may output very different clusters, each having a different desirable property of a dense or well separated network substructure [29]. Therefore, multiobjective optimisation was successfully applied to network community detection [30]. However, assessing quality of a clustering of a biological network [31] remains hard and often requires to fall back to usage of a reference solution [18,30] or simply requesting verification from a biologist. Additionally, clustering or partitioning of a network [32] might often lead to NP-hard combinatorial optimisation problems [33], which generally require specific attention [4,5].

Protein folding (PF) network beta3s. This network represents conformation space of a 20 residue antiparallel β-sheet peptide investigated by NMR spectroscopy. Vertices represent conformations and edges represent transitions. The network seems to represent a complex system, in which spontaneous folding of protein is modelled as a (weighted) random walk on the conformation space network. Due to space and methods being used, we only consider the structure of the network and omit the weights [34].

PPI and PF networks have also been previously studied in the context of centrality metrics and their stability and potential decomposition [35]. Enumerative and spectral analytical methodologies were also used to study their structure [24]. Statistical analysis of complex networks helps in understanding of the large-scale properties of these networks, too [36].

Combinatorial optimisation problems in networks. We investigate five different classical NP-hard combinatorial optimisation problems [4,5]. For simplicity, we describe these problems only less formally.

Maximum clique is the largest subgraph, in which each pair of vertices is adjacent. In the context of PPI networks, it is the largest group of proteins, in which all proteins mutually interact. Maximum clique size is denoted by ω. There is a spectrum of algorithms for this problem [37].

Graph colouring is an assignment of colours to vertices such that each for each edge, its vertices are differently coloured. Minimum number of colours needed to obtain a graph colouring is called chromatic number and is denoted by χ. Chromatic number is useful, since for each graph, it holds that $\omega \leq \chi$ [38], i.e. maximum clique and chromatic number represent bounds for each other. Randomised algorithms are frequently used to solve graph colouring, too [39].

Maximum independent set is the largest subgraph, in which no pair of vertices is adjacent. Maximum independent set size is denoted by α. In a PPI network, independent set is the largest set of mutually non-interacting proteins.

Minimum vertex clique covering is a partitioning of the network into as few non-overlapping cliques as possible. In PPI networks, it represents a problem of finding the minimum number of clusters such that within each cluster, all proteins must be mutually interacting. The number of cliques in a minimum vertex clique covering is denoted by ϑ. Similarly to maximum clique and graph colouring, it holds that $\alpha \leq \vartheta$ [8]. Hence, maximum independent set and minimum vertex clique covering represent bounds for each other, too.

The last studied problem is the *minimum dominating set problem*. Minimum dominating set is the smallest subset of vertices such that each vertex is either in the dominating set or has a neighbour in it. Minimum dominating set size is denoted by γ. For PPI networks, dominating set represents a set of "central" proteins such that all other proteins interact with at least one protein of the dominating set.

3   Our Experimental Approach

Graph-theoretical approaches represent a vital part of the tools used to analyse biological networks [43]. We aim to provide an approach for their exploration, which ensures solid generalisation and computes properties, which are naturally related to functional module identification. Indeed, large cliques, independent sets and dominating sets represent such properties. Additionally, these problems have clear definitions and approaches, which can easily be applied to previously unexplored PPI or possibly other biological networks. The aim is to provide a hybrid technique, providing bounds for several well-defined valuable properties of an unknown network, which lead to NP-hard combinatorial optimisation problems.

Algorithm 1. Experimental Approach for Analysis of Combinatorial Optimisation Problems in Large Networks

Input: network modelled as a graph $G = [V, E]$
Output: maximum clique and chromatic number bound interval $[\omega_L, \chi_U]$
maximum independent set and clique covering number bound interval $[\alpha_L, \vartheta_U]$
minimum dominating set interval $[\gamma_L, \gamma_U]$

1 find a lower bound $\omega_L \leq \omega$ using the following greedy algorithm
  for construction of a clique Q
2     $Q = \emptyset$
3   order vertices in G from the largest degree to the smallest
4   for vertex v in this ordering
5   if v is adjacent to all vertices in C
6           $Q = Q \cup \{v\}$
7     $\omega_L = |Q|$
8 find a colouring C and an upper bound $\chi_U \geq \chi$ using Brélaz's heuristic [9]
  with binary heap [40,41]
9 if $\omega_L \neq \chi_U$
10   use IG graph colouring heuristic [42,10] starting with C,
      combined with RLS for maximum clique starting with Q
      to compute new bounds ω_L and χ_u
11 find an independent set I and a lower bound $\alpha_L \leq \alpha$ using the greedy
   approximation algorithm for maximum independent set in sparse graphs [7]
12 use IG heuristic for minimum vertex clique covering, combined with RLS
   for maximum independent set [8], starting with independent set I,
   to compute bounds α_L and ϑ_U
13 use the greedy approximation algorithm for minimum dominating set,
   based on minimum set covering [11] to compute upper bound γ_U
14 compute the number of connected components c
15 compute the lower bound $\gamma_L = \min \left\{ c, \dfrac{|V|}{\Delta + 1}, \dfrac{\gamma_U}{\ln \Delta + 1} \right\}$,
   where Δ is the maximum degree of a vertex

This way, we are able to characterise the structure of the networks using cliques, independence and domination and avoid the broad notion of general clustering.

To carry out our investigations, we use a collection of classical heuristics, as well as order-based stochastic algorithms to find high-quality solutions to our combinatorial optimisation problems. The main process of mining from the network data is characterised by the pseudocode of Algorithm 1.

Let us now describe the steps in a slightly more detailed way. Due to lack of space, we are not able to review all aspects of the algorithms we used. However, an interested reader may refer to the referenced work.

In steps 1–7, we use a simple greedy clique algorithm. It starts with an empty clique and orders vertices from largest degree to the smallest. It puts the current vertex to the clique if and only if the clique property is not violated by adding the new vertex. In fact, this approach is equivalent to use of greedy algorithm for independent set [7] for the complement of our graph.

In step 8, we use Brélaz's heuristic implemented with binary heap to find a colouring of the network in $\mathcal{O}(m \log n)$ time, where n is the number of vertices and m is the number of edges.

If maximum clique from steps 1–7 and number of colours used in step 8 are not equal, we use iterated greedy (IG) graph colouring search heuristic [10, 42], combined with randomised local search (RLS) for maximum clique. This is represented by steps 9–10 of Algorithm 1. We start with clique and colouring found in steps 1–8. IG uses randomised block-based moves to possibly reduce the colouring. RLS for maximum clique has not previously been used. Therefore, we describe it in more detail.

RLS for maximum clique uses the same algorithm for clique construction as in steps 1–7. However, it works with a predefined permutation instead of ordering the vertices by their degrees. In the beginning, vertices of clique Q are put into a permutation first and other vertices are ordered at random after that. In each time step of RLS, *jump* move is attempted. The *jump* move simply takes a uniformly random vertex from the permutation and puts it to the first position in the permutation. The other vertices are then shifted to the right. Resulting permutation is used to construct a new clique and is accepted if the new clique is at least as large as the current one.

In step 11, we use the greedy approximation algorithm for maximum independent set in sparse and bounded degree graphs [7]. We use binary heap as a priority queue.

In step 12, we apply the recently proposed IG heuristic for minimum vertex clique covering with RLS for maximum independent set [8].

In step 13, the greedy approximation algorithm for dominating set is used to compute an upper bound for minimum dominating set size [11]. Additionally, a lower bound for the size of minimum dominating set is computed in steps 14–15. This lower bound represents a maximum of three different lower bounds. One of them is the number of components c, the second bound is a general bound derived from maximum degree Δ and the third bound is implied by logarithmic approximation guarantee of the greedy algorithm.

Note that our approach is not specifically restricted to PPI and PF networks. It can easily be applied to social networks or other complex network data. However, for the purpose of this study, we focus specifically on its suitability to explore biological network data.

4   Experimental Results and Discussion

We performed the evaluation in two parts. We first used the approach without the stochastic techniques based on IG and RLS (i.e. we omitted steps 10 and 12). Hence, we used only greedy algorithms. To provide an upper bound for ϑ, we used Brélaz's heuristic applied to complementary graph $\overline{G}$. $\overline{G}$ contains edges between pairs of vertices, which are not adjacent in G and vice versa. In Table 1, we present the best results obtained by this approach in 20 independent runs.

For evaluation of the impact of stochastic components of the approach, we then used the full approach, as specified by Algorithm 1. These results are presented in Table 2. Similarly, we performed 20 independent runs for PPI networks and the reduced PF network $beta3s.reduced$ and present the best results obtained. For the large PF network $beta3s$, we performed only one long run.

The stochastic subroutines of our approach were parameterised as follows. For IG for graph colouring and RLS for maximum clique, we used a simultaneous implementation with 5 iterations of RLS per one iteration of IG. Stochastic optimisation was stopped when $100n$ iterations without improvement of neither clique nor colouring were encountered. Similarly, IG for minimum vertex clique covering and RLS for maximum independent set were used in an implementation with 5 iterations of RLS per one iteration of IG. Stopping criterion was similar, too. Optimisation was stopped when $100n$ iterations without improvement of neither clique covering nor independent set were encountered. Interestingly, these stopping criteria led to results with good quality and solid scalability for all four of these problems.

Both Tables 1 and 2 have the following structure. The first column contains the name of the network. Its number of vertices n, number of edges m, number of connected components c and the number of triangles τ are specified along with the name. The next columns present the maximum clique size ω, chromatic number χ, maximum independent set size α, minimum clique covering size ϑ and minimum dominating set size γ. If a cell contains only one value, it means that the value is a numerically proven optimum for the particular characteristic. If it contains two values separated by $-$, it means that the value is located within the interval specified by presented values. Symbol n in the table means that the value is upper bounded only by the number of vertices n. Bold numbers in Table 2 represent values, which were obtained only by the stochastic approach, i.e. randomised search techniques were beneficial for these instances.

Additionally, we also performed listing of maximal cliques for each network [46]. A clique is maximal if it is not a subgraph of some other clique. The reason is to confront of the number of maximal cliques and maximum (i.e. largest) cliques and to further analyse the cliques as building blocks of the networks.

Network $ecoli$ contains a maximum clique of 6 mutually interacting proteins. Using emumeration based on triangles, we found that there are 657 maximal cliques of size at least 3. There are 5 of these cliques, which consist of 6 proteins. The network $ecoli$ can be partitioned into $\vartheta = 161$ non-overlapping cliques, with an average size of such a clique being 1.678. There also is a dominating set of 69 proteins, for which it holds that all other proteins interact with at least protein of this set.

For network $elegans$, we have that its maximum clique size is 3 and there are 39 triangles representing maximum cliques. It can be partitioned into 294 non-overlapping cliques. The average size of such a clique is 1.276, which makes it the network with the smallest average clique size in minimum vertex clique

Table 1. Experimental results obtained for PPI and PF networks by using only greedy algorithms (i.e. without steps 10 and 12 in Algorithm 1).

G	ω	χ	α	ϑ	γ
PPI networks					
ecoli [44]	5–6	5–6	160–161	160–161	20–69
$n = 270$, $m = 716$, $c = 20$, $\tau = 478$					
elegans [45]	3	3	293–294	293–294	20–71
$n = 375$, $m = 405$, $c = 20$, $\tau = 13$					
helico [45]	*3*	3–4	521–528	521–528	33–163
$n = 732$, $m = 1403$, $c = 16$, $\tau = 76$					
yeast [45]	3–8	3–8	2641–2673	2641–2673	146–959
$n = 4142$, $m = 7839$, $c = 99$, $\tau = 1562$					
PF networks					
beta3s.reduced [34]	37–39	37–39	229–301	229–301	11–70
$n = 1287$, $m = 23948$, $c = 1$, $\tau = 219165$					
beta3s [34]		37–39	37–39	64053-n	64053-n	5375–40323
$n = 132168$, $m = 228967$, $c = 2$, $\tau = 241209$					

Table 2. Experimental results obtained for PPI and PF networks by using the full stochastic approach, including IG and RLS algorithms (i.e. full Algorithm 1, including steps 10 and 12). Bold values represent instances, for which IG and RLS provided improved results compared to purely greedy algorithms.

G	ω	χ	α	ϑ	γ
PF networks					
ecoli [44]	**6**	**6**	**161**	**161**	20–69
$n = 270$, $m = 716$, $c = 20$, $\tau = 478$					
elegans [45]	3	3	**294**	**294**	20–71
$n = 375$, $m = 405$, $c = 20$, $\tau = 13$					
helico [45]	*3*	3–4	**528**	**528**	33–163
$n = 732$, $m = 1403$, $c = 16$, $\tau = 76$					
yeast [45]	**7**	**7**	**2673**	**2673**	146–959
$n = 4142$, $m = 7839$, $c = 99$, $\tau = 1562$					
PF networks					
beta3s.reduced [34]	**38**–39	**38**–39	**259–282**	**259–282**	11–70
$n = 1287$, $m = 23948$, $c = 1$, $\tau = 219165$					
beta3s [34]		**38**–39	**38**–39	**64497–69667**	**64497–69667**	5375–40323
$n = 132168$, $m = 228967$, $c = 2$, $\tau = 241209$					

covering. This is understandable, since this network is the sparsest. It contains a dominating set consisting of 71 vertices.

For network *helico*, we obtained a clique of size 3, while we were only able to find a 4-colouring. This is the only PPI network, for which we obtained a gap between an estimate for maximum clique size and chromatic number. Using enumeration, we found that there is no 4-clique and the number of triangles of mutually interacting proteins is 76. However, this confirms that while chromatic number can be used as a good upper bound on the size of the maximum clique of mutually interacting proteins, it seems that one cannot guarantee that these values for PPI networks will be equal. Network *helico* can be partitioned into 528 non-overlapping cliques of average size 1.386. It also contains a dominating set of size 164.

Instance *yeast* contains 1872 maximal cliques, which is the largest number of maximal cliques among the studied PPI networks. However, only 12 of them are also maximum cliques, which contain 7 vertices. These will shortly be discussed below. Network *yeast* can be partitioned into 2673 non-overlapping cliques, which have average size 1.550. Dominating set on 959 vertices for this network is the largest among the PPI networks, too.

It is not surprising that numbers of maximal and maximum cliques, as well as the properties of non-overlapping and overlapping cliques seem to vary between different networks. Hence, it might be interesting to discuss the properties of large clique a bit further.

Table 3 presents a listing of 12 maximum cliques of size 7 in the *yeast* PPI network. One can notice that the first clique and the last two cliques consist of proteins, which are not present in other cliques. On the other hand, all other cliques represent extensions of clique $CEF1, SEC28, SET1, SFA1, SFB2$. This indicates that some interesting substructures might be relatively isolated, while other substructures form larger clusters. These structures can be modelled

Table 3. Listing of 12 maximum cliques of size 7 in *yeast* PPI network.

$ALD3, ALG2, ALG9, ANC1, ANP1, AOS1, APA1$
$CEF1, LIP5, MKK2, SEC28, SET1, SFA1, SFB2$
$CEF1, LIP5, MKK2, SEC28, SET2, SFA1, SFB2$
$CEF1, LIP5, PUB1, SEC28, SET1, SFA1, SFB2$
$CEF1, LIP5, PUB1, SEC28, SET2, SFA1, SFB2$
$CEF1, LSM3, MKK2, SEC28, SET1, SFA1, SFB2$
$CEF1, LSM3, MKK2, SEC28, SET2, SFA1, SFB2$
$CEF1, LSM3, PUB1, SEC28, SET1, SFA1, SFB2$
$CEF1, LSM3, PUB1, SEC28, SET2, SFA1, SFB2$
$CEF1, PUB1, SEC28, SET1, SFA1, SFB2, SIN3$
$GRE2, HIR2, HIR3, HIS6, HIS7, HIT1, HMG1$
$SPA2, YAP1802, YAP3, YAP5, YAP6, YAR003W, YAR009C$

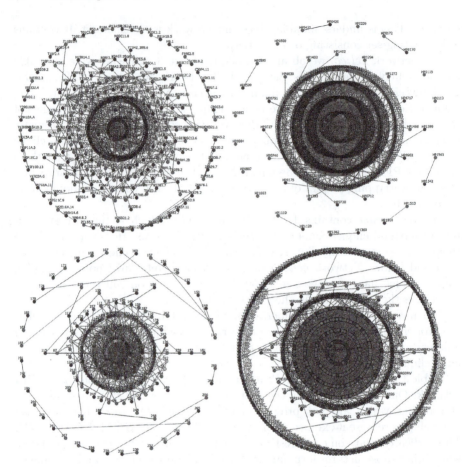

Fig. 1. Visualisation of colourings found for protein-protein interaction networks *elegans* (upper, on the left-hand side), *helico* (upper, on the right-hand side), *ecoli* (lower, on the left-hand side) and *yeast* (lower, on the right-hand side). These colourings represent good upper bounds for the size of maximum clique of mutually interacting proteins for these PPI networks. Based on the availability of protein labels and expected visual quality, labels or indices of vertices are presented for some of the networks and vertices.

e.g. by merging cliques [21]. Additionally, large cliques seem to comprise smaller cliques. This suggests that some PPI networks might have a hierarchical structure [47]. While functional modules are formed by groups of cliques, it seems that one can even identify smaller cliques as low-level building blocks of the network. Interestingly, while labels of proteins are naturally dependent on conventions of biologists, some of the identified maximum cliques seem to consist of proteins with lexicographically similar labels.

PF networks have slightly different characteristics. Network *beta3s.reduced* is atypical due to its reduced representation, which features a drastic cutoff in

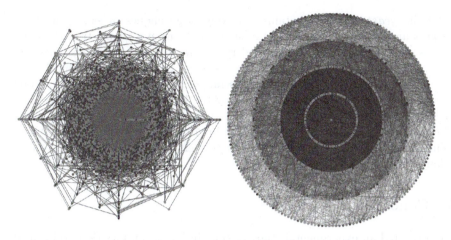

Fig. 2. Visualisation of *beta3s* protein folding network (on the left-hand side), which is the largest of studied networks, with over 10^5 vertices and its "core" *beta3s.reduced* (on the right-hand side). One can easily see that *beta3s.reduced* is the subgraph, which requires a high number of colours, while the visualisation of *beta3s* highlights mostly the three colours, which are used to colour most of the vertices in the "outer layer" of the network.

vertices with low degree. As a consequence, both *beta3s* and *beta3s.reduced* contain maximum clique of size 38–39, while the average size of a clique needed to partition *beta3s.reduced* into non-overlapping cliques is 4.564–4.969. This is a value range previously observed in variations of random graphs and graphs with planted cliques [8]. The original network *beta3s* requires cliques of size 1.897–2.049 to obtain a minimum vertex clique covering. However, it is worth mentioning that this value is still somewhat higher than the values obtained for PPI networks. This indicates a denser structure of PF networks with some large embedded cliques found in the "core" of the network. Such a phenomenon has not been observed in the studied PPI networks.

Summarising the above results, combinatorial optimisation properties seem to vary between different PPI networks. Comparison between a purely greedy and a stochastic approach confirms that stochastic optimisation techniques help in combinatorial optimisation for PPI and PF networks. Figure 1 indicates that to a certain extent, PPI networks seem to have similar structure. This figure shows colourings obtained for the PPI networks and groups vertices to layers, based on distance from a vertex with maximum degree. Visualisations reveal dense subgraphs in the proximity of the vertex with maximum degree, while this property seems to be more accentuated for large networks. Figure 2 presents a similar visualisation for the PF networks. This visualisation reveals slightly "layered" structure of *beta3s.reduced*. In this context, it is not surprising that large cliques are located within *beta3s.reduced*, while the "outer" layer of *beta3s* is sparser and seems to contain much smaller cliques.

While *yeast* network contains relatively large cliques of size 7, networks *elegans* and *helico* do not contain a clique larger than a simple triangle. Large cliques may both heavily overlap and represent relatively "isolated" substructures. Properties of cliques in network *yeast* seem to indicate hierarchical structure. For this purpose, data reductions might represent a promising research direction. A specific case is represented by the large *beta3s* PF network, which might further be studied in this context, too. Dominating sets were explored using an approximation algorithm. More interesting results might be obtained using a nature-inspired heuristic for this problem, e.g. algorithms based on ant colony optimisation [48].

5   Conclusions

We presented an experimental study of combinatorial optimisation problems in protein-protein interaction (PPI) and protein folding (PF) networks. Studied problems included maximum clique, chromatic number, maximum independent set, minimum vertex clique covering and minimum dominating set. We presented a unified technique to estimate these properties of large networks, which lead to NP-hard problems in general. Our experimental approach revealed several interesting properties of four PPI networks of the European COSIN project, as well as PF network *beta3s* and its reduced version. Even though the approach was applied to biological networks, its ideas are general and can also be used to analyse other complex networks, such as social networks or research citation networks.

Our investigation found maximum cliques for all PPI networks and provided a very small interval for the maximum clique of PF network *beta3s*. For all four PPI networks, we found the optimal solution to the problem of their partitioning into non-overlapping cliques. We confronted our method with the use of stochastic elements of iterated greedy (IG) and randomised local search (RLS) algorithms to its variant without the elements of stochastic optimisation. This confrontation revealed that stochastic optimisation approaches provide results of better quality for maximum clique, chromatic number, maximum independent set and minimum vertex clique covering.

Overlapping cliques were investigated using enumerative methods, too. This investigation suggests that some of the studied PPI networks have a hierarchical structure, with large overlapping cliques possibly consisting of smaller cliques. We also identified the dominating sets of these networks. In the context of PPI networks, these are the sets of "central" proteins such that all other proteins interact with at least one protein of the dominating set.

We believe that this approach might be beneficial especially in exploration of new biological networks. Most of the studied problems are closely related to functional module detection. However, unlike network clustering, studied characteristics are clearly defined and can be used as a systematic basis for further investigations.

References

1. Boyer, F., Morgat, A., Labarre, L., Pothier, J., Viari, A.: Syntons, metabolons and interactons: an exact graph-theoretical approach for exploring neighbourhood between genomic and functional data. Bioinformatics **21**(23), 4209–4215 (2005)
2. Gao, L., Sun, P., Song, J.: Clustering algorithms for detecting functional modules in protein interaction networks. J. Bioinform. Comput. Biol. **7**(1), 217–242 (2009)
3. Cohen, J.: Bioinformatics - an introduction for computer scientists. ACM Comput. Surv. **36**(2), 122–158 (2004)
4. Cormen, T.H., Leiserson, C.E., Rivest, R.L., Stein, C.: Introduction to Algorithms, 3rd edn. MIT Press, Cambridge (2009)
5. Karp, R.M.: Reducibility among combinatorial problems. In: Miller, R., Thatcher, J. (eds.) Proceedings of a Symposium on the Complexity of Computer Computations, pp. 85–103. Plenum Press, New York (1972)
6. COSIN: Coevolution and Self-organization in Dynamical Networks. http://www.cosinproject.org/
7. Halldórsson, M.M., Radhakrishnan, J.: Greed is good: approximating independent sets in sparse and bounded-degree graphs. Algorithmica **18**(1), 145–163 (1997)
8. Chalupa, D.: Construction of near-optimal vertex clique covering for real-world networks. Computing and Informatics (to appear)
9. Brélaz, D.: New methods to color vertices of a graph. Commun. ACM **22**(4), 251–256 (1979)
10. Culberson, J.C., Luo, F.: Exploring the k-colorable landscape with iterated greedy. In: Johnson, D.S., Trick, M. (eds.) Cliques, Coloring and Satisfiability: Second DIMACS Implementation Challenge, pp. 245–284. American Mathematical Society, RI (1995)
11. Chvátal, V.: A greedy heuristic for the set-covering problem. Math. Oper. Res. **4**(3), 233–235 (1979)
12. Atias, N., Sharan, R.: Comparative analysis of protein networks: hard problems, practical solutions. Commun. ACM **55**(5), 88–97 (2012)
13. Kuchaiev, O., Stevanović, A., Hayes, W., Pržulj, N.: Graphcrunch 2: software tool for network modeling, alignment and clustering. BMC Bioinf. **12**(1), 24 (2011)
14. Gavin, A.C., Aloy, P., Grandi, P., Krause, R., Boesche, M., Marzioch, M., Jensen, C.R.L.J., Bastuck, S., Dümpelfeld, B., et al.: Proteome survey reveals modularity of the yeast cell machinery. Nature **440**(7084), 631–636 (2006)
15. Zaki, N., Berengueres, J., Efimov, D.: Prorank: a method for detecting protein complexes. In: Proceedings of the 14th Annual Conference on Genetic and Evolutionary Computation, pp. 209–216. ACM (2012)
16. Li, X., Wu, M., Kwoh, C.K., Ng, S.K.: Computational approaches for detecting protein complexes from protein interaction networks: a survey. BMC Genomics **11**(Suppl. 1), S3 (2010)
17. Pizzuti, C., Rombo, S.E., Marchiori, E.: Complex detection in protein-protein interaction networks: a compact overview for researchers and practitioners. In: Giacobini, M., Vanneschi, L., Bush, W.S. (eds.) EvoBIO 2012. LNCS, vol. 7246, pp. 211–223. Springer, Heidelberg (2012)
18. Pizzuti, C., Rombo, S.E.: Algorithms and tools for protein-protein interaction networks clustering, with a special focus on population-based stochastic methods. Bioinformatics **30**(10), 1343–1352 (2014)
19. Becker, E., Robisson, B., Chapple, C.E., Guénoche, A., Brun, C.: Multifunctional proteins revealed by overlapping clustering in protein interaction network. Bioinformatics **28**(1), 84–90 (2006)

20. Adamcsek, B., Palla, G., Farkas, I.J., Derényi, I., Vicsek, T.: CFinder: locating cliques and overlapping modules in biological networks. Bioinformatics **22**(8), 1021–1023 (2012)
21. Li, X.L., Tan, S.H., Foo, C.S., Ng, S.K.: Interaction graph mining for protein complexes using local clique merging. Genome Inf. **16**(2), 260–269 (2005)
22. Girvan, M., Newman, M.E.J.: Community structure in social and biological networks. Proc. Nat. Acad. Sci. **99**(12), 7821–7826 (2002)
23. Cho, Y.R., Hwang, W., Zhang, A.: Identification of overlapping functional modules in protein interaction networks: information flow-based approach. In: Sixth IEEE International Conference on Data Mining Workshops, ICDM Workshops 2006, pp. 147–152 (2006)
24. Hawick, K.A.: Applying enumerative, spectral and hybrid graph analyses to biological network data. In: International Conference on Computational Intelligence and Bioinformatics (CIB 2011), pp. 89–96. IASTED, Pittsburgh, USA, 7–9 November 2011
25. Sun, J., Xie, Y., Zhang, H., Faloutsos, C.: Less is more: sparse graph mining with compact matrix decomposition. Stat. Anal. Data Min. **1**(1), 6–22 (2008)
26. King, A.D., Pržulj, N., Jurisica, I.: Protein complex prediction via cost-based clustering. Bioinformatics **20**(17), 3013–3020 (2004)
27. Pizzuti, C., Rombo, S.E.: Experimental evaluation of topological-based fitness functions to detect complexes in PPI networks. In: Proceedings of the 14th Annual Conference on Genetic and Evolutionary Computation, pp. 193–200. ACM (2012)
28. Pizzuti, C., Rombo, S.E.: A coclustering approach for mining large protein-protein interaction networks. IEEE/ACM Trans. Comput. Biol. Bioinf. (TCBB) **9**(3), 717–730 (2012)
29. Leskovec, J., Lang, K.J., Mahoney, M.W.: Empirical comparison of algorithms for network community detection. In Rappa, M., Jones, P., Freire, J., Chakrabarti, S. (eds.) Proceedings of the 19th International Conference on World Wide Web, WWW 2010, pp. 631–640. ACM, New York, NY (2010)
30. Pizzuti, C.: A multiobjective genetic algorithm to find communities in complex networks. IEEE Trans. Evol. Comput. **16**(3), 418–430 (2012)
31. Brohee, S., Helden, J.V.: Evaluation of clustering algorithms for protein-protein interaction networks. BMC Bioinf. **7**(1), 488 (2006)
32. Schaeffer, S.E.: Graph clustering. Comput. Sci. Rev. **1**(1), 27–64 (2007)
33. Šíma, J., Schaeffer, S.E.: On the NP-completeness of some graph cluster measures. In: Wiedermann, J., Tel, G., Pokorný, J., Bieliková, M., Štuller, J. (eds.) SOFSEM 2006. LNCS, vol. 3831, pp. 530–537. Springer, Heidelberg (2006)
34. Rao, F., Caflisch, A.: The protein folding network. J. Mol. Biol. **342**(1), 299–306 (2004)
35. Hawick, K.A.: Centrality metrics for comparing protein-protein interaction networks with synthesized NK systems. In: Proceedings of the IASTED International Conference on Biomedical Engineering, pp. 1–8. IASTED, Zurich, Switzerland, 23–25 June 2014
36. Albert, R., Barabási, A.L.: Statistical mechanics of complex networks. Rev. Mod. Phys. **74**(1), 47–97 (2002)
37. Wu, Q., Hao, J.K.: A review on algorithms for maximum clique problems. Eur. J. Oper. Res. **242**(3), 693–709 (2015)
38. Welsh, D.J.A., Powell, M.B.: An upper bound for the chromatic number of a graph and its application to timetabling problems. Comput. J. **10**(1), 85–86 (1967)

39. Galinier, P., Hamiez, J.-P., Hao, J.-K., Porumbel, D.: Recent advances in graph vertex coloring. In: Zelinka, I., Snasel, V., Abraham, A. (eds.) Handbook of Optimization: From Classical to Modern Approach. ISRL, vol. 38, pp. 505–528. Springer, Heidelberg (2013)

40. Morgenstern, C.: Improved implementations of dynamic sequential coloring algorithms. Technical report CoSc-91-4, Department of Computer Science, Texas Christian University (1991)

41. Turner, J.S.: Almost all k-colorable graphs are easy to color. J. Algorithms **9**(1), 63–82 (1988)

42. Culberson, J.C.: Iterated greedy graph coloring and the difficulty landscape. Technical report TR92-07, University of Alberta (1992)

43. Pavlopoulos, G.A., Secrier, M., Moschopoulos, C.N., Soldatos, T.G., Kossida, S., Aerts, J., Schneider, R., Bagos, P.G., et al.: Using graph theory to analyze biological networks. BioData Min. **4**(10), 1–27 (2011)

44. Butland, G., Peregrín-Alvarez, J.M., Li, J., Yang, W., Yang, X., Canadien, V., Starostine, A., Richards, D., Beattie, B., Krogan, N., Davey, M., Parkinson, J., Greenblatt, J., Emili, A.: Interaction network containing conserved and essential protein complexes in Escherichia coli. Nature **433**, 531–537 (2005)

45. UCLA: Database of Interacting Proteins. http://dip.doe-mbi.ucla.edu/dip/Main.cgi

46. Eppstein, D., Löffler, M., Strash, D.: Listing all maximal cliques in sparse graphs in near-optimal time. In: Cheong, O., Chwa, K.-Y., Park, K. (eds.) ISAAC 2010, Part I. LNCS, vol. 6506, pp. 403–414. Springer, Heidelberg (2010)

47. Kovács, I.A., Palotai, R., Szalay, M.S., Csermely, P.: Community landscapes: an integrative approach to determine overlapping network module hierarchy, identify key nodes and predict network dynamics. PloS ONE **5**(9), e12528 (2010)

48. Potluri, A., Singh, A.: Two hybrid meta-heuristic approaches for minimum dominating set problem. In: Panigrahi, B.K., Suganthan, P.N., Das, S., Satapathy, S.C. (eds.) SEMCCO 2011, Part II. LNCS, vol. 7077, pp. 97–104. Springer, Heidelberg (2011)

A Multi-objective Genetic Programming Biomarker Detection Approach in Mass Spectrometry Data

Soha Ahmed[1], Mengjie Zhang[1], Lifeng Peng[2], and Bing Xue[1(✉)]

[1] School of Engineering and Computer Science, Victoria University of Wellington, PO Box 600, Wellington 6140, New Zealand
{soha.ahmed,mengjie.zhang,bing.xue}@ecs.vuw.ac.nz
[2] School of Biological Sciences, Victoria University of Wellington, PO Box 600, Wellington 6140, New Zealand
lifeng.peng@vuw.ac.nz

Abstract. Mass spectrometry is currently the most commonly used technology in biochemical research for proteomic analysis. The main goal of proteomic profiling using mass spectrometry is the classification of samples from different clinical states. This requires the identification of proteins or peptides (biomarkers) that are expressed differentially between different clinical states. However, due to the high dimensionality of the data and the small number of samples, classification of mass spectrometry data is a challenging task. Therefore, an effective feature manipulation algorithm either through feature selection or construction is needed to enhance the classification performance and at the same time minimise the number of features. Most of the feature manipulation methods for mass spectrometry data treat this problem as a single objective task which focuses on improving the classification performance. This paper presents two new methods for biomarker detection through multi-objective feature selection and feature construction. The results show that the proposed multi-objective feature selection method can obtain better subsets of features than the single-objective algorithm and two traditional multi-objective approaches for feature selection. Moreover, the multi-objective feature construction algorithm further improves the performance over the multi-objective feature selection algorithm. This paper is the first multi-objective genetic programming approach for biomarker detection in mass spectrometry data.

1 Introduction

Nowadays, much attention is given to the high-throughput mass spectrometry (MS) technology in proteomics. MS enables the detection and the discrimination of patterns between diseased and healthy samples of complex mixtures of proteins [1,2]. MS datasets typically consist of tens or thousands of mass to charge (m/z) ratios. Each m/z value corresponds to a mass of a certain peptide and reflects the abundance of this peptide through an intensity value [3]. From

© Springer International Publishing Switzerland 2016
G. Squillero and P. Burelli (Eds.): EvoApplications 2016, Part I, LNCS 9597, pp. 106–122, 2016.
DOI: 10.1007/978-3-319-31204-0_8

machine learning perspective, each of the abundances of the peptides is a feature for classification. This causes the critical issue of *curse of dimensionality*, which leads to the degradation of classification performance due to the large number of features and the small number of examples.

Feature manipulation can help solving the biomarker detection problem [1]. It provides means to transform the representation of the input to a classification algorithm to improve its performance [4]. Feature manipulation consists of *feature selection* and *feature construction*.

While feature selection aims at selecting a subset of relevant original features, feature construction aims at generating new high-level features. Generally, feature selection and construction methods can be divided into filter, wrapper or embedded approaches [4,5].

In filter approaches, features are evaluated with some relevance measure such as, t-statistics [6] and mutual information [7]. Although the filter approach is efficient in terms of computational cost, most of the features selected by the filter approach are still correlated [3]. Therefore, features are mostly redundant and include some sort of data noise, which leads to the reduction of their effectiveness in terms of classification accuracy. In wrapper approaches, an inductive algorithm (mostly a classifier) is wrapped as an evaluation criterion to the selected features. Although a wrapper approach is more effective than a filter approach, its computational cost is a major obstacle to the use of this approach. Moreover, in high feature-to-sample ratio data, the wrapper approach may face the problem of overfitting [3]. The embedded approach also uses an inductive algorithm but the main difference from the wrapper approach is that the inductive algorithm is used for both feature selection and classification. Therefore, embedded approaches can overcome the disadvantages of wrapper approaches.

Genetic programming (GP) is an algorithm which, inspired by natural evolution, searches for good solutions in a population of programs. GP proved to be an effective technique for feature selection, feature construction and classification especially for high dimensional data [8].

Many feature selection techniques have been proposed to detect the potential biomarkers in MS data [9,10]. Despite the great promise of the previously proposed methods, none of these methods considered the number of features as an important objective to optimise. Although some studies considered the relative importance of the number of features to classification accuracy [11,12], the major limitation of these approaches is the prior specification of the weighting factor of each objective. More related work can be seen from [13–15], which are not detailed here due to the page limit.

Multi-objective optimisation offers solutions to the optimisation of different conflicting objectives. Multi-objective optimization is evaluated in terms of the trade-off between the conflicting objectives, which have to be minimised or maximised.

Biomarker detection must consider the trade-off between the classification performance and the number of features. The number of features should be as small as possible to be able to pass them to experimental validation. Therefore, for evaluation of biomarker detection, two objectives should be considered, which are maximizing the classification performance and at the same time minimising the number of features. This paper represents the first attempt to use GP as a multi-objective approach to biomarker detection.

1.1   Goals:

The overall goal of this paper is to develop GP-based multi-objective feature selection and construction approaches to classification of MS data. In feature selection, the proposed GP method uses the ideas of NSGAII [16] and SPEA2 [17] to evolve models that keep the balance between the conflicting objectives. We notate these methods as NS-$GPMOFS$ and SP-$GPMOFS$. The main goal here is to evolve a Pareto front of non-dominated solutions, which include a small number of selected original features and achieve a better classification accuracy than using the whole set of features.

In feature construction, a single evolved tree is used to construct multiple features to replace the original features by combining them using the GP functions. Multi-objective optimisation is used to reduce the number of constructed features while keeping the high classification accuracy. We notate these methods as NS-$GPMOFC$ and SP-$GPMOFC$.

In both approaches, an embedded approach is used to take advantages of the low computational cost and better classification accuracy.

Precisely, we will investigate the followings:

- whether using GP as a multi-objective approach to feature selection can produce better solutions than using the single objective GP algorithm,
- whether using multi-objective GP feature selection methods (NS-$GPMOFS$ and SP-$GPMOFS$) can select feature subsets that improve the classification performance and reduce the number of features than using the traditional multi-objective algorithms ($NSGAII$ and $SPEA2$), and
- whether the GP-based multi-objective feature construction methods (NS-$GPMOFC$ and SP-$GPMOFC$) can further improve the performance over the multi-objective GP feature selection methods (NS-$GPMOFS$ and SP-$GPMOFS$).

1.2   Organisation

The rest of the paper is organised as follows. Section 2 describes the GP-based multi-objective feature selection and the GP-based multi-objective feature construction approaches. Section 3 describes the experimental design that includes the settings and the MS datasets used. Section 4 presents the experimental results and discussions. Section 5 concludes the paper.

2   New GP Multi-objective Approaches

This section describes the two multi-objective GP approaches.

2.1   The GP Multi-objective Feature Selection Approach

In this section, we propose a new approach to feature selection for MS data with the aim of biomarker detection using multi-objective GP, with two main objectives to explore the Pareto front of feature subsets. The objectives here are maximising the classification accuracy and minimising the number of features used in each individual of the population. As mentioned earlier, an embedded approach is taken in the proposed algorithm. GP is employed here as a classifier as well, and the number of correctly classified instances in the training set is stored in an external archive. The classification accuracy is used to assess the first objective. The second objective here is to minimise the cardinality of the selected features (number of features selected automatically in the GP tree). When a new solution is evolved, it is compared to the other solutions stored in the archive. If the evolved solution is not worse in both objectives and it is better than a solution in the list in at least one of the objectives, it will dominate that solution. Pareto optimal contains the set of non-dominated solutions where a specific solution can not improve any of the objectives without degrading at least one of the other conflicting objectives [18]. The non-dominated solution forms the Pareto front in which no solution can be judged better than the others.

2.1.1   Pareto Fitness Assignment in *NS-GPMOFS* and *SP-GPMOFS*

In evolutionary multi-objective optimisation, solutions are usually ranked according to their performance on the different objectives to measure the Pareto dominance. The Pareto dominance is measured through the dominance rank or dominance count [17] (or both) of a certain solution. Dominance rank of a solution is the number of solutions that dominates this solution, while the dominance count is the number of solutions that a given solution dominates. A solution with a smaller number of solutions that dominate it (lower rank) and a higher count is a better solution.

We investigate two mechanisms to measure the Pareto fitness. The first uses the dominance rank of a solution S_i for evaluating the fitness which is similar to the idea of NSGAII [16], i.e., the number of other solutions in the population that dominate S_i, and we call this method as *NS-GPMOFS*.

Similar to SPEA2 [17], the second mechanism uses both dominance rank and dominance count in the Pareto refined fitness, and this method is named as *SP-GPMOFS*.

2.1.2   Crowding Distance

In addition to the previously mentioned Pareto dominance measures used in the fitness, a crowding distance measure is used to generate more diversity among the population [19]. The crowding distance used is the Manhattan distance between the solutions. This distance measure is used only when two or more solutions have the same Pareto dominance measures, which means that if solutions have an equal rank, then the solution with smaller crowding distance is selected. The crowding distance is the average distance between the two solutions with each of the objectives, where a lower distance indicates a better result.

2.1.3 *NS-GPMOFS* and *SP-GPMOFS* Algorithms

Algorithm 1 shows the pseudocode of *GPMOFS* algorithms.

The input is D, the dataset, and the output is the Pareto front archive of solutions (PF). At each generation, the parent and offspring populations are merged. The fittest individuals (according to the two objectives) in this merged population acts as the new population ($CHILD$) in the next generation. The population is reduced to size N (original size of the population) using dominance rank and crowding distance for *NS-GPMOFS*. While for *SP-GPMOFS* dominance rank, dominance count and the crowding distance are measured. The size of $CHILD$ is the same as the size of the original population and it is produced using the traditional genetic operators (crossover and mutation operators). In case of *SP-GPMOFS*, the size of PF (Pareto front solutions) is kept fixed while in *NS-GPMOFS* it does not have a specific size. Another difference between using *NS-GPMOFS* and *SP-GPMOFS* is the use of elitism in *SP-GPMOFS*, which is not used in *NS-GPMOFS*. The non-dominated solutions in $CHILD$ are identified and copied to PF. These steps are repeated until the maximum number of generations is reached. At the end of the evolutionary search, the solutions of PF are used to project the datasets and passed for evaluation. The evaluation is done through both classification accuracy and the number of features used in each solution in the archive.

Algorithm 1. Pseudo-Code of *NS-GPMOFS* and *SP-GPMOFS*

Require: D, a dataset that contains a vector of instances with m original features.
Ensure: PF, a Pareto front (PF) of a set of solutions (low-level features).
    begin
    Divide D into training and test sets.
    Initialise the population (P)
  while Maximum generation is not reached **do**
        Evaluate the two objectives of each individual   { // Acc, $|F|$}
        Select the individuals using the selection method
        Generate new population ($CHILD$) using the genetic operators
        if *NS-GPMOFS* is used **then**
            Non-dominated sorting of the individuals based on ranking and the crowding distance
      else if *SP-GPMOFS* **then**
            evaluate the individuals based on ranking, count, and the crowding distance
      end if
        Copy both $CHILD$ and P to *Archive*
        Identify the individuals who have non-dominated solutions in *Archive* and add to Pareto front (PF)
        Select a population of size N based upon ranking and crowding distance
        Generate new population ($CHILD$) using the genetic operators
  end while
    Use the solutions in PF to project the test set
    Calculate the test set classification accuracy of the different solutions
    Calculate the number of selected features in each solution in PF
    return a vector S that contain the number of features and classification accuracy of each solution
  in PF

2.2 The GP Multi-objective Feature Construction Approach

GPMOFC constructs new high-level features from the original features [20] (resulted features from the tree branches). In addition to the features constructed from the branches (all the internal function nodes), the final feature constructed from the root node of the tree is also used.

2.2.1 *SP-GPMOFC* and *NS-GPMOFC* Algorithm

Algorithm 2 describes the pseudocode of *SP-GPMOFC* and *NS-GPMOFC*. The two algorithms are similar to the feature selection algorithms (*SP-GPMOFS* and *NS-GPMOFS*) except for the feature sets. The difference between the two algorithms for feature selection and construction is that instead of using the original features selected in *SP-GPMOFS* and *NS-GPMOFS*, high-level features are constructed to optimise the two objectives in *SP-GPMOFC* and *NS-GPMOFC*.

Algorithm 2. Algorithm of *NS-GPMOFC* and *SP-GPMOFC*

Require: D, a dataset that contains a vector of instances with m original features.
Ensure: PF, A Pareto front (PF) (solutions with high-level features).
    begin
    Divide D into 50 % for training and 50 % testing.
    Randomly Initialise the population (P)
    Save the high-level features resulting from the branches and the root of the individual tree
  while Maximum generation is not reached **do do**
        Evaluate the number of constructed features and *Acc* of each individual
        Select the individuals using the selection method
        Generate new population ($CHILD$) using the genetic operators
      if *NS-GPMOFC* **then**
            Non-dominated sorting of the individuals based on ranking and the crowding distance
      else if *SP-GPMOFC* **then**
            Non-dominated sorting of the individuals based on ranking, count and the crowding distance
      end if
        Copy both $CHILD$ and P to $Archive$
        Identify the individuals who have non-dominated solutions in $Archive$ and add to Pareto front (PF)
        Select a population of size N based upon ranking and crowding distance
        Generate new population ($CHILD$) using the genetic operators
  end while
    Use the solutions in PF to project test set
    Calculate the test set classification accuracy of the different solutions
    Calculate the number of high-level features in each solution in PF
    return a vector S that contain the number of high-level features and classification accuracy of each solution in PF
    end

2.3 Summary of the Two Systems

As shown in Fig. 1, after preprocessing of the MS spectra datasets, the system for *GPMOFS* or *GPMOFC* starts by dividing the dataset into training and test sets. Each program in the population uses a subset of features in its tree terminal nodes and generates the objective value. The objective value (classification accuracy) is measured by GP individual classifier's accuracy that is passed as a

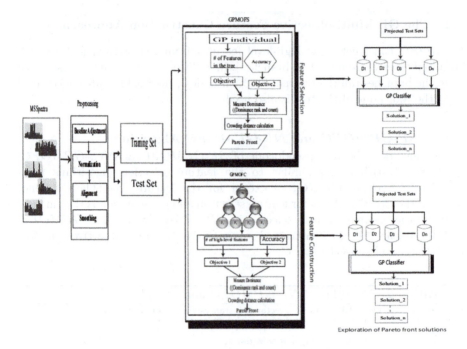

Fig. 1. General overview of the multi-objective approaches

objective value to measure the pareto dominance. Dominance rank, dominance count and crowding distance are used to measure the fitness of the solutions. After the objective calculation, the objective value of each solution is compared to the Pareto front archive. If the solution in the archive is dominated by the new solution, the new solution will replace it in the archive. Each solution in the Pareto front has a subset of features that were selected in the terminal nodes. The Pareto front solutions are used to project the datasets, therefore if the size of the archive is n, there will be n projected datasets. To test the subsets of features, the test set is evaluated using GP classifier. As explained earlier, the main difference between $GPMOFS$ and $GPMOFC$ is the use of the low-level or the high-level features.

2.3.1 Objective Functions

For both $GPMOFS$ and $GPMOFC$, the first objective is to maximise the classification accuracy (Acc). The second objective used is to minimise the number of features selected or constructed by each GP tree in the terminal nodes $|F|$.

Acc is defined as:

$$Acc = \frac{TP + TN}{TP + TN + FP + FN}$$

where TP, TN, FP and FN are the true positives, true negatives, false positives, and false negatives, respectively. For each instance of the training set, if the output of the program is less than or equal to zero then the instance is classified as class 1, otherwise it is classified as class 2.

3 Experiment Design

This section explains the MS datasets used to test *GPMOFS* and *GPMOFC*, GP operators and parameters, benchmark algorithms used for comparison, and the evaluation criteria.

3.1 MS Datasets

To test the effectiveness of the proposed GP multi-objective approaches, eight different MS datasets are used.

- OVA1 and OVA2 [21]: OVA1 is composed of 216 spectra where 121 spectra are cancerous samples and 95 spectra are healthy ones, while OVA2 consists of 253 spectra with 162 spectrum are cancerous samples and 91 are healthy samples. The number of features is 15000 and 15154 in OVA1 and OVA 2, respectively.
- PAN [22]: The dataset has 181 spectral examples, where 80 are in the affected class and 101 are in the healthy class. The number of features in each spectrum is 6771.
- TOX [23]: The dataset consists of 62 spectra (28 in the positive and 34 in the negative class) and each spectrum has 45200 features.
- HCC [24]: HCC has 150 spectra (78 affected and 72 non-affected) with 36802 features in each spectrum.
- DGB [25]. This dataset contains three groups of samples (78 healthy, 25 hepatocellular carcinoma and 25 chronic liver diseased). The total number of features is 16075.
- Pros dataset [26]: This dataset is composed of four classes which are: Healthy (63 samples), Benign stage$_1$ (190 samples), Prostate Cancer stage$_2$ (26 samples) and Prostate Cancer stage$_3$ (43 samples). The number of features in Pros is 15000. For DGB and Pros datasets, we used only two classes of instances.
- Appleminus: This dataset is composed of 365 features with ten instances of each class. Three classes contain five predefined biomarkers, and the last class is not spiked-in. Only one of the spiked-in classes and the non-spiked class are used in our algorithms.

Several preprocessing steps were applied to each of the datasets. The preprocessing of MS data is important to convert the data to a homogeneous matrix which can be used for feature selection and classification of the data [25]. The preprocessing steps used in our experiments include baseline adjustment, spectrum normalisation, alignment and filtering with different parameters for each dataset. The baseline removal is used to remove the low-range noise. The baseline is estimated by passing a window on the spectra and the minimum m/z values

Table 1. Preprocessing parameters

	OVA1 and OVA2	TOX	PAN	HCC	DGB	Pros
Window size for baseline removal	500	-	200	50	200	-
Smoothing frame size	5	6	3	6	6	3
Maximum intensity after normalisation	300					

are calculated. A piecewise linear interpolation method is used for the regression of the baseline. To make the intensity values range the same, normalisation is performed. The normalisation of the spectra is done by calculating the area under the curve [23] and rescaling the spectra to have a maximum intensity value of 300. This is done by using the *msnorm* function in the Matlab toolbox [27]. After normalisation is performed, alignment of the peaks is performed to match the similar peaks across all the spectra. Finally, smoothing of the spectra is done to remove the low signal fluctuation. Smoothing is done via a Savitzky-Golay filter. Pros and TOX datasets were already baseline adjusted. Therefore, both of the datasets were only filtered and normalised. Table 1 shows the running parameters of the preprocessing steps used with each of the datasets. The parameters are selected based on the original papers of the datasets [22–25]. The spike-in Appleminus dataset is available in NetCDF format, and it is preprocessed using XCMS [28] with the settings described in [29].

3.2   Performance Evaluation

GP as a classifier is used to test the selected features in each solution in the archive on the test sets. The performance is evaluated according to both the classification accuracy of the test set and the number of features.

3.3   Terminal Set, Function Set, Genetic Operators and Parameters

In the experiments, we adopt standard tree-based GP, which produces a single floating point number [30] for each instance in the dataset. Each of the output values is then used to determine the classification accuracy of the genetic program. The initial population is generated using the ramped-half-and-half method [31]. The function set consists of the four standard elementary mathematical operators $\{+, -, \%, \times\}$ and also a square root $\sqrt{}$ operator. The $\%$ and $\sqrt{}$ are "protected" where $\%$ returns zero for division by zero and $\sqrt{}$ returns zero for negative numbers. The terminal set has only variable terminals that are the feature values. The population size is set to 1024. Crossover and mutation probabilities are 0.8, and 0.2, respectively, and tournament selection is used with the size of 7. The GP, NSGAII and SPEA2 implementations used in the experiments are based on the Evolutionary Computing Java-based (ECJ) package [32]. Other parameters for NSGAII and SPEA2 are set as the default values in the ECJ library. The evolution terminates at a maximum number of 20 generations.

For each dataset, the experiment is repeated for 30 independent runs with 30 different random seeds. Each run outputs a set of non-dominated solutions in the Pareto front. The 30 sets of non-dominated solutions from the 30 runs are combined to one set by removing the dominated solutions from the different sets.

3.4 Benchmark Algorithms

GPMOFS and *GPMOFC* are compared to the following benchmark algorithms:

1. Standard (Single-Objective) GP is the standard GP classification framework using the overall classification accuracy as a single objective to maximise. The features selected in the terminal nodes of the tree are treated as the selected features.
2. NSGAII: Multi-objective optimisation using NSGAII and Fisher criterion based class separability for feature selection [33]. The evaluation is done through both the higher Fisher criterion and the smaller number of features. The first objective which is maximising the Fisher criterion or the class separability, that is defined as,

$$\text{Fitness function} = \text{Fisher criterion} = \sum_{n=1}^{N} |\frac{\mu_i - \mu_j}{\sigma_i^2 - \sigma_j^2}|$$

where μ_i and μ_j are the means, σ_i^2 and σ_j^2 are the variances of the samples which belong to class i and class j, respectively. N is the number of samples in the training set. The second objective is minimising the number of features.
3. SPEA2: Multi-objective optimisation using SPEA2 where the first objective is to maximise Fisher criterion and the second objective is to minimise the number of features.

Similar to *GPMOFS* and *GPMOFC*, the population size is set to 1024 and the number of generations is 20. For both NSGAII and SPEA2, each individual is encoded as a binary vector. The length of the vector is equal to the total number of features in the dataset. Hence, if the bit is 1, this means that the feature is selected and if the bit is 0 the feature is not selected.

4   Results and Discussions

Figure 2 shows the results of *GPMOFS* compared to using the single objective GP method, and the SPEA2 and NSGAII, while Fig. 3 shows the results of *GPMOFC* compared to *GPMOFS*. The multi-objective methods have different numbers of non-dominated solutions. The results are the non-dominated solutions obtained from the 30 independent runs. The x-axis refers to the number of features selected by each method, whereas the y-axis indicates the classification accuracy. Each figure is divided into a number of sub-figures where each sub-figure represents the results of each dataset.

4.1   Performance of *GPMOFS*

It can be noticed from Fig. 2 that using *SP-GPMOFS* has the potential to evolve solutions, which have better classification performance and a smaller number of features than using *NS-GPMOFS* in seven out of the eight datasets. The proposed method also outperformed the single objective GP approach and the two benchmark multi-objective methods SPEA2 and NSGAII, on all the eight datasets. This supports our hypothesis that using multi-objective GP can improve the feature selection performance from both the classification accuracy and the number of features points of view.

In some cases, *NS-GPMOFS* and *SP-GPMOFS* have common solutions such as in the TOX, and HCC datasets during the left region of the front. Only in the TOX dataset, *NS-GPMOFS* evolves solutions at the right region of the frontier which have better accuracy, but the number of features in these solutions are larger. In the Appleminus dataset, *NS-GPMOFS* is the best followed by *SP-GPMOFS*. The single-objective GP method for the Appleminus dataset has evolved solutions with a large number of features and lower accuracy compared to the multi-objective approaches.

The multi-objective approaches SPEA2 and NSGAII for feature selection are both used with Fisher criterion for comparison to the proposed method. Comparing *NS-GPMOFS* and *SP-GPMOFS* with SPEA2 and NSGAII, it is clear that GP has improved the performance of both NSGAII and SPEA2 for feature selection. This can be explained by the GP capability to select the subsets of features that are more relevant to classification. Using multi-objective optimisation along with GP improves both objectives of reducing the number of features and having a better performance. This suggests that GP improves the capability of the multi-objective approaches through its ability to select the better subsets of features.

4.2   Comparison of *GPMOFS* and *GPMOFC*

Considering the experimental results of *GPMOFC* that are shown in Fig. 3, it can be noticed that the multi-objective feature construction is better than the multi-objective feature selection in most cases. For OVA1, *SP-GPMOFC* is the best with a smaller number of features. For PAN and HCC, feature construction approaches evolve better solutions than the feature selection algorithms. In dataset OVA2, *SP-GPMOFC* is equivalent to *SP-GPMOFS* and it outperforms *NS-GPMOFS*.

The results suggest that multi-objective feature construction tends to achieve the balance between reducing the dimensionality and improving the performance better than multi-objective feature selection. This supports our first hypothesis that feature construction can further improve the multi-objective feature manipulation performance through the construction of high-level features that identify the interactions and relations between the original low-level features.

The exceptions to the observaton mentioned above that multi-objective feature construction can achieve better results than the multi-objective feature

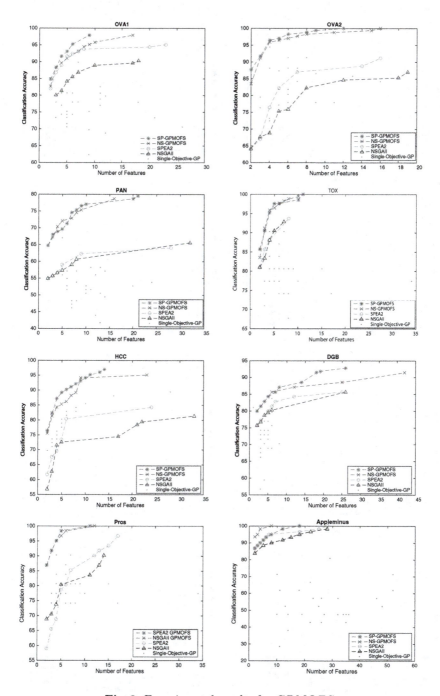

Fig. 2. Experimental results for *GPMOFS*.

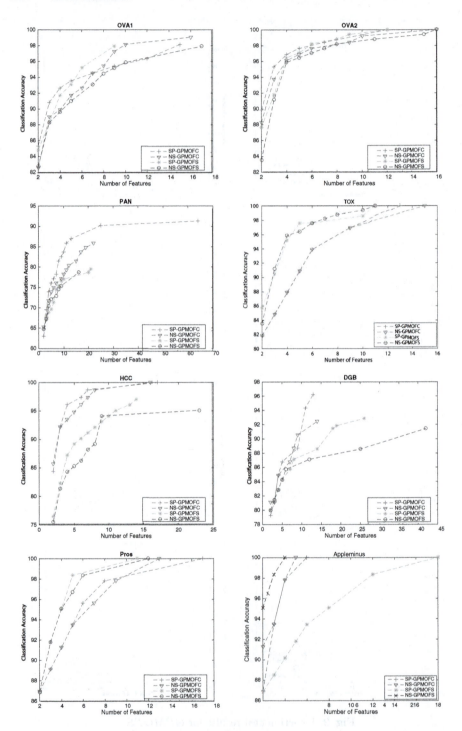

Fig. 3. Experimental results for *GPMOFC*.

selection on these datasets are the TOX and Pros datasets. For Appleminus dataset, NS-$GPMOFS$ has the best set of solutions and the two feature construction methods come next. For these two datasets, $GPMOFS$ is better than $GPMOFC$. $GPMOFC$ tries to reduce the number of constructed features and decreases the dimensionality better than $GPMOFS$ in these two datasets, but this came on the account of the classification performance. However, the gap between the selection and construction is very small. Both selection and construction can achieve 100 % accuracy with a number of features of 10–12 for feature selection and 12–16 for feature construction, from over 15,000 features in Pros and 45,000 in TOX.

4.3 Comparison of $GPMOFC$ to Single Objective GP, SPEA2 and NSGAII Approaches

Comparing Figs. 2 and 3, $GPMOFC$ is outperforming both SPEA2 and NSGAII in all the cases. If the results of $GPMOFC$ and the single objective GP are compared, it is also clear the multi-objective construction is better on all the tasks.

This indicates the increased effectiveness of using the high-level features over the selected original features, and gives more credibility to GP as a feature construction approach.

4.4 Biomarker Detection

We tested the features selected from the Appleminus dataset to check the number of detected predefined biomarkers by each method. Table 2 shows the selection status of the biomarker by each of the multi-objective feature selection methods. It can be noticed from Table 2 that SP-$GPMOFS$ has outperformed the other three methods and managed to detect the five predefined biomarkers. SPEA2 detected four predefined biomarkers while both NS-$GPMOFS$ and $NSGAII$ detected three out of the five predefined biomarkers. This suggests that SP-$GPMOFS$ has better performance in terms of biomarker detection as well as higher accuracy solutions with a smaller number of features

Table 2. Identified spike-in biomarkers by SP-$GPMOFS$, NS-$GPMOFS$, $SPEA2$ and $NSGAII$

m/z value	SP-$GPMOFS$	NS-$GPMOFS$	$SPEA2$	$NSGAII$
(5 Biomarkers) 463.0	✓	✓	✓	✗
447.09	✓	✓	✓	✓
273.03	✓	✓	✓	✓
435.13	✓	✗	✗	✗
227.07	✓	✗	✓	✓

5 Conclusions

This paper proposes the first multi-objective biomarker detection approach for MS data. Moreover, the paper also presents the first multi-objective feature construction algorithm that is applied to MS data.

In Sect. 2 of the paper, *GPMOFS*, the proposed GP multi-objective feature selection method manages the trade-off between the classification accuracy and the cardinality of features. According to the results, *GPMOFS* evolves non-dominated solutions, which has the potential to solve the problem of high dimensionality and a small number of examples in MS data. The method outperforms the single-objective feature selection GP method in terms of both objectives. The method uses the embedded capability of GP to select features with the dominance rank, dominance count and crowding distance to evaluate the solutions. The proposed method also outperforms both SPEA2 and NSGAII multi-objective feature selection approaches using Fisher criterion.

The second part of the paper presents *GPMOFC*, the first multi-objective feature construction method on MS data. For the construction of multiple high-level features, the features generated from the branches of the evolved GP tree in addition to the root features are used. This generates a number of new high-level features, which has the potential to improve the classification performance. To reduce the dimensionality by generating a smaller number of features, *GPMOFC* uses ideas from SPEA2 and NSGAII to keep the trade-off between the number of features and the classification performance. The results show that *GPMOFC* outperformed *GPMOFS* in almost all the cases, and that it was also better than SPEA2 and NSGAII approaches and the single objective GP feature selection method.

References

1. Morris, J.S., Coombes, K.R., Koomen, J., Baggerly, K.A., Kobayashi, R.: Feature extraction and quantification for mass spectrometry in biomedical applications using the mean spectrum. Bioinformatics **21**(9), 1764–1775 (2005)
2. Ahmed, S., Zhang, M., Peng, L., Xue, B.: Genetic programming for measuring peptide detectability. In: Dick, G., et al. (eds.) SEAL 2014. LNCS, vol. 8886, pp. 593–604. Springer, Heidelberg (2014)
3. Yang, P., Zhang, Z.: A clustering based hybrid system for mass spectrometry data analysis. In: Chetty, M., Ngom, A., Ahmad, S. (eds.) PRIB 2008. LNCS (LNBI), vol. 5265, pp. 98–109. Springer, Heidelberg (2008)
4. Liu, H., Motoda, H.: Feature Selection for Knowledge Discovery and Data Mining. Kluwer Academic Publishers, Norwell (1998)
5. Xue, B., Fu, W., Zhang, M.: Differential evolution (de) for multi-objective feature selection in classification. In: Proceedings of the 2014 Conference Companion on Genetic and Evolutionary Computation Companion, GECCO Comp 2014, pp. 83–84. ACM, New York (2014)
6. Golub, T.R., Slonim, D.K., Tamayo, P., Huard, C., Gaasenbeek, M., Mesirov, J.P., Coller, H., Loh, M.L., Downing, J.R., Caligiuri, M.A., Bloomfield, C.D., Lander, E.S.: Molecular classification of cancer: class discovery and class prediction by gene expression monitoring. Science **286**(5439), 531–537 (1999)

7. Peng, H., Long, F., Ding, C.: Feature selection based on mutual information criteria of max-dependency, max-relevance, and min-redundancy. IEEE Trans. Pattern Anal. Mach. Intell. **27**(8), 1226–1238 (2005)

8. Neshatian, K., Zhang, M.: Unsupervised elimination of redundant features using genetic programming. In: Nicholson, A., Li, X. (eds.) AI 2009. LNCS, vol. 5866, pp. 432–442. Springer, Heidelberg (2009)

9. Gertheiss, J., Tutz, G.: Supervised feature selection in mass spectrometry-based proteomic profiling by blockwise boosting. Bioinformatics **25**(8), 1076–1077 (2009)

10. Somnath, D.: Classification of breast cancer versus normal samples from mass spectrometry profiles using linear discriminant analysis of important features selected by random forest. Stat. Appl. Genet. Mol. Biol. **7**(2), 1–14 (2008)

11. Muni, D., Pal, N., Das, J.: Genetic programming for simultaneous feature selection and classifier design. IEEE Trans. Syst. Man Cybern. Part B: Cybern. **36**(1), 106–117 (2006)

12. Ahmed, S., Zhang, M., Peng, L.: Improving feature ranking for biomarker discovery in proteomics mass spectrometry data using genetic programming. Connection Sci., 1-29 (2014). doi:10.1080/09540091.2014.906388

13. Kourid, A., Batouche, M.: Biomarker discovery based on large-scale feature selection and MapReduce. In: Amine, A., Bellatreche, L., Elberrichi, Z., Neuhold, E.J., Wrembel, R. (eds.) Computer Science and Its Applications. IFIP AICT, vol. 456, pp. 81–92. Springer, Heidelberg (2015)

14. Duval, B., Hao, J.K.: Advances in metaheuristics for gene selection and classification of microarray data. Briefings Bioinform. **11**(1), 127–141 (2010)

15. Xue, B., Cervante, L., Shang, L., Browne, W.N., Zhang, M.: Binary PSO and rough set theory for feature selection: a multi-objective filter based approach. Int. J. Comput. Intell. Appl. **13**(2), 1450009 (2014)

16. Deb, K., Pratap, A., Agarwal, S., Meyarivan, T.: A fast elitist multi-objective genetic algorithm: NSGA-II. IEEE Trans. Evol. Comput. **6**, 182–197 (2000)

17. Zitzler, E., Laumanns, M., Thiele, L.: SPEA2: improving the strength pareto evolutionary algorithm for multiobjective optimization. In: Evolutionary Methods for Design, Optimisation, and Control, CIMNE, Barcelona, Spain, pp. 95–100 (2002)

18. Ngatchou, P., Zarei, A., El-Sharkawi, M.: Pareto multi objective optimization. In: Proceedings of the 13th International Conference on Intelligent Systems Application to Power Systems, pp. 84–91 (2005)

19. Bhowan, U., Johnston, M., Zhang, M., Yao, X.: Evolving diverse ensembles using genetic programming for classification with unbalanced data. IEEE Trans. Evol. Comput. **17**(3), 368–386 (2013)

20. Ahmed, S., Zhang, M., Peng, L., Xue, B.: Multiple feature construction for effective biomarker identification and classification using genetic programming. In: Proceedings of the 2014 Conference on Genetic and Evolutionary Computation, GECCO 2014, pp. 249–256. ACM, New York (2014)

21. Petricoin, E.F., Ardekani, A.M., Hitt, B.A., Levine, P.J., Fusaro, V.A., Steinberg, S.M., Mills, G.B., Simone, C., Fishman, D.A., Kohn, E.C., Liotta, L.A.: Use of proteomic patterns in serum to identify ovarian cancer. Lancet **359**, 572–577 (2002)

22. Hingorani, S.R., Petricoin III, E.F., Maitra, A., Rajapakse, V., King, C., Jacobetz, M.A., Ross, S., Conrads, T.P., Veenstra, T.D., Hitt, B.A., Kawaguchi, Y., Johann, D., Liotta, L.A., Crawford, H.C., Putt, M.E., Jacks, T., Wright, C.V., Hruban, R.H., Lowy, A.M., Tuveson, D.A.: Preinvasive and invasive ductal pancreatic cancer and its early detection in the mouse. Cancer Cell **4**(6), 437–450 (2003)

23. Petricoin, E.F., Rajapaske, V., Herman, E.H., Arekani, A.M., Ross, S., Johann, D., Knapton, A., Zhang, J., Hitt, B.A., Conrads, T.P., Veenstra, T.D., Liotta, L.A., Sistare, F.D.: Toxicoproteomics: serum proteomic pattern diagnostics for early detection of drug induced cardiac toxicities and cardioprotection. Toxicol. Pathol. **32**, 122–130 (2004)

24. Ressom, H., Varghese, R.S., Orvisky, E., Drake, S., Hortin, G., Abdel-Hamid, M., Loffredo, C.A., Goldman, R.: Ant colony optimization for biomarker identification from MALDI-TOF mass spectra. In: Proceedings ofthe 28th IEEE Annual International Conference in Engineering in Medicine and Biology Society, pp. 4560–4563 (2006)

25. Armañanzas, R., Saeys, Y., Inza, I., García-Torres, M., Bielza, C., Larranaga, P., van de Peer, Y.: Peakbin selection in mass spectrometry data using a consensus approach with estimation of distribution algorithms. IEEE/ACM Trans. Comput. Biol. Bioinform. **8**(3), 760–774 (2011)

26. Petricoin, E.F., Ornstein, D.K., Paweletz, C.P., Ardekani, A., Hackett, P.S., Hitt, B.A., Velassco, A., Trucco, C., Wiegand, L., Wood, K., Simone, C.B., Levine, P.J., Linehan, W.M., Emmert-Buck, M.R., Steinberg, S.M., Kohn, E.C., Liotta, L.A.: Serum proteomic patterns for detection of prostate cancer. J. Nat. Cancer Institute **94**(20), 1576–1578 (2002)

27. MATLAB: version 7.10.0 (R2010a). The MathWorks Inc., Natick, Massachusetts (2010)

28. Smith, C., Want, E., O'Maille, G., Abagyan, R., Siuzdak, G.: XCMS: processing mass spectrometry data for metabolite profiling using nonlinear peak alignment, matching, and identification. Anal. Chem. **78**, 779–787 (2006)

29. Datta, S.: Feature selection and machine learning with mass spectrometry data. In: Matthiesen, R. (ed.) Mass Spectrometry Data Analysis in Proteomics. Methods in Molecular Biology, vol. 1007, pp. 237–262. Humana Press (2013)

30. Koza, J.: Genetic Programming III: Darwinian Invention and Problem Solving. A Bradford book, Elsevier Science & Tech, Massachusetts, Philadelphia (1999)

31. Neshatian, K., Zhang, M., Johnston, M.: Feature construction and dimension reduction using genetic programming. In: Orgun, M.A., Thornton, J. (eds.) AI 2007. LNCS (LNAI), vol. 4830, pp. 160–170. Springer, Heidelberg (2007)

32. Luke, S.: Essentials of Metaheuristics, 2nd edn. Lulu (2013). http://cs.gmu.edu/sean/book/metaheuristics/

33. Soyel, H., Tekguc, U., Demirel, H.: Application of NSGA-II to feature selection for facial expression recognition. Comput. Electr. Eng. **37**(6), 1232–1240 (2011)

Automating Biomedical Data Science Through Tree-Based Pipeline Optimization

Randal S. Olson[1]([✉]), Ryan J. Urbanowicz[1], Peter C. Andrews[1],
Nicole A. Lavender[2], La Creis Kidd[2], and Jason H. Moore[1]

[1] Institute for Biomedical Informatics, University of Pennsylvania,
3700 Hamilton Walk, Philadelphia, PA 19104, USA
olsonran@upenn.edu
[2] University of Louisville, 505 S. Hancock St., Louisville, KY 40202, USA

Abstract. Over the past decade, data science and machine learning
has grown from a mysterious art form to a staple tool across a variety of fields in academia, business, and government. In this paper, we
introduce the concept of tree-based pipeline optimization for automating
one of the most tedious parts of machine learning—pipeline design. We
implement a Tree-based Pipeline Optimization Tool (TPOT) and demonstrate its effectiveness on a series of simulated and real-world genetic
data sets. In particular, we show that TPOT can build machine learning
pipelines that achieve competitive classification accuracy and discover
novel pipeline operators—such as synthetic feature constructors—that
significantly improve classification accuracy on these data sets. We also
highlight the current challenges to pipeline optimization, such as the
tendency to produce pipelines that overfit the data, and suggest future
research paths to overcome these challenges. As such, this work represents an early step toward fully automating machine learning pipeline
design.

Keywords: Pipeline optimization · Hyperparameter optimization ·
Data science · Machine learning · Genetic programming

1   Introduction

Data science is a fast-growing field: Between 2011 and 2015, the number of self-
reported data scientists has more than doubled [1]. At the same time, machine
learning—one of the primary tools of the modern data scientist—has experienced
a revitalization as academics, businesses, and governments alike discovered new
applications for automated algorithms that can learn from data. As a consequence, there has been a growing demand for tools that make machine learning
more accessible, scalable, and flexible. Unfortunately, the successful application
of these machine learning tools often requires expert knowledge of the tool as
well as the target problem, an awareness of all assumptions involved in the
analysis, and/or the application of simple exhaustive, brute force search. These

© Springer International Publishing Switzerland 2016
G. Squillero and P. Burelli (Eds.): EvoApplications 2016, Part I, LNCS 9597, pp. 123–137, 2016.
DOI: 10.1007/978-3-319-31204-0_9

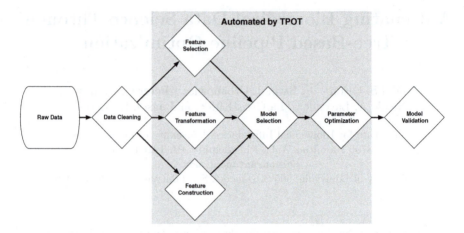

Fig. 1. A depiction of a typical machine learning pipeline. Before building a model of the data, the practitioner must ensure that the data is ready for modeling by performing an initial exploratory analysis (e.g., looking for missing or mislabeled data) and either correct or remove the offending records (i.e., data cleaning). Next, the practitioner may transform the data in some way to make it more suitable for modeling, e.g., by normalizing the features (i.e., feature transformation), removing features that do not seem useful for modeling (i.e., feature selection), and/or creating new features from the existing data (i.e., feature construction). Afterward, the practitioner must select a machine learning model to fit to the data (i.e., model selection) and select the model parameters that allow the model to make the most accurate classification from the data (i.e., parameter optimization). Lastly, the practitioner must validate the model in some way to ensure that the model's predictions generalize to data sets that it was not fitted on (i.e., model validation), for example, by testing the model's performance on a holdout data set that was excluded from the earlier phases of the pipeline. The light grey area indicates the steps in the pipeline that are automated by the tree-based pipeline optimization tool (TPOT).

requirements can make the application of many machine learning approaches a time-consuming and computationally-demanding endeavor.

As an example, a typical machine learning practitioner may build a pipeline as shown in Fig. 1. At each step, there are dozens of possible choices to make: How to preprocess the data (e.g., what feature selector? what feature constructor? etc.), what model to use (e.g., support vector machine (SVM)? artificial neural network (ANN)? random forest (RF)? etc.), what parameters to use (e.g., how many decision trees in a RF? how many hidden layers in an ANN? etc.), and so on. Experienced machine learning practitioners often have good intuitions on what choices are appropriate for the problem domain, but some practitioners can easily spend several weeks tinkering with model parameters and data transformations until the pipeline achieves an acceptable level of performance.

In the past two decades, we have seen the growth of intelligent systems such as evolutionary algorithms that are capable of outperforming humans in a wide variety of tasks, such as antenna design for space missions [2], discovering and patching bugs in large software projects [3], and even in the study of finite algebras [4]. Considering the creative power of these intelligent systems, we must ask ourselves: Can intelligent systems automatically design machine learning pipelines?

In this paper, we report on the early development of an algorithm that automatically constructs and optimizes machine learning pipelines through a Tree-based Pipeline Optimization Tool (TPOT). We use a well-known evolutionary computation technique called genetic programming [5] to automatically construct a series of data transformations and machine learning models that act on the data set with the goal of maximizing classification accuracy. We demonstrate TPOT's capabilities on an array of simulated data sets to explore the limits of the algorithm, then apply TPOT to a genetic analysis of prostate cancer. In particular, we show that TPOT can construct machine learning pipelines that achieve competitive classification accuracy, and that TPOT can discover novel pipeline operators—such as synthetic feature constructors—that significantly improve accuracy when added to the pipeline.

2   Related Work

Historically, automated machine learning pipeline optimization has focused on optimizing specific elements of the pipeline [6]. For example, grid search is the most commonly-used form of hyperparameter optimization, where practitioners apply brute force to explore a broad ranged sweep of model parameter combinations in search of the parameter set that allows the machine learning model to perform best. Recent research has shown that randomly exploring parameter sets within the grid search often discovers high-performing parameter sets faster than an exhaustive search [7], suggesting that there is promise for intelligent search in the hyperparameter space. Bayesian optimization of model hyperparameters, in particular, has been effective in this realm and has even outperformed manual hyperparameter tuning by expert practitioners [8].

Another focus of automated machine learning has been feature construction. One recent example of automated feature construction is "The Data Science Machine," which automatically constructs features from relational data sets via deep feature synthesis [9]. In their work, Kanter *et al.* demonstrated the critical role of automated feature construction in machine learning pipelines by entering their Data Science Machine into three machine learning competitions and achieving expert-level performance in all of them.

All of these findings point to one take-away message: Intelligent systems are capable of automatically designing portions of machine learning pipelines, which can save practitioners considerable amounts of time by automating one of the most laborious parts of machine learning. To our knowledge, there have been no published attempts at automatically optimizing entire machine learning pipelines to date. Thus, the work presented in this paper establishes a blueprint for future research on the automation of machine learning pipeline design.

3   Methods

In this section, we describe tree-based pipeline optimization in detail, including the tools and concepts that underlie the Tree-based Pipeline Optimization Tool (TPOT). We begin this section by describing the basic pipeline operators that are currently implemented in TPOT. Next, we describe how the operators are combined together into a tree-based pipeline, and show how tree-based pipelines can be evolved via genetic programming. Finally, we end this section by providing an overview of the data sets that we used to evaluate TPOT.

3.1   Decision Trees and Random Forests

In the version of TPOT presented here, we only used decision tree and random forest machine learning models as they are implemented in scikit-learn [10], a general-purpose Python machine learning library. For both models, we used versions that perform binary classification, i.e., splitting the records into two pre-defined groups based on their feature vectors.

A *decision tree* model is a flowchart-like structure that asks a series of binary questions about each record's features, all the while attempting to differentiate the two groups as much as possible. An example question may be, "Height <= 182 cm?", where the decision tree proceeds down the path to the right if the inequality is true and left otherwise. Generally, decision trees will select a question based on its ability to divide the records at each split by their group the most (i.e., maximizing the "purity" of the remaining records at that node), and will continue attempting to divide the remaining records down each path until either the maximum tree depth is reached or the path reaches a state where the remaining records completely belong to one group.

A *random forest* model uses several decision trees in an ensemble to make the same classification, where each decision tree is trained on a random sample (with replacement) of the training data. Once all of the decision trees in a random forest are constructed, their aggregate "vote" is used as the classification for the random forest. For further reading on decision trees and random forests, see [11].

3.2   Synthetic Feature Construction

In some cases, building new features from the existing feature set can prove useful for extracting vital information from data sets, especially when the features interact in some important way that would not be captured by methods that analyze only one feature at a time. In previous work, synthetic features constructed by random forests proved effective for combining genetic markers into constructed features that could then be used for classification [12]. In this paper, we allowed such synthetic features to be constructed by both decision trees and random forests. By default, whenever a decision tree or random forest was used to perform a classification in a TPOT pipeline, the classifications were also added as a constructed synthetic feature to the resulting data set.

3.3 Decision Tree-Based Feature Selection

Often times, it is necessary to reduce the number of features in large data sets to improve classification accuracy—especially in genetic analyses, where it is not uncommon for the number of genetic features to number in the thousands or more. In this paper, we created a custom implementation of a decision tree-based feature selection method that reduces the feature set down to a parameterized number of feature pairs. To evaluate the feature pairs, we exhaustively constructed every possible two-feature combination from the feature set. These feature pairs were then ranked based on the training classification accuracy of a decision tree that was provided only those two features, where the feature pairs that resulted in higher training classification accuracy were selected first. This method allowed for the detection of epistatic interactions between features, which is typically overlooked by traditional machine learning methods that only consider the interaction between the endpoint and one feature at a time.

3.4 Tree-Based Pipelines

To combine all of these operators into a flexible pipeline structure, we implemented the pipelines as trees as shown in Fig. 2. Every tree-based pipeline began with one or more copies of the input data set at the bottom of the tree, which was then fed into one of the many available pipeline operators: feature construction, feature selection, or classification by decision tree or random forest. These operators modified the provided data set then passed the resulting data set to the next operator as the data proceeded up the tree. In cases where multiple copies of the data set were being processed, it was also possible for the two data sets to be combined into a single data set via a data set combination operator.

Each time a data set passed through a decision tree or random forest operator, the resulting classifications were stored in a *guess* column, such that the most recent classifier to process the data would have its classifications in that column. Once the data set was fully processed by the pipeline—e.g., when the data set passed through the *Decision Tree Classifier* operator in Fig. 2—the values in the *guess* column were used to determine the classification accuracy of the pipeline. In this paper, we divided the data into stratified 75 % training and 25 % testing sets—such that the pipeline never trained but only predicted on the testing set—and each pipeline's accuracy was reported only on the testing set.

This tree-based pipeline structure allowed for arbitrary pipeline representations: For example, one pipeline could only apply operations in serial on a single copy of the data set, whereas another pipeline could just as easily work on several copies of the data set and combine them at the end before making a final classification.

3.5 Genetic Programming

To automatically generate these tree-based pipelines, we used a well-known evolutionary computation technique called genetic programming (GP) as implemented in the Python package DEAP [13]. Traditionally, GP builds trees of

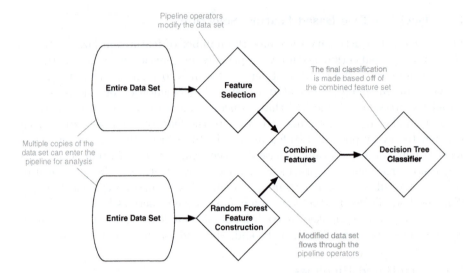

Fig. 2. An example tree-based machine learning pipeline. The data set flows through the pipeline operators, which add, remove, and modify the features in a successive manner. Combination operators allow separate copies of the data set to be combined, which can then be provided to a classifier to make the final classification.

Table 1. Genetic programming and experiment settings.

GP Parameter	Value
Selection	10 % elitism, rest 3-way tournament (2-way parsimony)
Population size	100
Per-individual mutation rate	90 %
Per-individual crossover rate	5 %
Generations	100
Replicates	30

mathematical functions to optimize toward a given criteria; in this case, by the same token, GP builds trees of pipeline operators to maximize the final classification accuracy of the pipeline. Here, we used GP to evolve the sequence of pipeline operators that acted on the data set as well as the parameters of these operators, e.g., the number of trees in a random forest or the number of feature pairs to select during feature selection.

In this paper, the TPOT GP algorithm followed a standard evolutionary algorithm procedure. At the beginning of every TPOT run, we randomly generated a fixed number of tree-based pipelines to constitute what we call the *population*. These pipelines were then evaluated based on their classification accuracy, which we used as the pipeline's *fitness*.

Once all of the pipelines were evaluated, we proceeded to the next iteration (i.e., *generation*) of the GP algorithm. To generate the next generation's population, we first created exact copies of the pipeline with the highest fitness and placed them into the new population until they represented 10 % of the maximum population size (i.e., elitism of 10). To construct the remainder of the new population, we randomly selected three pipelines from the existing population then placed them in a tournament to decide which pipeline reproduces. In this tournament, the pipeline with the lowest fitness was removed, then the least complex pipeline of the remaining two (i.e., the pipeline with the fewest pipeline operators) was chosen as the winner to be copied into the new population (i.e., 3-way tournament selection with 2-way parsimony). This tournament selection procedure was repeated until the remaining 90 % of the new population was created.

With the new population created, we then applied a *one-point crossover* operator to a fixed percentage of the copied pipelines, where two pipelines are selected at random, split at a random point in the tree, then have their contents swapped between each other. Following that, a fixed percentage of the remaining unaffected pipelines had a random change (i.e., a *mutation*) applied to them:

- **Uniform mutation**: A random operator in the pipeline was replaced with a new randomly-generated sequence of pipeline operators.
- **Insert mutation**: A new randomly-generated sequence of pipeline operators was inserted into a random place in the pipeline.
- **Shrink mutation**: A random subset of the pipeline operators were removed from the pipeline.

where each mutation operator had a $\frac{1}{3}$ chance of occurring when a pipeline was mutated. In all crossover and mutation operations, creating an invalid pipeline was disallowed, e.g., a pipeline that attempts to pass a data set into a parameter that expects a single integer would not be allowed.

Once the crossover and mutation operations completed, the previous generation's pipelines were deleted and this evaluate-select-crossover-mutate process was repeated for a fixed number of generations. In this manner, TPOT's GP algorithm continually tinkered with the pipelines—adding new pipeline operators that improve fitness and removing redundant or detrimental operators—in an intelligent, guided search for high-performing pipelines. At all times, the single best-performing pipeline ever discovered during the TPOT run was tracked and stored in a separate location and used as the representative pipeline at the completion of the run.

Table 1 describes the specific GP settings used in this paper. Every TPOT replicate was seeded with a unique random number generator seed to ensure that the runs were distinct.

3.6   GAMETES Simulated Data Sets

In order to evaluate TPOT, we adopted a diverse, complex simulation study design. We generated a total of 12 models and 360 associated data sets using

GAMETES [14], an open source software package designed to generate a diverse spectrum of pure, strict epistatic models. GAMETES generated random, biallelic, n-locus single nucleotide polymorphism (SNP) models with a precise form of epistasis, which we refer to as pure. An n-locus model is purely epistatic if all n loci, but no fewer, are predictive of disease status. Models were precisely generated with the desired heritabilities, SNP minor allele frequencies, and population prevalences.

In this study, all data sets included 100 SNP attributes—8 that were predictive of a binary case/control endpoint, and 92 that were randomly generated using an allele frequency between 0.05 and 0.5. The 8 predictive SNPs were simulated as four separate purely epistatic models, additively combined using the newly-added "hierarchical" data simulation feature in GAMETES. In doing so, each separate interaction model additively contributed to the determination of the endpoint, but the overall data set did not include main effects (i.e., direct associations between single SNP variables and the endpoint).

We simulated two-locus epistatic genetic models with heritabilities of (0.1, 0.2, or 0.4) and attribute minor allele frequencies of 0.2 in GAMETES and selected the model with median difficulty from all those generated [15]. Data sets with a sample size of either 200, 400, 800, or 1600 were generated, within which each of the four underlying two-locus epistatic models carried an equal additive weight. 30 replicates of each model and data set combination were generated, yielding a total of 360 data sets (i.e., 3 heritabilities * 4 sample sizes * 30 replicates). Together, this simulation study design allowed us to evaluate TPOT across a range of data sets with varying difficulties and sample sizes to explore the limits of TPOT's modeling capabilities.

3.7   CGEMS Prostate Cancer Data Set

To demonstrate TPOT's performance on a real-world data set, we applied TPOT to a genetic analysis of a nationally available genetic data set from 2,286 men of European descent (488 non-aggressive and 687 aggressive cases, 1,111 controls) collected through the Prostate, Lung, Colon, and Ovarian (PLCO) Cancer Screening Trial, a randomized, well-designed, multi-center investigation sponsored and coordinated by the National Cancer Institute (NCI) and their Cancer Genetic Markers of Susceptibility (CGEMS) program. We focused here on prostate cancer aggressiveness as the endpoint, where the prostate cancer is considered aggressive if it was assigned a Gleason score ≥ 7 and was in tumor stages III/IV. Between 1993 and 2001, the PLCO Trial recruited men ages 55–74 years to evaluate the effect of screening on disease specific mortality, relative to standard care. All participants signed informed consent documents approved by both the NCI and local institutional review boards. Access to clinical and background data collected through examinations and questionnaires was approved for use by the PLCO. Men were included in the current analysis if they had a baseline PSA measurement before October 1, 2003, completed a baseline questionnaire, returned at least one Annual Study Update (ASU), and had available SNP profile data through the CGEMS data portal (http://cgems.cancer.gov/).

We used a biological filter to reduce the set of genes to just those involved in apoptosis (programmed cell death), DNA repair, and antioxidation/carcinogen metabolism. These biological processes are hypothesized to play an important role in prostate cancer. This report evaluated a total of 219 SNPs in the afore-mentioned biological pathways in relation to aggressive prostate cancer.

4   Results

In this section, we present the performance of the tree-based pipeline optimiza-tion tool (TPOT) on the GAMETES simulated data sets and the real-world CGEMS prostate cancer data set.

4.1   GAMETES Simulated Data Sets

To begin, we tested TPOT on a broad range of heritability settings and data set sizes from GAMETES to explore TPOT's limits. Figure 3 shows that in the higher heritability setting of 0.4—with the least amount of noise in the data set—TPOT achieved >80 % testing accuracy even with only 200 records to train on. Even at the more difficult heritability setting of 0.1—with a large amount of noise in the data set—TPOT achieved >65 % testing accuracy with only 800 records for training. However, at data set sizes of <=200 records and 0.1 heritability, the pipelines discovered by TPOT didn't perform much better than chance (i.e., 50 % testing accuracy due to the balanced classes). Generally, this array of tests demonstrated that TPOT performs best when provided with (a) larger data sets to train on and/or (b) models with higher heritability (i.e., less noise), which is to be expected since both configurations entail that the signal in the data set is easier to detect.

We also compared TPOT's classification performance to that of a basic ran-dom forest with 100 decision trees and a version of TPOT that optimized only model selection and parameters (i.e., neither versions had access to the fea-ture selection nor feature construction operators). As shown in Fig. 3, TPOT performed significantly better with the feature selection and construction oper-ators included in all but the most difficult data sets. This finding showed that TPOT can discover useful feature construction and selection operators that that improve classification accuracy over solely optimizing the model parameters.

Lastly, we compared TPOT's classification performance with selection (i.e., genetic programming optimizing for classification accuracy) to a version of TPOT that randomly generated the same number of pipelines (i.e., randomly generates population size * generations number of pipelines, or 10,000 pipelines in this case). Figure 3 shows that TPOT with selection did not perform signif-icantly differently than TPOT via random search. This finding suggests that randomly generating pipelines eventually discovers a top-performing pipeline, and selection may not have performed a vital role on the GAMETES data sets.

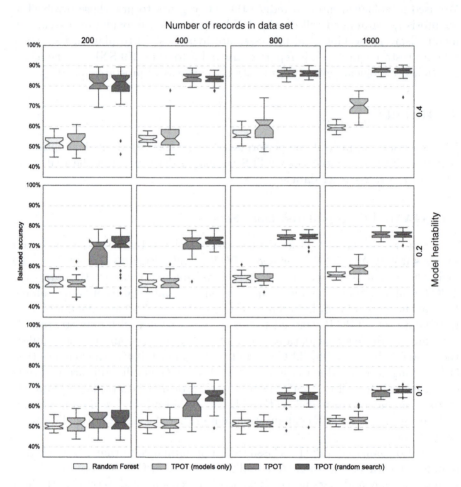

Fig. 3. Tree-based pipeline optimization tool (TPOT) performance comparison across a range of data set sizes and difficulties. Each subplot on the grid shows the distribution of balanced 10-fold cross validation accuracies, where each notched box plot represents a sample of 30 data sets. (Note: the notches in the box plots indicate 95 % confidence intervals of the median.) The experiments compared include a random forest with 100 decision trees ("Random Forest"), a version of TPOT that only automates model selection and model parameters ("TPOT (models only)"), standard TPOT with genetic programming ("TPOT"), and standard TPOT with random generation of pipelines ("TPOT (random search)"). The subplots correspond to varying GAMETES configurations, where the x-axis modifies the number of records in the data set and the y-axis modifies the heritability in the model (where higher heritability reduces the amount of noise and vice versa). The grid ranges from easy configurations in the top right (larger data sets generated from higher heritability models) to difficult configurations in the bottom left (smaller data sets generated from low heritability models).

Fig. 4. Balanced testing accuracy scores on the CGEMS prostate cancer data set as the data set progresses through the pipeline. Each step in this diagram corresponds to a pipeline operator in the final TPOT pipeline. Starting from the raw data set, the pipeline applies a series of three feature construction steps (a random forest feature constructor, a decision tree feature constructor, then another random forest feature constructor), adding a new synthetic feature to the data set each time before proceeding to the next step. At the final classification step, the random forest classifier has access to a data set (E_4) containing all the original features along with the three new synthetic features, which allows it to accurately classify significantly more patients than without the synthetic features.

4.2   CGEMS Prostate Cancer Data Set

Next, we applied TPOT to the CGEMS prostate cancer data set to demonstrate its performance on a real-world genetic analysis. As shown in Fig. 4, TPOT discovered a pipeline that achieved a balanced testing accuracy of 60.8 %, which is competitive with previous accuracy of 59.8 % on the same data set with the Computational Evolution System (CES) [16]. However, the balanced testing accuracy of the final TPOT pipeline dropped significantly to 51.7 % with 10-fold cross-validation, suggesting that there was overfitting occurring with the pipelines despite our use of a 75 %/25 % training/testing validation split during the optimization process.

In addition, Fig. 4 shows the progression of balanced testing accuracy as the pipeline added constructed features to the data set. Starting with the raw data set (E_1), a random forest is only capable of achieving a 54.7 % balanced testing accuracy on E_1. However, when the pipeline used that same random forest to construct and add a new synthetic feature to the data set—creating a new data set, E_2—a decision tree achieves a significantly-improved balanced testing accuracy of 58.7 %. At the final random forest classifier, the pipeline had added three synthetic features in total to the data set (E_4), enabling the random forest to achieve a much-improved 60.8 % balanced testing accuracy. These findings further support the theory that TPOT can discover novel feature construction methods that improve the pipeline's classification accuracy.

To provide a more detailed view of the pipeline, we looked at the top 10 features (according to their feature importance scores) that were used during the synthetic feature construction and classification steps. When constructing the first synthetic feature ("Synthetic Feature 1", Fig. 5), the random forest primarily used the *NAT2* and *BCL2* SNPs to classify the patients. Interestingly, CES also discovered that these SNPs play an important role in determining the aggressiveness of prostate cancer [16], which lends support to their importance in the biological functions of prostate cancer aggressiveness. During the

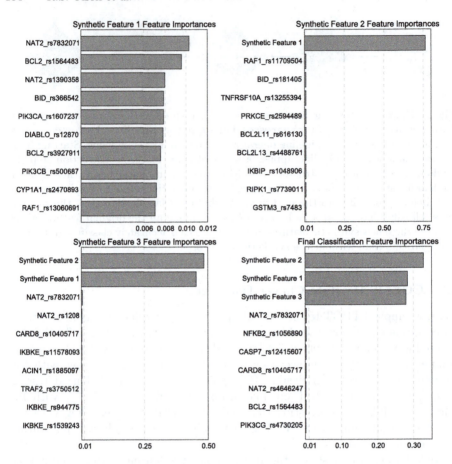

Fig. 5. Top 10 features for each pipeline step for the CGEMS prostate cancer data set. Each bar chart shows the feature importance scores of the top 10 features for each pipeline step, where the feature importance scores are determined by the feature's Gini importance [17], i.e., the sum of the feature's contributions to dividing the two classes. Synthetic features 1, 2, and 3 are constructed features built by the pipeline operators. Each feature is named after the corresponding SNP in the CGEMS prostate cancer data set.

construction of the second synthetic feature ("Synthetic Feature 2", Fig. 5), Synthetic Feature 1 played a much larger role in classifying the patients, further showing the synthesizing power of synthetic features. Despite the major role that Synthetic Feature 1 played, the decision tree was also able to integrate information from other SNPs such as *RAF1* and *BID* into Synthetic Feature 2. The third and final synthetic feature ("Synthetic Feature 3", Fig. 5) similarly relied on the previous synthetic features to classify the patients yet integrated new information from the remaining SNPs, which allowed the final random forest classifier to make a significantly improved classification based on the three synthetic features ("Final Classification", Fig. 5).

5   Discussion

It is important to note that the goal of pipeline optimization is not to replace data scientists nor machine learning practitioners. Rather, we aim for the tree-based pipeline optimization tool (TPOT) to be a "Data Science Assistant" that can explore the data, discover novel features, and recommend pipelines to the practitioner. From there, the practitioner is free to build on the automated pipelines and integrate their domain knowledge as they see fit. To aid in this goal, we have released TPOT as an open source Python package that provides a flexible implementation of the concepts introduced in this paper. We encourage interested practitioners to involve themselves in the project on GitHub (https://github.com/rhiever/tpot/).

In this paper, we showed that TPOT is capable of building machine learning pipelines that achieve competitive classification accuracy and discovering novel pipeline operators—such as synthetic feature constructors—that significantly improve classification accuracy. However, TPOT is not without its drawbacks. For example, TPOT with guided search did not perform significantly differently than TPOT with randomly-generated pipelines (Fig. 3), which suggests that the additional layer of guided search (i.e., genetic programming) is not useful in the current version of TPOT. We believe that guided search did not outperform random search in this case because TPOT currently lacks clear building blocks [18]—i.e., small combinations of pipeline operators that can be combined to improve accuracy—for evolution to act on, and instead builds pipelines from powerful pipeline operators such as random forest classifiers. In future work, we will explore methods to divide the pipeline operators into better building blocks for TPOT to work with.

Furthermore, in Sect. 4.2 we showed that TPOT built a pipeline that suffered significantly from overfitting on the CGEMS prostate cancer data set. In this work, we attempted to prevent overfitting by evaluating the TPOT pipelines by dividing the data set into a 75 %/25 % training/testing validation set, where the pipelines were trained on the training set and assigned a fitness according to their balanced accuracy on the testing set. In place of overfitting on the training data, the pipeline overfit on the testing data instead, which suggests that alternative methods are required to prevent TPOT from overfitting. In future work, we will explore methods such as multi-objective [19] and Pareto optimization [20], where one of the objectives/Pareto fronts can be pipeline generalization performance or pipeline complexity (assuming less-complex pipelines will generalize better).

In the GAMETES simulated data sets, the random forest and decision tree classifiers failed to detect the epistatic pairs in the data sets on their own, which resulted in poor classification accuracy (Fig. 3, "Random Forest" and "TPOT (models only)"). This failure likely occurred because traditional machine learning algorithms only look at single feature-class correlations, which will necessarily overlook any interactions between the features. We believe that this result points to the need for machine learning algorithms such as Spatially Uniform ReliefF (SURF) [21] that can reliably and efficiently discover important features that have epistatic interactions with other features.

When applied to the CGEMS prostate cancer data set, TPOT discovered several synthetic features that significantly contributed to the classification accuracy of the pipeline. Notably, several of the SNPs that TPOT used to classify the patients—*NAT2* and *BCL2*, in particular—were also found to play an important role in the prediction prostate cancer aggressiveness in a previous study [16], which highlights TPOT's ability to contribute to knowledge discovery. In future work, we will continue to develop methods to extract new knowledge such as feature importances from TPOT pipelines.

Beyond the work discussed above, in the near future we will continue developing TPOT to integrate more feature selection (e.g., SURF [21]) and construction operators, more machine learning models (e.g., support vector machines, logistic regression, K-nearest neighbor classifiers, etc.), and better hyperparameter optimization (e.g., Bayesian optimization methods [8]). In general, we seek to integrate as many pipeline operators as possible into TPOT, with the assumption that genetic programming will discover the best operators (or set of operators) for the classification task at hand.

6   Conclusions

Machine learning pipeline optimization is poised to transform data science by automating one of the most tedious parts of machine learning. In this paper, we introduced a new method for automatically creating and optimizing machine learning pipelines, tree-based pipeline optimization, and demonstrated its effectiveness on a series of simulated and real-world genetic data sets. In particular, we showed that such a system can automatically build machine learning pipelines that achieve competitive classification accuracy as well as discover novel pipeline operators—such as synthetic feature constructors—that significantly improve classification accuracy on these data sets. We also highlighted the current challenges to pipeline optimization, such as the tendency to produce pipelines that overfit the data, and suggested future research paths to overcome these challenges. As such, this work represents an early step toward fully automating machine learning pipeline design.

Acknowledgments. We thank Sebastian Raschka for his valuable input during the development of this project. We also thank the Michigan State University High Performance Computing Center for the use of their computing resources. This work was supported by National Institutes of Health grants LM009012, LM010098, and EY022300.

References

1. RJMetrics: The State of Data Science, November 2015. https://rjmetrics.com/resources/reports/the-state-of-data-science/
2. Hornby, G.S., Lohn, J.D., Linden, D.S.: Computer-automated evolution of an X-band antenna for NASA's space technology 5 mission. Evol. Comput. **19**(1), 1–23 (2011)

3. Forrest, S., Nguyen, T., Weimer, W., Le Goues, C.: A genetic programming approach to automated software repair. In: Proceedings of the 11th Annual Conference on Genetic and Evolutionary Computation, GECCO 2009, pp. 947–954. ACM, New York (2009)

4. Spector, L., Clark, D.M., Lindsay, I., Barr, B., Klein, J.: Genetic programming for finite algebras. In: Proceedings of the 10th Annual Conference on Genetic and Evolutionary Computation, GECCO 2008, pp. 1291–1298. ACM, New York (2008)

5. Banzhaf, W., Nordin, P., Keller, R.E., Francone, F.D.: Genetic Programming: An Introduction. Morgan Kaufmann, San Meateo (1998)

6. Hutter, F., Lücke, J., Schmidt-Thieme, L.: Beyond manual tuning of hyperparameters. KI - Künstliche Intelligenz 29(4), 329–337 (2015)

7. Bergstra, J., Bengio, Y.: Random search for hyper-parameter optimization. J. Mach. Learn. Res. 13, 281–305 (2012)

8. Snoek, J., Larochelle, H., Adams, R.P.: Practical bayesian optimization of machine learning algorithms. In: Pereira, F., Burges, C.J.C., Bottou, L., Weinberger, K.Q. (eds.) Advances in Neural Information Processing Systems, vol. 25, pp. 2951–2959. Curran Associates, Inc. (2012)

9. Kanter, J.M., Veeramachaneni, K.: Deep feature synthesis: towards automating data science endeavors. In: Proceedings of the International Conference on Data Science and Advance Analytics. IEEE (2015)

10. Pedregosa, F., Varoquaux, G., Gramfort, A., Michel, V., Thirion, B., Grisel, O., Blondel, M., Prettenhofer, P., et al.: Scikit-learn: machine learning in python. J. Mach. Learn. Res. 12, 2825–2830 (2011)

11. Hastie, T.J., Tibshirani, R.J., Friedman, J.H.: The Elements of Statistical Learning: Data Mining, Inference, and Prediction. Springer, New York (2009)

12. Pan, Q., Hu, T., Malley, J.D., Andrew, A.S., Karagas, M.R., Moore, J.H.: A system-level pathway-phenotype association analysis using synthetic feature random forest. Genet. Epidemiol. 38(3), 209–219 (2014)

13. Fortin, F.A., Gardner, M.A., Parizeau, M., Gagne, C., de Rainville, F.M.: DEAP: evolutionary algorithms made easy. J. Mach. Learn. Res. 13, 2171–2175 (2012)

14. Urbanowicz, R.J., Kiralis, J., Fisher, J.M., Moore, J.H.: Predicting the difficulty of pure, strict, epistatic models: metrics for simulated model selection. BioData Min. 5(1), 1–13 (2012)

15. Urbanowicz, R.J., Kiralis, J., Sinnott-Armstrong, N.A., Heberling, T., Fisher, J.M., Moore, J.H.: GAMETES: a fast, direct algorithm for generating pure, strict, epistatic models with random architectures. BioData Min. 5(1), 1–14 (2012)

16. Moore, J.H., Hill, D.P., Sulovari, A., Kidd, L.C.: Genetic analysis of prostate cancer using computational evolution, pareto-optimization and post-processing. In: Riolo, R., Vladislavleva, E., Ritchie, M.D., Moore, J.H. (eds.) Genetic Programming Theory and Practice X, pp. 87–101. Springer, New York (2013)

17. Breiman, L., Cutler, A.: Random forests - classification description, November 2015. http://www.stat.berkeley.edu/breiman/RandomForests/cc_home.htm

18. Goldberg, D.E.: The Design of Innovation: Lessons from and for Competent Genetic Algorithms. Kluwer Academic Publishers, Norwell (2002)

19. Konak, A., Coit, D.W., Smith, A.E.: Multi-objective optimization using genetic algorithms: a tutorial. Reliab. Eng. Syst. Saf. 91(9), 992–1007 (2006)

20. Deb, K., Pratap, A., Agarwal, S., Meyarivan, T.: A fast and elitist multiobjective genetic algorithm: NSGA-II. IEEE Trans. Evol. Comput. 6(2), 182–197 (2002)

21. Greene, C.S., Penrod, N.M., Kiralis, J., Moore, J.H.: Spatially Uniform ReliefF (SURF) for computationally-efficient filtering of gene-gene interactions. BioData Min. 2(1), 1 (2009)

Bicliques in Graphs with Correlated Edges: From Artificial to Biological Networks

Aaron Kershenbaum[1], Alicia Cutillo[1], Christian Darabos[1,3], Keitha Murray[2], Robert Schiaffino[2], and Jason H. Moore[1 (✉)]

[1] Institute for Biomedical Informatics, The Perelman School of Medicine, University of Pennsylvania, Philadelphia, PA 19104, USA
jhmoore@upenn.edu
[2] Department of Computer Science, Iona College, New Rochelle, NY 10801, USA
[3] Research Computing, Dartmouth College, Hanover, NH 03755, USA

Abstract. Networks representing complex biological interactions are often very intricate and rely on algorithmic tools for thorough quantitative analysis. In bi-layered graphs, identifying subgraphs of potential biological meaning relies on identifying bicliques between two sets of associated nodes, or variables – for example, diseases and genetic variants. Researchers have developed multiple approaches for forming bicliques and it is important to understand the features of these models and their applicability to real-life problems. We introduce a novel algorithm specifically designed for finding maximal bicliques in large datasets. In this study, we applied this algorithm to a variety of networks, including artificially generated networks as well as biological networks based on phenotype-genotype and phenotype-pathway interactions. We analyzed performance with respect to network features including density, node degree distribution, and correlation between nodes, with density being the major contributor to computational complexity. We also examined sample bicliques and postulate that these bicliques could be useful in elucidating the genetic and biological underpinnings of shared disease etiologies and in guiding hypothesis generation. Moving forward, we propose additional features, such as weighted edges between nodes, that could enhance our study of biological networks.

1 Introduction

Relationships of many types can be modeled as graphs, where the related entities are the nodes of the graph and the relationships between them are the edges. In order to make the model more complete and solutions more realistic, properties relevant to the specific problem can be associated with the nodes and edges, transforming the graph into a network [20]. One type of problem which has received a great deal of interest is to identify sets of nodes which are closely related, e.g., diseases with common etiologies. In the context of graph theory, this involves a search for collections of nodes with many connections between them. In the ideal case, edges between all pairs of nodes in the subset would exist. Such graphs are called cliques.

© Springer International Publishing Switzerland 2016
G. Squillero and P. Burelli (Eds.): EvoApplications 2016, Part I, LNCS 9597, pp. 138–155, 2016.
DOI: 10.1007/978-3-319-31204-0_10

It is often true that the nodes are of two distinct types, say A and B, and that the edges can be drawn between nodes of one type and the other [19]. For example, we might be interested in relationships between diseases and genes. The problem now becomes one of finding a collection of nodes in A that are all connected to all nodes in B. These nodes form bipartite graphs, or simply bigraphs, and are known as bicliques [11]. They may represent significant relationships between the two sets of nodes. A maximal biclique is one that is not contained in any other one; i.e., at least one of its component sets is not properly contained in the corresponding set in any other biclique.

We present an algorithm for solving the problem of finding all maximum bicliques of a bigraph. We discuss the algorithm's runtime and memory requirements as a function not only of the number of nodes and edges in the network, but also of the density of the graph (fraction of edges present), the distribution of node degrees, and the correlation among the edges.

We apply our algorithm to the analysis of two Human Phenotype Networks (HPN) [7], bigraphs models of human phenotype interactions based on shared biology. The first uses biological pathways, and the second genes, to identify common etiology in disease phenotypes. Through the use of Genome-Wide Association Study (GWAS), we have amassed a wealth of knowledge linking genetic variants to human disease phenotypes. However, our understanding remains limited by the complexity of human disease and by the epistatic, polygenetic and pleiotropic effects of genotype-phenotype interactions. Bicliques and other phenotype networking tools allow for a systematic examination of shared biology between diseases. These tools can be used to (1) identify diseases previously believed to be unrelated, (2) pinpoint genetic or biochemical pathway associations that underlie common disease etiologies and (3) identify therapeutic targets that may be applicable to multiple diseases.

2   Background

2.1   Bicliques in Research

Due to the large number of applications they model, bigraphs, bicliques, quasi-bicliques, and maximal bicliques have been studied extensively in the context of human genotype-to-phenotype relationships. Cheng and Church [4] introduced the concepts of "biclustering" both genes and conditions and illustrated their technique by identifying co-regulation patterns in yeast and human gene expression data. Tanay *et al.* [26], Wang *et al.* [27], and Liu *et al.* [15] address the same problem using bipartite graphs to model relationships between genes and conditions and identified bicliques in the bipartite graphs. Sanderson *et al.* [24] use bipartite graphs and bicliques in phylogenetics to improve the accuracy of tree construction. Zhang *et al.* [28] use maximal bicliques to identify groups of genes that are associated with related biological functions.

With the generalizability of these tools, many other applications exist both inside and outside of biology. Bicliques can be applied to the study of other factors influencing human disease, such as environmental exposures, or to the study various other biological networks. For instance, Maulik *et al.* [18] use quasi-bicliques to study viral-host protein-protein interaction networks. Liu *et al.* [16] use quasi-bicliques to find interacting protein group pairs in protein-protein interactions. Beyond biology, Sim *et al.* [25] use quasi-bicliques to cluster groups of stocks and financial ratios to identify investment opportunities.

2.2 Some Technical Considerations

This range of applications has resulted in significant research to find efficient algorithms to identify maximal bicliques. Prisner [22] demonstrated that the number of maximal bicliques in a bipartite graph with n vertices varies according to the formula $2^{n/2}$. Zhang *et al.* [28] noted that algorithms for finding maximal bicliques follow one of three approaches: exhaustive search, reduction to the clique enumeration problem in a general graph, and reduction to the frequent itemset mining problem in a transaction database. Alexe *et al.* [1] enumerate all maximal bicliques using a consensus algorithm similar to the consensus method for finding prime implicants of a Boolean function. Makino and Uno [17] use the second approach to convert a bigraph into a general graph by adding all edges to connect all vertices within the same partition. In this case, algorithms for enumerating maximal cliques in a general graph can be applied. However, Zhang *et al.* [28] noted that this approach is neither practical or scalable. Using the third approach, the adjacency matrix of a graph can be viewed as a transaction database, and a biclique corresponds to a pair of frequent closed itemsets in the transaction database.

Li *et al.* [14] suggest using frequent itemset mining techniques to mine maximal bicliques. They show that the problem of enumerating all maximal complete bipartite subgraphs is equivalent to mining frequent closed itemsets in the adjacency matrix. That is, a closed itemset and the set of transactions containing the closed itemset form a biclique. Liu *et al.* [25] point out that the large maximal biclique mining problem has size constraints on both vertex sets, while the frequent itemset mining problem puts size constraints on only one side of the transaction set. Their approach uses a divide-and-conquer technique to prune the search space.

The number of bicliques may be limited based on specific properties of the nodes and edges. For example, if the maximum node degree can be limited, the number of bicliques can become a polynomial in the largest node degree. Thus, the problem of finding all maximal bicliques is intractable for graphs allowing nodes of very high degree. It is therefore important to better understand the features that separate tractable instances of the problem from intractable ones and to better understand which of these factors are intrinsically associated with real world problems. This knowledge can help determine which types of problems are approachable by analysis using bicliques and which are not.

2.3   The Human Phenotype Network

Human Phenotype Networks (HPNs) [7,9] are mathematical graph models where the nodes represent human phenotypic traits. Edges represent genetic connections between traits, the granularity of which can be modulated from shared genetic variants (Single Nucleotide Polymorphisms, or SNPs), genes, or pathways, to name only a few. Because HPNs rely on Genome-Wide Association Study (GWAS) data, they incorporate diseases, physical and behavioral traits. HPNs are bigraphs by nature, before being projected onto the space of phenotype nodes only. In their projected form (single set of phenotype nodes), they have been successfully used to study diverse aspects of human traits and disease interactions, from pleiotropy and epistasis [8], to Type 2 Diabetes in East Asian populations [23], to environmental effects [7]. In this work we utilize two different HPNs in their bigraph (pre-projection) form. The first one is based on shared pathways between diseases and the second relies on shared genes between the traits [8].

Biological networks are generally expected to have heterogeneous connectivity placing them in the scale-free family. This means that the degree distribution follows a power-law, or exponential decay. Within the network, this translates into the presence of a minority of highly connected nodes (i.e. hubs). When the degree distribution of a scale-free network is plotted on a logarithmic scale, the resulting curve is approximately linear across the top [20].

3   Methods

In the present work, we utilize both pathway and gene-centric HPNs, using data from the May 2014 version of the NHGRI GWAS catalog [6]. To maximize coverage, HPNs included genotype-phenotype associations found in the database of Genotypes and Phenotypes (dbGaP). The pathway HPN incorporates genetic pathways information for Kegg and Reactome to build association between phenotypes/traits. The GWAS catalog and dbGaP report 1,252 traits combined, annotated with 37,681 SNPs in 16,411 loci. Bipartite HPNs rely on almost 1,000 phenotypic traits, and over 10,000 genes and almost 1,500 pathways, respectively. Throughout the analysis HPNs remain bipartite, where phenotypes and genes/pathways each represent a set of nodes. The resulting bipartite network can be projected in the space of phenotype vertices to obtain the HPN found in literature.

3.1   Definition of Terms

A **graph** $G = (N, E)$ [12] is defined by a set of nodes NN and a set of edges NE, where

$$N = \{n_i | i = 1, 2, ...NN\} \ , \quad E = \{e_k = (n_i, n_j) | k = 1, 2, ...NE\}$$

and n_i, and n_j are contained in N. Graphs are used to model relationships. The nodes are the objects being related and the edges model the link between them. In the present work, we will only consider undirected graphs.

A **clique**, $C = (NC, EC)$ is a subgraph of a graph where NC and EC are subsets of N and E.

A **bigraph** (or bipartite network) $G = (U, V, E)$ is a graph whose node set is partitioned into disjoint sets of nodes, U and V, and edges E connect nodes in U to nodes in V; i.e., there are no edges between nodes in U and also no edges between nodes in V.

$$E = \{e_k | k = 1, 2, ...NE\}$$

where $e_k = (u_i, v_j)$ for u_i, contained in U and v_j contained in V.

A **biclique** $CB = (UC, VC, EC)$ is a subgraph of a bigraph $B = (U, V, E)$ where UC, VC, and EC are subsets of U, V and E, respectively and for all u_i, in UC and v_j in VC there is an edge $e_k = (u_i, v_j)$ for pairs u_i, in UC and v_j in VC. Bicliques model relationships between nodes in U and nodes in V. For example, U may be a set of diseases, V may be a list of genes and E may be a set of relationships between the diseases and the genes.

A **maximal biclique** is one containing nodes that are not proper subsets of the nodes in any other biclique. Maximal bicliques represent strongly related groups of nodes in G and represent groupings of the nodes into categories. Note that maximal bicliques need not be disjoint; nodes may overlap.

Networks are graphs where properties are associated with the nodes and edges. We therefore use these terms interchangeably. The search for bicliques is motivated by the desire to find groups of nodes that share one or more properties. Indeed, the problems we solve are often aimed at identifying relevant properties relating nodes to one another. In this paper, we restrict the discussion to unweighted (or all edges have the same weight), undirected edges but the algorithm we present extends naturally to the weighted, directed problems.

3.2   Biclique Detection Algorithm

We present and analyze an algorithm for finding maximal bicliques. A bigraph is defined as a triple, $G = (U, V, E)$. U and V are disjoint sets, not necessarily of equal cardinality. We will refer to a typical member of U as u_i and to a typical member of V as v_j. The members of E, $e = (u_i, v_j)$ are undirected edges so that U and V are interchangeable.

In the algorithm, we represent a biclique as a triple (L, R, A) where L is a subset of U, R is a subset of V, and A is the set of nodes, u_m in U that are adjacent to u_i, a given node in L (think of L and R as simply being left and right sets of nodes, with no particular significance to left and right.) For a biclique, $c_i = (L_i, R_i, A_i)$, we refer to L_i, R_i, and A_i as the biclique's L-set, R-set and A-set, respectively.

We say two nodes, u_i and u_m, in L are adjacent if they share at least one neighbor, v_k, in R; i.e., if there exist one or more pairs of edges (u_i, v_k) and

(u_m, v_k) that are members of E in G. We define a **singleton biclique**, s_i, as a biclique where L contains a single node.

As an example, consider Fig. 1. Let the nodes in the top row be U and the nodes in the bottom row be V. As shown in red, nodes 8, 9, 3 and 4 form a biclique (which is not a singleton biclique), C, because both 8 and 9 are connected to both 3 and 4. Note that for nodes to form a biclique all the left nodes have to be directly connected to (not just adjacent to them by the definition above) all the right nodes. Nodes 3 and 4 are both directly connected to nodes 0, 5, 8 and 9.

We thus have:

$$L_c = \{8, 9\} \quad ; \quad R_c = \{3, 4\} \quad ; \quad A_c = \{0, 5\}$$

Note that we do not include 8 and 9 in AC because they are already in L_c.

A singleton biclique, s_i, is a biclique where L contains a single node. In green in Fig. 1, we show an example of a singleton biclique: [{14} ; {7, 10, 12} ; {5, 9, 11, 13, 15, 17, 18}] where the 3 sets inside the square brackets are respectively L_c, R_c and A_c.

The algorithm maintains a collection of maximal bicliques found so far and a collection of candidate bicliques to be expanded into larger bicliques. The expansion of biclique is a set of bicliques formed by adding each member of its A-set to its L-set.

The collection of maximal bicliques is managed as a map where each key is the R-set of a biclique and the associated value is the biclique itself. The candidate collection is managed as a priority queue. In order for this to work, sets of nodes have to be comparable. Since they are sets, we can test for inclusion and equality. In addition to this, the algorithm also has to know when two sets are not comparable. Two sets are said to be not comparable if they are not equal and neither includes the other. We will refer to the map of bicliques as MB and the queue of candidate bicliques as QC.

MB contains all bicliques that are currently maximal. As the algorithm proceeds, new bicliques are added and to MB and bicliques in MB may be replaced by bicliques that dominate them. A biclique $c_i = (L_i, R_i, A_i)$ dominates biclique

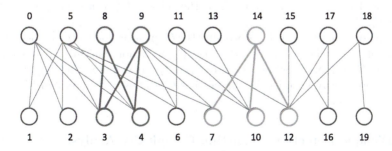

Fig. 1. Biclique schematic. L nodes 8 and 9 form a biclique with R nodes 3 and 4, with adjacent nodes 0 and 5. A singleton biclique is formed between L node 14 and R nodes 7, 10, and 12 (Color figure online).

$c_j = (L_j, R_j, A_j)$ if L_j and R_j, are contained in L_i ,and R_i. If $L_i = L_j$, and $R_i = R_j$, then $c_i = c_j$. Note that since the A-sets are derivable from the R-sets, if R_j is contained in (or equal to) R_i then A_j is contained in (or equal to) A_i.

The algorithm is implemented from the following basic operations.

Merging Two Bicliques: The merger, M_{ij}, of two bicliques, c_i and c_j is defined as

$$M_{ij} = c_k = (L_k, R_k, A_k)$$

where

$$L_k = L_i \cup L_j \quad ; \quad R_k = R_i \cap R_j \quad ; \quad A_k = A_i \cap A_j$$

Expansion of a Biclique: The expansion, EXP_i of a biclique c_i is the set of bicliques M_{ij} formed by merging $c_i s_j$ with each singleton biclique, s_i, in the A-set of c_i. Thus,

$$EXP_i = \{M_{ik} | k \in A_i\}$$

In its simplest form, the algorithm begins by putting all the singleton bicliques into QC and also into MB. It then pops c_i from the front of QC and expands it. Each M_{ik} in EXP_i is tested. Specifically, the biclique, c_m, (if any) whose key equals the R-set of M_{ik} is retrieved from MB. If is null, M_{ik} is put into MB. If is not *null*, it is merged with M_{ik}, forming M_{im} , and M_{im} replaces c_m in MB.

As the algorithm proceeds, new biclique are added to MB and previously found bicliques are replaced by ones that dominated them. As a biclique is expanded, it is removed from QC. When newly found bicliques (either new ones or merged ones that dominate previously found ones) are placed in MB they are also added to QC. The process terminates when QC is empty.

Given the original graph $G = (U, V, E)$, we form a set, $T = U$; i.e. $t_i = u_i$. We then order the t_i based on their degrees in G, smallest first. Other ordering are possible; e.g. by degree, largest first. For each t_i we will form a graph $G_i = (U_i, V_i, E_i)$ which we will use to find all the maximal bicliques in G_i.

We form U_i first by removing from U all nodes, u_j not adjacent to t_i in G, i.e., where adjacency is defined as above by the existence of a path through some v_j in V_i. We then remove from U_i all t_j for $j < i$. V_i contains all v_j in V for which an edge (u_i, v_j) exists in G. E_i is the subset of E whose endpoints are in U_i and V_j.

The algorithm proceeds by processing the G_i in succession as described above, finding all the bicliques in G_i. Because U_i does not contain any previously processed G_i removed previously processed t_j we will not find any maximal bicliques previously found when processing G_j. Thus each maximal biclique is found exactly once.

3.3   Biclique Detection Algorithm Complexity Analysis

Our algorithm, as described above, starts with singleton nodes and expands each one by adding adjacent nodes to them one at a time via a depth first search. Its runtime complexity and memory requirements can be modeled as

$O(N_{gc} \times E_{pc})$,where N_{gc} is the number of bicliques generated and E_{pc} is the effort required to generate a biclique and check it for dominance. E_{pc} can be seen to be linear in the size of the biclique, specifically the size of the L-set, R-set and A-set. These are in the worst case linear in the number of nodes in the network. N_{gc} on the other hand could grow exponentially with the number of nodes in the network and is therefore our principal concern. N_{gc} must grow at least linearly with Nc, since every maximal clique must be generated and checked. Ideally, N_{gc} would grow only linearly with N_c.

4   Results

4.1   Artificial Networks

We ran experiments on artificially generated networks using a desktop computer with 16 GB of memory and a 2.4 GHz. processor using CentOS Linux. Our principal goal was to determine the functional dependence of the complexity on the size of the network, the average nodal degree, the variation in nodal degree and the amount of correlation among the edges.

Experiment 1 examined the effect of network density on the number of maximal bicliques and the complexity of our algorithm. Figure 2 summarizes the results of this experiment for 100 and 200 node networks. Within each set of networks we varied the density between 0.05 and 0.5 for the 100 node network (Fig. 2a) and between 0.025 and 0.25 for the 200 node network (Fig. 2b). We recorded the number of maximal bicliques, N_c, in the network, and the algorithm's running time.

Both time and maximal bicliques increased exponentially with density and at comparable rate. This indicates that the algorithm is not expanding many nodes that are not producing maximal bicliques. The number of generated bicliques also converges to a rate that is roughly the same as that of the maximal bicliques

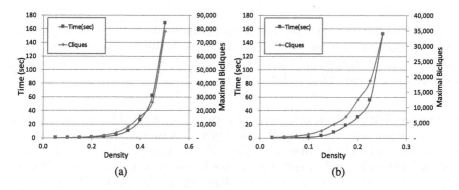

Fig. 2. Effect of network density on complexity. Time and number of maximal bicliques as a function of graph density for network sizes (a) N = 100 and (b) N = 200.

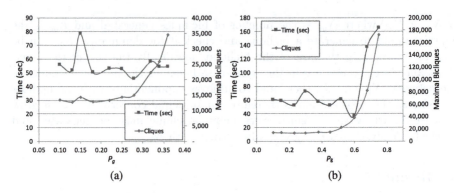

Fig. 3. Effect of correlation on complexity. Time and number of maximal bicliques as a function of P_g, the probability of an edge between nodes within a group, for network size N = 400 and groups (a) NG = 4 and (b) NG = 8.

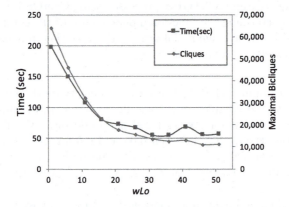

Fig. 4. Effect of node weight distribution on complexity. Time and number of maximal bicliques as a function of wLo, the minimum node weight in the network, for network sizes (a) N = 200 and (b) N = 2000. Note that as wLo increases, weight distribution narrows and degrees of nodes becomes more uniform.

N_c, but its value is approximately N times the number of cliques expanded, which is approximately twice N_c. This difference is due to the fact that in its current form, when the algorithm expands a node, n, it generates a new node corresponding to each node adjacent to n.

Experiment 2 explored the effects of correlation. We divided the nodes into groups of equal sizes and varied the probabilities of edges between nodes in the same group and nodes in different groups. We generated random networks with 200 and 400 nodes and in each case divided into groups of nodes of different sizes ($NG = 4$, $NG = 8$). P_g, the probability of an edge existing between two nodes in the same group was varied from 0.1 to 0.8, while the probability of an edge existing between nodes in different groups was set to a value that made the overall density 0.2.

This simulated the situation where the network contained groups of nodes with different properties (e.g., different types of diseases) and where the edges between nodes in the same group were correlated. Figure 3 summarizes the results of Experiment 2. We see overall complexities similar to those observed in Experiment 1, and while the correlation does have some effect, we did see some increase in the rate of maximal clique generation with the smaller sample size ($N = 200$). In the larger sample size ($N = 400$), however, we observed in Fig. 3 an exponential increase in the number of maximal bicliques generated, with the rate of growth increasingly sharply around $P_g = 0.3$ ($NG = 4$, Fig. 3a) and $P_g = 0.5$ of 0.5 ($NG = 8$, Fig. 3b).

Experiment 3 examines the effect of the distribution of node weights varied linearly between wLo and wHi. We varied wLo from a value of 1 to 50 and varied wHi to maintain a network density of 0.2 in all cases.

We generated two sets of networks, one with 100 nodes and the other with 200. Figure 4 summarizes the results for $N = 200$. As the degree distributions becomes wider, the number of maximal bicliques becomes larger, increasing by more than a factor of 4. This effect was more pronounced in the larger network and thus is shown here.

In summary, we found that the density of the network is the factor most seriously affecting complexity and by working with smaller networks we were able to examine denser networks without consuming an inordinate amount of computer time and memory. It is our intent to optimize the algorithm implementation and run it on larger problems in the near future.

4.2   Bicliques and Correlation in the HPNs

We ran our algorithm on both versions of the HPN. The pathways network containing 807 phenotypes, 1444 pathways and 82,855 edges, with a density of 0.07 and the gene HPN containing 916 phenotypes, 10,011 genes and 55,223 edges, and a density of 0.006. In both cases, the degrees of the nodes vary widely. However, there are hundreds of phenotypes associated with more than 200 pathways/genes. As discussed in Sect. 4.1, this gives rise to a large number of maximal bicliques and both the runtime and memory requirements of the algorithm rise correspondingly.

We ran an experiment in which we trimmed the network by removing all nodes exceeding a maximum degree. Table 1 and Fig. 5a report results for the pathways HPN, where the maximum degree varied between 100 and 200. Results for the gene HPN follow the same trend and are only presented in Fig. 5b, for which we varied the maximum degree between 100 and 500.

It can be seen in Table 1 that as the maximum degree increases, the fraction of phenotype nodes retained increases from 73 % to 84 % of the total number of phenotypes while the number of retained edges increases from 8 % to 28 %.

Table 1. Effects of varying maximum node degree on HPN pathways network.

Maximum degree	100	125	150	175	200
# of phenotypes	589	593	633	654	676
# of edges	6,259	9,837	14,264	17,820	23,068
# of generated bicliques	96,391	498,802	2,063,365	7,075,168	26,550,050
# of expanded bicliques	5,693	12,690	28,618	63,892	180,303
# of maximum bicliques	2,001	4,944	11,705	27,728	85,481
Running time:	2.845	16.843	90.436	413.109	9,020.638

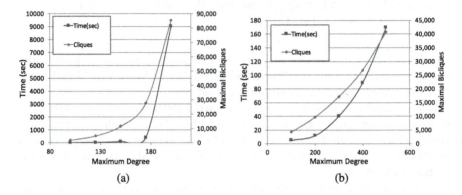

(a) (b)

Fig. 5. Effect of maximum node degree on network complexity in pathway- and gene-centric HPN. Time and number of maximal bicliques as a function of the networks maximum allowed degree for (a) HPN pathway network and (b) HPN gene network.

Thus, most of the nodes have degrees less than 100 while most of the edges are associated with nodes with degree over 200. The number of expanded and maximal bicliques increases multiplicatively as the maximum degree increases by a constant. This is exponential behavior.

Note that the number of expanded bicliques and the number of maximal bicliques as seen in Table 1 grow at the same rate with the number of expanded bicliques, with their ratio converging towards 2. This shows that the algorithm expands only about 2 bicliques for every maximal biclique found. Thus, the complexity of the algorithm in terms of the number of bicliques expanded is linear in the number of maximal bicliques, but both are exponential in the number of nodes. The algorithm's complexity cannot decrease below linear in the number of maximal bicliques since it generates all of them. Therefore, the algorithm's complexity is minimal.

The number of generated bicliques, however, increases at a higher rate than the expanded or maximal bicliques. This is due to the implementation of the algorithm, which in expanding bicliques considers all adjacent nodes and eliminates most of the generated bicliques because they are dominated.

The running time increases even more rapidly than the number of generated bicliques. This is most probably a consequence of Java's Virtual Machine Garbage Collecting technology that automatically releases memory used by the software. As the number of objects in memory increases, in our case, the number of bicliques, so is the time necessary to free unused memory blocks. This effect is most evident for the case where the maximum degree is 200. Thus, to truly reduce the complexity and make more problems tractable, even after optimizing the implementation, we must reduce the density of the graph. This is treated in more detail in the Conclusions section below.

5   Biomedical Insights into Genotype-Phenotype Bicliques

We performed a preliminary biomedical analysis of bicliques from the gene HPN. Phenotypic data includes disease-specific and non-disease phenotypes, such as body mass index and blood glucose levels.

Though comprehensive analysis of phenotypes and genes yielded over 80,000 edges, various methods can be used to filter this input in future studies. Trimming nodes with degree 100 or more, for instance, eliminates many non-specific phenotypes such as body weight and creatinine, which have been included in various GWAS studies and which may impart little useful information in elucidating the basis for human disease. On the other hand, trimming these nodes can eliminate important diseases such as multiple sclerosis and coronary artery disease, which have been extensively studied and, due in part to their complex etiologies, have a large number of genetic and pathway associations. It may therefore be advantageous to rely upon clinical expertise to eliminate phenotypes extraneous to a study, or to focus studies on a predetermined subset of phenotypes and therefore limit the number of nodes analyzed, i.e. knowledge-driven filtering.

5.1   Assessment of the Relevance of Biclique Output

For a large set of data, output can be restricted to bicliques containing a specified number of phenotype, gene or pathway-specific nodes. For instance, we restricted output to bicliques containing at least 40 (Table 2a) or 50 (Table 2b) genes and 2 phenotypes to generate 72 and 24 bicliques, respectively. We further trimmed the output to bicliques containing at least one disease-specific phenotype – i.e. we eliminated bicliques containing only non-disease phenotypes, such as gambling and heart rate. We hypothesized that bicliques containing a large number of genes would match phenotypes with known clinical, biological, or genetic overlap and examined the bicliques to assess our hypothesis.

To summarize the findings, the phenotypes in each biclique were divided into three different categories in Table 2: redundant, meaning phenotypes are clinically equivalent or one is a subtype of another; directly correlated, meaning phenotypes are pathologically or physiologically linked and thus expected to occur in the same patient; or distinct phenotypes, meaning phenotypes may share underlying genetic or molecular pathways but are not expected to occur in the same patient.

Note that in future studies, disease ontology can be used to filter these redundant or directly correlated phenotypes. In the present study, we chose not to filter these phenotypes in order to assess the validity of the algorithm in matching these phenotypes. As such, there are several noted redundancies seen between matched phenotypes in these analyses.

More interesting for clinical research are connections found between the distinct phenotypes in Table 2. Most of these phenotypes do in fact have a biochemical, pathological or clinical correlation. For instance, Crohn's disease and Ulcerative Colitis are clinical subtypes of inflammatory bowel disease but have overlapping symptomatology and are often misdiagnosed; genetic markers may have some diagnostic utility in distinguishing these diseases, and our algorithm could be useful in identifying markers that are shared versus distinct [5].

Beyond the ability of the algorithm to replicate known clinical and biological correlation, the algorithm provides an efficient tool for investigating candidate genes and pathways that underlie phenotypes with shared etiologies. For example, the algorithm can be used to identify candidate genes underlying etiological overlaps between multiple sclerosis (MS) and type I diabetes (TIDM), which tend to co-occur in families, though their genetic makeup (HLA patterns) are thought to be mutually exclusive, with some variants being a risk factor for one disease and protective against the other [3,13]. As shown in Table 3, our algorithm matched MS and TIDM to several genes already known to influence the risk of both diseases, including HLA DRB1 and HLA DRQ1. Our algorithm also matched the diseases with nine other HLA loci that could be used as candidates for future analysis of variants.

5.2   Assessment of a Disease-Limited Data Set and Hypothesis Generation

In addition to TIDM and MS, there is a high degree of connectivity between various types of autoimmune disease, as can be seen in Table 2. The development of autoimmune disease is genetically and environmentally complex and much is yet unknown. We can use bicliques and past research to draw hypotheses regarding candidate genes and pathways involved in these diseases.

For instance, recent research has shown that abnormal dopamine levels are observed within brain cells of mouse models and within tissues of human patients with autoimmune disease [21]. To identify candidate genes involved in this dopamine dysregulation, it may be useful to examine bicliques that contain autoimmune diseases as well as psychiatric disorders, which are largely driven by dopamine dysregulation.

Examination of several bicliques in Table 3 reveals a gene called NOTCH4 in several bicliques containing autoimmune disease as well as schizophrenia. As reflected in additional bicliques that are not shown, GWAS has implicated NOTCH4 in autoimmune diseases including multiple sclerosis, asthma, systemic sclerosis, lupus, ulcerative colitis and rheumatoid arthritis. In various studies,

Table 2. Clinical assessment of phenotypes pairs matched by bicliques. Phenotypes matched in bicliques containing two phenotypes and a large number of genes: a minimum of fifty (a) or forty (b). The phenotypes are divided into categories based on clinical characteristics. Redundancy indicates that the phenotype pair is the same clinical entity or that one phenotype is a subtype of the other. Direct correlation indicates that the phenotypes are patho-physiologically related and therefore expected to occur in the same patients. Distinct phenotypes indicates separate clinical entities that may be biologically or genetically associated and have been paired in bicliques according to the algorithm. Note: IBD stands for inflammatory bowel disease and MS stands for multiple sclerosis.

(a) Bicliques,containing 50 or more genes and 2 phenotypes.

Redundancy (4)	Direct correlation (3)	Distinct phenotypes (4)
IBD - Crohn's disease	Osteoporosis - Bone density	ADD/ADHD - Gambling
AIDS - HIV-1	Asthma - QT interval	Schizophrenia - Gambling
Coronary dx - Restenosis	Sudden death - Resting HR	Breast neoplasms - Breast size
IBD - Urcerative colitis		Crohn's - Ulcerative colitis

(b) Bicliques containing 40 or more genes and 2 phenotypes.

Redundancy (7)	Direct correlation (7)	Distinct phenotypes (11)
IBD - Crohn's disease	Osteoporosis - Bone density	ADD/ADHD - Gambling
AIDS - HIV-1	Asthma - Respiratory function	Schizophrenia - Gambling
Coronary dx - Restenosis	Asthma - Heart rate	Breast neoplasms - Breast size
IBD - Ulcerative colitis	Asthma - QT interval	Crohn's dx - Ulcerative colitis
Cardiac hypertrophy - Cardiomegaly	Sudden death - Resting heart rate	Alzheimer's - IgG glycosylation
Pancreas cancer - Pancreas neoplasms	Dialysis mortality - Type 2 Diabetes	Alzheimer's - lipids
Biliary cirrhosis - Primary BC	Prostate cancer - Erectile dysfunction	Type 1 Diabetes - MS
		Behcet Syndrome - MS
		Crohn's disease - MS
		Lupus - MS
		IBD - MS

NOTCH4 has also been established as a candidate gene for schizophrenia and bipolar disorder [10], and is thought to influence the age of onset of disease as well as response to antipsychotic treatment [2]. Drawing on this previous knowledge and the results of our data analysis, we hypothesize that NOTCH4 may

Table 3. Example maximal bicliques containing two phenotypes and corresponding genes. Note: maximal bicliques contain variable numbers of phenotypes and genes.

Left nodes (phenotypes)	Right nodes (genes)			
Multiple sclerosis type I diabetes	HLA-C	BAT1	PPIAP9	ZBTB12
	HLA-S	BTNL2	RPL3P2	RPL32P23
	HLA-DQA2	MICA	TAP2	FCRL3
	HLA-DRB1	NOTCH4	PRM3	SLC44A4
	HLA-DRB5	TNXB	TNP2	CFB
	HLA-DRA	CLEC16A	BAT3	TYK2
	HLA-DRB9	HNRNPA1P2	CLECL1	C2
	HLA-DOB	IL2RA	KIAA0350	C16orf75
	HLA-DQB2	BACH2	ZFP36L1	IL7R
	HLA-DMB	HCG26	BAT2	MICB
Rheumatoid arthritis schizophrenia	BDP1P	GUSBP1	HLA-DRB5	HLA-DRB9
	CDH12	HLA-DQA1	LASS6	NOTCH4
	DKFZp667F0711		PRKCQ	SALL3
Multiple sclerosis lupus	BTNL2	RNF39	ZBTB12	MTCO3P1
	DDR1	TNXB	WASF5P	PVT1
	DHFRP2	HLA-DRA	CBLN2	HLA-DRB1
	HLA-B	BACH2	HLA-S	HLA-DRB9
	HLA-X	HCG18	RPS2P6	FENDRR
	IER3	HLA-DOB	FOXF1	HCG26
	MICA	HLA-L	IRF8	TCF19
	MICB	HLA-DQB2	CLEC16A	TNFAIP3
	NOTCH4	EHMT2	HLA-DQA2	TRIM26
	PSORS1C1	BAT2	HLA-DQB1	AIF1
	BAT3	CFB		

also play a role in dopaminergic pathway dysregulation underlying autoimmune etiology. Investigation of genetic and pathway overlap between distinct diseases can be used in biomedical studies to develop such hypotheses.

6   Conclusions and Future Work

We found that the complexity of our algorithm is the product of the number of nodes and the number of generated bicliques. As hypothesized, network density is the major determinant of computational complexity. Node degree distribution and correlation among edges are secondary factors, except when resulting in a significant number of nodes of high degree.

To address complexity in large data sets, it may seem beneficial to trim the network by removing nodes of high degree. However, it is important to assess the significance of these nodes in the context of the investigation, in our case the biological and genetic underpinnings of disease. It may also be beneficial to partition the network into distinct parts; for instance, to analyze one set of diseases and their pathway and gene associations.

In phenotype-pathway analysis, we found that few of the bicliques contained a large number of phenotypes and pathways. Many of the bicliques contained a small number of phenotypes links to a large number of pathways, or vice versa. Many bicliques included redundancies between similar phenotypes and pathways, as expected, and such nodes could be eliminated or combined to reduce the size and complexity of the problem, if clinically or scientifically appropriate.

In the phenotype-gene analysis, the algorithm linked many phenotypes that were redundant, closely related, or have been associated in clinical or biological studies. The algorithm further reveals genes that may underlie these associations and also provides a tool for genotype-phenotype hypothesis generation.

As we apply the algorithm to additional investigations in biology and medicine, we will evaluate the techniques that can be guided by biological properties. The maximal biclique problem described above was defined on an unweighted graph. The problem can be extended naturally to one where an edge is given an associated weight, indicating the strength of the relationship between nodes or where a node is given a weight based on its biological significance.

Potentially, these insights can be used to enhance our understanding of the biochemical basis for disease and to identify useful targets for drug repositioning between diseases. Besides phenotype-gene interactions, the algorithm can be used to investigate single nucleotide polymorphisms (SNPs), as well as environmental and other risk factors associated with disease.

Acknowledgments. This work was supported by National Institutes of Health grants LM009012, LM010098, and EY022300.

References

1. Alexe, G., Alexe, S., Crama, Y., Foldes, S., Hammer, P.L., Simeone, B.: Consensus algorithms for the generation of all maximal bicliques. Discrete Appl. Math. **145**(1), 11–21 (2004). Graph Optimization {IV}
2. Anttila, S., Illi, A., Kampman, O., Mattila, K.M., Lehtimaki, T., Leinonen, E.: Interaction between NOTCH4 and catechol-O-methyltransferase genotypes in schizophrenia patients with poor response to typical neuroleptics. Pharmacogenetics **14**(5), 303–307 (2004)
3. Atkinson, M.A., Eisenbarth, G.S.: Type 1 diabetes: new perspectives on disease pathogenesis and treatment. Lancet **358**(9277), 221–229 (2001)
4. Cheng, Y., Church, G.M.: Biclustering of expression data. In: Proceedings of the International Conference Intelligent Systems for Molecular Biology, vol. 8, pp. 93–103 (2000)

5. Cleynen, I., Boucher, G., Jostins, L., Schumm, L.P., Zeissig, S., Ahmad, T., Andersen, V., Andrews, J.M., Annese, V., Brand, S., Brant, S.R., Cho, J.H., Daly, M.J., Dubinsky, M., Duerr, R.H., Ferguson, L.R., Franke, A., Gearry, R.B., Goyette, P., Hakonarson, H., Halfvarson, J., Hov, J.R., Huang, H., Kennedy, N.A., Kupcinskas, L., Lawrance, I.C., Lee, J.C., Satsangi, J., Schreiber, S., Théâtre, E., van der Meulen-de Jong, A.E., Weersma, R.K., Wilson, D.C., Parkes, M., Vermeire, S., Rioux, J.D., Mansfield, J., Silverberg, M.S., Radford-Smith, G., McGovern, D.P.B., Barrett, J.C., Lees, C.W.: Inherited determinants of Crohn's disease, ulcerative colitis phenotypes: a genetic association study. Lancet (2015)

6. Darabos, C., Desai, K., Cowper-Sal·lari, R., Giacobini, M., Graham, B.E., Lupien, M., Moore, J.H.: Inferring human phenotype networks from genome-wide genetic associations. In: Vanneschi, L., Bush, W.S., Giacobini, M. (eds.) EvoBIO 2013. LNCS, vol. 7833, pp. 23–34. Springer, Heidelberg (2013)

7. Darabos, C., Grussing, E.D., Cricco, M.E., Clark, K.A., Moore, J.H.: A bipartite network approach to inferring interactions between environmental exposures and human diseases. In: Pacific Symposium on Biocomputing, pp. 171–182 (2015)

8. Darabos, C., Harmon, S.H., Moore, J.H.: Using the bipartite human phenotype network to reveal pleiotropy and epistasis beyond the gene. In: Pacific Symposium on Biocomputing, pp. 188–199 (2014)

9. Darabos, C., White, M.J., Graham, B.E., Leung, D.N., Williams, S.M., Moore, J.H.: The multiscale backbone of the human phenotype network based on biological pathways. BioData Min. **7**(1), 1 (2014)

10. Dieset, I., Djurovic, S., Tesli, M.: Up-regulation of NOTCH4 gene expression in bipolar disorder. Am. J. Psychiatry **169**, 1292–1300 (2012)

11. Gaspers, S., Kratsch, D., Liedloff, M.: On independent sets and bicliques in graphs. In: Broersma, H., Erlebach, T., Friedetzky, T., Paulusma, D. (eds.) WG 2008. LNCS, vol. 5344, pp. 171–182. Springer, Heidelberg (2008)

12. Gondran, M., Minoux, M., Vajda, S.: Graphs and Algorithms. Wiley, New York (1984)

13. Lernmark, A.: Multiple sclerosis and type 1 diabetes: an unlikely alliance. Lancet **359**(9316), 1450–1451 (2002)

14. Li, J., Li, H., Soh, D., Wong, L.: A correspondence between maximal complete bipartite subgraphs and closed patterns. In: Jorge, A.M., Torgo, L., Brazdil, P.B., Camacho, R., Gama, J. (eds.) PKDD 2005. LNCS (LNAI), vol. 3721, pp. 146–156. Springer, Heidelberg (2005)

15. Liu, J., Wang, W.: Op-cluster: clustering by tendency in high dimensional space. In: 2003 Third IEEE International Conference on Data Mining, ICDM 2003, pp. 187–194, November 2003

16. Liu, X., Li, J., Wang, L.: Modeling protein interacting groups by quasi-bicliques: complexity, algorithm, and application. IEEE/ACM Trans. Comput. Biol. Bioinform. **7**(2), 354–364 (2010)

17. Makino, K., Uno, T.: New algorithms for enumerating all maximal cliques. In: Hagerup, T., Katajainen, J. (eds.) SWAT 2004. LNCS, vol. 3111, pp. 260–272. Springer, Heidelberg (2004)

18. Maulik, U., Mukhopadhyay, A., Bhattacharyya, M., Kaderali, L., Brors, B., Bandyopadhyay, S., Eils, R.: Mining quasi-bicliques from HIV-1-human protein interaction network: a multiobjective biclustering approach. IEEE/ACM Trans. Comput. Biol. Bioinform. **10**(2), 423–435 (2013)

19. Milner, R.: Bigraphs and their algebra. Electron. Notes Theor. Comput. Sci. **209**, 5–19 (2008)

20. Newman, M.: Networks: An Introduction. Oxford University Press Inc., New York (2010)
21. Pacheco, R., Contreras, F., Zouali, M.: The dopaminergic system in autoimmune diseases. Front. Immunol. **5**, 1–17 (2014)
22. Prisner, E.: Bicliques in graphs I: bounds on their number. Combinatorica **20**(1), 109–117 (2000)
23. Qiu, J., Darabos, C., Moore, J.H.: Studying the genetics of complex diseases with ethnicity-specific human phenotype networks: the case of type 2 diabetes in east asian populations. In: 5th Translational Bioinformatics Conference (2014)
24. Sanderson, M.J., Driskell, A.C., Ree, R.H., Eulenstein, O., Langley, S.: Obtaining maximal concatenated phylogenetic data sets from large sequence databases. Mol. Biol. Evol. **20**(7), 1036–1042 (2003)
25. Sim, K., Li, J., Gopalkrishnan, V., Liu, G.: Mining maximal quasi-bicliques to co-cluster stocks and financial ratios for value investment. In 2006 Sixth International Conference on Data Mining, ICDM 2006, pp. 1059–1063, December 2006
26. Tanay, A., Sharan, R., Shamir, R.: Discovering statistically significant biclusters in gene expression data. Bioinformatics **18**(Suppl 1), S136–44 (2002)
27. Wang, H., Wang, W., Yang, J., Yu, P.S.: Clustering by pattern similarity in large data sets. In: Proceedings of the ACM SIGMOD International Conference on Management of Data, SIGMOD 2002, pp. 394–405. ACM, New York (2002)
28. Zhang, Y., Phillips, C.A., Rogers, G.L., Baker, E.J., Chesler, E.J., Langston, M.A.: On finding bicliques in bipartite graphs: a novel algorithm and its application to the integration of diverse biological data types. BMC Bioinform. **15**, 110 (2014)

Hybrid Biclustering Algorithms for Data Mining

Patryk Orzechowski[(⊠)] and Krzysztof Boryczko

Department of Automatics and Bioengineering,
AGH University of Science and Technology, al. Mickiewicza 30,
30-059 Cracow, Poland
{patrick,boryczko}@agh.edu.pl
http://www.agh.edu.pl

Abstract. Hybrid methods are a branch of biclustering algorithms that emerge from combining selected aspects of pre-existing approaches. The syncretic nature of their construction enriches the existing methods providing them with new properties. In this paper the concept of hybrid biclustering algorithms is explained. A representative hybrid biclustering algorithm, inspired by neural networks and associative artificial intelligence, is introduced and the results of its application to microarray data are presented. Finally, the scope and application potential for hybrid biclustering algorithms is discussed.

Keywords: Data mining · Biclustering techniques · Gene expression data · Microarray analysis

1 Introduction

Biclustering is one of very popular recent data mining techniques intended to detect local similarities within a dataset. Biclustering (also called co-clustering or two-mode clustering) algorithms, in opposition to classic clustering approaches, take into account rows and columns of the input matrix simultaneously. A variety of biclustering methods has already gained recognition in multiple domains, such as biology and biomedicine, genetics, recommendation systems, marketing, text mining and pattern recognition [1–13].

It has already been acknowledged that no single biclustering algorithm is able to outperform others in every competition. As universal solution is not available, any algorithm fitted to specific type of data yields advantage [14]. Hence arises the necessity of reviewing and classifying the most established methods and indicating their strengths [15]. This pushes forward the foundations of new algorithms, especially those that could be considered as more universal.

P. Orzechowski—AGH University of Science and Technology, Faculty of Electrical Engineering, Automatics, Computer Science and Biomedical Engineering, Department of Automatics and Bioengineering.

K. Boryczko—AGH University of Science and Technology, Faculty of Computer Science, Electronics and Telecommunications, Department of Computer Science.

G. Squillero and P. Burelli (Eds.): EvoApplications 2016, Part I, LNCS 9597, pp. 156–168, 2016.
DOI: 10.1007/978-3-319-31204-0_11

Combining the strengths of multiple approaches has been known as ensemble approaches. Such techniques are believed to provide more stable and effective results [7,16,17]. Those algorithm sub-sample the data, apply various learning methods (sometimes with varying parameters, too), gather, and in their final stage combine the results, for example by means of voting.

In this paper a different mechanism of incorporating multiple biclustering algorithms is presented. In the following sections we explain the motivation and properties of hybrid methods, provide an exemplary algorithm and assess its effectiveness on four GEO Series datasets.

2 Methods

Out of the multiplicity of biclustering approaches [18], six well recognized algorithms have been selected for comparison with the hybrid Propagation-Based Biclustering Algorithm (PBBA) introduced in the present paper: Cheng-Church (CC) [19], Plaid Models [20], xMotifs [21], BiMax [22], CPB [23] and QUBIC [24].

2.1 Cheng-Church (CC)

The Cheng and Church algorithm uses the Mean Square Residue $H(I, J)$ at each iteration as presented in (1) to determine the low variance bicluster.

$$H(I, J) = \frac{1}{|I||J|} \sum_{i \in I, j \in J} (a_{ij} - a_{iJ} - a_{Ij} + a_{IJ})^2 \tag{1}$$

where a_{ij} is the element in i-th row and j-th column of the input matrix, a_{iJ} and a_{Ij} are row and column means, respectively, and a_{IJ} is the mean of submatrix (I, J). Averages by rows and columns are used in order to delete nodes (columns or nodes) exceeding the scaled $H(I, J)$ and later to add nodes lower than $H(I, J)$ [19]. Multiple bicluster detection is possible by masking the previously obtained bicluster with random values.

2.2 Bimax

Bimax algorithm locates inclusion-maximal biclusters in binarized data by following a divide-and-conquer approach [22]. A bicluster is called inclusion-maximal, when it may not be included in any other bicluster, thus its consistency is lost when adding any row or column. After discretizing the dataset by applying a threshold (preferably: the mean of the data) based on a selected pattern, the input matrix is resorted and divided into three possibly overlapping submatrices. The first one contains elements concordant with the pattern, the second one elements that are partially concordant or non-concordant with the pattern and the third one only null elements. The clever resorting of the rows allows disregarding the third submatrix and executing the algorithm recursively on the remaining two submatrices, with special attention paid to their overlapping part.

2.3   XMotifs

The xMotifs algorithm uses a probabilistic approach to determine a *conserved gene expression motif*, also referred to as the xMotif [21]. The following conditions need to be met for a bicluster to be an xMotif: size (at least $1/\alpha$ columns need to be contained in a bicluster B, where $\alpha > 0$), conservation (every row in a bicluster needs to have the same value in each of the bicluster columns) and maximality (the values in each row outside the bicluster may share at most $1/\gamma$ of values inside the bicluster, where $\gamma < 1$).

The algorithm starts with determining intervals between ordered values for each row. As normal distribution of data is assumed, all rows that were unlikely to have been generated by an uniform distribution are considered as significant. A seed column (randomly selected) and a discriminating set of columns (defined uniformly at random) are selected. All rows with the same states in the seed columns and the discriminating set are added to the bicluster.

2.4   QUBIC

The QUBIC algorithm looks for heavy subgraphs in a bipartite graph [24]. Each row in the input matrix is sorted and ranked according to the position of its values placed in an ascending order. The weighted complete graph is built from the data, with rows are represented as vertices and edges assigned weights that reflect similarity between the vertices (i.e. the number of identical values in the same columns). By browsing the sorted list of edges in descending order, bicluster seeds are determined. The algorithm iteratively expands seed edges aiming at finding biclusters $B(I,J)$ with $min\{|I|,|J|\}$ as large as possible. Consistency relaxation is used to include rows that are not entirely concordant with the rest of the bicluster as well. In its final step, QUBIC includes oppositely regulated values in the bicluster.

2.5   Plaid Models

Plaid merges two-sided clustering with ANOVA methods to generate so-called layers [20]. A general additive model of overlapping layers is used to model the values a_{ij} of the input matrix (2):

$$a_{ij} = \mu_0 + \sum_{k=1}^{K} (\mu_k + \alpha_{ik} + \beta_{jk})\rho_{ik}\kappa_{jk} \tag{2}$$

where: K represents the number of layers (i.e. biclusters), ρ_{ik} and κ_{jk} are binary values specifying if row i and column j belong to bicluster k, μ_k is the background effect and α_{ik} and β_{jk} represent layer k orderings of the effects upon the rows and columns.

The Plaid approach resembles the Expectation Maximization (EM) algorithm or unconstrained Singular Value Decomposition (SVD) [20,25].

2.6   CPB

Correlated Pattern Biclustering (CPB) algorithm [23] uses Pearson Correlation Coefficient with regard to a randomly selected set of input matrix columns to determine similarity of rows. An initial bicluster is improved by modifying in turns the set of rows and set of columns with a technique similar to mean-shift [26]. At each iteration the bicluster center (analogical to the center of mass) is determined and updated by calculating correlations between the center and each of the rows. Bicluster columns are updated in a different manner, using the root mean squared error (RMSE) value. The procedure is repeated until convergence or the 20th iteration is reached.

3   Hybrid Biclustering Algorithms

The advantage of combining multiple strategies is the fact that each technique provides a different tangent of insight into the data. Thus, performing an analysis with a single (bi)clustering algorithm provides only partial information about the dataset. Thus ensemble and aggregated approaches emerged as multi-objective criterion selection tool. They usually include partial solution by means of voting, applying a consensus function (for example Normalized Mutual Information [27]) or a dedicated technique, such as bagging (for example random forests [28]) or boosting [29]. The consolidated methods (for example consensus clustering) aim to determine a median partition (also referred to as consensus) representing on average the highest similarity among all stochastic methods applied. Typically, the consensus matrix is built and an extra algorithm is used to provide more stable results [30]. Consensus and ensemble clustering approaches are said to improve quality of the results, robustness, multi-view perspective and considering the match across the base solutions. Ensembles are also easily separable and enable distributed computing [27].

Unfortunately there are certain problematic issues concerning ensemble clustering algorithms as well. The most important drawback of ensembles is lack of transparency. Even though the mechanism of reaching agreement between different clustering results is overt (rules of voting), the various enseble members may provide inconsistent class predictions. To avoid this, independent training of the classifiers needs to be performed. Otherwise, some of the techniques may become overrepresented, leading to impediment of the voting reliability. Independent training is also one of the issues that impedes the extensibility of an ensemble.

Distributed processing allows cutting execution time at cost of increased communication and greater memory consumption. The results of each of the classifiers need to be computed, stored and merged into a single assembled answer, resulting in increased latency.

The aforementioned drawbacks of ensemble algorithms motivated the introduction of *biclustering algorithms*. These are methods combining the mechanisms underpinning various algorithms, such as classifiers, techniques or structures from various existing solutions. What makes hybrid algorithms distinctive is the fact that the result (i.e. a specific bicluster) is determined *during* the classification instead of combining the results of various classifiers naively at the end, as it takes place in ensemble methods. Thus, the hybrid algorithms may be considered cooperative, whereas ensemble methods would be perceived as competitive. We have observed that intelligent synergy of the mechanism coming from different algorithms allows the solution obtained to generate more reliable and transparent results in comparison to frequently unclear manner of merging various sets of previously generated results.

Admittedly, both hybrid and ensemble algorithms share some similarities. They are both data mining methods wherein multiple techniques are applied to draw conclusions about the data structure. The complexity of both methods is determined by the complexity of their members, but ensemble methods require extra voting procedure as a final stage. Both algorithms are difficult to extend. Adding another ensemble method reasonably requires insight into classifiers, so that each classifier is assigned an appropriate voting strength. On the other hand, devising a hybrid method requires proper selection of techniques or metrics. Sometimes this requires reconsidering the whole algorithm and relations between various measures, which may be hardly possible.

The major advantages of hybrid approaches include increased transparency, decreased storage requirements and computation times. In hybrid biclustering algorithms, the resulting biclusters are determined as conjunction of existing methods. This allows precise determination of what and how affects the formation of a bicluster. Overall efficiency of hybrid methods is increased as they share common data types and structures. Extra overhead required by the voting procedure is not necessary either.

3.1 Propagation-Based Biclustering Algorithm (PBBA)

One of hybrid biclustering algorithm representatives is the Propagation-Based Biclustering Algorithm (PBBA). The method is inspired by neural networks and associative artificial intelligence [31,32] in its essence, but was developed as a hybrid of two biclustering algorithms. The algorithm combines the concept of incremental version of BiMax [22] with a conserved expression concept of xMotifs [21]. BiMax applies a binary threshold on the dataset and considers values above the threshold only while PBBA takes into consideration discrete values in each row. For floating point numbers a multi-levelled threshold is applied, in a way similar to ranking values in QUBIC [24].

The general concept of the algorithm is presented in Fig. 1. The algorithm iterates consecutive rows and propagates each row's motif in order to find similarities in other rows. Different subsets of the seed row are detected by finding the nearest row with the exact value appearing in the particular column.

The details of the algorithm have been presented in the Algorithm 1 section. The version of the PBBA algorithm presented herein was implemented in C++.

Each row, beginning with the last one, is depicted as a new seed, treated as a conserved sequence motif. In every iteration the set of biclusters corresponding to the motif is expanded, either by appending existing biclusters or intersecting

Fig. 1. The mechanism of PBBA - the algorithm seeks the nearest row with the same value in a particular column

---

Algorithm 1. Propagation-based biclustering algorithm (PBBA)

1: **procedure** PBBA(matrix A)
2:     $M_{all} \leftarrow \emptyset$ ▷ set with all biclusters
3:     $R_{all} \leftarrow \emptyset$ ▷ set with restricted motifs
4:     **for** $i \leftarrow n \dots 1$ **do**   ▷ set each row as seed
5:         $M_i \leftarrow \emptyset$ ▷ store all biclusters common with i-th pattern
6:         $M_i \leftarrow \text{insert}(M_i, R_i, \ B(i, A_{i*}))$ ▷ add seed to retrieved biclusters
7:         $mask \leftarrow A_{i*}$ ▷ propagate the motif to further rows
8:         $\{lev, pat\} \leftarrow \text{next\_level}(mask)$
9:         **while** $pat \neq \emptyset$ **do**   ▷ proceed through all rows similar to $seed$
10:            $M_i \leftarrow \text{insert}(M_i, R_i, \ B(lev \cup i, \ pat))$ ▷ intersect lev with M_i
11:            $R_{lev} \leftarrow R_{lev} \cup \{pat\}$ ▷ forbid addition of any subset of pat
12:            $mask(\{j : mask(j) = lev\}) \leftarrow v(mask(j))$ ▷ proceed to next row
13:            $\{lev, pat\} \leftarrow \text{next\_level}(mask)$
14:        **end while**
15:        $M_{all} \leftarrow M_{all} \cup M_i;$
16:    **end for**
17:    $\text{print}(M_{all})$
18: **end procedure**

with new patterns. This is performed within the *insert* procedure, where biclusters originating from i-th row and contained in M_i are step by step extended to their maximality. A verification takes place if an intersection has not been previously discovered. The algorithm also maintains a list of restrictions for each row (denoted as R_i) in order to skip previously detected motifs.

PBBA may be considered a generalization of BiMax as it applies a multi-threshold on the data where BiMax uses a binary one.

4   Results

In this section different areas of hybrid method applicationare presented. The validation of correctness of hybrid PBBA algorithm was performed on a synthetic dataset. The validation of biological significance was performed on microarray gene expression datasets.

4.1   Synthetic Databases

The correctness of the PBBA algorithm has been verified with two well established benchmarks provided by Prelic [22] and Li [24]. For the verification of the algorithm correctness, the data has been binarized with various thresholds. As expected for this type of data, the results of PBBA and BiMax have been exact. This proves that PBBA is capable of providing the results of BiMax, therefore it may be considered more general.

4.2   Microarray Gene Expression Data

For verification of biological significance of the results four Gene Series (GSE) datasets representing human tissues have been used (see Table 1). The datasets were previously used in [33]. An R package called *GEOquery* [34] has been used to acquire the data from NCBI Gene Expression Omnibus (GEO). Raw data has been read with *affy* R package [35] and preprocessed with a standard technique for Affymetrix gene chips (MAS 5.0). Similarly to Eren et al. [14], the biclusters were considered enriched if the P-value after Benjamini-Hochberg correction for any gene ontology term was smaller than 0.05.

The histograms of the datasets obtained are presented in Fig. 2.

The input parameters for the algorithms were based on remarks of the authors. All the algorithms were run with a grid of input parameters. Thus, for Bimax multiple different binarization thresholds were checked, ranging from 1 % of the data to the data mean. Different values of minimum number of rows and columns were also considered (generally speaking: at least 9 columns and at least 15 rows were required). For CC α varied from 1.2 to 1.5 and δ from 1 to 1500. Both input parameters of the Plaid algorithm were taken from the grid of values ranging from 0.5 to 0.7. For xMotifs, the following settings were taken into account: n_s=10, n_d=1000, α=0.05, with β ranging from 10 to 100 and s_d

Table 1. Datasets used for verification of biological enrichments. The table includes the experiment name, number of genes and samples, microarray platform and tissue description.

GEO ID	Genes	Samples	Platform	Description
GSE3585	22215	12	U133A	Heart
GSE5090	22215	17	U133A	Omental adipose tissue
GSE5390	22215	15	U133A	Dorsolateral prefrontal cortex
GSE7148	22215	14	U133A	Peripheral blood leukocytes

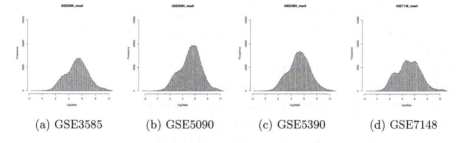

(a) GSE3585   (b) GSE5090   (c) GSE5390   (d) GSE7148

Fig. 2. Histograms of the analyzed datasets.

from 5 to 7. For PBBA, the discretization threshold varied from 20 to 100. The minimum numbers of columns and rows considered were similar to BiMax.

Terms from Biological Processes Ontology have been selected for validation. Genes that were overrepresented have been filtered out using the *genefilter* R library [36] with an arbitrary threshold of non-specific filtering (IQR) set to 0.5. Additionally, probes with small variation, no GO annotation or no Entrez Gene ID have been disregarded. The significance of each bicluster has been verified with the hypergeometric test constituting a part of the *GOstats* R package [37], applied to each GO term. Following the comparison of Eren et al. [14], the biclusters were considered significantly enriched if the adjusted p-value of any term after Benjamini-Hochberg's correction [38] was smaller than 0.05.

Biclusters generated by all input parameter settings for each algorithm have been taken into account. Overlapping biclusters with more than 25 % in common with the most enriched ones have been disregarded. The most enriched bicluster found for each dataset by two best biclustering algorithms has been shown in Table 2.

The proportions of the most enriched biclusters averaged for four datasets are presented in Fig. 3. Notice that in our tests the Bimax algorithm was unable to detect any significantly enriched biclusters for the GSE3585 dataset.

It needs to be mentioned that one of the input parameters of various algorithms concerned the expected number of biclusters, possibly impeding the presented statistics. Thus all biclusters detected by the algorithms were taken into

Table 2. Biclusters containing the most enriched terms for each dataset. Note that Bimax did not manage to find significantly enriched term for the GSE3585 dataset.

Method	p-value	GO term	Enrichment
GSE3585			
PBBA	6.34e-06	GO:0006614	SRP-dependent cotranslational protein targeting to membrane
CC	1.76e-05	GO:0006879	cellular iron ion homeostasis
xMotifs	0.00278402	GO:0019405	alditol catabolic process
CPB	0.00133498	GO:0018394	peptidyl-lysine acetylation
QUBIC	0.0028911	GO:0018917	fluorene metabolic process
Plaid	0.0306778	GO:0003357	noradrenergic neuron differentiation
GSE5090			
Bimax	9.17e-82	GO:0006415	translational termination
PBBA	3.97e-80	GO:0006415	translational termination
Plaid	1.1857e-60	GO:0006415	translational termination
CPB	3.94261e-39	GO:0006614	SRP-dependent cotranslational protein targeting to membrane
xMotifs	2.4337e-23	GO:0006414	translational elongation
QUBIC	1.29138e-07	GO:0071294	cellular response to zinc ion
CC	8.1269e-05	GO:0030317	sperm motility
GSE5390			
Bimax	7.67e-20	GO:0006414	translational elongation
PBBA	7.21863e-15	GO:0006414	translational elongation
CPB	1.85827e-11	GO:0002480	antigen processing and presentation of exogenous peptide antigen
QUBIC	4.49754e-07	GO:0035637	multicellular organismal signaling
xMotifs	3.49099e-06	GO:0002480	antigen processing and presentation of exogenous peptide antigen
CC	0.000197919	GO:0000718	nucleotide-excision repair, DNA damage removal
Plaid	0.00346504	GO:0051084	'de novo' posttranslational protein folding
GSE7148			
CPB	2.44e-21	GO:0034340	response to type I interferon
CC	2.72041e-13	GO:0032501	multicellular organismal process
QUBIC	2.01813e-09	GO:0002576	platelet degranulation
Bimax	9.14979e-07	GO:0019884	antigen processing and presentation of exogenous
PBBA	8.96482e-07	GO:0019884	antigen processing and presentation of exogenous antigen
xMotifs	0.000304073	GO:0006297	nucleotide-excision repair, DNA gap filling
Plaid	0.00899831	GO:0005999	xylulose biosynthetic process

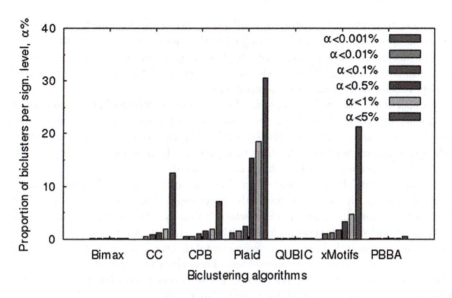

Fig. 3. The proportion of significantly enriched biclusters per significance level

account and those that overlapped with previously detected biclusters have been disregarded.

It may be noticed that for three of the datasets taken into account, the PBBA results placed the algorithm among the higher ranking tier. In one case PBBA found the most enriched biclusters and twice it was second to BiMax. On the other hand, applying a technique other than that of BiMax allowed the hybrid algorithm to find more enriched terms.

5   Summary

The present paper outline a new branch of biclustering algorithms called *hybrid biclustering algorithms*. These are algorithms combining various measures, techniques or structures of the pre-existing biclustering algorithms. An example of hybrid biclustering algorithms called the Propagation-Based Biclustering Algorithm (PBBA) has been presented. It has been proved that the method may be successfully applied not only to gene expression data, but also towards other areas such as text mining or analysis of symbolic datasets within knowledge spreadsheets.

At this stage, the formal definition of what may or may not be considered as a feature for hybrid algorithms requires further investigations. Developing a parallel hybrid biclustering algorithm, especially on GPU [39] or Intel MIC or may be worthwhile.

The hybrid methods should not be treated as "yet another biclustering technique". Their foundation is based on a completely new concept which utilizes hitherto achievements in an original manner. The algorithms aggregate various techniques to perform biclustering, unlike consensus functions for biclustering in ensemble approaches aiming at aggregating the biclustering results [40].

The major advantages exhibited by hybrid methods in comparison to the well-established ensemble methods include increased transparency, decreased storage requirements and lower computation time.

Furthermore, the paper suggests that it may be worthwhile to reconsider previously designed algorithms, focus on the techniques used and adapt the very components constituting the strengths of particular methods. Combining different approaches opens the door to a completely new result quality, at the same time rendering hybrid methods more general.

In the authors' belief hybrid algorithms are more advanced than ensemble methods. Firstly, the ensemble methods involve no recipe or agreement as to how the optimum solution should be selected. Secondly, the internal structures of methods included in the ensemble algorithm is ignored. Adding another method may always cause imbalance in assessing capabilities of the ensemble as some of the set features become overrepresented. Thus the disturbance may cause the method to decrease its objectivity level. Those two reasons explain the need for developing more sophisticated data mining methods.

Unquestionably, developing a valuable representative of hybrid biclustering algorithms remains an effort. Additionally, a mixture of different features does not guarantee improved results. The design of a successful algorithm sometimes requires a combination of vision and luck on the part of the designer.

Acknowledgements. This research was funded by the Polish National Science Center (NCN), grant No. 2013/11/N/ST6/03204. This research was supported in part by PL-Grid Infrastructure.

References

1. Broder, A., Fontoura, M., Josifovski, V., Riedel, L.: A semantic approach to contextual advertising. In: Proceedings of the 30th Annual International ACM SIGIR Conference on Research and Development in Information Retrieval, pp. 559–566. ACM (2007)
2. Busygin, S., Prokopyev, O., Pardalos, P.M.: Biclustering in data mining. Comput. Oper. Res. **35**(9), 2964–2987 (2008)
3. de Castro, P.A.D., de França, F.O., Ferreira, H.M., Von Zuben, F.J.: Applying biclustering to text mining: an immune-inspired approach. In: de Castro, L.N., Von Zuben, F.J., Knidel, H. (eds.) ICARIS 2007. LNCS, vol. 4628, pp. 83–94. Springer, Heidelberg (2007)
4. Franca, F.O.D.: Scalable overlapping co-clustering of word-document data, pp. 464–467. IEEE, December 2012
5. Henriques, R., Madeira, S.: Biclustering with flexible plaid models to unravel interactions between biological processes. IEEE/ACM Trans. Comput. Biol. Bioinf. **12**, 738–752 (2015)

6. Hussain, S.F., Bisson, G., Grimal, C.: An improved co-similarity measure for document clustering. In: Proceedings of the 2010 Ninth International Conference on Machine Learning and Applications, ICMLA 2010, pp. 190–197. IEEE Computer Society, Washington, DC (2010)

7. Kaiser, S.: Biclustering: methods, software and application. PhD thesis, Ludwig-Maximilians-Universitt Mnchen (2011)

8. Liang, T.P., Lai, H.J., Ku, Y.C.: Personalized content recommendation and user satisfaction: theoretical synthesis and empirical findings. J. Manag. Inf. Syst. **23**(3), 45–70 (2006)

9. Mimaroglu, S., Uehara, K.: Bit sequences and biclustering of text documents. In: ICDMW, pp. 51–56. IEEE (2007)

10. Stawarz, M., Michalak, M.: eBi – the algorithm for exact biclustering. In: Rutkowski, L., Korytkowski, M., Scherer, R., Tadeusiewicz, R., Zadeh, L.A., Zurada, J.M. (eds.) ICAISC 2012, Part II. LNCS, vol. 7268, pp. 327–334. Springer, Heidelberg (2012)

11. Stawarz, M., Michalak, M.: HRoBi – the algorithm for hierarchical rough biclustering. In: Rutkowski, L., Korytkowski, M., Scherer, R., Tadeusiewicz, R., Zadeh, L.A., Zurada, J.M. (eds.) ICAISC 2013, Part II. LNCS, vol. 7895, pp. 194–205. Springer, Heidelberg (2013)

12. Zhang, K., Katona, Z.: Contextual advertising. Mark. Sci. **31**(6), 980–994 (2012)

13. Zhao, H., Wee-Chung Liew, A., Wang, D.Z., Yan, H.: Biclustering analysis for pattern discovery: current techniques, comparative studies and applications. Curr. Bioinf. **7**(1), 43–55 (2012)

14. Eren, K., Deveci, M., et al.: Bbiclustering algorithms for gene expression data. Briefings Bioinf. **14**, 279–292 (2012)

15. Orzechowski, P.: Proximity measures and results validation in biclustering – a survey. In: Rutkowski, L., Korytkowski, M., Scherer, R., Tadeusiewicz, R., Zadeh, L.A., Zurada, J.M. (eds.) ICAISC 2013, Part II. LNCS, vol. 7895, pp. 206–217. Springer, Heidelberg (2013)

16. Hanczar, B., Nadif, M.: Ensemble methods for biclustering tasks. Pattern Recogn. **45**(11), 3938–3949 (2012)

17. Kotsiantis, S., Pintelas, P.: Combining bagging and boosting. Int. J. Comput. Intell. **1**(4), 324–333 (2004)

18. Pontes, B., Girldez, R., Aguilar-Ruiz, J.S.: Biclustering on expression data: a review. J. Biomed. Inform. **57**, 163–180 (2015)

19. Cheng, Y., Church, G.M.: Biclustering of expression data. Proceedings of the Eighth International Conference on Intelligent Systems for Molecular Biology, vol. 8, pp. 93–103 (2000)

20. Lazzeroni, L., Owen, A., et al.: Plaid models for gene expression data. Statistica Sinica **12**(1), 61–86 (2002)

21. Murali, T., Kasif, S.: Extracting conserved gene expression motifs from gene expression data. In: Proceedings of Pacific Symposium on Biocomputing, vol. 3, pp. 77–88 (2003)

22. Prelić, A., Bleuler, S., Zimmermann, P., Wille, A., Bühlmann, P., Gruissem, W., Hennig, L., Thiele, L., Zitzler, E.: A systematic comparison and evaluation of biclustering methods for gene expression data. Bioinformatics **22**(9), 1122–1129 (2006)

23. Bozdağ, D., Parvin, J.D., Catalyurek, U.V.: A biclustering method to discover co-regulated genes using diverse gene expression datasets. In: Rajasekaran, S. (ed.) BICoB 2009. LNCS, vol. 5462, pp. 151–163. Springer, Heidelberg (2009)

24. Li, G., Ma, Q., Tang, H., Paterson, A.H., Xu, Y.: QUBIC: a qualitative biclustering algorithm for analyses of gene expression data. Nucleic Acids Res. **37**(15), e101–e101 (2009)
25. Madeira, S.C., Oliveira, A.L.: Biclustering algorithms for biological data analysis: a survey. IEEE/ACM Trans. Comput. Biol. Bioinf. **1**(1), 24–45 (2004)
26. Comaniciu, D., Meer, P.: Mean shift: a robust approach toward feature space analysis. IEEE Trans. Pattern Anal. Mach. Intell. **24**(5), 603–619 (2002)
27. Ghosh, J., Acharya, A.: Cluster ensembles. Wiley Interdiscip. Rev. Data Min. Knowl. Discov. **1**(4), 305–315 (2011)
28. Breiman, L.: Random forests. Mach. Learn. **45**(1), 5–32 (2001)
29. Parvin, H., Minaei-Bidgoli, B., Alinejad-Rokny, H., Punch, W.F.: Data weighing mechanisms for clustering ensembles. Comput. Electr. Eng. **39**(5), 1433–1450 (2013)
30. Lancichinetti, A., Fortunato, S.: Consensus clustering in complex networks. Scientific reports 2 (2012)
31. Horzyk, A.: Information freedom and associative artificial intelligence. In: Rutkowski, L., Korytkowski, M., Scherer, R., Tadeusiewicz, R., Zadeh, L.A., Zurada, J.M. (eds.) ICAISC 2012, Part I. LNCS, vol. 7267, pp. 81–89. Springer, Heidelberg (2012)
32. Horzyk, A.: How does human-like knowledge come into being in artificial associative systems? In: Proceedings of the 8-th International Conference on Knowledge, Information and Creativity Support Systems, Krakow, Poland (2013)
33. McCall, M.N., Almudevar, A.: Affymetrix GeneChip microarray preprocessing for multivariate analyses. Brief. Bioinf. **13**(5), 536–546 (2012)
34. Davis, S., Meltzer, P.: GEOquery: a bridge between the Gene Expression Omnibus (GEO) and BioConductor. Bioinformatics **14**, 1846–1847 (2007)
35. Gautier, L., Cope, L., Bolstad, B.M., Irizarry, R.A.: Affy–analysis of Affymetrix GeneChip data at the probe level. Bioinformatics **20**(3), 307–315 (2004)
36. Gentleman, R., Carey, V., Huber, W., Hahne, F.: Genefilter: methods for filtering genes from microarray experiments. R package version 1(0) R package version 1.42.0. (2011)
37. Falcon, S., Gentleman, R.: Using GOstats to test gene lists for GO term association. Bioinformatics **23**(2), 257–8 (2007)
38. Benjamini, Y., Hochberg, Y.: Controlling the false discovery rate: a practical and powerful approach to multiple testing. J. Roy. Stat. Soc. Series B (Methodol.) **57**, 289–300 (1995)
39. Orzechowski, P., Boryczko, K.: Effective biclustering on gpu-capabilities and constraints. Przeglad Elektrotechniczny **1**, 133–6 (2015)
40. Hanczar, B., Nadif, M.: Study of consensus functions in the context of ensemble methods for biclustering (2013). http://cap2013.sciencesconf.org/21492/document

Discovering Potential Clinical Profiles of Multiple Sclerosis from Clinical and Pathological Free Text Data with Constrained Non-negative Matrix Factorization

Jacopo Acquarelli[1], The Netherlands Brain Bank[2], Monica Bianchini[3],
and Elena Marchiori[1(✉)]

[1] Institute for Computing and Information Sciences (ICIS), Radboud University,
Nijmegen, The Netherlands
{j.acquarelli,elenam}@cs.ru.nl
[2] Netherlands Institute for Neuroscience, Amsterdam, The Netherlands
[3] Dipartimento di Ingegneria dell'Informazione e Scienze Matematiche,
Università Degli Studi di Siena, Siena, Italy
monica@diism.unisi.it

Abstract. Constrained non-negative matrix factorization (CNMF) is an effective machine learning technique to cluster documents in the presence of class label constraints. In this work, we provide a novel application of this technique in research on neuro-degenerative diseases. Specifically, we consider a dataset of documents from the Netherlands Brain Bank containing free text describing clinical and pathological information about donors affected by Multiple Sclerosis. The goal is to use CNMF for identifying clinical profiles with pathological information as constraints. After pre-processing the documents by means of standard filtering techniques, a feature representation of the documents in terms of bi-grams is constructed. The high dimensional feature space is reduced by applying a trimming procedure. The resulting datasets of clinical and pathological bi-grams are then clustered using non-negative matrix factorization (NMF) and, next, clinical data are clustered using CNMF with constraints induced by the clustering of pathological data. Results indicate the presence of interesting clinical profiles, for instance related to vision or movement problems. In particular, the use of CNMF leads to the identification of a clinical profile related to diabetes mellitus. Pathological characteristics and duration of disease of the identified profiles are analysed. Although highly promising, results of this investigation should be interpreted with care due to the relatively small size of the considered datasets.

Introduction

Due to the high cost and large amount of under-utilized data, health care needs computational methods to boost the benefits of available data, in order to yield

© Springer International Publishing Switzerland 2016
G. Squillero and P. Burelli (Eds.): EvoApplications 2016, Part I, LNCS 9597, pp. 169–183, 2016.
DOI: 10.1007/978-3-319-31204-0_12

more efficient practice and research [1]. Application of such methods to data can help to guide and facilitate research, leading to new knowledge and improved clinical standards [2].

The situation applies prominently to the Netherlands Brain Bank (NBB). The NBB supplies the international scientific community with clinical and neuro-pathological well-documented brain tissue from donors [3]. It contains also a host of yet un-utilized data, describing anonymous clinical information collected from the donors, including the medical and family history, the type and course of the diseases, and the clinical diagnosis. Moreover the NBB contains pathological post-mortem information about the donors including the true (pathological) diagnosis and various pathological hallmarks.

Pathological parameters collected post-mortem cannot be used as features for performing diagnosis, treatment and prognosis. Nevertheless, such valuable side information can be exploited for discovering clinical disease profiles characterized by few clinical features linked to pathological parameters of interest. These profiles could be used by domain experts as a guidance for performing targeted studies, in order to improve diagnosis, treatment and prognosis of neuro-degenerative diseases. In machine learning, data that can be used for both constructing and testing a model (such as the clinical data of the NBB) are also called *technical*, while data that cannot be used for testing (such as the pathological data of the NBB) are called *privileged* [4].

In this paper we investigate a methodology based on clustering with free text technical and privileged data from the NBB to identify clinical profiles of donors affected by Multiple Sclerosis (MS), a chronic inflammatory disease of the central nervous system.

In our analysis of NBB documents we do not take into account medical knowledge, but only use the text as it is, except for the application of standard lexical and statistical filtering methods. We perform cluster analysis of these data using Non-Negative Matrix Factorization (NMF), a state-of-the-art machine learning technique for clustering documents [5]. First, a clustering of the pathological data using NMF is computed. The resulting clusters induce a partition of the data in classes. Next, the clinical data are clustered using Constrained NMF (CNMF) [6], by integrating the class information of the donors as an additional constraint. In this way, donors from the same pathological cluster will be merged together in the new (clinical) representation space.

Results of this clustering methodology indicate the presence of clinical profiles describing subtypes of the MS disease. In particular, we show that the obtained clusters correspond to groups of donors with different duration of disease.

1   Related Work

Medical documents, such as clinical or pathological annotations, contain valuable information about patients, such as their medical history (diseases, injuries, medical symptoms) and responses (diagnoses, prescriptions, and drugs) [7]. These data have a huge potential in order to build profiles for individual patients or classes of patients sharing a similar disease course, to discover disease correlations [8], and enhance patient care [9]. Moreover, also in medical error detection a number of tools have emerged, from medical informatics and computer science — natural language processing, visualization, and machine learning tools — as well as methods for understanding cognitive processes, which can collect a large amount of important clinical information, that normally lie locked in narrative reports, unavailable to automated decision support systems [10]. Whatever the particular problem, large volumes of medical documents have been recently generated by electronic health record systems, whose nature is normally unstructured or semi-structured, making the information extraction procedure a very difficult task. Due to the intrinsic diversity among medical documents, it is just a challenge to discover the underlying patterns from a corpus. Thus, document clustering techniques, being an efficient way of navigating and summarizing documents, have been intensively investigated in biomedical research. As a dimension reduction method, non–negative matrix factorization [11] has been widely applied to medical document clustering [12,13]. By imposing non–negativity constraints in both basis and weight factorization matrices, NMF guarantees to preserve the local structure of the original data. Moreover, the resulting latent semantic space of the clustered documents produced by NMF may be explained in a very intuitive way. Specifically, each axis in the semantic space represents the basic topic of a particular cluster, whilst each document in a collection is viewed as the additive combination of the basic topics. Therefore, a particular document is grouped into the cluster where it has the largest projection value. Many extensions of the basic NMF method have also been explored for clustering biomedical documents. For instance, in [13], Multi–view NMF, which can integrate different data sources, was applied for clustering clinical document, based on medication/symptom names, whereas, in [12], ensemble NMF, able to achieve a consensus solution from a set of runs with different initial conditions, was tested on the TREC genomic 2004 track. Finally, also more complex techniques were recently introduced in order to cope with graph representations of medical documents [14]. Actually, Subgraph Augmented Non–negative Tensor Factorization (SANTF), in addition to relying on atomic features (e.g., words in clinical narrative text), automatically mines higher–order features by converting sentences into a graph representation and identifying important subgraphs. Latent groups of atomic features were shown to help in better correlate latent groups of higher–order features. Moreover, feature analysis also identified latent groups of higher–order features that lead to interesting medical insights.

2   Multiple Sclerosis

Multiple sclerosis (MS) is a complex, chronic inflammatory disease that affect the central nervous system and whose causes are largely unknown. The disease usually causes relapsing-remitting attacks of inflammation, demyelination and axonal damage, leading to various degrees and spectra of neurological symptoms and disability (see [15,16]). The main types of MS[1] are listed below.

---

[1] See http://www.multiplesclerosis.com/us/treatment.php.

- *Relapsing-Remitting* MS (RR-MS) is the most diffuse type. People affected by RR-MS have temporarily periods called relapse when the symptoms appear.
- *Secondarily Progressive* MS (SP-MS) is considered as an advanced state of RR-MS, with symptoms that go worse steadily without relapses.
- *Primarily Progressive* MS (PP-MS) is characterized by slowly worsening of the neurological functions without any remission from the beginning.
- *Progressive-Relapsing* MS (PR-MS) is a rare form of MS characterized by a steadily worsening disease state from the beginning, with acute relapses but no remissions, with or without recovery.

Main MS symptoms include numbness or weakness in the arms and legs, blurred or double vision, or pain during eye movement, partial or total vision loss, pain or tingling in different areas of the body, lack of coordination or unsteady walking, tremors, dizziness, and extreme fatigue.

3   Methods

We begin by describing the data and techniques used in our analysis.

3.1   Data Pre-processing

We consider 149 records of MS donors, which are free text documents composed by three parts:

- *General information:* contains gender, age, etc.;
- *Clinical information:* contains the clinical history of the patient including prescriptions and diagnosis;
- *Pathological information:* contains information including the presence of diverse types of brain lesions and the pathological diagnosis.

The pathological and clinical parts of the records are used in our analysis as privileged and technical data, respectively.

A free text is a mixture of words and signs so the first pre-processing step is devoted to put it in the form of '*word space word*'. Given that it is a medical text, there are a lot of acronyms and nomenclatures typical of this area. Also a free text generally contains noise from uninformative words to be filtered out. We choose to consider only words that belong to the adjective, adverb, noun and verb classes, with a length greater than two.

A Python script was used for filtering, taking advantage of the WordNet library, a large lexical database, containing tools to process English text, including grammatical analysis and stemming, and stop-word removing procedures. Stemming transforms a generic word into its basic form (for example 'chronically' will be transformed into 'chronic'). It is an important operation because it serves to merge semantically similar words. Instead, stop-word removing discards uninformative words contained in the text.

After filtering, bi-grams are generated, yielding 64858 bi-gram features, whose values are represented using the term frequency-inverse document frequency

(TF-IDF), a score estimating how important a bi-gram is to a document within a collection [17]. For a bi-gram t, in a record x of the (filtered) data X, term frequency and inverse document frequency are defined as follows.

$$TF(t,x) = 0.5 + 0.5 f(t,x)/\max(\{f(d,x) \mid d \in X\}),$$

where $f(t,x)$ is the frequency of t in x;

$$IDF(t,X) = log(|X|/|\{x \in X \text{s.t. } t \text{ occurs in } x\}|),$$

where $|\ |$ denotes the size of a set.

The TF-IDF of a bi-gram t is the product of $TF(t,x)$ and $IDF(t,X)$.

In order to further reduce the input dimension of the data, *trimming* is applied to the distribution of TF-IDF values: 2-grams with TF-IDF value on the tail of the distribution are removed from the data. Specifically, we remove features that appear in more than 50 % of the records, as well as those that appear in less than three documents. In this way a dataset with 1084 features is obtained.

3.2 Non-Negative Matrix Factorization

NMF approximates the input matrix X of dimension $m \times n$ (in our context $m = 1084$, $n = 149$) as the product of two matrices W and H, of dimension $m \times k$ and $k \times n$ respectively, with k much smaller than m and n.

The problem can be formulated as finding W and H that minimize the objective function

$$||X - WH||^2,$$

with the constraint that both W and H are non-negative. Here $||\ ||$ denotes the Frobenius norm. Since this problem is intractable [18], heuristic methods are used to find locally optimal solutions. Here, to find W and H that locally optimize the objective function, the following multiplicative update rules [19] are used:

$$H_{\alpha\mu} = H_{\alpha\mu} \frac{\sum_i W_{i\alpha} X_{i\mu}/(WH)_{i\mu}}{\sum_k W_{k\alpha}},$$

$$W_{\alpha\mu} = W_{\alpha\mu} \frac{\sum_i H_{i\alpha} X_{i\mu}/(WH)_{i\mu}}{\sum_k H_{k\alpha}}.$$

H represents a soft clustering, where the entry (i,j) can be interpreted as the membership degree with which record j belongs to cluster i. W provides information about the relevance of terms for the clusters: each element w_{ij} of W represents the degree with which term (in our context bi-gram) t_i belongs to cluster j.

Crisp clusters can be generated from H by assigning each donor to the cluster having the highest membership degree (where ties are broken randomly).

3.3 Constrained Non-negative Matrix Factorization

The resulting clusters form disjoint classes of donors, which can be used as privileged information to cluster the clinical data by means of CNMF.

CNMF [6] is a kind of semi-supervised clustering method. It considers an input dataset for which the class label of some (l) elements is given. From these class labels, an $l \times k$ indicator matrix C is constructed, whose (i,j) entry contains a 1 if the i-th point of the dataset X is labelled with the j-th class. C is used to define the constraint matrix A:

$$A = \begin{bmatrix} C_{l \times k} & 0 \\ 0 & I_{n-l} \end{bmatrix},$$

where I_{n-l} is a $n-l \times n-l$ identity matrix. Then, CNMF approximates the input matrix X by the product WAZ, with the constraint that W and Z are both non-negative. This corresponds to find W and Z that minimize the objective

$$||X - WAZ||^2,$$

under the constraint that $W \geq 0$, $H \geq 0$. To find W and Z that locally optimize this function, the multiplicative update rules given in [6] are used.

4 Experiments

In our experiments we set the parameters, namely the number k of clusters and the number l of labelled donors, to the somewhat arbitrary values of 3 and 120 (about 80 % of the donors randomly selected from the dataset). Therefore, we aim at identifying three main clinical MS profiles, using a large amount of privileged information provided by the clustering on the pathological data. The following experiments were carried out on the pre-processed data:

1. cluster pathological data using NMF;
2. cluster clinical data using NMF;
3. cluster clinical data using CNMF, using the clustering of pathological data as privileged knowledge.

A set matching algorithm [20] was used to align clusters from two clusterings of the donors, for instance the two clustering obtained by using the clinical and pathological data. For each cluster in one clustering C, a best match in the other clustering C' is found. This is done by processing the elements n_{ij} of the contingency table in decreasing order, where

$$n_{ij} = |C_i \cap C'_j|$$

is the number of donors in the intersection of the cluster i of the clustering set C and the cluster j in the clustering set C'. The largest number, say n_{ab}, entails a match between the two clusters $a \in C$ and $b \in C'$, whereas the second largest number entails the second match, and so on.

5   Results

We first analyze results of clustering pathological and clinical documents separately, and then by using CNMF. Specifically, for each cluster we report the 10 words considered as most important by the clustering algorithms, and analyze the composition of the clusters with respect to two external metrics: duration of disease and type of MS.

5.1   Clustering Pathological Data

The top 10 relevant bi-grams for the pathological clusters are shown in Table 1.

Table 1. Top 10 relevant bi-grams for clusters of the pre-processed pathological dataset clustered using NMF.

Cluster 1	Cluster 2	Cluster 3
inactive-plaque	cortex-white	spinal-marrow
chronically-inactive	matter-ependyma	abnormally-spinal
chronic-inactive	chronic-active	thoracic-spinal
lateral-string	sclerosis-plaque	brain-laminate
sclerosis-plaque	abnormality-spc	cord-section
posterior-string	waes-contains	abnormality-conclusion
small-plaque	section-contain	nucleus-putamen
axonal-density	matter-cortex	brain-stem
decrease-myeline	diagnosis-multiple	plaque-find
myelinated-axon	cord-abnormality	visible-brain

Clustering pathological data yields interesting results. As shown in Table 1, the top 10 relevant bi-grams reveal three pathological profiles characterized by different brain lesion types:

- (chronic-inactive): this profile contains as top terms 'chronic-inactive', but also terms such as 'small-plaque', 'decrease-myeline' and 'myelinated-axon' which provide further characteristics of brain lesions in this profile.
- (chronic-active): this profile contains as top terms 'chronic-active', but also terms such as 'abnormality-spc' and 'matter-ependyma' which provide further characteristics of brain lesions in this profile.
- (spinal-cord): this profile contains as top terms concerning spinal cord injuries, but also terms such as 'brain-laminate' and 'nucleus-putamen' which provide further characteristics of brain lesions in this profile.

Table 2. Top 10 relevant bi-grams for the clusters of the pre-processed clinical dataset clustered using NMF.

Cluster 1	Cluster 2	Cluster 3
patient-suffer	right-arm	relapsive-progressive
patient-complain	lesion-visible	start-special
patient-underwent	arm-leg	phase-start
admit-hospital	focal-lesion	neuritis-subsequent
right-leg	left-arm	optic-neuritis
physical-examination	left-side	sympt-optic
complain-pain	right-side	subsequent-relapse
tension-mmhg	right-leg	eds-die
get-worse	raise-signal	iggindex-elevate
situation-get	periventricular-lesion	remark-prominent

5.2   Clustering Clinical Data

The top 10 relevant bi-grams for the clinical clusters are shown in Table 2.

The main characteristics of the three clinical profiles stemming from the clustering can be summarized as follows.

– Cluster 1 appears to contain terms related to **pain and complaints** from the patient (patient-suffer, patient-complain, complain-pain, get-worse, situation-get) and other patient-focused outcomes (patient-underwent, admit-hospital, physical examination).
– Cluster 2 seems to have a preponderance of symptoms related to **movement disabilities**.
– Cluster 3 seems to emphasize disease progression and state (start-special, phase-start, relapsive-progressive, subsequent-relapse, eds-die, neuritis-subsequent) and contains symptoms related to **vision** ('optic-neuritis' and 'sympt-optic').

Table 3 shows the size of the crisp clusters obtained by clustering clinical and pathological data separately using NMF: there are two somewhat large clusters and a small one.

5.3   Clustering Clinical Data with Privileged Pathological Information

The three pathological clusters indicate the presence of pathological profiles of donors with diverse types of brain lesions. These clusters are used in the sequel as privileged information to identify profiles from the clinical data using CNMF.

The alignment between the clinical clustering generated by NMF and CNMF leads to same correspondence between clusters as that indirectly obtained from

Table 3. Number of records in the aligned clusters obtained by applying NMF to the two pre-processed clinical and pathological datasets independently.

Cluster no.	Clinical Docs	Pathological Docs
1	50	40
2	88	97
3	11	12

the separate alignment of the NMF and CNMF clusterings with the pathological one. Therefore, for instance, NMF cluster 1 of the clinical data, CNMF cluster 1 of the clinical data and NMF cluster 1 of the pathological data are all aligned with each other.

Table 4 shows the top 10 bi-grams of these new clinical clusters.

Table 4. Top 10 relevant bi-grams for the clusters of the pre-processed clinical dataset clustered with CNMF.

Cluster 1	Cluster 2	Cluster 3
patient-suffer	right-arm	periventricular-lesion
patient-complain	arm-leg	lesion-visible
progressive-phase	left-arm	lateral-ventricle
phase-start	left-side	patient-suffer
right-leg	patient-underwent	signal-intensity
eds-die	muscle-strength	matter-lesion
optic-neuritis	patient-suffer	raise-signal
start-special	left-leg	white-matter
examination-reveal	right-leg	mellitus-type
right-side	paresis-right	diabetes-mellitus

From this table three clinical profiles emerge, whose properties can be summarized as follows.

– The top 10 most relevant bi-grams in cluster 1 represent symptoms about pain and complaints by the patient. Moreover the presence of the bi-gram 'progressive-phase' indicates the presence of a progressive phase of MS, and 'phase-start' could indicate that a new phase of MS is starting (relapse) after remitting. This could mean an SP type of MS. Cluster 1 contains also as top term 'optic-neuritis', a symptom involving vision problems.
– Cluster 2 top bi-grams refer mainly to symptoms related to movement disabilities, involving arms and legs, as well as paresis.
– Cluster 3 contains two bi-grams referring to diabetes mellitus. These could indicate a specific characteristic of a disease sub-type. Top bi-grams in cluster

3 refer also to brain lesions detected by magnetic resonance imaging (bi-grams: periventricular-lesion, lesion-visible, lateral-ventricle, matter-lesion).

The size of the clusters obtained by applying CNMF to the clinical data, and the size of the corresponding best match pathological clusters are shown in Table 5.

6   Interpretation of Results

We now use an external cluster validation metric, namely the duration of disease (DOD), to analyze clinical clusters. DOD measures the number of years from the moment a clinical diagnosis of the disease was performed to the death of the patient.

Table 6 shows the average DOD per type of MS and the corresponding number of records contained in the dataset.

In the clustering of pathological data the majority of donors with SP-MS belongs to cluster 1 and 3 while the majority of PP-MS donors is in cluster 2. Information about the type of MS is only partially available (see Table 7).

6.1   Clinical Clusters Generated by NMF

Table 8 shows the average DOD of the three clinical clusters produced by NMF.

In order to assess whether clusters have significantly different DOD, the Rank Sum Wilcoxon test is applied. Results of the test show that the three NMF clinical clusters differ significantly one from each other with respect to DOD. Composition of clinical clusters with respect to the type of MS is shown in Table 9.

Table 5. Size of clusters of the clinical dataset partitioned with CNMF and of their best match pathological clusters.

Cluster no.	Clinical Docs	Pathological Docs
1	109	97
2	32	12
3	8	40

Table 6. Average DOD for the different MS types and number of donors of that type in the dataset.

MS type	# Records	Average DOD
RR	0	-
SP	33	22.21
PP	12	25.92
PR	0	-

Table 7. Composition of pathological clusters wrt type of MS.

Cluster no.	Available/Total	# SP	# PP
1	26/97	20 (60.61 %)	6 (50.00 %)
2	5/12	2 (6.06 %)	3 (25.00 %)
3	14/40	11 (33.33 %)	3 (25.00 %)
Total		33 (100.00 %)	12 (100.00 %)

Table 8. Average DOD values (Average DOD) and standard deviation (STD) of the clusters of the clinical data obtained using NMF.

Cluster no.	Available/Total	Average DOD	STD
1	66/88	27.97	±14.15
2	11/11	24.45	±11.80
3	38/50	27.21	±12.78

Table 9. Composition of clinical clusters generated by NMF with respect to the type of MS.

Cluster no.	Available/Total	# SP	# PP
1	27/88	18 (54.55 %)	9 (75.00 %)
2	7/11	7 (21.21 %)	0 (0.00 %)
3	11/50	8 (24.24 %)	3 (25.00 %)
Total		33 (100.00 %)	12 (100.00 %)

In summary, the clustering analysis using NMF identifies clinical profiles which can be summarized as follows:

- (pain-complaints-relative-long-DOD): this clinical profile involves symptoms involving pain and complaints. This cluster contains donors with a relatively long DOD.
- (movement-disabilities-short-DOD): this clinical profile involves symptoms concerning movement disabilities. This cluster contains donors with a relatively short DOD.
- (vision-MS-phases-SP/PP-relative-long-DOD): this clinical profile concerns symptoms involving vision and phases of the MS. This cluster contains donors with a relatively long DOD.

6.2   Clinical Clusters Generated by CNMF

Table 10 shows the average DOD of the three clinical clusters produced by CNMF.

In order to assess whether clusters have significantly different DOD, the Rank Sum Wilcoxon test is applied. Results of the test show that cluster 1 and 2 differ

Table 10. Average DOD values (Average DOD) and standard deviation (STD) of the clusters of the clinical data obtained using CNMF.

Cluster no.	Available/Total	Average DOD	STD
1	87/109	26.52	±12.89
2	25/32	30.24	±13.73
3	3/8	28.67	±23.30

significantly, as well as cluster 1 and 3. Cluster 3 has few available records so it is not possible to use its DOD value to link it to its corresponding pathological cluster.

The composition of the clusters with respect to the type of MS is shown in Table 11.

Table 11. Composition of clinical clusters generated by CNMF with respect to the type of MS.

Cluster no.	Available/Total	# SP	# PP
1	32/109	26 (78.79 %)	6 (50.00 %)
2	12/32	6 (18.18 %)	6 (50.00 %)
3	1/8	1 (3.03 %)	0 (0.00 %)
Total		33 (100.00 %)	12 (100.00 %)

The clustering analysis using CNMF identifies clinical profiles which can be summarized as follows:

- (pain-complaints-short-DOD): this clinical profile involves symptoms involving pain and vision complaints of the patient, and symptoms indicating advanced RR-MS. This cluster contains donors with a relatively short DOD.
- (movement-disabilities-long-DOD): this clinical profile involves symptoms involving movement disabilities of arms and legs, and paresis. This cluster contains donors with a long DOD and PP-MS type of disease.
- (diabetes-mellitus-long-DOD): this clinical profile is related to symptoms involving lesions and diabetes mellitus. This cluster contains donors with a relatively long DOD.

As expected, in this case the composition of clusters with respect to different MS types identified by the clinic CNMF clustering corresponds to that produced by clustering pathological data.

6.3   Comparison Between NMF and CNMF Clusters

Clinical records mainly report symptoms or treatments and describe the medical history of the patients. Thus the lexicon is quite different between the pathological and the clinic parts of the patient record: that explains the importance of bi-grams in the NMF clustering such as 'admit-hospital', 'complain-pain' or bi-grams describing parts of the patients' body. CNMF clusters contains less of these somewhat generic bi-grams which are replaced by more specific ones in terms of information from the medical literature, for example terms referring to diabetes. Also as expected CNMF clusters are more faithful to the pathological ones with respect to MS type composition.

The use of privileged information from the pathological clustering changes the rank of important bi-grams in the clinical clusters; in some cases bi-grams disappear or migrate from one cluster to another one.

Table 12 report words which appear in both the NMF and CNMF clusterings (column 'Stay'), which are in CNMF but not NMF (column 'In'), and which disappear (column 'Out').

From this table we can see that while for cluster 1 and 2 some of the 10 most important bi-grams of the NMF clustering remain most important also in the CNMF clustering (3/10 for cluster 1 and 5/10 for cluster 2), no bi-grams for NMF cluster 3 are kept in the corresponding CNMF cluster.

Table 12. Changes in top 10 relevant bi-grams from clinical clustering with NMF to that with CNMF.

Cluster	In	Out	Stay
1	progressive-phase phase-start eds-die optic-neuritis start-special esamination-reveal right-side	patient-underwent admit-hospital physical-examination complain-pain tension-mmhg get-worse situation-get	patient-suffer patient-complain right-leg
2	patient-underwent muscle-strength patient-suffer left-leg paresis-right	lesion-visible focal-lesion right-side raise-signal periventricular-lesion	right-arm arm-leg left-arm left-side right-leg
3	periventricular-lesion lesion-visible lateral-ventricle patient-suffer signal-intensity matter-lesion raise-signal white-matter mellitus-type diabetes-mellitus	relapsive-progressive start-special phase-start neuritis-subsequent optic-neuritis simpt-optic subsequent-relapse eds-die iggindex-elevate remark-prominent	

Bi-grams which appear in both the clusterings (column 'Stay') mainly refer to pain and complaints in parts of the patients' body. For cluster 1 top bi-grams which are in CNMF but not NMF (column 'In') are mainly associated with the phase of the disease, while for cluster 2 such bi-grams are related to symptoms related to movement, and for cluster 3 they mainly refer to lesions and diabetes.

In general, results indicate that the information provided by the pathological clustering affect the characteristics of clinical profiles by favoring the emergence of a small cluster associated to diabetes mellitus. The co-occurrence of MS and diabetes mellitus has been reported by a number of studies. In particular as mentioned in [21] for the combined effect of diabetes mellitus type 1 and type 2 there is evidence that this is associated with a worse progression of disability compared to MS patients without type 1 or type 2 diabetes. This is in accordance with clinical profile we identified using privileged information which indicates that the neuropathology associated with this form of diabetes might influence the disease course (symptoms related to lesions, spinal cord brain lesion) and contribute to the severity of MS (short DOD).

7   Conclusion

In this paper, we investigated a methodology to identify clinical MS profiles using pathological information as privileged data to guide the identification process. To this aim, data from the NBB, consisting of free text documents containing clinical and pathological records, were used. The data were first pre-processed and NMF was used to cluster both clinical and pathological data independently. CNMF was then employed to cluster clinical data, using constraints inferred from the pathological data. The obtained results indicate the presence of profiles with characteristics which reflect underlying neuropathological differences, and differing DOD outcomes. In particular, the use of privileged information lead to the identification of a clinical profile related to diabetes mellitus.

Although potentially interesting, these results should be interpreted with care because of the number of DOD missing values and the small size of the data. The proposed analysis applied to a larger dataset would provide deeper understanding and confidence on the potential relevance of these profiles.

It is an interesting and relevant matter of future research to provide stronger ties between the observed clusterings, in particular the final ones, and the neuropathology of MS from literature studies, as well as to use a corpus of documents from the literature as external prior knowledge to enhance the clustering process.

Acknowledgments. This work has been partially funded by the Netherlands Organization for Scientific Research (NWO) within the NWO project 612.001.119.

References

1. Urbach, D., Moore, J.H.: Data mining and the evolution of biological complexity. BioData Min. **4** (2011)
2. Davis, D., Chawla, N.V.: Exploring and exploiting disease interactions from multi-relational gene and phenotype networks. PloS ONE **6**(7), e22670 (2011)

3. Bell, J.E., et al.: Management of a twenty-first century brain bank: experience in the BrainNet Europe consortium. Acta Neuropathol. **115**(5), 497–507 (2008)
4. Vapnik, V., Vashist, A.: A new learning paradigm: learning using privileged information. Neural Netw. **22**(5–6), 544–557 (2009)
5. Lee, D.D., Seung, H.S.: Learning the parts of objects by non-negative matrix factorization. Nature **401**(6755), 788–791 (1999)
6. Wu, H., Liu, Z.: Non-negative matrix factorization with constraints. In: Proceedings of the 24th AAAI Conference on Artificial Intelligence, pp. 506–511 (2010)
7. Roberts, K., Harabagiu, S.M.: A flexible framework for deriving assertions from electronic medical records. J. Am. Med. Inform. Assoc. **18**(5), 568–573 (2011)
8. Roque, F.S., et al.: Using electronic patient records to discover disease correlations and stratify patient cohorts. PLoS Comput. Biol. **7**(8), E1002141 (2011)
9. Hripcsak, G., et al.: Mining complex clinical data for patient safety research: a framework for event discovery. J. Biomed. Inform. **36**(1), 120–130 (2003)
10. Melton, G.B., Hripcsak, G.: Automated detection of adverse events using natural language processing of discharge summaries. J. Am. Med. Inform. Assoc. **12**, 448–457 (2005)
11. Xu, W., Liu, X., Gong, Y.: Document clustering based on non-negative matrix factorization. In: Proceedings of the 26th Annual International ACM SIGIR Conference, pp. 267–273. ACM (2003)
12. Huang, X., Zheng, X., Yuan, W., Zhu, S.: Enhanced clustering of biomedical documents using ensemble non-negative matrix factorization. Inf. Sci. **181**, 2293–2302 (2012)
13. Ling, Y., Pan, X., Li, G., Hu, X.: Clinical documents clustering based on medication/symptom names using multi-view nonnegative matrix factorization. IEEE Trans. Nanobiosci. **14**(5), 500–504 (2015)
14. Luo, Y., et al.: Subgraph augmented non-negative tensor factorization (SANTF) for modeling clinical narrative text. J. Am. Med. Inform. Assoc. **22**(5), 1009–1019 (2015)
15. Bö, L., Geurts, J.J.G., Mörk, S.J., Van der Valk, P.: Grey matter pathology in multiple sclerosis. Acta Neurol. Scand. **113**, 48–50 (2006)
16. Van der Valk, P., De Groot, C.J.A.: Staging of multiple sclerosis (MS) lesions: pathology of the time frame of MS. Neuropathol. Appl. Neurobiol. **26**, 2–10 (2000)
17. Feldman, R., Fresko, M., Kinar, Y., Lindell, Y., Liphstat, O., Rajman, M., Schler, Y., Zamir, O.: Text mining at the term level. In: Żytkow, J.M. (ed.) PKDD 1998. LNCS, vol. 1510, pp. 65–73. Springer, Heidelberg (1998)
18. Vavasis, S.A.: On the complexity of nonnegative matrix factorization. SIAM J. Optim. **20**(3), 1364–1377 (2009)
19. Lee, D.D., Seung, H.S.: Algorithms for non-negative matrix factorization. In: NIPS, pp. 556–562. MIT Press (2000)
20. Meilǎ, M., Heckerman, D.: An experimental comparison of model-based clustering methods. Mach. Learn. **42**(1–2), 9–29 (2001)
21. Tettey, P., Simpson, S., Taylor, B.V., van der Mei, I.A.F.: The co-occurrence of multiple sclerosis and type 1 diabetes: shared aetiologic features and clinical implication for MS aetiology. J. Neurol. Sci. **348**(1), 126–131 (2015)

Application of Evolutionary Algorithms for the Optimization of Genetic Regulatory Networks

Elise Rosati[✉], Morgan Madec, Abir Rezgui, Quentin Colman,
Nicolas Toussaint, Christophe Lallement, and Pierre Collet

ICube Laboratory (Engineering Sciences, Computer Sciences and Imaging
Laboratory, UMR 7357), University of Strasbourg/CNRS,
300 boulevard Sébastien Brandt, 67412 Illkirch Cedex 02, France
erosati@unistra.fr

Abstract. Synthetic biology aims at reinvesting theoretical knowledge from various do-mains (biology, engineering, microelectronics) for the development of new bio-logical functions. Concerning the design of such functions, the classical trial-error approach is expensive and time consuming. Computer-aided design is therefore of key interest in this field. As for other domains, such as microelectronics or robotics, evolutionary algo-rithms can be used to this end. This article is a first step in this direction: it describes the optimization of an existing artificial gene regulatory network using evolutionary algorithms. Evolutionary algorithms successfully find a good set of parameters (the simu-lated response of the system which fits at 99 % the expected response) in about 200 s (corresponding to 5000 generations) on a standard computer. This is the proof of concept of our approach. Moreover, results analysis allows the biologist not only to save time during the design process but also to study the specificity of a system.

Keywords: Synthetic biology · EASEA · Gene regulatory networks · Design automation · Biosystems modeling

1 Introduction

Synthetic biology is an emerging field which aims at reinvesting theoretical knowledge acquired during the past decades in biology and the know-how in the design of systems, such as large-scale integrated circuits, autonomous embedded systems (e.g. smartphone) or heterogeneous macro-systems (automotive, robotics, ...). As a consequence, this science is at the interface between biology, biotechnologies, microelectronics and computer science. Synthetic biology is coming of age as it now has many applications in several domains, such as the manufacturing of new low-cost drugs [1], the implementation of Boolean functions with biological material for cancer detection purpose [2], biological sensing [3] or the synthesis and the optimization of bio-fuels [4].

© Springer International Publishing Switzerland 2016
G. Squillero and P. Burelli (Eds.): EvoApplications 2016, Part I, LNCS 9597, pp. 184–200, 2016.
DOI: 10.1007/978-3-319-31204-0_13

Synthetic biology involves several aspects. One of the most interesting for us is not *what you can do* but *how you can do it*. More specifically, the question we try to answer is the following: *is it possible to switch from trial-error design process to virtual prototyping*. In this context, evolutionary algorithms can play an important role for design automation and system-level optimization.

In this paper, focus is put on a specific field of synthetic biology which consists in the design of artificial gene regulatory networks that can be integrated in a living cell (bacteria for instance) so that this network implements a new functionality. Design automation for gene regulatory networks has been widely investigated over the past decade [5,6] and remains a hot topic [7]. One of the most interesting features of such biological systems lies in the fact that they can be described by Boolean relationships at a high level of abstraction [8]. Thus, design methods and associated tools used for digital electronics can be directly applied to synthetic biology. GeNeDA (GEne NEtwork Design Automation) is an example of tools developed upon this principle [9,10]. GeNeDA is based on a digital synthesizer and a technological mapper, initially developed for the implementation of digital function in FPGAs (Field-Programmable Gate Array). These tools have been adapted to the biological context. In the same vein, several alternative tools have been developed [11,12].

Nevertheless, by opposition to microelectronics, there can be a huge gap between the Boolean abstraction of a gene regulatory network and its actual behavior. In particular, connexions between genes (when the protein synthesized by the expression of a gene #1 regulates the expression of another one, *e.g.* gene #2) are not so obvious. For instance, even if gene #1 is active, the amount of synthesized protein may be not sufficient in order to activate gene #2. In addition, several other mechanisms or functions can not be described by Boolean abstraction (*e.g.* the *amplification function* described in [2]). Another example is Basu's band detector [13] for which the output reporter protein (green fluorescent protein in this case) is synthesized only for an intermediate concentration of input protein (acyl-homoserin lactone). For such circuits, two alternatives exist. Firstly, the development of new tools based on mathematics that provide an intermediate level of abstraction. Secondly, the adaptation of *analog synthesis* methods from microelectronics, that is to say the design automation of analog circuits that are described by transfer function and Kirchhoff's networks.

On intermediate levels of abstraction, René Thomas' investigations deserve to be highlighted [14]. Indeed, he developed a formalism for the modeling of dynamical behaviour of biological regulatory network through multivalued logic variables, rules, graphs and graphical representation of the state space. A couple of years later, Gilles Bernot extended this approach to include temporal properties of gene regulatory networks [15]. An alternative has been recently investigated based on fuzzy logic which is used to describe the rules that govern the relations between protein concentration and gene states [16]. The main asset of fuzzy logic in comparison with standard multivalued logic is that the link between fuzzy value and actual concentration of protein is never lost. Fuzzy logic can be used to describe systems at an intermediate level of abstraction, but also for design purposes, as it has been shown in [16]. In this case, an algorithm iteratively tests several combinations of rule matrices (picked up from a library) and finds the one that best meets the expected behaviour. Each rules matrix corresponds to a gene-protein interaction and its content provides the designer with important clues about the choice of the protein-gene interaction to implement.

On the other hand, research on *analog synthesis* started in the beginning of the 80's and this topic remains worth investigation in electronics. Several tools and methods have been demonstrated using specific formalisms and formal computation [17,18]. Nevertheless, these developments have not led to a generic tool which would have been widespread in analog designers community. The main reason is that they were too complex and too specific to be used for a large range of circuits. In addition, they require libraries and/or artificial learning methods or the translation of designer experience to formal rules, which is not straightforward. In a more general way, it was observed that the ratio between the implementation complexity of such algorithms and complexity of circuits that can be synthesized were poor in comparison to a hand-made design.

One of the most outstanding breakthrough in the domain of *analog synthesis* has been made by Koza in 1997 [19] based on genetic programming algorithms. He demonstrates the potential of his method on a large set of electronic circuits (filters, amplifiers, controllers) for which genetic algorithms provided solutions (circuit topology and component dimensioning) very competitive in comparison with human intelligence [20]. At the time, the main shortcoming of evolutionary algorithms were that they required computing power that could only be provided by supercomputers. This is no longer true with current technologies: the exploitation of the parallelized computation over small networks and/or the exploitation of the performance of the Graphical Processor Units (GPU) [21] makes it possible to obtain such results with reasonably priced computing systems.

The topic of this paper is to introduce the basic principles of the use of genetic algorithms in genetic regulatory networks design automation. Designing a genetic regulatory network requires a first step of architecture design where abstract biological parts are assembled together. In a second step, these parts need to be actuated: modeling and optimization of the parameters of the system are required to choose which actual parts will be used (for example to choose between a strong or weak promoter). The next section of this paper describes the above-mentioned biological system that is used as case study to illustrate these principles. Then, focus is put on the settings used for the evolutionary algorithm. Results are given in Sect. 4 and discussed in Sect. 5. In this last section, the opportunity to use genetic programming, which is the next step toward gene regulatory network design automation, is also discussed.

2   Description of the Case Study

As a proof of concept, we choose to model and simulate a modified version of a biological band-pass system developed by Basu et al. [23] (cf. Fig. 1). This system allows the detection of an intermediate concentration of acyl-homoserine lactone (AHL). It consists in two populations of cells: senders and receivers. Senders synthesize and emit isotropic AHL (the signal) inside a Petri dish. AHL is a small molecule able to diffuse in the gelose of the Petri dish and to enter cells. Receivers react to the concentration of AHL and produce GFP, a Green Fluorescent Protein (the output in this system) if the concentration in AHL ([AHL]) is comprised within a specific interval (around $5.10^{-2}\,\mu$M). In practice, one or many groups of senders are laid on specific spots on a Petri dish covered with receiver cells.

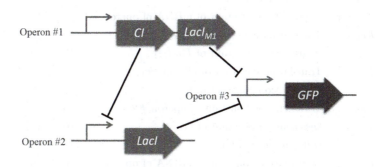

Fig. 1. Simplified Basu system

Our focus was put on the biological core that is computing the output, namely the receiver cells. They are composed of three "operons" (cf. Fig. 1). Operon #1 is positively regulated by AHL (via LuxR, a constitutively expressed protein) and expresses a modified LacI ($LacI_{M1}$). $LacI_{M1}$ inhibits the expression of operon #3 GFP. Operon #1 also produces CI, which in turn inhibits the expression of operon #2 LacI. LacI inhibits the epxression of GFP (operon #3). Each gene transcription into mRNA and subsequent translation into protein can be modeled by differential equations parametrized by the parameters given in Table 1. These 22 values (that constitute the genome of the individuals to be evolved) correspond to the different kind of regulators and promoters (regulated part of the operon). At the steady-state and for given parameters, GFP concentration only depends on [AHL], as described by Eqs. 1, 2, 3 and 4:

$$[LacI_{M1}] = \frac{k_{TL_1}}{d_{LacI_{M1}}} \cdot \frac{k_{TR_1}}{d_{mRNA_1}} \cdot \frac{1}{1 + \left(\frac{K_{A_1}}{[AHL]}\right)^{n_{A_1}}} \tag{1}$$

$$[CI] = \frac{k'_{TL_1}}{d_{CI}} \cdot \frac{k_{TR_1}}{d_{mRNA_1}} \cdot \frac{1}{1 + \left(\frac{K_{A_1}}{[AHL]}\right)^{n_{A_1}}} \tag{2}$$

Table 1. Parameters table. The boundaries correspond to the limits in which the parameters were constrained during initialization (as well as crossover and mutation if relevant). Dissociation constants are given in μM and transcription constants in $\mu M \cdot s^{-1}$.

algogene	Description	Boundaries	
		Min	Max
k_{TR_1}	Transcription constant of operon #1	10^{-3}	10^3
K_{A_1}	Dissociation constant of AHL	10^{-4}	10^3
n_{A_1}	Hill's number of AHL	1	4
d_{mRNA_1}	Degradation constant of mRNA of operon #1	10^{-3}	10^{-1}
k_{TL_1}	Translation constant of $LacI_{M1}$ (op. #1)	10^{-7}	10^{-5}
$d_{LacI_{M1}}$	Degradation constant of $LacI_{M1}$	10^{-4}	10^{-2}
k'_{TL_1}	Translation constant of CI (op. #1)	10^{-7}	10^{-5}
d_{CI}	Degradation constant of CI	10^{-4}	10^{-2}
k_{TR_2}	Transcription constant of operon #2	10^{-3}	10^3
K_{R_2}	Dissociation constant of CI	10^{-4}	10^3
n_{R_2}	Hill's number of CI	1	4
d_{mRNA_2}	Degradation constant of mRNA of operon #2	10^{-3}	10^{-1}
k_{TL_2}	Translation constant of operon #2	10^{-7}	10^{-5}
d_{LacI}	Degradation constant of LacI	10^{-4}	10^{-2}
k_{TR_3}	Transcription constant of operon #3	10^{-3}	10^3
K_{R_3}	Dissociation constant of $LacI_{M1}$	10^{-4}	10^3
n_{R_3}	Hill's number of $LacI_{M1}$	1	4
K'_{R_3}'	Dissociation constant of LacI	10^{-4}	10^3
n'_{R_3}	Hill's number of LacI	1	4
d_{mRNA_3}	Degradation constant of mRNA of operon #3	10^{-3}	10^{-1}
k_{TL_3}	Translation constant of operon #3	10^{-7}	10^{-5}
d_{GFP}	Degradation constant of GFP	10^{-4}	10^{-2}

$$[LacI] = \frac{k_{TL_2}}{d_{LacI}} \cdot \frac{k_{TR_2}}{d_{mRNA_2}} \cdot \frac{1}{1 + \left(\frac{[cI]}{K_{R_2}}\right)^{n_{R_2}}} \tag{3}$$

$$[GFP] = \frac{k_{TL_3}}{d_{GFP}} \cdot \frac{k_{TR_3}}{d_{mRNA_3}} \cdot \frac{1}{1 + \left(\frac{[LacI_{M1}]}{K_{R_3}}\right)^{n_{R_3}} + \left(\frac{[LacI]}{K'_{R_3}}\right)^{n'_{R_3}}} \tag{4}$$

3   Setup of the Evolutionary Algorithm

This section deals with the settings used for the evolutionary algorithm, *i.e.* population size, initialiser, crossover function, mutator function and evaluator.

In the following, to avoid confusion, *algogene* refers to the genes in the algorithm and *biogene* to the biological genes (composing the Basu system).

Population Size and Genetic Engine — 1000 individuals. At each generation, 1000 offspring were generated but to the difference of a generational engine, individuals for the next generation were selected *via* a binary tournament with weak elitism (the best of parents+offspring survives). Parents selection also uses a binary tournament.

Initialiser — Each parameter is initialized randomly within its respective range (see Table 1). For the four Hill's numbers, a random number is drawn between 1.0 and 4.0. Because we are in a biological system, for all the other parameters, a random value is drawn between the two logarithmic values of the interval (the log of min and max values of the "Boundaries" in Table 1); the parameter is then initialized to the power of this value. For example, for k_{TR_1} a random value x is drawn between -3 and 3 so that $k_{TR_1} = 10^x$.

Crossover Function — Different crossover functions were tested: a simple replacement (a tosscoin decides whether the child will be a copy of parent 1 or 2), a barycentric crossover, a BLX-α [22] crossover and an SBX [23] crossover (cf. Table 2). The crossover function is called for all children creation and involves 2 parents for the creation of 1 child.

Mutation Function — Different mutation functions were tested: a simple random draw in the previously defined range, relative mutation (addition of a gaussian noise) and auto-adaptive mutation (from Evolutionary Strategies, see [24]). The mutation operator is applied to each *algogene* with a probability of 0.05 %. If involved, the parameter sigma of the auto-adaptive mutation is mutated whenever its corresponding *algogene* is.

Evaluator — To evaluate the individuals, a first step was to select N absciss points (spread uniformly in the logarithmic scale) from 10^{-4} to 10^1. Then on each of these points the genome parameters are used to compare the value of the Basu function (Eqs. 1, 2, 3 and 4) to the target value. To obtain the target values, we approximated the band-pass system described by Basu et al. by a gaussian curve centered on $10^{-1.5}$, of maximal height 20 and of standard deviation 0.2 as follows:

$$f(x) = GFP_{max} \cdot e^{-\frac{(x-\mu)^2}{2 \cdot \sigma^2}} \tag{5}$$

The Mean Square Error (MSE) is calculated on all these points. The aim of the algorithm is to minimize the error. If the operators allowed the parameters to cross the boundaries used for initialization, an additional limiting step was used.

Table 2 shows the operators that were used in the results analysis: barycentric crossover operator, auto-adaptative mutation or gaussian noise, gaussian curve with 40 points for the evaluator. Stopping criteria was the number of generations. Analysis of the biological parameters is performed on the genome of the best individual after 5000 generations.

Table 2. Summary of the preliminary tests. When not indicated, mutator operator was applied with a probability of 0.05 %. Each score corresponds to the mean score of three to six runs after 1000 generations, with their standard deviation value. BLXα-x (respectively SBXν-x): x corresponds to the value of α (respectively ν). Evaluation was performed with 40 samples taken uniformly in the log domain.

Crossover		Mutator	Bound.	Average time (s)	Average score of the best
Operator	Domain				
Par.Replacement	lin	simple bounded	No	35.0	0.02 ± 0.007
Par.Replacement	lin	simple bounded 0.1 %	No	34.8	0.04 ± 0.013
Par.Replacement	lin	relative sig0.2	Yes	30.3	0.012 ± 0.005
Par.Replacement	lin	relative sig0.1	Yes	32.4	0.006 ± 0.001
Par.Replacement	lin	relative sig0.3	Yes	30.8	0.011 ± 0.003
Par.Replacement	lin	mut_autoad noise 1.0	Yes	32.5	0.05 ± 0.031
Par.Replacement	lin	mut_autoad noise sqrt(L)	Yes	28.5	0.02 ± 0.008
Barycentric	log	simple bounded 0.01 %	No	39.9	0.08 ± 0.004
Barycentric	log	simple bounded	No	40.1	0.07 ± 0.002
Barycentric	log	simple bounded 0.1 %	No	40.5	0.10 ± 0.009
Barycentric	log	simple bounded 0.2 %	No	40.6	1.27 ± 0.555
Barycentric	log	mut_autoad noise sqrt(L)	Yes	41.8	0.04 ± 0.002
BLXα-0.1	lin	simple bounded	No	36.4	12.23 ± 10.56
BLXα-0.1	log	simple bounded	No	40.2	0.05 ± 0.005
BLXα-0.1	log	simple bounded	Yes	40.3	0.05 ± 0.006
BLXα-0.2	log	simple bounded	No	40.3	0.03 ± 0.001
BLXα-0.2	log	simple bounded	Yes	40.0	0.04 ± 0.005
SBXν-1	log	simple bounded 0.005 %	Yes	35.2	4.15 ± 0.057
SBXν-1	log	simple bounded	Yes	35.3	2.66 ± 0.609
SBXν-2	log	simple bounded 0.005 %	Yes	38.7	1.00 ± 0.867
SBXν-2	log	Relative (sig = 0.1)	Yes	39.1	1.19 ± 1.309

4   Results

The evolutionary algorithm has been implemented on the EASEA platform [25,26]. Computations have been performed on a standard computer without GPU acceleration. Results are summarized in Table 2 and discussed in the following subsections. The score reflects how far the function fed by the 22 parameters found by the algorithm is from the target function (precisions are given in the previous section).

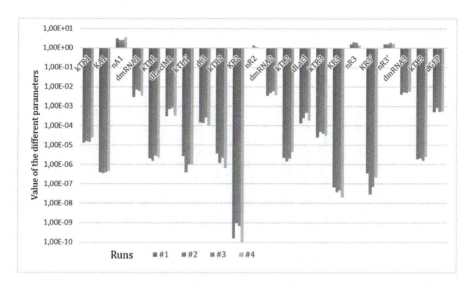

Fig. 2. Value of the *algogenes* for four runs after 5000 generations. The dissociation constants are given in M, the transcription constants in $M \cdot s^{-1}$, the translation and the degradation constants in s^{-1}.

In the following subsections, scores and computing time correspond to the mean of these values over 30 runs if not indicated otherwise. For clarity, only the first four runs are shown on Figs. 2 and 3.

4.1 Is Parental Replacement Relevant on a Biological Point of View?

When using parental replacement as crossover operator, the results are the best in terms of computing time (158 s) and score of the best individual ($9.91 \cdot 10^{-3}$ corresponding to a relative error (score of the best individual over the height h of the peak) of 0.05 %). Out of five runs, four parameters (k_{TR_1}, d_{CI}, K_{R_2}, n'_{R_3}) systematically reached their lower boundary and one (nA_1) reached its upper one. Decreasing the mutator's gaussian noise (variance is no longer the range of the interval for the parameter in the log domain but 1.0) did not alter significantly this tendency (out of four runs, the same parameters but k_{TR_1} reached systematically one of their boundary). A Covariance Matrix Adaptation Evolution Strategy (CMA-ES) [27] algorithm could be tested.

4.2 Validity of the Results

As shown on Fig. 2, the algorithm solutions always requires for K_{R_3} (the repression constant of $LacI_{M1}$) to be lower than K_{R_2} by at least one order of magnitude. In the original work [13], this is also a requirement, which they solve by using CI as a strong repressor (to have a low K_{R_2} constant), and $LacI_{M1}$ as a

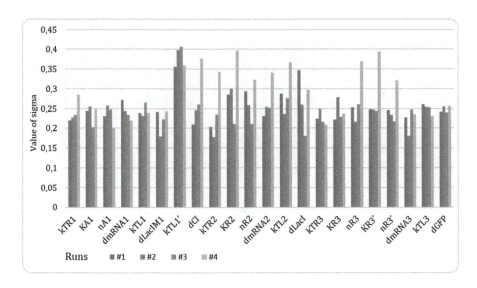

Fig. 3. Values of the auto-adaptive mutator sigma values for four runs after 5000 generations.

weaker repressor. Similarly, we observe that the constants related to LacI and $LacI_{M1}$ (K_{R_3} and $d_{LacI_{M1}}$ and respectively K'_{R_3} and d_{LacI}) are close for the two proteins. This is also the case in the original work. It is to be noted that the degradation constants of the proteins obtained by our algorithm are in accordance with the parameters used by Basu *et al.* to represent their actual system. As for the global synthesis constants (for a given protein, it corresponds to $\frac{k_{TR} \cdot k_{TL}}{d_{mRNA}}$), they are in the range of $0.1 \, \mu M \cdot min^{-1}$ in our results and of $1 \, \mu M \cdot min^{-1}$ in the original work.

4.3 The Assets of Auto-adaptive Mutation

Auto-adaptive mutation offers the possibility to retrieve information about key parameters of the system. After the algorithm reached its stopping criteria, we retrieved the sigma values of the best individual. Out of 30 runs, less than 5 % *algogenes* showed a sigma value lower than 0.2 whereas more than 17 % had sigma values greater than 0.3 (see Fig. 3). We observed that no *algogene* has a particularly low value of sigma. On the contrary, parameter k'_{TL1} has an average sigma value of 0.38, showing that it is less sensitive to variation.

4.4 Validation on Different Targeted Response

We tried to see whether the algorithm could solve similar problems with the same set-up by shifting the target function and increasing its peak (Fig. 4). Three other gaussian curves were tested: one with $\mu = 1$ and $GFP_{max} = 5 \, \mu M$, another one with $\mu = 10^{-1}$ and $GFP_{max} = 10 \, \mu M$ and a last one with $\mu = 10^{-3}$

and $GFP_{max} = 40\,\mu M$. In each case, relative error was under 0.3 %, in 200 s on average (see Fig. 4). For each peak, we took the average of each parameter out of three runs. It appears that most parameters are similar. Major differences can be observed for the dissociation constants of the regulators, namely activation constant K_{A_1} and repression constants K_{R_2}, K_{R_3} and K'_{R_3}. Apart for K'_{R_3} where the tendency is opposed, they decrease with the height of the peak. The same is observed for n_{R_3}. To observe whether this difference was due to the positional shift or the height difference, a new set of runs were performed with the same position for each peak ($10^{-1}\,\mu M$). The results showed that the tendency observed previously is absent. Moreover, the differences between the dissociation constants of each setup are negligible (they are non-existent or spread over less than one order of magnitude). No difference could be observed for the values of k_{TL_3}, the translation constant of GFP. Only minor differences could be observed (when observed).

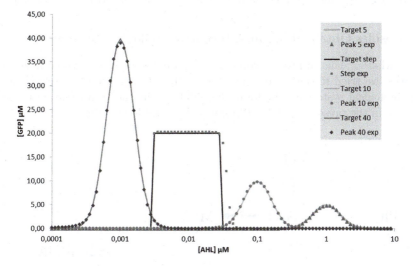

Fig. 4. Representation of the best individual's GFP response in function of [AHL] for different targets. The continuous lines show the targets and the points the corresponding results for three runs. "Target x" corresponds to a gaussian curve with height $h = x$ and "Peak x exp" the cognate results. For $h = 40$ (resp. 10, 5), the gaussian is centered on $\mu = 10^{-3}$ (resp. 10^{-1}, 10^0). "Target step" corresponds to a step function centered on 10^{-2} with a plateau value of $20\,\mu M$. The algorithm was run for 5000 generations with barycentric crossover and gaussian noise mutator.

We also wanted to see whether this biological function was able to return a step-shaped response. The target function was changed to a step function with a value of $20\,\mu M$ between $10^{-2.5}\,\mu M$ and $10^{-1.5}\,\mu M$ and $0\,\mu M$ everywhere else. The algorithm found decent solutions (average relative error inferior to 2 %) in 200 s on average (see Fig. 4).

4.5 Grouped Evolution of *algogenes*

In nature, some *biogenes* coevolve: they are constrained by the same rules during evolution [28,29]. We tried to reproduce this phenomenon by allowing the algorithm to divide the genome in n groups of *algogenes* (n varying from 1 to 22). In a group, the *algogenes* are crossed with the same parameters. From two to five groups, the scores are higher than the one group-algorithm. The algorithm is converging prematurely. Above five groups, compared to the one group-algorithm the scores are similar. However, the algorithm requires 1.15 more time when using groups.

We also tried to run the algorithm with autoadaptive groups. Each individual had a random number of groups, composed of *algogenes* randomly picked. Group setup of each individual was allowed to mutate. Crossover of two individual creates a child with a number of groups of one of his two parents. Results were not conclusive in terms of score, nor in terms of biological interest: it was not observed that specific *algogenes* had a higher tendency to form a group together.

4.6 Results Obtained on an Alternative Biological System

To see whether this approach could be generalized to other types of genetic networks, we tried to optimize a bio-logic XOR gate. The complete description of the biological system can be found here [30]. A simplified illustration is given on Fig. 5. In the presence of both Phloretin (Ph) and Erythromycin (Er), mRNA

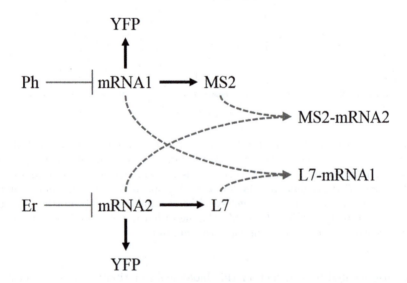

Fig. 5. Simplified XOR bio-logic gate. Barred red lines indicate a repression on the operon coding for the corresponding mRNA; red dashed arrows indicate a binding reaction; black heavy arrows indicate production. Ph: Phloretin. Er: Erythromycin. YFP: Yellow Fluorescent Protein (Color figure online).

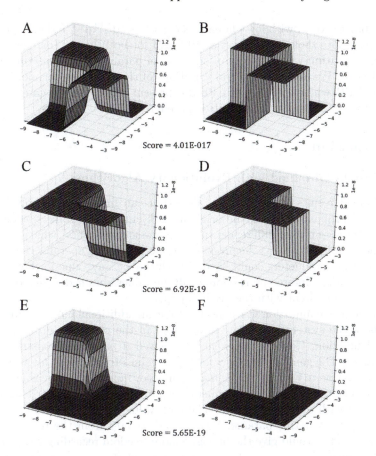

Fig. 6. [YFP] in function of [Er] and [Ph] for three different targets. B (resp. D and F) shows the target function for a XOR (resp. OR and INH) gate. A (resp. C and E) corresponds to the results for the target B (resp. D and F). Score corresponds to the fitness score of the best individual for each target.

are not synthesized. Thus Yellow Fluorescent Protein (YFP) is not produced. If there is no Ph nor Er, both mRNA are synthesized but inhibited because they are bound with L7 (for mRNA1) and MS2 (for mRNA2). Finally, when Ph (resp. Er) is present alone, mRNA2 (resp. mRNA1) is produced but not MS2 (resp. L7). As a consequence, mRNA2 or mRNA1 can be translated and YFP is synthesized.

The equation set that models this system is composed of 15 parameters (2 transcription rates, 1 translation rate, 2 dissociation constant and 2 Hill's numbers for Ph and Er repression, 2 dissociation constants for MS2-mRNA2 and L7-mRNA1 binding reaction and 3 translation rate and 4 decay rates).

Evaluation is performed by taking the MSE between the value of this function and the target function on 10 values of [Ph] times 10 values of [Er], spread uniformly on the log domain between 10^{-9} and $10^{-3}\,\mu$M. The target function is a classic binary XOR, with low plateaus being set at $10^{-12}\,\mu$M and high plateaus at $10^{-8}\,\mu$M. Threshold value is set at $10^{-6}\,\mu$M for both inputs (namely Er and Ph). The algorithm was able to produce a set of parameter with a score around 10^{-17} (Fig. 6). We also tried several variations of the XOR gate, namely an OR gate (no YFP when both Er and Ph are present) and an INH gate (YFP is produced when Er is present and pH is absent).

5   Discussion

5.1   Results Obtained with Evolutionary Algorithms

The obtained results show that the algorithm can find coherent solutions to a biological problem. Indeed, the original work emphasizes some specificity of their system (significant differences in the repressor constants) which the algorithm successfully shows. As the process contains modifying operators (crossover, mutation), the parameters could have virtually taken any value in their respective allowed range. Interestingly enough, all the values obtained were biologically relevant and consistent with the original paper.

Auto-adaptive mutation was expected to be an additional hint towards which elements of the system are of key importance in the correct realization of the expected function. However, no *algogene* showed a particular constraint (low sigma value). The general tendency for sigma values to be rather above average than below is due to the way the sigma is mutated. Indeed, its value is multiplied by the exponential of a random number drawn from a Gaussian distribution centered on 0.0 with a variance of 1.0. The high value of k'_{TL_1} (translation of CI) sigma suggests that this parameter is allowed to vary from the returned optimized value. This would give the biologist more freedom regarding the promoter he needs to use.

Running the algorithm with other targets revealed key parameters to set the desired position of the peak. These parameters are dissociation constants of regulators, which makes sense since these parameters control the sensitivity of a promoter towards its regulator. Two groups can be distinguished. K_{A_1} and K_{R_2} allow to shift the peak, and K_{R_3} and $K_{(R_3}$ to sharpen it. Indeed, a decrease on K_{A_1} will lead to a higher sensitivity of operon #1 promoters towards AHL, so that a lower [AHL] will be required to initiate $LacI_{M1}$ and CI expression. This leads to an increase of GFP at lower [AHL], namely a shift of the rising edge of the peak to the lower concentrations. Similarly, a decrease on K_{R_2} will increase the sensitivity of operon #2 towards CI, resulting in a stronger repression of LacI by CI: the falling edge of the peak will shift to lower [AHL]. Finally, K_{R_3} and $K_{R_{3'}}$ control directly GFP expression and therefore act on the thinness of the peak. Indeed an increase on $K_{R_{3'}}$ decreases the sensitivity of GFP's promoter towards CI. The rising edge of the peak will therefore be shifted towards lower [AHL]. Seemingly, a decrease on K_{R_3} will result in a shift of the falling edge of the peak towards lower [AHL] as well. This can be verified by simulating GFP levels in function of [AHL] while varying the above mentionned constants (data not shown).

We can expect that a change in GFP peak height only would involve a similar change in GFP translating constant. As it is not the case, we investigated the product $\frac{k_{TL_3} \cdot k_{TR_3}}{d_{mRNA_3} \cdot d_{GFP}}$ (see Eq. 4), which indeed grows proportionally to the height of the peak. The algorithm is therefore capable of finding non trivial solutions regarding this biological problem.

With the possibility for the *algogenes* to evolve in groups, we expected to observe the appearance of groups of linked parameters. Indeed, as described above, the maximal [GFP] depends of the constants k_{TL_3} , k_{TR_3}, d_{GFP} and d_{mRNA_3}. When the algorithm found a good individual, if k_{TL_3} and k_{TR_3} (respectively d_{GFP} and d_{mRNA_3}) evolve in the same direction (or not at all), the individual should keep its good score. We therefore expected for such parameters to tend to gather in the same group. This tendency was not observed.

To further validate the strength of this approach, a comparison with other methods (such as particle swarm or simulated annealing) could be carried on.

5.2   From an Evolution Strategy to Genetic Programming

The algorithm used to obtain the presented results is close to an Evolution Strategy to optimize a highly dimensional biological system (22 parameters). What would be a following step is to create an algorithm able to find a relevant biological system to answer the biologists' needs. Indeed, a typical approach in synthetic biology begins with the specifications describing a given problem. The solution is often a biological system, that fulfills the specification requirements. The next step for the biologist is to design the said biological system, and finally optimize its components. The last step is realized by optimizing the constants, as shown above. A still missing link is to find, in a library (typically the Biobricks library [31]), the closest components to the returned parameters. However, the biologist still has to imagine the biological system's architecture before 'feeding' it to the algorithm. That is why an algorithm able to create a biological function fulfilling a biological set of requirements is needed.

Genetic Programming could be used for this. As a biological system can be abstracted in a biological function as seen above, we first tried to evolve a function with the same evaluation as previously. Preliminary results showed that the returned equations were too complicated in order to find the biological system associated to the obtained equation. In synthetic biology, systems are made of so called parts, DNA sequences coding for promoters, proteins, terminators, among others. A good idea would be to use the standard decomposition of systems into blocks to evolve a graph, composed of such elements. A new formalism must be imagined for mutation, crossover and evaluation. To what corresponds the nodes and edges is still unclear. Ideally, the algorithm would be able to build the most compact biological system which, given the right constants, would fulfill the requirements. This hints towards an evaluation process taking into account the size of the system (it is easier to produce a promoter sensitive towards one or two regulators than towards four or five). Because some systems' behavior depends heavily on their parameters (*e.g.* the system presented here) the graph should not only produce an architecture (which product regulates the expression of which other product) but also include quantitative or semi-quantitative elements. Of course, as an optimization step is included afterwards, this step should be quite quick. Taking inspiration from fuzzy logic which introduces an intermediate level between binary and continuous elements, a further step would be to introduce parts with three different level of sensitivity, expression, etc.

6   Conclusion and Future Work

This work demonstrates the relevance of evolutionary algorithms in synthetic biology, in particular in the field of biological networks. Being able to optimize the parameters of a defined system before going to the bench is an important speedup in the process of biological networks design. Indeed, biologists typically have to test various sets of components before finding the most suitable. Such an algorithm gives hints on which parameters are of key importance and which kind of components should be used (*e.g.* a strong or weak repressor). Moreover, as this tool is capable of generating sets of parameters realizing a large variety of functions, it can also help to understand the specifics of a system, by varying the target function used in evaluation and analyzing the returned parameters. This modularity enlarges designers' horizon by giving them the possibility to preliminary test in silico any (crazy) idea they might have.

This optimizing step is to be preceded by a network conception step, also ideally automated. Genetic programming is a promising field in this regard. Formalism of the networks components, the operators to use and the evaluator is what comes next.

References

1. Ro, D.K., Paradise, E.M., Ouellet, M., Fisher, K.J., Newman, K.L., Ndungu, J.M., Ho, K.A., Eachus, R.A., Ham, T.S., Kirby, J., et al.: Production of the antimalarial drug precursor artemisinic acid in engineered yeast. Nature **440**(7086), 940–943 (2006)
2. Xie, Z., Wroblewska, L., Prochazka, L., Weiss, R., Benenson, Y.: Multi-input RNAi-based logic circuit for identification of specific cancer cells. Science **333**(6047), 1307–1311 (2011)
3. Levskaya, A., Chevalier, A.A., Tabor, J.J., Simpson, Z.B., Lavery, L.A., Levy, M., Davidson, E.A., Scouras, A., Ellington, A.D., Marcotte, E.M., et al.: Synthetic biology: engineering escherichia coli to see light. Nature **438**(7067), 441–442 (2005)
4. Peralta-Yahya, P.P., Zhang, F., Del Cardayre, S.B., Keasling, J.D.: Microbial engineering for the production of advanced biofuels. Nature **488**(7411), 320–328 (2012)
5. Beal, J., Weiss, R., Densmore, D., Adler, A., Appleton, E., Babb, J., Bhatia, S., Davidsohn, N., Haddock, T., Loyall, J., et al.: An end-to-end workflow for engineering of biological networks from high-level specifications. ACS Synth. Biol. **1**(8), 317–331 (2012)
6. Marchisio, M.A. (ed.): Computational Methods in Synthetic Biology, Methods in Molecular Biology, vol. 1244. Springer, New York (2015)

7. Myers, C.J.: Computational synthetic biology: progress and the road ahead. IEEE Trans. Multi-scale Comput. Syst. **1**(1), 19–32 (2015)
8. Bhatia, S., Roehner, N., Silva, R., Voigt, C.A., Densmore, D.: A framework for genetic logic synthesis **103**(11) (2015)
9. Icube laboratory - genetic netowrk design automation (2015). http://geneda.fr
10. Madec, M., Pecheux, F., Gendrault, Y., Bauer, L., Haiech, J., Lallement, C.: EDA inspired open-source framework for synthetic biology. In: 2013 IEEE Biomedical Circuits and Systems Conference (BioCAS), pp. 374–377. IEEE (2013)
11. Bilitchenko, L., Liu, A., Cheung, S., Weeding, E., Xia, B., Leguia, M., Anderson, J.C., Densmore, D.: Eugene-a domain specific language for specifying and constraining synthetic biological parts, devices, and systems. PloS ONE **6**(4), e18882 (2011)
12. Yaman, F., Bhatia, S., Adler, A., Densmore, D., Beal, J.: Automated selection of synthetic biology parts for genetic regulatory networks. ACS Synth. Biol. **1**(8), 332–344 (2012)
13. Basu, S., Gerchman, Y., Collins, C.H., Arnold, F.H., Weiss, R.: A synthetic multicellular system for programmed pattern formation. Nature **434**(7037), 1130–1134 (2005)
14. Thomas, R., Thieffry, D., Kaufman, M.: Dynamical behaviour of biological regulatory networks—i. Biological role of feedback loops and practical use of the concept of the loop-characteristic state. Bull. Math. Biol. **57**(2), 247–276 (1995)
15. Bernot, G., Comet, J.P., Richard, A., Guespin, J.: Application of formal methods to biological regulatory networks: extending thomas' asynchronous logical approach with temporal logic. J. Theor. Biol. **229**(3), 339–347 (2004)
16. Gendrault, Y., Madec, M., Lemaire, M., Lallement, C., Haiech, J.: Automated design of artificial biological functions based on fuzzy logic. In: 2014 IEEE Biomedical Circuits and Systems Conference (BioCAS), pp. 85–88. IEEE (2014)
17. Doboli, A., Vemuri, R.: Exploration-based high-level synthesis of linear analog systems operating at low/medium frequencies. IEEE Trans. Comput. Aided Des. Integr. Circ. Syst. **22**(11), 1556–1568 (2003)
18. Lohn, J.D., Colombano, S.P.: Automated analog circuit synthesis using a linear representation. In: Sipper, M., Mange, D., Pérez-Uribe, A. (eds.) ICES 1998. LNCS, vol. 1478, pp. 125–133. Springer, Heidelberg (1998)
19. Koza, J.R., Bennett, F.H., Andre, D., Keane, M.A., Dunlap, F.: Automated synthesis of analog electrical circuits by means of genetic programming. IEEE Trans. Evol. Comput. **1**(2), 109–128 (1997)
20. Koza, J.R.: Genetic Programming IV: Routine Human-Competitive Machine Intelligence. Kluwer Academic Publishers, Norwell (2003)
21. Maitre, O., Baumes, L.A., Lachiche, N., Corma, A., Collet, P.: Coarse grain parallelization of evolutionary algorithms on GPGPU cards with EASEA. In: Proceedings of the 11th Annual Conference on Genetic and Evolutionary Computation, pp. 1403–1410. ACM (2009)
22. Eshelman, L.J., Schaffer, J.D.: Real-coded genetic algorithms and interval-schemata. In: Whitley, L.D. (ed.) FOGA, pp. 187–202. Morgan Kaufmann, San Mateo (1992)
23. Deb, K., Agrawal, R.W.: Simulated binary crossover for continuous search space. Complex Syst. **9**, 115–148 (1995)
24. Schwefel, H.P.: Adaptive Mechanismen in der biologischen Evolution und ihr Einfluß auf die Evolutionsgeschwindigkeit

25. Collet, P., Lutton, E., Schoenauer, M., Louchet, J.: Take it EASEA. In: Deb, K., Rudolph, G., Lutton, E., Merelo, J.J., Schoenauer, M., Schwefel, H.-P., Yao, X. (eds.) PPSN 2000. LNCS, vol. 1917, pp. 891–901. Springer, Heidelberg (2000)
26. Collet, P., Krüger, F., Maitre, O.: Automatic parallelization of EC on GPGPUs and clusters of GPGPU machines with EASEA and EASEA-CLOUD. In: Tsutsui, S., Collet, P. (eds.) Massively Parallel Evolutionary Computation on GPGPUs, pp. 15–34. Springer, Heidelberg (2013)
27. Hansen, N., Ostermeier, A.: Completely derandomized self-adaptation in evolution strategies. Evol. Comput. **9**(2), 159–195 (2001)
28. Ehrlich, P., Raven, P.: Butterflies and plants: a study in coevolution. Evolution **18**(4), 586–608 (1964)
29. Goh, C.S., Bogan, A.A., Joachimiak, M., Walther, D., Cohen, F.E.: Co-evolution of proteins with their interaction partners. J. Mol. Biol. **299**(2), 283–293 (2000)
30. Ausländer, S., Ausländer, D., Müller, M., Wieland, M., Fussenegger, M.: Programmable single-cell mammalian biocomputers. Nature **487**(7405), 123–127 (2012)
31. Registry of standard biological parts. http://parts.igem.org/Main_Page

EvoCOMNET

A Hybrid Discrete Artificial Bee Colony Algorithm for the Multicast Routing Problem

Yannis Marinakis[1]([✉]), Magdalene Marinaki[1], and Athanasios Migdalas[2,3]

[1] School of Production Engineering and Management,
Technical University of Crete, Chania, Greece
marinakis@ergasya.tuc.gr, magda@dssl.tuc.gr
[2] Department of Civil Engineering, Aristotle University of Thessalonike,
54124 Thessalonike, Greece
samig@civil.auth.gr, athmig@ltu.se
[3] Industrial Logistics, Department of Business Administration,
Technology and Social Sciences, Luleå Technical University, 97187 Luleå, Sweden

Abstract. In this paper, a new algorithm is proposed for the solution of the Multicast Routing Problem. The algorithm is based on the Artificial Bee Colony approach hybridized with Variable Neighborhood Search. The quality of the algorithm is evaluated with experiments conducted on suitably modified benchmark instances of the Euclidean Traveling Salesman Problem from the TSP library. The results of the algorithm are compared to results obtained by several versions of the Particle Swarm Optimization algorithm. The comparisons indicated the effectiveness of the new approach.

Keywords: Artificial bee colony · Variable neighborhood search · Multicast routing problem

1 Introduction

In the last few years, a number of swarm intelligence algorithms based on the behavior of the bees has been presented [1–3]. These algorithms are mainly divided in two categories according to the behavior in the nature which they simulate: the foraging behavior, and the mating behavior. The main algorithms based on the mating behavior are the Honey Bees Mating Optimization Algorithm (HBMO) [4–6], the Bumble Bees Mating Optimization Algorithm (BBMO) [7] and the Bees Life Algorithm (BLA) [8]. There is a large number of algorithms that simulate the foraging behavior of the bees, the Virtual Bee Algorithm [9], the BeeHive [10] algorithm, the Artificial beehive algorithm (ABHA) [11], the Bees Algorithm (BA) [12], the Bee Swarm Optimization (BSO) Algorithm [13], the Bumblebees algorithm [14], the Bee Colony Optimization (BCO) algorithm [15], the Bee Colony inspired Algorithm (BCiA) [16], the Bee System (BS) [17], the Honeybee Social Foraging (HBSF) [18], the OptBees algorithm [19], the Simulated Bee Colony (SBC) [20] and the Wasp Swarm Optimization (WSO) [21,22].

© Springer International Publishing Switzerland 2016
G. Squillero and P. Burelli (Eds.): EvoApplications 2016, Part I, LNCS 9597, pp. 203–218, 2016.
DOI: 10.1007/978-3-319-31204-0_14

The most important algorithm that simulates the foraging behavior of the bees is the Artificial Bee Colony (ABC) algorithm proposed by [23,24]. This algorithm has been applied to many problems with very good results [25]. In this paper, a hybridized version of the Artificial Bee Colony algorithm, the Hybrid Discrete Artificial Bee Colony (HDABC), is proposed for the solution of the multicast routing problem. From all algorithms that simulate the behavior (foraging or mating) of the bees, a hybridization of the Artificial Bee Colony algorithm is selected, implemented and used in this paper as it seems to be the most effective of all bee inspired algorithms and an algorithm that gives comparative and in some problems better results from other evolutionary algorithms. Thus, as the Multicast Routing Problem is a difficult combinatorial optimization problem we would like to use for its solution the most effective bee inspired algorithm. Routing has always been of immense importance in communication networks due to its impact on the network performance. During the last years the increasing demand for internet and mobile communications has increased the significance of scalable and adaptive routing. A routing algorithm selects one or more paths over which devices communicate with each other. The selection of these paths is based on the status of the network and optimization criteria aiming at maximizing network performance by pursuing e.g. optimization of resource utilization, or/and minimization of congestion (packet delay), or/and minimization of packet loss, etc. Due to these facts, routing problems are combinatorial optimization problems which differ with respect to type, i.e., a routing may be of unicast, multicast or anycast type depending on the number of communicating devices, with respect to the network topology which may be of fixed infrastructure or of ad-hoc wireless infrastructure.

The rest of the paper is organized as follows: In the next section, a short description of the Multicast Routing Problem is presented while in Sect. 3 the proposed algorithm is described and analyzed in detail. In Sect. 4, the computational results of the algorithm are given. Finally, in the last section conclusions and possible future directions are given.

2   The Multicast Routing Problem

Multicast routing is a technique to simultaneously transfer information from a source to a set of destinations in a communication network. Compared to **unicast routing** which is based on a point-to-point transmission, multicast is more efficient since it utilizes the inherent network parallelism, that is, shares resources, such as links and forwarding nodes, in order to efficiently deliver data from a source to destination nodes. Due to the increasing development of multimedia applications concerning video and audio transmissions but also the increased importance of tele-conferencing, collaborative environments, distance learning, and e-commerce, the importance of multicast routing and the importance of the development of efficient multicast routing algorithms have increased significantly [26–28].

Most multicast routing algorithms have the goal of minimizing the cost of the constructed multicast tree. However, such a tree is a (**rooted**) **Steiner tree** and it is well known that the problem of finding it is NP-hard [29]. Therefore, much attention has been directed towards developing heuristics of polynomial complexity that produce near optimal results and often guarantee that the produced solutions are within twice the cost of the optimum one [27,28,30–32]. Moreover, the multicast routing has also been formulated as **constrained Steiner tree** problems. Thus, Kompleea et al. [33] require that each path in the tree must satisfy an end-to-end delay bound, Tode et al. [34] restrict the number of packet copies per network node, Hwang et al. [35] minimize the cost of the multicast tree subject to law path delay and Wu and Hwang [36] consider the minimization of the cost of the tree subject to multiple constraints concerning such issues as end-to-end delay and end-to-end loss probability. In [37] the authors propose the first robust optimization model for jointly optimizing the topology and the routing in body area networks under traffic uncertainty. In [38] the authors investigate the robust multiperiod network design problem and propose, as a remedy against traffic volume uncertainty, a robust optimization model based on multiband robustness.

The **multicast routing problem** (**MRP**) can be stated as follows [26]:

Given a network $\mathcal{G} = (\mathcal{N}, \mathcal{E}, c)$ where $\mathcal{N} \neq \emptyset$ is the set of nodes, $\mathcal{E} \subset \mathcal{N} \times \mathcal{N}$ is the set of edges connecting pairs of nodes. Let $|\mathcal{N}| = n$ and $|\mathcal{E}| = m$. Moreover, let $s \in \mathcal{N}$ be a distinct node, called the source node, and let $\mathcal{D} \subseteq \mathcal{N} \backslash \{s\}$ be the set of destination nodes. Let $c : \mathcal{E} \to \mathcal{R}_0^+$ be a cost function that assigns a real value to each edge.

When a **Single Flow routing** problem is considered:

$$y_e^k = \begin{cases} 1 \text{ if link e is used to transmit flow to destination k,} \\ 0 \text{ otherwise,} \end{cases} \forall k \in \mathcal{D}. \quad (1)$$

When a **Multicommodity Flow routing** is considered, let $\mathcal{F} = \{1, 2, \ldots, \nu\}$ be the set of different flows emanating from the source s and let $\mathcal{D}_f \subseteq \mathcal{N} \backslash \{s\}$, for $f \in \mathcal{F}$, be the corresponding set of destination nodes, then:

$$y_e^{fk} = \begin{cases} 1 \text{ if link } e \text{ is used to transmit} \\ \quad \text{flow f to destination k,} \quad \forall k \in \mathcal{D}_f, \forall f \in \mathcal{F} \\ 0 \text{ otherwise,} \end{cases} \quad (2)$$

Consider the single objective problem flow case. It is sought a subnetwork $T(\{s\} \cup \mathcal{D}) = (\mathcal{N}_T, \mathcal{E}_T, c_T)$ of $\mathcal{G}$ such that:

- $\{s\} \cup \mathcal{D} \subseteq \mathcal{N}_T$.
- $\mathcal{E}_T \subset \mathcal{N}_T \times \subset \mathcal{N}_T$.
- c_T is the restriction of c to $\mathcal{E}_T$.
- there is a path from s to every node $v \in \mathcal{D}$.
- the total cost of $T(\{s\} \cup \mathcal{D})$, i.e. $c(T(\{s\} \cup \mathcal{D})) = \sum_{e \in \mathcal{E}_T} c(e)$, is minimized.

The cost in this problem setting may refer to **various metrics** of network resource utilization such as delay, bandwidth, number of links. The approach has received criticism because performance factors, such as delay and bandwidth utilization, do not in general get optimized by minimizing the sum of the edge costs [39].

The two most common and important requirements when designing multicast trees are **delay** and **bandwidth utilization**. A cost occurs from using and/or reserving network resources such as bandwidth. The **end-to-end delay** is the sum of the total delays encountered along the paths from source to each destination. Typically, the delay should be within a certain bound in real time communications.

Such additional requirements must be imposed separately resulting in constrained versions of the problem [35, 36, 40]. Let $\mathcal{P}_v \subset \mathcal{E}_T$ denote, in the form of a sequence of edges, the path in $\mathcal{T}(\{s\} \cup \mathcal{D})$ from the source s to the destination $v \in \mathcal{D}$.

If besides c we, also, introduce the delay function $d : \mathcal{E} \rightarrow \mathcal{R}_0^+$, then, the delay $d(\mathcal{P}_v)$ along the path $\mathcal{P}_v$ is the sum of the delays on all links along the path, i.e. $d(\mathcal{P}_v) = \sum_{e \in \mathcal{P}_v} d(e)$.

The delay of the multicast tree $\mathcal{T}(\{s\} \cup \mathcal{D})$ is, then, defined as the maximum delay among all such paths, that is, $d(\mathcal{T}(\{s\} \cup \mathcal{D})) = \max_{v \in \mathcal{D}}\{d(\mathcal{P}_v)\}$.

Hence, the **delay-constrained least cost routing tree** must satisfy all the previous requirements plus the additional constraints: $d(\mathcal{P}_v) \leq b, \; \forall v \in \mathcal{D}$, where b is a specified delay bound.

A number of evolutionary algorithms (based mainly on Genetic Algorithms and Ant Colony Optimization) have been proposed for solving the single objective and multiobjective routing problems. For an analytical presentation of these algorithms please see [26].

3   Hybrid Discrete Artificial Bee Colony

In this paper, an extended version of the Artificial Bee Colony optimization algorithm [23, 24] in the discrete space, the Hybrid Discrete Artificial Bee Colony (HDABC), is proposed for the Multicast Routing Problem. The algorithm used in this paper is based on the algorithm proposed in [41] for the solution of a clustering problem where instead of a Greedy Randomized Adaptive Search Procedure (GRASP) algorithm [42] for the creation of the initial population, a random process is used, and a local search based on Variable Neighborhood Search (VNS) [43] is utilized for the improvement of the solutions. For a recent complete survey about hybridization techniques in metaheuristics please see [44].

In the Artificial Bee Colony optimization algorithm [23, 24], there are three kind of artificial bees in the colony, the employed bees, the onlooker bees and the scouts. Initially, a set of food source positions (possible solutions) are randomly generated and, then, are randomly selected by the employed bees and their nectar amounts (fitness functions) are determined. The solutions should consist of values equal to 0 or to 1. In our algorithm, the food sources are generated exactly as in the original algorithm (i.e. randomly produced continuous values) and subsequently these values are transformed by using a sigmoid function:

$$sig(x_{ij}) = \frac{1}{1 + exp(-x_{ij})}. \tag{3}$$

Then the food sources are calculated as follows:

$$y_{ij} = \begin{cases} 1, & \text{if } rand_1 < sig(x_{ij}) \\ 0, & \text{if } rand_1 \geq sig(x_{ij}), \end{cases} \tag{4}$$

where x_{ij} is the solution (food source), $i = 1, \ldots, N$ (N is the number of food sources), $j = 1, \ldots, d$ (d is the dimension of the problem), y_{ij} is the transformed solution and $rand_1$ is a random number in the interval (0,1). Subsequently the fitness of each food source is calculated and an employed bee is attached to each food source. These equations have, also, been used for the Discrete Particle Swarm Optimization [45]. The employed bees return in the hive and perform the waggle dance in order to inform the other bees (onlooker bees) about the food sources. Then, the onlooker bees choose the food source which they will visit based on the nectar information taken from the waggle dance of the employed bees. The probability of choosing a food source is given by [23, 24]:

$$p_i = \frac{fit_i}{\sum_{n=1}^{N} fit_n}. \tag{5}$$

where fit_i is the fitness function of each food source.

The nectar information corresponds to the fitness function value of each food source. Since the multicast routing problem is a minimization problem, if a solution has a high value in the cost function, then, it is not a good solution and its fitness value must be small. So, a high fitness value must correspond to a solution with a small cost function. A way to accomplish this is to find initially the solution in the population with the maximum cost and to subtract from this value the cost of each of the other solutions. Now, the higher fitness value corresponds to the solution with the shorter cost [41].

Since the probability of selecting a food source by the onlooker bees is related to its fitness, and since the food source with the worst cost has fitness equal to zero, it will never be selected for food gathering. To avoid this possibility the fitness of all food sources is incremented by one [41]. Afterwards, the employed and the onlooker bees are placed in the selected food sources. In order to produce a new food source position from the old one, the Artificial Bee Colony algorithm uses the equation [23, 24]:

$$x'_{ij} = x_{ij} + rand_2(x_{ij} - x_{kj}) \tag{6}$$

where x'_{ij} is the candidate food source, k is a different from i food source, and $rand_2$ is a random number in the interval (0,1). As the values of the candidate food source are not suitable for the multicast routing problem they are transformed to the y'_{ij} using the Eqs. (3) and (4), and the fitness of each food source is then calculated.

Next, a local search algorithm (Variable Neighborhood Search) is applied in each food source. It should be noted that if there is a large number of bees in a food source, then, due to the local search moves that each bee performs, this food source provides greater exploitation abilities in each iteration. If a better food source is found in an iteration, this food source replaces the old one. If for a number of iterations a solution is not improved, then, this solution is assumed to be abandoned and a scouter bee is placed in a new random position (a new food source).

3.1   Path Representation

All the solutions (food sources) are represented by vectors of length equal to the number of nodes and of integer values $\{0, 1, 2, \cdots\}$, where a zero value means that the corresponding node does not belong to the path, the value 1 means that the node belongs to all paths, the value 2 means that the node belongs only to the path number 1, the value 3 means that the node belongs only to the path number 2, the value 4 means that the node belongs only to the path number 3, and so on. If the node belongs two more than one paths but not in all paths and we have 3 paths, then, the number 5 means that the node belongs to the paths number 1 and 2, the number 6 means that the node belongs to the paths number 1 and 3 and number 7 means that the node belongs to the paths number 2 and 3. For example, if we have a food source with ten nodes and starting node the node 1, then, a unicast routing with end node the node 10 can be represented by the vector

$$
\begin{array}{cccccccccc}
1 & 2 & 3 & 4 & 5 & 6 & 7 & 8 & 9 & 10 \\
\hline
1 & 0 & 1 & 0 & 0 & 1 & 1 & 0 & 1 & 1 \\
\hline
\end{array}
$$

which corresponds to the following path:

$$
1 \longrightarrow 3 \longrightarrow 6 \longrightarrow 7 \longrightarrow 9 \longrightarrow 10.
$$

For a food source with ten nodes and starting node the node 1, a multicast routing with two end nodes, the nodes 9 and 10, a vector representation of the form

$$
\begin{array}{cccccccccc}
1 & 2 & 3 & 4 & 5 & 6 & 7 & 8 & 9 & 10 \\
\hline
1 & 2 & 1 & 3 & 2 & 0 & 1 & 0 & 2 & 3 \\
\hline
\end{array}
$$

means that the paths are as follows:

$$
1 \longrightarrow 2 \longrightarrow 3 \longrightarrow 5 \longrightarrow 7 \longrightarrow 9
$$
$$
1 \longrightarrow 3 \longrightarrow 4 \longrightarrow 7 \longrightarrow 10.
$$

3.2   Variable Neighborhood Search

A Variable Neighborhood Search (VNS) [43, 46] algorithm is applied in order to optimize the food sources. Both combinatorial optimization local search algorithms ($2-$opt, $1-0$ relocate, and $1-1$ exchange) and continuous optimization local search algorithms are utilized. The latter are denoted by LS1, $\cdots$, LS6. All the algorithms are applied for a number of iterations (ls_{num}). The first one (LS1) uses a transformation of the solution inside the solution space. The second one (LS2) combines the current solution with the best food source ($bfsource$) and the third one (LS3) combines the current solution with its personal best ($pfsource$). The fourth one (LS4) is a combination of the current solution, the personal best and the best food source. The fifth one (LS5) and the sixth one (LS6) are crossovers of the food source with the personal best and the best food source, respectively. The methods used are described by the following equations [26]:

$$\text{LS1:} \quad x_{ij}(t_1 + 1) = rand_5 \cdot x_{ij}(t_1) \tag{7}$$

$$\text{LS2:} \quad x_{ij}(t_1 + 1) = rand_6 \cdot bfsource_j + (1 - rand_6) \cdot x_{ij}(t_1) \tag{8}$$

$$\text{LS3:} \quad x_{ij}(t_1 + 1) = rand_7 \cdot pfsource_{ij} + (1 - rand_7) \cdot x_{ij}(t_1) \tag{9}$$

$$\text{LS4:} \quad x_{ij}(t_1 + 1) = rand_8 \cdot rand_9 \cdot bfsource_j$$
$$+rand_8 \cdot (1 - rand_9) \cdot pfsource_{ij} + (1 - rand_8) \cdot x_{ij}(t_1) \tag{10}$$

$$\text{LS5:} \quad x_{ij}(t_1 + 1) = \begin{cases} pfsource_{ij}, & \text{if } rand_{10} \le 0.5 \\ x_{ij}(t_1), & \text{otherwise} \end{cases} \tag{11}$$

$$\text{LS6:} \quad x_{ij}(t_1 + 1) = \begin{cases} bfsource_j, & \text{if } rand_{11} \le 0.5 \\ x_{ij}(t_1), & \text{otherwise,} \end{cases} \tag{12}$$

where t_1 is the local search iteration number, $rand_6, \cdots, rand_{11}$ are random numbers in the interval [0,1] and $rand_5$ is a random number in the interval [-1,1].

Two different versions of the VNS algorithm are considered. The first one is the classic version. It starts with a certain neighborhood and when a local optimum is found with respect to that one, the algorithm proceeds with the next (enlarged) neighborhood in turn. In the second version, called sequential VNS (SVNS), all selected neighborhoods are applied consecutively in each iteration. These algorithms are utilized both with continuous neighborhoods (VNS1 and SVNS1) and with discrete ones (VNS2 and SVNS2).

4   Computational Results

The algorithm was implemented in modern Fortran and tested on five modified benchmark instances from the TSPLIB: Eil51, Eil76, pr264, A280 and pr439. The Traveling Salesman Problem (TSP) instances were modified to suitable instances for the multicast routing problem with the number of nodes ranging from 51 to 439, a single source (the first node) and two, three or five destination nodes corresponding to highest indexed nodes. There is no connection between the destination nodes and the links between the other nodes are uni-directional. We do not allow using the nodes in different order from their labels. The link cost is the distance between the two end nodes while the link delay is chosen randomly. Finally, a different delay bound b is associated with each instance. The value of b depends on the number of nodes and the random delay values.

In order to test the efficiency of the proposed algorithm, the algorithm is compared with three versions of the Particle Swarm Optimization (PSO) algorithm [26] where in each version a different velocity equation and a different neighborhood topology is proposed. The parameters of the proposed algorithm were selected after thorough testing several alternative values. Those which gave the best results with respect to the solution quality and the computational time are listed in Table 1. In this table, the parameters of the PSO algorithms from [26] have been included. The parameters it_{num}, neighbors, c_1, c_2 and c_3 are parameters of the PSO algorithms, where the first two are used in the neighborhood and the last three are used in the velocity equations.

Table 1. Parameters for all algorithms

	HDABC	PSOLGNT	PSOLGENT1	PSOLGENT2
Food sources/particles	100	100	100	100
Iterations	1000	1000	1000	1000
Employed bees	20	-	-	-
Onlookers bees	80	-	-	-
ls_{num}	10	10	10	10
it_{num}	-	-	10	10
Neighbors	-	5	3 to 99	3 to 99
c_1	-	1.35	1.35	1.35
c_2	-	1.35	1.35	1.35
c_3	-	1.40	1.40	1.40

The results given in Table 2 are for the new algorithm, for the three versions of PSO that do not employ local search (WLS), and for those PSO variants that use four versions of VNS: VNS1, VNS2, SVNS1 and SVNS2, respectively.

In all tables and figures of this section, PSOLGNT denotes the variant of the PSO algorithm that uses a different equation of velocities than the one that is most commonly used and a static number of local neighborhoods, PSOLGENT1 denotes the version of PSO which employs two features, the different equation of velocities and the expanding neighborhood procedure based on the best particle while PSOLGENT2 is the variant with the expanding neighborhood procedure based on the personal best position of each particle [26].

All figures of this section show the improvement progress of the solution associated with the best food source (for the proposed algorithm) and the best particle (for the PSO implementations) during the iterations.

The analysis of the Table 2 is performed using two different directions. Initially, we would like to see which of the VNS variant performs better. In order to achieve this goal, we divided the Table in 60 different run sets. Each run set contains five runs (the results for each instance (eil51, eil76, ...), for each algorithm used in the comparisons (HDABC, PSOLGNT, PSOLGENT1 and PSOLGENT2) and for each number of end nodes (2, 3 or 5 end nodes)) with

Table 2. Results of the proposed algorithm and of the PSO versions in all benchmark instances

	2 end nodes				3 end nodes				5 end nodes			
	HDABC	PSO LGNT	PSO LGENT1	PSO LGENT2	HDABC	PSO LGNT	PSO LGENT1	PSO LGENT2	HDABC	PSO LGNT	PSO LGENT1	PSO LGENT2
					eil51							
WLS	1159.59	1138.33	1134.48	1111.48	1152.83	1156.55	1173.3	1195.24	1119.44	1198.88	1168.61	1192.67
VNS1	650.17	714.86	620.2	609.68	595.21	716.81	766.61	787.13	652.99	624.53	663.89	746.67
VNS2	40.32	40.32	40.32	40.32	70.77	70.93	66.99	72.95	208.67	217.68	228.38	263.41
SVNS1	400.6	514.93	494.6	480.86	496.58	486.91	529.3	523.66	419.87	436.05	585.88	575.8
SVNS2	40.32	40.32	40.32	40.32	123.93	184.04	158.1	298.49	260.81	301.92	347.92	341.17
					eil76							
WLS	2070.07	2045.34	2070.07	1978.26	1882.36	2081.62	1965.78	1819.01	1932.19	1962.04	1932.19	1900.05
VNS1	1235.01	1286.94	1270.51	1287.74	1266.49	1357.52	1333.8	1313.87	1371.41	1257.16	1368.58	1460.93
VNS2	49.26	49.26	49.68	58.27	81.68	100.76	224.78	184.67	270.67	350.27	372.89	297.01
SVNS1	1003.3	982.36	967.59	1025.45	911.52	883.63	889.59	886.02	798.11	839.79	821.76	859.54
SVNS2	49.38	75.16	72.31	113.51	323.08	351.18	400.24	402.32	586.98	690.67	554.06	519.18
					pr264							
WLS	97394.81	108571.8	111604.2	107558.2	124733.5	120118.6	125187.9	120216.8	157758	154892.8	156176.4	183879.7
VNS1	65285.36	75691.6	64693.97	66538.3	81091.98	80139.88	75062.42	74962.39	102113.1	99901.38	97441.84	97458.36
VNS2	29768.99	40796.37	23029.15	25133.66	34095.24	59726.99	56202.67	59568.67	65416.32	67859.46	73662.71	73985.34
SVNS1	76252.24	67842.44	70433.64	71303.2	78879.08	82955.7	79964.22	81332.49	99202.7	93282.33	97972.02	97342.74
SVNS2	36603.97	63308.6	48494.66	44260.11	67224.04	78409.79	67230.2	72943.94	86926.66	92508.23	91243.62	92957.12
					A280							
WLS	4194.44	4291.98	4470.23	3830.7	4702.43	4613.91	4856.83	4609.35	5712.92	6124.67	6064.63	6436.14
VNS1	3758.41	4126.98	4146.05	4167.34	3234.03	2888.7	2866.08	2888.35	3217.92	3934.39	3396.08	3777.71
VNS2	980.37	1508.22	1423.46	1391.02	1429.69	1466.97	1573.11	1413.36	1883.80	2222.28	2687.92	2530.96
SVNS1	2868.46	2772.25	3016.28	2827.1	3296.04	3156.02	3152.75	3153.15	4227.6	3130.2	3094.09	3149.1
SVNS2	1131.5	1542.63	1221.8	1444.07	1999.18	2194.93	2059.23	2087.41	2274.11	2473.8	2518.66	2557.19
					pr439							
WLS	320463.2	335412.5	292549.1	352364.2	380014.6	421655.5	379063.6	410003	598892.5	483956.7	552124.5	471173.5
VNS1	263379.4	290356.9	167689.7	174279.3	352901.5	176416.7	301053.1	198538	200809.5	244557.2	482690.6	200821.6
VNS2	74639.68	76115.86	73313.09	73458.76	87811.39	73632.12	96066.55	79018	97035.61	104698.6	135617.2	134594.3
SVNS1	163445.4	166381.6	142149.6	137240.6	164394.8	177385.7	170619.3	178557	196953.4	194711.4	206554.3	200500.3
SVNS2	71395.54	142266.1	105677	99727.28	95290.95	147354.8	157639.9	115567.7	153649.7	159415.4	176267	179654.2

different variants of VNS (WLS, VNS1, VNS2, SVNS1 and SVNS2). Thus, the run set 1 contains the results of HDABC with 2 end nodes for the instance eil51 using for local search algorithms WLS, VNS1, VNS2, SVNS1 and SVNS2, respectively. For this run set we would like to see which local search performs better. The run set 2 contains the results of PSOLGNT with 2 end nodes for the instance eil51 using for local search algorithms WLS, VNS1, VNS2, SVNS1 and SVNS2, respectively and so on. It can be observed that for all run sets, the variants that do not employ local search (WLS) perform worse than all the other variants of all algorithms. It can also be noted that the HDABC and the versions of PSO with VNS2 and SVNS2 perform better than those employing VNS1 and SVNS1. Moreover, VNS2 performs in most cases better than SVNS2. In total, VNS2 performs best in 53 of the run sets, SVNS2 performs best in 3 sets while in the remaining four run sets, VNS2 and SVNS2 have equal performance (Table 3).

Afterwards, we would like to see which of the algorithms (HDABC, PSOL-GNT, PSOLGENT1 and PSOLGENT2) performs better. A number of 75 run sets were created. Each run set contains four runs (the results for each instance (eil51, eil76, ...), for each local search algorithm (WLS, VNS1, VNS2, SVNS1 and SVNS2) and for each number of end nodes (2, 3 or 5 end nodes)) with different algorithms (HDABC, PSOLGNT, PSOLGENT1 and PSOLGENT2). Thus, in this case, the run set 1 contains the results of WLS with 2 end nodes for the instance eil51 for all algorithms HDABC, PSOLGNT, PSOLGENT1 and PSOLGENT2, respectively. For this run set we would like to see which algorithm performs better.

Table 3. Number of run sets in which each local search algorithm performs better for the algorithms used

	Total	HDABC	PSOLGNT	PSOLGENT1	PSOLGENT2
WLS	0	0	0	0	0
VNS1	0	0	0	0	0
VNS2	53	13	14	12	14
SVNS1	0	0	0	0	0
SVNS2	3	1	0	2	0
VNS2 and SVNS2	4	1	1	1	1
	60	15	15	15	15

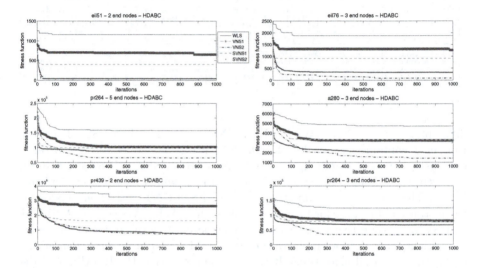

Fig. 1. Performance comparisons of the proposed algorithm without local search (WLS) and with all VNS variants for a number of instances and destination nodes.

However, comparing HDABC, PSOLGNT, PSOLGENT1 and PSOLGENT2, it is clear that none of the variants dominates the others for all problem instances. More precisely, when no local search is employed (WLS), HDABC performs best in 4 run sets, PSOLGNT in 2, PSOLGENT1 in 2, and PSOLGENT2 in 7. When VNS1 is used, HDABC performs best in 6 run sets, PSOLGNT in 3, PSOL-GENT1 in 4, and PSOLGENT2 in 2. If VNS2 is used, then, all variants find the same solution for one instance, in one instance HDABC and PSOLGNT find the same solution, HDABC performs best in 8 run sets, PSOLGNT in 1, PSOLGENT1 in 3, and PSOLGENT2 in 1. When SVNS1 is used, HDABC performs best in 5 run sets, PSOLGNT in 5, PSOLGENT1 in 4, and PSOLGENT2 in 1. Finally, in the case of SVNS2, all variants find the same solution for one instance, HDABC performs best in 13 run sets, and PSOLGENT2 in 1. PSOL-

GNT performs best in 5 run sets, PSOLGENT1 in 6, PSOLGENT2 in 3. All in all, in two cases all the variants find the same solutions, in one case HDABC and PSOLGNT find the same solution, HDABC performs best in 36 run sets, PSOLGNT in 11, PSOLGENT1 in 13, and PSOLGENT2 in 12 (Table 4). In general, we can say that the proposed algorithm gave the best results for almost half of the instances.

In Fig. 1, a graphical presentation is given for 6 out of the first 60 run sets discussed previously. In these figures, six representative run sets are depicted, taking care to show runs for all instances and for different number of destination nodes every time. Thus, in Fig. 1 the reduction of the objective function of the best food source is demonstrated for runs of HDABC without employing any local search algorithm and with all VNS variants. It can be seen that

Table 4. Number of run sets in which each algorithm performs better based on the local search algorithm used

	Total	WLS	VNS1	VNS2	SVNS1	SVNS2
HDABC	36	4	6	8	5	13
PSOLGNT	11	2	3	1	5	0
PSOLGENT1	13	2	4	3	4	0
PSOLGENT2	12	7	2	1	1	1
All algorithms	2	0	0	1	0	1
HDABC and PSOLGNT	1	0	0	1	0	0
	75	15	15	15	15	15

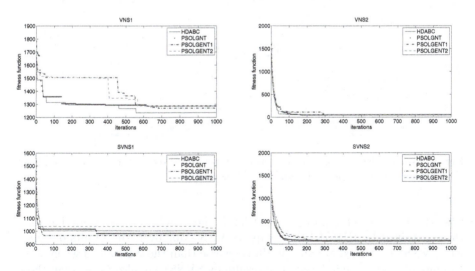

Fig. 2. Results of the proposed algorithm and of the three variants of PSO with all VNS versions using 2 end nodes for the instance eil76.

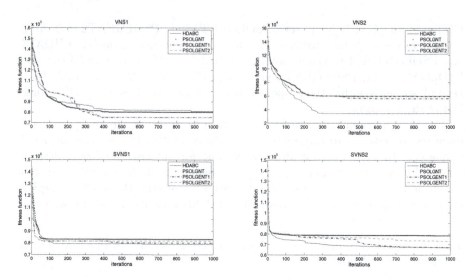

Fig. 3. Results of the proposed algorithm and of the three variants of PSO with all VNS versions using 3 end nodes for the instance pr264.

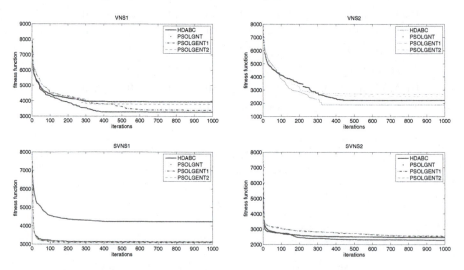

Fig. 4. Results of the proposed algorithm and of the three variants of PSO with all VNS versions using 5 end nodes for the instance a280.

when no local search algorithm is utilized, the results are less favorable than those obtained by HDABC that employs VNS. Moreover, the results obtained when VNS1 is utilized are almost always worse than those obtained when other variants of VNS are employed. The results by SVNS1 are also worse than those obtained by employing either VNS2 or SVNS2. The two best performing variants of VNS are VNS2 and SVNS2. However, the VNS2 converges to better values than SVNS2 in most cases.

In Figs. 2, 3 and 4, 12 of the second 75 run sets are depicted graphically. Figure 2 is divided in four subfigures where the results obtained for the instance eil76 with two end nodes by the proposed algorithm using VNS1, VNS2, SVNS1 and SVNS2 are depicted. Similarly Fig. 3 depicts the results for the instance pr264 with three end nodes, and Fig. 4 shows the results for the instance a280 with five end nodes. From these results it is obvious that no algorithm performs better than the others for all problem instances. It can be seen, however, that the new algorithm performs in most cases better than the other algorithms. Also, the proposed algorithm using VNS2 and SVNS2 performs better as it converges to better values than those found when the other two VNS versions are utilized. It should be noted that when the proposed algorithm uses either VNS2 or SVNS2 it converges to better objective function values than those found by the PSO implementations using the same VNS variants.

5   Conclusions and Future Research

In this paper, a new hybridized version of the Artificial Bee Colony algorithm with the Variable Neighborhood Search was presented for solving the Multicast Routing Problem. Several variants of the algorithm were tested on a number of modified instances from the TSPLIB and the importance of employing VNS was verified. The algorithm was compared with three versions of Particle Swarm Optimization algorithm and it performed better as in a total of 75 different run sets the algorithm gave superior results for half of the run sets. The new algorithm compared to variants of PSO has the advantage that it needs fewer parameters and in addition produces better results than those of the tested PSO variants. We therefore conclude that the HDABC algorithm can be applied effectively to the specific problem. Future research will be focused on the application of the methodology to other difficult problems.

References

1. Baykasoglu, A., Ozbakir, L., Tapkan, P.: Artificial bee colony algorithm and its application to generalized assignment problem. In: Chan, F.T.S., Tiwari, M.K. (eds.) Swarm Intelligence, Focus on Ant and Particle Swarm Optimization, pp. 113–144. I-Tech Education and Publishing, Vienna (2007)
2. Karaboga, D., Akay, B.: A survey: algorithms simulating bee swarm intelligence. Artif. Intell. Rev. **31**, 61–85 (2009)
3. Xing, B., Gao, W.J.: Innovative Computational Intelligence: A Rough Guide to 134 Clever Algorithms. Intelligent Systems Reference Library, vol. 62. Springer International Publishing, Cham (2014)
4. Abbass, H.A.: A monogenous mbo approach to satisfiability. In: International Conference on Computational Intelligence for Modeling, Control and Automation, CIMCA 2001, Las Vegas, NV, USA (2001)

5. Abbass, H.A.: Marriage in honey-bee optimization (MBO): a haplometrosis polygynous swarming approach. In: The Congress on Evolutionary Computation (CEC 2001), Seoul, Korea, pp. 207–214 (2001)

6. Marinakis, Y., Marinaki, M., Dounias, G.: Honey bees mating optimization algorithm for the vehicle routing problem. In: Krasnogor, N., Nicosia, G., Pavone, M., Pelta, D. (eds.) NICSO 2007. SIC, vol. 129, pp. 139–148. Springer, Heidelberg (2008)

7. Marinakis, Y., Marinaki, M.: Bumble bees mating optimization algorithm for the vehicle routing problem. In: Panigrahi, B.K., Shi, Y., Lim, M.-H. (eds.) Handbook of Swarm Intelligence. ALO, vol. 8, pp. 347–369. Springer, Heidelberg (2011)

8. Bitam, S., Mellouk, A.: Bee life-based multi constraints multicast routing optimization for vehicular ad hoc networks. J. Netw. Comput. Appl. **36**, 981–991 (2013)

9. Yang, X.-S.: Engineering optimizations via nature-inspired virtual bee algorithms. In: Mira, J., Álvarez, J.R. (eds.) IWINAC 2005. LNCS, vol. 3562, pp. 317–323. Springer, Heidelberg (2005)

10. Wedde, H.F., Farooq, M., Zhang, Y.: Beehive: an efficient fault-tolerant routing algorithm inspired by honey bee behavior. In: Dorigo, M., Birattari, M., Blum, C., Gambardella, L.M., Mondada, F., Stützle, T. (eds.) ANTS 2004. LNCS, vol. 3172, pp. 83–94. Springer, Heidelberg (2004)

11. Munoz, M.A., Lopez, J.A., Caicedo, E.: An artificial beehive algorithm for continuous optimization. Int. J. Intell. Syst. **24**, 1080–1093 (2009)

12. Pham, D.T., Ghanbarzadeh, A., Koc, E., Otri, S., Rahim, S., Zaidi, M.: The bees algorithm - a novel tool for complex optimization problems. In: IPROMS 2006 Proceeding 2nd International Virtual Conference on Intelligent Production Machines and Systems. Elsevier, Oxford (2006)

13. Drias, H., Sadeg, S., Yahi, S.: Cooperative bees swarm for solving the maximum weighted satisfiability problem. In: Cabestany, J., Prieto, A.G., Sandoval, F. (eds.) IWANN 2005. LNCS, vol. 3512, pp. 318–325. Springer, Heidelberg (2005)

14. Comellas, F., Martinez-Navarro, J.: Bumblebees: a multiagent combinatorial optimization algorithm inspired by social insect behavior. In: Proceedings of First ACM/ SIGEVO Summit on Genetic and Evolutionary Computation (GECCO), pp. 811–814 (2009)

15. Teodorovic, D., Dell'Orco, M.: Bee colony optimization - a cooperative learning approach to complex transportation problems. In: Advanced OR and AI Methods in Transportation, Proceedings of the 16th Mini - EURO Conference and 10th Meeting of EWGT, pp. 51–60 (2005)

16. Hackel, S., Dippold, P.: The bee colony-inspired algorithm (BCiA)-a two stage approach for solving the vehicle routing problem with time windows. In: Proceedings of GECCO 2009, Montreal, Quebec, Canada, pp. 25–32 (2009)

17. Sato, T., Hagiwara, M.: Bee system: finding solution by a concentrated search. In: IEEE International Conference on Systems, Man, and Cybernetics (SMC), pp. 3954–3959 (1997)

18. Quijano, N., Passino, K.M.: Honey bee social foraging algorithms for resource allocation: theory and application. Eng. Appl. Artif. Intell. **23**, 845–861 (2010)

19. Maia, R.D., Castro, L.N.D., Caminhas, W.M.: Bee colonies as model for multimodal continuous optimization: the optbees algorithm. In: IEEE World Congress on Computational Intelligence (WCCI), Brisbane, Australia, pp. 1–8 (2012)

20. McCaffrey, J.D., Dierking, H.: An empirical study of unsupervised rule set extraction of clustered categorical data using a simulated bee colony algorithm. In: Governatori, G., Hall, J., Paschke, A. (eds.) RuleML 2009. LNCS, vol. 5858, pp. 182–192. Springer, Heidelberg (2009)

21. Fan, H., Zhong, Y.: A rough set approach to feature selection based on wasp swarm optimization. J. Comput. Inf. Syst. **8**, 1037–1045 (2012)
22. Theraulaz, G., Goss, S., Gervet, J., Deneubourg, J.L.: Task differentiation in polistes wasps colonies: a model for self-organizing groups of robots. In: First International Conference on Simulation of Adaptive Behavior, pp. 346–355. MIT Press, Cambridge (1991)
23. Karaboga, D., Basturk, B.: A powerful and efficient algorithm for numerical function optimization: artificial bee colony (ABC) algorithm. J. Global Optim. **39**, 459–471 (2007)
24. Karaboga, D., Basturk, B.: On the performance of artificial bee colony (ABC) algorithm. Appl. Soft Comput. **8**, 687–697 (2008)
25. Karaboga, D., Gorkemli, B., Ozturk, C., Karaboga, N.: A comprehensive survey: artificial bee colony (ABC) algorithm and applications. Artif. Intell. Rev. **42**, 21–57 (2014)
26. Marinakis, Y., Migdalas, A.: A particle swarm optimization algorithm for the multicast routing problem. In: Batsyn, M.V., Kalyagin, V.A., Pardalos, P.M. (eds.) Models, Algorithms and Technologies for Network Analysis: Third International Conference on Network Analysis. Springer Proceedings in Mathematics & Statistics, vol. 104, pp. 69–91. Springer International Publishing, Switzerland (2014)
27. Oliveira, C.A.S., Pardalos, P.M.: A survey of combinatorial optimization problems in multicast routing. Comput. Oper. Res. **32**, 1953–1981 (2005)
28. Oliveira, C.A.S., Pardalos, P.M., Resende, M.G.C.: Optimization problems in multicast tree construction. In: Resende, M.G.C., Pardalos, P.M. (eds.) Handbook of Optimization in Telecommunications, pp. 701–731. Springer, New York (2006)
29. Garey, M.R., Johnson, D.S.: Computers and Intractability: A Guide to the Theory of NP-Completeness. W. H. Freeman and Company, New York (1979)
30. Chow, C.H.: On multicast path finding algorithms. In: IEEE INFOCOM 1991, pp. 1974–1283. IEEE (1991)
31. Takahashi, H., Mutsuyama, A.: An approximate solution for the steiner problem in graphs. Math. Jpn. **6**, 573–577 (1980)
32. Waxman, B.M.: Routing of multipoint connections. IEEE J. Sel. Areas Commun. **1**, 286–292 (1988)
33. Kompleea, V.P., Pasquale, J.C., Polyzos, G.C.: Multicast routing for multimedia communication. IEEE/ACM Trans. Networking **1**, 286–292 (1993)
34. Tode, H., Sakai, Y., Yamamoto, M., Okada, H., Tezuka, Y.: Multicast routing algorithm for nodal load balancing. In: IEEE INFOCOM 1992, pp. 2086–2095. IEEE (1992)
35. Hwang, R.H., Do, W.Y., Yang, S.C.: Multicast routing based on genetic algorithms. In: WiOpt 2003: Modeling and Optimization in Mobile, Ad Hoc and Wireless Networks. INRIA Sophia-Antipolis, France (2003)
36. Wu, J.J., Hwang, R.H.: Multicast routing with multiple constraints. Inf. Sci. **124**, 29–57 (2000)
37. D'Andreagiovanni, F., Nardin, A.: Towards the fast and robust optimal design of wireless body area networks. Appl. Soft Comput. **37**, 971–982 (2015)
38. D'Andreagiovanni, F., Krolikowski, J., Pulaj, J.: A fast hybrid primal heuristic for multiband robust capacitated network design with multiple time periods. Appl. Soft Comput. **26**, 497–507 (2015)
39. Doar, M., Leslie, I.: How bad is naive multicast routing. In: Twelfth Annual Joint Conference of the IEEE Computer and Communications Societies. Networking: Foundation for the Future, INFOCOM 1993, Proceedings, pp. 82–89. IEEE (1993)

40. Oh, J., Pyo, I., Pedram, M.: Constructing minimal spanning/steiner trees with bounded path length. In: European Design and Test Conference, pp. 244–249 (1996)
41. Marinakis, Y., Marinaki, M., Matsatsinis, M.: A hybrid discrete artificial bee colony - grasp algorithm for clustering. In: 39th International Conference on Computers and Industrial Engineering, Troyes, France (2009)
42. Feo, T.A., Resende, M.G.C.: Greedy randomized adaptive search procedure. J. Global Optim. **6**, 109–133 (1995)
43. Hansen, P., Mladenovic, N.: Variable neighborhood search: principles and applications. Eu. J. Oper. Res. **130**, 449–467 (2001)
44. Blum, C., Puchinger, J., Raidl, G.R., Roli, A.: Hybrid metaheuristics in combinatorial optimization: a survey. Appl. Soft Comput. **11**, 4135–4151 (2011)
45. Shi, Y., Eberhart, R.: A modified particle swarm optimizer. In: Proceedings of 1998 IEEE World Congress on Computational Intelligence, pp. 69–73 (1998)
46. Hansen, P., Mladenović, N., Moreno-Pérez, J.A.: Variable neighborhood search: methods and applications. Ann. Oper. Res. **175**, 367–407 (2010)

Evolving Coverage Optimisation Functions for Heterogeneous Networks Using Grammatical Genetic Programming

Michael Fenton[1(✉)], David Lynch[1], Stepan Kucera[2], Holger Claussen[2], and Michael O'Neill[1]

[1] Natural Computing Research and Applications Group, UCD, Dublin, Ireland
michaelfenton1@gmail.com
[2] Bell Laboratories, Alcatel-Lucent, Dublin, Ireland

Abstract. Heterogeneous Cellular Networks are multi-tiered cellular networks comprised of Macro Cells and Small Cells in which all cells occupy the same bandwidth. User Equipments greedily attach to whichever cell provides the best signal strength. While Macro Cells are invariant, the power and selection bias for each Small Cell can be increased or decreased (subject to pre-defined limits) such that more or fewer UEs attach to that cell. Setting optimal power and selection bias levels for Small Cells is key for good network performance. The application of Genetic Programming techniques has been proven to produce good results in the control of Heterogenous Networks. Expanding on previous works, this paper uses grammatical GP to evolve distributed control functions for Small Cells in order to vary their power and bias settings. The objective of these control functions is to evolve control functions that maximise a proportional fair utility of UE throughputs.

1 Introduction

Recent technological advancements have created a paradigm shift in the way that mobile phones are used. The advent of large screens on handheld devices has prompted a shift from voice traffic to video and data streaming [1]. Until recently network operators have prioritised power management and cost minimisation over capacity maximisation [2,3]. However, the recent surge in high data-rate traffic [1] has prompted a switch from cost minimisation to capacity maximisation as carriers and operators struggle to cope with the increased demand.

Traditional network deployments are characterised by a distribution of high-powered transmitters (known as Macro Cells) placed on a hexagonal grid pattern to maximise coverage [4]. User Equipments (UEs) greedily attach to whichever cell provides the strongest signal. Cells then transmit data to all attached UEs by sub-dividing available bandwidth amongst them. A higher number of attached UEs per cell results in higher congestion, thus reducing the bandwidth. Consequently, each UE receives less data as congestion increases [4]. A standard

G. Squillero and P. Burelli (Eds.): EvoApplications 2016, Part I, LNCS 9597, pp. 219–234, 2016.
DOI: 10.1007/978-3-319-31204-0_15

method for dealing with increased demand is to increase the density of cells in the network.

Heterogenous Networks (HetNets) are multi-tiered cellular network deployments comprised of Macro Cells (MCs) and Small Cells (SCs) where both cell tiers operate on the same bandwidth. SCs transmit at a lower power and have a smaller operational range than MCs. They are employed to supplement the MC tier by offloading User Equipments (UEs), thus easing network congestion. SCs are often deployed in an ad-hoc manner by business owners in high-traffic areas such as restaurants, cafés, and shopping malls. As such, network operators may not necessarily have control over their placement [5].

Since SCs will be sub-optimally distributed, network operators seek to vary cell parameters in order to optimise the overall state of the network. One method of network optimisation is the notion of control algorithms [2,3,6–8]. These control algorithms either operate locally on individual cells or globally through a central server to manage the state of the network. In this study we expand on previous works by the authors [9] which used Grammatical Evolution (GE) [10,11] to evolve optimal SC settings for network load balancing. While results were highly optimal, the major limitation of this work was that it was necessary to do a full evolutionary run to find good settings. In this paper we adopt a symbolic regression approach to search the space of SC control functions, allowing for optimisation of SC settings on the fly, in a fraction of the time.

The remainder of this paper is structured as follows. A detailed definition of the problem is given in Sect. 2. Section 3 will give an overview of HetNet optimisation under the 3GPP standard, including a description of grammatical GP. Our approach summary is detailed in Sect. 4, including descriptions of our simulation environment (Sect. 4.1), grammatical representation (Sect. 4.2) and our fitness function (Sect. 4.3). Experimental studies are described in Sect. 4.4, and the results are discussed in Sect. 5. Finally, our conclusions and recommendations for future work are given in Sect. 6.

2   Problem Definition

Optimization of HetNets involves varying parameters of the network such that some objective is satisfied, usually the maximisation of overall UE throughput (with fairness). While MCs are invariant, SCs have adjustable parameters that can affect UE attachment [5]. Each SC s can vary its power P_s in order to modulate its operational range (the area in which it is the strongest serving cell). However, under the 3rd Generation Partnership Project - Long Term Evolution (3GPP-LTE) framework [4], SCs have an additional variable parameter that affects UE attachment, namely the Cell Selection Bias (CSB), β_s.

The CSB is a mechanism which artificially increases the effective range of the SC. UEs in this "expanded region" of the SC will attach to the cell in deference to their better serving MC for the global good of the network. A UE u will therefore attach to a cell k in accordance with the attachment rule in Eq. 1:

$$k = \arg\max_i(S_{ui} + \beta_i). \tag{1}$$

where S_{ui} is the perceived signal strength of cell i for UE u, $\beta_i = 0, \forall i \in \mathcal{M}$, the set of all MCs, and $\beta_i \geq 0, \forall i \in \mathcal{S}$, the set of all SCs [8]. Note that P_i (the transmitting power of cell i) is subject to path loss such that the signal strength perceived by u is given by:

$$S_{ui} = P_i[\text{dBm}] + G_{ui}[\text{dB}], \tag{2}$$

such that G_{ui} is the signal gain from cell i to u, see Sect. 4.1.

Cells transmit data during 1 ms intervals with each interval referred to as a subframe (f). During each subframe, a cell will transmit data across the available bandwidth to all attached UEs. Transmitted data primarily consists of packets of data, along with some minimal control signals. A full frame $\mathcal{F}$ consists of 40 subframes (i.e. 40 ms of network run-time).

The performance of UE u in any given subframe f is quantified by the down-link rate $R_{u,f}$ (in bits/sec, bps) from a cell i to u. Shannon's formula gives the downlink rate for wireless transmission in the presence of noise as [12]:

$$R_{u,f} = \frac{B_i}{N_{i,f}} * \log_2(1 + SINR_{ui,f}) \tag{3}$$

where B_i is the available bandwidth, $N_{i,f}$ is the total number of scheduled UEs attached to cell i for subframe f, and $SINR_{ui,f}$ is the Signal to Interference and Noise Ratio (the ratio of the received signal strength to the sum of all interfering signal strengths from all other cells in the network including background thermal noise) from source cell i to UE u in subframe f. Note that the available band-width across which the cell can transmit is divided by the total number of UEs attached to that particular cell. Therefore, the greater the number of attached UEs to a cell, the less bandwidth will be available to each individual UE [13].

Since by definition any UE within the expanded region of a SC (the additional area served by the SC due to its non-zero CSB) must experience significant interference from their strongest serving MC, provision has been made in recent 3GPP releases [4] for an enhanced Inter-Cell Interference Coordination (eICIC) mechanism for HetNets implementing CSBs. This mechanism mitigates inter-tier cell edge interference by employing Almost Blank Subframes (ABSs) [8]. With ABSs, MCs periodically mute across their entire bandwidth (save for some neg-ligible but necessary control signals), thus giving nearby SCs quiet subframes in which they can transmit with greatly reduced interference. UEs in the expanded region of SCs (those UEs who are most vulnerable to interference from neighbor-ing MCs) experience greatly improved $SINR$ during ABSs and, therefore, receive greater throughput. Unfortunately, UEs attached to MCs that are implementing ABSs will receive no data transmissions during muted SFs.

Control algorithms which adjust SC power and bias settings have been used to optimise HetNets [6]. Increasing or decreasing the operational range of individual SCs can change the number of attached UEs, thereby affecting global network performance. Control algorithms can be designed to operate centrally across an entire network or independently on individual cells [6,7]. This study will focus on generating control algorithms for individual SCs. Each SC will run the same algorithm that is capable of adjusting both power and bias settings based measurement reports collected by the SCs.

3   Previous Work

Release 10 of 3GPP [4] describes eICIC conceptually but does not specify methods for configuring ABS patterns, setting SC powers and CSBs or scheduling UEs, as these are non-trivial tasks. Deb *et al.* prove (Sect. 4-A) that optimising ABS patterns alone is an NP-hard problem, even for minimal networks with a single MC and multiple SCs [8]. Huge growth forecasts [14] for the SC market motivate algorithms which can maximise the benefits from eICIC. The literature describes three resource allocation problems which jointly determine the performance of HetNets implementing eICIC [15]. They are:

1. setting SC powers and CSBs to ensure optimal offloading from the MC tier,
2. setting ABSrs to protect UEs at SC edges, and
3. scheduling SC attached UEs so that protected resources (ABSs) are optimally utilised.

A number of contributions address one or multiple components of this joint optimisation problem.

Tall *et al.* proposed a stochastic approximation technique to optimise SC CSBs and the ABSrs of MCs [3]. They first derived a Self Organising Network (SON) load balancing update function that minimised the load imbalance between MCs and SCs. It computed CSB adjustments based on averaged load statistics and operated in a distributed manner across SCs. Stochastic Approximation theorems from Combes *et al.* [16] proved that their SON converges to the set of optimal CSBs. The authors also derived update equations from a proportional fair utility of UE throughputs in order to optimise MC ABSrs. Simulations showed that the load balancing SON, in combination with the ABSr optimisation SON from the second implementation above, achieved the best tradeoff between overall network throughput and cell edge throughput (i.e. fairness).

Deb *et al.* formulated the eICIC optimisation problem as a non-linear programming (NLP) problem [8]. They adopted the proportional fair sum logarithm of UE throughputs used by [3]. This utility function negotiates a tradeoff between fairness for cell edge UEs and maximisation of overall network throughput [17]. The authors simulated their algorithm using a realistic HetNet deployment in Manhattan. The 5th percentile of UE throughputs was improved by more than 50 % under eICIC without significant throughput losses for MC attached UEs. However, their algorithm requires measurement statistics from each UE's best serving SC and MC, but in reality UEs only attach to one cell [18].

As the problem is NP-hard and the structure of the solution is unclear it presents an opportunity for Genetic Programming (GP), a heuristic technique in Evolutionary Computation (EC).

3.1 Grammatical Genetic Programming

Grammatical Genetic Programming is a subset of GP techniques which use a formal grammar to define the terminal sets [19]. The use of a grammar means that programs can be generated in an arbitrary language [10,11,20]. Grammatical GP methods draw metaphorical inspiration from the principles of evolutionary and molecular biology to create machine executable solutions for a diverse spectrum of problems [20]. In contrast to canonical GP [21], where solutions are represented directed by parse trees, Grammatical GP techniques use a formal grammar to map from genotype to phenotype. Solutions can be generated using derivation trees [19,22] or using variable-length integer strings (chromosomes) [10,11] to map to programs (phenotypes) using a Backus-Naur Form (BNF) [23] grammar definition [24]. A key strength of these grammar-based techniques is that bias can be incorporated into the grammar to guide the search towards more desired solutions.

Grammatical GP methods such as Context-Free Grammatical GP (CFG-GP) [19,22] and Grammatical Evolution (GE) [10,11] have been successfully applied to financial modelling [25], structural engineering [26–28] and indeed HetNet optimisation [2,9,29,30]. Such flexibility is possible because problem specific domain knowledge can be incorporated into the grammars. This heuristic approach is appropriate for problems that do not easily admit analytic treatment, i.e. those where complete domain knowledge is lacking, or for dynamic environments [20].

3.2 Coverage Optimisation

A number of EC techniques have been applied in the field of telecommunications networks [31], but there have been relatively few in the area of coverage optimisation. Ho *et al.* applied GP to optimise the coverage of femtocell deployments (SCs with a range of several meters that are designed to support plug-and-play deployment) in enterprise environments [6]. Cell powers must be set in order to achieve load balancing and minimisation of coverage gaps and signal leakage. This problem is multi-objective with conflicting objectives since, for example, increasing power to reduce coverage gaps may increase leakage. The authors evolved programs that adjusted the power on individual femtocells based on local measurement statistics. Solutions responded sensibly to network conditions. This study represents a proof of concept that controllers can be automatically generated for wireless networks.

Hemberg *et al.* also examined a variety of different grammars on the related HetNet coverage optimisation problem [2,29,30]. In these instances, the three conflicting objectives of mobility minimisation (number of UE hand-overs), load balancing, and cell power minimisation (leakage) were jointly optimised for various indoor femtocell deployment scenarios using the multi-objective optimisation algorithm NSGA-II [32]. The authors found in [29] that the weighted fitness function used in [6] caused premature convergence to local optima, and they employed a symbolic regression approach in [2] to evolve femtocell power control equations. The grammar combined smooth and non-linear functions so that a wide range of non-trivial behaviours were accessible to evolved solutions. In [30] the authors compared a symbolic regression grammar, a grammar consisting exclusively of conditional statements and a hybrid combining both conditionals and functions. The purely conditional grammar allowed discrete power changes and was found to converge faster than the less constrained symbolic regression grammar. The combined grammar was slowest to converge and evolved solutions exhibited significantly worse fitness over all scenarios. It was noted that less domain knowledge is required for symbolic regression grammars but engineers favour the easily interpretable conditional solutions [6]. Finally, the utilities of control programs evolved using GE were found to match and sometimes exceed those achieved by partial enumeration of the search spaces.

There are two main differences between coverage optimisation and eICIC optimisation problems. Firstly, the objective function is univariate in eICIC because our goal is simply to maximise network capacity with fairness, while coverage optimisation observes a multivariate objective. Secondly, we currently have three degrees of freedom in eICIC: SC powers, SC CSBs and MC ABS patterns, as opposed to the single variable of SC powers. With this in mind, we now describe our simulation environment for a HetNet that implements eICIC.

4 Experimental Setup

4.1 Simulation

The simulation environment covers a $3.61\,\mathrm{km}^2$ area of Dublin City Centre (Fig. 1), with a resolution of $2\,\mathrm{m}^2$. A total of 21 MCs are placed on a hexagonal grid, with 79 SCs scattered randomly across the map. The random placement of SCs accounts for their ad-hoc deployment, since the manufacturer (or indeed the carrier) may have little control over their placement. UEs are distributed on the map with an average density of 60 UEs per MC sector, giving a total of 1,260 UEs. UE hotspots are modeled as dense congregations of between 5 to 25 UEs, such that 20 % of all UEs are located in hotspots. Hotspots are distributed at randomly selected SC locations with a probability of 90 %. Otherwise they are placed randomly on the map, thus simulating the tendency of business owners to deploy SCs in high-traffic areas.

An environmental encoding is generated from a Google Maps [33] screenshot of the region served by the network. The encoding recognises four environmental categories: buildings, bodies of water, parks, and roads/footpaths. UEs and SCs are not placed in bodies of water, but their placement in all other locations respects a uniform distribution (subject to the distribution of hotspots). A 2-dimensional signal gain path loss matrix G_i is then calculated for each cell i. Path loss is based on cell location, cell gain, shadow fading, and environmental obstacles such as buildings. The signal gain from a cell i to a location $[x, y]$ is thus indexed by $G[i, x, y]$. UE locations do not change throughout the optimisation procedure, with all UEs requesting data constantly (a "full-buffer" model).

The power range for a SC is 23–35 dBm, while the bias β_s can vary from 0–15 dBm. No cell can be completely turned off/muted. In this study we allow

Fig. 1. Simulated coverage area of network deployment.

SCs to adopt non-zero CSB only if their power is already at a maximum, thus minimising the number of SC expanded regions, more formally defined in Eq. 4:

$$\beta_s = \begin{cases} 0 & \text{if } P_s \leq P_{s\_max} \\ \geq 0 & \text{if } P_s = P_{s\_\max} \end{cases}, \forall s \in \mathcal{S}. \tag{4}$$

The set of UEs attached to a cell i is denoted by $\mathcal{A}_i$. The set of UEs for whom cell i is the strongest serving cell of its tier based solely on received power (those UEs that have the *potential* to attach to i, i.e. Eq. 1 when $\beta_i = 0$) is denoted by $\mathcal{P}_i$. The set of UEs attached to SC s who are in the expanded region of s is denoted by $\mathcal{E}_s$. Note that $\mathcal{A}_i \subset \mathcal{P}_i$ and $\mathcal{E}_s \subset \mathcal{A}_s$. We denote by $N_{\mathcal{X}}$ the cardinality of the set $\mathcal{X}$.

ABS. The ratio of the number of ABSs to non-ABSs in a full frame is known as the ABS ratio (ABSr) [3]. This is defined as:

$$ABSr_m = \frac{\left[\dfrac{\sum\limits_{s \in S} N_{(\mathcal{E}_s \cap \mathcal{P}_m)}}{\left(\sum\limits_{s \in S} N_{(\mathcal{E}_s \cap \mathcal{P}_m)} \right) + N_{\mathcal{A}_m}} \times 8 \right]}{8}. \tag{5}$$

Note that this ratio is sensitive only to the ratio of UEs within a single MC sector.

Release 10 of the 3GPP-LTE framework [4] cites eight distinct ABS patterns that can be used (shown in Table 1). While these patterns can be combined in any fashion to suit given ABSrs, since there are only eight patterns ABS ratios must consequently be multiples of 0.125. A MC can never be completely blanked for a full frame (i.e. no MC can run an ABSr of 1; the maximum ABSr is 0.875), and for this study a minimum ABSr of 0.125 is set for all MCs (i.e. no MC can run an ABSr of 0).

Table 1. There are eight possible ABS patterns [4]. The patterns are isomorphic to a full frame of 40 SFs. 1 indicates MC transmission, 0 indicates an ABS.

ABS pattern 1	0111111101	1111110111	1111011111	1101111111
ABS pattern 2	1011111110	1111111011	1111101111	1110111111
ABS pattern 3	1101111111	0111111101	1111110111	1111011111
ABS pattern 4	1110111111	1011111110	1111111011	1111101111
ABS pattern 5	1111011111	1101111111	0111111101	1111110111
ABS pattern 6	1111101111	1110111111	1011111110	1111111011
ABS pattern 7	1111110111	1111011111	1101111111	0111111101
ABS pattern 8	1111111011	1111101111	1110111111	1011111110

For the purposes of this study, each UE u attached to cell i is scheduled to receive data transmissions in all subframes within a full frame for which they have an $SINR_{ui,f} > 1$ for every subframe $f \in \mathcal{F}$, the full frame (i.e. they are receiving more signal from their host cell than interference from the rest of the network). Note that muting of MCs during ABSs affects $SINR$ for all UEs as the sum of all interfering signals diminishes. A UE u can not be scheduled to receive data transmissions from a cell i during a subframe f if their $SINR_{ui,f}$ is less than 1 as this would result in a transmission outage and their data packets would be dropped. Since intelligent scheduling can have a significant effect on UE downlink performance [34], we wish to remove any variability in the simulation so that any performance improvement may be definitively ascribed to the evolved optimisation algorithm.

Finally, since the Shannon formula defined in Eq. 3 gives the downlink rate (in *bits/sec*)for a UE for one particular subframe, the total received downlink for a UE across a full frame $\mathcal{F}$ is averaged across all 40 subframes in the frame:

$$R_{u\_avg} = \frac{\sum\limits_{f \in \mathcal{F}} R_{u,f}}{40}. \tag{6}$$

4.2 Grammar

Since MC powers are invariant, it is desirable that SCs should operate autonomously and regulate their output in accordance with some measurements about their environment. SCs obtain this environmental information through reports from both attached UEs and local MCs. Each UE u attached to a SC s reports specific information back to the SC:

– $SINR_{us,\mathcal{F}}$, the array of all SINRs across the full frame for UE u
– the id of the UE's strongest serving MC
– R_{u_avg} (Eq. 6)

With this information the SC has a profile of its surrounding environment. The SC collates a list of the nearby MCs from all attached UEs. The SC area is then sub-divided into individual MC sectors (i.e. areas where individual MCs are the strongest serving MCs).

Each MC m within the SC s region is then queried for further information:

$$N_{\mathcal{A}_m}, \tag{7}$$

the number of m attached UEs,

$$N_{(\mathcal{P}_m \cap \mathcal{A}_s)}, \tag{8}$$

the number of s attached UEs who are in the MC m sector,

$$avg\_R_m, \tag{9}$$

the average downlink rates of all m attached UEs,

$$\sum_{u \in \mathcal{A}_m} \log(R_{u\_avg}), \tag{10}$$

the sum of the log of the downlink rates of all m attached UEs (a logarithmic scale is used to reward solutions that achieve fairness; changes in the throughput of the worst performing UEs will be highlighted but decreases in performance of the best UEs are less critical),

$$avg\_R_{(\mathcal{P}_m \in \mathcal{A}_s)}, \tag{11}$$

the average downlink rates of all s attached UEs who are in the MC m sector, and

$$\sum_{u \in (\mathcal{P}_m \cap \mathcal{A}_s)} \log(R_{u\_avg}), \tag{12}$$

the sum of the log of the downlink rates of all s attached UEs who are in the MC m sector.

Furthermore, each SC s can provide further information about itself:

$$N_{\mathcal{A}_s}, \tag{13}$$

the number of s attached UEs,

$$avg\_R_s, \tag{14}$$

the average downlink rates of all s attached UEs, and

$$\sum_{u \in \mathcal{A}_s} \log(R_{u\_avg}), \tag{15}$$

the sum of the log of the downlink rates of all s attached UEs.

A grammar (as shown in Fig. 2) was then written for a typical symbolic regression application [21]. The grammar is capable of producing formulae using basic mathematical expressions: $+, -, \times, \%, sin, cos, tan, log, sqrt$. These are used to build arithmetic compositions of the SC-specific values defined in Eqs. 7 to 15, as outlined in Table 2.

Since the values given in Table 2 are defined per MC in a SC sector, evaluation of the function derived from the grammar generates a numerical value for each MC sector within the SC region. The sum of all such generated values per SC is then used as an update value U_s for the power and bias of SC s (as defined in Eq. 16, subject to Eq. 4).

```
<expr> ::= <reg> | <reg> | <val> <op> <val> | <val>
<reg> ::= (<expr> <op> <expr>) | <fn>(<expr>)
<op> ::= + | - | * | %
<fn> ::= sin | cos | tan | log | sqrt
<val> ::= <epsilon> | N_m | N_s | N_ms |
          s_log_R | m_log_R | ms_log_R |
          avg_R_s | avg_R_m | avg_R_ms
<epsilon> ::= 0.000<n><n><n>
<n> ::= 0|1|2|3|4|5|6|7|8|9
```

Fig. 2. Grammar for evolution of HetNet control algorithms.

Table 2. Cell-dependent grammar elements

Grammar element	Meaning
N_m	$N_{\mathcal{A}_m}$
N_s	$N_{\mathcal{A}_s}$
N_ms	$N_{(\mathcal{P}_m \cap \mathcal{A}_s)}$
avg_R_m	avg_R_m
avg_R_s	avg_R_s
avg_R_ms	$avg\_R_{(\mathcal{P}_m \cap \mathcal{A}_s)}$
s_log_R	$\sum_{u \in \mathcal{A}_s} \log(R_{u\_avg})$
m_log_R	$\sum_{u \in \mathcal{A}_m} \log(R_{u\_avg})$
ms_log_R	$\sum_{u \in (\mathcal{P}_m \cap \mathcal{A}_s)} \log(R_{u\_avg})$
epsilon	Constant, from 0.000000 to 0.000999

$$U_s = \sum_{m \in \mathcal{M}} \text{eval}(expr_m). \tag{16}$$

All SCs are initialised with minimum power levels and bias of 0 in order to fairly compare solutions between individuals. Note from Eq. 4 that the power and bias of a SC can be treated as a single entity, as any increase in power past the upper power limit of a SC is translated as an increase in bias.

The objective of this grammar is to provide a single optimization algorithm which can be applied to all SCs in the network deployment (rather than individual algorithms tailored for specific cells). This control algorithm will set the power or bias of the SC based on the evolved parameters. The same algorithm will be run on all SCs across the entire network (i.e. solutions must be generalized and not tailored to individual cells).

4.3   Fitness Function

The performance of a network instance (i.e. a network state with fixed powers, biases, and UE attachments, running for a single full frame) is calculated from the data throughputs of the UEs via:

$$Performance = \sum_{u \in \mathcal{U}} \log(R_{u\_avg}). \tag{17}$$

A logarithmic scale is used in order to magnify the changes in R_{u_avg} of the worst performing UEs. Any decrease in throughput of the best performing UEs is deemed relatively unimportant by this fitness metric, as the focus is on improving the performance of those worst performing UEs. The objective of evolved control algorithms is to maximise this performance metric. The fitness of an individual solution is given by the change in the performance of the network from its initial state to its optimized state after the algorithm is run:

$$Fitness = avg\left(\sum_{u \in \mathcal{U}} \log(R_{u\_avg})_{post\ opt} - \sum_{u \in \mathcal{U}} \log(R_{u\_avg})_{pre\ opt}\right) \tag{18}$$

This fitness is averaged across multiple scenarios in order to evolve solutions which are generalisable.

4.4   Evolutionary Setup

Experiments in this paper adhere to the recommendations of Hemberg et al. [2]. Each individual is evaluated across 10 different scenarios in order to ensure generalizability of solutions. Each scenario is characterized by a different UE distribution (subject to the distribution of hotspots, as described in Sect. 4.1). Evolutionary parameters are shown in Table 3. SC powers and biases for each scenario were initialized at 23 dBm and 0 dBm respectively.

Table 3. Evolutionary parameter settings

Number of runs:	50
Pop. Size:	100
Generations:	100
Initialisation:	Sensible
Max tree depth:	10
Crossover type:	Subtree
Crossover probability:	70%
Mutation types:	Subtree & Point
Selection:	Tournament
Tournament size:	2
Replacement:	Generational with Elites
Elite size:	1

A list of all previously evaluated phenotypes per evolutionary run is retained. If a phenotype has already been evaluated in the evolutionary process then it is discarded and a new randomly initialised individual is inserted into the population in its place. This ensures that all 10,000 evaluations in a single run are unique. The net result of this is that a much wider search area is covered, since the evolutionary process does not waste time re-evaluating known solutions as the best of these exist in the form of elites [29, 35].

5   Results and Discussion

The results of the evolutionary runs are shown in the graph in Fig. 3, which displays the average of the best fitnesses across each run as generations progress.

The best evolved algorithm is described in Eq. 19. Note that the control algorithm is run independently on each SC $s \in \mathcal{S}$, and that it must be run once for each MC $m \in \mathcal{M}$ which intersects with the area of influence of s (as described in Sect. 4.2).

$$U_s = \sum_{m \in \mathcal{M}} \frac{\left(\sum_{u \in (\mathcal{P}_m \cap \mathcal{A}_s)} \log(R_{u\_avg}) \right) + N_{(\mathcal{P}_m \cap \mathcal{A}_s)}}{N_{\mathcal{A}_s}} \tag{19}$$

The primary method to evaluate success of a strategy is through analysis of a Cumulative Distribution Function (CDF) graph of the log of the downlink of all UEs in a scenario of the network. Figure 4 compares the evolved strategy against a number of benchmark methods of setting SC powers and biases. A number of observations can be made from Fig. 4:

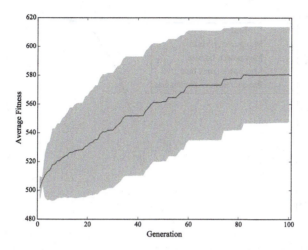

Fig. 3. Average best fitness across all runs, including standard deviation of average best fitness across all runs.

- The height of the lowest portion of any curve represents the total number of UEs in the network for whom $SINR_{max} < 1$ and, thus, cannot be scheduled for any data transmissions.
- The red curve (minimum power, and bias of 0 dBm) represents the CDF of pre-optimised $\log R_{u\_avg}, \forall\, u \in \mathcal{U}$. It can be seen to begin not at the origin of the y-axis, but at a level of $y = 0.0722$. This indicates that there are 91 UEs in the original pre-optimised network for whom $SINR_{max} < 1$ and, thus, cannot be scheduled for any data transmissions.
- Since $\log 0$ is undefined, the performance metric of a network state (as described in Eq. 17) does not account for those UEs who cannot be scheduled to receive any data transmissions (i.e. those UEs that have a data throughput of 0 bps). Thus, easy improvements can be made to the fitness of the network by increasing the number of scheduled UEs. The evolved algorithm generates significant improvement in fitness by increasing the overall number of scheduled UEs, more than any of the benchmark methods.
- Since the fitness function (as defined in Eq. 18) operates on a logarithmic scale, the greatest gains can be made by increasing the throughput of the worst performing UEs in the network. It can be seen that the black curve lies to the right of and below the majority of the other curves, indicating a significant increase in performance for all UEs. This desirable behaviour emerges automatically and is not explicitly rewarded by the fitness function.
- While setting all powers and biases to maximum levels (as shown by the dashed green line) results in marginally improved performance for the majority of UEs over the evolved method, the evolved method can schedule more UEs overall, leading to a better fitness.

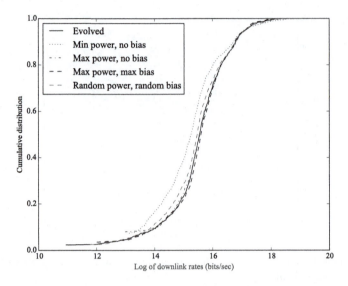

Fig. 4. Cumulative Distribution Function graph of $\log R_{u\_avg}, \forall u \in \mathcal{U}$, the set of all UEs. The black line represents the performance of all UEs under the evolved algorithm. Red, orange, green, and blue lines represent various baseline methods of setting powers and biases(Color figure online).

6 Conclusions and Future Work

Grammatical GP has been shown to be capable of successfully evolving control algorithms for HetNets which give acceptable performance. Analysis of the performance of best evolved algorithm shows that not only does it improve the throughput of the worst performing UEs in the network, but that it also reconfigures the network such that the number of UEs who cannot be scheduled (due to $SINR_{u,max} < 1$) is minimised. This is a critical issue, as fairness is a desirable quality in cellular network performance [4].

A number of areas for further study have been identified as a result of this work:

- At present, all UEs are scheduled to receive data transmissions from their hosting cell in every subframe f where they have an $SINR_{ui,f} > 1$. Significant improvements in UE throughput are expected by more intelligent scheduling methods.
- The function for describing ABS is a simple model. It is possible that further network improvements can be made by evolving optimal ABS functions.
- Optimisation of networks for minimal transmission outages (i.e. minimise the number of UEs for whom $SINR_{max} < 1$) has been described in this paper, but has not been fully realised here as this was not the main objective of the fitness function. The evolution of control algorithms for minimisation of transmission outages would be a worthwhile study.

Acknowledgments. This research is based upon works supported by the Science Foundation Ireland under grant 13/IA/1850.

References

1. Cisco: Cisco visual networking index: global mobile data traffic forecast update, 2014 2019. Technical report (2015). http://www.cisco.com/c/en/us/solutions/collateral/service-provider/visual-networking-index-vni/white_paper_c11-520862.html
2. Hemberg, E., Ho, L., O'Neill, M., Claussen, H.: Evolving femtocell algorithms with dynamic and stationary training scenarios. In: Coello, C.A.C., Cutello, V., Deb, K., Forrest, S., Nicosia, G., Pavone, M. (eds.) PPSN 2012, Part II. LNCS, vol. 7492, pp. 518–527. Springer, Heidelberg (2012)
3. Tall, A., Altman, Z., Altman, E.: Self organizing strategies for enhanced ICIC (eICIC). arXiv preprint arXiv:1401.2369 (2014)
4. 3gpp (2014). http://www.3gpp.org/
5. Chandrasekhar, V., Andrews, J., Gatherer, A.: Femtocell networks: a survey. Commun. Mag. **46**, 59–67 (2008)
6. Ho, L.T., Ashraf, I., Claussen, H.: Evolving femtocell coverage optimization algorithms using genetic programming. In: 20th International Symposium on Personal, Indoor and Mobile Radio Communications (IEEE PIMRC), pp. 2132–2136. IEEE (2009)
7. Fagen, D., Vicharelli, P.A., Weitzen, J.A.: Automated wireless coverage optimization with controlled overlap. IEEE Trans. Veh. Technol. **57**, 2395–2403 (2008)
8. Deb, S., Monogioudis, P., Miernik, J., Seymour, J.P.: Algorithms for enhanced inter-cell interference coordination (eICIC) in LTE HetNets. IEEE/ACM Trans. Networking (TON) **22**, 137–150 (2014)
9. Fenton, M., Lynch, D., Kucera, S., Claussen, H., O'Neill, M.: Load balancing in heterogeneous networks using an evolutionary algorithm. In: Proceedings of IEEE Conference on Evolutionary Computation, pp. 70–76, Sendai, Japan (2015)
10. Ryan, C., Collins, J.J., Neill, M.O.: Grammatical evolution: evolving programs for an arbitrary language. In: Banzhaf, W., Poli, R., Schoenauer, M., Fogarty, T.C. (eds.) EuroGP 1998. LNCS, vol. 1391, pp. 83–96. Springer, Heidelberg (1998)
11. O'Neill, M., Ryan, C.: Grammatical evolution. IEEE Trans. Evol. Comput. **5**, 349–358 (2001)
12. Shannon, C.E.: Communication in the presence of noise. Proc. Inst. Radio Eng. **37**, 10–21 (1949)
13. M. R1-060877: Frequency domain scheduling for E-UTRA. Technical report TSG RAN1#44bis, March (2006)
14. Bian, Y.Q., Rao, D.: Small Cells big opportunities (2014). www.huawei.com/ilink/en/download/HW_330984
15. Bedekar, A., Agrawal, R.: Optimal muting and load balancing for eICIC. In: 11th International Symposium on Modeling & Optimization in Mobile, Ad Hoc & Wireless Networks (WiOpt), pp. 280–287. IEEE (2013)
16. Combes, R., Altman, Z., Altman, E.: Self-organization in wireless networks: a flow-level perspective. In: Proceedings of INFOCOM, pp. 2946–2950. IEEE (2012)
17. Wu, Q., Esteves, E., Proprietary, Q.: The CDMA2000 High Rate Packet Data System. Artech House, Norwood, MA, USA (2002)

18. López-Pérez, D., Claussen, H.: Duty cycles and load balancing in HetNets with eICIC almost blank subframes. In: 24th International Symposium on Personal, Indoor and Mobile Radio Communications (IEEE PIMRC Workshops), pp. 173–178. IEEE (2013)
19. Whigham, P.A.: Grammatically-based genetic programming. In: Proceedings of the Workshop on Genetic Programming: From Theory to Real-World Applications. vol. 16, no. 3, pp. 33–41 (1995)
20. Dempsey, I., O'Neill, M., Brabazon, A.: Foundations in Grammatical Evolution for Dynamic Environments. SCI, vol. 194. Springer, Heidelberg (2009)
21. Koza, J.R.: Genetic Programming: On the Programming of Computers by Means of Natural Selection. MIT Press, Cambridge (1992)
22. Whigham, P.A.: Inductive bias and genetic programming. In: 1st International Conference on Genetic Algoritms in Engineering Systems: Innovations and Applications (GALESIA), pp. 461–466. IET (1995)
23. Backus, J.W., Nauer, F.L., Green, J., Katz, C., McCarthy, J., Naur, P., Perlis, A.J., Rutishauser, H., Samelson, K., Vauquois, B.: Revised report on the algorithmic language ALGOL 60. Comput. J. **5**, 349–367 (1963)
24. McKay, R.I., Hoai, N.X., Whigham, P.A., Shan, Y., O'Neill, M.: Grammar-based genetic programming: a survey. Genet. Program. Evolvable Mach. **11**, 365–396 (2010)
25. Brabazon, A., O'Neill, M.: Biologically Inspired Algorithms for Financial Modelling. Springer, Berlin (2006)
26. Byrne, J., Fenton, M., Hemberg, E., McDermott, J., O'Neill, M., Shotton, E., Nally, C.: Combining structural analysis and multi-objective criteria for evolutionary architectural design. In: Di Chio, C., Brabazon, A., Di Caro, G.A., Drechsler, R., Farooq, M., Grahl, J., Greenfield, G., Prins, C., Romero, J., Squillero, G., Tarantino, E., Tettamanzi, A.G.B., Urquhart, N., Uyar, A.Ş. (eds.) EvoApplications 2011, Part II. LNCS, vol. 6625, pp. 204–213. Springer, Heidelberg (2011)
27. Fenton, M., McNally, C., Byrne, J., Hemberg, E., McDermott, J., O'Neill, M.: Automatic innovative truss design using grammatical evolution. Autom. Constr. **39**, 59–69 (2014)
28. Fenton, M., McNally, C., Byrne, J., Hemberg, E., McDermott, J., O'Neill, M.: Discrete planar truss optimization by node position variation using grammatical evolution. In: IEEE Transactions on Evolutionary Computation (2016, in press)
29. Hemberg, E., Ho, L., O'Neill, M., Claussen, H.: A symbolic regression approach to manage femtocell coverage using grammatical genetic programming. In: Proceedings of the 13th Annual Conference Companion on Genetic and Evolutionary Computation, pp. 639–646. ACM (2011)
30. Hemberg, E., Ho, L., O'Neill, M., Claussen, H.: A comparison of grammatical genetic programming grammars for controlling femtocell network coverage. Genet. Program. Evolvable Mach. **14**, 65–93 (2013)
31. Alba, E., Chicano, J.F.: Evolutionary algorithms in telecommunications. In: Electrotechnical Conference (MELECON), pp. 795–798. IEEE (2006)
32. Deb, K., Pratap, A., Agarwal, S., Meyarivan, T.: A fast and elitist multiobjective genetic algorithm: NSGA-II. IEEE Trans. Evol. Comput. **6**, 182–197 (2002)
33. Google maps (2014). https://maps.google.com
34. Weber, A., Stanze, O.: Scheduling strategies for HetNets using eICIC. In: International Conference on Communications (IEEE ICC), pp. 6787–6791. IEEE (2012)
35. Glover, F.: Future paths for integer programming and links to artificial intelligence. Comput. Oper. Res. **13**, 533–549 (1986)

Joint Topology Optimization, Power Control and Spectrum Allocation for Intra-Vehicular Multi-hop Sensor Networks Using Dandelion-Encoded Heuristics

Javier Del Ser[1,2(✉)], Miren Nekane Bilbao[2], Cristina Perfecto[2],
Antonio Gonzalez-Pardo[3], and Sergio Campos-Cordobes[1]

[1] TECNALIA. OPTIMA Unit, 48160 Derio, Spain
{javier.delser,sergio.campos}@tecnalia.com
[2] University of the Basque Country UPV/EHU, 48013 Bilbao, Spain
{javier.delser,nekane.bilbao,cristina.perfecto}@ehu.eus
[3] Basque Center for Applied Mathematics (BCAM), 48009 Bilbao, Spain
agonzalezp@bcamath.org

Abstract. In the last years the interest in multi-hop communications has gained momentum within the research community due to the challenging characteristics of the intra-vehicular radio environment and the stringent robustness imposed on critical sensors within the vehicle. As opposed to point-to-point network topologies, multi-hop networking allows for an enhanced communication reliability at the cost of an additional processing overhead. In this context this manuscript poses a novel bi-objective optimization problem aimed at jointly minimizing (1) the average Bit Error Rate (BER) of sensing nodes under a majority fusion rule at the central data collection unit; and (2) the mean delay experienced by packets forwarded by such nodes due to multi-hop networking, frequency channel switching time multiplexing at intermediate nodes. The formulated paradigm is shown to be computationally tractable via a combination of evolutionary meta-heuristic algorithms and Dandelion codes, the latter capable of representing tree-like structures like those modeling the multi-hop routing approach. Simulations are carried out for realistic values of intra-vehicular radio channels and co-channel interference due to nearby IEEE 802.11 signals. The obtained results are promising and pave the way towards assessing the practical performance of the proposed scheme in real setups.

Keywords: Intra-vehicular networks · Routing · Spectrum allocation · Dandelion encoding · Evolutionary meta-heuristics

1 Introduction

The last decade has witnessed a number of technological advances in short-range wireless networking as a consequence of impending research efforts towards

© Springer International Publishing Switzerland 2016
G. Squillero and P. Burelli (Eds.): EvoApplications 2016, Part I, LNCS 9597, pp. 235–250, 2016.
DOI: 10.1007/978-3-319-31204-0_16

improving their efficiency in terms of energy consumption, resource usage and self-organization capabilities. Evidences abound, ranging from newly developed protocols and stacks to the practical assessment of their performance in different communication environments. Consequently, a plethora of application scenarios have been shown to potentially benefit from the inherent advantages of wireless networks (i.e. ease of deployment, less infrastructure needed, increased mobility and lower costs [1]).

The automotive sector is one of the application scenarios where wireless technologies have been extensively studied as a lightweight alternative to sense different constituent parts of the vehicle [2–5]. Surprisingly it has not been until recently [6,7] when multi-hop wireless networking has been hypothesized as an effective communication option to cope with the stringent fading statistics of the intra-vehicular channel, whose lack of frequency selectivity impedes the adoption of adaptive spectrum shaping techniques [8,9]. Such challenging radio conditions get even more involved with the presence of outer, non-controllable interfering sources whose transmitted signals collide with that forwarded by the deployed vehicular sensors to the central on-board unit. For instance, many contributions have analyzed the behavior and resilience of IEEE 802.15.4 (Zigbee) sensors under interfering IEEE 802.11 signals in both general (i.e. application agnostic) setups [10,11] and specific vehicular environments [12,13].

Indeed, multi-hop networking has attracted most of the recent literature on intra-car wireless communications. To the knowledge of the authors studies so far have gravitated on proving the feasibility of multi-hop networks by applying well-established protocols to controlled intra-vehicular communication setups. The most notable example is the work in [14], where the so-called Collection Tree Protocol was utilized in its naïve form to verify the predicted gains of multi-hop intra-vehicular wireless networking. Although resource allocation has been recently studied for the intra-vehicular environment [15], spectrum coordination, topology and power control has not been yet jointly tackled for this specific communication scenario, even though the underlying radio particularities – i.e. strong fading characteristics, eventual spectrum interferers – and the spectrum flexibility of avant-garde short-range wireless sensors [16] call for further investigation on this topic.

This manuscript elaborates on the above noted lack of research by proposing a centralized meta-heuristic scheme for optimally (1) routing the information captured by the compounding nodes of an intra-vehicular wireless network; (2) allocating the power utilized by each of such nodes; and (3) selecting the spectrum channel between each pair of sensors. This work builds upon the number of multi-channel short-range wireless transceivers made commercially available in recent times, such as the GP712 single radio multi-protocol chipset developed by GreenPeak Technologies [17]. By exploiting the a priori knowledge of the propagation statistics between sensors (which can be obtained by aside channel estimation techniques), the proposed scheme resorts to a combination of evolutionary meta-heuristics and a tree encoding approach to balance two conflicting objectives: the overall delay of the captured information due to propagation, multiplexing and channel access, and the bit error rate averaged over all sensors when their information is sent over multiple paths and fused at the on-board central unit (OBU). Experiments over realistically modeled intra-vehicular communication scenarios are discussed towards evincing the satisfactory performance of the proposed resource allocation algorithm.

The remainder of the manuscript is structured as follows: Sect. 2 poses the optimization problem to be tackled. Section 3 delves into the design of the proposed resource allocation algorithm. Experimental results are analyzed in Sect. 4 and finally, Sect. 5 concludes the paper.

2   System Model and Problem Formulation

As often made in the literature an intra-vehicular network can be conceptualized as a directed graph defined by the ordered pair $\mathcal{G} \doteq (\mathcal{V}, \mathcal{E})$, where $\mathcal{V}$ denotes the set of vertices or nodes compounding the network and $\mathcal{E}$ represents the set of edges or links, each connecting a given pair of nodes. Let $N \doteq |\mathcal{E}|$ and $M \doteq |\mathcal{V}|$, with $|\cdot|$ standing for set cardinality. Point-to-point communication over each link $e \in \mathcal{E}$ is established over a certain frequency channel $f(e) \in \mathcal{F} \doteq \{f_1, \ldots, f_F\}$, where F is the number of available radio channels (e.g. $F = 15$ for Zigbee [18]). On the other hand, each node $v \in \mathcal{V}$ transmits at a certain power $p(v) \in \mathbb{R}[0, P_{max}]$, where it is assumed that the radio transceiver is capable of adjusting its utilized power at any intermediate value within that given range.

The link between node v_i and v_j (with $j \in \{1, \ldots, i-1, i+1, \ldots, N\}$) undergoes degradation due to the fading statistics between different constituent parts of the network. In this context, measurements reported and analyzed in [8] showed that the whole intra-vehicular wireless medium features frequency-flat signal fading, with different Line-Of-Sight (LOS) and Non-Line-Of-Sight (NLOS) propagation characteristics depending on the physical location of the transmitting sensor and the receiving node. As a consequence, the degradation suffered by the above link – hereafter denoted as $e_{i,j}$ – imprints a Bit Error Rate BER(v_i, v_j) on the transmitted signal. This error rate can be expressed as a closed-form expression depending on the Signal to Interference and Noise Ratio SINR(v_i, v_j) over such a link, which in turn is fixed by the transmit power $p(v_i)$ and the signals interfering on receiver v_j over the utilized frequency $f(e_{i,j})$. For instance, in the IEEE 802.15.4-2006 standard utilized by Zigbee [18, Sect. E.4.1.8] BER(v_i, v_j) can be approximated by

$$\text{BER}(v_i, v_j) \doteq \text{BER}(e_{i,j}) \approx \frac{1}{30} \sum_{n=2}^{16} (-1)^n \binom{16}{n} e^{\frac{20(1-n)\text{SINR}(v_i,v_j)}{n}}, \tag{1}$$

where effects from fading and interfering sources are usually included as summing contributions in SINR(v_i, v_j), i.e.

$$\text{SINR}(v_i, v_j) = \frac{p(v_i) \cdot L(v_i, v_j)}{\sum\limits_{\substack{i'=1 \\ i' \neq i,j}}^{M} p(v_{i'}) L(v_{i'}, v_j) \mathbb{I}\left[f(e_{i',j}) = f(e_{i,j})\right] + (N_0/2)\Delta_{f(e_{i,j})}},$$

with $L(v_i, v_j)$ comprising fading/scattering channel losses from node v_i to node v_j; Δ_f represents the bandwidth of channel $f \in \mathcal{F}$; $\mathbb{I}[\cdot]$ is an auxiliary function taking value 1 if its argument is true (and 0 otherwise); and N_0 is the noise power spectral density measured in watts per hertz. Loss factors also depend logarithmically on the distance between nodes v_i and v_j via a loss exponent $\gamma_{i,j}$.

Following the diagram depicted in Fig. 1, each compounding sensor of the intra-vehicular network forwards their captured information to an on-board central receiver (coined previously as OBU) through a multi-hop tree-like route. By a slight notational abuse, each of such routes will be denoted as a subset $\mathcal{E}_i = \{e_1^i, e_2^i, \ldots, e_{N_i}^i\} \subseteq \mathcal{E}$, which includes all such $N_i \doteq |\mathcal{E}_i|$ edges in the network graph that participate in the route from node v_i to the OBU. Since this work focuses on tree topologies, it should be obvious that $|\cup_{i=1}^M \mathcal{E}_i| = M$, i.e. the number of total links compounding the multi-hop route is equal to the overall number of nodes in the network. If a decode-and-forward relaying approach is assumed, the bit error rate at the OBU for the information sent by node v_i will be given, for odd N_i, by

$$\mathrm{BER}_{\circledast}(v_i) = \sum_{x=1}^{N_i} BER(e_x^i) \prod_{\substack{y=1 \\ y \neq x}}^{N_i} \left(1 - BER(e_y^i)\right)$$

$$+ \sum_{x=1}^{N_i} BER(e_x^i) \sum_{\substack{y=1 \\ y \neq x}}^{N_i} BER(e_y^i) \sum_{\substack{z=1 \\ z \neq x,y}}^{N_i} BER(e_z^i) \prod_{\substack{k=1 \\ k \neq x,y,z}}^{N_i} \left(1 - BER(e_k^i)\right)$$

$$+ \ldots + \sum_{x=1}^{N_i} (1 - BER(e_x^i)) \prod_{\substack{y=1 \\ y \neq x}}^{N_i} BER(e_y^i), \qquad (2)$$

i.e. as the probability that an odd number of links within $\mathcal{E}_i$ incurs in error (a similar expression can be obtained for even N_i). The average Bit Error Rate of the intra-vehicular sensor network when operating on the relaying routes specified by $\{\mathcal{E}_i\}_{i=1}^M$ will be hence expressed as

$$\mathrm{BER}_{\circledast}^{avg} \doteq \frac{1}{M} \sum_{i=1}^M \mathrm{BER}_{\circledast}(v_i), \qquad (3)$$

which can be conceived as an overall measure of the communication quality of the sensor network deployed inside the vehicle.

The interference caused among nodes within the network and the eventual presence of external, non-controllable radio sources capable of creating co-channel interference in the same spectral band (e.g. IEEE 802.11 and Zigbee) can be mitigated by overlapping Ψ redundant multi-hop routes that transmit the same information to the central OBU, which fuses the received flows under a given fusion criteria. One of such criteria is the so-called majority voting, which decides for the value of the k-th bit b_k^i sent by node i over each overlaid multi-hop route $\{\mathcal{E}_i^\psi\}_{i=1}^M$ as

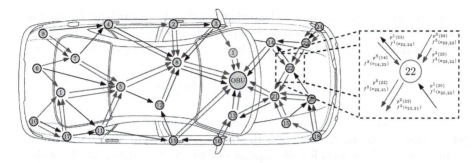

Fig. 1. Diagram representing the scenario tackled in this manuscript for $\Psi = 3$ multi-hop routes overlaid over $M = 24$ nodes. Each colored route can be modeled as a tree graph whose error and delay performance for every compounding node depends not only on the topology itself, but also on interfering signals from nearby nodes.

$$
b_k^i = \begin{cases} 1 \text{ if } \sum\limits_{\psi=1}^{\Psi} b_k^\psi \geq \Psi/2, \\ 0 \text{ if } \sum\limits_{\psi=1}^{\Psi} b_k^\psi < \Psi/2. \end{cases} \tag{4}
$$

If by a similar development to Expression (2) we denote the Bit Error Rate for sensor i over multi-hop route $\{\mathcal{E}_i^\psi\}_{i=1}^M$ as $\mathrm{BER}_\circledast^\psi(v_i)$, the overall error obtained for sensor i at the OBU under majority voting (mv) is given, for odd Ψ, by

$$
\mathrm{BER}_\circledast^{mv}(v_i) \doteq \sum_{z=\lceil \Psi/2 \rceil}^{\Psi} \left(\sum_{\mathcal{M} \in \mathcal{M}^z} \left(\prod_{i \in \mathcal{M}} \mathrm{BER}_\circledast^\psi(v_i) \prod_{i' \in \mathcal{V}/\mathcal{M}} (1 - \mathrm{BER}_\circledast^\psi(v_{i'})) \right) \right), \tag{5}
$$

where $\mathcal{M}^z$ denotes the set of all z-length combinations drawn from the integer set $\{1, \ldots, M\}$. By averaging over the whole set of nodes an overall measure of the communication robustness of the whole network can be obtained as

$$
\mathrm{BER}_\circledast^{avg,mv} = \frac{1}{M} \sum_{i=1}^M \mathrm{BER}_\circledast^{mv}(v_i), \tag{6}
$$

which should be minimized by properly optimizing the values of its controlling variables $\{p^\psi(v_i)\}_{i=1}^M$ (power), $\{f^\psi(e_{i,j})\}_{i,j=1}^M$ (frequency channel between nodes) and the topology of each of the different multi-hop routes. For notational convenience edges belonging to the tree graph modeling each of the overlaid multi-hop routes connecting all nodes to the OBU will be denoted as $\{e_{i,j}^\psi\}_{i,j=1}^N$.

Unfortunately the minimization of the above error comes along with a delay penalty due to (1) the switch of frequency channels at every transmitting node; and (2) the time multiplexing of these flows in intermediate nodes receiving data from different nodes and forwarded over redundant routes. As for the former the incurred delay is assumed to increase linearly with the different number of channels to be heard by every node. Specifically, if the time taken by the radio interface to shift from one channel to another is denoted by T_{sw} and no parallelization between reception and transmission tasks is assumed, the worst-case switching delay at node v_j and multi-hop route $\{\mathcal{E}_i^{\psi}\}_{i=1}^{M}$ will be given by

$$D_{sw}^{\psi}(v_j) = T_{sw} \left(\left| \{f(e_{i,j}^{\psi})\}_{\substack{i=1 \\ i \neq j}}^{N} \right| + \left| \{f(e_{j,n}^{\psi})\}_{\substack{n=1 \\ n \neq j}}^{N} \right| - 2 \right), \tag{7}$$

i.e. as the switching delay times the number of unique frequency channels used for the links incoming at – and outgoing from – node v_j. The second term of the overall delay will assume all nodes operate in saturation regime, which yields a multiplexing time linearly dependent on the number of links to be relayed and the estimated time of the utilized communication stack for transmitting a packet. In Zigbee this time reduces to a sum of the time-on-air of the packet at hand and the time for CSMA-CA channel access and eventual retries. For instance, for the XBee commercial radio module [19], 16-bit addressing, 127-byte payload and clear channel access, this latency is

$$D_{mx}^{\psi}(v_i) = 0.001 \cdot (0.544 + (0.032 \cdot 127 \cdot |\{e_{i,j}^{\psi}\}_{\substack{j=1 \\ j \neq i}}^{N}|)) \text{ (in seconds)}. \tag{8}$$

The information sent by every node undergoes accumulated delays at intermediate nodes due to the above two terms. As a result the overall, worst-case delay for the information sent by node v_i will be given by

$$D_{\circledast}^{\psi}(v_i) = \sum_{j \in \mathcal{M}_{i \to OBU}^{\psi}} \left(D_{mx}^{\psi}(v_j) + D_{sw}^{\psi}(v_j) \right), \tag{9}$$

where $\mathcal{M}_{i \to OBU}^{\psi}$ denotes the subset of nodes that participate in the path from node v_i (included) to the OBU. From this expression the overall delay metric averaged over all nodes in the network can be computed under two different assumptions: if all nodes await for the completion of all transmissions over a certain multi-hop route $\{\mathcal{E}_i^{\psi}\}_{i=1}^{M}$ before proceeding with the next one, the overall delay will be fixed for all nodes and given by

$$D_{\circledast}^{\text{avg}} = \sum_{\psi=1}^{\Psi} \max_{i \in \{1,\dots,M\}} D_{\circledast}^{\psi}(v_i), \tag{10}$$

which, in what regards to the allocation of resources, can be conceived as a set of independent routes. On the contrary, if transmissions over different routes are allowed to hold concurrently in the network, nodes will require switching frequency channels and relay links belonging to different routes. This implies that the overall worst-case delays in Expressions (7) and (8) are modified to

$$D_{sw}(v_j) = T_{sw} \left(\left| \{\{f(e_{i,j}^{\psi})\}_{\substack{i=1 \\ i \neq j}}^{N}\}_{\psi=1}^{\Psi} \right| + \left| \{\{f(e_{j,n}^{\psi})\}_{\substack{n=1 \\ n \neq j}}^{N}\}_{\psi=1}^{\Psi} \right| - 2 \right), \tag{11}$$

$$D_{mx}(v_i) = 0.001 \cdot (0.544 + (0.032 \cdot 127 \cdot |\{\{e_{i,j}^{\psi}\}_{\substack{j=1 \\ j \neq i}}^{N}\}_{\psi=1}^{\Psi}|)), \tag{12}$$

giving rise to an overall average delay given by

$$D_\circledast^{avg} = \frac{1}{M} \sum_{i=1}^{M} \max_{\psi \in \{1,\dots,\Psi\}} \left(\sum_{j \in \mathcal{M}_{i \rightarrow OBU}^{\psi}} (D_{mx}(v_j) + D_{sw}(v_j)) \right), \qquad (13)$$

which depends roughly on the transmitted power, frequency and topology of the constructed overlay routes. With these definitions in mind, the problem tackled in this paper can be formulated as finding an optimal allocation $\{p^\psi(v_i)\}_{i=1}^{M}$ (power), $\{f^\psi(e_{i,j})\}_{i,j=1}^{M}$ (frequency channel between nodes) and $\{e_{i,j}^\psi\}_{i,j=1}^{N}$ (tree topology) for a given number of overlaid routes Ψ such as the conflicting objectives defined in Expressions (6) (overall average BER) and (10) or (13) (overall delay) are simultaneously optimized, i.e.

$$\left[\{p^\psi(v_i)\}_{i=1}^{M}, \{f^\psi(e_{i,j})\}_{\substack{i,j=1 \\ i \neq j}}^{M}, \{e_{i,j}^\psi\}_{i,j=1}^{N} \right]_{\psi=1}^{\Psi} = \arg \min \left[\text{BER}_\circledast^{avg,mv}, D_\circledast^{avg} \right],$$

$$(14)$$

subject to

$$p^\psi(v_i) \in [0, P_{max}] \ \forall i \in \{1, \dots, M\}, \qquad (15)$$

$$f^\psi(e_{i,j}) \in \mathcal{F} = \{f_1, \dots, f_F\} \ \forall e_{i,j} \in \mathcal{E}, \qquad (16)$$

and the additional constraint that routes $\{e_{i,j}^\psi\}_{i,j=1}^{N} \ \forall \psi \in \{1, \dots, \Psi\}$ can be modeled as a tree graph spanning the whole set of nodes within the network and rooted on the OBU.

Due to the complexity of the problem and the conflicting nature of both objectives, this work proposes to resort to multi-objective evolutionary meta-heuristic algorithms aimed at estimating the Pareto front differently trading one objective for the other. This approach requires efficient solution encoding strategies capable of jointly representing the tree topology of the multi-hop routes, the power and the frequency channel utilized by the nodes for each route as will be explained in detail in the following section.

3   Proposed Resource Allocation Algorithm

The ultimate purpose of the resource allocation algorithm proposed in this paper is to estimate the set of values for the optimization problem posed in Expressions (14) to (16) optimally trading bit error rate performance for average delay from the nodes to the central OBU. To this end a population-based multi-objective meta-heuristic approach simultaneously comprising power, frequency and topology optimization will be next described, including the solution encoding, its constituent operators and overall operation strategy.

3.1   Solution Encoding

All variables are represented by a vector $\mathbf{X}(k)$, with $k \in \{1, \dots, K\}$ denoting the index of the solution within the population. This vector will contain the power (one value per node and route, $M\Psi$ in total), frequency channel (one value per link and route, and since in every tree the number of links coincides with the number of sensors, $M\Psi$ in total) and an integer representation of the tree topology of every route. As for the latter our approach will resort to the family of Dandelion-like codes, a subclass of the general category known as Cayley codes. A Cayley code is essentially a bijection between the set of all labeled unrooted trees on M nodes and $M - 2$ tuples of node labels, i.e. each tree corresponds to a unique Cayley code and vice versa. Therefore, the overall solution vector will be given by $\Psi \cdot (3 \cdot M - 2)$ integer values.

While the variable encoding for the transmit power and frequency per link is straightforward, the rationale for selecting Dandelion codes for representing the multi-hop routes lies on their full coverage, zero-bias and perfect closure, all desirable properties for ensuring efficient encoding/decoding and a bounded locality when perturbed via evolutionary operators. In fact, Dandelion codes have been shown to possess better locality and heritability characteristics than any other Cayley codes when undergoing different evolutionary operators [20,21]. This superior performance of Dandelion codes has unleashed a flurry of research gravitating on their application in different scenarios and sectors, such as Telecommunications [22] and Energy [23].

3.2   Meta-Heuristic Solver

With the previous solution encoding approach in mind, a meta-heuristic algorithm is required to evolve the solution population towards regions of progressively increased Pareto optimality. This will be accomplished by the constituent operators of the so-called Harmony Search (HS) meta-heuristic algorithm, first presented in [24] and since then proven to perform statistically better than other meta-heuristic schemes in a wide variety of applications [25]. This algorithm inspires from the collaborative behavior of musicians when improvising aesthetically good harmonies; in fact the compounding operators of HS can be regarded as computationally modeled behavioral patterns commonly observed in music composition. At this point the authors would like to point out that despite the controversy around this algorithm in regards to its similarity to a special case of $(\mu + 1)$ Evolutionary Strategies [26,27], in this paper we will use the HS notation in impartial conformity with the majority of the related literature (i.e. *note* $\rightarrow$ variable, *harmony* $\rightarrow$ solution).

Following the above simile solutions are referred to as *harmonies* in the context of HS. Likewise, the population of potential solutions as harmony memory or HM. This memory undergoes a set of intelligent operators repeatedly until a stop criterion is satisfied. HS applies these operators on a per-note basis with statistical independence between nodes, which ultimately permits to balance more effectively the intensification and diversification of the underlying search procedure. When particularized to the problem tackled in this paper, the nominal search process of the HS algorithm breaks down into four steps, schematically depicted in Fig. 2 and described as follows:

1. Initialization of the HM: since no a priori information will be assumed, the HM is populated with harmonies whose individual elements are drawn uniformly at random from their corresponding alphabets. Without loss of generality other initialization criteria can be used such as e.g. a distance-based minimum spanning tree for the notes representing the topology of the routes.

2. Improvisation: after initializing the set of stored solutions, a new harmony memory $\{\widehat{\mathbf{X}}(k)\}_{k=1}^{K}$ is improvised. For each note in $\{\mathbf{X}(k)\}_{k=1}^{K}$ the following operators are subsequently applied:
   - The Harmony Memory Considering Rate (HMCR) is driven by the probabilistic parameter $\vartheta \in [0, 1]$, and establishes the probability that the new improvised value for a given note is drawn uniformly from the values taken by the same note in the $K - 1$ remaining harmonies. Otherwise (with probability 1-ϑ) the value of the note is kept unaltered.
   - The Pitch Adjustment Rate, controlled by the parameter $\varphi \in [0, 1]$, sets the probability that the value for a given note is replaced with any of its neighborhood in the corresponding alphabet. Here the notion of neighborhood depends roughly on how the value of the fitness functions behaves over the alphabet with respect to its value for the note value of reference. In this work the alphabet for all variables will be sorted in natural ordering; while this criterion can be intuitively aligned with the expected impact of the transmit power on the objective functions, this intuition may not hold for those variables representing the frequency assignment and the topology of the multi-hop routes. Hence, in such variables this operator reduces to a uniform randomization of the value, similar to the naïve mutation operator in Genetic Algorithms.

3. Evaluation and update of the Harmony Memory: once the fitness values for every candidate harmony in the newly produced memory have been computed, this bi-objective solver selects the prevailing set of K harmonies under a criterion based on rank and crowding distance: each improvised harmony $\widehat{\mathbf{X}}(k)$ is labeled with a numerical score depending on its Pareto dominance level with respect to the rest of individuals (both the other new harmonies and those remaining from the previous iteration). Once all individuals have been ranked in these terms, their crowding distance is computed and utilized as a secondary score to evaluate individuals within the same Pareto dominance level. Finally, those K harmonies first ranked in terms of Pareto optimality (primary criterion) and crowding distance (secondary criterion) are kept for subsequent iterations.

4. Termination: a stopping criterion based on a maximum number of iterations is imposed to declare the HM as the best Pareto solution attained by the proposed scheme. If the criterion has not been met steps 2 to 4 are repeated.

As for the implementation of the above meta-heuristic algorithm, it is important to note that a high number of variables are optimized via a single fitness function. For instance, when dealing with $\Psi = 3$ routes and $M = 50$ nodes a total of $\Psi \cdot (3 \cdot M - 2) = 444$ variables are optimized by the above solver. In order to handle efficiently the possible counteractions between the variables over

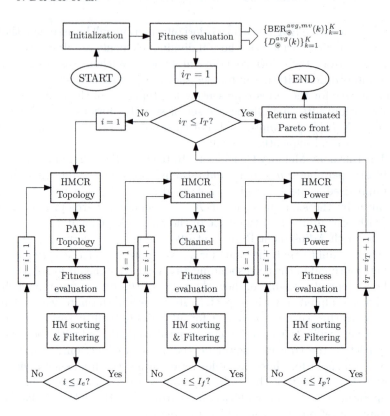

Fig. 2. Flow diagram of the proposed meta-heuristic resource allocation algorithm. HM sorting is performed based on Pareto rank and crowding distance criteria. It is important to note that the application of the HS operators is iterated over the resources to be allocated – topology, channel or power – during a number of iterations each.

the considered objective functions, the application of the constituent harmony search operators is iterated over each variable set – topology, spectrum channel and transmit power – during a certain number of iterations I_e, I_f and I_p, respectively. This block-wise iterative procedure is then globally repeated for I_T iterations, accounting up to a total of $I_T \cdot (I_e + I_f + I_p)$ fitness evaluations.

4   Experiments and Results

In order to assess the performance of the proposed resource allocation algorithm under different operational radio conditions, several scenarios have been designed and simulated by considering an intra-vehicular mesh of $M = 75$ wireless sensors uniformly deployed at random inside the car. The central OBU has been assumed to be located in the front part of the vehicle. The maximum transmit power of every compounding node is $P_{max} = 0$ dBm, whereas noise power spectral density is fixed to $N_0 = -134$ dBm/Hz. Communications take place over a spectrum band divided in $|\mathcal{F}|$ channels of equal bandwidth (5 MHz each).

Radio links established inside the vehicle are modeled as Rayleigh or Rician fading channel depending on whether source and destination nodes are located (Line Of Sight, LOS) or not (NLOS) inside the same vehicle compartment (i.e. trunk, cabin, in engine, under engine). The path loss exponent driving the exponential dependence of the transmit power on distance is set to $\gamma_{i,j} = 3$ $\forall i, j \in M \times M$. An additional log-normal shadowing component with standard deviation equal to 8 dB is included in the large-scale fading model. Values for the ratio between the direct component and the variance of the multi-hop signal are set so as to reflect (1) the radio propagation complexity of each section of the car due to e.g. metallic parts; and (2) the boundaries between the afore-mentioned vehicle compartments. Figure 3a and b exemplify this channel modeling by depicting the average channel gain for a source sensor located in the trunk and inside the engine, respectively. The values adopted for the above radio parameters are suitable for modeling intra-vehicular wireless networks according to recent literature [8, 28].

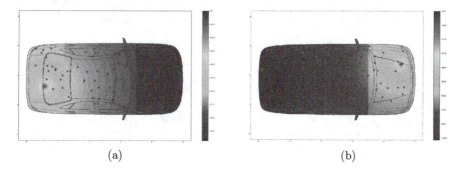

| (a) | (b) |

Fig. 3. Heatmap representing the average channel gain (in dB) inside the vehicle for a sensor node transmitting (a) from the trunk; (b) from the outer part of the engine. Deployed nodes are depicted as red circles. The channels utilized in the simulation experiments incorporate log-normal shadowing. Color scales vary between figures (Color figure online).

Switching between two frequency channels is assumed to take $T_{sw} = 2 \cdot 10^{-4}$ seconds. The parameters controlling the resource allocation algorithm are fixed to $K = 100$, $\vartheta = 0.7$, $\varphi = 0.1$, $I_e = I_f = I_p = 10$ and $I_T = 10$, values that have been optimized via an off-line grid search (results not included in the manuscript for the sake of space). The algorithm has been simulated over two different scenarios so as to shed light on the single-shot performance of the proposed algorithm, which are discussed in what follows.

Scenario A: Full Spectrum Availability

In this case a total of $|\mathcal{F}| = 16$ channels (as in physical stacks such as IEEE 802.15.4) are available. It is expected that the resource allocation algorithm has

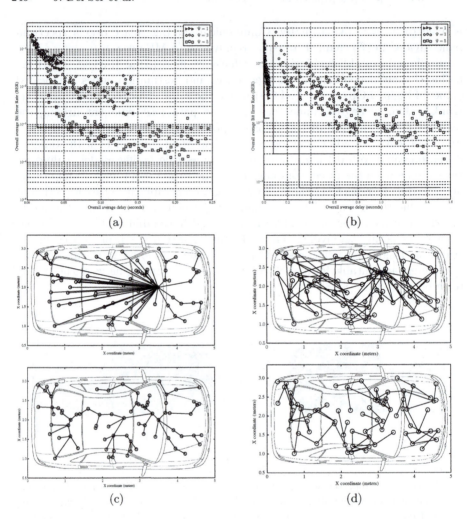

Fig. 4. Approximate Pareto fronts produced by the proposed resource allocation algorithm for different values of Ψ and the delay model in Expressions (10) (a) and (13) (b). Figures (c) and (d) depict the network topology of a multi-hop route selected at random in the case with $\Psi = 3$ for the extreme points of the corresponding fronts.

enough channel diversity to achieve the lowest interference level within the set of simulated scenarios. When combined with a suitable, interference-aware transmit power assignment over nodes, the Pareto front optimally balancing error rate and delay should connect maximal points characterized by (1) an underlying star network topology (minimum delay, maximum error rate); and (2) a tree-like network layout with short-range links (maximum delay, minimum error rate).

The obtained results verify our prior intuitions on the behavior of the algorithm under the two assumptions taken for the modeling of the overall delay. When nodes await for the completion of all transmissions within a multi-hop route

$\{\mathcal{E}_i^\psi\}_{i=1}^M$, the results in Fig. 4a evince that the Pareto front approximation produced by the algorithm approaches several performance asymptotes estimated based on:

- The majority fusion of Ψ flows characterized by individual, interference-free error rates for each node.
- The minimum delay per node attainable by star network topologies with no need for channel switching, i.e. $\min D_\circledast^{\mathrm{avg}} = \Psi \cdot D_{mx}^\psi(v_i)$.

The good agreement of the estimated resource allocation algorithm with its expected behavior is also buttressed by the plots in Fig. 4c, which depicts the topology of a randomly chosen route produced by the algorithm at both extremes of the approximated Pareto front for $\Psi = 3$. On the contrary, when modeling the delay under the assumption that transmissions belonging to different routes are concurrently routed at intermediate nodes, the objectives to be optimized undergo the effects of interference, frequency switching and multiplexing. The corresponding approximated Pareto fronts degrade accordingly, as Fig. 4b clearly shows. In regards to Fig. 4d, topologies associated to the extreme Pareto points in the $\Psi = 3$ case of Fig. 4b do not resemble any longer a tree-like or star network, the reason being that the allocation of resources becomes more involved due to the interactions between different multi-hop routes.

Scenario B: Restricted Spectrum Availability

In this second scenario the vehicle at hand is assumed to suffer from the interfering effects of an external IEEE 802.11 access point transmitting at 20 dBm 7 m away from the trunk. The access point is assumed to occupy different percentages Δ of the set of $|\mathcal{F}|$ spectrum channels available for intra-vehicular networking. We will restrict this second analysis to the results obtained for the simplistic delay model.

Results in Fig. 5a and b reveal that the proposed algorithm reacts satisfactorily against the in-band interference imposed by the external access point. When focusing on low-delay allocation policies, the assigned resources do not vary significantly due to the localized effect of the interference at the receiving end which, in the case of star-like topologies, is mainly located at the OBU. By contrast, when focusing on resilient topologies, the network is built by short-length links which, for the sake of minimizing inter-node interference, demand as many frequency channels as possible to provide enough spectral diversity among transmissions concurring at the same node.

As the interference from the external access point becomes dominant over the transmission band (i.e. as Δ increases) nodes located inside the trunk undergo a increasingly severe spectrum scarcity, thus their achievable error and delay figures degrade accordingly. In Fig. 5b it is important to observe that as the number of interference-free spectrum channel reduces, nodes inside the trunk are connected via physically longer radio links, which can be justified by the fact that the algorithm finds it *better* (in the strict sense of Pareto optimality) to

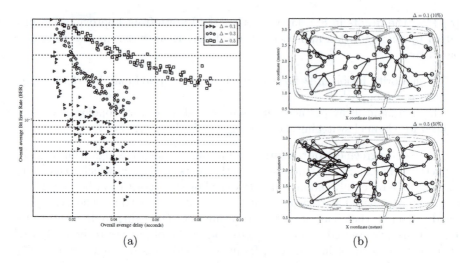

Fig. 5. (a) Approximate pareto front and (b) network topology associated to the points with minimum overall bit error rate for the simplistic delay model in presence of a external IEEE 802.11 interferer over Δ % of the utilized spectrum band.

minimize the number of intermediate hence relaying nodes. This is accomplished by making the network layout less deep and more *central*.

5   Conclusions and Future Research Lines

This manuscript has elaborated on the simultaneous management of spectrum, topology and power in an intra-vehicular wireless network. The stringent radio propagation and interference characteristics of this radio environment and their impact on the optimal allocation of the aforementioned resources are mathematically formulated as a constrained bi-objective optimization problem jointly considering two conflicting criteria: bit error rate after majority data fusion and overall average latency undergone by the compounding nodes of the network. Intuitively, error rate performance is strongly affected by co-channel interference and fading statistics, which impose short-range communications and spectrum coordination between nearby nodes. On the other hand, latency is mainly due to switching among different spectrum channels and the service (queuing) time at intermediate nodes, which in contrast call for star-like networks with long-range links and scarce spectrum diversity.

The formulated problem aims at finding the set of Pareto-optimal resource allocation strategies that best trade one objective for the other. In order to cope with the inherent complexity of the problem evolutionary meta-heuristics have been utilized. The proposed approach resorts to the so-called Dandelion solution encoding scheme, which permits representing tree-like routes with maximum locality and heritability under evolutionary operators. In order to enhance the convergence speed of the overall solver, three separate yet interrelated search procedures are scheduled in practice, each operating on a single resource for a given number of iterations. The performance of the allocation method is assessed and discussed through experiments over a number of simulation scenarios, from which it is concluded that the proposed meta-heuristic scheme is capable to optimally allocate resources under distinct spectrum availability levels and external interfering signals.

Future research will be conducted towards including energy efficiency (i.e. minimization of the power consumption) as an objective in the formulated optimization problem. In addition, more realistic latency models will be considered by discarding any assumption on the transmission regime of the nodes. Finally, a performance benchmark between different evolutionary solvers (e.g. genetic algorithms) will be done in terms of Pareto optimality.

Acknowledgments. This work has been supported by the ARTEMIS Joint Undertaking and the Spanish *Ministerio de Economia y Competitividad* (DEWI project, ref. 621353).

References

1. Kraemer, R., Katz, M.D.: Short-Range Wireless Communications: Emerging Technologies and Applications. John Wiley & Sons, Hoboken (2009)
2. Lin, J.R., Talty, T., Tonguz, O.K.: On the potential of bluetooth low energy technology in vehicular applications. IEEE Commun. Mag. **53**(1), 267–275 (2015)
3. Lu, N., Cheng, N., Zhang, N., Shen, X., Mark, J.W.: Connected vehicles: solutions and challenges. IEEE Internet Things J. **1**(4), 289–299 (2014)
4. Bas, C.U., Ergen, S.C.: Ultra-wideband channel model for intravehicular wireless sensor networks beneath the chassis: from statistical model to simulations. IEEE Trans. Veh. Technol. **62**(1), 14–25 (2013)
5. Wu, X., Subramanian, S., Guha, R., White, R.G., Li, J., Lu, K.W., Bucceri, A., Zhang, T.: Vehicular communications using DSRC: challenges, enhancements, and evolution. IEEE J. Sel. Areas Commun. **31**(9), 399–408 (2013)
6. Hashemi, M., Si, W., Laifenfeld, M., Starobinski, D., Trachtenberg, A.: Intra-Car multihop wireless sensor networking: a case study. IEEE Commun. Mag. **52**(12), 183–191 (2014)
7. Ben Jaballah, W., Conti, M., Mosbah, M., Palazzi, C.E.: Fast and secure multihop broadcast solutions for intervehicular communication. IEEE Trans. Intell. Transp. Syst. **15**(1), 433–450 (2014)
8. Moghimi, A.R., Tsai, H.-M., Saraydar, C., Tonguz, O.K.: Characterizing Intra-car wireless channels. IEEE Trans. Veh. Technol. **58**(9), 5299–5305 (2009)
9. De Francisco, R., Huang, L., Dolmans, G., de Groot, H.: Coexistence of zigbee wireless sensor networks and bluetooth inside a vehicle. In: IEEE International Symposium on Personal, Indoor and Mobile Radio Communications, pp. 2700–2704 (2009)
10. Zhang, X., Shin, K.G.: Enabling coexistence of heterogeneous wireless systems: case for zigbee and WiFi. In: Proceedings of the Twelfth ACM International Symposium on Mobile Ad Hoc Networking and Computing, p. 6 (2011)
11. Thonet, G., Allard-Jacquin, P., Colle, P.: ZigBee-WiFi coexistence. Schneider Electr. White Pap. Test Rep. **1**, 1–38 (2008)

12. Lin, J.-R., Talty, T., Tonguz, O.K.: An empirical performance study of intra-vehicular wireless sensor networks under WiFi and bluetooth interference. In: Proceedings of the IEEE Global Communications Conference (GLOBECOM), pp. 581–586 (2013)

13. Tsai, H.-M., Tonguz, O.K., Saraydar, C., Talty, T., Ames, M., Macdonald, A.: Zigbee-based intra-car wireless sensor networks: a case study. IEEE Wireless Commun. **14**(6), 67–77 (2007)

14. Hashemi, M., Si, W., Laifenfeld, M., Starobinski, D., Trachtenberg, A.: Intra-car wireless sensors data collection: a multi-hop approach. In: 2013 IEEE 77th Vehicular Technology Conference (VTC Spring), pp. 1–5 (2013)

15. Sadi, Y., Ergen, S.C.: Optimal power control, rate adaptation and scheduling for uwb-based intra-vehicular wireless sensor networks. IEEE Trans. Veh. Technol. **62**(1), 219–234 (2013)

16. Xu, R., Shi, G., Luo, J., Zhao, Z., Shu, Y.: Muzi: multi-channel zigbee networks for avoiding wifi interference. In: International Conference on Internet of Things and 4th International Conference on Cyber, Physical and Social Computing (iThings/CPSCom), pp. 323–329 (2011)

17. GreenPeak Technologies press release: GreenPeak's new multi-channel chipset will simultaneously support ZigBee and Thread networks. http://www.greenpeak.com/Press/. Retrieved on October 2015

18. IEEE Standard 802.15.4-2006: Wireless Medium Access Control (MAC) and Physical Layer (PHY) Specifications for Low-Rate Wireless Personal Area Networks (WPANs) (2006)

19. Xbee module. Digi International Inc. http://www.digi.com/lp/xbee. Retrieved on October 2015

20. Thompson, E., Paulden, T., Smith, D.K.: The dandelion code: a new coding of spanning trees for genetic algorithms. IEEE Trans. Evol. Comput. **11**(1), 91–100 (2007)

21. Perfecto, C., Bilbao, M.N., Del Ser, J., Ferro, A.: On the heritability of dandelion-encoded harmony search heuristics for tree optimization problems. In: International Symposium on Innovations in Intelligent SysTems and Applications, pp. 1–8 (2015)

22. Jafari, A.H., Lopez-Perez, D., Song, H., Claussen, H., Ho, L., Zhang, J.: Small cell backhaul: challenges and prospective solutions. EURASIP J. Wireless Commun. networking **2015**(1), 1–18 (2015)

23. Sabattin, J., Bolton, C.C., Arias, M., Parada, V.: Evolutionary optimization of electric power distribution using the dandelion code. J. Electr. Comput. Eng. **2012**, Article ID 738409, 5 (2012)

24. Geem, Z.W., Kim, J.H., Loganathan, G.V.: A new heuristic optimization algorithm: harmony search. Simulation **76**(2), 60–68 (2001)

25. Manjarres, D., Landa-Torres, I., Gil-Lopez, S., Del Ser, J., Bilbao, M.N., Salcedo-Sanz, S., Geem, Z.W.: A survey on applications of the harmony search algorithm. Eng. Appl. Artif. Intell. **26**(8), 1818–1831 (2013)

26. Weyland, D.: A critical analysis of the harmony search algorithm - how not to solve sudoku. Oper. Res. Perspect. **2**, 97–105 (2015)

27. Geem, Z.W.: Research commentary: survival of the fittest algorithm or the novelest algorithm? Int. J. Appl. Metaheuristic Comput. **1**(4), 75–79 (2010)

28. Rahman, M.A.: Reliability analysis of ZigBee based intra-vehicle wireless sensor networks. In: Sikora, A., Berbineau, M., Vinel, A., Jonsson, M., Pirovano, A., Aguado, M. (eds.) Nets4Cars/Nets4Trains 2014. LNCS, vol. 8435, pp. 103–112. Springer, Heidelberg (2014)

A Heuristic Crossover Enhanced Evolutionary Algorithm for Clustering Wireless Sensor Network

Muyiwa Olakanmi Oladimeji$^{(\boxtimes)}$, Mikdam Turkey, and Sandra Dudley

School of Engineering, London South Bank University,
103 Borough Road, Newington, London SE1 0AA, UK
{oladimm2,turkeym,dudleyms}@lsbu.ac.uk

Abstract. In this paper, a Heuristic-Crossover Enhanced Evolutionary Algorithm for Cluster Head Selection is proposed. The algorithm uses a novel heuristic crossover operator to combine two different solutions in order to achieve a high quality solution that distributes the energy load evenly among the sensor nodes and enhances the distribution of cluster head nodes in a network. Additionally, we propose the Stochastic Selection of Inactive Nodes, a mechanism inspired by the Boltzmann Selection process in genetic algorithms. This mechanism stochastically considers coverage effect in the selection of nodes that are required to go into sleep mode in order to conserve energy of sensor nodes. The proposed selection of inactive node mechanisms and cluster head selections protocol are performed sequentially at every round and are part of the main algorithm proposed, namely the Heuristic Algorithm for Clustering Hierarchy (HACH). The main goal of HACH is to extend network lifetime of wireless sensor networks by reducing and balancing the energy consumption among sensor nodes during communication processes. Our protocol shows improved performance compared with state-of-the-art protocols like LEACH, TCAC and SEECH in terms of improved network lifetime for wireless sensor networks deployments.

Keywords: Evolutionary algorithm · Heuristic crossover · Wireless sensors networks · Clustering · Network lifetime

1   Introduction

Recent advances in Micro-Electro-Mechanical Systems (MEMS) have led to the development of autonomous and self-configurable wireless sensor nodes that are densely deployed throughout a spatial region [1]. These sensor nodes have the capability to sense any event or abnormal environmental conditions such as motion, moisture, heat, smoke, pressure etc. in form of data. A large number of these sensors can be networked and deployed in remote and inaccessible areas, hence providing a wireless sensor network (WSN) connectivity. WSNs

S. Dudley—Member of the Institute of Electrical and Electronics Engineers(MIEEE).

© Springer International Publishing Switzerland 2016
G. Squillero and P. Burelli (Eds.): EvoApplications 2016, Part I, LNCS 9597, pp. 251–266, 2016.
DOI: 10.1007/978-3-319-31204-0_17

are usually composed of small, inexpensive and battery-powered wireless sensor nodes. In fact, WSNs have contributed tremendously to a number of military and civil applications such as target field imaging, event detection, weather monitoring, security and tactical surveillance. However, sensor nodes are constrained by energy supply and bandwidth. Such constraints combined with a typical deployment of large number of sensor nodes pose many challenges to the design and management of WSNs and necessitate energy-awareness at all layers of the networking protocol stack [2]. Thus, *innovative techniques that eliminate energy inefficiencies that would shorten the lifetime of the network are required.* In this work, we present a solution that balances the consuming energy among sensor nodes in order to prolong the network lifetime. The energy balance is done using two mechanisms; firstly, a genetic algorithm (GA) for cluster heads (CHs) selection was used to ensure that well distributed nodes with higher energy will be selected as CHs. Secondly, a mechanism inspired by Boltzmann selection was utilized to select nodes to send into sleep mode without causing an adverse effect on the coverage.

Most existing routing protocols designed to tackle the above challenges are broadly classified into two classes, namely flat and hierarchical. Flat protocols include the old-fashion Direct Transmission (DT) and Minimum Transmission Energy (MTE), which cannot promise a balanced energy distribution among sensors in a WSN. The drawback of the MTE is that a remote sensor uses a relay sensor for data transmission to the sink and this makes the relay sensor the first to die. In the DT, sensors communicate directly with the sink and this causes the remote sensor to die first. Therefore designing energy-efficient clustering protocols becomes a major factor for sensor lifetime extension. Generally, clustering protocols can outperform flat protocols in balancing energy consumption and network lifetime prolongation by adopting data aggregation mechanisms [3,4]. Theoretically, there are three types of nodes: the cluster-head (CH), member node (MN) and sink node (SN). The member node is responsible for sensing the raw data and employs Time Domain Multiple Access (TDMA) scheduling to send the raw data to the cluster head. The main role of the CH is to aggregate data received from MN and thereby forwards the aggregated data to the SN through single-hop or multi-hop. CHs selection can either be done by the sensors themselves, by the SN or pre-determined by the wireless network designer. In this paper, CH selection is done by the SN because the SN has enough energy and can perform complex calculations. The CH selection problem can be viewed as an optimisation problem where the methods have used GA to solve. From that end, we define an objective function to access the individual solution and propose a novel heuristic crossover that is enhanced by the knowledge of our problem.

In this paper, a Heuristic Algorithm for Clustering Hierarchy (HACH) protocol has been developed and it has two mechanisms namely the Heuristic-Crossover Enhanced Evolutionary Algorithm for Cluster Head Selection (HEECHS) and Stochastic Selection of Inactive Nodes (SSIN) mechanisms. HEECHS exploits our knowledge of the problem to develop an effective heuristic crossover that combines genetic material in a unique way to produce improved CHs configuration.

This process bears some similarity to a form of optimisation algorithm called Memetic Algorithms (MAs). MAs are a class of stochastic global search heuristics in which Evolutionary Algorithms-based approaches are combined with local search techniques to improve the quality of the solutions created by evolutions [5]. The (SSIN), a mechanism inspired by Boltzmann selection process in GA was used to reduces the number of active nodes in each round by sending some nodes to sleep or inactive mode to conserve energy and extend the network lifetime without adversely affecting coverage. These two mechanisms work collaboratively to maximize network lifetime by balancing energy consumption amongst sensor nodes during the communication process. The balance in energy consumption is achieved by selecting spatially distributed nodes with higher energy as CHs and also sending some nodes into sleep mode without causing an adverse effect on coverage. Protocol presented is a more energy efficient protocol compared with other protocols that employ GA in the sense that it integrates knowledge of the problem into the GA crossover operator.

The remainder of this study is organised as follows. Section 2 discusses related work on the application of genetic algorithms in energy-efficient wireless sensor networks. Section 3 describes the network and radio model underlying the protocol presented. In Sect. 4 we describe our proposed algorithm under three major stages namely the proposed selection mechanism, clustering algorithm and the energy consumption calculation. Section 5 presents our experimental settings, performance measures, result and discussion. Finally, a conclusion is provided in Sect. 6.

2   Related Work

Recently, research interests in the area of energy-efficient wireless sensor network has tends towards energy efficient clustering protocols [6–9]. The Low-Energy Adaptive Clustering Hierarchy (LEACH) assumes that the energy of each sensor node is the same at the time of CH selection and non-CH selection. The selection process is carried out probabilistically. The role of the CH is to aggregate data received from its members within its cluster and pass the aggregated data directly to the sink. Problems with this protocol are that the location of the selected CH might be far away from the sink, so it consumes more energy when transmitting to the sink and this can lead to CH nodes to dying faster than other nodes [4]. A two-level LEACH (TL-LEACH) proposed in [10], adds another level to the cluster whereas LEACH that has only one level. This extra level reduces energy consumption most especially for CH far away from the sink. The hybrid energy efficient distributive (HEED) protocol [11] selects CHs using residual energy and the least amount of energy dissipated for communication between the CHs and non-CHs. Data are sent to the sink using a multi-hop communication approach.

In Topology-Controlled Adaptive Clustering (TCAC) protocol [12], a large number of nodes consider themselves candidates for CH nodes and inform other nodes. Each of the candidate CH nodes checks whether the other candidate CH

nodes have a higher residual energy level or not. If there are none with higher
residual energy, the highest announces itself as the CH. Non-CH nodes select
the CH that has the minimum cost distance between itself and the CH to the
sink. The TCAC balances the size of the cluster and send the data directly
to the sink from the CH. In the scalable energy efficient clustering hierarchy
(SEECH) protocol [13], the network nodes are divided into three layers namely
the member nodes, CH nodes and relays. Clusters are formed based on centrality
of the CH node with minimum intra-cluster energy distribution. A closer node
to the sink in a cluster is often selected as the relay node. The relay node helps
the CH node to transmit aggregated data to the sink through hop or multi-hop
communication.

In [14] a genetic algorithm based energy efficient cluster (GABEEC) protocol
that uses static clustering with dynamic CH selection was described. At the end
of each round, an associate member node becomes a CH and this decision is
based on the residual energy of the current CHs and the average energy of all
members in the cluster. A Genetic algorithm approach was used to minimize
the communication distance and maximize the lifetime of the network. In [15],
a centralized energy-aware cluster-based protocol was discussed to extend the
sensor network lifetime by using Optimization (PSO) algorithms. A new cost
function that simultaneously takes into account the maximum distance between
the non-CH node and its CH, and the remaining energy of CH candidates in the
CH selection algorithm was defined.

3   Network and Radio Model

In the proposed HACH protocol, the following assumptions of the network model
are assumed:

- The data sink is a stationary and resource-rich device that is placed far away
  from the sensing field.
- All sensor nodes are homogeneous in terms of energy and are stationary after
  deployment.
- All sensors have GPS or other location determination devices
- Nodes located close to each other have correlated data.
- Nodes are capable of acting in inactive mode or a low power sleeping mode.
- The communication channel is symmetric (i.e., the energy required to transmit
  a message from sensor node s1 to sensor node s2 is the same as energy required
  to transmit a message from node s2 to node s1 for a given signal to noise ratio).

For fair comparison with previous protocols [4,16,17], we assume the simple
model shown in Fig. 1 for the radio hardware energy dissipation where the transmitter dissipates energy $E_{Tx}(k, d)$ to run the radio electronics and the power
amplifier, and the receiver dissipates energy $E_{Rx}(k)$ to run the radio electronics.
For the experiments described here, both the free space (d^2 power loss) and the
multipath fading (d^4 power loss) channel models were used, depending on the

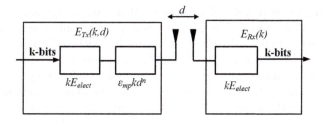

Fig. 1. Radio Energy Dissipation Model

Algorithm 1. HACH Protocol
___

Let *AliveNodes* be the total number of sensor nodes
Compute the network total coverage.
while (*AliveNodes* > 0) **do**
    Use algorithm SSIN to select inactive nodes. (See Algorithm 2)
    Put selected nodes into sleep mode.
    Apply the proposed HEECHS algorithm for CHs configuration. (See Algorithm 3)
    Compute the energy values of E_{CH}, E_{Mem} and E_{Res}. (refer to Sect. 4.3)
    Find out the number of dead node (node with energy equal or less than 0).
    Update AliveNodes
end while
___

distance (d) between the transmitter and receiver. The power amplifier is appropriately set in such a way that if the distance is less than a threshold distance, the free space (fs) model is used; otherwise, the multipath (mp) model is used. Thus, to transmit a k-bit message at distance d, the radio expends:

$$E_{Tx}(k, d) = \begin{cases} kE_{elect} + \varepsilon_{mp}kd^4, & \text{if } d > d_0 \\ kE_{elect} + \varepsilon_{fs}kd^2, & \text{if } d < d_0 \end{cases} \tag{1}$$

And to receive k-bit message, the radio expends:

$$E_{Rx}(k) = kEelect \tag{2}$$

where $d_0 = \sqrt{\varepsilon_{fs}/\varepsilon_{mp}}$ denotes the threshold distance and the electronics energy, E_{elect} depends on factors such as the digital coding, modulation, filtering, and spreading of the signal, whereas the amplifier energy, ε_{mp} or ε_{fs} depends on the distance to the receiver and the acceptable bit-error rate.

4   The Proposed Protocol: HACH

This protocol performs three sequential operations: sending nodes to sleep, clustering and network operations. At the set-up phase of the algorithm, the sink transmits control packets in order to receive node information in terms of the nodes ID, location and energy. The proposed SSIN protocol helps to dynamically select which nodes to send to sleep by initially generating a candidate list.

The candidates list is populated with nodes that have energy smaller than the average energy of all nodes. Using a stochastic process, only a small number of nodes is sent into sleep mode in the candidate list without causing an adverse on the coverage. CH selection using HEECHS is then performed on the nodes that are kept active. In this work, the authors formulated clustering as an optimisation problem that is best addressed using GA. The genetic operators used in this approach are the tournament selection, heuristic crossover and mutation operator. The best CH configuration that ensures balanced energy consumption across the networks is selected at every network operation round. At the end of each round, the residual energy of each node is computed. This value is then used to compute the average energy for the next round and this cycle continues until all nodes are dead as shown in Algorithm 1.

4.1   Inactive Node Selection Using SSIN

The SSIN makes decisions on which nodes to send into inactive mode at the beginning of each network operation round. The candidate list of sleeping nodes is built up by checking the residual energy of nodes that are less than the computed average energy. This inactive node selection mechanism is synonymous to the Boltzmann selection process whereby a method is adopted for controlling the selection pressure [18]. In Boltzmann selection, the temperature parameter is varied in order to control the selection pressure. In this work, the maximum coverage effect, Max_{eff} is used to control the effect of putting nodes to sleep on the WSNs, which is defined as:

$$Max_{eff} = 2 \times \pi \times R_s^2 \tag{3}$$

where R_s is the sensing range of a sensor node (taking the coverage area as a circle with radius R_s), $(pi \times R_s^2)$ signifies the coverage of one node and the value $'2'$ represents coverage of two nodes.

The effect of putting a node to sleep based on the coverage is defined by the coverage effect C_{eff} as shown in Fig. 2. The total coverage effect is calculated by invoking a matrix called the Coverage Matrix. The coverage matrix captures

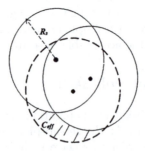

Fig. 2. Illustration of nodes to sleep on coverage area

Algorithm 2. Proposed SSIN protocol

$Acc_{eff} = 0$;
Compute the residual energy, E_{Res} of each node. (refer to Sect. 4.3)
Compute the average energy of all nodes, E_{Avg}.
Generate a candidate list for nodes that satisfies the condition $E_{Res} < E_{Avg}$.
Compute Max_{eff}. (refer to Eq. 3)
while $(Acc_{eff} < Max_{eff})$ **do**
    Compute probability, P of adding nodes to the sleeping list. (See Eq. 4)
    if $(rand() < P)$ **then**
        Create list of sleeping node from the candidate list.
        Compute the coverage effect, C_{eff}.
        $Acc_{eff} = Acc_{eff} + C_{eff}$
    end if
end while

---

the overlapped areas of nodes coverage which allows the identification of nodes that can be sent into sleep mode without affecting coverage as there will be other nodes covering part of the selected node's area. The accumulated Coverage effect Acc_{eff} is defined as the total effect on the coverage as a result of allowing some nodes to sleep. Our algorithm is designed in such a way that the Acc_{eff} value is expected to be less than the Max_{eff} for optimum coverage ($Acc_{eff} < Max_{eff}$). The probability that a node will be added to the sleeping node list can be computed using:

$$P = e^{(-C_{eff}/Max_{eff})/(1-(Acc_{eff}/Max_{eff})))^2} \tag{4}$$

A randomly generated number is compared with the computed probability, P. A candidate list of inactive node is created if the random number is less than the probability, P. The accumulated frequency Acc_{eff} is computed by adding its current value to coverage effect C_{eff} value. The operations of SSIN continues until the Acc_{eff} is greater than the maximum acceptable coverage effect, Max_{eff} as shown in Algorithm 2.

4.2 Cluster Heads Selection Using HEECHS

Our clustering algorithm involves the task of CH selection and membership association. The proposed HEECHS protocol is developed for the CH selection task using a heuristic-based GA. Similar to conventional GA, it runs through a cycle of tasks such as the creation of population strings, evaluation of each string, selection of the best strings and reproduction to create a new population. The significant difference is that it uses a problem-dependent knowledge-based heuristic crossover to find the best CH configuration with the optimum number of well distributed CH nodes. The population, $P(t)$ comprises of individuals with binary representation of 0s or 1s. This represents CH individuals with size p_s and each individual CH has length N_s that is evaluated by computing the fitness value using Eq. 6. Individual CHs with the best fitness value are selected from two randomly selected parent population, $P_x(t)$ and $P_y(t)$ and this continues until the mating pool is filled. Our proposed heuristic crossover is applied to mate the individuals in the pool and subsequently produce a new population of $P(t+1)$. Again, the fitness value of each individual in this new population is computed using Eq. 6 and the whole cycle continues until the stopping criterion is achieved. Stopping criterion is achieved when the average fitness of the population does not change further.

Proposed Heuristic Crossover. The core component of the proposed HEECHS is the heuristic crossover and this term simply means a unique form of crossover that uses the knowledge of a problem to combine two candidate solution in order to produce an offspring. A heuristic crossover operator was proposed by Lixin Tang [19] as an operator which utilizes the parents implicit information to produce offspring. In a conventional GA, the selected members of a population are paired, and some two parents undergo crossover to produces pairs of offspring that tends to substitute the parents. Heuristic crossover operator produces only one child from two or more parents, which differs from the canonical approach. Also in the canonical approach, there is no guarantee that an offspring produced would be better than the mating parent [20].

The proposed crossover produces individuals with CH configurations that are well distributed across the sensing field and favors the ones with higher energy. We can see from the Algorithm 3 that the proposed crossover does not allow two CHs to be selected within the same region and gives selection priority to the CH with higher energy. Two individuals are selected from the parent pool and the position of CH genes inside each individual is computed. Each individual has its own array set of CH positions, which is denoted by CH_1 and CH_2. We choose to define the threshold distance between any two neighbouring CH position as $\frac{\sqrt{(x_{max}-x_{min})^2+(y_{max}-y_{min})^2}}{n \times 0.04}$, where the (x_{min}, y_{min}) and (x_{max}, y_{max}) are minimum and maximum xy-coordinates of the sensor fields respectively, $(n \times 0.04)$ represent 4% of the total sensor nodes. The union of CH_1 and CH_2 is represented by $CH_{all} = CH_1 \cup CH_2$. By default, the first CH position $CH_{all}(1)$ in the set CH_{all} is transferred to a newly created set CH_{new}. Each subsequent CHs position in the CH_{all} is compared with the CH_{new} array set in order to ensure decisions based on distance between the CHs and their residual energy (E) as shown in Algorithm 3.

Proposed Objective Functions. In order to solve the CH selection problem, there is a need to develop objective functions because CH selection is dealt with as an optimisation problem. This objective function returns a fitness value which is used to access the quality of a candidate solution. The objective functions is derived by considering the parameters such as the energy of all sensor nodes and the Risk penalty R. The reason for considering the energy parameter of sensor nodes is to ensure that nodes with greater energy are given higher priority in CHs selection.

Algorithm 3. Proposed Heuristic Crossover

Select two chromosomes from the parent pool
Compute and store the cluster head position in each chromosomes in sets CH_1 and CH_2.
Compute the threshold distance, T (refer to Sect. 4.2)
Compute the union set CH_{all}. (refer to Sect. 4.2)
Obtain the first cluster position $CH_{all}(1)$ in the set CH_{all}.
Create a new set CH_{new} and transfer the $CH_{all}(1)$ to it.
Compute the distance, D between cluster head positions in the sets CH_1 and CH_2.
while $(D < T)$ **do**
    if (Energy in CH_{all} node is less than CH_{new} node) **then**
        Discard the cluster head node. (i.e. do not add to CH_{new} set)
    end if
    Replace the cluster head node in the CH_{new} set
end while
Add to the cluster head node in the set CH_{all} into the CH_{new} set.

The Risk penalty, R for the CH selection is defined as:

$$R = \begin{cases} Lower - L, & \text{if } L < Lower \\ L - Upper, & \text{if } L > Upper \\ 0, & \text{otherwise} \end{cases} \tag{5}$$

where $Lower$ and $Upper$ are calculated as the 4 % and 6 % of total number all sensor nodes in the field (n) respectively. R imposes restrictions on the number of CHs (L).

Therefore, the objective function is computed as:

$$F(X) = w_1 * \frac{AvgENCH}{AvgECH} + w_2 * R \tag{6}$$

Where w_1 and w_2 are the weighting factors; the average energy of non-CHs and CHs is given by $AvgENCH = \frac{\sum_{i \varepsilon NCH} E_i}{n - L}$ and $AvgECH = \frac{\sum_{i \varepsilon CH} E_i}{L}$ respectively. The ratio $\frac{AvgENCH}{AvgECH}$ is assigned a higher weighting factor ($w_1=0.9$) than the Risk penalty, R ($w_2=0.1$) because much emphasis is placed upon it. (Note: CH and NCH represent the set of all CHs and non-CHs respectively).

Other Operators. Crossover and mutation provide exploration, compared with the exploitation provided by selection. The effectiveness of GA depends on the trade-off between exploitation and exploration [21,22].

The *tournament selection operator* is used to randomly choose a group of individuals from the current population, and the one with the best fitness is selected. The tournament size determines the extent of selection pressure. In this work, a tournament size of two is used in order to reduce the selection pressure. This process is repeated as often as desired (usually until the mating pool is filled).

The *mutation operator* alters one individual (parent), to produce a single new individual (child) with a probability of mutation (p_m).

The initial population pool (parents) and offspring individuals that was generated in the previous step are sorted in ascending order based on the objective function values. Then the first CH chromosomes with minimum objective function values are selected to form the population pool for the next generation.

The *stopping criterion* is met when the objective function value of the population does not change further.

4.3   Network Operations and Energy Consumption Computation

Energy consumption at each round can be explained with regards to the two phases in the algorithm such as the set-up phase and the steady state phase.

Set-up Phase. The set-up phase involves the transmission and reception of control packets k_{CP} from the sink to all nodes to initialise inter- and intra-communication. In the set-up phase, the sink sends a control packet which contains a short message to wake up and request IDs, positions and energy levels of all sensor nodes in the sensor field. As in Eq. 2, the energy $E_{Rx}(k_{CP})$ is dissipated when control packets are received from the sink. All sensors report their IDs, positions and energy levels back to the sink. Also, as in Eq. 1, energy $E_{Tx}(k_{CP}, d)$ is consumed when transmitting control packets k_{CP} to the sink. The control packets received from all sensor nodes is processed by the sink in order to make the following vital decision; which nodes to keep active, CH selection, and the associated CH membership. Energy $E_{Rx}(k_{CP})$ is also dissipated for receiving the membership status information from the sink. The energy required for all CHs to transmit the TDMA schedules to their members can be computed using:

$$E_{Tx(ch_i)}(k_{CP}, d_{i-toMem}) = \sum_{i=1} ch_i * \begin{cases} k_{CP}E_{elect} + \varepsilon_{mp}k_{CP}d_{i-toMem}^4, & \text{if } d < d_0 \\ k_{CP}E_{elect} + \varepsilon_{fs}k_{CP}d_{i-toMem}^2, & \text{if } d > d_0 \end{cases}$$
$$(7)$$

Also the energy spent by each member node to receive the TDMA schedule from the CH can also be computed using Eq. 2.

Steady Phase. During the steady phase, the active sensor nodes start sending data packets k. Each node sends the sensed data to its CH according to the TDMA schedule received. The CH nodes receiver must always be ready to receive packets from its node within its cluster. Data aggregation is performed on all received data at the CH and data are converted into a single data stream. This aggregated data is transmitted from the CH to the sink. During the process the sensor node transceiver consumes energy calculated by Eq. 9. The total amount of energy spent by all member nodes to transmit to their respective CHs is computed using:

$$E_{Rx(m_i)}(k) = \sum_{i=1} m_i k E_{elec} \tag{8}$$

Where m_i denotes the member nodes, which ranges from $i = 1, 2, 3, ..., n - L$. Also, the energy dissipated by the CH for aggregating data received from all its members and itself can be calculated using:

$$E_{DA(m_i+1)}(k) = k E_{DA} * \left(\sum_{i=1} m_i + 1\right) \tag{9}$$

Finally, the amount of energy spent by the CH node to transmit to the sink is computed using:

$$E_{Tx(ch_i)}(k_{CP}, d_{i-toSink}) = \sum_{i=1} ch_i * \begin{cases} k_{CP} E_{elect} + \varepsilon_{mp} k_{CP} d_{i-toSink}^4, & \text{if } d > d_0 \\ k_{CP} E_{elect} + \varepsilon_{fs} k_{CP} d_{i-toSink}^2, & \text{if } d < d_0 \end{cases} \tag{10}$$

Total Energy Dissipated. In our algorithm, the total energy dissipated by all the CHs can be computed using:

$$\begin{aligned} E_{CHs} = 2 * E_{Rx}(k_{CP}) + E_{Tx}(k_{CP}, d_{i-toSink}) + E_{Tx}(k_{CP}, d_{i-toMem}) \\ + E_{Rx(m_1)}(k) + E_{DA(m_i+1)}(k) \end{aligned} \tag{11}$$

And the energy dissipated by the member nodes is computed as:

$$E_{Mem} = E_{Tx}(k_{CP}, d_{i-toSink}) + E_{Tx}(k_{CP}, d_{i-toCH}) + 3 * E_{Rx}(k_{CP}) \tag{12}$$

Therefore, the overall energy dissipated by all nodes is represented by $E_{TOTAL} = E_{CHs} + E_{Mem}$. (Note: Residual energy E_{Res} of individual node at each round can be computed by subtracting the energy consumption from the current residual energy.

5   Experimental Results

In this work, we evaluate protocols from an energy efficiency perspective by examining the number of nodes alive versus rounds. With the aid of MATLAB tools, a simulation model was developed to test the proposed algorithm and graph results are generated to evaluate the lifetime of the sensor nodes. The proposed technique is scalable and it may lead to energy efficiency improvements across various network topologies. To assess this claim the performance of HACH is compared to three other protocols LEACH, TCAC and SEECH across three scenarios with different network sizes; small, medium and large. Table 1 describes the parameters of proposed scenes in details.

Table 1. Parameter values for each experiment

Experiment	Parameter settings			
	Dimension	Sink coordinates	Number of sensors	Initial energy
Experiment I	100×100	(50,175)	100	0.5J
Experiment II	100×100	(50,200)	400	0.5J
Experiment III	200×200	(50,350)	1000	1.0J

5.1   Experimental Settings

The common communication parameters used for all three experiments presented in Table 1 below are the electronic energy, E_{elect} =50 nJ/bit, free space loss ε_{fs}=10 pJ/bit/m^2, multipath loss, ε_{mp}=0.0013 pJ/bit/m^4, threshold distance, d_0=87m, data aggregation energy E_{DA}= 5 nJ/ bit/signal, packet size k=4000, and control packet size, k_{CP}=50. The GA parameters are set as population size, p_s=100 and mutation rate, p_m= 0.05.

5.2   Performance Measures

There are many metrics used to evaluate the performance of clustering protocols [19]. These measures again are used in this paper to evaluate the performance of HACH protocol:

1. **First Dead Node (FDN)**: This is the number of rounds after which the first node dies. It can also refer to the operational lifetime or stability period of the network.
2. **Last Dead Node (LDN)**: This is the number of rounds from the start of network operation until the last node dies.
3. **Instability Period Length (IPL)**: The round difference between the round at which the last node dies and the first node dies. (i.e. $IPL = LND - FND$).
4. **Average Energy at First Node Dies (AEFND)**: The average energy of all sensor nodes when the first node dies.

Clearly, the longer the stability period and the shorter the instability period are, the more reliable the clustering process of the wireless sensor network.

5.3   Results and Discussion

The values presented in this section are the averaged results obtained from 100 simulation runs. In each simulation run, a new set of sensor nodes was distributed to a network area. Experimental results shown in Fig. 3 depicts the number of alive nodes after each round. For a summarized comparison FND, LND, and IPL measures belonging to the graphs are presented in Table 2. Experiment I (100 nodes) shows that the HACH protocol maintains the network operational lifetime of 36, 131 and 338 more than the SEECH, TCAC and LEACH respectively.

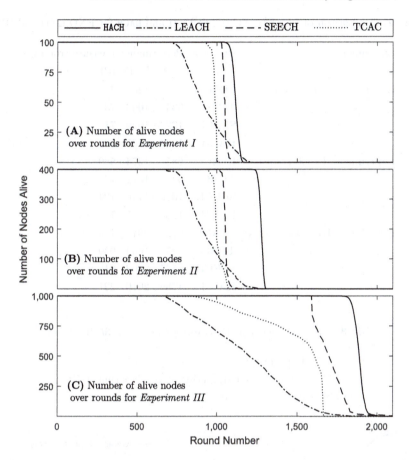

Fig. 3. Network Lifetime Comparison of HACH with LEACH, SEECH, TCAC

The energy of the sensor nodes are balanced until the FND times and this is observed on the graph by a sharp decrease in the number of alive nodes for the TCAC, SEECH and HACH.

When the FND is achieved or during the IPL, lots of nodes begin to die due to the limited residual energy. The IPL values of various protocols are shown in Table 2 with HACH having a very low IPL values in Experiment II(400 nodes) and III (1000 nodes) except in Experiment I which has 30 rounds more than TCAC. This shows that our algorithm performs better in a dense network. Using our proposed HACH protocol, the AEFND values for Experiments I,II and III are shown in Table 3. It can be observed that the AEFND value for each of the three experiments is approximately zero at the time the first node dies. For example, Fig. 4 shows that at FND time of 1064, the AEFND of sensor nodes is 0.0232J: which means we were able to used almost all of our initial energy till the death of first node.

From the above, it can be deduced that the proposed HACH decrease in energy consumption and energy balancing can increase the network lifetime.

Table 2. Comparison of LEACH, TCAC, SEECH and HACH for FND,LND and IPL

Experiment	Protocol	Performance measure (round)		
		FND	LND	IPL
Experiment I (100 Nodes)	LEACH	726	1209	483
	TCAC	933	1006	73
	SEECH	1028	1099	71
	HACH	1064	1167	103
Experiment II (400 Nodes)	LEACH	685	1274	589
	TCAC	948	1071	123
	SEECH	1016	1140	124
	HACH	1235	1307	72
Experiment III (1000 Nodes)	LEACH	672	2014	1342
	TCAC	725	1664	939
	SEECH	1587	2202	615
	HACH	1789	2010	221

Table 3. Average Energy at FND for our proposed HACH protocol

	Experiments		
	Experiment I	Experiment II	Experiment III
AEFND	0.0232	0.0164	0.0650

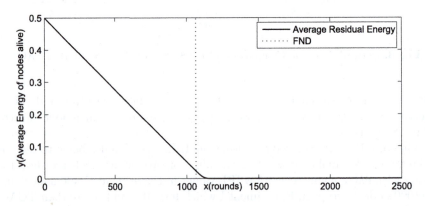

Fig. 4. Average Energy of alive node versus Rounds (refer to Experiment I)

At each round, our proposed HACH protocol conserves energy by selectively allowing some nodes to become inactive before network operation. Also, it uses an adaptive heuristic approach to select the best CHs configuration by ensuring that CH are optimally distributed over the sensor field.

6   Conclusion

In this paper, a Heuristic Algorithm for Clustering Hierarchy (HACH) protocol was proposed where the major two operations included sending some nodes into sleep mode and the cluster head selection. We proposed the Heuristic-Crossover Enhanced Evolutionary Algorithm for Cluster Head Selection protocol (HEECHS), a mechanism that employs genetic algorithms for a cluster head selection that balance energy consumption and enhances better distribution of cluster heads. To achieve the selection process, an objective function to access the quality of each solutions was designed. A problem-dependent heuristic crossover to produce a better cluster head configuration was implemented. Also, we employed a new mechanism called Stochastic Selection of Inactive Node (SSIN) inspired by Boltzmann selection process to stochastically select some nodes to send into sleep without causing an adverse effect on coverage. The two proposed mechanisms work collaboratively to reduce and balance energy consumption by selecting distributed nodes with high energy as cluster heads in order to prolong network lifetime. From the three experiments, simulation results demonstrated that our proposed HACH outperforms other protocol in terms of energy consumption, network lifetime and stability period in all the three experiments.

References

1. Naeimi, S., Ghafghazi, H., Chow, C.-O., Ishii, H.: A survey on the taxonomy of cluster-based routing protocols for homogeneous wireless sensor networks. Sensors **12**(6), 7350–7409 (2012)
2. Chakraborty, A., Mitra, S.K., Naskar, M.K.: Energy efficient routing in wireless sensor networks: A genetic approach. CoRR abs/1105.2090 (2011)
3. Abbasi, A.A., Younis, M.: A survey on clustering algorithms for wireless sensor networks. Comput. commun. **30**(14), 2826–2841 (2007)
4. Heinzelman, W.B., Chandrakasan, A.P., Balakrishnan, H.: An application-specific protocol architecture for wireless microsensor networks. IEEE Trans. Wirel. Commun. **1**(4), 660–670 (2002)
5. Hart, W.E., Krasnogor, N., Smith, J.E.: Recent Advances in Memetic Algorithms, vol. 166. Springer Science & Business Media, Heidelberg (2005)
6. Kang, S.H., Nguyen, T.: Distance based thresholds for cluster head selection in wireless sensor networks. IEEE Commun. Lett. **16**(9), 1396–1399 (2012)
7. YeMao, L., Fa, C., Hai, W.: An energy efficient clustering scheme in wireless sensor networks. Ad Hoc & Sensor Wireless Networks (to be published)
8. Dimokas, N., Katsaros, D., Manolopoulos, Y.: Energy-efficient distributed clustering in wireless sensor networks. J. parallel Distrib. Comput. **70**(4), 371–383 (2010)
9. Lin, S., Zhang, J., Zhou, G., Lin, G., Stankovic, J.A., He, T.: Atpc: adaptive transmission power control for wireless sensor networks. In: Proceedings of the 4th international conference on Embedded networked sensor systems, pp. 223–236 (2006)
10. Loscri, V., Morabito, G., Marano, S.: A two-levels hierarchy for low-energy adaptive clustering hierarchy (tl-leach). In: IEEE Vehicular Technology Conference, vol. 62, pp. 1809. IEEE; 1999 (2005)

11. Younis, O., Fahmy, S.: Heed: a hybrid, energy-efficient, distributed clustering approach for ad hoc sensor networks. IEEE Trans. Mobile Comput. **3**(4), 366–379 (2004)
12. Dahnil, D.P., Singh, Y.P., Ho, C.K.: Topology-controlled adaptive clustering for uniformity, increased lifetime in wireless sensor networks. IET Wirel. Sens. Syst. **2**(4), 318–327 (2012)
13. Tarhani, M., Kavian, Y.S., Siavoshi, S.: Seech: scalable energy efficient clustering hierarchy protocol in wireless sensor networks. IEEE Sens. J. **14**(11), 3944–3954 (2014)
14. Bayrakli, S., Erdogan, S.Z.: Genetic algorithm based energy efficient clusters (gabeec) in wireless sensor networks. Procedia Comput. Sci. **10**, 247–254 (2012)
15. Latiff, N.M., Tsimenidis, C.C., Sharif, B.S.: Energy-aware clustering for wireless sensor networks using particle swarm optimization. In: Personal, Indoor and Mobile Radio Communications, PIMRC 2007. IEEE 18th International Symposium on, pp. 1–5. IEEE (2007)
16. Liu, J.-L., Ravishankar, C.V., et al.: Leach-ga: genetic algorithm-based energy-efficient adaptive clustering protocol for wireless sensor networks. Int. J. Mach. Learn. Comput. **1**(1), 79–85 (2011)
17. Go, K.: An amend implementation on leach protocol based on energy hierarchy. Int. J. Curr. Eng. Technol. **2**(4), 427–431 (2012)
18. Dumitrescu, D., Lazzerini, B., Jain, L.C., Dumitrescu, A.: Evolutionary Computation. International Series on Computational Intelligence. Taylor & Francis, New York (2000)
19. Lixin, T.: Improved genetic algorithms for tsp. J. Northeastern Univ. (Nat. Sci.), p. 01 (1999)
20. Hasan, B.S., Khamees, M., Mahmoud, A.S.H., et al.: A heuristic genetic algorithm for the single source shortest path problem. In: Computer Systems and Applications, AICCSA 2007. IEEE/ACS International Conference on, pp. 187–194 (2007)
21. Halke, R., Kulkarni, V.A.: En-leach routing protocol for wireless sensor network. Int. J. Eng. Res. Appl. **2**(4), 2099–2102 (2012)
22. Brunda, J.S., Manjunath, B.S., Savitha, B.R., Ullas, P.: Energy aware threshold based efficient clustering (eatec) for wireless sensor networks. Energy, 2(4) (2012)

A Variable Local Search Based Memetic Algorithm for the Load Balancing Problem in Cloud Computing

Nasser R. Sabar[1]([✉]), Andy Song[1], and Mengjie Zhang[2]

[1] School of Computer Science and I.T., RMIT University, Melbourne, Australia
{nasser.sabar,andy.song}@rmit.edu.au
[2] Victoria University of Wellington, Wellington, New Zealand
mengjie.zhang@ecs.vuw.ac.nz

Abstract. Load balancing (LB) is an important and challenging optimisation problem in cloud computing. LB involves assigning a set of services into a set of machines for which the goal is to optimise machine usages. This study presents a memetic algorithm (MA) for the LB problem. MA is a hybrid method that combines the strength of population based evolutionary algorithms with local search. However the effectiveness of MA mainly depends on the local search method chosen for MA. This is because local search methods perform differently for different instances and under different stages of search. In addition, invoking local search at every generation can be computationally expensive and compromise the exploration capacity of search. To address these issues, this study proposes a variable local search based MA in the context of LB problem. The proposed MA uses multiple local search mechanisms. Each one navigates a different area in search space using a different search mechanism which can leads to a different search path with distinct local optima. This will not only help the search to avoid being trap in a local optima point, but can also effectively deal with various landscape search characteristics and dynamic changes of the problem. In addition, a diversity indicator is adopted to control the local search processes to encourage solution diversity. Our MA method is evaluated on instances of the Google machine reassignment problem proposed for the ROADEF/EURO 2012 challenge. Compared with the state of the art methods, our method achieved the best performance on most of instances, showing the effectiveness of variable local search based MA for the Load Balancing problem.

Keywords: Local search · Memetic algorithms · Load balancing · Cloud computing · Meta-heuristics

1 Introduction

Cloud computing is a fast growing technology that provides on-demand computing services over the Internet [3,6]. It offers network access to a various shared pool of configurable computing resources including storage, processing,

© Springer International Publishing Switzerland 2016
G. Squillero and P. Burelli (Eds.): EvoApplications 2016, Part I, LNCS 9597, pp. 267–282, 2016.
DOI: 10.1007/978-3-319-31204-0_18

bandwidth and memory. A cloud provider, such as Google and Amazon, manages a data centre of which the computing resources are to be shared by end users. With the rapid growth of the demand in cloud services, optimal resources allocation becomes one the most important targets in cloud computing [6].

Load balancing (LB) is one of the cloud resource allocation tasks seeking for the best arrangement of services into a set of machines so the usage of these machines can be improved. In this paper, we consider the LB problem introduced by Google for the ROADEF/EURO 2012 challenge [1]. The task is named as *Machine Reassignment Problem* (MRP). The goal of MRP is to improve the usage of resources by reassigning a set of processes into a set of machine, while all problem constraints must be satisfied. A range of methods have been proposed to solve MRP. These include variable neighbourhood search [8], constraint programming-based large neighbourhood search [15], large neighbourhood search [4], multi-start iterated local search [14], simulated annealing [20] and restricted iterated local search [13].

In this study, we propose a memetic algorithm (MA) based method for this load balancing problem. MA is a stochastic optimisation search method which combines population based algorithm with local search. The rationale of MA is to synergise the exploration power of population based algorithms with the exploitation capability of local search [16]. MA has been proven very successful in solving various difficult optimisation problems [17]. However the success of MA is not automatic [21, 22, 25]. There are two important aspects that have to be considered when designing MA for a particular problem [18]. Firstly, the choice of local search is important. The performance of MA heavily depends on the selected local search algorithm. It is difficult for one local search to fit with diverse features of different instances of different problems. Even for the same instance, the characteristic of search space under different stages may vary significantly [24]. That makes the choice of local search method difficult and critical. Secondly, MA often faces the challenge of how to preserve the diversity of a search process [23]. Excessive use of local search may consume more computation on exploitation compromising the effort on exploration. To address these two issues, we propose a variable local search based memetic algorithm. It combines genetic algorithm (GA) with multiple local search algorithms in which each one can navigate a different area in the search space. Different search mechanisms can lead to a different search path with distinct local optima. Furthermore a diversity indicator is adopted to control the invocation of local search to prevent lost of diversity in the population of solutions. With the proposed method, there is no need to examine the nature of a load balancing problem and to choose an appropriate local search for the problem. The need of tuning the local search is also unnecessary in the proposed MA approach.

The proposed algorithm are evaluated on small and large scale instances of the machine reassignment problem from ROADED/EURO 2012 challenge. For comparison purposes, the state of the art algorithms for this challenge are included as well. In Sect. 2, this challenge is described in details. The proposed variable local search based MA is presented in Sect. 3. Section 4 shows the experiment settings while the results are listed in Sect. 5. Section 6 concludes this study.

2   Problem Description

The so-called machine reassignment problem (MRP) introduced by Google [1] for the ROADEF/EURO 2012 challenge is load balancing problem. It is a challenging combinatorial optimisation problem which is to find the optimal way to assign processes to machines in order to improve the usage of a given set of machines. One machine consists of a set of resources such as RAM and CPUs. One process can be moved from one machine to another to improve overall machine usage. The allocation of processes must not violate the following hard constraints:

- *Capacity constraints*: the sum of requirements of resource of all processes should be less than or equal to the capacity of the allocated machine.

- *Conflict constraints*: processes of the same service should be allocated into different machines.

- *Transient usage constraints*: if a process is moved from one machine to another, it requires adequate amount of capacity on both machines.

- *Spread constraints*: the set of machines is partitioned into locations and processes of the same service should be allocated to machines in a number of distinct locations.

- *Dependency constraints*: the set of machines are partitioned into neighbourhoods. Then, if there is a service depends on another service, then the process of first one should be assigned to the neighbouring machine of second one or vice versa.

A solution to MRP is a process-machine assignment which satisfies all hard constraints and minimises the weighted cost function as much as possible which is calculated as follows:

$$f = \sum_{r \in R} weight_{loadCost}(r) \times loadCost(r)$$
$$+ \sum_{b \in B} weight_{balanceCost}(b) \times balanceCost(b)$$
$$+ weight_{processMoveCost} \times processMoveCost$$
$$+ weight_{serviceMoveCost} \times serviceMoveCost$$
$$+ weight_{machineMoveCost} \times machineMoveCost \qquad (1)$$

where R is a set of resources, *loadCost* represents the used capacity by resource r which exceeds the safety capacity, *balanceCost* represents the use of available machine, *processMoveCost* is the cost of moving a process from its current machine to a new one, *serviceMoveCost* represents the maximum number of moved processes over services and *machineMoveCost* represents the sum of all moves weighted by relevant machine cost. $weight_{loadCost}$, $weight_{balanceCost}$, $weight_{processMoveCost}$, $weight_{serviceMoveCost}$ and $weight_{machineMoveCost}$ define the importance of each individual cost.

The detailed explanation of the constraints, the costs and their weights can be found on the challenge documentation [1]. Note that the quality of a solution is evaluated by the given solution checker, which returns fitness measure to the best solution generated by our MA. Another important aspect of this challenge is the time limit. It was stated that *"The maximum execution time will be fixed to 5 min by instance on a core2duo E8500 3.16 MHz with 4Go RAM on debian 64 or Win7 64 bits."* All methods have to finish within the 5 min timeframe to ensure the fairness of the comparison.

3   Methodology

Hybridised algorithms have recently received increased interest from the optimisation research community [17]. It is expected that integrating the components of multiple algorithms under one framework may result in a more effective and efficient optimisation method [19]. One of such hybridisation frameworks is the Memetic Algorithms (MAs) [16,17]. MA is a class of search methods that merge the strengths of population-based algorithms and local search algorithms. Local search is to improve the convergence of traditional population-based algorithms by exploiting the surrounding area of the evolved solutions in the search space. MAs can not only produce high quality solutions but also converge faster than other methods. However, as mentioned before the performance of a MA highly depends on its local search method of which the suitability is problem dependent. In addition the excessive invocation of local search may harm the exploration. The balance between exploration and exploitation, that is the balance between population-based search versus local search, should be carefully maintained.

Our proposed MA is to address these aforementioned issues. It introduces a set of local search algorithms, which are invoked according to the search process, and a diversity indicator to balance the exploration and exploitation. Figure 1 shows the overall flowchart of the proposed MA, which is based on steady-state genetic algorithms (GA). This choice is mainly due to the 5 min time limit imposed on this machine reassignment challenge. The process shown in the figure is actually similar to that of classic steady-state GA. The main different is the addition of the diversity control after mutation. The diversity condition of each new solution will be checked at this step. If the condition is satisfied then the solution will be sent to the variable local search component for improvement. Basically solutions are to be improved by a sequence of local search. A solution that can not be improved will be abandoned. Otherwise it will be added back to the population.

A main purpose of the variable local search is to address the question of *"which local search should be used on which solution?"*. Multiple local search strategies are to be applied depending on the individual situation. The detail of the process is explained in Sect. 3.3. Other main components of the proposed MA are also presented in details in the following subsections.

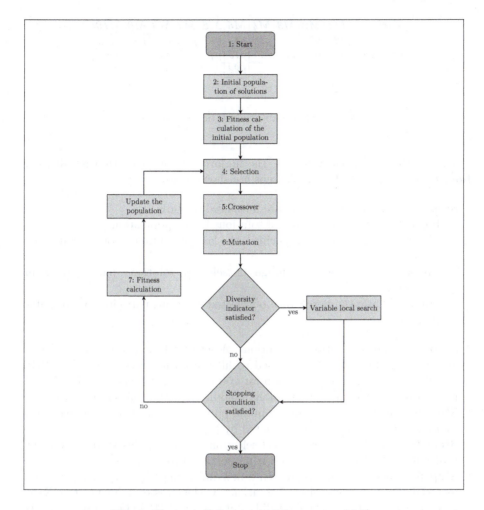

Fig. 1. The overall flowchart of the proposed MA

3.1 Genetic Algorithm

The proposed MA method is based on steady-state genetic algorithm which is a variation of GA, the well-known nature inspired population based meta-heuristic that mimics the process of natural selection [9].

A solution for one MRP instance is represented as a chromosome as shown in Fig. 2. The number of alleles is the number of available machines. Each allele encodes the processes that are to be executed by the corresponding machine. For example machine $M1$ on the figure will run processes $p1, p9, p3, p20$ in sequence. In this example there are 22 processes in total. A valid solution should contain all of these process while there is no duplicates. Furthermore the solution must satisfy the constraints that mentioned in Sect. 2.

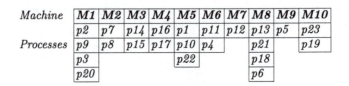

Machine	M1	M2	M3	M4	M5	M6	M7	M8	M9	M10
Processes	p2	p7	p14	p16	p1	p11	p12	p13	p5	p23
	p9	p8	p15	p17	p10	p4		p21		p19
	p3				p22			p18		
	p20							p6		

Fig. 2. Solution representation for MRPs

As for the evolutionary aspect, the majority of the algorithm is inherited from steady-state GA. The main steps are shown below:

* **Step 1:** Start. GA parameters are initialised including:
  - Population size: the number of solutions in the population;
  - Maximum generations: unspecified, the last generation before reaching the 5 min limit;
  - Crossover rate: the percentage of selected solutions participating in crossover;
  - Mutation rate: the percentage of selected solutions participating in mutation.

* **Step 2:** Initial population of solutions. A set of solutions are randomly created and the feasible ones are used to fill the initial population. For MRP, Google provides an initial solution for every problem instance [1]. The initial population are generated by randomly modifying this solution many times.
* **Step 3:** Fitness calculation of the initial population. The fitness value of each solution is its weighted cost calculated by Eq. (1).
* **Step 4:** Selection. Roulette wheel selection is used in this study where the probability of selecting a solution is proportional to its fitness.
* **Step 5:** Crossover. This is to create new individuals by exchanging genetic materials between two selected solutions. Order crossover operator (OX) is used here to avoid generating invalid solutions for MRP. In OX, two cut points are randomly selected. The alleles between the cut points on the first parent are copied to one child. These identical alleles are removed from the second parent which will contribute its remaining alleles to fill the other parts of the child. The same process can repeat by copying alleles between cut points from the second parent and other alleles from the first parent to generate another child. The individuals generated by OX has no missing alleles or duplications. They are the rearrangement of their parents. This is particular suitable for MRP.
* **Step 6:** Mutation. The mutation operator modifies one solution by randomly selecting two different alleles from the gene and then swap their positions, for example a solution of $M2$ running $\{p7, p8\}$ and $M7$ running $p12$ becomes a new solution that allocates $\{p7, p8\}$ to $M7$ and $p12$ to $M2$ after the mutation.
* **Step 7:** Fitness calculation. The fitness value of the generated solutions are again calculated using Eq. (1).

3.2   Diversity Indicator

An important aspect of MA is to avoid exploitation dominating exploration. To attain high quality solutions, balance should be maintained between the exploration from GA and the exploitation from local search. Excessive invocation of local search, for example running it at every generation, is costly and may increases the risk of the premature convergence, due to bias towards exploitation. Therefore local search should be applied only when it is necessary. In this study a diversity indicator is introduced to control when local search is needed by measuring the diversity of the current population. The measurement is from the work of Neri and Cotta [18]:

$$\text{Diversity} = 1 - \left| \frac{f_{\text{avg}} - f_{\text{best}}}{f_{\text{worst}} - f_{\text{best}}} \right| \tag{2}$$

where f_{avg} is the average fitness of the population of solutions, f_{best} is the fitness of the best solution and f_{worst} is the fitness of the worst solution in that population. If the *Diversity* is greater than a pre-defined value *threshold*, local search will be invoked to improve the solution generated after crossover and mutation.

3.3   Variable Local Search

Local search starts with a single solution and attempts to improve it by exploiting local solution space until the predefined termination criterion is reached [2,27]. In other words, to improve a given solution, local search will try to find the best neighbour. One major drawback of traditional local search is being trapped in the so-called local minima. In the literature there are different mechanisms to guide search away from non-improving areas and continue the search beyond the local optima. Examples of the well-known local search include simulated annealing [11], great deluge [7] and the late acceptance hill climbing [5].

It is not known in advance which local search would be the most suited to perform well on all problem instances for a memetic algorithm [12]. Therefore we propose a new approach in which multiple local search will be executed in sequence. Each one can utilise different rules to navigate different area in the search space. The integration of these local search can help avoid local optima and lead to high quality solutions for majority of instances.

Four local search algorithms are used in our variable local search mechanism. Each solution that is sent to this component for improvement will be assigned to a sequence of local search algorithms. This sequence consists of all four algorithms but in a random order. If the first local search resulted in an improved solution then the search would stop. The improved solution will be added into the population. The index of this local search is also recorded. If the stopping condition of the first search algorithm is met but no improvement can be made, then the second local search will be applied. This process repeats until an improved solution is found or the last local search in the sequence is terminated. Note the sequence of local search is circular. For example if a solution starts with local search No.3, then the sequence of execution is local search No. 4, No.1 and No.2, if no improvement can not found early.

In the case of no better solution found after trying all four local search mechanisms, the original solution will return to population with no further modification. The index of the last successful local search will remain unchanged. Then the local search process will start over again on the next solution.

The four local search algorithms to construct the sequence are:

– **Steepest Descent (SD)** [26]. Given an initial solution, S_0, generates a neighbourhood solution, S_1, by randomly selecting one process and then move it into a different machine. Replace S_1 with the S_0 if the quality of S_1 is better than S_0. If not, S_1 will be rejected and a new iteration will be started. The search process will be terminated after a fixed number of non-improving iterations. In our preliminary experiment that number is set as 10.

– **Simulated Annealing (SA)**. SA was introduced by [11]. It tries to escape local optima by accepting worse solutions. Given an initial solution, S_0, generates a neighbourhood solution, S_1, by randomly selecting one process and then move it into a different machine. Replace S_1 with the S_0 if the quality of S_1 is better than S_0 or satisfying the probability condition, $R < PA$, where R is uniform random number in [0,1] interval and PA is calculated as follows:

$$PA = \exp(-\delta/t) \tag{3}$$

where δ is the change in the fitness values of S_0 and S_1, t is the temperature which is set to 50 % of the fitness value of S_0. The t value controls the acceptance ration of worse solutions and its gradually decreases by α ($\alpha = 0.85$) during the search process. The search process will terminate when $t=0$.

– **Great Deluge (GD)**. GD was introduced by [7]. Similar to SA, GD also accepts worse solutions in order to get out of a local optima point but using different rules. Given an initial solution, S_0, generates a neighbourhood solution, S_1, by randomly selecting one process and then move it into a different machine. Replace S_1 with the S_0 if the quality of S_1 is better than S_0 or lower than the *level*. Initially, the value of *level* is set equal to the fitness value of the initial solution. At each iteration, the value of *level* is decreased by ε as follows:

$$level = level - \varepsilon \tag{4}$$

and

$$\varepsilon = (f(S_1) - f(best_{sol}))/NI \tag{5}$$

where NI is the number of iterations which is fixed to 1000 (the 1000 was determined based on a preliminary test). The search process will stop when the *level* value is lower than the best solution found so far.

- **Late Acceptance Hill Climbing.** This is an improved variation of hill climbing [5]. It also attempts to escape from local optima point by accepting worse solutions. Its main idea is to accept the generated solution if it is not worse than the quality of a saved solution which was the current one several steps before. Given an initial solution, S_0, generates a neighboured solution, S_1, by randomly selecting one process and then move it into a different machine. Replace S_1 with the S_0 if the quality of S_1 is better than S_0 or better than f_v where f_v is the quality of v^{it} solution saved in list L that contains qualities of current solutions during a number of recent iterations. v is calculated as follows:

$$v = I \mod L_{size} \tag{6}$$

where L_{size} is the size of the list L and I is the iteration counter. At each iteration, LA will insert the quality of the current solution into the beginning of L and removes the last one from the end. In this work, L_{size} size was set to 20 and the search will terminate after 10 consecutive none improving iterations.

4   Experimental Settings

This section briefly introduces the MRP instances from the ROADEF/EURO 2012 challenge and presents the parameter settings of our proposed MA.

4.1   Problem Instances

Two groups of instances from the ROADEF/EURO 2012 challenge are named as a and b. Group a has two subgroups $a1$ and $a2$. Each group contains 10 instances with diverse characteristics in terms of the number of machines, the number of processes, neighbourhood, and so on. Table 1 shows the main characteristics of these instances. In the table, R is the number of resources; TR is the number of resources that need transient usage; M is the number of machines; P is the number of processes; S is the number of services; L is the number of locations; N is the number of neighbourhoods; B is number of triples and SD is the number of service dependencies.

4.2   Parameters Settings

The proposed MA contains mainly six parameters, as shown in Table 2. The setting of these parameters was determined by our preliminary experiments. In particular, for each parameter, different values were tested and the one that leads to the best trade-off between the computational time and solution quality is selected [10].

Table 1. The characteristics of the problem instances

Instance	R	TR	M	P	S	L	N	B	SD
a1_1	2	0	4	100	79	4	1	1	0
a1_2	4	1	100	1000	980	4	2	0	40
a1_3	3	1	100	1000	216	25	5	0	342
a1_4	3	1	50	1000	142	50	50	1	297
a1_5	4	1	12	1000	981	4	2	1	32
a2_1	3	0	100	1000	1000	1	1	0	0
a2_2	12	4	100	1000	170	25	5	0	0
a2_3	12	4	100	1000	129	25	5	0	577
a2_4	12	0	50	1000	180	25	5	1	397
a2_5	12	0	50	1000	153	25	5	0	506
b_1	12	4	100	5000	2512	10	5	0	4412
b_2	12	0	100	5000	2462	10	5	1	3617
b_3	6	2	100	20000	15025	10	5	0	16560
b_4	6	0	500	20000	1732	50	5	1	40485
b_5	6	2	100	40000	35082	10	5	0	14515
b_6	6	0	200	40000	14680	50	5	1	42081
b_7	6	0	4000	40000	15050	50	5	1	43873
b_8	3	1	100	50000	45030	10	5	0	15145
b_9	3	0	1000	50000	4609	100	5	1	43437
b_10	3	0	5000	50000	4896	100	5	1	47260

Table 2. Parameter settings of MA

Parameter	Tested range	Suggested value
Population size	5–50	30
Crossover rate	0.1–0.9	0.7
Mutation rate	0.1–0.9	0.2
threshold value th	0.1–0.99	0.5
Maximum number of non-improving iterations for SD and LA	5–50	10
Number of iterations for GD	100–1500	1000

5   Results and Comparison

In this section, we first evaluate the effectiveness of the proposed MA by comparing it with basic GA and its counterpart that did not use the proposed variable local search enhancement. Secondly, the performance of the proposed MA is compared with the state of the art algorithms for MRP.

Table 3. The p-values of comparing MA with GA, MA1-5 on all instances

MA vs. Instance	GA p-value	MA1 p-value	MA2 p-value	MA3 p-value	MA4 p-value	MA5 p-value
a1_1	0	0.01	0.06	0.03	0.05	0.07
a1_2	0	0.02	0.02	0.01	0	0.06
a1_3	0	0	0.01	0.02	0.01	0.03
a1_4	0	0.04	0	0	0	0.04
a1_5	0	0.04	0.01	0.01	0.01	0.5
a2_1	0	0	0	0	0	0
a2_2	0	0	0.01	0	0	0
a2_3	0	0	0	0	0	0.03
a2_4	0	0.01	0.04	0	0	0
a2_5	0	0.02	0	0	0.03	0.02
b_1	0	0	0	0	0	0.01
b_2	0	0	0	0	0	0
b_3	0	0	0	0	0	0
b_4	0	0	0	0	0	0
b_5	0	0	0	0	0	0
b_6	0	0	0	0	0	0
b_7	0	0	0	0	0	0
b_8	0	0	0	0	0	0
b_9	0	0	0	0	0	0
b_10	0	0	0	0	0	0

5.1 Comparing the Proposed MA with GA and Other MAs

To evaluate the effectiveness of the proposed MA, it is compared with following algorithms which are similar to our method but with missing parts:

- **GA:** steady-state GA as described in Sect. 3.1.

- **MA1:** uses Steepest Descent only.

- **MA2:** uses Simulated Annealing only.

- **MA3:** uses Great Deluge only.

- **MA4:** uses Late Acceptance Hill Climbing only.

- **MA5:** no diversity indicator scheme described in Sect. 3.2.

The proposed MA and all the methods for comparison, including GA, MA1, MA2, MA3, MA4 and MA5, were tested with 31 independent runs on all instances from both group a and group b. All runs have the same computational resources, terminating within 5 min.

The final results of each method over these 31 runs are statistically compared using the Wilcoxon statistical test with a significance level of 0.05. The p-values of MA versus all of these methods are shown in Table 3. In the table, a p-value less than 0.05 means MA is statistically better than the compared algorithm. A value greater than 0.05 indicates the good performance of our proposed MA is not so significant. This table does not include the actual cost achieved by these methods. The costs obtained by our proposed MA can be found in the Sect. 5.2.

As can be seen from the table, MA is statistically better than GA with no local search on all 20 instances. MA is also significantly better than other memetic algorithms. It outperformed MA1 (steepest descent only) and MA3 (great deluge only) on all 20 instances. Comparing with MA2 (simulated annealing only) and MA4 (late acceptance hill climbing), there is only one instance ($a1\_1$) that our MA is not significantly better. Comparing with MA5, which is the most similar to the proposed MA just without the diversity indicator, the proposed MA achieved significant better results on 17 instances.

This positive result clearly justifies the benefits of the proposed variable local search enhancements. In particular, the proposed MA outperformed the basic GA. That demonstrates the benefit of local search on evolutionary search, the combination of exploitation with exploration. The comparison with MA5 shows the benefit of the diversity indicator method.

5.2   Comparing with the State of the Art Methods

In this section, the results obtained by the proposed MA are compared with those obtained by the state of the art algorithms. The six algorithms are:

1. **VNS:** Variable neighbourhood search [8].
2. **CLNS:** CP-based large neighbourhood search [15].
3. **LNS:** Large neighbourhood search [4].
4. **MILS:** Multi-start iterated local search [14].
5. **SA:** Simulated annealing [20].
6. **RILS:** Restricted iterated local search [13].

Table 4 presents the comparison results which are the cost of the best solution obtained by our proposed MA and that of the other six algorithms (VNS, CLNS, LNS, MILS, SA) on all 20 instances. Note that the computational resources of these seven methods are identical as all of them have to complete the search within 5 min. For MRPs the lower the cost the better the solution. The best solution for one instance is highlighted in bold. There might be multiple results shown in bold for the same instance as all of them reached the same best value. The results of VNS, CLNS, LNS, MILS, SA are reported by the authors. Among all these methods, RILS did not have results on group a instances.

Table 4. The results MA compared to the state of the art algorithms

Instance	MA	VNS	CLNS	LNS	MILS	SA	RILS
a1.1	44,306,501	44,306,501	44,306,501	44,306,575	44,306,501	44,306,935	–
a1.2	777,533,308	777,536,907	778,654,204	788,074,333	780,499,081	777,533,311	–
a1.3	583,005,810	583,005,818	583,005,829	583,006,204	583,006,015	583,009,439	–
a1.4	250,866,958	251,524,763	251,189,168	278,114,660	258,024,574	260,693,258	–
a1.5	727,578,310	727,578,310	727,578,311	727,578,362	727,578,412	727,578,311	–
a2.1	164	199	196	1,869,113	167	222	–
a2.2	720,671,537	720,671,548	803,092,387	858,367,123	970,536,821	877,905,951	–
a2.3	1,193,311,432	1,190,713,414	1,302,235,463	1,349,029,713	1,452,810,819	1,380,612,398	–
a2.4	1,680,596,746	1,680,615,425	1,683,530,845	1,689,370,535	1,695,897,404	1,680,587,608	–
a2.5	312,124,226	309,714,522	331,901,091	385,272,187	412,613,505	310,243,809	–
b.1	3,302,947,648	3,307,124,603	3,337,329,571	3,421,883,971	3,516,215,073	3,455,971,935	3,511,150,815
b.2	1,011,789,473	1,015,517,386	1,022,043,596	1,031,415,191	1,027,393,159	1,015,763,028	1,017,134,891
b.3	158,102,214	156,978,411	157,273,705	163,547,097	158,027,548	215,060,097	161,557,602
b.4	4,677,819,137	4,677,961,007	4,677,817,475	4,677,869,484	4,677,940,074	4,677,985,338	4,677,999,380
b.5	923,311,250	923,610,156	923,335,604	940,312,257	923,857,499	923,299,310	923,732,659
b.6	9,525,857,758	9,525,900,218	9,525,867,169	9,525,862,018	9,525,913,044	9,525,861,951	9,525,937,918
b.7	14,836,237,140	14,835,031,813	14,838,521,000	14,868,550,671	15,244,960,848	14,836,763,304	14,835,597,627
b.8	1,214,411,947	1,214,416,705	1,214,524,845	1,219,238,781	1,214,930,327	1,214,563,084	1,214,900,909
b.9	15,885,546,811	15,885,548,612	15,885,734,072	15,887,269,801	15,885,617,841	15,886,083,835	15,885,632,605
b.10	18,051,241,638	18,048,499,616	18,049,556,324	18,092,883,448	18,093,202,104	18,049,089,128	18,052,239,907

It can observed that the proposed MA is very competitive in comparison with these state-of-the-art methods. It achieved the lowest cost on 12 out of 20 instances. Among these 12 cases, 11 of them are the new best, meaning being higher than all other methods. On instance $a1\_1$, the proposed MA achieved the best result equivalent to that of VNS, CNS and MILS.

The second best is the VNS method which is the leader on 6 instances. Method SA championed 3 instances. Method CLNS was the best on 2 instances while method MILS achieve the best cost on instance $a1\_1$. Method RILS did not lead on any of the instances. Our proposed MA outperformed VNS on 14 instances, SA on 17 instances, CLNS on 18 instances, MILS on 19 instances, LNS and RILS on all tested instances. Overall, the comparison results clearly show that the proposed MA is an effective method for the MRP.

6   Conclusions

This study proposed a memetic algorithm for the load balancing problems. It combines the strengths of genetic algorithm and local search. The proposed algorithm integrates two important aspects to improve the performance of traditional memetic algorithms. Firstly, it uses multiple local search mechanisms to avoid getting stuck in a local optimum and to effectively deal with various types of search space. Four different local search algorithms were used in a sequential manner to improve the solutions evolved by genetic algorithm. Secondly, a diversity indicator was used to control the invocation of local search in order to avoid losing diversity caused by excessive exploitation. The performance of the proposed algorithm was evaluated using the Google machine reassignment benchmark instances that were used during the ROADEF/EURO 2012 challenge. The proposed MA method outperformed general GA, GA with single local search and memetic algorithm without diversity indicator. In comparison with the state-of-the-art algorithms, our proposed MA method is also very competitive achieving the best performance on most instances. We conclude that the variable local search based memetic algorithm is a good method for solving load balancing problem.

References

1. Roadef/euro challenge 2012: Machine reassignment. http://challenge.roadef.org/2012/en/
2. Emile Aarts, H.L., Lenstra, J.K.: Local Search in Combinatorial Optimization. Princeton University Press, Princeton (2003)
3. Armbrust, M., Fox, A., Griffith, R., Joseph, A.D., Katz, R., Konwinski, A., Lee, G., Patterson, D., Rabkin, A., Stoica, I., et al.: A view of cloud computing. Commun. ACM **53**(4), 50–58 (2010)
4. Brandt, F., Speck, J., Völker, M.: Constraint-based large neighborhood search for machine reassignment. Ann. Oper. Res., 1–29 (2012)
5. Burke, E.K., Bykov, Y.: A late acceptance strategy in hill-climbing for exam timetabling problems. In: PATAT 2008 Conference, Montreal, Canada (2008)

6. Calheiros, R.N., Ranjan, R., Beloglazov, A., De Rose, C.A.F., Buyya, R.: Cloudsim: a toolkit for modeling and simulation of cloud computing environments and evaluation of resource provisioning algorithms. Softw. Pract. Experience **41**(1), 23–50 (2011)
7. Dueck, G.: New optimization heuristics: the great deluge algorithm and the record-to-record travel. J. Comput. Phys. **104**(1), 86–92 (1993)
8. Gavranović, H., Buljubašić, M., Demirović, E.: Variable neighborhood search for google machine reassignment problem. Electron. Notes Discrete Math. **39**, 209–216 (2012)
9. Holland, J.H.: Adaptation in Natural and Artificial Systems. The University of Michigan Press, Ann Arbor (1975)
10. Kendall, G., Bai, R., Błazewicz, J., De Causmaecker, P., Gendreau, M., John, R., Li, J., McCollum, B., Pesch, E., Qu, R., et al.: Good laboratory practice for optimization research. J. Oper. Res. Soc. (2015)
11. Kirkpatrick, S., Daniel Gelatt, C., Vecchi, M.P., et al.: Optimization by simulated annealing. Science **220**(4598), 671–680 (1983)
12. Krasnogor, N., Smith, J.: A memetic algorithm with self-adaptive local search: tsp as a case study. In: GECCO, pp. 987–994 (2000)
13. Lopes, R., Morais, V.W.C., Noronha, T.F., Souza, V.A.A.: Heuristics and matheuristics for a real-life machine reassignment problem. Int. Trans. Oper. Res. **22**(1), 77–95 (2015)
14. Masson, R., Vidal, T., Michallet, J., Penna, P.H.V., Petrucci, V., Subramanian, A., Dubedout, H.: An iterated local search heuristic for multi-capacity bin packing and machine reassignment problems. Expert Syst. Appl. **40**(13), 5266–5275 (2013)
15. Mehta, D., O'Sullivan, B., Simonis, H.: Comparing solution methods for the machine reassignment problem. In: Milano, M. (ed.) CP 2012. LNCS, vol. 7514, pp. 782–797. Springer, Heidelberg (2012)
16. Moscato, P., et al.: On evolution, search, optimization, genetic algorithms and martial arts: towards memetic algorithms. Caltech concurrent computation program, C3P Report, 826:1989 (1989)
17. Neri, F., Cotta, C.: Memetic algorithms and memetic computing optimization: a literature review. Swarm Evol. Comput. **2**, 1–14 (2012)
18. Neri, F., Tirronen, V., Karkkainen, T., Rossi, T.: Fitness diversity based adaptation in multimeme algorithms: a comparative study. In: IEEE Congress on Evolutionary Computation, CEC 2007, pp. 2374–2381. IEEE (2007)
19. Nguyen, S., Zhang, M., Johnston, M., Tan, K.C.: Hybrid evolutionary computation methods for quay crane scheduling problems. Comput. Oper. Res. **40**(8), 2083–2093 (2013)
20. Ritt, M.R.P.: An Algorithmic Study of the Machine Reassignment Problem. Ph.D. thesis, Universidade Federal do Rio Grande do Sul (2012)
21. Sabar, N.R., Ayob, M.: Examination timetabling using scatter search hyper-heuristic. In: 2nd Conference on Data Mining and Optimization, DMO 2009, pp. 127–131. IEEE (2009)
22. Sabar, N.R., Ayob, M., Kendall, G., Qu, R.: A dynamic multiarmed bandit-gene expression programming hyper-heuristic for combinatorial optimization problems. IEEE Trans. Cybern. **45**(2), 217–228 (2015)
23. Sabar, N.R., Song, A.: Dual population genetic algorithm for the cardinality constrained portfolio selection problem. In: Dick, G., Browne, W.N., Whigham, P., Zhang, M., Bui, L.T., Ishibuchi, H., Jin, Y., Li, X., Shi, Y., Singh, P., Tan, K.C., Tang, K. (eds.) SEAL 2014. LNCS, vol. 8886, pp. 703–712. Springer, Heidelberg (2014)

24. Sabar, N.R., Zhang, X.J., Song, A.: A math-hyper-heuristic approach for large-scale vehicle routing problems with time windows. In: 2015 IEEE Congress on Evolutionary Computation (CEC), pp. 830–837. IEEE (2015)
25. Sabar, N.R., Ayob, M., Kendall, G., Qu, R.: Automatic design of a hyper-heuristic framework with gene expression programming for combinatorial optimization problems. IEEE Trans. Evol. Comput. **19**(3), 309–325 (2015)
26. Talbi, E.-G.: Metaheuristics: From Design to Implementation, vol. 74. John Wiley and Sons, Hoboken (2009)
27. Xie, J., Mei, Y., Song, A.: Evolving self-adaptive tabu search algorithm for storage location assignment problems. In: Proceedings of the Companion Publication of the 2015 on Genetic and Evolutionary Computation Conference, pp. 779–780. ACM (2015)

An (MI)LP-Based Primal Heuristic for 3-Architecture Connected Facility Location in Urban Access Network Design

Fabio D'Andreagiovanni[1,2]([✉]), Fabian Mett[1], and Jonad Pulaj[1]

[1] Department of Mathematical Optimization, Zuse Institute Berlin (ZIB),
Takustr. 7, 14195 Berlin, Germany
{d.andreagiovanni,mett,pulaj}@zib.de
[2] Institute for System Analysis and Computer Science,
National Research Council of Italy (IASI-CNR),
via Dei Taurini 19, 00185 Roma, Italy

Abstract. We investigate the 3-architecture Connected Facility Location Problem arising in the design of urban telecommunication access networks integrating wired and wireless technologies. We propose an original optimization model for the problem that includes additional variables and constraints to take into account wireless signal coverage represented through signal-to-interference ratios. Since the problem can prove very challenging even for modern state-of-the art optimization solvers, we propose to solve it by an original primal heuristic that combines a probabilistic fixing procedure, guided by peculiar Linear Programming relaxations, with an exact MIP heuristic, based on a very large neighborhood search. Computational experiments on a set of realistic instances show that our heuristic can find solutions associated with much lower optimality gaps than a state-of-the-art solver.

Keywords: Telecommunications · FTTX access networks · Connected facility location · Mixed integer linear programming · Tight linear relaxations · MIP heuristics

1 Introduction

In the last two decades, telecommunications have increasingly assumed a major role in our everyday life and the volume of traffic exchanged over wired and wireless networks has enormously increased. Major telecommunications companies forecast that such growth will powerfully continue, thus requiring the need for more technologically complex networks. In this context, *telecommunication access networks*, which connects users to service providers, have become a vital part of urban

The work of Fabio D'Andreagiovanni and Jonad Pulaj was partially supported by the *Einstein Center for Mathematics Berlin* (ECMath) through Project MI4 (ROUAN) and by the *German Federal Ministry of Education and Research* (BMBF) through Project VINO (Grant 05M13ZAC) and Project *ROBUKOM* (Grant 05M10ZAA) [1].

G. Squillero and P. Burelli (Eds.): EvoApplications 2016, Part I, LNCS 9597, pp. 283–298, 2016.
DOI: 10.1007/978-3-319-31204-0_19

metropolitan infrastructures. A critical component of such networks is represented by optical fiber connections, which provide higher capacity and better transmission rates than the old copper-based connections. In the last years, the trend in access networks has been to provide broadband internet access through different types of optical fiber deployments. These several deployments, usually called *architectures*, are denoted as a whole by the acronym *FTTX* (Fiber-To-The-X), where the *X* is specified on the basis of where the optical fiber granting access is terminated: major examples of architectures are *Fiber-To-The-Home (FTTH)*, bringing a fiber directly to the final user, and *Fiber-To-The-Cabinet (FFTC)* and *Fiber-To-The-Building* (FTTB), respectively bringing a fiber to a street cabinet or to the building of the user and then typically connecting the fiber termination point to the user through a copper-based connection. We refer the reader to [2] for an exhaustive introduction to FTTX network design and to [3] for a thorough discussion about the features of FTTH network design. Nowadays, an access network implementing a *full* FTTH architecture seems impractical, because of its extremely high deployment costs and since not all users are willing to pay higher fees for faster connections. As a consequence, in recent times higher attention has been given to deployments mixing *two architectures* like FTTH and FTTC/FTTB (e.g., [4]). An even more recent and promising trend has been represented by the integration of wired and wireless connections, providing service to users also through wireless links and leading to *3-architecture* networks that includes also the so-called *Fiber-To-The-Air* (FTTA) architecture [2,5]. This integration aims to get the best of both worlds: the high capacity offered by optical fiber networks and the mobility and ubiquity offered by wireless networks [5]. Moreover, it aims at getting a critical cost advantage, since the deployment of wireless transmitters is generally simpler and less expensive than that of optical fibers.

In this paper, we provide an original optimization model for the design of 3-architecture urban access networks integrating wireless and wired connections. A distinctive feature of our model w.r.t. state-of-the-art literature available on the topic (see [2,4] for an overview) is to include the mathematical expressions that model wireless signal coverage and that evaluates the relation between useful and interfering signals. The inclusion of such expressions is critical in any wireless network design problem considering wireless signal coverage, since the exclusion may lead to wrong design decisions (see [6,7] for a discussion). The resulting problem has proved to be very difficult to solve even for a state-of-the-art commercial MIP solver like IBM ILOG CPLEX [8].

In this work, our main original contributions are:

1. we propose the first optimization model for the problem of optimally designing a 3-architecture access network, explicitly modelling the *signal-to-interference formulas* that express wireless signal coverage. Specifically, we trace back the design problem to a 3-architecture variant of the *Connected Facility Location Problem* that includes additional variables and constraints for modelling the service coverage of the wireless architecture;

2. in order to strengthen the mathematical formulation of the problem, we propose to include two families of valid inequalities that model conflicts between

variables representing the activation of wireless transmitters and the assignment of users to the transmitters;

3. we develop a new primal heuristic for solving the problem. The heuristic is based on the combination of a probabilistic variable fixing procedure, guided by suitable *Linear Programming* (LP) relaxations of the problem, with an exact *Mixed Integer Programming* (MIP) heuristic, which provides for executing a *very large neighborhood search* formulated as a *Mixed Integer Linear Program* (MILP) and solved exactly by a state-of-the-art MIP solver;

4. we present computational experiments over a set of realistic network instances, showing that our new algorithm is able to produce solutions of much higher quality than those returned by a state-of-the-art MIP solver.

The remainder of this paper is organized as follows: in Sect. 2 we review the 2-architecture Connected Facility Location Problem; in Sect. 3, we introduce the new formulation for 3-architecture network design; in Sects. 4 and 5, we present our new metaheuristic and discuss computational results.

2   2-Architecture Connected Facility Location

We start our modeling considerations by taking into account a generalization of a *Connected Facility Location Problem* (ConFL) including two types of architectures. For an exhaustive introduction to foundations of network flow theory on graphs and to the ConFL, we refer the reader to [9] and [10]. The ConFL can be essentially described as the problem of (a) deciding the assignment of a set of served users to a set of open facilities and (b) how to connect open facilities through a Steiner tree, in order to minimize the total cost deriving from opening and connecting facilities and the assignment of facilities to users. The ConFL has been introduced and proven to be NP-Hard in [11]. A hop-constrained version of ConFL that is related to the design of single-architecture access network has been studied in [12].

The canonical ConFL considers a *single architecture* and can be associated to the design of an urban access network using a single *technology* (i.e., either optical fiber or copper connections). However, as we highlighted in the introduction, a new modern trend is to integrate two architectures and mix optical fiber and copper connection technologies. This leads to an extension of the ConFL that has been first considered and modeled in [4] and that we denote by 2-ConFL. We now proceed to define an optimization model for 2-ConFL that we use as basis for introducing our new model for a 3-architecture ConFL including wireless technology.

The 2-ConFL in access network design involves a set of potential facilities that can install one among two technologies and that provide a telecommunication service to a set of potential users. A served user must be assigned to exactly one open facility and each open facility must be connected to a central office. The aim is to guarantee a minimum coverage of users by each technology, while minimizing the cost of deployment of the network.

To formally define the 2-ConFL, it is useful to consider a modeling of the network as a directed graph $G(V, A)$ where:

1. the set of nodes V is the (disjoint) union of (i) a set of users U associated with a weight $w_u \geq 0$ representing the importance of each user $u \in U$, (ii) a set of facilities F with opening cost $c_f^t \geq 0, \forall f \in F$ that depends upon the technology $t \in T$ used by f, (iii) a set of central offices Γ, with opening cost $c_\gamma \geq 0, \forall \gamma \in \Gamma$, (iv) a set of Steiner nodes S. We call *core nodes* the subset of nodes $V^C = F \cup \Gamma \cup S$ that does not include the user nodes. Additionally, we denote by F_u^t the subset of facilities using technology t that may serve user u and by U_f^t the subset of users that may be served by facility f when using technology t.

2. the set of arcs A is the (disjoint) union of (i) a set of *core arcs* $A^C = \{(i, j) : i, j \in V^C\}$ that represent connections only between core nodes and are associated with a cost of realization $c_{ij} \geq 0$; (iii) a set of *assignment arcs* $A^{ASS} = \{(f, u) \in A : u \in U, f \in F_u\}$ representing connection of facilities to users and associated with a cost of realization c_{fu}^t that depends upon the used technology.

We call *core graph* the subgraph $G^C(V^C, A^C)$ of $G(V, A)$ representing the potential topology of the optical fiber deployment (*core network*) that has the *core nodes* as set of nodes and the *core arcs* as set of arcs. To take into account the opening cost of central offices, we use the common trick to add an artificial root node r to $G(V, A)$ that is connected to each central office $\gamma \in \Gamma$ by an arc (r, γ) associated with cost $c_{r\gamma}$ that is set equal to the cost c_γ of opening γ. This entails the inclusion in $G(V, A)$ of an additional set of (artificial) arcs $A^R = \{(r, \gamma) : \gamma \in \Gamma\}$. In what follows, we will use the notation $A^{R\text{-}C} = A^R \cup A^C$ to denote the union of the root and the core arcs. The total cost of deployment of the access network is obtained by summing the cost of opening central offices and facilities, the cost of connections established within the core graph and the cost of connecting open facilities to served users.

For each architecture, it is necessary to ensure a minimum weighted coverage of users. Given the total weight of users $W = \sum_{u \in U} w_u$, we express the coverage requirement for the architecture corresponding to technology $t \in T$ by introducing thresholds $W_t \in [0, W], t \in T$. We assume that $W_1 \leq W_2$, i.e. the coverage requirement of the more performing and costly technology $t = 1$ is not higher than that of the lower class technology $t = 2$. We base this on the realistic assumption that just a part of the users is willing to pay more for getting a higher quality of service.

On the basis of the previous formalization of the problem, we can finally introduce a *mixed integer linear program* to model the 2-ConFL. To this end, we introduce the following family of variables: (1) facility opening variables $z_f^t \in \{0, 1\} \; \forall f \in F, t \in T$ - the generic z_f^t is equal to 1 if facility f is open and uses technology t and is 0 otherwise; (2) arc installation variables $x_{ij} \in \{0, 1\}$ $\forall (i, j) \in A^{R\text{-}C}$ - the generic x_{ij} is equal to 1 if the root or core arc (i, j) is installed and is 0 otherwise; (3) assignment arc variables $y_{fu}^t \in \{0, 1\} \; \forall u \in U,$

$t \in T$, $f \in F_u^t$ - the generic y_{fu}^t is equal to 1 if facility f is connected to user u by technology t and is 0 otherwise; (4) user variables $v_u^t \in \{0,1\}$, $\forall u \in U$, $t \in T$ - the generic v_u^t is equal to 1 if user u is served by technology t and is 0 otherwise; (5) flow variables ϕ_{ij}^f, $\forall (i,j) \in A^{\text{R-C}}$, $f \in F$ representing the amount of flow sent on a root or core arc (i,j) for facility f. The Mixed Integer Linear Program for 2-ConFL (2-ConFL-MILP) is then:

$$\min \sum_{(i,j) \in A^{\text{R-C}}} c_{ij}\, x_{ij} + \sum_{f \in F} \sum_{t \in T} c_f^t\, z_f^t + \sum_{u \in U} \sum_{t \in T} \sum_{f \in F_u^t} c_{fu}^t\, y_{fu}^t \qquad \text{(2-ConFL-MILP)}$$

$$\sum_{t \in T} z_f^t \leq 1 \qquad\qquad\qquad\qquad f \in F \qquad\qquad (1)$$

$$\sum_{f \in F_u^t} y_{fu}^t = v_u^t \qquad\qquad\qquad u \in U, t \in T \qquad (2)$$

$$y_{fu}^t \leq z_f^t \qquad\qquad\qquad\qquad u \in U, f \in F, t \in T \quad (3)$$

$$\sum_{u \in U} \sum_{\tau=1}^{t} w_u\, v_u^t \geq W_t \qquad\qquad t \in T \qquad\qquad\quad (4)$$

$$\sum_{(j,i) \in A^{\text{R-C}}} \phi_{ji}^f - \sum_{(i,j) \in A^{\text{R-C}}} \phi_{ij}^f = \begin{cases} -\sum_{t \in T} z_f^t & \text{if } i = r \\ 0 & \text{if } i \neq r, f \\ +\sum_{t \in T} z_f^t & \text{if } i = f \end{cases} \quad i \in V^C \cup \{r\}, f \in F \quad (5)$$

$$0 \leq \phi_{ij}^f \leq x_{ij} \qquad\qquad\qquad\qquad (i,j) \in A^{\text{R-C}}, f \in F \quad (6)$$

$$v_u^t,\ z_f^t,\ x_{ij}^t,\ y_{fu}^t \in \{0,1\} \qquad\qquad (i,j) \in A, u \in U, f \in F, t \in T$$

The objective function aims at minimizing the total cost, expressed as the sum of the cost of activating root and core arcs (note that the corresponding summation includes the cost of activated central offices, opened facilities and of activated assignment arcs). The constraints (1) impose that each facility is activated on a single technology, whereas constraints (2) impose that if a user u is served by technology t, exactly one of the assignment arcs coming from a facility that can serve u is activated on technology t. The constraints (3) link the opening of a facility f on technology t to the activation of assignment arcs involving f and t. The constraints (4) impose the coverage requirement for each user (note that here the weighted sum of users getting the better technology $t = 1$ contributes to satisfying the requirement for the coverage of the worse technology).

The constraints (5) and (6) jointly model the fiber connectivity within the core network as a multicommodity flow problem that includes one commodity per facility. Specifically, (5) represents flow conservation in root and core nodes, while (6) are variable upper bound constraints that express the linking between the activation of a root or core arc and the activation of the arc.

We note that in contrast to the formulation proposed in [4], which models connectivity within the core network by cut-set inequalities and whose size is thus potentially exponential in the size of the problem input, we adopt a *compact formulation* based on multicommodity flows that is polynomial in the size of the problem input. The compact formulation is indeed more suitable for being used in our new heuristic, not requiring the execution of additional time consuming separation routines.

3   3-Architecture Connected Facility Location

We now proceed to introduce our new original generalization of the 2-ConFL problem, which additionally considers wireless *FTTA architecture* and explicitly embed the formulas expressing wireless coverage for a user.

As first step, we need to add an additional element to the set of available technologies, i.e. $T := T \cup \{3\}$ with index $t = 3$ denoting the wireless technology. We then assume that each facility $f \in F$ can also accommodate a *wireless transmitter*, which may provide service connection *without need of cables* to a subset of users. Transmitters are characterized by a number of radio-electrical parameters to set (e.g., the power emission, the frequency channel used to transmit, the modulation and coding scheme - see [13,14]). In principle, all these parameters can be set in an optimal way, by expressing their setting through a suitable mathematical optimization problem. However, just a (small) subset of parameters are typically optimized in a wireless network design problem [6,13]. A decision that is included in practically every design problem is the setting of power emissions. This is indeed a crucial decision that deeply influences the possibility of covering users with service [15].

In order to model the power emission of a wireless facility $f \in F$, we introduce a semi-continuous power variable $p_f \in [P_{\min}, P_{\max}] \, \forall \, f \in F$. A user u receives power from each wireless facility $f \in F$ and the power $P_f(u)$ that u receives from f is proportional to the power emitted by f by a factor $a_{fu} \in [0,1]$, i.e. $P_f(u) = a_{fu} \cdot p_f$. The factor a_{fu} is called *fading coefficient* and expresses the reduction in power that a signal propagating from f to u experiences [14]. We say that a user $u \in U$ is *covered* or *served* if it receives the wireless service signal within a minimum level of quality. The service is provided by one single transmitter, chosen as *server* of the user, while all the other transmitters interfere with the server and reduce the quality of service. The minimum quality condition can be expressed through the *Signal-to-Interference Ratio (SIR)*, a measure comparing the power received from the server with the sum of the power received by the interfering transmitters [14]:

$$\frac{a_{fu}\, p_f}{\eta + \sum_{k \in F \setminus \{f\}} a_{ku}\, p_k} \geq \delta . \tag{7}$$

The user is served if the SIR is above a threshold $\delta > 0$ that depends upon the wanted quality of service. We remark that in the denominator we must also include a constant $\eta > 0$ representing the noise of the system. By simple linear algebra operations, inequality (7) can be transformed in the so-called *SIR inequality*: $a_{fu}\, p_f - \delta \sum_{k \in F \setminus \{f\}} a_{ku}\, p_k \geq \delta\, \eta$. Since we do not know in advance which wireless facility $f \in F$ will be the server of user $u \in U$ (establishing the assignment facility-user is part of the decision process), given a user $u \in U$ we have one SIR inequality for each potential server $f \in F$, which must be activated or deactivated depending upon the assignment. In order to ensure that u is served through a wireless connection, at least one SIR inequality must be satisfied. We are thus actually facing a disjunction of constraints, which, according to a standard approach of Mixed Integer Programming (see [15,16]), can be represented by a variant of the SIR inequality that includes a sufficiently large positive constant M (the so-called *big-M coefficient*) and the assignment variable y_{fu}^3 representing the service connection of u through facility f by technology $t = 3$, namely:

$$a_{fu}p_f - \delta \sum_{k \in F \setminus \{f\}} a_{ku}p_k + M(1 - y_{fu}^3) \geq \delta N \qquad (8)$$

It is immediate to check that if $y_{fu}^3 = 1$, then u is wirelessly served by f and (8) reduces to a SIR inequality to be satisfied. If instead $y_{fu}^3 = 0$, then u is not wirelessly served by f and M activates, thus making (8) redundant and satisfied by any power vector $(p_1, p_2, \ldots, p_{|F|})$. The MILP for 3-ConFL is then:

$$\min \sum_{(i,j) \in A^{\text{R-C}}} c_{ij}\, x_{ij} + \sum_{f \in F} \sum_{t \in T} c_f^t\, z_f^t + \sum_{u \in U} \sum_{t \in T} \sum_{f \in F_u^t} c_{fu}^t\, y_{fu}^t \qquad \text{(3-ConFL-MILP)}$$

$$\sum_{t \in T} z_f^t \leq 1 \qquad\qquad f \in F$$

$$\sum_{f \in F_u^t} y_{fu}^t = v_u^t \qquad\qquad u \in U, t \in T$$

$$y_{fu}^t \leq z_f^t \qquad\qquad u \in U, f \in F, t \in T$$

$$\sum_{u \in U} \sum_{\tau = 1}^{t} w_u\, v_u^t \geq W_t \qquad\qquad t \in T$$

$$\sum_{(j,i) \in A^{\text{R-C}}} \phi_{ji}^f - \sum_{(i,j) \in A^{\text{R-C}}} \phi_{ij}^f = \begin{cases} -\sum_{t \in T} z_f^t & \text{if } i = r \\ 0 & \text{if } i \neq r, f \\ +\sum_{t \in T} z_f^t & \text{if } i = f \end{cases} \qquad i \in V^C \cup \{r\}, f \in F$$

$$0 \leq \phi_{ij}^f \leq x_{ij} \qquad\qquad (i,j) \in A^{\text{R-C}}, f \in F$$

$$a_{fu}p_f - \delta \sum_{k \in F \setminus \{f\}} a_{ku}p_k + M(1 - y_{fu}^3) \geq \delta N \qquad\qquad f \in F, u \in U \qquad (9)$$

$$0 \leq P^{\min} z_f^3 \leq p_f \leq P^{\max} z_f^3 \qquad\qquad f \in F$$

$$v_u^t,\; z_f^t,\; x_{ij}^t,\; y_{fu}^t \in \{0, 1\} \qquad\qquad (i,j) \in A, u \in U, f \in F, t \in T \qquad (10)$$

The major modifications w.r.t. the formulation (2-ConFL-MILP) concern: (1) the introduction of the variable bound constraints (10) that express the semi-continuous nature of variables p_f (when $z_f^3 = 0$, facility f does not install a wireless transmitter and the power p_f is thus forced to 0; when instead $z_f^3 = 1$, the transmitter is installed and its power must lie in $[P^{\min}, P^{\max}]$); (2) the introduction of the SIR constraints (9) for expressing the wireless coverage conditions.

Strengthening 3-ConFL-MILP. A key ingredient of the probabilistic fixing that we adopt in our new heuristic is represented by the combination of an a-priori and an a-posteriori measure of fixing attractiveness based on linear relaxations of 3-ConFL-MILP. In particular, we obtain the a-priori measure considering a *tighter formulation* (informally speaking, a problem with a "mathematically stronger" structure) defined by adding two class of valid inequalities to 3-ConFL-MILP: (1) superinterferer inequalities; (2) conflict inequalities. These two families of inequalities were respectively introduced in [6,17] and we refer the reader to these works for a detailed description. Here, we just provide a concise introduction to them.

The first class of inequalities captures the existence of so-called *superinterferers*: a superinterferer is an interfering transmitter that alone can deny service coverage to a user even when it emits at minimum power and the serving transmitter emits at maximum power. The corresponding valid inequalities are logical constraints of the form:

$$y_{fu}^3 \leq 1 - z_k^3 \quad \forall k \in K \setminus \{f\} : k \text{ is superinterfer for } u \text{ served by } f \qquad (11)$$

expressing that if k is a superinterferer facility and is activated, then the variable assigning user u to f installing wireless technology is forced to 0, since the corresponding SIR constraint cannot be satisfied (notice that the set of superinterferers depends upon the considered user and the user serving facility).

The second class of valid inequalities captures the existence of couples of SIR constraints that involve just two wireless facilities and that cannot be satisfied at the same time. More formally, consider the two SIR constraints corresponding to two users u_1, u_2 served by two distinct wireless facilities f_1, f_2, namely: (1) $a_{f_1 u_1} p_{f_1} - \delta a_{f_2 u_1} p_{f_2} \geq \delta N$; (2) $a_{f_1 u_1} p_{f_1} - \delta a_{f_2 u_1} p_{f_2} \geq \delta N$ (respectively representing u_1 served by f_1 and interfered by f_2 and u_2 served by f_2 and interfered by f_1). If there is no power vector (p_1, p_2) that satisfies the power bounds (10) and the two SIR constraints, then the following is a *valid inequality* for 3-ConFL-MILP stating that both SIR constraints cannot be activated simultaneously:

$$y_{f_1 u_1}^3 + y_{f_2 u_2}^3 \leq 1 \qquad (12)$$

Such valid inequalities can be easily identified in a pre-processing phase and can be added to the formulation to get remarkable strengthening (see [17]). In the next section, we denote by *Strong-3-ConFL-MILP*, the problem 3-ConFL-MILP suitably strengthened by inequalities (11) and (12).

4   A Primal Heuristic for the 3-ConFL-MILP

Being a mixed integer linear program, the problem 3-ConFL-MILP could in principle be solved by using a commercial MIP solver, such as IBM ILOG CPLEX [8]. However, the introduction of the wireless technology and of constraints (9) make 3-ConFL-MILP a very challenging extension of the 2-ConFL problem that result very difficult even for state-of-the-art solvers. According to our direct experience on realistic instances, in many cases CPLEX had big difficulties in finding good quality solutions even after several hours of computations. Such computational difficulties appear also in analogue optimization problems where flow models are combined with SIR constraints (e.g., [18]).

As an alternative to the direct use of a MIP commercial solver, we thus propose a new heuristic that combines a *probabilistic fixing procedure*, guided by the solution of peculiar linear relaxations of 3-ConFL-MILP, with an MIP heuristic, based on an *exact very large neighborhood search*. The probabilistic fixing is partially inspired by the algorithm ANTS (*Approximate Nondeterministic Tree Search*) [19] a refined version of an ant colony algorithm that tries to exploit information about bounds available for the optimization problem. More precisely, our new heuristic is based on considerations about the use of linear relaxations in place of generic bounds that have been first made in [20,21].

Since we exploit linear relaxations, in contrast to "simple" heuristics, we can provide a certificate of quality for the best solution produced by our heuristic. The certificate assumes the form of an *optimality gap*, measuring how far the best solution is from the best lower bound given by Strong-3-ConFL-MILP.

It is nowadays widely known that Ant Colony Optimization (ACO) is a meta-heuristic inspired by the behaviour of ants looking for food, initially proposed by Dorigo and colleagues for combinatorial optimization and then extended and refined in many works (e.g., [19,22–24] ACO is essentially centered on the execution of a cycle where a number of feasible solutions are iteratively built, using information about solution construction executed in previous runs of the cycle. An ACO algorithm (ACO-alg) presents the general structure of Algorithm 1.

Algorithm 1. General ACO Algorithm (ACO-alg)

1: **while** an arrest condition is *not* satisfied **do**
2:     ant-based solution construction
3:     pheromone trail update
4: **end while**
5: local search

In the step 2 of the while-cycle, a number of *ants* are defined and each ant builds a feasible solution in an iterative way. At every iteration, the ant is in a *state* that corresponds with a *partial solution* and can further complete the solution by making a *move*. The move corresponds to fixing the value of a not-yet-fixed variable and is chosen in a probabilistic way, evaluating a measure that combines an *a-priori* and an *a-posteriori* measure of fixing attractiveness. The a-priori attractiveness measure is called *pheromone trail value* in an ACO-alg context and is updated at the end of the construction phase, in the attempt of rewarding good fixing and penalizing bad fixing. Once that an arrest condition is met (typically, reaching a time limit), a local search is executed to improve the quality of the produced solutions and possibly identify a local optimum.

We stress that the algorithm that we propose is *not* an ACO-alg, but is rather an evolution and refinement of the ANTS algorithms that we strengthen by the use of peculiar linear relaxations. Specifically, in our case, the a-priori measure is provided by a strengthened linear relaxation of the problem - we use Strong-3-ConFL-MILP, namely problem 3-ConFL-MILP strengthened by inequalities (11) and (12) - whereas the a-posteriori measure is provided by the linear relaxation of 3-ConFL-MILP for partial fixing of the facility opening variables.

We now proceed to describe in detail our new primal heuristic.

Feasible Solution Construction. To explain how we build a feasible solution for the 3-ConFL-MILP, we first introduce the concept of *Facility Opening state*:

Definition 1. *Facility opening state (FOS): let $F \times T$ be the set of couples (f, t) that represent the activation of a facility f on a technology t. An FOS specifies an opening of a subset of facilities $\bar{F} \subseteq F$ on some technologies and excludes that the same facility is opened on more than one technology (i.e., $FOS \subseteq F \times T$: $\nexists (f_1, t_1), (f_2, t_2) \in FOS: f_1 = f_2 \text{ and } t_1 \neq t_2$).*

Given a FOS and a facility-technology couple $(f, t) \in FOS$, we denote by W_{ft}^{POT} the total weight of users that can be potentially served by f activated on technology t, i.e. $W_{ft}^{POT} = \sum_{u \in U_f^t} w_u$. We introduce this measure to distinguish between a partial and complete FOS for a technology $t \in T$. We say that a FOS is *partial for technology t* when the total weight of potential users that can be served by facilities appearing in the FOS using technology t does not reach the minimum coverage requirements W_t for t, i.e.:

$$\sum_{f \in F: (f,t) \in FOS} \sum_{u \in U_f^t} w_u < W_t . \tag{13}$$

On the contrary, we say that a FOS is *complete for technology t* when the total weight is not lower than W_t. Additionally, we call *fully complete* a FOS that is complete for all technologies $t \in T$. We introduce the concept of completeness and the formula (13) in order to guide and limit the probabilistic fixing of facility opening variables during the construction phase of feasible solutions.

Given a *partial* FOS for technology t, the probability p_{ft}^{FOS} of operating an additional fixing $(f, t) \notin FOS$, thus making a further step towards reaching a complete FOS, is set according to the formula:

$$p_{ft}^{FOS} = \frac{\alpha \, \tau_{ft} + (1 - \alpha) \, \eta_{ft}}{\sum_{(k,t) \notin FOS} \alpha \, \tau_{kt} + (1 - \alpha) \, \eta_{kt}} , \tag{14}$$

which provides for a convex combination of the a-priori attractiveness measure τ_{ft} and the a-posteriori attractiveness measure η_{ft} through factor $\alpha \in [0, 1]$. In our specific case, τ_{ft} is provided by the optimal value of the linear relaxation Strong-3-ConFL-MILP including the additional fixing $z_f^t = 1$, whereas η_{kt} is the value of the linear relaxation of 3-ConFL-MILP obtained for a specified partial fixing of the facility opening variables z. We remark that (14) is a revised formula that was proposed in [19] to improve the computationally inefficient canonical formula of ACO, which includes products and powers of measures and depends upon a higher number of parameters.

At the end of a solution construction phase, the a-priori measures τ are updated, evaluating how good the fixing resulted in the obtained solutions. We stress that for the update we do not rely on the canonical ACO formula including the pheromone evaporation parameter, whose setting may result very tricky, but we use a revised version of the improved formula proposed for ANTS in [19]. To define the new formula, we first introduce the concept of *optimality gap* (*OGap*) for a feasible solution of value v and a lower bound L that is available on the optimal value v^* of the problem (note that it holds $L \leq v^* \leq v$): the *OGap* provides a measure of the quality of the feasible solution, comparing its value to the lower bound and is formally defined as $OGap(v, L) = (v - L)/v$. The a-priori attractiveness measure that we use is:

$$\tau_{ft}(h) = \tau_{ft}(h-1) + \sum_{\sigma=1}^{\Sigma} \Delta\tau_{ft}^{\sigma} \text{ with } \Delta\tau_{ft}^{\sigma} = \tau_{ft}(0) \cdot \left(\frac{OGap(\bar{v}, L) - OGap(v_{\sigma}, L)}{OGap(\bar{v}, L)} \right)$$
(15)

where $\tau_{ft}(h)$ is the a-priori attractiveness of fixing (f, t) at fixing iteration h, L is a lower bound on the optimal value of the problem (we remember that as lower bound we use the optimal value of the strengthened formulation Strong-3-ConFL-MILP), v_{σ} is the value of the σ-th feasible solution built in the last construction cycle and $\bar{v}$ is the (moving) average of the values of the Σ solutions produced in the previous construction phase. $\Delta\tau_{ft}^{\sigma}$ represents the penalization/reward factor for a fixing and depends upon the initialization value $\tau_{ft}(0)$ of τ (in our case, based upon the linear relaxation of Strong-3-ConFL-MILP), combined with the relative variation in the optimality gap that v_{σ} implies w.r.t. $\bar{v}$. We note that the use of a relative gap difference in (15) allows us to reward or penalize fixing adopted in the last solution making a comparison with the average quality of the last Σ solutions constructed.

Once that a *fully complete FOS* is built, we have characterized an opening of facilities that can *potentially* satisfy the requirements on the weighted coverage for each technology. We use the term "potentially", since the activation of facilities specified by the FOS does not necessarily admit a feasible completion in terms of connectivity variables and assignment of users of facilities: it is indeed likely that not all the SIR constraints (9) corresponding to wireless facilities can be activated simultaneously because of interference effects. It is thus possible that a complete FOS will result infeasible. Since a risk of infeasibility is present, after the construction of a complete FOS, we execute a *check-and-repair phase*, in which the feasibility of the FOS is checked and, if not verified, we make an attempt to repair and make it feasible. The reparation attempt is based on the same MIP heuristic based on an exact very large neighborhood search that we adopt at the end of the construction phase to possibly improve a feasible solution (see the next subsection for details).

Given a FOS that is complete for all technologies, we check its feasibility and attempt at finding a feasible solution for the complete problem 3-ConFL-MILP by defining a restricted version of 3-ConFL-MILP, where we set $z_f^t = 1$ if

$(f, t) \in FOS$. We solve this restricted problem through the MIP solver with a time limit: if this problem is recognized as infeasible by the solver, we run the MIP heuristic for reparation. Otherwise, we run the solver to possibly find a solution that is better than the best incumbent solution.

MIP-VLNS - An Exact MIP Repair/Improvement Heuristic. To repair an infeasible partial fixing of the variables z induced by a complete FOS or to improve an incumbent feasible solution, we rely on an MIP heuristic that operates a very large neighborhood search *exactly*, by formulating the search as a mixed integer linear program solved through an MIP solver [24]. Specifically, given a (feasible or infeasible) and possibly not complete fixing $\bar{z}$ of variables, we define the neighborhood $\mathcal{N}$ including all the feasible solutions of 3-ConFL-MILP that can be obtained by modifying at most $n > 0$ components of $\bar{z}$ and leaving the remaining variables free to vary. This condition can be expressed in 3-ConFL-MILP by adding an *hamming distance constraint* imposing an upper limit n on the number of variables in z that change their value w.r.t. $\bar{z}$:

$$\sum_{(f,t) \in F \times T:\ \bar{z}_f^t = 0} z_f^t + \sum_{(f,t) \in F \times T:\ \bar{z}_f^t = 1} (1 - z_f^t) \leq n$$

The modified problem is then solved through an MIP solver like CPLEX, running with a time limit. Imposing a time limit is essential from a practical point of view: optimally solving the exact search can take a very high amount of time to close the optimality gap; additionally, a state-of-the-art MIP solver is usually able to quickly find solutions of good quality for large problems whose size has been conveniently reduced by fixing. In what follows, we denote the overall procedure for repair/improvement that we have discussed by MIP-VLNS.

The Complete Algorithm. The complete algorithm for solving the 3-ConFL-MILP is presented in Algorithm 2. We base the algorithm on the execution of two nested loops: the outer loop runs until reaching a global time limit and contains an inner loop inside which we define Σ feasible solutions, by first defining a complete FOS and then executing the MIP heuristic to repair or complete the fixing associated with the FOS. More in detail, the first algorithmic task is to solve the linear relaxation of Strong-3-ConFL-MILP for each fixing $z_f^t = 1$, getting the corresponding optimal value and using it to initialize the a-priori measure of attractiveness $\tau_{ft}(0)$. This is followed by the definition of a solution X^* that represents the best solution found during the execution of the algorithm. Each run of the inner loop provides for building a complete FOS by considering, in order, fiber, copper and wireless technology. The complete FOS is built according to the procedure using the probability measures (14) and update formulas (15) that we have discussed before. The complete FOS provides a (partial) fixing of the facility opening variables $\bar{z}$ and the MIP solver uses it as a basis for finding a complete feasible solution X^* to the problem. If $\bar{z}$ is recognized as an infeasible fixing by the MIP solver, then we run the MIP-VLNS in a reparation mode. If instead $\bar{z}$ is feasible and leads to find a feasible solution to 3-ConFL-MILP that is better than the best solution found X^B in the current run of the inner loop, then X^B is updated. Then the inner loop is iterated. After that the execution of the inner loop is concluded, the a-priori measures τ are updated according to formula (15), considering the quality of the produced solutions, and we check the necessity of updating the global best solution X^*. After having reached the global time limit, the heuristic MIP-VLNS is eventually run with the aim of improving the best solution found X^*.

Algorithm 2. - Heuristic for 3-ConFL-MILP

1: compute the linear relaxation of Strong-3-ConFL-MILP for all $z_f^t = 1$ and initialize the values $\tau_{ft}(0)$ with the corresponding optimal values
2: let X^* be the best feasible solution found
3: **while** a global time limit is not reached **do**
4:     let X^B be the best solution found in the inner loop
5:     **for** $\sigma := 1$ to Σ **do**
6:         build a complete FOS
7:         solve 3-ConFL-MILP imposing the fixing $\bar{z}$ specified by the FOS
8:         **if** 3-ConFL-MILP with fixing $\bar{z}$ is infeasible **then**
9:             run MIP-VLNS for repairing the fixing $\bar{z}$
10:        **end if**
11:        **if** a feasible solution $\bar{X}$ is found by the MIP solver and $c(\bar{X}) < c(X^B)$ **then**
12:            update the best solution found $X^B := \bar{X}$
13:        **end if**
14:    **end for**
15:    update τ according to (15)
16:    **if** $c(X^B) < c(X^*)$ **then**
17:        update the best solution found $X^* := X^B$
18:    **end if**
19: **end while**
20: run MIP-VLNS for improving X^*
21: return X^*

---

5   Computational Results

We tested the performance of our algorithm on 15 instances based on realistic network data defined within past consulting and industrial projects for a major telecommunication company. The experiments were performed on a 2.70 GHz Windows machine with 8 GB of RAM and using IBM ILOG CPLEX 12.5 as MIP solver. The code was written in C/C++ and is interfaced with CPLEX through Concert Technology. The experiments ran with a time limit of 3600 s. All the instances refers to a urban district in the metropolitan area of Rome (Italy) and considers different traffic generation and user location scenarios. The considered area has been discretized into a grid of about 450 pixels, following the *testpoint model* recommended by international telecommunications regulatory bodies for wireless signal evaluation (see [6,25]). We considered 30 potential

Table 1. Experimental results

ID	Gap-CPLEX %	Gap-Heu %	ΔGap%
I1	148.57	131.23	-11.67
I2	136.74	106.16	-22.36
I3	99.46	72.96	-26.64
I4	156.47	123.73	-20.92
I5	78.86	49.98	-36.62
I6	93.42	64.04	-31.44
I7	117.00	82.05	-29.48
I8	95.21	59.73	-37.26
I9	178.94	119.62	-33.15
I10	98.80	77.66	-21.39
I11	89.13	66.17	-25.76
I12	104.11	71.23	-31.58
I13	95.20	52.08	-45.29
I14	112.44	82.48	-26.64
I15	103.00	74.30	-27.86

facility locations that can accommodate any of the 3 technology considered in the study and can be connected to 5 potential central offices. On the basis of past experience and preliminary tests, we imposed the following setting of the parameters of the heuristic: $\alpha = 0.5$ (a-priori and a-posteriori attractiveness are balanced), $\Sigma = 5$ (number of solutions built in the inner loop before updating the a-priori measure and width of the moving average). Additionally, we imposed a time limit of 3000 s to the execution of the outer loop of Algorithm 2 and a limit of 600 s to the execution of the improvement heuristic MIP-VLNS.

The computational results are presented in Table 1: here, for each instance, we report its ID, the best percentage optimality gap *Gap-CPLEX%* reached by CPLEX within the time limit, the best percentage optimality gap reached by our heuristic within the time limit *Gap-Heu%*. In the case of the heuristic, we note that the gap is obtained combining the best feasible solution found by Algorithm 2 with the best known lower bound obtained by CPLEX using the strengthened formulation Strong-3-ConFL-MILP.

Concerning the results, the first critical observation to be made is that 3-ConFL-MILP results very challenging even for a modern state-of-the-art solver like CPLEX: the minimum gap obtained for the majority of instances results far beyond 90 %. We believe that such difficulty in solving the problem is particularly due to the presence of the SIR constraints (9) associated with wireless coverage: pure wireless coverage problem constitutes indeed already very challenging optimization problems, as discussed in [6]. In comparison to CPLEX, our heuristic is able to get always (much) better optimality gaps, that on average are

28 % lower than those produced by CPLEX and can reach even reductions over 35 % (we remind that decreasing the optimality gap is crucial to "move towards" identifying the optimal value of an optimization problem, see [16]). We believe that this is a very promising performance and that the heuristic deserves further investigations to be enhanced.

6 Conclusion and Future Work

We considered the design of 3-architecture urban access networks, combining the use of wired optical fiber- and copper-based connections with wireless connections. In literature, it has been suggested that this problem can be modeled as a simple generalization of 2-architecture Connected Facility Location Problems, by including an additional technology index. However, this is a simplistic generalization that neglects the interaction between signals emitted by distinct wireless transmitters and that may possibly lead to service coverage plans not implementable in practice. As a remedy, we have proposed a new optimization model that also includes the variables and constraints modeling the power emissions of wireless transmitters and the signal-to-interference formulas that are recommended for evaluating wireless service coverage. The resulting MILP proves very challenging even for a state-of-the-art commercial MIP solver like CPLEX, so we have proposed a new heuristic based on the combination of a probabilistic variable fixing procedure, guided by suitable linear relaxations of the problem, and an exact very large neighborhood search. Computational experiments on a set of realistic instances indicate that our heuristic can provide solutions associated with much lower optimality gaps than those returned by CPLEX. As future work, we plan to integrate heuristic with a branch-and-cut algorithm to improve the overall performance. Also, we will address data-uncertain versions of the problem, using Multiband Robust Optimization (e.g., [7,26]).

References

1. Bauschert, T., Büsing, C., D'Andreagiovanni, F., Koster, A.M.C.A., Kutschka, M., Steglich, U.: Network planning under demand uncertainty with robust optimization. IEEE Comm. Mag. **52**, 178–185 (2014)
2. Grötschel, M., Raack, C., Werner, A.: Towards optimizing the deployment of optical access networks. EURO J. Comp. Opt. **2**, 17–53 (2014)
3. Zotkiewicz, M., Mycek, M., Tomaszewski, A.: Profitable areas in large-scale FTTH network optimization. Telecommun. Syst. **61**(3), 591–608 (2015). doi:10.1007/s11235-015-0016-7
4. Leitner, M., Ljubic, I., Sinnl, M., Werner, A.: On the two-architecture connected facility location problem. ENDM **41**, 359–366 (2013)
5. Ghazisaidi, N., Maier, M., Assi, C.: Fiber-wireless (FiWi) access networks: a survey. IEEE Comm. Mag. **47**, 160–167 (2009)
6. D'Andreagiovanni, F.: Pure 0–1 programming approaches to wireless network design. 4OR-Q: J. Oper. Res. **10**, 211–212 (2012)

7. D'Andreagiovanni, F.: Revisiting wireless network jamming by SIR-based considerations and multiband robust optimization. Optim. Lett. **9**, 1495–1510 (2015)

8. IBM ILOG CPLEX. http://www-01.ibm.com/software/integration/optimization/cplex-optimizer

9. Ahuja, R., Magnanti, T., Orlin, J.: Network Flows: Theory, Algorithms, and Applications. Prentice Hall, Upper Saddle River (1993)

10. Gollowitzer, S., Ljubic, I.: MIP models for connected facility location: A theoretical and computational study. Comput. & OR **38**, 435–449 (2011)

11. Gupta, A., Kleinberg, J., Kumar, A., Rastogi, R., Yener, B.: Provisioning a virtual private network: a network design problem for multicommodity flow. In Harrison, M., et al. (eds.) Proceedings of the STOC 2001, 389–398. ACM, New York (2001)

12. Ljubic, I., Gollowitzer, S.: Layered graph approaches to the hop constrained connected facility location problem. INFORMS J. Comp. **25**, 256–270 (2013)

13. D'Andreagiovanni, F.: On improving the capacity of solving large-scale wireless network design problems by genetic algorithms. In: Di Chio, C., et al. (eds.) EvoApplications 2011, Part II. LNCS, vol. 6625, pp. 11–20. Springer, Heidelberg (2011)

14. Rappaport, T.: Wirel. Commun.: Princ. Pract. Prentice Hall, Upper Saddle River (2001)

15. D'Andreagiovanni, F., Mannino, C., Sassano, A.: GUB covers and power-indexed formulations for wireless network design. Manag. Sci. **59**, 142–156 (2013)

16. Nehmauser, G., Wolsey, L.: Integer and Combinatorial Optimization. John Wiley & Sons, Hoboken (1988)

17. D'Andreagiovanni, F., Mannino, C., Sassano, A.: Negative cycle separation in wireless network design. In: Pahl, J., Reiners, T., Voß, S. (eds.) INOC 2011. LNCS, vol. 6701, pp. 51–56. Springer, Heidelberg (2011)

18. Dely, P., D'Andreagiovanni, F., Kassler, A.: Fair optimization of mesh-connected WLAN hotspots. Wirel. Commun. Mob. Comput. **15**, 924–946 (2015)

19. Maniezzo, V.: Exact and approximate nondeterministic tree-search procedures for the quadratic assignment problem. INFORMS J. Comp. **11**, 358–369 (1999)

20. D'Andreagiovanni, F., Krolikowski, J., Pulaj, J.: A fast hybrid primal heuristic for multiband robust capacitated network design with multiple time periods. App. Soft Comp. **26**, 497–507 (2015)

21. D'Andreagiovanni, F., Nardin, A.: Towards the fast and robust optimal design of wireless body area networks. App. Soft Comp. **37**, 971–982 (2015)

22. Dorigo, M., Maniezzo, V., Colorni, A.: Ant system: Optimization by a colony of cooperating agents. IEEE Trans. Syst. Man Cybern. B **26**, 29–41 (1996)

23. Gambardella, L.M., Montemanni, R., Weyland, D.: Coupling ant colony systems with strong local searches. Europ. J. Oper. Res. **220**, 831–843 (2012)

24. Blum, C., Puchinger, J., Raidl, G., Roli, A.: Hybrid metaheuristics in combinatorial optimization: a survey. Appl. Soft Comp. **11**, 4135–4151 (2011)

25. Italian Authority for Telecommunications (AGCOM): Specifications for a dvb-t planning software tool. http://www.agcom.it/Default.aspx?message=downloaddocument&DocID=3365 Accessed: 15 October 2015

26. Büsing, C., D'Andreagiovanni, F.: New results about multi-band uncertainty in robust optimization. In: Klasing, R. (ed.) SEA 2012. LNCS, vol. 7276, pp. 63–74. Springer, Heidelberg (2012)

Reducing Efficiency of Connectivity-Splitting Attack on Newscast via Limited Gossip

Jakub Muszyński, Sébastien Varrette$^{(\boxtimes)}$, and Pascal Bouvry

Computer Science and Communications (CSC) Research Unit,
University of Luxembourg, 6, rue Richard Coudenhove-Kalergi,
1359 Luxembourg, Luxembourg
{jakub.muszynski,sebastien.varrette,pascal.bouvry}@uni.lu

Abstract. Newscast is a Peer-to-Peer, nature-inspired gossip-based data exchange protocol used for information dissemination and membership management in large-scale, agent-based distributed systems. The model follows a probabilistic scheme able to keep a self-organised, small-world equilibrium featuring a complex, spatially structured and dynamically changing environment. Newscast gained popularity since the early 2000s thanks to its inherent resilience to node volatility as the protocol exhibits strong self-healing properties. However, the original design proved to be surprisingly fragile in a byzantine environment subjected to cheating faults. Indeed, a set of recent studies emphasized the hardwired vulnerabilities of the protocol, leading to an efficient implementation of a malicious client, where a few naive cheaters are able to break the network connectivity in a very short time. Extending these previous works, we propose in this paper a modification of the seminal protocol with embedded counter-measures, improving the resilience of the scheme against malicious acts without significantly affecting the original Newscast's properties nor its inherent performance. Concrete experiments were performed to support these claims, using a framework implementing all the solutions discussed in this work.

Keywords: Newscast · Gossip protocol · Peer-to-Peer · Byzantine faults · Cheating faults · Fault tolerance · Security

1 Introduction

Peer-to-Peer (P2P) overlay networks were created to address scalability problems in large, distributed applications. Such systems consist of equally privileged, equipotent participants called peers. The main characteristics of classic P2P architecture are heterogeneity, volatility of resources, and a symmetric distribution of data and control across the nodes. Thus, a well-designed platform can easily scale to millions of processes where each of which can join or leave whenever it pleases without seriously disrupting the Quality of Service (QoS). A large part of research on P2P systems focuses on routing protocols and self-management to handle group membership and communications in an automated

© Springer International Publishing Switzerland 2016
G. Squillero and P. Burelli (Eds.): EvoApplications 2016, Part I, LNCS 9597, pp. 299–314, 2016.
DOI: 10.1007/978-3-319-31204-0_20

and distributed way. The existing approaches may be classified according to the topology of the peers into structured and unstructured. In the first case, the participants are linked in a tightly controlled and organised way. Each joining peer has an assigned identifier mapped into a specific position in the overlay network, supporting efficient routing of messages, for instance: Chord [18], Tapestry [21] or Pastry [16]. In contrast to the previous model, unstructured P2P networks exploit randomness to organise the participants and to share information. Such solutions are mainly nature-inspired and fall into the category of gossip (or epidemic) protocols. They rely on random and self-organised topologies emerging from a viral dissemination of messages. This comes with a bigger overhead in terms of routing performances compared to the structured P2P overlays. However, this disadvantage is counterbalanced by a high resilience to the volatility of resources. A well-known example implementing the above paradigm is Newscast [7] which displays a scalable and robust behaviour. Due to the simplicity of the protocol, it has been rapidly adopted in academic research: as a platform for distributed optimisation [10], as a framework for peer-sampling services [5] or as a way for monitoring the status of large-scale decentralised systems [6].

The robustness of Newscast against the volatility of resources (modelled as *crash faults*) has been demonstrated successfully in many previous works [7,20]. This paper extends the study of the protocol to executions in byzantine environments and more precisely with malicious users (*cheaters* and *cheating faults* caused by them) present in the network. The initial data flow analysis conducted in [13] revealed a high sensitivity of Newscast to evil-intentioned tampering with its internal mechanisms. Subsequently, the findings were exploited to develop an efficient attack on the protocol able to break the connectivity between the peers in a relatively short time [14]. Here, we propose and analyse an extension of Newscast with an embedded counter-measure based on a limited gossip. The approach notably reduces the impact of the malicious clients (following the scheme proposed in [14]) without significantly affecting the original small-world properties of the protocol.

This paper is organized as follows: Sect. 2 details the background of this work and introduces the underlying notions. The proposed counter-measure of the seminal protocol is explained in Sect. 3. Implementation details of the proposed framework are provided in Sect. 4. The validation of the approach on concrete applications is expounded in Sect. 5, where the experimental results are reviewed. Section 6 reviews the related works. Finally, Sect. 7 concludes the paper and provides some future directions and perspectives opened by this study.

2   Context and Motivations

Before proceeding to a detailed analysis of the topic proposed in this paper, we introduce here key concepts and motivation of this work.

Algorithm 1. Newscast in pseudo-code.

```
Active thread
    while true do
        wait t_r;
        node_j ← select a node from cache_i;
        send cache_i and data_i to node_j;
        receive cache_j and data_j from node_j;
        cache_j ← Update(node_i, node_j, cache_j, data_j);
        cache_i ← Aggregate(cache_i, cache_j);

Passive thread
    while true do
        wait until cache_k and data_k is received from node_k;
        send cache_i and data_i to node_k;
        cache_k ← Update(node_i, node_k, cache_k, data_k);
        cache_i ← Aggregate(cache_i, cache_k);

function Update(node_a, node_b, cache_b, data_b)
    newEntry ← create a new entry using the address of node_b and data_b;
    return cache_b \ {entry reffering to node_a} ∪ {newEntry};

function Aggregate(cache_a, cache_b)
    return a cache with the c freshest items from cache_a ∪ cache_b;
```

2.1 Newscast

Newscast is a gossip protocol proposed by Jelasity and van Steen in [7] for interconnecting large-scale distributed systems. Without any central services or servers, the protocol differs from other similar approaches by its simplicity. Membership management follows a simple protocol. In order to join the system, a node needs to contact an already connected node from which it gets a list of neighbours. Whereas to leave, it requires to stop communicating for a predefined time. The dynamics of the system follow a probabilistic scheme able to keep a self-organized equilibrium. Such an equilibrium emerges from the loosely-coupled and decentralized run of the protocol within the different and independent nodes. The emerging graph behaves as a small-world topology allowing a scalable information dissemination. Despite the simplicity of the scheme, Newscast is particularly fault-tolerant against crash failures and exhibits a graceful degradation without requiring any extra mechanism other than its own emergent behaviour [20]. Each Newscast node keeps its own set of neighbours in a cache (*i.e.* a view) containing $c \in \mathbb{N}$ entries, referring to c other nodes in the network without duplicates. Each entry provides a reference to a node, a time-stamp of its creation (allowing replacement of old items) and optionally application-specific data.

Algorithm 1 shows the pseudo-code of the protocol. There are two different tasks carried out within each node. The active thread (**Active**) which proactively initiates a cache exchange once every cycle (one cycle takes t_r time units) and the passive thread (**Passive**) that waits for data-exchange requests. Every cycle, each $node_i$ initiates a cache exchange. It selects randomly a neighbour $node_j$ from its $cache_i$ with uniform probability. Then, the nodes $node_i$ and $node_j$ exchange their caches and data (respectively $data_i$ and $data_j$). At $node_i$, the entry from $cache_j$ referring to $node_i$ is substituted with a new item containing the address of $node_j$ and its data $data_j$ (**Update**). Conversely, the same operation is performed at $node_j$. Eventually, the caches are merged at both nodes following the aggregation function (**Aggregate**). It consists of picking the freshest c items from both caches to form a single cache. Since this function is applied at both nodes (the one initiating the request and the one serving the request), the result is that $node_i$ and $node_j$ will have some entries in common in their respective caches.

2.2   Faults, Fault Tolerance and Robustness

Fault tolerance can be defined as an ability of a system to behave in a well-defined manner once a failure occurs. A failure is due to an error of the system which is a consequence of a fault. Different kinds of faults are usually distinguished in function of their origin and their temporal duration [1]. They could be introduced intentionally or not, caused by software or hardware. When a given system is resilient to a given type of fault, one generally claims that this system is *robust*.

The most relevant fault models in distributed computing are: crash, omission, duplication, timing, and byzantine. A crash failure occurs when a computing node is unresponsive temporarily or permanently, with a possible reset of its state (partial or complete) to the initial conditions. Omission and duplication failures are usually linked with problems in communication. The first type occurs when a message is not sent or not received, the second — when a message is sent or received more than once. Timing failures are associated with not met constraints on the time required for execution or data delivery. The last model — byzantine (or arbitrary) failures — covers unexpected and inconsistent responses of a system at arbitrary times. They are the hardest to detect, as the behavior of the system is often similar to the expected one.

Malicious Users in Distributed Systems. In modern distributed systems, "crackers" or "black hat" hackers "*violate computer security for little reason beyond maliciousness, personal gain or satisfaction*" [12]. Generally, they can simulate or cause any failure described in the previous section, most of which are easily detectable with their effect significantly limited in the scope. However, byzantine failures are the most flexible and hardest to detect. As formalised in [19], an intentional introduction of byzantine failures into the system may be done by altering (*i.e.* corrupting) results or messages being sent. Such approach promotes further propagation of the problem to the other parts of the system.

As the altered data is not true (in the meaning of not genuine or not authentic), it may be defined as *cheated*, the process of its production and transfer as *cheating*, and the user sending it as a *cheater*.

Finally, cheaters may organise themselves into *colluding* groups with the aim to feed incorrect results [3]. It might be voluntary or resulting from a virus or a bug in the code [3].

Fault Tolerance of Newscast. One of the important issues regarding P2P computing is the robustness of the underlying protocols, as they need to provide a coherent view of a large-scale — and potentially — unreliable distributed system. In that sense, Newscast establishes the dynamics of a changing communication graph which requires to persistently maintain a small-world connectivity. As a first assessment of its robustness, Jelasity and van Steen [7] showed that the protocol maintains such property regardless of the scale or the initial state of the system (*i.e.* an execution of Newscast is able to bootstrap from any graph structure and consistently converges to a small-world graph). Nevertheless, two additional issues still need to be considered for characterizing a protocol as robust. The first is related to the spontaneous partitioning of the communication graph and the second to the resilience of the protocol to failures.

The spontaneous partitioning refers to the probability of a subgraph to become disconnected from the system as a consequence of the protocol dynamics. In Newscast, the probability of such event is mainly influenced by the cache size c. Jelasity and van Steen [7] conclude that the probability of the spontaneous partitioning is almost negligible for $c \geq 20$ regardless of the size of the network. However, the previous results do not take into account node failures, which is an inherent feature of large-scale distributed systems. In that sense, Voulgaris et al. [20] analyse the robustness of Newscast in a failure-prone scenario in which the nodes are removed until none is left. The study leads to the following conclusions: (1) the communication graph remains connected despite failures until a large percentage of nodes are removed (*e.g.* for $c = 40$, almost 90 % of the nodes have to be removed to split the graph); (2) when a partition finally happens, most of the nodes still remain connected in a large cluster. These findings allow stating that Newscast is robust against crash faults: the protocol consistently maintains the desired connectivity, even if a high percentage of nodes crash. However, the large-scale distributed systems are also subject to other potential sources of risks. Specifically, malicious users may pose a threat to the system, finding its vulnerabilities and exploiting them to disrupt the execution. This type of failures (*i.e. cheating faults*) also need to be considered for defining a protocol as fully robust. Therefore, assessing and eventually enabling security in Newscast is a necessary step to promote the uses of the protocol beyond the academic environment.

Muszyński et al. [13] highlighted possible vulnerabilities in Newscast through a data flow analysis in fault-free and byzantine environments. In the first scenario, the study demonstrated that most of the cache entries exchanged during the protocol execution are hardly ever used to establish a connection (resulting in a verification of the contained address). Additionally, new (fresh) cache entries spread freely to almost half of the cache size neighbours of a given node.

The analysis in a byzantine environment covered the worst case scenarios with the corrupted entries being the freshest possible, containing random addresses (most possibly pointing to non-existing nodes). They conclude that a one-time corruption of a single cache does not pose a serious threat to the protocol execution. In most of the tests the network recovered to a fully functional state (sifting incorrect data, *i.e.* self-healing) without any nodes losing the connectivity. However, a constant corruption of the cache always resulted in a disconnection of the affected node and in most of the cases along with at least one other client.

In [14], the authors continued the study using previously gathered data to develop a non-cooperating malicious client for Newscast. As they pointed out, blocking a given node from making any valid outgoing connection is relatively straightforward: an evil-intentioned user has to send a cache consisting of the freshest possible entries, containing random addresses (as previously). However, the self-healing property of the protocol makes the effect temporary and to overcome this, the cheater has to send the (refreshed) corrupted cache to the victim node more than once. Therefore, to disconnect the whole network, the key was the proper choice of the targeted nodes. In [14], it is thus proposed to target peers appearing frequently during the cache-exchanges. Such approach allows decreasing the density of valid links within the network and maximises the spread of the corrupted entries. The proposed scheme proved to be effective and scalable. Prioritizing the targets by their frequency of appearance during the cache-exchanges is the only mechanism required for the effectiveness of the solution. The scalability was achieved by ignoring the cache entries from incoming connections. With such settings, non-cooperating cheaters do not disturb each other much and do not get stuck on cheating one another.

The studies mentioned in the previous section demonstrated that the uncoordinated, independent cheaters are a serious threat for Newscast. Even a few malicious clients can break the connectivity in the network really fast. Moreover, the model is hard to detect at the level of each node as it does not lead to the formation of any graph anomalies easily detectable in the neighbourhood. The only symptom is a decreasing global number of (working) links between the nodes, which is similar to churn (volatility of resources). In case when the random addresses can not be used, they might be substituted with the ones discovered during the execution. The nodes with the lowest frequency of appearance are most likely inactive.

3   Countering Non-cooperating Cheaters in Newscast

Our main contribution in this article is to propose an extension for the seminal Newscast protocol that would defeat the effectiveness of the identified attack recalled in the previous section. In order to prevent malicious clients *i.e.* cheaters, to break the connectivity in the network according to the attack detailed previously, a step-wise approach was performed to define the set of effective countermeasures. In the end, it appeared that limiting the gossip *e.g* the number of cache entries exchanged at each cycle was the most effective (and sufficient) solution to sustain cheating faults.

The approach to limit the gossip is motivated by the fact that plain tampering with the time-stamps of the cache entries allows them to spread rapidly in the network. As cheating clients actively initiate the outgoing connections, they can choose a specific node and overwrite its whole cache (all $c \in \mathbb{N}$ entries). The same happens when any of the cheaters responds to the incoming request. However, during the connection between non-malicious peers, the number of updated items is significantly lower (approximately $c/2$) [13]. Therefore, we propose to limit the number of exchanged entries between the Newscast clients. Such change immediately prevents overwriting the full content of the cache in a single connection. Establishing the overall efficiency of the solution and its influence on the original properties of the protocol requires some formalisation that is now recalled from [14] with minor improvements.

First of all, the overlay network formed during the execution of Newscast can be modelled at a given time t as a directed (connection) graph $G_t = (V_t, A_t)$, consisting of the set of vertices V_t and the set of arcs A_t between them. V_t corresponds to the set of nodes in the network at the time t. An arc $a = \langle \overrightarrow{v_i, v_j} \rangle \in A_t$ connecting the vertex (or the node) $v_i \in V_t$ with the node $v_j \in V_t$, reflects the fact that v_j is in the cache (i.e. the view) of v_i. The cache size of each node remains constant within the graph G_t, i.e. $|cache_i| = c \ \forall_{v_i \in V_t \wedge t \geq 0}$. Execution of the protocol leads to a series of graphs G_t, given an initial graph G_0. For the sake of simplicity, we will assume that the number of non-malicious nodes in the network is constant and equal to $n_{\text{honest}} \ \forall_{t \geq 0}$. In addition, there are $n_{\text{cheating}} \ \forall_{t \geq t_c}$ malicious nodes i.e. cheaters, joining the network at the time t_c (all at the same simulation step). This leads to a partition of the set of vertices in V_t between non-malicious (i.e. honest) nodes and cheaters. Thus, $V_t = V_t^{\text{honest}} \cup V_t^{\text{cheating}}$, where $|V_t^{\text{honest}}| = n_{\text{honest}}$ and $|V_t^{\text{cheating}}| = n_{\text{cheating}}$.

An important concept to measure the robustness of the protocol against cheaters is the size of the connected components of non-malicious nodes in G_t. Let G_t^{honest} be a subgraph of G_t induced by the vertex set V_t^{honest}. A j-th (weakly) connected component $\hat{\mathcal{C}}_j$ in G_t^{honest} is a maximal subgraph of G_t^{honest} such that every node in $\hat{\mathcal{C}}_j$ is reachable from every other node in $\hat{\mathcal{C}}_j$. In the sequel, $\hat{\mathcal{C}}_t^{\text{max}}$ will denote the connected component of maximum size (i.e. the giant component) at a given time t. Obviously $0 < |\hat{\mathcal{C}}_t^{\text{max}}| \leq n_{\text{honest}}$. Now we can formalize the impact of cheating failures on the connection graph.

Definition 1 (Impact of cheating failures). *Let $\hat{\mathcal{C}}_t^{max}$ be the (weakly) connected component of maximum size in the subgraph G_t^{honest} of G_t induced by the set of honest nodes V_t^{honest}, at a given time t. We say that the protocol was subject to cheating iff there exist a sequence of points in time (of the size greater than one) such that $|\hat{\mathcal{C}}_t^{max}|$ and the number of edges in G_t^{honest} are strictly decreasing.*

Definition 1 excludes a possibility of a *spontaneous partitioning* of the connection graph. It may happen during a fault-free execution of a given protocol due to its probabilistic nature and the limited knowledge about the network at each node. However, in such case, the number of edges in G_t^{honest} would stay constant.

The above formalisation allows to precisely measure the efficiency of malicious clients and proposed solutions countering their activities. Furthermore, normalisation of the $|\hat{\mathcal{C}}_t^{max}|$ yields the probability that a given node $v \in V_t^{honest}$ belongs to the $\hat{\mathcal{C}}_t^{max}$, aiding comparison of obtained results for various parameter settings.

4   Implementation and Experimental Setup

In order to perform the analysis presented in this paper, we implemented in Java a simulator for the protocol execution, including all cheating models presented in the following sections. The resulting framework contains a set of monitoring sensors able to track the complete state of the network — from the nodes to the individual cache entries. This allows gathering various network statistics (e.g. connectivity, clusters, etc.). Thanks to the GRAPHSTREAM [4] Java library, the tool is able to graphically display the dynamics of the system (e.g. active connections, link updates, etc.). During the execution, the whole simulation is divided into steps (called *simulation/synchronization steps*). In each of them, every node is selected once with a uniform probability to initiate a cache-exchange according to the protocol specification (see Sect. 2.1). That is: establishing an outgoing connection, sending own cache, receiving a cache from the destination and performing the cache merge. At the onset of every experiment (step 0), the network is initialized as a bidirectional grid lattice. Then we let the protocol run for 50 steps (i.e. *bootstrap*), which enables the network to converge into a self-organized equilibrium. The actual time required to reach this desired state is usually below 30 steps and depends greatly on the actual configuration (the number of nodes, the cache size, etc.). The external interactions (all malicious activities) are made after the bootstrap, i.e. starting at the 51st step.

All simulations were executed assuming idealised environment, where all other faults than cheating can not occur. The nodes and links are fully operational throughout the whole simulation. In particular, establishing a connection from one node to another is always successful provided that the destination address exists in the network.

5   Validation and Experimental Results

The results are split in two groups. First, the influence of the cache-exchange limit on the network disconnecting speed is analysed. The probability that a given node $v \in V_t^{honest}$ belongs to the $\hat{\mathcal{C}}_t^{max}$ (defined in Sect. 3, averaged from 100 executions for each combination of the parameters) is used as the measure of robustness. Then, in the second group, the impact of the limit on the original (small-world) properties of Newscast is examined.

Limiting the Merge Against Uncoordinated Malicious Clients. Figure 1 presents the obtained results. As visible, limiting the number of exchanged cache entries to at most $c/2$ of the freshest items rapidly decreases the effectiveness of

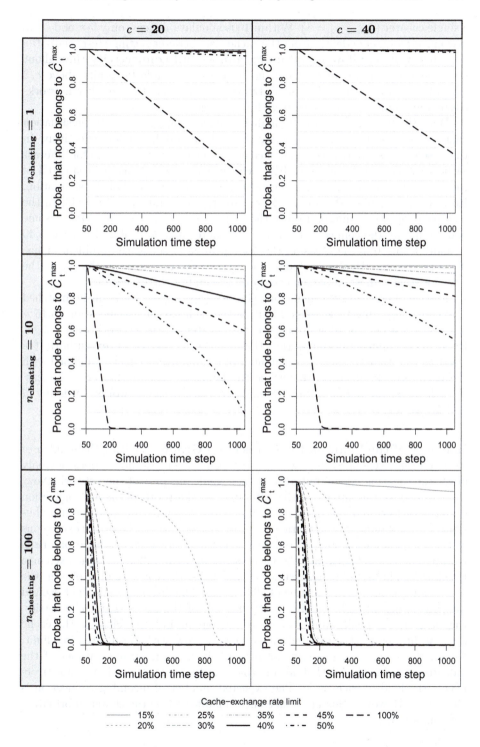

Fig. 1. Influence of the cache-exchange limit on the disconnecting speed for 1, 10 and 100 cheaters, for a Newscast network consisting of 1000 nodes with the cache containing 20 and 40 entries.

a single cheater ($n_{\text{cheating}} = 1$). Within 1000 simulation steps, only few nodes are disconnected independently from the cache size. Stricter settings (especially the rates below 40 %) **allow maintaining almost full connectivity** throughout the execution. On the other hand, no benefit was visible for limits above $c/2$ (not plotted for clarity). The same trends were observed for smaller networks (results are omitted due to the page limit restrictions).

With the increased number of malicious clients ($n_{\text{cheating}} = 10$) the situation changes. The performance of cheating is still impaired, but less effectively. In this case there is a noticeable difference between the results obtained for different cache sizes. For $c = 20$ and the 50 % cache-exchange rate limit, less than 10 % of the nodes stay connected in the biggest component at the end of the simulation. However, more than 50 % of the peers can still cooperate when $c = 40$. In the previously optimal configuration — the cache-exchange limit set to 40 % — some nodes still lose the connectivity (for $c = 20$ more than 20 % of nodes, for $c = 40$ — more than 10 %). Decreasing the setting further, solves the problem. Thus, if the merge operation is limited to the 30 % of all the entries in the cache, the network stays almost fully connected and the end of the execution.

Further increase in the malicious activity ($n_{\text{cheating}} = 100$) can be matched again with appropriate settings. Limiting the cache-exchange operation to 15 % of the freshest entries allows tolerating the cheaters in the network with a slight loss of the computing resources (less than 10 % at the end of the execution). It is worth noting that the increasing number of the malicious clients combined with the decreasing limit starts to reverse the benefits from the bigger cache size. In such cases, the influence of cheaters is greater for the networks consisting of nodes with larger c.

Influence of the Limit on the Connection Graph. Limiting the number of cache entries exchanged during a connection between two nodes does not only affect the cheating clients but also the non-malicious part of the network. Now, we will analyse the impact of this alteration in the protocol on the connection graph. The first and the most noticeable effect concerns the indegree distributions. Figure 2 presents the average value of this statistic through 100 executions at the 50th simulation step. The data was gathered for each combination of the parameters: the network and the cache sizes, and the different cache-exchange rate limits. The trend is identical for all the cases, therefore we only present the results for the network consisting of 1000 nodes. The distribution is right-skewed for all the settings. The smaller is the number of the exchanged cache entries, the closer is the peak number of nodes having the indegree close to the cache size. Additionally, the distribution is more narrow with the decreasing limit. Hence, the number of peers in the network having the extreme values of the indegree is reduced. All that means that the introduced restriction positively affects the load of each node. The expected number of incoming connection is more even. However, this has a side effect manifested by the slower information dissemination.

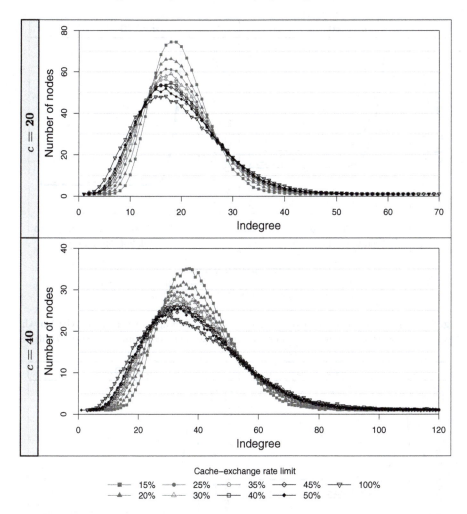

Fig. 2. Influence of the cache-exchange limit on the average indegree in the Newscast network consisting of 1000 nodes with the cache containing 20 and 40 entries.

The small-world properties of the Newscast network are also affected by the introduced limit. Figure 3 presents the results of the average path length (APL) and Fig. 4 — the average clustering coefficient (CC). The values were obtained from 100 executions, measured for networks consisting of 1000 nodes, with different cache-exchange limits and the cache sizes set to 20 and 40. As previously, the measurements for smaller networks are characterised by the similar trends, therefore they are omitted from the discussion. The initial values of the APL are high due to the initialisation phase (described in Sect. 4). The average path length in a grid consisting of 1000 nodes is approximately 17,88. This value drops to the lower levels with the progress in the execution of the protocol. Hence, the range

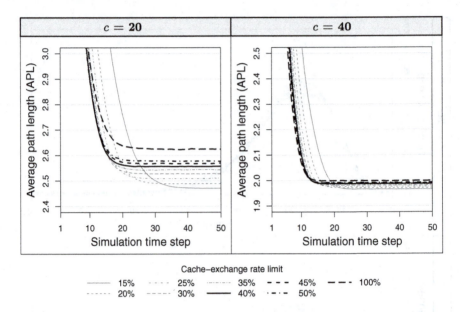

Fig. 3. Influence of the cache-exchange limit on the average path length (APL) in the Newscast network consisting of 1000 nodes with the cache containing 20 and 40 entries.

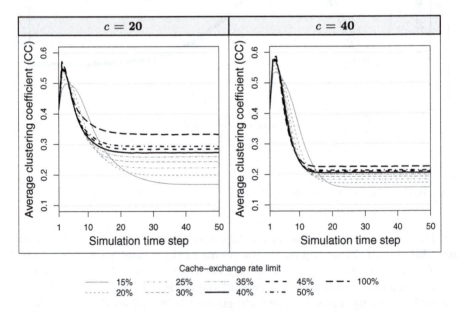

Fig. 4. Influence of the cache-exchange rate limit on the average clustering coefficient (CC) in the Newscast network consisting of 1000 nodes with the cache containing 20 and 40 entries.

of the vertical axis was limited to increase the readability of the relevant values. As visible on Figs. 3 and 4, the time required to reach the self-organised equilibrium in the network extends with the decreasing limit on the number of exchanged entries. Both statistics also converge to lower values. As previously, these trends are amplified with the decreasing number of exchanged entries. Generally, the networks with a bigger cache sizes are less affected by the change in the protocol mechanics. It is worth noting that the equilibrium is reached in all the cases within the bootstrap time (50 simulation steps) set for the experiments presented before.

Summary. Limiting the number of exchanged cache entries during a connection proved to be an effective method against the cheating model introduced in [14]. The increasing number of malicious clients in the network may be matched by the bigger cache sizes combined with the limitation on the exchanged items. Even at the 50 % limit, the solution gives satisfactory results. The nodes stay connected in the biggest component much longer without significantly affecting the initial (small-world) properties of Newscast. Yet, further decrease of the limit, allowing tolerating greater number of the malicious clients, has more influence on the protocol operation. Limiting the cache-exchange to the 30 % of the freshest entries proves to tolerate almost completely up to 1 % of the malicious users during 1000 simulation steps. On the other hand, decreasing the limit to 15 % allows resisting the attack from 10 % of the cheating clients. However, such serious change in the protocol mechanics affects the small-world properties of the connection graph. Therefore, the adjustment should be made dynamically during the execution of the protocol, minimising the impact of the approach on the executed computation.

6   Related Work

The problem of malicious users is not new in the context of P2P networks and in particular — gossip protocols. There are many existing attempts to solve it. Cheating-tolerance (or more general — byzantine fault tolerance) is usually achieved at the cost of limited gossip performance or loss of other (desirable) properties, like decentralisation or scalability. Generally, solutions are based on *cryptography* or *operational constraints*. In the first case either the centralisation is re-introduced into the scheme or the local view of each peer becomes only a simulated feature, as members of the network have to possess and maintain the full, global view (i.e. information about all active nodes). When it comes to introducing the operational constants, the available solutions are simply focused on limiting the number of gossips (send and received). We now proceed to a short overview of the concrete solutions found in the literature of the subject.

Minsky and Schneider in [11] proposed *path verification protocols*, based on the assumptions that it is possible to determine a sender of each message and that no data alteration is possible during its transit over the network. The solution allows tolerating fixed amount (t) of faults. JetStream [15], Fireflies [9] and Veracity [17] are based on principals found in social networks. The first two utilise mechanisms for predictable, pseudo-random, and verifiable gossiping. Such approach combined with ignoring unexpected connections, limits the amount of gossips between the nodes. Additionally, each peer in Fireflies stores the global view on the network and uses only a small part of it for the protocol execution. Despite the embedded cryptographic solutions, scalability and reliability of gossip is maintained up to a certain point. Frequent malicious activities increase the computational effort associated with communication. In Veracity on the other hand, a slightly different approach is used. The protocol is executed on top of a fully functional Distributed Hash Table DHT service, increasing the complexity of the solution. Messages are accepted, when they are checked by a (small) set of other nodes from the DHT network. However, the base of the protocol opens additional directions of attacks [8].

An interesting implementation of gossiping was proposed by Bortnikov et al. in [2]. They performed a theoretical analysis of a bias in the local view of each peer caused by the malicious activities. This lead to creation of Brahms protocol, in which an additional sample of the nodes was introduced (comparing to the standard scheme). This set is built from all descriptors received during the view exchanges, using min-wise independent permutations [2]. Such construct ensures uniform distribution of the sample regardless of the frequency and order in which the addresses are collected. Random elements from this set are added to the view during its update. The approach proved to be effective in a testing environment, however in a real-world setting, when the churn is present, the uniform random sample is impossible to achieve [8]. This is caused by the method of its construction, which heavily depends on a stable environment.

The last protocol described here, is a prestige-based solution proposed by Jesi et al. in [8]. It has a unique feature in comparison to the previously described schemes — an *exploratory view exchange*. Each node, during its synchronisation cycle, additionally to following the standard send-receive scheme, sends a predefined number of *requests for gossip* to randomly chosen peers. If B receives such message from A, it has to respond with its view, for which A sends its own. Lists of descriptors collected this way are used to compute and maintain local statistics on neighbourhoods of each node. This allows identification of structural abnormalities in the neighbourhoods (e.g. hubs are *"over-represented in the partial views"* [8]) and blocking responsible peers from communication. As suggested by authors, like in the previous case (Brahms), it is possible to build a sample of network members, which in case of disconnection or view pollution may be used to restore full connectivity and performance. However, the solution might be broken by sending the views containing old or random addresses (and in a realistic scenario — most probably belonging to inactive/non-existing nodes). In such case, the efficiency of the communication would be significantly crippled, increasing the overhead associated with maintenance of the statistics and the local repository of known nodes.

7   Conclusion

Newscast is a Peer-to-Peer, nature-inspired gossip-based data exchange protocol used for information dissemination and membership management in large-scale distributed systems. The model follows a probabilistic scheme able to keep a self-organised, small-world equilibrium featuring a complex, spatially structured and dynamically changing environment. While Newscast has been proven resilient to crash faults [7,20] (*i.e.* when nodes simply fail or stop working), recent advances in the analysis of the gossiping mechanism [13,14] reveal hard-wired vulnerabilities of the protocol, leading to an efficient attack where a few malicious clients are able to break the network connectivity in a very short time. In this paper, we propose and analyse an extension of Newscast with an embedded counter-measure able to defeat this attack while preserving the original protocol properties and its inherent performance. Indeed, the most effective counter-measure appeared to be the limiting in the number of cache entries exchanged at each connection. The implementation of a Java-based simulator (able to model the protocol behaviour in a byzantine environment) permitted to tune the parameters of our enhanced version of the seminal protocol while establishing its resilience against cheating faults. For instance, limiting the cache-exchange to the 30 % of the freshest entries proved to tolerate almost completely up to 1 % of the malicious users.

The future work induced by this study includes more large-scale experiments, in particular to tackle and tune our proposed extension of Newscast against coordinated attacks. One ambition is also to define efficient detection techniques to identify the misbehaving peers, *i.e.* cheaters, to embed accurate blacklisting capabilities to the Newscast protocol.

Acknowledgments. The experiments presented in this paper were carried out using the HPC facility of the University of Luxembourg.

References

1. Avizienis, A., Laprie, J.C., Randell, B., Landwehr, C.E.: Basic concepts and taxonomy of dependable and secure computing. IEEE Trans. Dependable Secure Comput. **1**, 11–33 (2004)
2. Bortnikov, E., Gurevich, M., Keidar, I., Kliot, G., Shraer, A.: Brahms: Byzantine resilient random membership sampling. Comput. Netw. **53**, 2340–2359 (2009)
3. Canon, L., Jeannot, E., Weissman, J.: A scheduling and certification algorithm for defeating collusion in desktop grids. In: 2011 31st International Conference on Distributed Computing Systems (ICDCS), pp. 343–352, June 2011
4. Dutot, A., Guinand, F., Olivier, D., Pigné, Y.: GraphStream: A Tool for bridging the gap between Complex Systems and Dynamic Graphs. In: Emergent Properties in Natural and Artificial Complex Systems. Satellite Conference within the 4th European Conference on Complex Systems (ECCS'2007), Dresden, Allemagne (2007–2010). http://graphstream-project.org/
5. Jelasity, M., Guerraoui, R., Kermarrec, A.-M., van Steen, M.: The peer sampling service: experimental evaluation of unstructured gossip-based implementations. In: Jacobsen, H.-A. (ed.) Middleware 2004. LNCS, vol. 3231, pp. 79–98. Springer, Heidelberg (2004)
6. Jelasity, M., Montresor, A., Babaoglu, O.: Gossip-based aggregation in large dynamic networks. ACM Trans. Comput. Syst. **23**(3), 219–252 (2005)

7. Jelasity, M., van Steen, M.: Large-Scale newscast computing on the internet. Technical Report IR-503, Vrije Universiteit Amsterdam, Department of Computer Science, Amsterdam, The Netherlands, October 2002

8. Jesi, G.P., Montresor, A., Van Steen, M.: Secure peer sampling. Comput. Netw. **54**, 2086–2098 (2010)

9. Johansen, H., Allavena, A., van Renesse, R.: Fireflies: Scalable support for intrusion-tolerant network overlays. SIGOPS Oper. Syst. Rev. **40**(4), 3–13 (2006)

10. Laredo, J., Eiben, A., Steen, M., Merelo, J.: Evag: a scalable peer-to-peer evolutionary algorithm. Genet. Program Evolvable Mach. **11**(2), 227–246 (2010)

11. Minsky, Y.M., Schneider, F.B.: Tolerating malicious gossip. Distrib. Comput. **16**(1), 49–68 (2003)

12. Moore, R.: Cybercrime: Investigating high-technology computer crime. LexisNexis/ Matthew Bender (2005)

13. Muszyński, J., Varrette, S., Laredo, J.L.J., Bouvry, P.: Analysis of the data flow in the newscast protocol for possible vulnerabilities. In: Kotulski, Z., Księżopolski, B., Mazur, K. (eds.) CSS 2014. CCIS, vol. 448, pp. 89–99. Springer, Heidelberg (2014)

14. Muszyński, J., Varrette, S., Laredo, J.L.J., Bouvry, P.: Exploiting the hard-wired vulnerabilities of newscast via connectivity-splitting attack. In: Au, M.H., Carminati, B., Kuo, C.-C.J. (eds.) NSS 2014. LNCS, vol. 8792, pp. 152–165. Springer, Heidelberg (2014)

15. Patel, J., Gupta, I., Contractor, N.: Jetstream: Achieving predictable gossip dissemination by leveraging social network principles. In: Fifth IEEE International Symposium on Network Computing and Applications, 2006, NCA 2006 pp. 32–39, July 2006

16. Rowstron, A., Druschel, P.: Pastry: scalable, decentralized object location, and routing for large-scale peer-to-peer systems. In: Guerraoui, R. (ed.) Middleware 2001. LNCS, vol. 2218, pp. 329–350. Springer, Heidelberg (2001)

17. Sherr, M., Loo, B.T., Blaze, M.: Veracity: A fully decentralized service for securing network coordinate systems. In: Proceedings of the 7th International Conference on Peer-to-peer Systems, IPTPS 2008, p. 15. USENIX Association, Berkeley (2008)

18. Stoica, I., Morris, R., Liben-Nowell, D., Karger, D., Kaashoek, M., Dabek, F., Balakrishnan, H.: Chord: a scalable peer-to-peer lookup protocol for internet applications. IEEE/ACM Trans. Netw. **11**(1), 17–32 (2003)

19. Varrette, S., Tantar, E., Bouvry, P.: On the resilience of [distributed] evolutionary algorithms against cheaters in global computing platforms. In: Proceedings of the 14th International Workshop on Nature Inspired Distributed Computing (NIDISC 2011), part of the 25th IEEE/ACM International Parallel and Distributed Processing Symposium (IPDpPS 2011). IEEE Computer Society, Anchorage (Alaska), USA, 16–20 May 2011

20. Voulgaris, S., Jelasity, M., van Steen, M.: A robust and scalable peer-to-peer gossiping protocol. In: Moro, G., Sartori, C., Singh, M.P. (eds.) AP2PC 2003. LNCS (LNAI), vol. 2872, pp. 47–58. Springer, Heidelberg (2004)

21. Zhao, B., Huang, L., Stribling, J., Rhea, S.C., Joseph, A.D., Kubiatowicz, J.D.: Tapestry: A resilient global-scale overlay for service deployment. IEEE J. Sel. Areas Commun. **22**(1), 41–53 (2004)

A Distributed Intrusion Detection Framework Based on Evolved Specialized Ensembles of Classifiers

Gianluigi Folino$^{(\boxtimes)}$, Francesco Sergio Pisani, and Pietro Sabatino

Institute of High Performance Computing and Networking (ICAR-CNR),
Rende, Italy
{folino,fspisani,pietro.sabatino}@icar.cnr.it

Abstract. Modern intrusion detection systems must handle many complicated issues in real-time, as they have to cope with a real data stream; indeed, for the task of classification, typically the classes are unbalanced and, in addition, they have to cope with distributed attacks and they have to quickly react to changes in the data. Data mining techniques and, in particular, ensemble of classifiers permit to combine different classifiers that together provide complementary information and can be built in an incremental way. This paper introduces the architecture of a distributed intrusion detection framework and in particular, the detector module based on a meta-ensemble, which is used to cope with the problem of detecting intrusions, in which typically the number of attacks is minor than the number of normal connections. To this aim, we explore the usage of ensembles specialized to detect particular types of attack or normal connections, and Genetic Programming is adopted to generate a non-trainable function to combine each specialized ensemble. Non-trainable functions can be evolved without any extra phase of training and, therefore, they are particularly apt to handle concept drifts, also in the case of real-time constraints. Preliminary experiments, conducted on the well-known KDD dataset and on a more up-to-date dataset, ISCX IDS, show the effectiveness of the approach.

1 Introduction

As the number of network connections and also the speed of these networks is increasing, the problem of analyzing large streams of data in real time for detecting possible attacks gains relevance in the scientific community of cybersecurity. Typically, *Intrusion Detection Systems* (IDS) [1] are used to detect unauthorized accesses to computer systems and networks (in this case, they are named *Network Intrusion Detection Systems*, NIDS). Data mining techniques are largely used to support the detection phase of the NIDS, as witnessed by a large number of papers published in recent years. Despite this, most of the data mining algorithms are not suitable to capture in real time new trends and changes (often denoting a network intrusion) in streaming data, as they assume that data are static and not changing due to external modifications.

© Springer International Publishing Switzerland 2016
G. Squillero and P. Burelli (Eds.): EvoApplications 2016, Part I, LNCS 9597, pp. 315–331, 2016.
DOI: 10.1007/978-3-319-31204-0_21

Ensemble [2–4] is a learning paradigm where multiple component learners are trained for the same task by a learning algorithm, and the predictions of the component learners are combined for dealing with new unseen instances. Among the advantages in using ensemble of classifiers, we would like to remind that they help to reduce the variance of the error, the bias, and the dependence from a single dataset; furthermore, they can be build in an incremental way and they are apt to distributed implementations. Finally, for the particular task of intrusion detection, they are able to combine different classifiers that together provide complementary information and also are particularly apt to handle unbalanced classes.

For the above considerations, this work proposes a distributed intrusion detection framework, which includes a detection module based on a meta-ensemble algorithm. In this module, the different ensembles are specialized to detect particular types of attack and normal connections; in addition, a distributed GP tool [5] is used to generate a non-trainable function to combine each specialized ensemble. The main characteristic of the non-trainable functions is that it can be evolved without any extra phase of training and, therefore, they are particularly apt to handle concept drifts, also in the case of real-time constraints.

The distributed framework includes also a module for detecting change in the data and for consequently updating the ensemble and an alert correlation module. However, in this paper, we are only interested in studying the behavior of the specialized ensembles in detecting the intrusions; therefore the above-cited modules are not analyzed in this paper. More details concerning these modules are supplied in the paper [6].

Due to the problems in obtaining a fully labeled dataset in this particular domain, the experiments were conducted on the ISCX IDS dataset from the Information Security Centre of Excellence of the University of New Brunswick [7]. This dataset was generated by capturing seven days of network traffic transiting in a controlled testbed made of a subnetwork behind a firewall. Normal traffic was generated with the help of agents that simulated requests of human users following probability distributions extrapolated from real traffic captured during a week. Attack were generated with the aid of human operators following formal descriptions of the steps to be performed.

In [8], specialized detectors were successful employed to improve the classification accuracy of malware detection. Differently from our approach, the detectors are combined in an ensemble using standard pre-defined functions, i.e., majority voting, stacking, etc. Therefore, the approach is not easily adaptable to the case of data stream and when the data change and this is confirmed by the experiments contained in the paper. The adoption of non-trainable functions for cybersecurity problems in general, and also for the intrusion detection, is also explored in [9] in which, differently from this paper, the non-trainable functions were used for handling missing data. In [10], the authors develop a GP-based framework to evolve the fusion function of the ensemble both for heterogenous and homogeneous ensemble. The main aim of the paper is to improve the accuracy of the generated ensemble, while distributed implementations and the problems concerning incomplete and unbalanced datasets are not explored.

In addition, differently from our approach, the authors do not consider weights depending from the performance of the classifiers on the datasets.

The rest of the paper is structured as follows. The main dataset used in the experiment is presented and discussed in Sect. 2. Section 3 illustrates the overall distributed framework for the intrusion detection task. In Sect. 4, the meta-ensemble approach is described in detail. Section 5 show some experiments conducted to verify the effectiveness of the approach on two interesting datasets. Finally, Sect. 6 concludes the work and addresses future research directions.

2   A Realistic Dataset for Testing Intrusion Detection Systems

The main problem in testing algorithms for IDSs is to find a suitable testbed. There are mainly two possibilities.

The first one is to use a publicly available dataset such as for instance CAIDA[1] from The Cooperative Association for Internet Data Analysis, The Internet Traffic Archive[2] from Lawrence Berkeley National Laboratory, LBNL/ICSI Enterprise Tracing Project[3], DARPA[4] from Lincoln Laboratory and KDD[5] Cup 1999 from the University of California, see [1] for a short discussion on these datasets. DARPA and KDD are a popular choice in the research community, see Travallaee et al. [11]; however, they have been thoroughly criticized for being unable to provide a realistic scenario. These critics are carried out in a number of papers, see Mahoney and Chan [12], McHugh [13] and Brown [14] among the others; in particular, the conclusions contained in Tavallaee et al. [15] regarding the issues of the KDD dataset lead the authors to produce a new dataset based on it, the so called NSL-KDD. Even for this last dataset, doubts still remain on the ability to provide up to date attack scenarios.

A second option, is to generate an ad hoc dataset by capturing traffic directly from an operational subnetwork and to inject attacks by employing tools openly available on the web. A drawback of this approach is that obtaining a fully labeled dataset is usually a difficult operation that requires a certain amount of time and resources to be performed. Moreover, because of privacy concerns, these datasets are not publicly supplied, and in the case they are made so, they are anonymized, e.g. the entire payload may be removed resulting in a far less effective dataset. On the other side, as pointed out by many authors, Paxson [16], Tavallaee et al. [11] and Shiravi et al. [7] only to cite a few, it is of great importance to have a common testbed for the validation and comparison of different solutions.

For the reasons outlined in the above discussion, in this paper we choose to conduct our experimental testing on the ISCX IDS dataset from the Information

---

[1] https://www.caida.org/data/.

[2] http://ita.ee.lbl.gov/index.html.

[3] http://www.icir.org/enterprise-tracing/.

[4] http://www.ll.mit.edu/ideval/data/.

[5] http://kdd.ics.uci.edu/databases/kddcup99/kddcup99.html.

Table 1. Main characteristics of the ISCX IDS dataset.

Day	Description	Size of the pcap file (GB)	Number of Flows	Percentage of Attacks
Day 1	*Normal traffic without malicious activities*	16.1	359,673	0.000 %
Day 2	*Normal traffic with some malicious activities*	4.22	134,752	1.545 %
Day 3	*Infiltrating the network from the inside &* Normal traffic	3.95	153,409	6.395 %
Day 4	*HTTP Denial of Service &Normal traffic*	6.85	178,825	1.855 %
Day 5	*Distributed Denial of Service using an IRC Botnet*	23.4	554,659	6.686 %
Day 6	*Normal traffic without malicious activities*	17.6	505,057	0.000 %
Day 7	*Brute Force SSH + Normal activities*	12.3	344,245	1.435 %

Security Centre of Excellence of the University of New Brunswick [7]. This dataset is the result of capturing seven days of network traffic in a controlled testbed made of a subnetwork placed behind a firewall. Normal traffic was generated with the aid of agents that simulated normal requests of human users following some probability distributions extrapolated from real traffic. Attack were generated with the aid of human operators. The result is a fully labelled dataset containing realistic traffic scenarios. Indeed, the dataset consists of standards `pcap` (packet capture) files one for each day containing the relative network traffic, as illustrated in Table 1. Different days contain different attack scenarios, ranging from HTTP Denial of Service, DDos, Brute Force SSH and attempts of infiltrating the subnetwork from the inside.

3   A Distributed Framework for Intrusion Detection

In this section, we outlined a general architecture for network intrusion detection and illustrate the particular preprocessing phase adopted. More details about the architecture and the different solutions, which can be used, are shown in [6].

3.1   Preprocessing

Usually network traffic is acquired with the aid of software based on `libpcap`, such as `tcpdump` or `Gulp` (more suitable to an high speed network environment), and stored in the standard `pcap` (packet capture) format.

The ISCX IDS 2012 dataset consists of standards `pcap` (packet capture) files, one for each day containing the relative network traffic. Network data streams acquired in form of pcap files require a preprocessing phase in order to extract features needed in the subsequent analysis. With respect to this task, network traffic may be modeled at the level of flows, i.e., first the connections are aggregated into flows and then suitable features are extracted and aggregated. Beside the level of flows, data streams may be analyzed at the level of packet payloads, using the so-named *deep packet inspection* technique. Indeed, the payload of `HTTP` packets is made up of textual content and then it may be analyzed by using the technique of n-grams, usually employed for the analysis of textual content. It is worth to notice that n-grams based techniques have the drawback of presenting an exponential grow in n of the feature space and consequently they are computationally intensive. In any case, note that the choice of techniques used to analyze the data streams depends clearly on the type of analysis performed in the subsequent stages. In order to extract a relatively compact set of features, our choice was to model the traffic at the level of flows; in practice, the `pcap` files are aggregated in flows and preprocessed with the aid of the `flowcalc`[6] tool from the MuTriCs project, Fomerski et al. [17]. We preferred this tool over more common choices such as `tcptrace` since it is more apt to a streaming environment. Moreover, it can be extended with a number of plugins; for instance, we used the plugins, which permit to process the payload and to extract basic statistics from it. In particular we enabled the following plugins: `basic`, `counters`, `pktsize`, `lpi`, `web`. These modules provide respectively basic statistics on packets payload such as size and inter arrival times, number of packets and bytes, sizes of the first packets in the flow, payload protocol and finally various statistics on the web traffic content.

3.2   The Distributed Architecture and the Modules

The proposed architecture is build around the Collaborative Intrusion Detection paradigm, i.e. it implements a mechanism that correlates information coming from different nodes or sensors in which network data stream is collected. This mechanism permits the discovery of more complex attacks and may work either at the level of alarms or at the level of models generated for a particular type of attack. Sharing models of malicious activity has the side effect of spreading the knowledge acquired locally to other nodes; in practice, a node can receive models for attacks observed elsewhere, but not locally and use them in a proactive way to inspect the local traffic. Moreover, working at the level of alarms, has the advantage of enabling collaboration among nodes of different nature, i.e. running commercial IDSs as well as open source solutions such as Snort and BroIDS to share meaningful information, see Fig. 1.

For the sake of scalability, a distributed alert correlation mechanism that involves each single node as active part is illustrated in the architecture.

---

[6] http://mutrics.iitis.pl/flowcalc.

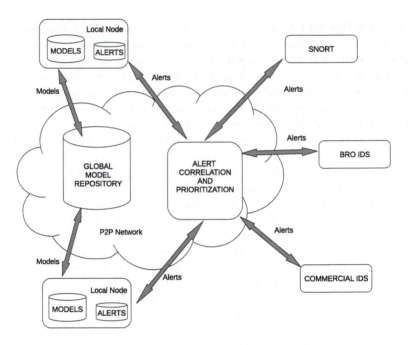

Fig. 1. Ensemble-based NIDS general Architecture

In order to accomplish the above tasks, and also to cope with the fast changing nature of network traffic and attacks, the local components of the architecture contains some specialized interacting modules. In fact, the module Change
Detector is responsible for a large scale analysis of the data stream that flows
through a local node in search of significant deviations from the normal behavior. In order to guarantee real time responses, the Change Detector analyzes the
data stream in time windows and subsequently it computes suitable functions
on values obtained from aggregated features extracted from each time window;
we call the set of values of these functions, obtained from a window, a "concept"
(relative to that time window).

Many algorithms, based for instance on time series or fractal dimension, can
be employed to detect anomalies in the sequence of "concepts", and, in the
case anomalies are detected, the Model Generator is activated consequently. It
builds new models by classification, clustering and by using other statistical
methods and maintains the model together with the relative "concept". The
overall ensemble is employed to monitor the network data stream, see Fig. 2.
Since it is not plausible to add models indefinitely to the ensemble, a module
called Update Ensemble is responsible for managing the active models of the
ensemble; a strategy, currently in development, permits to compare "concepts"
stored as metadata in the models with that coming from the current data stream,
and consequently activate/deactivate the models.

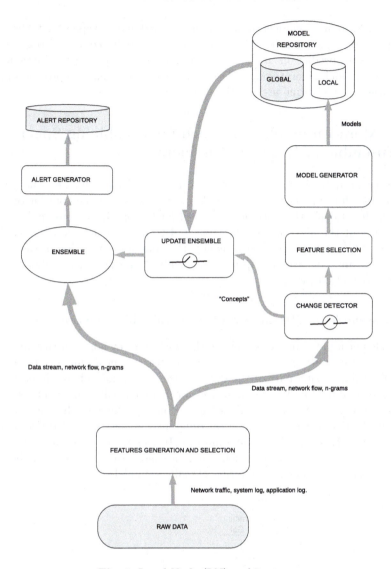

Fig. 2. Local Node (LN) architecture

As already said, the ensemble incorporates models of traffic and it is responsible for the analysis of the incoming data stream in search of malicious activities.

IDSs benefit from using algorithms based on the ensemble paradigm for a number of reasons. Indeed, given the speed of modern networks and the consequent increase in the volume of data that need to be processed, solutions based on algorithms that can be easily implemented in parallel/distributed environment (multi-core CPUs and GPGPU hardware) should be preferred, and clearly solutions based on the ensemble paradigm can be easily implemented on parallel hardware. Moreover, models may be trained on different type of attacks or on some parts or on some levels of the network and finally combined together, to assure a better prediction. In the following section, we present the implementation of the model generator module based on a evolutionary approach and the meta-ensemble technique used in the framework.

4   A Meta-Ensemble Approach for Combining Specialized Ensemble of Evolved Classifiers

In this section, we introduce some background information, i.e., the general schema used for combining an ensemble of classifiers and the concept of "non-trainable functions", that can be used in order to combine an ensemble of classifiers without the need of a further phase of training. Then, we illustrated the distributed GP framework used to evolve the combining function of the ensemble and the meta-ensemble approach to combine specialized classifiers for the intrusion detection task.

4.1   Ensemble of Classifiers and Non-trainable Functions

Ensemble permits to combine multiple (heterogenous or homogenous) models in order to classify new unseen instances. In practice, after a number of classifiers are built usually using part of the dataset, the predictions of the different classifiers are combined and a common decision is taken. Different schemas can be considered to generate the classifiers and to combine the ensemble, i.e. the same learning algorithm can be trained on different datasets or/and different algorithms can be trained on the same dataset. In this work, we follow the general approach shown in Fig. 3, in which different algorithms are used on the same dataset in order to build the different classifiers/models.

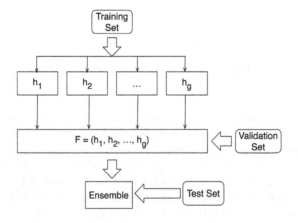

Fig. 3. A general schema for combining ensemble of classifiers

Let $S = \{(x_i, y_i) | i = 1, \ldots, N\}$ be a training set where x_i, called example or tuple or instance, is an attribute vector with m attributes and y_i is the class label associated with x_i. A predictor (classifier), given a new example, has the task to predict the class label for it.

Ensemble techniques build g predictors $(h_1, h_2, \ldots, h_g)$ using different algorithms on the same training set, then combine them together to classify the test set. As an alternative, the g predictors could be built using the same algorithm on different training sets.

The largely used boosting algorithm, introduced by Schapire [18] and Freund [19], follows a different schema; in order to boost the performance of any "weak" learning algorithm, i.e. an algorithm that "generates classifiers which need only be a little bit better than random guessing" [19], the method adaptively changes the distribution of the training set depending on how difficult each example is to classify.

This approach was successfully applied to a large number and types of datasets; however, it has the drawback of needing to repeat the training phase for a number of rounds and that could be really time-consuming for large datasets. The applications and the datasets in hard domains, as cyber security, have real-time requirements, which do not permit to re-train again the base models. On the contrary, ensemble strategies following the schema shown in Fig. 3 do not need any further phase of training, whether the functions used can be combined without using the original training set. The majority vote is a classical example of this kind of combiner function. Some types of combiner has no extra parameters that need to be trained and consequently, the ensemble is ready for operation as soon as the base classifiers are trained. These are named non-trainable combiners [20] and could be used as functions in a genetic programming tree.

Before describing the GP framework used, here, we introduce some definitions useful to understand how the algorithm works.

Let $x \in R^N$ be a feature vector and $\Omega = \{\omega_1, \omega_2 \ldots, \omega_c\}$ be the set of the possible class labels. Each classifier h_i in the ensemble outputs c degrees of support, i.e., for each class, it will give the probability that the tuple belong to that class. Without loss of generality, we can assume that all the c degrees are in the interval $[0, 1]$, that is, $h_i : R^N \rightarrow [0, 1]^c$. Denote by $H_{i,j}(x)$ the support that classifier h_i gives to the hypothesis that x comes from class ω_j. The larger the support, the more likely the class label ω_j. A non-trainable combiner calculates the support for a class combining the support values of all the classifiers. For each tuple x of the training set, and considering g classifiers and c classes, a Decision Profile matrix DP can be build as follow:

$$DP(x) = \begin{bmatrix} H_{1,1}(x) \ldots H_{1,j}(x) \ldots H_{1,c}(x) \\ H_{i,1}(x) \ldots H_{i,j}(x) \ldots H_{i,c}(x) \\ H_{g,1}(x) \ldots H_{g,j}(x) \ldots H_{g,c}(x) \end{bmatrix}$$

where the element $H_{i,j}(x)$ is the support for j-th class of i-th classifier.

The functions used in our approach simply combine the values of a single column to compute the support for $j - th$ class and can be defined as follow:

$$\mu_j(x) = F[H_{1,j}(x), H_{2,j}(x), ..., H_{g,j}(x)]$$

For instance, the most simple function we can consider is the average, which can be computed as:

$$\mu_j(x) = \frac{1}{g} \sum_{i=1}^{g} H_{i,j}(x)$$

The class label of x is the class with maximum support μ.

4.2   The Pseudocode and the Software Architecture of the Detection Algorithm

The model generator of the distributed architecture shown in Sect. 3 is built on the well-known Massive Online Analysis (MOA) toolbox[7], adopted to handle the data streams and also used for the implementation of the classifier algorithms and on the distributed GP tool, CellulAr GEnetic programming (CAGE) [5], which is used to evolve the combining function of the ensemble. The latter is based on the fine-grained cellular model and runs both on distributed-memory parallel computers and on distributed environments, by partitioning the overall population of the GP algorithm into subpopulations, one for computation node; then a standard (panmictic) GP algorithm is executed on each node.

As the streams flows, we suppose that part of it is labelled. When a sufficient number of labelled tuples is collected, they are divided into training set and validation set and used to train the ensemble, as described in the following. The base classifiers, chosen among the best performing on a set of benchmarks of the MOA toolbox, are trained on the training set; then, a weight, proportional to the error on the training set, is associated to each classifier; at the same time a decision support matrix is built. This phase could be computationally expensive, but it could be performed in parallel, as the different algorithms are independent from each other.

After that, the combiner function of the ensemble is evolved by using the distributed GP tool, CAGE, on the validation set. As nodes of the GP tree, some non-trainable functions are chosen, better specified in the experimental section, while the leafs of the tree are the different classifiers selected in the previous phase (see Fig. 4) and the fitness function is simply the accuracy of the ensemble computed on the validation set. It is worth to remember that no extra computation on the data is necessary, as the validation set is only used to verify the correct class is assigned and consequently to compute the fitness function.

Finally, the overall combiner function is used to classify the new coming tuples of the stream. Implicitly, the function selects the classifiers/models more apt to the particular datasets considered. The final function is used to combine the base classifiers and classify the incoming stream. Also this phase can be

---

[7] MOA prerelease 2014.01; http://moa.cms.waikato.ac.nz/overview/.

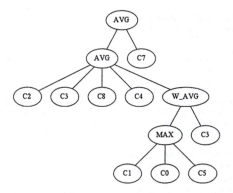

Fig. 4. An example of GP tree generated from the tool.

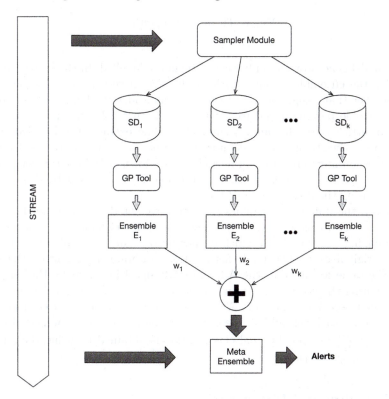

Fig. 5. The software architecture of the meta-ensemble architecture.

performed in parallel, by partitioning the stream among different nodes and applying the function to each partition.

The entire process is better illustrated in Fig. 5 and then specified in the pseudocode of Fig. 6. An infinite stream of tuples is supposed to flow as input

Let α be the total number of base classifiers algorithms.
Let l be the number of base classifiers algorithms effectively used.
Given a set of k datasets $SD_1, SD_2, \ldots, SD_k$.
For each D_i
    Consider the dataset SD_i partitioned in train, validation and test set: $SDtrain_i$, $SDvalid_i$ and $SDtest_i$
    Train α different classification algorithms on $SDtrain_i$
    Maintain the l classifiers obtaining the best accuracy on the training set.
    Build l decision profile matrixes, one for each classifiers, $DP_1, DP_2, \ldots, DP_l$ using the respective validation set,
    one for each classifier of dimension $n \times c$, where n is the number of tuples and c the number of classes.
    Run the distributed GP tool on the validation set $SDvalid_i$ in order to obtain the combiner function of the ensemble.
    Obtain an Ensemble E_i, a combiner function FC_i and a weight W_i,
    computed on the basis of the error of the given ensemble on the validation set.
end for each
Build the decision profile matrix DP of the entire ensemble E,
where each element $H_{i,j}(x)$ is the support that classifier h_i gives to the hypothesis that the tuple x comes from class ω_j.
Compute the weighted mean for each class j: $\mu_j(x) = \frac{\sum w_i * H_{ij}(x)}{\sum w_i}$ on the test set.
Compute the class by using the formula $class(x) = argmax_j(\mu_j(x))$.

Fig. 6. The overall algorithm

to the model generator and only part of them are labelled. In the startup phase, in which the ensemble is built, each newly arrived tuple, i.e. T_i, is passed to the classifier that assigns a class to it. The main idea of this paper is adopt a meta-ensemble paradigm and to specialize each ensemble on a particular type or set of attacks. To this aim, a sampler module extracts the labelled tuples coming from the streams and assign them to the appropriate datasets (together to a corresponding percentage of normal connections, 50 % in our experiments), named $SD_1, SD_2, \ldots, SD_k$. Then, an ensemble for each specialized dataset is built, by using the distributed GP tool.

Finally, the different ensembles perform a weighted vote in order to decide the correct class. It is worth to remember that each ensemble evolves a function for combining the classifiers, which do not need of any extra phase of training on the original data. The final classification is obtained computing the error using the same formulae of the Adaboost.M2 algorithm used by the GP tool, by computing the error of the entire ensemble instead of a single classifier as in the original boosting algorithm. Afterwards, for each tuple x, for each possible class j and for each ensemble i, the errors are computed using a weighted mean: $\mu_j(x) = \frac{\sum w_i * E_{ij}(x)}{\sum w_i}$ an the final classification is obtained by using the formula $class(x) = argmax_j(\mu_j(x))$

5   Experimental Investigation

In this Section, we illustrate the preliminary experiments performed to evaluate the effectiveness of the ensembles of specialized detectors using the parameters and the datasets described in the next subsection. To this aim, we compared our approach with other related approaches using the well-known KDD dataset and we analyzed the performance of the specialized ensembles on a a really interesting dataset in the field of intrusion detection, presented in the Sect. 2, which overcomes the drawbacks of the KDD dataset. It is worth noticing that these preliminary experiments only aim to evaluate the ensemble module of a local node of the distributed framework, illustrated in Sect. 3 and the cluster cited in next subsection is used only to run the distributed GP tool.

5.1 Datasets and Experiment Setup

All the experiments were performed on a Linux cluster with 16 Itanium2 1.4 GHz nodes, each having 2 GBytes of main memory and connected by a Myrinet high performance network. No tuning phase has been conducted for the GP tool, but the same parameters used in the original paper [5] were used, listed in the following: a probability of crossover equal to 0.7 and of mutation equal to 0.1, a maximum depth equal to 7, a population of 120 individuals and 500 as number of generations. All the results were obtained by averaging 30 runs.

The non-trainable functions chosen to combine the classifiers are listed in the following: average, weighted average, multiplication, maximum and median. Each function is replicated with a different arity, typically from 2 to 5, how specified in the following.

The **average** function, used with an arity of 2, 3 and 5, is defined as: $\mu_j(x) = \frac{1}{g}\sum_{i=1}^{g} H_{i,j}(x)$.

The **multiplication** function (arity 2, 3 and 5) is defined as: $\mu_j(x) = \prod_{i=1}^{g} H_{i,j}(x)$.

The **maximum** function returns the maximum support for 2, 3 and 5 classifiers and can be computed as: $\mu_j(x) = \max_i \{H_{i,j}(x)\}$.

The **median** function (arity 3 and 5) can be computed as: $\mu_j(x) = median_i \{H_{i,j}(x)\}$.

Finally, the **weighted** version of the **average** function uses the weights computed during the training phase to give a different importance to the models on the basis of the performance on the training set, and can be computed as: $\mu_j(x) = \frac{1}{\sum_{i=1}^{g} w_{i,j}} \sum_{i=1}^{g} w_{i,j} * H_{i,j}(x)$. For this function the values of 2, 3 and 5 are chosen for the arity.

5.2 Experiments on the Two Datasets: KDD 99 and ISCX IDS

To evaluate the system proposed, we choose two datasets: the first, the KDD Cup 1999[8] is one of the most used in this field, but present some drawbacks, remarked in Sect. 2. However, we choose to include it in the experiments for the sake of comparison with other similar approaches using this dataset. KDD contains 494,020 records, representing normal connections and 24 different attack types. Each attack is clustered into four main categories, so each connection belongs to the following classes: normal (normal, i.e., no attack), DoS (Denial of Service connections), R2L (Remote to Local, remote attacks addressed to gain local access), U2R (User to Root, exploits used to gain root access) or Probe (probing attack to discover known vulnerabilities). The second dataset is ISCX

---

[8] http://www.sigkdd.org/kdd-cup-1999-computer-network-intrusion-detection.

IDS, consisting of 2,230,620 records, fully labelled, representing realistic traffic scenarios of 7 days, containing different types of attack, i.e., HTTP Denial of Service, DDos, Brute Force SSH and attempts of infiltrating the subnetwork from the inside (see Table 1).

Among the many metrics for evaluating classifier systems, in this paper we choose recall and precision, because they give an idea of the capacity of the system in individuating the attacks and in reducing the number of false alarms; indeed, recall represents the proportion of correctly predicted attack cases to the actual size of the attack class (a value of 100 % indicate we detect all the attacks, however, we can individuate also a large number of false attacks); precision represents the proportion of attack cases that were correctly predicted relative to the predicted size of the attack class (a value of 100 % indicates that no false alarms were signaled, however a large number of alarms could be not detected).

In order to analyze the behavior of our approach in recognizing the minority classes, i.e., the attacks, we compare our approach with that presented in [21], for the KDD Cup dataset. The authors propose a boosting approach, named Greedy-Boost, which builds an ensemble of classifier based on a linear combination of models, specifically designed to operate for the intrusion detection domain. The main idea of that algorithm is to extend the boosting process maintaining the models that behave better on the examples badly predicted in the previous round of the boosting algorithm (while the classical boosting algorithm adjusts only the weights and not the models).

In Table 2 and in Table 3, we reported the result of the comparison between CAGE-MetaCombiner and the Greedy-Boost algorithm on the KDDCup 99 dataset (precision and recall are reported for all the classes). It is evident that our approach performs better both for the precision and the recall measure when considering the minority classes, i.e., the attacks and it is comparable for the other classes.

As for the ISCX dataset, we want to verify the performance of the algorithm, trying to understand the benefits in using specialized classifier. It is worth to remember that an intrusion detection algorithm must able to detect all the types of attack. Therefore, we randomly sampled the labelled tuples of the dataset and

Table 2. Precision for different strategies for the KDD Cup dataset. In the first column, it is reported the class distribution for the dataset.

	Class distribution	Precision	
		Greedy-Boost	CMC specialized
DoS	79.28 %	100.0	99.99
Normal	19.86 %	99.1	99.83
Probe	0.84 %	99.0	99.43
R2L	0.023 %	93.2	98.29
U2R	1.06E-3 %	88.5	91.07

Table 3. Recall for different strategies for the KDD Cup dataset. In the first column, it is reported the class distribution for the dataset.

	Class distribution	Recall	
		Greedy-Boost	CMC specialized
DoS	79.28 %	100.0	99.99
Normal	19.86 %	100.0	99.97
Probe	0.84 %	97.1	97.31
R2L	0.023 %	71.9	86.43
U2R	1.06E-3 %	44.2	74.54

used them as training and validation datasets to train one ensemble for each class; for the normal class, a completely random subsample of the dataset is extracted; for a given attack, a subsample having 50 % of normal connections and 50 % of the attack is extracted. The overall sample extracted represents 2 % of the entire dataset. Finally, the entire dataset is classified with a weighted combination of the ensembles, as illustrated in Sect. 4.2.

In Table 4, we reported precision and recall for the CAGE-MetaCombiner algorithm, for the case of using specialized ensembles and for the case of using not specialized ensemble. The specialized ensembles behaves better in terms of precision for all the types of attack (every day presents a different type of attack) while, as for the precision metric, the two approaches are comparable, with the exception of day 4, in which the behavior of the specialized ensembles is worst; this could be due to the nature of the attacks: indeed, most connections are related to HTTP DDoS attacks and these are quite close to normal connections resulting in increasing of the false positive rate; however, this aspect must be further investigated in future works.

Table 4. Precision and Recall for different strategies (with and without specialized ensemble) for the ISCX dataset.

	Precision		Recall	
	Non specialized	Specialized	Non specialized	Specialized
Day 2	96.12	99.47	99.95	100.00
Day 3	56.67	67.00	92.56	91.82
Day 4	42.24	54.88	81.63	67.55
Day 5	91.50	96.89	97.83	97.49
Day 7	96.58	97.35	99.94	100.00

6   Conclusions and Future Work

A distributed framework for the problem of intrusion detection has been presented and the different components were illustrated. In particular, the meta-ensemble engine used to classify the attacks and the normal connections, based on a non-trainable function evolved by a well-known GP tool, is described in detail.

The experiments conducted on the KDD dataset and on a more up-to-date dataset demonstrate that the framework is able to cope with the intrusion detection task even in the case of really unbalanced classes, by exploiting the advantages of the specialized classifiers. The algorithm succeeds in minimizing the false alarms and generally recognizes well the attacks, however it needs some improvements to detect some particular kinds of attack. In future works, we want to explore the capacity of the algorithm to work in presence of concept drifts and to test the other modules of the distributed framework.

Acknowledgment. This work has been partially supported by MIUR-PON under project PON03PE_00032_2 within the framework of the Technological District on Cyber Security.

References

1. Bhuyan, M., Bhattacharyya, D., Kalita, J.: Network anomaly detection: methods, systems and tools. Commun. Surv. Tutorials IEEE **16**, 303–336 (2014)
2. Breiman, L.: Bagging predictors. Mach. Learn. **24**, 123–140 (1996)
3. Freund, Y., Shapire, R.: Experiments with a new boosting algorithm. In: Machine Learning, Proceedings of the Thirteenth International Conference (ICML 1996), Morgan Kaufmann, pp. 148–156 (1996)
4. Kuncheva, L.: Combining Pattern Classifiers: Methods and Algorithms. Wiley, Chichester (2004)
5. Folino, G., Pizzuti, C., Spezzano, G.: A scalable cellular implementation of parallel genetic programming. IEEE Trans. Evol. Comput. **7**, 37–53 (2003)
6. Cuzzocrea, A., Folino, G., Sabatino, P.: A distributed framework for supporting adaptive ensemble-based intrusion detection. In: 2015 IEEE International Conference on Big Data, Big Data 2015, Santa Clara, CA, USA, 29 October - 1 November 2015, pp. 1910–1916 (2015)
7. Shiravi, A., Shiravi, H., Tavallaee, M., Ghorbani, A.A.: Toward developing a systematic approach to generate benchmark datasets for intrusion detection. Comput. Secur. **31**, 357–374 (2012)
8. Khasawneh, K.N., Ozsoy, M., Donovick, C., Abu-Ghazaleh, N., Ponomarev, D.: Ensemble learning for low-level hardware-supported malware detection. In: Bos, H. (ed.) Raid 2015. LNCS, vol. 9404, pp. 3–25. Springer, Heidelberg (2015)
9. Folino, G., Pisani, F.S.: Combining ensemble of classifiers by using genetic programming for cyber security applications. In: Mora, A.M., Squillero, G. (eds.) EvoApplications 2015. LNCS, vol. 9028, pp. 54–66. Springer International Publishing, Switzerland (2015)
10. Acosta-Mendoza, N., Morales-Reyes, A., Escalante, H.J., Gago-Alonso, A.: Learning to assemble classifiers via genetic programming. IJPRAI **28**, 19 (2014)

11. Tavallaee, M., Stakhanova, N., Ghorbani, A.: Toward credible evaluation of anomaly-based intrusion-detection methods. IEEE Trans. Syst. Man Cybern. Part C Appl. Rev. **40**, 516–524 (2010)
12. Mahoney, M.V., Chan, P.K.: An analysis of the 1999 DARPA/Lincoln laboratory evaluation data for network anomaly detection. In: Vigna, G., Kruegel, C., Jonsson, E. (eds.) RAID 2003. LNCS, vol. 2820, pp. 220–237. Springer, Heidelberg (2003)
13. McHugh, J.: Testing intrusion detection systems: a critique of the 1998 and 1999 DARPA intrusion detection system evaluations as performed by lincoln laboratory. ACM Trans. Inf. Syst. Secur. **3**, 262–294 (2000)
14. Brown, C., Cowperthwaite, A., Hijazi, A., Somayaji, A.: Analysis of the 1999 DARPA/Lincoln laboratory IDS evaluation data with NetADHICT. In: Proceedings of the Second IEEE International Conference on Computational Intelligence for Security and Defense Applications. CISDA 2009, Piscataway, NJ, USA, pp. 67–73. IEEE Press (2009)
15. Tavallaee, M., Bagheri, E., Lu, W., Ghorbani, A.: A detailed analysis of the KDD CUP 99 data set. In: IEEE Symposium on Computational Intelligence for Security and Defense Applications, 2009. CISDA 2009, pp. 1–6 (2009)
16. Paxson, V.: Empirically derived analytic models of wide-area TCP connections. IEEE/ACM Trans. Netw. **2**, 316–336 (1994)
17. Foremski, P., Callegari, C., Pagano, M.: Waterfall: rapid identification of IP flows using cascade classification. In: Kwiecień, A., Gaj, P., Stera, P. (eds.) CN 2014. CCIS, vol. 431, pp. 14–23. Springer, Heidelberg (2014)
18. Schapire, R.E.: The strength of weak learnability. Mach. Learn. **5**, 197–227 (1990)
19. Schapire, R.E.: Boosting a weak learning by majority. Inf. Comput. **121**, 256–285 (1995)
20. Kuncheva, L.: Combining Pattern Classifiers: Methods and Algorithms. Wiley-Interscience, New York (2004)
21. Bahri, E., Harbi, N., Huu, H.N.: Approach based ensemble methods for better and faster intrusion detection. In: Herrero, A., Corchado, E. (eds.) CISIS 2011. LNCS, vol. 6694, pp. 17–24. Springer, Heidelberg (2011)

UAV Fleet Mobility Model
with Multiple Pheromones
for Tracking Moving Observation Targets

Christophe Atten[1], Loubna Channouf[2],
Grégoire Danoy[2(✉)], and Pascal Bouvry[2]

[1] SnT Interdisciplinary Centre, University of Luxembourg,
Luxembourg City, Luxembourg
christophe.atten.001@student.uni.lu
[2] CSC Research Unit, University of Luxembourg, Luxembourg City, Luxembourg
loubna.channouf.001@student.uni.lu, {gregoire.danoy,pascal.bouvry}@uni.lu

Abstract. The last years, UAVs have been developed to address a variety of applications ranging from searching and tracking to the surveillance of an area. However, using a single UAV limits the range of possible applications. Therefore, fleets of UAVs are nowadays considered to work together on a common goal which requires novel distributed mobility management models. This work proposes a novel nature-inspired mobility model for UAV fleets based on Ant Colony Optimisation approaches (ACO). It relies on two types of pheromones, a repulsive pheromone to cover the designated area in an efficient way, and an attractive pheromone to detect and to track the maximum number of targets. Furthermore, all decision takings are taken online by each UAV and are fully distributed. Experimental results demonstrate promising target tracking performances together with a small increase in the exhaustivity of the coverage.

Keywords: UAV · Mobility model · Pheromone · Target tracking · Ant Colony Optimisation

1 Introduction

Airborne surveillance of an area or tracking of some moving targets are typical military applications. However, these are also applicable in various civilian domains, from search and rescue to forest fire detection and water pollution detection. The usage of unmanned aerial vehicles (UAVs) for such applications has seen a dramatic increase in the past years. Current solutions typically rely on a single UAV, which restricts the observation capabilities and range of action due to a limited payload capacity and autonomy. In addition tracking can only be achieved on a single target.

Using several UAVs, referred to as a fleet, equipped with wireless communication capabilities permit to address these limitations. These have been recently

© Springer International Publishing Switzerland 2016
G. Squillero and P. Burelli (Eds.): EvoApplications 2016, Part I, LNCS 9597, pp. 332–347, 2016.
DOI: 10.1007/978-3-319-31204-0_22

considered for various applications, ranging from forest fires surveillance [1] to forest environment assessment [2] and emergency management in crisis zones [3]. Such communicating UAVs fleets constitute a specific type of mobile ad hoc network (MANET) referred to as Flying Ad Hoc Networks (FANETs). This work specifically considers a fleet of autonomous fixed-wings UAVs equipped with communication and sensing (e.g. radar) hardware. Operating fleets of autonomous UAVs raises novel issues such as mobility management and communication maintenance with the base station.

We here propose a novel nature-inspired mobility model, based on Ant Colony Optimisation (ACO), that extends the surveillance-oriented mobility model introduced by Schleich *et al.* in [4]. In the latter, the UAVs could perform an efficient scanning of an area by dropping repulsive pheromones on already scanned zones. The objective was to prevent other UAVs of the fleet to visit recently scanned areas. In order to fulfil the mission of target discovery and tracking, a second type of digital pheromones, i.e. attractive pheromones, is added in this contribution. These will help the UAVs to follow the targets by dropping such attractive pheromones where targets were discovered. As pheromones evaporate in time as in nature, UAVs can only use them during a limited amount of time to find back lost targets.

The remainder of this paper is organised as follows. In Sect. 2 the related work is presented, followed by a detailed and formalised problem description in Sect. 3. Then, in Sect. 5, the different quality metrics to assess the algorithm's performance are presented. The proposed novel mobility model with two types of pheromones is described in Sect. 4. Sections 6 and 7 respectively present the setup used for the evaluation of the different scenarios and the analysis of the numerical results. Finally, Sect. 8 provides some conclusions and future perspectives.

2   Related Work

The UAV field has seen a tremendous growth in the last years which raises new issues, from legal aspects such as flight regulations, to R&D perspectives such as autonomous swarms management [5].

This work focuses on mobility management of autonomous UAV swarms in the context of area surveillance, target detection and tracking. UAVs are equipped with wireless networking interfaces that allow them to communicate in an ad hoc fashion, forming so called Flying Ad hoc NETworks (FANETs).

In order to benefit from the full flexibility offered by autonomous UAV swarms, online and distributed path planning methods are desirable. However most existing works consider either centralised methods or distributed ones but using offline pre-computation and online decision taking [6,7].

Few works proposed both distributed and online methods. One promising approach is ant colony, a nature-inspired method that relies on stigmergy. Ants deposit virtual pheromones in the environment, e.g. to indicate recently explored regions. As in nature, these pheromones will then evaporate in time.

This work proposes to extend one such pheromone-based mobility model [8], that uses digital repulsive pheromones for surveillance. This way the UAVs perform an efficient coverage by scanning in priority areas that contain no or a small amount of pheromones.

Some evolutions of pheromone-based models were proposed, such as [9] which performance was demonstrated against a random mobility model for surveillance, target acquisition and tracking.

Nevertheless, network connectivity was still presented as a problem. This was addressed in [4] which focused on using multiple UAVs for area coverage with a connectivity constraint. The latter work however focused only on a surveillance scenario and used only one type of pheromone, called repulsive pheromone. A main characteristic of this pheromone is to lead other UAVs in the fleet to other places in the area which are not controlled yet. This strategy leads to a higher coverage of the aforementioned area. In addition, this model served as basis for the pheromone model proposed in this paper, which will use a supplementary type of digital pheromone in order to fulfill the mission of target discovery and tracking.

A general multiple pheromone approach is described in [10], in which different rules can be created for the different pheromones. However, this was used in a different context, i.e. for classification. The algorithm designed in this work is based on a second type of pheromone, an attractive one for target discovery and tracking, along with the already existing repulsive pheromone used for surveillance.

3   Problem Description

This work aims at optimising the mobility of swarms of autonomous UAVs in a distributed and online fashion for both area surveillance and targets tracking while maintaining the UAV network connectivity. The latter goal is important for safety reasons as it is generally required by air regulations to be able to bring a flying entity back to the base station at any time. The considered UAVs are fixed-wing units which payload embeds wireless communication interfaces, allowing to send and receive message from/to the base station and other UAVs in an ad hoc fashion, and a scanner that permits to cover some geographical area.

The focus is put on the following three main objectives to appraise the quality of the proposed mobility models:

- The exhaustivity of the coverage: covering a maximum of the surveilled area.
- The global connectivity between the UAVs and base station reachability.
- The observation of targets in the surveyed area are composed of target tracking, untracked targets and target detection.

4   Multiple Pheromone UAV Mobility Model

In the following section the main contribution of this work is provided, i.e. a UAV mobility model based on two different types of pheromones: *repulsive pheromones*

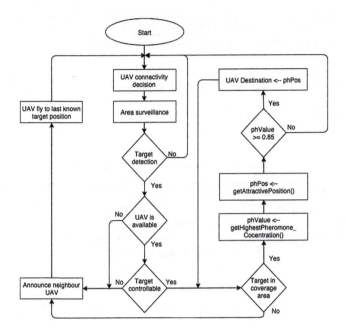

Fig. 1. Flowchart of the multiple pheromone mobility approach

to cover the area and *attractive pheromones* to follow targets. The repulsive pheromones allow UAVs to scan the area in an ant colony fashion avoiding to visit places that were covered recently, i.e. still containing repulsive pheromones. Attractive pheromones are deposited by UAVs that are tracking targets, so as to allow other UAVs to follow targets or to attract them in areas where the probability to find new targets is high.

This mobility model consists of the following sequential steps: connectivity approaches, area surveillance, target discovery, decision taking based on the UAV state and position, following best attractive positions and finally, target loss and recovery. The corresponding flowchart is presented in Fig. 1. Each step is described in detail in the following subsections.

4.1   Connectivity Approaches

The proposed mobility model is based on multiple pheromones. It thus uses the idea of stigmergy, i.e. the ability to coordinate agents in an indirect fashion based on the exchange of information (pheromones) in the environment. In our scenario, the environment, i.e. the area to be scanned, cannot be used for exchanging pheromones. As a result, we rely on a virtual and discrete representation of the territory via a map.

Furthermore, two different UAV connectivity approaches are considered. The first variant, the *UAV Inter-communication with usage of unavailable neighbour UAVs*, referred to as *maximum connectivity*, focuses on staying connected as

Algorithm 1. Available UAV discovers a Target

1: **if** *UAV is unused* **and** *target is in coverage range* **then**
2:    **if** *discovered target is untracked* **then**
3:       **if** *compare flying direction is acceptable* **then**
4:          *Execute Algorithm 3: Pheromone Decision Taking*
5:       **else**
6:          *Execute Algorithm 6: Notify Neighbour UAV*
7:       **end if**
8:    **end if**
9: **end if**

Algorithm 2. Used UAV discovers a Target

1: **if** *UAV is used* **and** *target is in coverage range* **then**
2:    **if** *discovered target is untracked* **then**
3:       **if** *discovered target != UAV's current target* **then**
4:          *Execute Algorithm 6: Notify Neighbour UAV*
5:       **end if**
6:    **end if**
7: **end if**

much as possible to cover the area, even if a neighbour UAV is currently following a target and not used to cover the area only. All the neighbour UAVS are used, which puts emphasis on the connectivity, but potentially to the expense of the area coverage. The second variant, the *UAV Inter-communication without usage of unavailable neighbour UAVs*, referred to as *low connectivity*, ignores the UAVs which are following targets for building a connected UAV fleet group. The latter could lead to a worse connectivity of the overall fleet, but might improve the coverage of the area and optimise the following of the targets, as the UAVs would be more spread. Both models have a so-called critical point that occurs when a UAV is following a target. The latter will choose at every time-step a new destination based on the attractive pheromones to optimise the following of the targets.

4.2   Area Surveillance

By default, all UAVs scan the area in order to detect targets. The UAVs scanning approach is similar to the one previously proposed for area surveillance [4]. It relies on repulsive pheromones, evaporating in time, that indicate recently visited geographical areas. Each UAV then moves left, right, or forward, with a probability that is inversely proportional to the amount of pheromones in these places.

4.3   Target Detection

In order to ensure target discovery and tracking, the mobility model was extended with a second type of pheromone, referred to as attractive pheromone.

Algorithm 3. Pheromone Decision Taking

1: **Part 1: Drop Attractive Pheromone**

2: **if** *target is in coverage range* **then**
3:     *pCell ← Determine the pheromone cell of the target position*
4:     *pCell.add(Attractive Pheromone)*
5: **end if**

6: **Part 2: Take decision on attractiveness**

7: **if** *UAV is used* **then**
8:     *pValue ← getHighestPheromoneConcentration()*
9:     *phPos ← getAttractivePosition()*
10:     **if** *phValue* ≥ 0.85 **then**
11:         *UAVDestination ← phPos*
12:     **else**
13:         *Proceed with covering the area*
14:     **end if**
15: **else**
16:     *Proceed with covering the area*
17: **end if**

The discovery of a target is described in Algorithms 1 and 2 for an available UAV or used UAV (i.e. already following a target) respectively. These are placed where and when a target is discovered and spread as long as it is followed by a UAV, as described in Algorithm 3.

4.4 Following a Target Based on Attractiveness

Target tracking is defined by how a UAV decides to follow or to cover the area for new targets. This is described in Algorithm 3. Part 1 of the algorithm describes the deposit of attractive pheromones. These are placed on cells that compose the discretised geographical area, a single cell being described as pCell. The second part of the algorithm describes the decision taking between either covering or tracking. When the concentration of attractive pheromones, phValue, in one or more neighbouring cells (front, left and right) is higher than 0.85 (85 %), the UAV chooses the position of the highest attractive pheromone concentration, phPos, as new destination. If the concentration is lower than 0.85 in the three neighbouring cells, the UAV will continue with covering the area.

4.5 Computing Best Decision

Fixed-wing UAVs cannot hover nor change instantly their direction to follow every discovered target. It might then happen that a UAV cannot follow one detected target. At each iteration a UAV has then to decide whether to keep following the target or to notify a neighbour UAV that is a good candidate for tracking the considered target. For this, three algorithms are presented to illustrate this decision taking.

Algorithm 4. Compare Flying Direction

1: $uFly \leftarrow$ UAV flying direction
2: $uPos \leftarrow$ UAV position
3: $tFly \leftarrow$ Target flying direction
4: $tPos \leftarrow$ Target position
5: **if** $uFly == tFly$ **and** $uPos$ is well positioned to $tPos$ **then**
6:     **return true**
7: **else**
8:     **return false**
9: **end if**

Algorithm 5. Target probably lost

1: *Initialise(targetProbablyLostTime)*
2: **if** $(getCurrentTime() - targetProbablyLostTime) > Shorty\ lost\ time$ **then**
3:     *Target is completely lost*
4: **else if** *UAV found target back* **then**
5:     *Reset(targetProbablyLostTime)*
6:     *Execute Algorithm 1: Pheromone Decision Taking*
7: **end if**

Algorithm 4 shortly describes the verification if a UAV is well-positioned to follow a discovered target. In more details, when a UAV discovers a target, it compares its position and flying direction to the target ones. For instance, when the UAV is behind the target and both the UAV and target are flying to the North, the UAV is well positioned to follow the target.

This algorithm will search for the closest neighbour UAV which is available, which means, when a UAV is not used to track a target. This UAV is then notified to fly to the last known position of the target.

4.6   Target Loss and Recovery

The last part of this new mobility model deals with the case when a UAV lost a target and wants to find it back. Algorithm 5 shortly describes the usage of a new parameter, i.e. the time a target is shortly lost. This parameter is added to give a UAV a chance to find the target back after loosing it. If the UAV finds the target back, the tracking is not interrupted, otherwise its is interrupted and Algorithm 6 is executed. When the UAV has definitively lost the target, the amount of time a UAV probably lost a target until the time the target is definitively lost, will be substracted from the average tracking time.

5   Quality Metrics

This section introduces the different metrics used to assess the performance of the proposed mobility models, both in terms of surveillance and target observation. The first three metrics are related to the observation and networking objectives,

---

Algorithm 6. Notify neighbour UAV

1: **Part 1: Notify neighbour UAV to try to follow a target**
2: $closestUAV \leftarrow null$
3: **for** UAV in $availableNeighbourUAVs$ **do**
4:    **if** $closestUAV == null$ **or** UAV is closer to myself than $closestUAV$ **then**
5:       $closestUAV \leftarrow UAV$
6:    **end if**
7: **end for**
8: *Notify closestUAV to try to follow the target*

9: **Part 2: Receiving order to control an unreachable target**
10: **if** *UAV got message from neighbour to control an unreachable target* **then**
11:    *Unused UAV will fly to the last known position of this target*
12:    **if** *UAV discovers a target* **then**
13:       *Execute Algorithm 1: Available UAV discovers a Target*
14:    **else**
15:       *Unused UAV will continue normal patrolling*
16:    **end if**
17: **end if**

---

namely the exhaustivity of the coverage, the global connectivity between the UAVs and the reachability of the base station. The other three metrics permit to evaluate the targets observation and tracking abilities, i.e. the average and best amount of time a target is tracked, the amount of time a target is untracked and the number of target detections.

5.1 Exhaustivity of the Coverage

The quality of the geographical area coverage is expressed by the exhaustivity. It describes the amount of area that has never been scanned during a simulation run. Literally, it is the percentage of never scanned cells in the discretised area. More information about the exhaustivity of the coverage can be found in [4]. As a complement, the percentage of scanned cells during the simulation is also measured.

5.2 Connectivity

Two types of connectivities are considered, i.e. between UAVs and between UAVs and the base station. The number of connected components grants information about how much disconnected is a graph, i.e. how many groups of UAVs exist. The optimal value for this metric is thus 1. Moreover, the number of connections to the base station is represented in percentage. This value provides information on how many UAVs can communicate over one or multiple hops to the base station.

5.3   Average and Longest Target Tracking Time

The target tracking time is the amount of time a UAV manages to follow a target it has detected. This metric is calculated for all targets and represented as the average in milliseconds. The longest tracking time metric complements the previous one. It measures the longest time a UAV was able to follow a target during a run. This metric shows the average of all the longest tracking times in milliseconds.

5.4   Target Untracked Time

While some UAVs successfully fulfil their mission of tracking targets, others may not be able to do so. Target unattainability is an inevitable consequence of a low amount of UAVs and respectively high amount of targets. The untracked time starts when a target status is undetected and ends when a UAV discovered it. This metric represents the average time during which targets are untracked by a UAV.

5.5   Number of Target Detections

The number of target detections is an important metric to consider in combination with the target untracked time and the tracking time. This metrics is represented as the total amount of detections in a run.

6   Experimental Setup

In this section, the simulation environment is presented along with the most important experimental parameters. Afterwards, the two variants of the mobility model for the UAVs are presented as well as the single pheromone mobility model these are compared to. Finally, the two mobility models for the targets are introduced.

6.1   Simulation Environment

The simulation environment is fully based on the dynamic graph library used in [4]. In this subsection we provide details concerning the implementation of the area, the autopilot of the UAVs and the main simulation parameters which values are summarised in Table 1.

The Area. The simulation area is a rectangle with a long side L_{long}, short side L_{short} and discretised in square cells of side L_{cell} as in [4]. Additional parameters are required when the Manhattan mobility model is used for the targets (described in Sect. 6.3). MB_{width} and MB_{height} describe the size of a Manhattan box, which is a square of dimension $MB_{width} * MB_{height}$. Hence, the quantity of Manhattan boxes in the simulation area L_{long} and L_{short} is:

$$MB_{quantity} = \frac{L_{long} * L_{short}}{MB_{width} * MB_{height}} \tag{1}$$

Table 1. Main simulation parameters

Simulation Area	Parameter Value	Target Autopilot	Parameter Value
L_{long}	2000	Speed	[0 .. 5]
L_{short}	1000	Acceleration	[-1 .. 1]
L_{cell}	20	Heading change	[0 .. 0.1]
Base station position	(1000,0)	Decision Frequency	[1, 30]
UAV Autopilot	**Parameter Value**	Shortly lost time	500ms
Speed	[0 .. 5]	**Manhattan Grid**	**Parameter Value**
Acceleration	[-1 .. 1]	MB_{width}	200
Heading change	[0 .. 0.1]	MB_{height}	100
Decision Frequency	[1, 30]	**Experiments**	**Parameter Value**
Future	30	Number of UAVs	[10 .. 50]
Wireless Range	400	Number of runs per experiment	30
Coverage Range	2	Number of Targets	10
Attractive Pheromone Threshold	0.85 (85%)		

UAV Autopilot. The implemented autopilot for the fixed-wing UAVs is similar to [4] in case of area surveillance, with a destination change every 30 time-steps. However, in order to ensure a higher accuracy when tracking targets, the current destination is updated every time-step. This is motivated by the criticality of this phase for the UAVs that need to react fast to target destination changes.

Simulation Parameters. As all the mobility models are stochastic, each set of experiments was repeated 30 times in order to obtain a good statistical confidence. For all quality metrics the average was computed. The results obtained on the 8 scenarios have been statistically compared according to the Wilcoxon unpaired signed rank test for the 8 metrics [11]. As it can be seen in Table 1, experiments were conducted with very different density conditions, ranging from 10 UAVs up to 50 with a wireless transmission range fixed at 400.

Limitations. As the UAV model is based on [4], the limitations are also similar. Additionally, it is assumed that every UAV knows if a neighbour UAV is busy or not. This information could get exchanged in a real-world execution through so-called beaconing.

6.2 Compared UAV Mobility Models

The multiple pheromone mobility model is compared to the repulsive pheromone mobility model from [4]. In this single pheromone mobility model, each UAV locally decides where to move based on different criterias, such as future position of one-hop neighbors, as well as the concentration of the repulsive pheromone around, to attract UAVs to places that are characterised by the lowest pheromone concentration. The target detection has no direct impact on the movement of the old mobility model proposed in [4] as long as there is no attractive pheromone around. Indeed, the attractive pheromone has a higher priority than the repulsive one.

Table 2. Possible scenarios with abbreviations

UAV connectivity		UAV model			Target model		*MC = Maximum Connectivity LC = Low Connectivity Abbreviation
MC	LC	Multiple pheromones	Single pheromone		Random	Manhattan	
X		X			X		MCMR
X		X				X	MCMM
X			X		X		MCSR
X			X			X	MCSM
	X	X			X		LCMR
	X	X				X	LCMM
	X		X		X		LCSR
	X		X			X	LCSM

6.3   Target Mobility Models

Two different target mobility models are used, a *random mobility model* and a *Manhattan mobility model*. The *random mobility model* decides uniformly at random between three possible directions with some fixed probabilities: front (60 %), right (20 %) and left (20 %). So each target randomly chooses a new direction (Left, Front, Right). This decision is taken every 30 time-steps when the target is untracked. If the target is tracked, it will switch to the critical mode, where the decision is made every time-step. The second mobility model, *Manhattan mobility model*, is based on a grid of "streets", that represents a simplified city road topology. Targets mobility is constrained on these streets. The dimension of the grid is expressed in Table 1. Targets have again three direction choices, i.e. left, front or right, with 25 %, 25 % and 50 % probabilities respectively. Decisions can only be made on crossroads. Furthermore, the critical mode is not available by the *Manhattan mobility model* through the fact that the targets can only decide on corner points.

7   Numerical Results

This section presents the evaluation and comparison of the different mobility models using the metrics presented in Sect. 5 and Table 2.

7.1   Coverage and Connectivity

Number of Scanned Cells. For the number of scanned cells, presented in Fig. 2, it can be observed that the new multiple pheromones UAV models, LCMM, LCMR, MCMM and MCMR are performing slightly better than the other ones, especially in the densest scenarios. This outlines that the usage of the two pheromone types permits the UAVs to reach unvisited places faster. Furthermore, we can claim with statistical confidence that the usage of two types of pheromones, MCMR and LCMR, provides the best results.

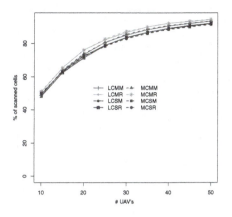

Fig. 2. Percentage of scanned cells

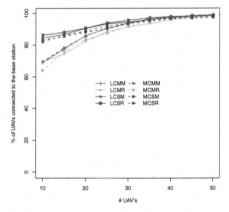

Fig. 3. Percentage of never scanned cells

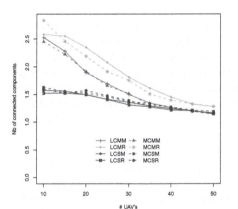

Fig. 4. Number of connected compo-nents

Fig. 5. UAVs connected to the base station

Number of Never Scanned Cells. The number of never scanned cells in Fig. 3 confirms the previous results. More scanned cells logically implies less never scanned cells. For the lowest density, i.e. 10 UAVs, LCMR and MCMR perform best with statistical confidence on the random mobility targets scenario with values ranging between 6 and 8 %, which is two times better than the previous approaches. Similarly, the multiple pheromone models, LCMR and MCMR, per-form better on the random mobility targets than the Manhattan approach with statistical confidence. This advantage decreases as the UAVs density increases, which logically means that the area is better covered.

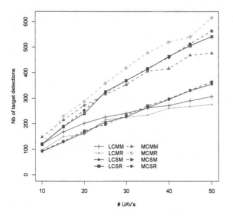

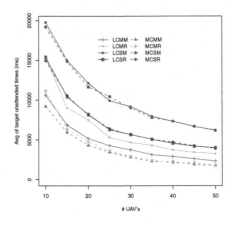

Fig. 6. Number of target detections   **Fig. 7.** Average untracked time

Number of Connected Components. This metrics represents the number of connected components (groups) during the simulation see Fig. 4. The optimal value is 1, which means that the base station can reach any UAV using multi-hop communication. At the connectivity level, we can say with statistical confidence, that the multiple pheromone approaches perform less good than the single pheromone approaches. This is not very surprising since the target following and the higher priority of the attractive pheromones imply that UAVs are more distributed in the simulation area.

Number of Components Connected to the Base Station. Similarly to the number of connected components, we can say with statistical confidence that the multiple pheromone approaches are performing less good at a low density of UAVs, which is not very surprising since the target following has a higher priority, as presented in Fig. 5. However, for all number of UAVs we can say that there is no statistical difference between all scenarios.

7.2   Tracking

Number of Detections. For the number of detections in Fig. 6 we can say that a lower number of detections might reflect to higher observation times while following a target. Nevertheless, it could also lead to high times for a few untracked targets, while some UAVs are actually following their target and the rest of the fleet is not able to track down the remaining targets.

Average Untracked Time of a Target. It clearly appears with statistical confidence that the multiple pheromone models always outperform the random ones on all scenarios, i.e. LCMM, MCMM, LCMR and MCMR, see Fig. 7. The maximum connectivity approaches, MCMM and MCMM provide better results

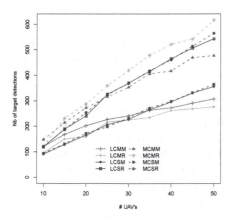
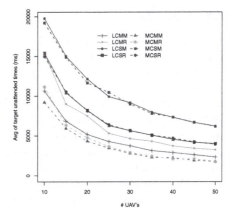

Fig. 8. Average tracking times

Fig. 9. Average of longest tracking times

with statistical confidence to other scenarios, which means that UAVs flying in a more connected way are able to better track targets back. This difference remains the same no matter the UAV density.

Average Tracking Time of a Target. In Fig. 8 we can observe with statistical confidence that the multiple pheromone models with the low connectivity approach, i.e. LCMM and LCMR provide the best results on the two scenarios, with an average of 7200 ms. This result is around 4100 ms better than by using the maximum connectivity variant, which is logical, since in that case UAVs that are tracking targets might loose them to remain connected with the swarm. A notifiable difference exists between the two target mobility models. The Manhattan model is more continuous and stable. This is explained by the fact that targets move on streets and not randomly in the whole area. With statistical confidence we can say that UAVs perform better on the Manhattan scenario model in comparison to the random one. This can be explained by the random model properties, that permits targets to move shortly out of the area covered by the UAVs. Targets will thus be lost since the UAVs cannot put attractive pheromones outside of the area, nor follow an attractive pheromone path.

Average of the Best Tracking Times of a Target. In Fig. 9, we can see the average of the best following times for a UAV. With statistical confidence, LCMM is in average performing better due to the fact of the higher distribution in the area, which leads to a possible higher number of detections and possible better positioning in the area to follow a target and the more continuous moving of the targets without leaving the area.

7.3 Summary

These experiments have provided information about the performance of the different UAV mobility models and associated connectivity approaches on two target models, i.e. random and Manhattan. For both coverage and tracking, the multiple pheromone models combined with the low connectivity strategy, LCMM and LCMR, provide the best results in most cases compared to the MCMM, MCMR or the random mobility model without tracking MCSM, MCSR, LCSM and LCSR. Mainly, LCMM and LCMR reach very good results for target tracking and even better results for coverage compared to their competitors. Nevertheless, these results induce a small degradation in terms of connectivity between the UAVs and the base station. This drawback is not a surprise, because the UAVs first priority is the tracking of targets. Therefore, this slightly worse connectivity is not too penalising since much better results are obtained in targets tracking.

8 Conclusion and Perspectives

This paper proposed a novel decentralised and online mobility model with multiple pheromones for surveillance and target tracking missions. The latter consists in the combination of a repulsive pheromone-based mobility model [4] with an attractive pheromone one. On the one hand, the repulsive pheromones are used to survey the area by repelling UAVs to less scanned cells, and on the other hand, the attractive pheromones are used to follow and track discovered target. Its performance is evaluated using a set of state-of-the-art and novel quality metrics. With this combination, the proposed mobility model was able to outperform a single pheromone mobility model extended with the proposed connectivity approaches in terms of coverage of the area (i.e. repulsive pheromone) and showed good target tracking abilities.

Future work will focus on extending the multiple pheromone mobility model with target path predictions in case of target loss as well as trying to adapt the threshold for the attractive pheromone. Moreover, we can implement a multi-level UAV swarm with different types of UAVs as in the article [12], to increase the connectivity stability and achieve possible better tracking times.

References

1. Casbeer, D.W., Kingston, D.B., Beard, R.W., McLain, T.W.: Cooperative forest fire surveillance using a team of small unmanned air vehicles. Int. J. Syst. Sci. **37**, 351–360 (2006)
2. Brust, M.R., Strimbu, B.M.: A networked swarm model for UAV deployment in the assessment of forest environments. In: Tenth IEEE International Conference on Intelligent Sensors, Sensor Networks and Information Processing, ISSNIP 2015, Singapore, 7–9 April 2015, pp. 1–6. IEEE (2015)
3. Bupe, P., Haddad, R., Rios-Gutierrez, F.: Relief and emergency communication network based on an autonomous decentralized UAV clustering network. In: SoutheastCon 2015, pp. 1–8 (2015)

4. Schleich, J., Panchapakesan, A., Danoy, G., Bouvry, P.: UAV fleet area coverage with network connectivity constraint. In: Proceedings of the 11th ACM International Symposium on Mobility Management and Wireless Access. MobiWac 2013, pp. 131–138. ACM, New York (2013)
5. Valavanis, K.P., Vachtsevanos, G.J.: Handbook of Unmanned Aerial Vehicles. Springer, Netherlands (2015)
6. Althoff, D., Althoff, M., Scherer, S.: Online safety verification of trajectories for unmanned flight with offline computed robust invariant sets. In: IEEE/RSJ International Conference on Intelligent Robots and Systems (2015)
7. de la Cruz, J.M., Besada-Portas, E., Torre-Cubillo, L., Andres-Toro, B., Lopez-Orozco, J.A.: Evolutionary path planner for UAVs in realistic environments. In: Proceedings of the 10th Annual Conference on Genetic and Evolutionary Computation, GECCO 2008, pp. 1477–1484. ACM, New York (2008)
8. Simon, D.: Evolutionary Optimization Algorithms. Wiley, New York (2013)
9. Kuiper, E., Nadjm-Tehrani, S.: Mobility models for UAV group reconnaissance applications. In: International Conference on Wireless and Mobile Communications, ICWMC 2006, p. 33 (2006)
10. Salama, K.M., Abdelbar, A.M., Otero, F.E.B., Freitas, A.A.: Utilizing multiple pheromones in an ant-based algorithm for continuous-attribute classification rule discovery. Appl. Soft Comput. **13**, 667–675 (2013)
11. Wilcoxon, F.: Individual comparisons by ranking methods. Biometrics Bull. **1**, 80–83 (1945)
12. Danoy, G., Brust, M.R., Bouvry, P.: Connectivity stability in autonomous multi-level UAV swarms for wide area monitoring. In: Proceedings of the 5th ACM Symposium on Development and Analysis of Intelligent Vehicular Networks and Applications, DIVANet 2015, pp. 1–8. ACM, New York (2015)

EvoCOMPLEX

Towards Intelligent Biological Control: Controlling Boolean Networks with Boolean Networks

Nadia S. Taou$^{(\boxtimes)}$, David W. Corne, and Michael A. Lones

School of Mathematical and Computer Sciences, Heriot-Watt University,
Edinburgh EH14 4AS, UK
{nt2,d.w.corne,m.lones}@hw.ac.uk

Abstract. Gene regulatory networks (GRNs) are the complex dynamical systems that orchestrate the activities of biological cells. In order to design effective therapeutic interventions for diseases such as cancer, there is a need to control GRNs in more sophisticated ways. Computational control methods offer the potential for discovering such interventions, but the difficulty of the control problem means that current methods can only be applied to GRNs that are either very small or that are topologically restricted. In this paper, we consider an alternative approach that uses evolutionary algorithms to design GRNs that can control other GRNs. This is motivated by previous work showing that computational models of GRNs can express complex control behaviours in a relatively compact fashion. As a first step towards this goal, we consider abstract Boolean network models of GRNs, demonstrating that Boolean networks can be evolved to control trajectories within other Boolean networks. The Boolean approach also has the advantage of a relatively easy mapping to synthetic biology implementations, offering a potential path to in vivo realisation of evolved controllers.

Keywords: Gene regulatory networks · Boolean networks · Control · Evolutionary algorithms

1 Introduction

Recently there has been a growing interest in methods that can be used to control the kind of complex processes that occur widely in the natural world [22]. Many of these processes can be thought of as dynamical systems that take place on networks: such as ecologies, societies and economies, and most of them are important to human life and livelihood. In this paper, we focus on a form of complex dynamical network that is found within all biological cells: gene regulatory networks (GRNs). These play a fundamental role within both the development and ongoing function of biological organisms, and effective mechanisms for controlling them could open up new forms of therapeutic disease interventions [16]. For instance, in [17,18], the authors discussed how a series of perturbations could be used to induce desired changes in a cell's behaviour, and this

© Springer International Publishing Switzerland 2016
G. Squillero and P. Burelli (Eds.): EvoApplications 2016, Part I, LNCS 9597, pp. 351–362, 2016.
DOI: 10.1007/978-3-319-31204-0_23

could form the basis for systems-based drug discovery to cure complex diseases such as cancer. However, it is difficult to find or implement effective control methods within highly non-linear biological systems such as these. Nevertheless, recent developments in control theory and the computational discovery of control interventions have shown that control of such complex dynamical systems is possible [4,21,25]. This makes us hopeful that useful control can be achieved in biological systems.

GRNs can be modelled in many different ways [8]. One of the best known and often used GRN models is the Boolean network (BN) [15]. Although deceptively simple, these discrete time, binary-state models have proved surprisingly effective at capturing the qualitative dynamics of many biological circuits. For this reason, they have been widely studied [2,3,9,11,13]. However, they have also been criticised for being too simple [14], particularly regarding their ability to accurately capture the quantitative dynamics of regulatory circuits. Nevertheless, they remain popular. They also have a simple mapping to the kind of digital circuit models that are often used in synthetic biology, providing a possible route to realising BNs *in vivo*.

Controlling BNs can hence be thought of as a model for the control of GRNs, where their ease of implementation and simulation makes them a convenient tool for studying potential control interventions in biological systems. This has led to significant interest in the problem of controlling BNs, particularly from the perspective of conventional control [1,5,7,19,20]. However, it has proved to be a very difficult problem to solve analytically [1], with feasible solutions only possible for restricted models [1].

The objective of our research is to develop general methods for designing control interventions in biological circuits, with an eventual aim of applying these to the control of real biological circuits *in vivo*. To this end, this paper describes initial work looking at how BNs can be controlled using other BNs. This approach is motivated by the observation that computational models of GRNs (including BNs) have useful computational properties, with previous work using evolutionary algorithms showing how they can be optimised to generate computational and control behaviours that are complex, robust and relatively compact [22,23,28]. The ease of mapping a BN to a synthetic biology implementation also motivates our interest, since it provides a potential future path for implementing evolved controllers *in vivo*.

In this paper, we show that it is possible to evolve BNs that can control other BNs. We also discuss the effect that the size of a BN has on both its ability to control and its ability to be controlled, and make some observations regarding the effects of coupling and connectivity. The paper is organised as follows: Sect. 2 presents a brief introduction to BNs. Sections 3 and 4 review previous work on controlling and evolving BNs, highlighting their limitations, constraints and the challenges the BN control problem involves. Section 5 describes our methodology. Section 6 presents results and analysis, and Sect. 7 concludes.

2 Boolean Networks

A Boolean network (BN) is a discrete-time non-linear dynamical system represented as a directed graph $G(V, E)$ composed of nodes, or vertices, V and edges E [9,11,15] (see Fig. 1). The time evolution of a BN is expressed by a set of Boolean functions f_i, $i = 1, 2, 3, ...$ Each BN node has a binary state s which is updated synchronously according to its Boolean function and the states of the k input nodes that are connected to it. Formally, $s(t + 1) = f_i(s(t))$, where s is a set of network states $s \in \{0, 1\}^N$, $t = 0, 1, 2, 3, 4, ...$ is the discrete time, $f_i : \{0, 1\}^N \rightarrow \{0, 1\}$ (see Fig. 3). Since a BN is deterministic $s(t + 1)$ is only determined by $s(t)$. The possible number of Boolean functions is 2^{2^k}, and the state space is finite and equal to 2^N. Since the state space is finite, states must eventually be repeated (see Fig. 2), leading to temporal structures called attractors. In particular, three regimes can be observed in BNs: *ordered, chaotic* and *critical*. Ordered BNs have attractors with a relatively short period, repeating the same series of states over and over again. Chaotic BNs have long periods; although deterministic, they appear random. Critical BNs also have attractors with long periods, but these appear to have a complex internal order. A random Boolean network (RBN) is a BN which is randomly sampled from the set of all possible BNs, i.e. node inter-connections and the Boolean functions associated with each node are randomly generated. RBNs with $k < 2$ tend to be ordered; those with $k > 2$ tend to be chaotic; critical dynamics tend to be found when $k = 2$.

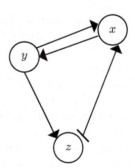

Fig. 1. Boolean network structure with three nodes

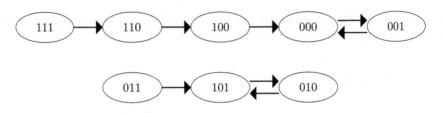

Fig. 2. State transition graph

xyz (t)	xyz (t+1)
000	001
001	001
010	101
011	101
100	000
101	010
110	100
111	110

(a) **Truth table**

$x^{(t+1)}$	y^t
$y^{(t+1)}$	$x^t \wedge z^t$
$z^{(t+1)}$	$\neg x^t$

(b) **Boolean functions**

Fig. 3. Boolean functions and truth table

3   Controlling Boolean Networks

The BN control problem involves guiding a BN's trajectory towards a particular part of its state space. Ideally this should be achieved by manipulating the states of a minimal group of nodes within the BN, and with the goal of achieving the target state in a minimal period of time. There have been a number of previous results concerning the control of BNs [1,7,16,19,29]. Akutsu et al. [1] demonstrated that the control of BNs is NP-complete. However they were able to find exponential time algorithms to control probabilistic BNs, and have also shown that the control problem can be solved in polynomial time in BNs that are restricted to a tree structure. Shi-Jian and Yi-Guang, in [29], developed a method for achieving chaos control in BNs. Cheng and Qi [7] and Kobayashi and Hiraishi [19] both developed optimal control techniques for BNs: the former using an algebraic approach, the latter applying integer programming to an asynchronous BN model. However, neither addressed the issue of computational complexity, and the scalability of these techniques to larger-scale networks. One potential mechanism for overcoming this problem can be found in the work of Kim et al. [16], who designed a general algorithm for identifying control kernels within GRNs. Control kernels are the minimal set of network nodes that need to be regulated to force the network to enter into a desired stable state. They have shown the applicability of their approach on a variety of biomolecular regulatory networks, such as the saccharomyces cerevisiae cell cycle network and the GRN underlying mouse myeloid development. This suggests that the BN control problem may be simplified by first identifying the control kernel and then applying control.

4   Evolving Boolean Networks

In addition to modelling biological GRNs, a number of studies have shown that GRN models can be used to carry out complex computational and control behaviours that are to some degree analogous to their biological activities [23]. Typically this is done by optimising the model using an evolutionary algorithm, and includes a number of approaches that have used BNs [10, 12, 28, 30]. For example, Bull and Preen evolved BNs to solve digital design problems such as multiplexing in synchronous and asynchronous systems [6]. Another notable work is that of Roli et al. [28], who evolved BNs to control robotic behaviours. In [24], Mesot and Teuscher demonstrated that BNs can achieve better performance than CAs on tasks that measure the capacity of distributed models to perform global density classification. In [12], the authors showed that BNs can be used to solve information processing problems, showing that network learning and generalisation can be optimised according to the complexity of the task and the quantity of information provided.

5   Methods

In this paper, we consider whether BNs can be optimised to control RBNs. We focus on the task of state space targeting, i.e. learning a control intervention that pushes a controlled RBN to a particular point in its state space. For simplicity, the target state is all-ones; however, this is no easier or harder to reach than any other arbitrary state for a particular sample of RBNs, and is not analogous to the max-ones problem in the genetic algorithms literature. At the start of the control task, the states of all the nodes in the controlled RBN are set to zero. Controller BNs are evolved using a conventional generational evolutionary algorithm, and RBNs are randomly sampled from the population of BNs of a particular size and connectivity. Figure 3 depicts the interaction between a controller BN and a controlled BN, with the grey lines illustrating coupling between the two.

An evolved controller comprises a BN, a set of coupling terms, and a value in the range [1,100] that indicates the speed of the controller. The BN is represented as an array of nodes, each comprising a Boolean function, an initial state, and a set of inputs, where each input is indicated by its position within the BN node array. The coupling terms indicate the nodes in the controlled RBN whose state will be changed by the controller BN; for an optimised controller, it is anticipated that this will be an indication of the control kernel of the controlled RBN. Inputs to the controller BN (i.e. feedback from the controlled BN) are always delivered by over-writing the states of nodes at the beginning of its node array, and outputs are always read from the states of nodes at the end of the array. The speed parameter determines how many time steps the controller BN will execute for each step of the controlled RBN, with values above 1 allowing the controller BN to execute faster than the controlled RBN (Fig. 4).

All of the controller's parameters can be optimised during the course of evolution. The evolutionary algorithm runs for 100 generations with a population size of 500 using tournament selection (size = 4), uniform crossover ($p = 0.15$), point mutation ($p = 0.06$), and elitism (size = 1), with crossover points always falling between node boundaries. These values were found to be appropriate during initial experiments. A controller's fitness is a measure of the distance between the controlled RBN's final state and the target state, after a control period of 20 time steps of the controlled RBN. The value is calculated by counting the number of 1 s in the controlled RBN's state at the final time step, and dividing by the size of the controlled RBN, i.e. a value in the interval $[0, 1]$ where 1 indicates the correct all-ones state was reached (Fig. 5).

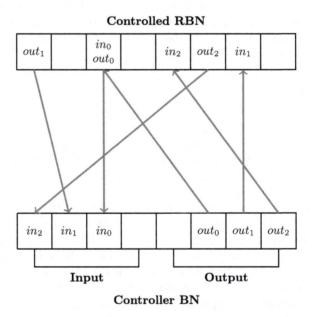

Fig. 4. Coupled Boolean networks, showing the linear encoding used by the evolutionary algorithm. Grey arrows indicate coupling between controller BN and controlled RBN.

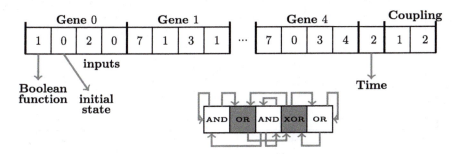

Fig. 5. Example of a Boolean network controller and its genetic representation. Since $k = 2$, functions are numbered between 0 and 15. The timing and coupling terms indicate that this network is executed twice for every step of the controlled network, that its control output is provided by Gene 1, and its feedback input from the controlled network is copied to Gene 2.

6   Results

Figures 6 shows the fitness distributions of controllers evolved to perform state space targeting in RBNs. Figure 7 shows the corresponding fitness for RBNs that are not controlled, i.e. that follow their natural dynamics. It is evident that fitnesses are much higher when a controller is used, indicating that BNs can be usefully evolved to carry out state space targeting in other BNs.

Figures 6 also shows the effect of network size, both for the controller and the controlled network. Perhaps unsurprisingly, it is evident that control becomes more difficult for larger RBNs, where the state space is much larger. However, optimal solutions are often found for RBNs up to size 10, and some optimal solutions are found beyond this. The size of the controller appears to be far less significant, though there is some indication that controllers are harder to find for the largest and smallest sizes we looked at. However, this is also unsurprising, since small BNs are less computationally expressive and large BNs have larger optimisation spaces.

Figure 8 shows the effect of the connectivity parameter k on controllability. This suggests relatively little effect up to $k = 5$ and a rapid fall in controllability after this, indicating that networks with a large degree of interconnectedness are hard to control. The figure also suggests that optimal controllers may be easier to find for the chaotic regime ($k > 2$), reflecting current understanding that chaotic dynamics are more controllable than ordered dynamics, due to their sensitivity to control signals. However, it should be noted that, in practice, GRNs have a distribution of connectivity that is scale-free rather than uniform, and in further work it would be interesting to consider how this pattern of connectivity affects controllability.

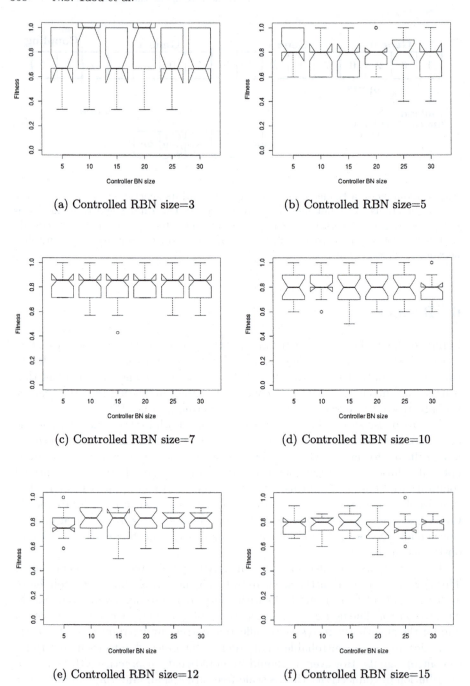

(a) Controlled RBN size=3

(b) Controlled RBN size=5

(c) Controlled RBN size=7

(d) Controlled RBN size=10

(e) Controlled RBN size=12

(f) Controlled RBN size=15

Fig. 6. Fitness distributions when controlling RBNs using evolved BNs with connectivity $k = 2$. High fitness values are better. Summary statistics of 20 evolutionary runs are shown as notched box plots, where overlapping notches indicate when median values (thick horizontal bars) are not significantly different at the 95 % confidence level.

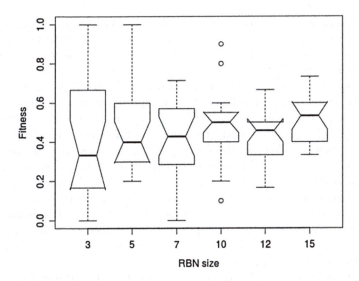

Fig. 7. Corresponding fitness distributions for uncontrolled RBNs ($k = 2$) that are following their natural dynamics.

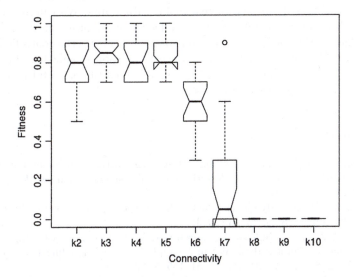

Fig. 8. Fitness distributions for different connectivity values.

7   Conclusions

In this paper we have proposed a novel method for controlling GRNs based on the optimisation of Boolean networks using evolutionary algorithms. Our results are promising, showing that Boolean networks can be evolved to carry out state space targeting in other Boolean networks. Whilst we use Boolean networks as a computationally efficient proxy for real biological GRNs, it is also possible to refine Boolean models into synthetic biology circuits. This provides a potential route to *in vivo* implementations. A potential use of these synthetic biology circuits would be to control disease processes caused by biological GRNs entering undesired states, using the synthetic circuit to control the GRN's trajectory back to a healthy state. Using artificial GRNs to do this would allow us to leverage the observed complexity, robustness and efficiency of these natural control systems.

However, achieving this would require a solution to the significant technological challenge of delivering the control system to diseased cells and tissues. In the meantime, an alternative route would be to study the evolved Boolean networks and attempt to understand their basis of control (something that has been done for other classes of artificial biochemical network [22]). This could then be mapped to more conventional drug delivery systems. For instance, such an analysis could indicate that certain target genes need to be modulated in a particular temporal pattern, and this could be implemented by a patient taking certain drugs at certain times.

In the near future, we plan to look more closely at how the evolved Boolean networks function, and whether there is a bias towards the control of certain topologies. In addition, we will look at controlled BNs whose topologies more closely match those seen in real biological circuits. We will also consider other models for both controller and controlled systems; for instance, continuous-state models for the controller, and agent-based models for the controlled. There has been considerable research effort in developing robust agent-based models of biochemical disease pathways [26,27], and these could be a useful platform for testing computational disease interventions.

References

1. Akutsu, T., Hayashida, M., Ching, W.K., Ng, M.K.: Control of boolean networks: hardness results and algorithms for tree structured networks. J. Theor. Biol. **244**(4), 670–679 (2007)
2. Aldana, M.: Boolean dynamics of networks with scale-free topology. Phys. D Nonlinear Phenom. **185**(1), 45–66 (2003)
3. Amaral, L.A., Díaz-Guilera, A., Moreira, A.A., Goldberger, A.L., Lipsitz, L.A.: Emergence of complex dynamics in a simple model of signaling networks. Proc. Natl. Acad. Sci. USA **101**(44), 15551–15555 (2004)
4. Azuma, S.-I., Imura, J.-I.: Synthesis of optimal controllers for piecewise affine systems with sampled-data switching. Automatica **42**(5), 697–710 (2006)
5. Ballesteros, F.J., Luque, B.: Random boolean networks response to external periodic signals. Phys. A Stat. Mech. Appl. **313**(3), 289–300 (2002)
6. Bull, L., Preen, R.: On dynamical genetic programming: random boolean networks in learning classifier systems. In: Vanneschi, L., Gustafson, S., Moraglio, A., De Falco, I., Ebner, M. (eds.) EuroGP 2009. LNCS, vol. 5481, pp. 37–48. Springer, Heidelberg (2009)
7. Cheng, D., Qi, H.: Controllability and observability of boolean control networks. Automatica **45**(7), 1659–1667 (2009)

8. De Jong, H.: Modeling and simulation of genetic regulatory systems: a literature review. J. comput. Biol. **9**(1), 67–103 (2002)

9. Drossel, B.: Random boolean networks. Rev. Nonlinear Dyn. Complex. **1**, 69–110 (2008)

10. Dubrova, E., Teslenko, M., Tenhunen, H.: A computational scheme based on random boolean networks. In: Priami, C., Dressler, F., Akan, O.B., Ngom, A. (eds.) Transactions on Computational Systems Biology X. LNCS (LNBI), vol. 5410, pp. 41–58. Springer, Heidelberg (2008)

11. Gershenson, C.: Introduction to random boolean networks. In: Bedau, M., Husbands, P., Hutton, T., Kumar, S., Suzuki, H. (eds.) Workshop and Tutorial Proceedings, Ninth International Conference on the Simulation and Synthesis of Living Systems (ALife IX), pp. 160–173 (2004)

12. Goudarzi, A., Teuscher, C., Gulbahce, N., Rohlf, T.: Emergent criticality through adaptive information processing in boolean networks. Phys. Rev. Lett. **108**(12), 128702 (2012)

13. Harris, S.E., Sawhill, B.K., Wuensche, A., Kauffman, S.: A model of transcriptional regulatory networks based on biases in the observed regulation rules. Complexity **7**(4), 23–40 (2002)

14. Harvey, I., Bossomaier, T.: Time out of joint: attractors in asynchronous random boolean networks. In: Proceedings of the Fourth European Conference on Artificial Life, pp. 67–75 (1997)

15. Kauffman, S.A.: The Origins of Order: Self Organization and Selection in Evolution. Oxford University Press, New York (1993)

16. Kim, J., Park, S.M., Cho, K.H.: Discovery of a kernel for controlling biomolecular regulatory networks. Sci. Rep. **3**, 1–9 (2013)

17. Kitano, H.: Computational systems biology. Nature **420**(6912), 206–210 (2002)

18. Kitano, H.: Cancer as a robust system: implications for anticancer therapy. Nat. Rev. Cancer **4**(3), 227–235 (2004)

19. Kobayashi, K., Hiraishi, K.: A petri net-based approach to control of boolean networks. In: 2012 Third International Conference on Networking and Computing (ICNC), pp. 399–403 (2012)

20. Kobayashi, K., Imura, J.I., Hiraishi, K.: Polynomial-time algorithm for controllability test of a class of boolean biological networks. EURASIP J. Bioinform. Syst. Biol. **2010**(1), 210685 (2010)

21. Lones, M., Turner, A.P., Caves, L.S., Stepney, S., Tyrrell, A.M.: Adaptive robotic gait control using coupled artificial signalling networks, hopf oscillators and inverse kinematics. In: 2013 IEEE Congress on Evolutionary Computation (CEC), pp. 1435–1442 (2013)

22. Lones, M., Turner, A.P., Caves, L.S., Stepney, S., Smith, S.L., Tyrrell, A.M.: Artificial biochemical networks: evolving dynamical systems to control dynamical systems. IEEE Trans. Evol. Comput. **18**(2), 145–166 (2014)

23. Lones, M.A.: Computing with artificial gene regulatory networks. In: Iba, H., Noman, N. (eds.) Evolutionary Algorithms in Gene Regulatory Network Research. Wiley, Hoboken (2016)

24. Mesot, B., Teuscher, C.: Deducing local rules for solving global tasks with random boolean networks. Physica D Nonlinear Phenom. **211**(1), 88–106 (2005)

25. Milias-Argeitis, A., Summers, S., Stewart-Ornstein, J., Zuleta, I., Pincus, D., El-Samad, H., Khammash, M., Lygeros, J.: In silico feedback for in vivo regulation of a gene expression circuit. Nat. Biotechnol. **29**(12), 1114–1116 (2011)

26. Read, M., Andrews, P.S., Timmis, J., Kumar, V.: A domain model of experimental autoimmune encephalomyelitis. In: CoSMoS, pp. 9–44 (2009)

27. Read, M., Andrews, P.S., Timmis, J., Kumar, V.: Techniques for grounding agent-based simulations in the real domain: a case study in experimental autoimmune encephalomyelitis. Math. Comput. Model. Dyn. Syst. **18**(1), 67–86 (2012)
28. Roli, A., Manfroni, M., Pinciroli, C., Birattari, M.: On the design of boolean network robots. In: Di Chio, C. (ed.) EvoApplications 2011, Part I. LNCS, vol. 6624, pp. 43–52. Springer, Heidelberg (2011)
29. Shi-Jian, C., Yi-Guang, H.: Control of random boolean networks via average sensitivity of boolean functions. Chin. Phys. B **20**(3), 036401 (2011)
30. Zanin, M., Pisarchik, A.N.: Boolean networks for cryptography and secure communication. Nonlinear Sci. Lett. B Chaos Fractal Synchronization **1**(1), 27–34 (2011)

The Emergence of Cooperation in Public Goods Games on Randomly Growing Dynamic Networks

Steve Miller[1]([⊠]) and Joshua Knowles[2]

[1] School of Computer Science, University of Manchester, Manchester, UK
stevemiller.gm@gmail.com
[2] School of Computer Science, University of Birmingham, Birmingham, UK

Abstract. According to evolutionary game theory, cooperation in public goods games is eliminated by free-riders, yet in nature, cooperation is ubiquitous. Artificial models resolve this contradiction via the mechanism of network reciprocity. However, existing research only addresses pre-existing networks and does not specifically consider their origins. Further, much work has focused on scale-free networks and so pre-supposes attachment mechanisms which may not exist in nature. We present a coevolutionary model of public goods games in networks, growing by random attachment, from small founding populations of simple agents. The model demonstrates the emergence of cooperation in moderately heterogeneous networks, regardless of original founders' behaviour, and absent higher cognitive abilities such as recognition or memory. It may thus illustrate a more general mechanism for the evolution of cooperation, from early origins, in minimally cognitive organisms. It is the first example of a model explaining cooperation in public goods games on growing networks.

Keywords: Evolution of cooperation · Evolutionary game theory · Public goods game · Complex networks

1 Introduction

The prisoner's dilemma has become a standard metaphor to represent cooperation in evolutionary game theory, however it only describes interactions between *pairs* of individuals. In nature, interactions are not necessarily constrained in this way and a broader representation of cooperation is useful, particularly in the case of social, economic and biological networks [1]. For such scenarios, the public goods game (PGG) offers a suitable alternative for groups of more than two members. Referred to variously, as the N-player prisoner's dilemma, the free-rider problem, or the tragedy of the commons [2], the PGG represents a group-based dilemma where there exists a tension between benefits to an individual following one (selfish) course of action versus benefits to the entire community if the individual chooses an alternative action.

© Springer International Publishing Switzerland 2016
G. Squillero and P. Burelli (Eds.): EvoApplications 2016, Part I, LNCS 9597, pp. 363–378, 2016.
DOI: 10.1007/978-3-319-31204-0_24

The formulation of the PGG is as follows: Each member of a group has the opportunity to contribute a 'cost' to a central 'pot'. They can choose to contribute, or not. The amount invested in the pot is then increased by a multiplier. The increased amount is divided amongst all members of the group, regardless of whether they contributed or not. Those contributing to the pot can be considered cooperators whilst those withholding, defectors (free-riders). As with the prisoner's dilemma, the choice which maximises payoff is to not contribute (to defect). Thus in the rational analysis, all individuals will choose to act selfishly, which will result in the worst case scenario for all: the minimisation of the public good. In nature however, the rational choice appears less appealing and communities are observed to cooperate, so as to preserve or maintain public goods.

Attempts to explain this apparent contradiction between theory and observed behaviour consider the importance of factors such as volunteering, reputation, punishment or reward [3–5]. Whilst it is easy to appreciate that such factors may affect the choices of, for example, humans, higher primates or birds, it is harder to extend such approaches to explaining cooperation in more primitive forms of life [6] such as microorganisms cooperating to establish protective shelters, forage for nutrients or aid dispersal [7]. In such cases, network reciprocity may offer an alternative explanation, requiring fewer assumptions.

The effect of spatial structure on cooperation was first highlighted in [8]. Whereas evolutionary game theory shows that cooperation cannot survive in evenly mixed populations, the presence of spatial relationships allows cooperators to cluster. This clustering increases their individual fitnesses and thus prevents extinction by defectors. Further research developed these findings to illustrate that heterogeneity of network structure promotes cooperation, in the case of *pair-wise* interactions modelled using the prisoner's dilemma [9]. A similar approach modelling *group-wise* interactions, using the PGG, was described in the work of [10]. Here, the mean field formulation of the PGG was spatially extended by mapping agents playing PGG to nodes of a network. The results of this work illustrated the emergence of cooperation on scale-free networks, thus reinforcing previous findings regarding pair-wise (prisoner's dilemma) cooperation [9].

Existing research has therefore established a consistent view of the positive role heterogeneous networks play in promoting cooperation, however the overwhelming majority of this work has focused on the pair-wise prisoner's dilemma and has primarily considered static networks. (A useful review of work focusing specifically on the PGG in networks may be found in [1]). Of the limited body of research that exists for cooperation in dynamic networks, most has focused on networks at some form of equilibrium, using approaches which involve modification of pre-existing (fully formed) networks (see reviews in [11,12]). A very limited number of publications consider network growth [13–15]; all of the latter focusing on prisoner's dilemma.

In this report we offer an initial attempt to fill this gap: We consider the growth of a population from its earliest origins and we ask how the social network affects and is affected by the *group* behaviour of the individuals within it.

Our aim is to establish a model based on group-wise cooperation which demonstrates the growth of networked populations of cooperative agents from original founder members. For such a model to be of value, it cannot be initially assumed that founder members are cooperators. Further, for the model to be broadly applicable, the sort of cognitive abilities (memory, recognition, reasoning) that are required for reciprocity or retaliation cannot be assumed. Finally for the model to be general, we make the simplest possible assumptions about the mechanism that new nodes use to attach to the existing network (i.e., we do not use preferential attachment).

2   Background

Here we discuss two models on which we have based our work and explain the rationale for the adaptations made in incorporating these into a single model. We provide this explanation in terms of the dynamic aspects of our model, divided into the two separate processes of attrition and growth.

2.1   An Existing Network Representation of the Public Goods Game

It has been demonstrated in [10] that heterogeneous network structure promotes cooperation in public goods games within static networks. The approach used represented a population in the form of multiple sub-groupings (neighbourhoods), each of which constitutes a PGG. More specifically, each node in the network initiates a single PGG and is also a participant in games initiated by its neighbours. Hence each node takes part in $g = k + 1$ games where k represents the degree of the node occupied by the agent. An agent x with direct connections to neighbours a and b therefore has a degree of 2 and takes part in 3 PGGs: the one initiated by itself and the ones initiated by its neighbours a and b. The total number of games in a population is therefore equal to N, the number of agents in the population. Within this work (ibid.), two variants of the PGG model were investigated, (i) where each agent had a fixed cost per game (FCPG) and therefore their overall contribution was proportional to g, and (ii) where each agent had a total fixed cost (fixed cost per individual, FCPI) and therefore their contribution was divided between all g games. The game-playing populations are incorporated into evolutionary simulations by means of a strategy updating process representing natural selection. Within this step, the strategies (behaviours) of fitter nodes probabilistically displace those of less fit neighbours.

We aim to use the above approach as a basis from which to develop an extended dynamic model that simulates *growth from founding members*. This naturalistic model is intended to explain the development of cooperation with respect to early origins of a population.

2.2   An Existing Model of Cooperative Network Growth

We take as our inspiration for developing a dynamic PGG model, the work of [13], who notably connected the dynamic structure of a network to the behaviour of agents within the network. In this approach, evolutionary processes, preferential attachment and agent behaviour were incorporated into a unified model of dynamic network-reciprocal cooperation (using the prisoner's dilemma), referred to as evolutionary preferential attachment (EPA).

2.3   Proposed Attrition Mechanism

In the EPA model, strategy updating still forms the primary evolutionary component, with selection acting on relative fitnesses resulting from *agent-agent* interactions. However, EPA also incorporates a secondary evolutionary mechanism into the *growth* processes of the social *network*. Within our model, we shift this secondary evolutionary component over to *shrinkage* of the network. Specifically, this *'global'* effect causes death of less fit *individual agents*. We consider this revision offers a model more analogous to the processes of selection in real world evolutionary situations. Such a shift separates evolutionary effects from the attachment processes responsible for network growth.

To implement such a culling mechanism, we impose a nominal maximum population size which is analogous to the concept of 'carrying capacity', as used in population biology. In this sense, the size of a population shrinks in response to extrinsic factors which are the result of environmental effects (such as predation, disease, food availability, many of which may be seasonal variations).

2.4   Proposed Growth Mechanism

The positive effect of scale-free degree distribution has featured significantly in research into the emergence of cooperation in networks [11]. However, we note that the fitness or degree-based mechanisms of preferential attachment, which are likely to be responsible for such structures, require underlying explanations for each occasion where they are found (for example, what specific process would enable a newcomer joining, or born into a population, to identify the fittest or most well-connected member in that situation). Clearly preferential mechanisms exist (although disagreements have arisen over claims in this respect [16,17]), however we suggest it is important that a general model for cooperation should be viable in the absence of mechanisms which require additional 'case-by-case' explanations or assumptions (even if when present they may further enhance cooperation).

To overcome these concerns, we implement the connection of new nodes to the existing network as an entirely random process. Such a mechanism does not cause the development of a simple Poisson degree distribution as would be found in a random network: chronological random attachment (CRA) results in older nodes having more connections. In the absence of other influences, the degree distribution in such a situation becomes exponential—giving a structure with heterogeneity somewhere between that of random and scale-free networks.

2.5   Summary: A Model of Population Fluctuation in Social Networks

The two processes described above, attrition of least fit nodes whenever a carrying capacity is reached, and growth of the network by random addition, continue until the simulation ends. We thus have a fluctuation system which (i) supports the growth of a network from founder members, and (ii) overcomes the unrealistic situation that a 'mature' network becomes fixed structurally. Further, as intended, this implementation gives us a minimal model which does not require assumption of higher cognitive abilities for its individual members and has no requirement regarding specific underlying mechanisms for the social network structure formation. This model, described in more detail below, is an extension of our earlier work [14], which considered preferential attachment and the prisoner's dilemma game.

3   Methods

Our model describes agents located at the nodes of networks. Interactions occur between agents on nodes that have connecting edges. Each node in the network has a 'neighbourhood', defined by the neighbours its edges connect to. A PGG occurs for each neighbourhood and hence a network of N nodes will result in N PGGs. Agents can contribute to a PGG (cooperate) or not (defect). Each agent in the network has a behaviour encoded by a 'strategy' variable representing either 'cooperate' or 'defect'. In a round robin fashion, each agent in turn initiates a PGG which involves their primary connected neighbours (their neighbourhood). Each agent in the population accumulates a fitness score which is the sum of its rewards from all the PGGs it participates in.

Within the evolutionary simulation, this process is repeated over generations. Agents are assessed at each generation, on the basis of their fitness score: Fitter agents' strategies remain unchanged; less fit agents are more likely to have strategies displaced by those of fitter neighbours. Fluctuation of the population occurs by repeated attrition and regrowth of the network.

The general outline of the evolutionary process, for one generation, is as follows:

1. *Play public goods games:* Each agent initiates a PGG involving its neighbours. Each agent will accumulate a fitness score that is the sum of payoffs from all the individual PGGs that it participates in.
2. *Update strategies:* Selection occurs. Agents with low scores will have their strategies replaced, on a probabilistic basis, by comparison with the fitness scores of randomly selected neighbours.
3. *Remove nodes:* If the network has reached the nominal maximum size, it is pruned by a tournament selection process that removes less fit agents.
4. *Grow network:* A specified number of new nodes are added to the network, each connecting to m randomly selected distinct existing nodes via m edges.

In the following, we provide more detail on the specifics of each of the four steps:

Play Public Goods Games. Each node of the network, in turn, initiates a PGG. Within a single PGG, all cooperator members of a neighbourhood contribute a cost c to 'the pot'. The resulting collective investment I is multiplied by r, and rI is then divided equally amongst all members of the neighbourhood, regardless of strategy.

In the FCPG variant of the PGG, each agent has a fixed cost *per game* and therefore their overall contribution, in one generation, is $c(k + 1)$ with contribution c to each game, and where k is the number of neighbours (degree). The single game individual payoffs of an agent x are given by the following equations, for scenarios where x is a defector (P_D) and a cooperator (P_C) respectively:

$$P_D = crn_c/(k_x + 1), \tag{1}$$

$$P_C = P_D - c, \tag{2}$$

where c is the cost contributed by each cooperator, r is the reward multiplier, n_c is the number of cooperators in the neighbourhood based around x, and k_x is the degree of x.

In the FCPI variant, each *individual* has a fixed cost c, i.e. their overall contribution is c and hence their contribution to each game is $c/(k + 1)$. The single game individual payoff for a node y having strategy s_y ($= 1$ if cooperator, $= 0$ if defector) present in the neighbourhood of x is given by:

$$P_{y,x} = \frac{r}{k_x + 1} \sum_{i=0}^{k_x} \frac{c}{k_i + 1} s_i - \frac{c}{k_y + 1} s_y, \tag{3}$$

where i is used to index each neighbour of x, and s_i is the strategy of neighbour i of x having degree k_i.

Update Strategies. Each node i selects a neighbour j at random. If the fitness of node i, f_i is greater or equal to the neighbour's fitness f_j, then i's strategy is unchanged. If the fitness of node i, f_i is less than the neighbour's fitness, f_j, then i's strategy is replaced by a copy of the neighbour j's strategy, according to a probability proportional to the difference between their fitness values. Thus poor scoring nodes have strategies displaced by those of more successful neighbours.

Hence, at generation t, if $f_i(t) \geq f_j(t)$ then i's strategy remains unchanged. If $f_i(t) < f_j(t)$ then i's strategy is replaced with that of the neighbour j with the following probability:

$$\Pi_{U_i}(t) = \frac{f_j(t) - f_i(t)}{\max(k_i(t), k_j(t))}, \tag{4}$$

where k_i and k_j are degrees of node i and its neighbour j respectively. The purpose of the denominator is to normalise the difference between the two nodes.

The term $max(k_i(t), k_j(t))$ represents the largest achievable fitness difference between the two nodes given their respective degrees.

Grow Network. New nodes, with randomly allocated strategies, are added to achieve a total of 10 at each generation. Each new node uses $m = 2$ edges to connect to existing nodes. Duplicate edges and self-edges are not allowed. The probability $\Pi_{G_i}(t)$ that an existing node i receives one of the m new edges is given by:

$$\Pi_{G_i}(t) = \frac{1}{N(t)}, \qquad (5)$$

where $N(t)$ is the number of nodes available to connect to at time t in the existing population. Given that in our model each new node extends $m = 2$ new edges, and multiple edges are not allowed, N is therefore sampled *without replacement*. Growth continues until a nominal maximum size (we used 1000 nodes) is achieved.

Remove Nodes (for Fluctuation Simulations). On achieving or exceeding the nominal maximum size, the network is pruned by a percentage X. This is achieved by repeated tournament selection using a tournament size equivalent to 1 % of the population. Tournament members are selected randomly from the population. The tournament member having the least fitness is the 'winner' and is added to a short list of nodes to be deleted. Tournament selection continues until the short list of $X\%$ nodes for deletion is fully populated.

   The nodes on the short list (and all of their edges) are removed from the network. Any nodes that become isolated from the network as a result of this process are also deleted. (Failure to do this would result in small numbers of single, disconnected, non-playing nodes, having static strategies and zero fitness values.) When there are multiple nodes of equivalent low fitness value, the selection is effectively random (on the basis that the members were originally picked from the population randomly). Where $X = 0$, no attrition occurs; in this case, on reaching maximum size, the network structure would become static.

General Simulation Conditions. Initial strategy types of founder nodes were specified in simulation setup (either 3 cooperators or 3 defectors). Strategy types of subsequently added nodes were allocated independently, uniformly, at random (cooperators and defectors with equal probability). All networks had an overall average degree of approximately $k = 4$, giving an average neighbourhood size of $g = 5$. Simulations were run until 20,000 generations. The final 'fraction of cooperators' values we use are means, averaged over the last 20 generations of each simulation, in order to compensate for variability that might occur from just using final generation values. Each simulation consisted of 25 replicates. We used shrinkage value of $X = 2.5\%$ for all fluctuation simulations. Simulation data is recorded after step 2 (*Update strategies*).

4 Results and Discussion

We now present the results of research investigating our model's ability to support cooperation in a range of simulations. Initially we consider its implementation in the type of scenarios that have dominated research into cooperation in dynamic networks, namely fully formed or 'pre-existing' networks. We then apply the model, as per our original motivation, to consider networks grown from a small number of founding members. In both types of investigation we have considered the two variants of PGG described in [10]: FCPG and FCPI.

4.1 Simulations Using Pre-existing Networks

We first consider the impact of our model in pre-existing networks. We initially consider the effect of the PGG variant (FCPG vs. FCPI) and subsequently we discuss specifically how the fluctuation mechanism achieves different outcomes to those seen for static networks.

Effect of PGG Variant in Simulations Using Pre-existing Networks.
It is established for the 2-player PGG (the prisoner's dilemma) that cooperation in pair-wise interactions on networks is promoted by the opportunity for cooperators to self-assort and form inter-connected groups (clusters) [8]. The larger the clusters which form, the greater the levels of cooperation which occur [9]. This effect is therefore enhanced by increased network degree heterogeneity, since greater heterogeneity allows for increasingly larger connected groups of cooperators within the network. Such findings for the prisoner's dilemma generalise to the PGG, however in the PGG there is also an opposing 'force' that limits cooperation, which we now explain. In the conventional FCPG representation of the PGG, an individual pays a cost for every single game they participate in. Since each individual in a population can initiate a PGG among their local neighbourhood, higher connectivity (more neighbours) means that an individual will participate in more games and will thus pay a penalty for their increased connectivity [18]. The classical result for the PGG in this case, is that the larger the neighbourhoods become, the less likely cooperation is. This finding makes intuitive sense, since the larger a PGG neighbourhood is, the closer it gets to representing a mean field scenario, where defection is the Nash equilibrium.

In Fig. 1a, (FCPG in static networks) we see that higher levels of cooperation are observed in static scale-free networks (green line with 'x' markers) than in networks of low or no heterogeneity—random and regular respectively. These results are consistent with the view that heterogeneity promotes cooperation. In the case of FCPI PGG (see Fig. 1b), the lack of any penalty on larger neighbourhood size weakens the dilemma i.e. it reduces the 'temptation to defect' and therefore increases levels of cooperation. By comparing corresponding lines for FCPG and FCPI PGG in Figs. 1a and b, we can see how FCPI causes different horizontal shifts in cooperation profiles for networks of differing heterogeneity. We thus see that the impact of FCPI is nonexistent for regular networks (no

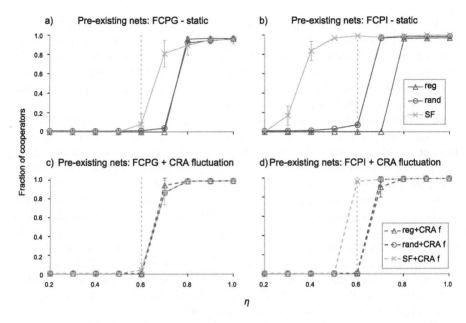

Fig. 1. Plots comparing the effect of network type for simulations (25 replicates) on pre-existing networks of increasing heterogeneity (regular, random and scale-free respectively). Final fraction of cooperators in population is plotted against η, the PGG reward multiplier r, normalised with respect to average neighbourhood size ($g = 5$). Variability is indicated by error bars showing 95 % confidence intervals. Simulation details are as described in Methods section. The dashed line at $\eta = 0.6$ is a reference for the eye (Color figure online).

visible shift, see blue lines with triangle markers) and becomes more relevant as increasing heterogeneity allows for increasing neighbourhood size (marked shift for scale-free networks, see green lines with 'x' markers).

Figures 1c and d illustrate the effect of PGG variant in fluctuating networks. Here we expect to see two general results: (i) Given that the fluctuation mechanism drives all networks to the same final degree distribution, we would expect similar result profiles, within each variant of the PGG, regardless of starting network topology; (ii) Further, we would expect these profiles, based on approximately exponentially distributed final networks, to lie somewhere between the two extremes of heterogeneity represented by scale-free (highly heterogeneous) and regular (non heterogeneous) networks, as observed in the static FCPI results (see Fig. 1b).

We find that the CRA-fluctuation profiles do indeed lie within the expected region of the graph, however we see an anomaly for the scale-free FCPI result (Fig. 1d) which achieves higher levels of cooperation than regular and random networks. This result is unexpected because given an assumption that cooperation is only dependent on the final degree distribution, we would expect to see the same result profiles for all network types. We have compared final degree distributions for all network types and find no discernible difference.

We propose that the explanation for the anomaly seen for initially scale-free networks lies in the differing challenges presented by the topology of the initial networks; specifically, the diameter of the network (rather than the degree distribution). We have measured the average shortest path length in our initial networks and find these to be approximately: 125 for regular networks, infinite (network disconnected) for random networks, and 4 for scale-free networks. Final networks have lengths of 6 in the case of FCPG and 7 in the case of FCPI. In order for cooperation to percolate through the network, sufficient reward (η) has to be present to drive assortativity by strategy, however assortativity will inevitably be impeded in those cases where the network is fragmented or does not have the small path lengths that are a defining characteristic of small-world networks. In such cases, cooperation cannot readily percolate, until the fluctuation mechanism has brought about sufficient changes to reduce the average path length and/or the number of network components. Whilst cooperation in scale-free networks is still dependent upon the value of η, such networks do not have to overcome the path length issues faced by random and regular networks. Thus, while all network topologies end up with the same final degree distributions, scale-free networks potentially start with a 'small-world' advantage which may support the emergence of cooperation at lower values of η.

This proposed explanation raises the question of why a difference exists between scale-free network results for FCPG and FCPI (see Figs. 1c and d). In response to this, our above explanation does indeed apply to fluctuating scale-free networks for both variants of the PGG, however in the case of FCPG, cooperation is limited by the additional constraint of neighbourhood size.

Effect of Static vs. Fluctuating Networks. As reported in the work of [15], a model based on fluctuating population size can promote cooperation in networks. This outcome arises from the greater opportunity for strategies to self-assort, given repeated perturbation of network structure, due to deletions of low fitness nodes.

From comparing Figs. 1a and c, we see how the incorporation of population fluctuation affects results for FCPG. Profiles for all network types are now super-imposed. We observe similar findings in the case of FCPI (compare Figs. 1b and d) except for in the case of scale-free networks—an anomaly which, as explained earlier, is believed to be due to the 'beneficial' impact of short average path lengths found in initially scale-free networks. The *general* consistency of cooperation profiles is because the CRA-fluctuation mechanism converts all networks to the same final degree distribution regardless of initial topology. Figure 2 illustrates this conversion by showing initial and final degree distributions in simulations starting from a scale-free network. We note that whilst the final degree distribution due to CRA alone would be exponential, the additional effect of node deletion compresses the exponential curve, giving a degree heterogeneity lying between that of a Gaussian and an exponential distribution.

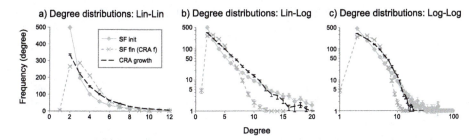

Fig. 2. (**a**) Graph illustrating initial and final degree distributions (25 replicates) for an initially scale-free network, of 1000 nodes, subjected to CRA-fluctuation. As a reference, the dashed black line represents the exponential distribution of a network 'built' by CRA—without any node deletion. To highlight characteristic distribution profiles, the same data is also presented on, (**b**) lin-log (exponential appears as a straight line), and (**c**) log-log (scale-free appears as a straight line) scales. For clarity, the x axis is truncated for plots **a** and **b**. Data is shown with error bars representing 95 % confidence intervals. (The presence of small numbers of nodes of degree 1 is an artefact of our implementation of fluctuation: Whilst all new nodes initially have $m = 2$ edges, the deletion process may leave limited numbers of nodes with a single edge.) (Color figure online)

We highlight an important point here. CRA-fluctuation converts *all* networks, regardless of initial type to this compressed exponential distribution which has moderate heterogeneity. In the case of scale-free networks, fluctuation therefore brings about a *decrease* in heterogeneity. Hence for initially scale-free networks, we should expect to see reduced cooperation in the fluctuation model compared to the static one. This is indeeed true for FCPI results (green lines with 'x' markers in Figs. 1b and d) but not the case for FCPG (Figs. 1a and c). Fluctuation does not cause the expected reduction in FCPG PGG because in the static FCPG implementation, scale-free networks are *already* constrained in their ability to cooperate i.e. they cannot achieve their full potential in supporting cooperation due to the penalties FCPG imposes on large neighbourhoods.

Figure 3 presents our results so as to separately illustrate the impact of fluctuation on each of the network types studied. Here we generally see (comparing dashed line for all plots) that CRA-fluctuation shows increased or similar levels of cooperation in comparison to results for static networks. We observe this effect in all cases except FCPI initially scale-free networks (Fig. 3f).

Our general observation here, that cooperation is increased or unchanged by fluctuation, is primarily a result of the CRA-fluctuation mechanism converting all networks to the same compressed exponential degree distribution, regardless of starting topology. Whilst the fluctuation model is in this way able to shift cooperation profiles to the left (increasing cooperation), such a change can only be achieved for networks having initially lower heterogeneity (random and regular networks). Hence, in the case of scale-free (highly heterogeneous) networks, levels of cooperation should be lower in the fluctuating network than in the static network. Our results for scale-free networks are however not as clear cut as

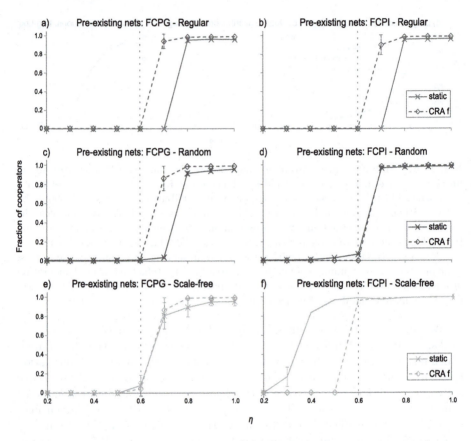

Fig. 3. Plots comparing results from networks with static population size to those where size fluctuates around a nominal maximum value. Three types of pre-existing networks were investigated: regular, random and scale-free. Final fraction of cooperators in population is plotted against η, the PGG reward multiplier r, normalised with respect to average neighbourhood size ($g = 5$). Variability (25 replicates) is indicated by error bars showing 95 % confidence intervals. Simulation details are as described in Methods section. The dashed line at $\eta = 0.6$ is a reference for the eye.

this simple explanation would suggest and we now provide further clarification of why this is so.

In Fig. 3e, for FCPG PGG on initially scale-free networks, we see no reduction in levels of cooperation when the fluctuation model is applied. The reason we do not see such a reduction is because, as highlighted earlier, FCPG restricts the levels of cooperation that are achievable in heterogeneous networks. This constraint limits levels of cooperation in the case of static networks and there is no reason to expect the fluctuation model to be able to overcome such a constraint. We hence see similar results for both fluctuation and static models in FCPG PGG. In Fig. 3f, we clearly see how the absence of such a constraint gives us the expected results. Here, for the fluctating network, we see reduced levels of cooperation due to CRA-fluctuation reducing the heterogeneity of the network as it converts it from scale-free to compressed exponential degree distribution.

In summary we make the general observation that fluctuation is more beneficial to cooperation in the case of FCPG. Whereas FCPI enables higher levels of cooperation to be achieved without the need for such 'further assistance'.

4.2   Simulations in Networks Grown from Founder Populations

We now report on results of simulations grown from founder populations of either 3 cooperators, or 3 defectors. We compare two implementations of our model. In the first ('non-fluctuating'), the population grows by a process of CRA until it reaches a maximum size, after which the network structure remains constant. In the second implementation ('fluctuating'), the network grows by means of CRA until it reaches a nominal maximum size, whereupon it is pruned and then allowed to regrow. This fluctuation cycle repeats thereafter, until the simulation ends. Strategy updating continues throughout the entirety of both implementations. As previously, we present results for both FCPG and FCPI variants of the PGG.

Effect of PGG Variant in Networks Grown from Founder Populations. In the case of cooperator-founded populations, by comparing corresponding curves (cyan lines) between Figs. 4a and b, it appears that FCPI may result in a marginal increase in levels of cooperation compared to FCPG. Small increases would be consistent with our understanding that FCPI can relax the penalty paid by cooperator clusters in heterogeneous networks. As described earlier, networks formed by CRA-fluctuation are only moderately heterogeneous, thus any increase afforded by FCPI over FCPG would be expected to be minor.

In the case of defector-founded populations (red lines in Figs. 4a and b), we see what may be a minor difference between FCPG and FCPI in fluctuating networks (dashed lines) and a more marked difference for non-fluctuating networks (solid lines). Curiously, FCPG *in the case of defector founders* appears to *promote* rather than restrict cooperation in non-fluctuating networks. Closer inspection of the data for these specific FCPG simulations shows very high variability for values of $\eta > 0.8$. Here individual replicate simulations proceeded to either high levels of cooperation or almost complete defection. We have been unable to establish a definitive explanation for this effect but strongly suspect that it is a result of FCPG favouring smaller neighbourhoods. In defector-founded populations, simulations where many cooperators connect to the large 'core' neighbourhood of defectors will only result in further growth of this defector core, since added cooperators are inevitably converted by the highly-connected defector core. However in simulations where larger numbers of cooperators randomly attach elsewhere, it may become possible for initially small groups of cooperators to get a 'toehold' in the population away from the main defector core. Growth of this type, where cooperation arises in smaller clusters away from the core, may be promoted by FCPG, which by penalising cooperators in larger neighbourhoods, favours those in smaller ones.

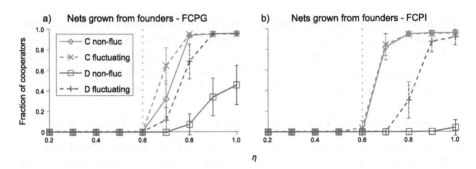

Fig. 4. Plots comparing results from fluctuating and non-fluctuating simulations, for networks grown from founder populations (25 replicates). We compare networks grown from 3 cooperators to those grown from 3 defectors. Final fraction of cooperators in population is plotted against η, the PGG reward multiplier r, normalised with respect to average neighbourhood size ($g = 5$). Populations were specified to have a nominal maximum size of 1000 individuals, at which point fluctuation is triggered. Variability is indicated by error bars showing 95 % confidence intervals. Further details on simulations are as described in Methods section. The dashed line at $\eta = 0.6$ is a reference for the eye (Color figure online).

Effect of Founder Population Strategy on Non-fluctuating and Fluctuating Networks. For cooperator-founded populations (blue lines in Figs. 4a and b), levels of cooperation are not markedly changed by incorporation of fluctuation once the network has reached its maximum size. We propose that the strategy assortativity that has taken place whilst the network was growing has maximised the amount of cooperation that can occur (for a given value of η). Cooperation levels thus do not rise beyond this limit when fluctuation is added.

For the more challenging, defector-founded scenario (red lines in Figs. 4a and b), fluctuation (dashed lines) brings about a marked increase in cooperation. To explain this effect we first consider the non-fluctuating system. Here, an initial defector population converts all nodes that attach to it to defection and hence significantly biases the growing population against cooperation. The founder defectors develop increased connectivity over subsequent generations and become a well-connected core of defectors, easily capable of converting any new cooperators that may attach. In the non-fluctuating model, the starting point of defector founders thus 'locks in' long term defector behaviour—to the extent that even where the dilemma collapses at $\eta = 1$, populations can still grow to be predominantly populated by defectors. In this scenario, rather like the growth of a coral reef, the core of defectors is 'dead'. 'Life' (positive fitness values) only occurs at the periphery of the core where new cooperators attach and hence allow defectors to gain positive fitness values. If the new nodes are defectors they instantly become part of the core; cooperators will do the same as soon as they have played PGG and been converted to defectors. Thus the defector core continues to grow. Whilst the core will primarily have zero fitness, it is still however impenetrable to invasion by cooperators, because they will be instantly converted as soon as they connect to the periphery of the core.

We now consider how the fluctuation model changes the above scenario to bring about such a marked increase in cooperation. Both fluctuating and non-fluctuating models operate identically until the maximum network size is reached. After this point, in the fluctuating model, the least fit individuals and the nodes they occupy are deleted from the population. New nodes are then added which link to randomly selected existing nodes and have randomly allocated strategies. Thus the defector core which is invulnerable in the non-fluctuating mode, contains many nodes which are highly vulnerable in the fluctuating model, due to their zero fitness. Our evolutionary model of fluctuating populations thus creates an escape from the domination of the defector core.

We highlight that a preferential (rather than random) attachment system for network growth in a defector-founded population as we have described, may potentially *reduce* the likelihood of cooperation emerging. This is because in preferential attachment, new nodes would be most likely to connect with existing nodes of highest degree, i.e. the defector-founders. As a result any cooperators would be highly likely to be immediately converted to defectors. Random attachment however allows for cooperators to connect elsewhere and hence allows for the development of cooperator clusters away from the founders where they are less likely to be converted to defection.

5   Conclusion

In this work we have developed a model of the coevolution of cooperative behaviour alongside the growth of a networked population. Importantly, our model demonstrates the emergence of such behaviour, in networks grown from non-cooperative founder members. Our results highlight that the absence of any perturbation of the system may potentially 'lock in' defector behaviour in the long term—a concept that appears to merit further investigation. We also note the possibility that, in certain circumstances, preferential attachment can impede the emergence of network-reciprocal cooperation. In addition to investigating growing networks, we have applied our model to pre-existing networks, populated with initially random strategies. Regardless of initial topology, such networks achieve a compressed exponential degree distribution and the emergence of cooperation is observed.

Our model has no requirements for agents to possess higher cognitive abilities such as memory or recognition. It also does not require underlying explanations to describe preferential attachment of nodes in network formation. Levels of network heterogeneity that are sufficient for cooperation to emerge, arise simply by random connections formed over time, combined with attrition of least fit members of the population. Finally, we highlight that the model supports cooperation in cases of costly interaction (FCPG) and also where costs are trivialised (FCPI), real world scenarios being likely to lie somewhere along a spectrum between these two extrema. The fluctuation mechanism proposed here is the first example of a model describing the emergence of group-based cooperation in both growing *and* dynamic-equilibrium networks. As such it forms an important step in understanding the origins of cooperation in networks.

Acknowledgements. This work has been funded by the Engineering and Physical Sciences Research Council (Grant reference number EP/I028099/1).

References

1. Perc, M., Gómez-Gardeñes, J., Szolnoki, A., Floría, L.M., Moreno, Y.: Evolutionary dynamics of group interactions on structured populations: a review. J. R. Soc. Interface **10**, 20120997 (2013)
2. Hardin, G.: The tragedy of the commons. Science **162**, 1243–1248 (1968)
3. Hauert, C., De Monte, S., Hofbauer, J., Sigmund, K.: Volunteering as red queen mechanism for cooperation in public goods games. Science **296**, 1129–1132 (2002)
4. Brandt, H., Hauert, C., Sigmund, K.: Punishment and reputation in spatial public goods games. Proc. R. Soc. Lond. B Biol. Sci. **270**, 1099–1104 (2003)
5. Szolnoki, A., Perc, M.: Reward and cooperation in the spatial public goods game. EPL (Europhys. Lett.) **92**, 38003 (2010)
6. Axelrod, R., Hamilton, W.D.: The evolution of cooperation. Science **211**, 1390–1396 (1981)
7. Crespi, B.J.: The evolution of social behavior in microorganisms. Trends Ecol. Evol. **16**, 178–183 (2001)
8. Nowak, M.A., May, R.M.: Evolutionary games and spatial chaos. Nature **359**, 826–829 (1992)
9. Santos, F.C., Pacheco, J.M.: A new route to the evolution of cooperation. J. Evol. Biol. **19**, 726–733 (2006)
10. Santos, F.C., Santos, M.D., Pacheco, J.M.: Social diversity promotes the emergence of cooperation in public goods games. Nature **454**, 213–216 (2008)
11. Szabó, G., Fáth, G.: Evolutionary games on graphs. Phys. Rep. **446**, 97–216 (2007)
12. Perc, M., Szolnoki, A.: Coevolutionary games: a mini review. BioSystems **99**, 109–125 (2010)
13. Poncela, J., Gómez-Gardeñes, J., Floría, L.M., Sánchez, A., Moreno, Y.: Complex cooperative networks from evolutionary preferential attachment. PLoS One **3**, e2449 (2008)
14. Miller, S., Knowles, J.: Population fluctuation promotes cooperation in networks. Sci. Rep. **5** (2015)
15. Miller, S., Knowles, J.: A minimal model for the emergence of cooperation in randomly growing networks. In: Proceedings of the European Conference on Artificial Life 2015 (ECAL 2015), vol. 13, pp. 114–121 (2015)
16. Clauset, A., Shalizi, C.R., Newman, M.E.: Power-law distributions in empirical data. SIAM Rev. **51**, 661–703 (2009)
17. Fox Keller, E.: Revisiting "scale-free" networks. BioEssays **27**, 1060–1068 (2005)
18. Boyd, R., Richerson, P.J.: The evolution of reciprocity in sizable groups. J. Theor. Biol. **132**, 337–356 (1988)

Influence Maximization in Social Networks with Genetic Algorithms

Doina Bucur[1]([⊠]) and Giovanni Iacca[2]

[1] Johann Bernoulli Institute, University of Groningen, Nijenborgh 9,
9747 AG Groningen, The Netherlands
d.bucur@rug.nl
[2] INCAS3, Dr. Nassaulaan 9, 9401 HJ Assen, The Netherlands
giovanniiacca@incas3.eu

Abstract. We live in a world of social networks. Our everyday choices are often influenced by social interactions. Word of mouth, meme diffusion on the Internet, and viral marketing are all examples of how social networks can affect our behaviour. In many practical applications, it is of great interest to determine which nodes have the highest influence over the network, i.e., which set of nodes will, indirectly, reach the largest audience when propagating information. These nodes might be, for instance, the target for early adopters of a product, the most influential endorsers in political elections, or the most important investors in financial operations, just to name a few examples. Here, we tackle the NP-hard problem of influence maximization on social networks by means of a Genetic Algorithm. We show that, by using simple genetic operators, it is possible to find in feasible runtime solutions of high-influence that are comparable, and occasionally better, than the solutions found by a number of known heuristics (one of which was previously proven to have the best possible approximation guarantee, in polynomial time, of the optimal solution). The advantages of Genetic Algorithms show, however, in them not requiring any assumptions about the graph underlying the network, and in them obtaining more diverse sets of feasible solutions than current heuristics.

Keywords: Social network · Influence maximization · Genetic algorithm · Graph theory · Combinatorial optimization

1 Introduction

Social networks are graphs of relationships natural to organized societies. Among humans, these relationships are the vehicle by which news, ideas, trends, advertising, or influence will spread, starting from an initial set of information owners. A process of influence spread is shown abstractly in Fig. 1. In a social network where a graph edge $a \rightarrow b$ signifies a likelihood that user a will support b in any election (e.g., for the purpose of work-related promotions), the long-term promotion outcome will depend on the initial set of network participants who cast

© Springer International Publishing Switzerland 2016
G. Squillero and P. Burelli (Eds.): EvoApplications 2016, Part I, LNCS 9597, pp. 379–392, 2016.
DOI: 10.1007/978-3-319-31204-0_25

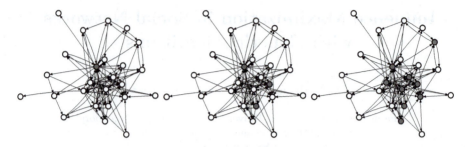

Fig. 1. Schematic spread of influence in a social network: three discrete steps in the state of the network starting from a single "seed" node. Nodes reached are drawn in red (grey in print). The information propagates via edges according to specific probabilistic spread models, each spread model particular to a type of network (Color figure online).

a vote. Also, in a product-marketing network where the nodes are products and the edges $a \rightarrow b$ model the likelihood of clients buying product b after product a was bought, an advertiser can increase sales by specifically promoting that set of products which trigger the largest co-buying effect upon the rest of the network.

The precise dynamics by which the graph structure enables new information to spread depends on the nature of the social network: a directed graph edge $a \rightarrow b$ may simply model a fixed probability that information will be adopted by b if a has just done so; alternatively, the likelihood of influence propagation across the edge depends on other features of node b, such as its number of direct relationships with other nodes. Social sciences have studied a number of such probabilistic *propagation models* [1].

The open problem of *influence maximization* in a social network is the following: given the network graph G, a discrete-time formal propagation model M, and a numerical "budget" $k \geq 1$ of network nodes to be initial "seeds" of influence, calculate that set of k seed nodes which will have the largest global influence upon the network. The problem was initially formulated in [2], and was proven to be NP-hard for most propagation models [1].

In this work, we tackle the influence maximization problem by means of a Genetic Algorithm (GA) which uses simple genetic operators typically used in discrete optimization. We evaluate the Genetic Algorithm on two large, real-world network datasets from the SNAP repository [3] modelling: (1) the who-voted-whom network of 7115 Wikipedia users through a number of years, and (2) a snapshot of the Amazon product co-purchasing network, of 262111 products. We compare the results against three existing heuristics (all of a greedy nature, either based on graph theory, or on exhaustive incremental search), and also with randomly sampled solutions. We show that, even on very large networks, the GA is at least as good as some of the known heuristics, without using any domain knowledge of the underlying graphs. Furthermore, the GA is able to find multiple diverse solutions with equally high network influence, suggesting that the fitness landscape of this problem may have a high level of multimodality.

The remainder of the paper is structured as follows. First, in the next section we briefly review the background concepts on influence propagation and the ways influence can be maximized, by using heuristics. Then, in Sect. 3 we describe the Genetic Algorithm used in our tests. Section 4 reports the experimental results and the related analysis. Finally, Sect. 5 concludes this work.

2   Background: Network and Propagation Models

A social network is modelled as a directed graph G. For the purpose of studying the propagation of influence, at a time t each node in G is either *active* (i.e., has adopted the new information) or *inactive*. As will be seen in Sect. 2.1 below, in the propagation models we use, the set of active nodes in G increases monotonously until the propagation process ends.

2.1   Modelling the Propagation of Influence: The Cascade Models

We consider a family of two classic, discrete-time influence-propagation models in social networks, known as "cascade" models [1], see Algorithm 1. In these discrete-time propagation models, the dynamics of information adoption is represented in individual steps. A given set A_0 of "seed" nodes start in the active state; these are the network entities which are originally targeted in the influence process. In the next time step $t \geq 1$, each node activated at time $t - 1$ may activate some other nodes according to a probability given by the propagation model; in the case of cascade models, this probability of activation is attached to every edge, and is an independent random variable. The propagation ends at the time t when no new nodes were activated, and the set containing all nodes made active before t is the end result of the propagation process.

The simplest cascade model is the Independent Cascade, first studied in the marketing domain, as an attempt to understand the effects that personal word-of-mouth communication has upon macro-level marketing [4]. In the Independent Cascade, each newly activated node n will succeed in activating each currently inactive neighbour m with a fixed probability p, which is a global property of the system, and is thus equal across all edges $n \rightarrow m$; when node n has more than one neighbour, the attempts at activation are sequenced in arbitrary order. As an example, for the simple network in Fig. 2, in which $A_0 = \{n\}$, the model has $p_1 = p_2 = p$ and, at time $t = 1$, nodes m_1 and m_2 are equally likely to become

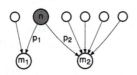

Fig. 2. Simple propagation example for the cascade models. With Independent Cascade, $p_1 = p_2$; with Weighted Cascade, $p_1 = 1/2$ and $p_2 = 1/5$.

active. Given $p = 0.5$, the expected size of the set of active nodes at the end of the propagation process is 2; this count includes the seed node n itself.

The second cascade model, named Weighted Cascade, differs from the Independent Cascade in that it assigns non-uniform probabilities of activation to edges: an edge $n \rightarrow m$ has probability $\frac{1}{\text{in-degree}(m)}$ of activating m when n is itself active[1]. It holds that, unlike for the Independent Cascade model, the expected number of neighbours which will succeed in activating any node equals 1.

Algorithm 1. The **Cascade** family of propagation models. G is the network graph, A_0 the set of "seed" nodes, and $p(n \rightarrow m)$ the probability that information will reach across a directed graph edge $n \rightarrow m$. In the **Independent Cascade** model, $p(n \rightarrow m)$ is constant and equal to the input parameter p for all edges $n \rightarrow m$ in G. In the **Weighted Cascade** model, $p(n \rightarrow m)$ is instead equal to $\frac{1}{\text{in-degree}(m)}$.

1: **procedure** CASCADE(G, A_0, p)
2:    $\tau \leftarrow$ *False*   ▷ τ: has the propagation ended?
3:    $A \leftarrow A_0$ ▷ A: the set of active nodes after the propagation ended
4:    $B \leftarrow A$ ▷ B: the set of nodes activated in the last time slot
5:    **while** not τ **do**
6:        $nextB \leftarrow \emptyset$
7:        **for** each $n \in B$ **do**   ▷ only nodes in B will activate new nodes
8:            **for** each direct neighbour m of n in G, where $m \notin A$, **do**
9:                with probability $p(n \rightarrow m)$, add m to $nextB$
10:        $B \leftarrow nextB$
11:        $A \leftarrow A \cup B$
12:        **if** B is empty **then**
13:            $\tau \leftarrow$ *True*
14:    **return** the size of A

2.2   Problem Statement

The influence-maximization problem optimizes the choice of the seed nodes in set A_0. The *influence* of a given seed set A_0, denoted $\sigma(A_0)$, is the expected size of the set A of active nodes, $\mathbb{E}[|A|]$, obtained by a propagation model from Algorithm 1 after completion. Given a number $k \geq 1$, the problem asks to compute the optimum set A_0, where $|A_0| = k$, such that $\sigma(A_0) = \mathbb{E}[|A|]$ is maximized over all possible sets A_0 in the graph G.

For both models, the problem of calculating the optimal "seed" set is NP-hard. Further, we also know an approximation hardness result: estimating the

---

[1] For any node n, we denote by *in-degree*(n) the number of edges incoming to n, and by *out-degree*(n) the number of edges outgoing from n. Unlike some of the related literature, which works with undirected rather than directed graphs, in our algorithms we make the distinction between the two degree counts explicit.

optimal solution by a factor better than $1 - \frac{1}{e}$ (where e is the base of the natural logarithm) is also NP-hard [1]; this was proven by showing that it is at least as hard as the classical NP-hard problem of maximum coverage (i.e., determining those k sets of elements whose union has the maximum size). This approximation factor amounts to an approximation guarantee just above 63 %, i.e., it is not possible to obtain in polynomial time a set A_0 whose influence is higher than 63 % of the true optimum.

2.3 Existing Heuristics. Their Complexity and Approximation Guarantees

Approximation algorithms are used to compute best-effort solutions. In this paper, we use three of the existing approximation heuristics as a basis for comparison of performance: *General greedy* (due to its high complexity, only when feasible computationally), *High degree*, and *Single discount*. The Kempe et al. [1] greedy hill-climbing algorithm (referred to in this paper as the *General greedy* heuristic) is proven to approximate the solution to within a factor arbitrarily close to the approximation guarantee; we thus use *General greedy* as an optimality benchmark. This heuristic has the following logic: it starts with $A_0 = \emptyset$ and the given problem size k, and adds one node at a time to A_0 until $|A_0| = k$. A new node n from G, with $n \notin A_0$, is chosen to be added to A_0 if n maximizes $\sigma(A_0 \cup \{n\})$.

This means that all nodes in G, but not yet in A_0 must have their added influence evaluated by computing their $\sigma(A_0 \cup \{\cdot\})$. Note that, since the propagation models are stochastic, the evaluation of $\sigma(A_0 \cup \{\cdot\})$ in polynomial time will not be exact; the problem of exactly computing $\sigma(A_0)$ for any A_0 under the Independent Cascade model was proven #P-complete in Wang et al. [5]. However, one can approximate $\sigma(A_0)$ by simulating the propagation process a number of times, as the computational budget allows.

The heuristics which followed the *General greedy* method fell into two categories: (a) heuristics which preserve the approximation guarantee of *General greedy* while lowering its average-case (but not worst-case) complexity, and (b) heuristics of better complexity, but either no optimality guarantees or much weaker ones.

From the latter category of heuristics without any guarantees attached, Kempe et al. [1] also tested experimentally the performance of a *High-degree* (or degree centrality) heuristic, which adds nodes n to A_0 in order of decreasing out-degrees. They motivate this heuristic as being a standard greedy heuristic for networks. Chen et al. [6] designs degree-discount heuristics based on *High-degree*: if a node n is already in A_0 and there exists an edge $m \to n$, then, when considering whether to add node m to A_0, this edge should not be counted towards the out-degree of m. This heuristic is applicable to all cascade models, and is denoted here as the *Single discount* heuristic. In our experiments, often we found that *Single discount* will only minimally improve over *High-degree*, in which case we compare only against *Single discount*. Many other heuristics than the ones described above are known. Among those without optimality guarantees, Jiang et al. [7] tests Simulated Annealing under Independent Cascade, and finds that it has a complexity advantage over greedy heuristics, and can also find narrowly better solutions.

Papers [1,6,7] evaluate the heuristics above comparatively on a small number of large social networks, and generally find that, for these datasets, the *General greedy* algorithm and its improved variants do find better solutions than the two degree-based heuristics, if in cases only by a small percentage. To concretely illustrate the fact that degree-based heuristics can underapproximate significantly, Fig. 3 gives two examples of small networks on which *High degree* is not effective. For the example on the left, *High degree* with $k = 2$ may select as optimal seeds the set $\{c, d\}$, from among other options of equal out-degrees. With Weighted Cascade, $\sigma(\{c, d\}) = 4$, while a better solution is the set $\{a, b\}$, with $\sigma(\{a, b\}) = 10$. On the right, under Independent Cascade with $p = 50\%$, the best single seed is $\{a\}$, with $\sigma(\{a\}) = 8.5$, rather than node b, which has the same degree, with $\sigma(\{b\}) = 6$.

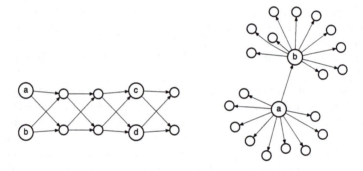

Fig. 3. Examples of networks executing cascade propagation models on which degree-based heuristics are ineffective. Left: a linear graph of nearly uniform node degrees, where *High degree* may choose $\{c, d\}$ as optimal seeds rather than the (here, optimal) $\{a, b\}$, under Weighted Cascade. Right: an extended star topology where a and b have equal degrees, but the seed set $\{a\}$ has higher influence, under Independent Cascade.

However, it is these degree heuristics without approximation guarantees which have the lowest computational complexity, rather than *General greedy*. Given a graph $G = (V, E)$, the problem size k, and number R of simulation repetitions to evaluate $\sigma(A_0)$, *High degree* has complexity $O(|V| \cdot \log(k))$ (with a min-heap implementation), while *General greedy* is $O(k \cdot |V| \cdot |E| \cdot R)$ [6]. In other words, for graphs that are large or dense, the *General-greedy* technique can be entirely infeasible.

3   Methodology: A Simple Genetic Algorithm

The application of Genetic Algorithms to the influence-maximization problem is straightforward: a candidate solution generated by the GA is encoded as a fixed-size sequence A_0 of k seed node (integer-valued) identifiers. For each solution A_0, we evaluate the fitness $\sigma(A_0)$ by running the Cascade Algorithm shown in Algorithm 1 (either according to the Independent Cascade or to the Weighted Cascade model). As the algorithm is stochastic, we run 100 simulation repetitions[2]. We then average, over the 100 repetitions, the size of A returned by the Cascade Algorithm. This average value is finally assigned to each candidate solution as its fitness.

The GA is configured to use 1-point crossover and a random mutation operator. The latter resets each node, with a given probability, to one of the possible node identifiers in G. Both genetic operators (crossover and mutation) are applied with probability one. Selection is performed using fixed-size tournament, with generational replacement and elitism (i.e., at each generation n_e best solutions in the population – the "elites" – are kept without being mutated). The complete list of parameters of the GA is reported in Table 1. We obtained these parameters through experimentation with different values; later in Sect. 4.3, we present the results of varying the population size, the number of elites, the tournament size, and the mutation rate.

Table 1. Parameters of the Genetic Algorithm

Parameter	Value
No. generations	$100 \times k$
No. elites	2 (for $k = \{10, 20\}$), 4 (for $k = \{30, 40, 50\}$)
Population size	100
Mutation rate (per node)	0.1
Crossover rate	1.0
Tournament size	5

The proposed parameters were empirically chosen by running preliminary experiments (see Sect. 4.3 on parameter analysis for further details). Overall, we observed that the performance of the GA is quite robust with different parameter settings, although we noted that some parameters (especially tournament size, mutation rate, and population size) can sensibly affect, as expected, the diversity of the solutions found at the end of the evolutionary process, as well as the optimization results. As for the number of elites, we found that maintaining a slightly higher number of unmutated individuals (4 instead of 2) is especially beneficial

---

[2] This number of repetitions was chosen as a practical compromise between the confidence interval that it affords, and the overall computational complexity of the Genetic Algorithm. With regards to the accuracy of the fitness estimation, 100 simulation repetitions give a 95 % confidence interval for the average in the approximate range of $[3, 10]$ nodes influenced, for all our experiments. Increasing the number of repetitions to 10000 would give a 95 % confidence interval for the average that is ≤ 1 nodes influenced in all experimental cases, but requires far longer runtimes.

for larger values of k. Indeed, as the search space dimensionality increases, keeping a larger group of diverse elites seems to prevent premature convergence and promote population diversity, especially in the later stages of evolution.

4   Experimental Results and Analysis

We test the proposed GA-based influence maximization algorithm on two large social network datasets, available from the SNAP repository [3]. A brief description of the datasets is reported in Table 2. It is to note that we chose networks with underlying graphs of strikingly different size and structure. The Wiki graph is relatively small at 7115 nodes, and very variate in terms of node degrees, with a large maximum out-degree. On the other hand, the Amazon graph (with over a quarter of a million nodes) is orders of magnitude larger, and has a flat degree landscape, with a maximum and median out-degree of 5.

For each dataset, we consider the Independent and Weighted Cascade model with values of k (the size of the seed node identifiers) ranging in $\{10, 20, 30, 40, 50\}$. In the Independent Cascade model, we fix $p = 1\%$; this low probability is realistic for practical social networks. The combination between the Amazon dataset and the Independent Cascade model is an exception, in that the low probability p, together with the consistently very low node degrees in the Amazon graph, yield a "flat" fitness landscape. Therefore, we do not present the results of this combination.

In order to evaluate the influence of the initial population and the robustness of the algorithm, we execute the GA three times on each experimental condition, with different random seeds. All the experiments are implemented in Python, by using the Genetic Algorithm provided by the Python package `inspyred` [8] (configured as in Sect. 3). Experiments are run on two Linux machines: a Ubuntu

Table 2. Large social networks from the Stanford large network dataset (SNAP [3]). The names in brackets indicate the dataset names used on the SNAP repository.

	Social network	Graph type and size	Node out-degrees
Wiki (wiki-Vote)	who-voted-whom in Wikipedia user elections (data collected on Jan 3 2008)	directed, 7115 nodes, 103689 edges	min 0, max 893, avg 14.57, stddev 42.28, median 2
Amazon (amazon0302)	Amazon product co-purchasing network (data collected on March 2 2003); if i is frequently co-purchased with product j, the graph has an edge $i \rightarrow j$	directed, 262111 nodes, 1234877 edges	min 0, max 5, avg 4.71, stddev 0.95, median 5

14.04 with 64 AMD Opteron 2.3 GHz cores and 256 GB RAM and a Ubuntu 12.04 with 32 Intel Xeon 2.0 GHz cores and 128 GB RAM. Computations are parallelized by running in multiple threads the evaluation of the candidate solutions generated by the GA.

A summary of the numerical results is shown in Fig. 4 for the Wiki and Amazon datasets. In the figures, we compare the highest influence (i.e., the number of active nodes) obtained by the GA with the results obtained by random sampling and with the heuristics. All algorithms evaluate the fitness of a seed set A_0 in exactly the same way, i.e., by simulating the propagation model, starting with A_0, 100 times, and reporting the mean number of influenced nodes as fitness. The figures report this mean value together with a measure of the confidence in the mean: this is the 95 % confidence interval, in the case of the algorithms which output a single solution (i.e., the degree heuristics *High degree* and *Single discount*, the greedy heuristic *General greedy*, and the Genetic algorithm), and the standard deviation in the case of random sampling, which samples 100 random values for A_0, each evaluated in 100 simulation repetitions. For the Wiki dataset, in the cases of both propagation protocols, the two degree-based heuristics, *High degree* and *Single discount*, yield results that are largely indistinguishable; because of this, we present only one in the figures, for clarity.

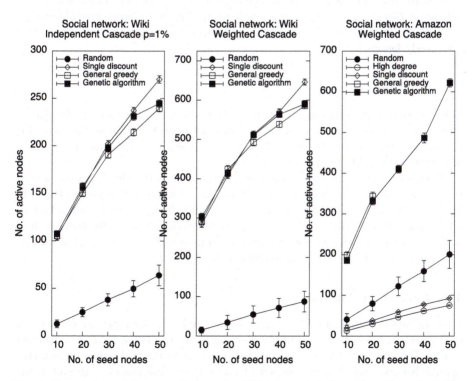

Fig. 4. Comparison of the numerical results: Wiki dataset, Independent Cascade (left); Wiki dataset, Weighted Cascade (middle); Amazon dataset, Weighted Cascade (right).

We also report, in Fig. 5, examples of the generational trends on some of the test cases. The evolutionary trends show on each test case little variation among the various repetitions. Additionally, the convergence rate of the GA is quite high, especially on the Wiki dataset.

4.1   Comparative Influence Results

The fact that the random sampling of seed nodes yields solutions of low network influence is expected, as is the fact that the computationally intensive *General greedy* heuristic, known to fulfill an approximation guarantee (as motivated in Sect. 2.3) computes significantly better solutions.

On the other hand, the performance of the degree-based heuristics is surprising: on the Wiki dataset, the degree-based heuristics find better solution than *General greedy* for the larger k values, while on the Amazon dataset their performance is outdone by all other algorithms, including random sampling. This fact can be explained by the different structure of the underlying graphs: essentially, the Amazon graph conforms to the general idea shown in the example from Fig. 3 (left), where we have shown that degree-based heuristics can be severely suboptimal. Overall, the degree heuristics have undependable performance.

Finally, the Genetic algorithm, although stochastic and not of an exhaustive nature (as *General greedy*), is consistently matching the *General greedy* heuristic; for some data points and the smaller Wiki dataset, the Genetic algorithm outdoes the greedy heuristic.

4.2   Runtimes

In Fig. 6 we report the average runtimes, in core hours, of the Genetic Algorithm and General greedy heuristic over the entire set of test scenarios. We omit, for brevity, the runtimes of the other heuristics as they are, in comparison, orders of magnitude smaller (< 1 core hour).

We observe that, for the Amazon dataset, the Genetic Algorithm has a runtime that is approximately half as big as the runtime of the General greedy heuristic. On the other hand, on the Wiki network the GA is more computationally expensive (up to 30 %) than the greedy heuristic, although for larger values of k the runtimes of the two algorithms are comparable, suggesting that on this particular dataset the GA is more efficient for larger k.

4.3   Parameter Analysis

We conclude our analysis by reporting some empirical observations we made in our experiments on the effect of the parameters of the GA on the optimization results. Due to limited computational resources, we performed only a preliminary qualitative analysis on a limited set of parameter configurations, to gain some general insight on the parametrization.

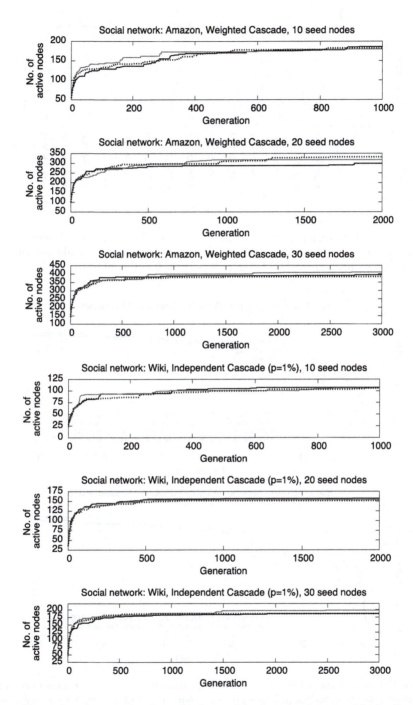

Fig. 5. Examples of generational trends observed in the experiments with the Genetic Algorithm. In each case, we report the trends of three repetitions of the GA.

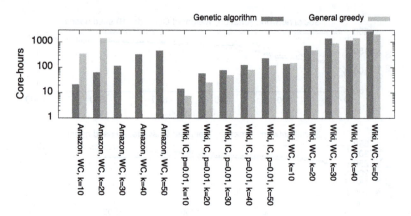

Fig. 6. Average runtimes, in core hours, of Genetic Algorithm and General greedy heuristic. Due to the limited computational resources available, we did not run the General greedy heuristic on the Amazon dataset for $k = \{30, 40, 50\}$.

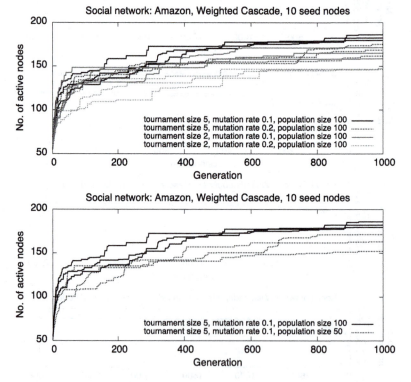

Fig. 7. Influence of the GA parameters on the optimization results: tournament size and mutation rate (top) and population size (bottom). For each parameter configuration, three repetitions of the GA are shown. The others parameters are kept constant, see Table 1 for details.

- **No. elites:** As reported in the previous section, elitism is beneficial as far as a small percentage ($2\% - 4\%$) of the best individuals in the population is maintained without undergoing mutation. Furthermore, for larger values of k there is a positive effect of higher numbers of elites on the diversity of individuals (particularly in the later stages of the optimization process).
- **Mutation rate:** We observed that smaller mutation rates (0.1) are overall beneficial in terms of optimization results, especially during the first stages of evolution. The other value we tested, 0.2, seems to produce disruptive mutations thus causing slower convergence. As shown in Fig. 7 (top) for the case ⟨Amazon, Weighted Cascade, $k = 10$⟩, this trend appears in both cases of tournament sizes 2 and 5.
- **Tournament size:** The effect of tournament size shows mostly in two aspects: one hand, increased tournament sizes increase the selection pressure [9], thus resulting in a higher convergence rate, see Fig. 7 (top). Also, tournaments of 5 individuals allowed us to find a more diverse set of solutions compared to binary tournaments, especially for larger values of k.
- **Population size:** We tested two values of population size, namely 50 and 100. Overall, we noted that bigger populations lead to better optimization results, as shown in Fig. 7 (bottom) for the case ⟨Amazon, Weighted Cascade, $k = 10$⟩. The intuitive explanation is that larger populations allow a higher diversity level, thus improving the exploration of the search space.

5   Conclusions

In this paper we made a first attempt to tackle the influence maximization problem in social networks by using Genetic algorithms. We performed an experimental campaign on two datasets taken from real-world applications, comparing the results of the GA against known heuristics based on either graph theory or a greedy, incremental exhaustive search. In short, our experiments showed that in each test case the performance of the GA is at least comparable to that of the best tested heuristic. However, as heuristics typically rely on specific network topology features, we observed a dramatic variation in their performance from one case to the other. On the other hand, since the GA is agnostic w.r.t. the properties of the networks, its results were more consistent across the whole range of test cases. In addition to that, we found, quite surprisingly, that on some test cases the GA was computationally cheaper than the greedy heuristic.

To the best of our knowledge, our study represents one of the very first examples of application of computational intelligence methods to influence maximization in social networks, whereas a number of heuristics have been designed and tested on this problem. Overall, our experiments revealed that Genetic algorithms are a viable tool to solve the influence maximization problem, especially in absence of any prior knowledge about the network topology and the characteristics of the graph underlying the social network. Other related studies have also shown that bio-inspired algorithms are suitable to compute solutions to computationally hard problems from the area of complex systems and networks in polynomial time: [10,11] applied natural selection as a means to stress-test wireless sensor networks and find vulnerable topologies; [12,13] optimized the path of strong attackers in large-scale mobile, wireless, urban delay-tolerant networks, and uncovered potential vulnerabilities in a message-routing protocol for such networks.

Considering the high relevance of social networks in all modern applications, we believe it is worth to further investigate the application of computational intelligence to this domain. In future research, we plan to analyze more thoroughly the influence the GA parametrization on this specific problem. Furthermore, we deem promising the possibility to combine GAs with graph-based heuristics, in order to create domain-specific memetic algorithms. Finally, it will be interesting to extend the experimental setup to other datasets.

References

1. Kempe, D., Kleinberg, J., Tardos, É.: Maximizing the spread of influence through a social network. Theory Comput. **11**(4), 105–147 (2015)
2. Richardson, M., Agrawal, R., Domingos, P.: Trust management for the semantic web. In: Fensel, D., Sycara, K., Mylopoulos, J. (eds.) ISWC 2003. LNCS, vol. 2870, pp. 351–368. Springer, Heidelberg (2003)
3. Leskovec, J., Krevl, A.: SNAP Datasets: Stanford large network dataset collection, October 2015. http://snap.stanford.edu/data
4. Goldenberg, J., Libai, B., Muller, E.: Talk of the network: a complex systems look at the underlying process of word-of-mouth. Mark. Lett. **12**(3), 211–223 (2001)
5. Wang, C., Chen, W., Wang, Y.: Scalable influence maximization for independent cascade model in large-scale social networks. Data Min. Knowl. Disc. **25**(3), 545–576 (2012)
6. Chen, W., Wang, Y., Yang, S.: Efficient influence maximization in social networks. In: Proceedings of the 15th ACM SIGKDD International Conference on Knowledge Discovery and Data Mining, KDD 2009, pp. 199–208. ACM, New York (2009)
7. Jiang, Q., Song, G., Cong, G., Wang, Y., Si, W., Xie, K.: Simulated annealing based influence maximization in social networks. In: Burgard, W., Roth, D. (eds.) AAAI. AAAI Press (2011)
8. Garret, A.L.: Inspyred: A framework for creating bio-inspired computational intelligence algorithms in Python, October 2015. https://pypi.python.org/pypi/inspyred
9. Miller, B.L., Goldberg, D.E.: Genetic algorithms, tournament selection, and the effects of noise. Complex Syst. **9**, 193–212 (1995)
10. Bucur, D., Iacca, G., Squillero, G., Tonda, A.: The impact of topology on energy consumption for collection tree protocols: an experimental assessment through evolutionary computation. Appl. Soft Comput. **16**, 210–222 (2014)
11. Bucur, D., Iacca, G., de Boer, P.T.: Characterizing topological bottlenecks for data delivery in CTP using simulation-based stress testing with natural selection. Ad Hoc Netw. **30**, 22–45 (2015)
12. Bucur, D., Iacca, G., Squillero, G., Tonda, A.: Black holes and revelations: using evolutionary algorithms to uncover vulnerabilities in disruption-tolerant networks. In: Mora, A.M., Squillero, G. (eds.) EvoApplications 2015. LNCS, vol. 9028, pp. 29–41. Springer, Switzerland (2015)
13. Bucur, D., Iacca, G., Gaudesi, M., Squillero, G., Tonda, A.: Optimizing groups of colluding strong attackers in mobile urban communication networks with evolutionary algorithms. Appl. Soft Comput. **40**, 416–426 (2016)

Measuring Diversity of Socio-Cognitively Inspired ACO Search

Ewelina Świderska[1], Jakub Łasisz[1], Aleksander Byrski[1(✉)], Tom Lenaerts[2],
Dana Samson[3], Bipin Indurkhya[4], Ann Nowé[5], and Marek Kisiel-Dorohinicki[1]

[1] Faculty of Computer Science, Electronics and Telecommunications,
AGH University of Science and Technology,
Al. Mickiewicza 30, 30-059 Kraków, Poland
{ewelina.swiderska,jakublasisz}@gmail.com, {olekb,doroh}@agh.edu.pl
[2] Université Libre de Bruxelles, Campus de la Plaine, ULB CP212,
Boulevard du Triomphe, 1050 Bruxelles, Belgium
tlenaert@ulb.ac.be
[3] Université catholique de Louvain, IPSY, Place Cardinal Mercier 10 Bte L3.05.01 à,
1348 Louvain-la-Neuve, Belgium
dana.samson@uclouvain.be
[4] Institute of Philosophy, Jagiellonian University,
ul. Grodzka 52, 31-044 Kraków, Poland
bipin@agh.edu.pl
[5] Vrije Universiteit Brussel, Boulevard de la Plaine 2, 1050 Ixelles, Belgium
anowe@como.vub.ac.be

Abstract. In our recent research, we implemented an enhancement of
Ant Colony Optimization incorporating the socio-cognitive dimension
of perspective taking. Our initial results suggested that increasing the
diversity of ant population — introducing different pheromones, different
species and dedicated inter-species relations — yielded better results. In
this paper, we explore the diversity issue by introducing novel diversity
measurement strategies for ACO. Based on these strategies we compare
both classic ACO and its socio-cognitive variation.

Keywords: Ant colony optimization · Metaheuristics · Diversity ·
Nature-inspired optimization · Discrete optimization

1 Introduction

In population-based metaheuristics, attaining certain balance between exploration of the search space in general, and exploitation of its most promising
parts is a major issue to achieve robust algorithms producing feasible solutions,
using rationally available resources (as computing power or time) [16]. Lack of
diversity may lead to stagnation and the system may focus on locally optimal
solutions (in other words—trapped in a local extremum), needing more randomness to escape [15]. This balance is usually attained by maintaining diversity

© Springer International Publishing Switzerland 2016
G. Squillero and P. Burelli (Eds.): EvoApplications 2016, Part I, LNCS 9597, pp. 393–408, 2016.
DOI: 10.1007/978-3-319-31204-0_26

in the population of individuals: different diversity-preserving mechanisms have been introduced to accomplish this, e.g. coevolution and speciation in Evolutionary Algorithms [16]. In this connection, it should be noted that diversity is shown to play a key role in creativity [9,21].

Ant systems have proven to be a popular tool for solving many discrete optimization problems, e.g. Traveling Salesman Problem (TSP), Quadratic Assignment Problem (QAD), Vehicle Routing Problem (VRP), Graph Coloring Problem (GCP) and others [8]. In our previous paper, we considered the ant system as a way to express socio-cognitive behaviors of a population of ants, differentiating them into species and defining their stigmergic interactions, following and enhancing the substantial results presented in [23], getting promising results in TSP optimization.

We claim that the promising results obtained by the socio-cognitive ants, as presented in [23], were, at least partially, due to an enhanced diversity of ants in the population (instead of a homogeneous population). However, to verify this claim we need to devise some way to measure and monitor diversity. One approach is to check the contents of the current solutions of the problem in the population of individuals. In particular, in the case of evolutionary algorithms, one can analyze the center of weight of the solutions, dispersion of genes etc. (see, e.g. [5,17]).

However, though measuring diversity in a real-valued space is relatively easy, it becomes a very difficult problem in a discrete-valued space. For example, looking for a center-of-weight of the strings (e.g. encoding TSP solutions) is an NP-hard problem *per se* [14]), so analysis of the search space is impossible in a feasible time period.

We introduce in this paper a new approach for measuring the diversity in ACO search and evaluate its feasibility on the socio-cognitive ant system compared to the classic ACO. Instead of analyzing solutions we focus on analyzing the pheromone table, treating the information contained therein as a kind of *derivative* of the information contained in the search space.

The rest of the paper is organized as follows. First, ACO and its selected variants including our socio-cognitive ones, are described. Later, dedicated diversity measurement techiques are introduced and appropriate results obtained from comparing the classic and the socio-cognitive ACO are shown. In the final section we present the conclusions and mention some future research issues.

2　Classic and Novel Approaches to Ant Colony Optimization

Ant System, introduced in 1991, applied to solve TSP, is considered to be a progenitor of all ant colony optimization (ACO) algorithms [6]. Because the action of a certain ant during one iteration is completely independent of the actions of other ants during any iteration, the sequential ant algorithm can easily be parallelized.

The ACO algorithm is an iterative process during which certain number of ants (agents) gradually create a solution [7,8]. The problem being solved is usually depicted as a graph, and the main goal of the ants is to traverse this graph in an optimal way. Each move of an ant consists in choosing a subsequent component of the solution (graph edge) with certain probability. This decision may be affected by interaction among the ants based on the levels of *pheromones*, which may be deposited into the environment (on the edges of the graph) by some ants and perceived by other ants. This interaction is guided by stigmergy (environment-mediated communication among individuals instead of direct contact) rules proposed in [6]). The iteration process is finished when a feasible solution is reached through the cooperative efforts of all the ants.

Recently, new interesting modifications of ACO-related techniques have been introduced. For example, multi-type ACO [19,24] define many species of ants and allow complex stigmergic interactions such as attraction to the pheromone of the same species and repulsion from that of the others. These algorithms have been successfully applied to problems such as edge disjoint path problem [19] and light path protection [24].

There are other modifications of the classic ACO, such as hierarchical ACO, where additional means of control are introduced to manage the output of particular ants or ant species [22]. In another approach, ants are endowed with different skills (e.g. sight, speed) in order to realize global path-planning for a mobile robot [13]. In a successfull approach to solve TSP, the authors propose to use two types of ants: classic and exploratory (creating 'short routes', moving according to some predefined conditions like near some selected cities, etc.) [12]. In [4], the authors introduce different ant sensitivity to pheromones such that ants with higher sensitivity follow stronger pheromone trails, while ants with lower sensitivity behave more randomly. This model strives to sustain a balance between exploration and exploitation.

Taking inspiration from these approaches, especially the ones proposed by Nowé et al. [19] (many species of ants with detailed stigmergic interactions), and by Chira et al. (different sensivity of the ants to the pheromones) [4], we proposed a novel method of simulation and analysis of socio-cognitive properties of individuals of a certain population, at the same time being an efficient optimization algorithm that already produced encouraging results [23].

Measuring of diversity in ACO has been tackled by Nakamichi et al. [18], who constructed elitist ACO and examined the number of paths found by elite ants. In other papers, one can find mostly visible enhancements of diversity (though not a measurement techniques themselves), see, e.g. [1,20]. Therefore, we chose our research goal to work out a universal diversity measurement techniques for ACO.

3   From Perspective Taking to Enhancing Diversity in ACO

In [23] we have explored the effect of incorporating socio-cognitive mechanism, namely perspective taking, on the search capabilities of ACO. We assumed that such an approach would promote the diversity of the ant colony, however we were not able to validate this claim: hence we undertook the research presented here. We briefly summarize below the main ideas from [23], which provide the context in which we address the problem of characterising diversity parameters.

Typically, perspective taking is seen as a one-dimensional ability: the degree to which an agent can take another one's perspective. But recent research has shown that human's variability in terms of perspective-taking performance can be better explained if one considers two dimensions [2]: the ability of an agent to handle conflict between its own and the other agent's perspectives, and the relative priority that an agent gives to his own perspective relative to the other's perspective. Individuals endowed with good cognitive skills to manage conflicting information are usually better perspective-takers [10]. In addition, the less a person focuses on her own perspective and the more that person will be motivated to engage in perspective taking [2]. Experimental research has also shown that situational factors such as someone's emotional state can selectively impact on one of the two perspective-taking dimensions (conflict handling or perspective priority), which further shows that both dimensions are important to characterize human diversity in perspective taking [3]. This two-dimensional approach to perspective taking inspired us to define four types of individuals.

Let us consider these four types of individuals and their possible interactions [23]:

- Egocentric individuals: Focus on their own perspective and can become creative by finding their own new solutions to a given task. They do not pay attention to others and do not get inspired by others' actions (or these inspirations do not become a main factor of their work).
- Altercentric individuals: Focus on the perspective of others and thus follow the mass of others. They are less creative but can end up supporting good solutions by simply following them.
- Good-at-conflict-handling individuals: Get inspired in a complex way by the actions of other individuals by considering different perspectives and choosing the best.
- Bad-at-conflict-handling individuals: Act purely randomly, following sometimes one perspective, sometimes another without any inner logic.

Now let us work on incorporating of these ideas by constructing a population consisting of different species of ants. These species will search for the solution not only using their own expertise, but also getting stigmergic inspirations from other ones (by the analysis and combination of different pheromones left by these species in the environment).

These four types of individuals are directly inspired by socio-cognitive phenomena. It is to note that a dedicated paper, reviewing different configurations of ACO populations (exceeding the above-mentioned inspirations) is under review.

4   Socio-Cognitive Ant Colony Optimization

In this section we recall the description of classic and socio-cognitive ACO (after [23]). The reason to do this is to prepare a simple formalism that will help in proper introducing different methods of diversity measuring in the next section.

Starting from the definition of classic ACO, we consider optimization of a combinatorial problem (e.g. to find a Hamiltonian cycle in a graph as in Travelling Salesman Problem). The method is based on agents, namely ants, that roam along the edges of a graph, searching for cycles and leaving trails of pheromones behind them.

4.1 Classic ACO

In the classical ACO algorithm, the ants are deployed in a graph consisting of vertices $V = \{i : i \in \mathbb{N}\}$ and edges $E = \{e_{ij} : i, j \in V\}$, where each edge is associated with the cost of moving along it. Each ant gets a randomly chosen starting graph node. Beginning from this node, the ant searches for a cycle, in a step-by-step manner, by moving from one node to another, choosing the next one and not coming back. While considering which node to visit next, the ant has to compute attractiveness for all possible paths that can be taken from the present node. The attractiveness n_{ij} of the edge ij starting from the node i where the ant is currently at is the basis for computing the probability of choosing a particular path:

$$\mu_{ij} = \frac{n_{ij}}{\sum_j n_{ij}} \tag{1}$$

where j is computed only for nodes that have not yet been visited by the ant.

The exact values of n_{ij}, which is the attractiveness computed for the next edges constituting the constructed path, for classic ants and all the introduced modifications are given below in details.

Finally, the ant randomly selects a path based on the previously computed probabilities: paths with higher attractiveness are more likely to be chosen. After visiting all the nodes exactly once, the ant finishes its trip and returns the found cycle as a proposed solution, and then retreats depositing certain amount of pheromone on the path of its current cycle. The amount of pheromone deposited on an edge e_{ij} is denoted by π_{ij}, and the deposition algorithm of ant a_k retreating along cycle c_{a_k} is as follows:

$$\pi'_{ij} \leftarrow \pi_{ij} + \frac{\pi_d}{\sum_{e \in c_{a_k}} cost(e)} \tag{2}$$

where the default pheromone deposit π_d is 1, e_{ij} denotes an edge in the cycle, and $cost(e_{ij}) : E \rightarrow \mathbb{R}$ is a function that assigns a cost to each edge.

The pheromone evaporates in each iteration (in each edge of the graph) according to this formula:

$$\pi'_{ij} = (1 - \pi_e) \cdot \pi_{ij} \tag{3}$$

Default pheromone evaporation coefficient π_e is 0.01.

Classic ants. They consider both pheromone and distance while choosing their direction by computing *path attractiveness* in order to complete the cycle. So an ant at node i will choose the next edge according to the following attractiveness:

$$n_{ij} = \frac{\pi_{ij}^{\alpha}}{cost(e_{ij})^{\beta}} \tag{4}$$

Default factors are, pheromone influence $\alpha = 2.0$, distance influence $\beta = 3.0$.

Each type of ant in Sect. 4.2 below uses Eq. 1 to calculate probabilities for the subsequent paths, differing only in **attractiveness** to various types of pheromones.

4.2   Multi-pheromone ACO

In socio-cognitive ACO, the idea of multiple pheromones is implemented by introducing different 'species' of ants and enabling their interactions (similar to the approach taken in [19]). The interaction is considered as a partial inspiration or perspective taking, realized by a particular ant reacting to the decisions taken by ants belonging to other species. This is made possible by having ants of different species leave different 'smells' (see Fig. 1). Different ants use different rules (consider different properties of the path) for computing attractiveness; and looking for inspirations or perspective taking, they utilize the smells of pheromones left by other species in a predefined way. Therefore different species may be treated as organisms with selective smelling capabilities (reacting to different combinations of the smells that are present).

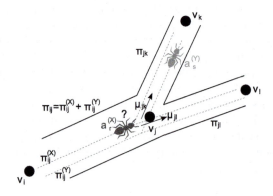

Fig. 1. Multi-pheromone ACO setting

Different ant species leave pheromones that 'smell' different, so the pheromone left at a particular edge is described as a sum of the following components:

$$\pi_{ij} = \pi_{ij}^{(EC)} + \pi_{ij}^{(AC)} + \pi_{ij}^{(GC)} + \pi_{ij}^{(BC)} \tag{5}$$

Other ants may react to different combinations of these pheromones. Of course, more species and more pheromones may be introduced into the system as needed.

Based on this framework, details of the actions undertaken by various ant species are described below.

Egocentric ants (EC). They are creative in trying to find a new solution and finding their own way. They care less about other ants and about the pheromone trail. Instead, they focus mostly on the distance as a way to determine their next directions. An ant at node i will choose the next edge with the following attractiveness:

$$\frac{1}{cost(e_{ij})^{\beta}} \tag{6}$$

Default distance influence $\beta = 3.0$, again.

Altercentric ants (AC). They follow the majority of other ants, thereby focusing on the pheromone, without caring for the distance. So an ant at node i will choose the next edge with the following attractiveness:

$$\pi_{ij}^{\alpha} \tag{7}$$

Default pheromone influence $\alpha = 2.0$.

Good-at-conflict-handling ants (GC). They wait and observe the others, thereby caring for all existing pheromones (the particular weights are to be determined experimentally). So an ant at node i will choose the next edge with the following attractiveness:

$$\left(14 \cdot \pi_{ij}^{(EC)} + 2 \cdot \pi_{ij}^{(AC)} + 2.5 \cdot \pi_{ij}^{(GC)} + 0.5 \cdot \pi_{ij}^{(BC)}\right)^{\alpha} \tag{8}$$

Default pheromone influence $\alpha = 2.0$.

Bad-at-conflict-handling ants (BC). They behave impulsively (in effect randomly), irrespective of the pheromone or the distance. So an ant at node i will choose the next edge with the following attractiveness:

$$\frac{1}{\sum_{e_{ik}, k \in V \setminus \{i\}}} \cdot 100\,\% \tag{9}$$

5   Measuring the Diversity of ACO Search

As it was mentioned in the introduction, direct measuring of the diversity in the discrete space is very difficult. However, a "derivative" of information contained in the search space, especially connected with the search abilities of the ants, resulting in their behavior (exploration—when they actively search for new solutions and exploitation—when they fine-tune the already found, good ones) is contained in the pheromone table. Let us consider the following measures based directly on the analysis of the pheromone table:

– The number of pheromone-marked edges of the graph should directly show the diversity of the search, as when a small number of the edges is marked, the ants will only travel using these edges, possibly getting stuck in a local extremum. Otherwise, when a large number of edges is marked, the ants will roam through the graph. Therefore we propose to treat pheromone dispersion as a first measure of diversity for ACO. In other words, this measure is based on the ratio of the pheromone-marked edges count to all edges count:

$$PR = 100\% \cdot \frac{\#\{e_{ij} : \pi_{ij} > 0\}}{\#\{e_{ij}\}}, \forall i, j \qquad (10)$$

PR standing for Pheromone Ratio.

– The second measure is based on the attractiveness, as the ants are directly driven by this parameter during their travels. In the extreme case, when the ants roam everywhere randomly, choosing directions with equal probability, the edges would have equal attractiveness. Therefore one can compute the attractiveness of each edge, and measure its dispersion throughout the whole graph. If the dispersion (measured e.g. by the means of standard deviation) is low, the attractiveness is equally distributed. If it is high, only part of the graph is marked with high attractiveness. Therefore the second measure of diversity, based on the attractivenesses of edges is as follows:

$$AD = \sigma(\{n_{ij}\}), \forall i, j \qquad (11)$$

AD standing for Attractiveness Dispersion.

– The third measure is also based on attractiveness, however the rationale for it is different compared to the second measure. If the ants have chosen only one solution (they have totally lost the diversity), they will travel only along one hamiltonian in the graph, and the edges belonging to this hamiltonian will have non-zero attractiveness, while the attractiveness of the other edges is zero. Therefore one can compute the sum of attractivenesses of the best edges belonging to the best solution, and divide it by the sum of attractivenesses of the other edges. Thus the third diversity measure, also based on attractiveness, is given as follows:

$$AR = 100\% \cdot \frac{\sum_{i,j} n_{ij}}{\sum_{k,l} n_{kl}}, k \neq i \wedge l \neq j, \forall i, j, k, l \qquad (12)$$

Where n_{ij} belong only to the currently best individual, and AR stands for Attractiveness Ratio.

6   Experimental Results

The experimental results were obtained from a dedicated software developed in Python[1], run on Zeus supercomputer[2] We considered the Travelling Salesman Problem: find a Hamiltonian in a graph defined by a network of cities, with the goal being a cycle with the least cost (distance) [11]. The instances used in the experiments were taken from TSPLIB library[3].

[1] www.python.org.

[2] http://plgrid.pl.

[3] http://www.iwr.uni-heidelberg.de/groups/comopt/software/TSPLIB95/.

6.1   Configuration and Infrastructure

Zeus cluster, which is a supercomputer consisting of different kinds of 2-processor servers with different processor frequencies, number of cores, number of cores per node and RAM memory per node. Experiments were run on machine with the following technical parameters: **Model**: HP BL2x220c G5, G6, G7, **Total number of cores**: 17516, **Processors**: 2x Intel Xeon L5420, L5640, X5650, E5645, **Number of cores per node**: 8–12, **Processor frequency**: 2,26-2,66 GHz, **RAM memory per node**: 16–24 GB.

The following platform configuration was assumed for each experimental run:

- Number of ants: 100.
- Number of iterations: 100.
- Number of trials for each experiment: 30. Final data is the avarage of these 30 trials.
- Tested data taken from TSPLIB: berlin52, rat195, ts225. These instances were taken to exemplify easy, medium and difficult TSP problems. Of course this evaluation is subjective, and a dedicated publication reviewing efficiency of socio-cognitive ACO for different TSP instances is under review.

During the experiment, the following compositions (with respect to proportions of different ant species) of the simulated population were considered:

- Classic Ant Population: Only ants acting as in classic ACO.
- Modification based on Human-inspired sample populations: Egocentrical without bad at conflict handling (**egoWithoutBad**) population: 60 % egocentric, 20 % altercentric, 20 % good at conflict handling, 0 % bad at conflict handling. The proportions were chosen arbitrarily as an exemplification of the socio-cognitive ACO algorithm. Publication of a paper focused on review of configurations of socio-cognitive ACO populations is pending.

PR diversity measure points out, that diversity of the ants belonging to socio-cognitive species is much higher than the one observed in ACO (see Figs. 2a, 3a, 4a). It is necessary to keep in mind, that the pheromone ratio was averaged for socio-cognitive ants, in order to be able to compare it with the classic ACO. Though it may seem that pheromone is located everywhere in the pheromone table, it is probably not equally distributed. Therefore the ants are able to explore the search space, by traveling from time to time along the edges with lower attractiveness, and exploit the search space by traveling along the highly attractive edges. It is also to note, that in the case of **ts225** problem (see Fig. 4a), classic ants performed so badly (cf. Fig. 5e), probably unable to mark any reasonable part of the graph with the pheromone, that the Pheromone Ration diversity measure outcome was zero.

In Figs. 2b, 3b, 4b presenting standard deviation of the average attractiveness depending on iteration. Its value becomes visibly bigger in **berlin52** and **rat195** (see Figs. 2b, 3b) problems for classic ACO when compared to socio-cognitive system. In the case of the experiment **ts225** (Fig. 4b) one should remember about the size of the problem (225 cities, it means maximally 25200 edges to be

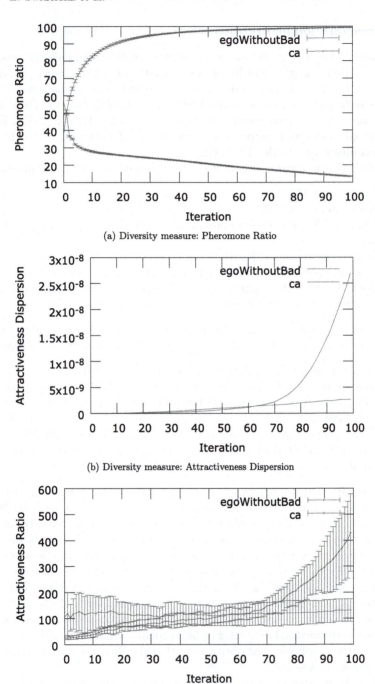

(a) Diversity measure: Pheromone Ratio

(b) Diversity measure: Attractiveness Dispersion

(c) Diversity measure: Attractiveness Ratio

Fig. 2. Problem: **berlin52**, iterations: **100**, ants: **100**, different diversity measures for classic and socio-cognitive ACO

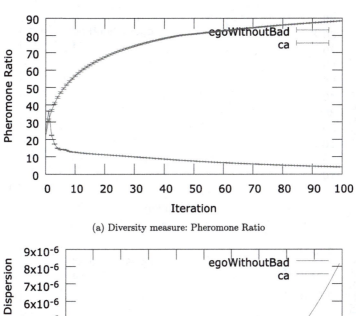

(a) Diversity measure: Pheromone Ratio

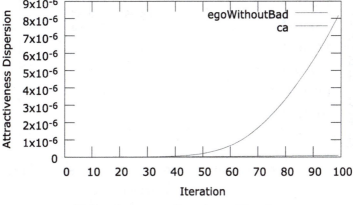

(b) Diversity measure: Attractiveness Dispersion

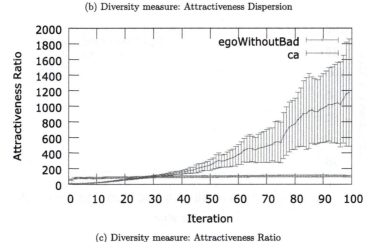

(c) Diversity measure: Attractiveness Ratio

Fig. 3. Problem: **rat195**, iterations: **100**, ants: **100**, different diversity measures for classic and socio-cognitive ACO

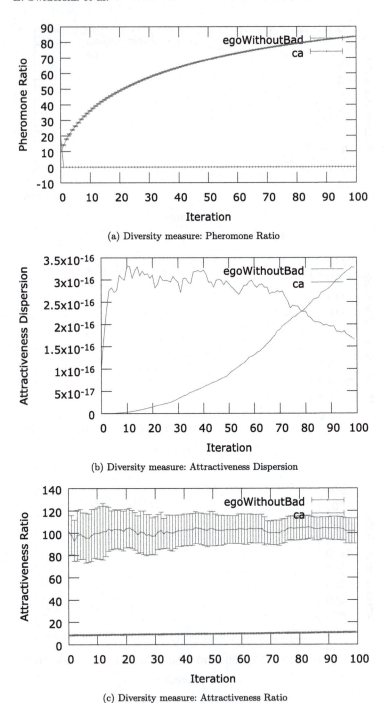

(a) Diversity measure: Pheromone Ratio

(b) Diversity measure: Attractiveness Dispersion

(c) Diversity measure: Attractiveness Ratio

Fig. 4. Problem: **ts225**, iterations: **100**, ants: **100**, different diversity measures for classic and socio-cognitive ACO

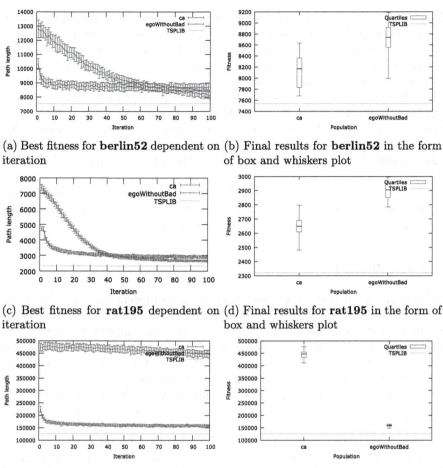

(a) Best fitness for **berlin52** dependent on iteration

(b) Final results for **berlin52** in the form of box and whiskers plot

(c) Best fitness for **rat195** dependent on iteration

(d) Final results for **rat195** in the form of box and whiskers plot

(e) Best fitness for **ts225** dependent on iteration

(f) Final results for **ts225** in the form of box and whiskers plot

Fig. 5. Comparison of **classic ACO** and **egoWithoutBad** populations fitnesses

marked by the pheromone, and only 100 ants in this experiment). The diversity measurement techniques based on attractiveness are not reliable here, as classic ants are unable to generate any reasonable solution in this case. The pheromone table for classic ants seems to be quite chaotic, thus the observation of the higher dispersion of their attractiveness. Anyway, this measure becomes the second one supporting the assumption of higher diversity in socio-cognitive ACO search.

Figures 2c, 3c, show clearly the dynamic nature of the search conducted in both systems. Classic ACO maintains high diversity in the beginning, exploring the space, later focusing on promising area, going into exploitation. At the same time, the socio-cognitive system maintains quite stable balance between exploitation and exploration. Figure 4c corresponds to the above-mentioned problems of classic ants in this setting—the diversity is practically zero, as the best solution in this case is probably not attractive enough in comparison with other edges marked with pheromones, to contribute to this fraction to a visible extent.

Finally one should check the actual outcome of the search, namely the evolution and final result obtained for **berlin52, rat195, ts225** problems in Fig. 5a, c, e. It is easy to see, that socio-cognitive system produced very quickly a very promising, sub-optimal result, while ACO needed at least 5 times more iterations to attain a similar result (Fig. 5a, c) (even though finally classic ACO was better than socio-cognitive one). However considering the hardest problem, **ts225**, classic ACO acted significantly poorer during all the observed iterations (Fig. 5e).

The curves of best fitness shown in Fig. 5a will behave in a similar way (reach suboptimal result of a similar quality as in Fig. 5c), as these two experiments were conducted with the same set of parameters.

7   Conclusions

We have presented here three novel diversity measurement strategies for ACO search: instead of using the search-space features, it is based on the information contained in the pheromone table. This strategies were applied to compare the diversity of socio-cognitive ant system and the classic one. The results obtained show that socio-cognitive ACO has a visibly higher diversity compared with the classic ACO in terms of the proposed diversity measures.

In the future, we plan to extend the analysis of the diversity measures and their impact on the ACO algorithms by getting a better insight on the pheromone table, distribution of attractiveness and pheromones, verification using other ACO-based metaheuristics and problems other than TSP. Moreover, the proposed measurement techniques will be used to automatically adapt the parameters of the ACO search. We would like also to further tune-up the parameters of our search in order to finally escape the local extrema that withhold us from reaching the global optima (found as best so far for TSPLIB), even though they are quite close to them.

Acknowledgments. This research received partial support from Polish-Belgian Joint Research Project (2015-2017), under agreement between the PAN and FNRS: "Modelling the emergence of social inequality in multi-agent systems". This research received partial support from AGH statutory fund, grant no. 11.11.230.124. This research was supported in part by PL-Grid Infrastructure.

References

1. Abdelbar, A.M., Wunsch II, D.C.: Promoting search diversity in ant colony optimization with stubborn ants. Procedia Comput. Sci. **12**, 456–462 (2012). Complex Adaptive Systems 2012

2. Bukowski, H.B.: What influences perspective taking. Ph.D. thesis, Catholic University of Louvain (2014)
3. Bukowski, H.B., Samson, D.: Can emotions affect level 1 visual perspective taking? Cogn. Neurosci. (in press)
4. Chira, C., Dumitrescu, D., Pintea, C.M.: Heterogeneous sensitive ant model for combinatorial optimization. In: Proceedings of the 10th Annual Conference on Genetic and Evolutionary Computation, GECCO 2008, pp. 163–164. ACM, New York (2008)
5. De Jong, K.: An analysis of the behavior of a class of genetic adaptive systems. Ph.D. thesis, University of Michigan, Ann Arbor (1975)
6. Dorigo, M., Stützle, T.: Ant Colony Optimization. Bradford Books, Scituate (2004)
7. Di Caro, G., Dorigo, M.: The ant colony optimization meta-heuristic. In: Corne, D., Dorigo, M., Glover, F. (eds.) New Ideas in Optimization, pp. 11–32. McGraw-Hill, Maidenhead (1999)
8. Di Caro, G., Gambardella, L.M., Dorigo M.: Ant algorithms for discrete optimization. Technical report, IRIDIA/98-10, Université Libre de Bruxelles, Belgium (1999)
9. Feitler, D.: The case for team diversity gets even better. Harvard Business Review (2014)
10. Fizke, E., Barthel, D., Peters, T., Rakoczy, H.: Executive function plays a role in coordinating different perspectives, particularly when ones own perspective is involved. Cognition **130**(3), 315–334 (2014)
11. Gutin, G.: Traveling salesman problem. In: Floudas, C.A., Pardalos, P.M. (eds.) Encyclopedia of Optimization, pp. 3935–3944. Springer US, New York (2009)
12. Hara, A., Matsushima, S., Ichimura, T., Takahama, T.: Ant colony optimization using exploratory ants for constructing partial solutions. In: 2010 IEEE Congress on Evolutionary Computation (CEC), pp. 1–7, July 2010
13. Lee, J.-W., Lee, J.-J.: Novel ant colony optimization algorithm with path crossover and heterogeneous ants for path planning. In: 2010 IEEE International Conference on Industrial Technology (ICIT), pp. 559–564, March 2010
14. Li, M., Ma, B., Wang, L.: On the closest string and substring problems. J. ACM **49**(2), 157–171 (2002)
15. McPhee, N.F.: Analysis of genetic diversity through population history. In: Proceedings of the Genetic and Evolutionary Computation Conference, pp. 1112–1120. Morgan Kaufmann (1999)
16. Michalewicz, Z.: Genetic Algorithms Plus Data Structures Equals Evolution Programs. Springer-Verlag New York, Inc., Secaucus (1994)
17. Morrison, R.W., De Jong, K.A.: Measurement of population diversity. In: Collet, P., Fonlupt, C., Hao, J.-K., Lutton, E., Schoenauer, M. (eds.) EA 2001. LNCS, vol. 2310, pp. 31–41. Springer, Heidelberg (2002)
18. Nakamichi, Y., Arita, T.: Diversity control in ant colony optimization. Artif. Life Robot. **7**(4), 198–204 (2004)
19. Nowé, A., Verbeeck, K., Vrancx, P.: Multi-type ant colony: the edge disjoint paths problem. In: Dorigo, M., Birattari, M., Blum, C., Gambardella, L.M., Mondada, F., Stützle, T. (eds.) ANTS 2004. LNCS, vol. 3172, pp. 202–213. Springer, Heidelberg (2004)
20. Nualhong, D., Chusanapiputt, S., Phomvuttisarn, S., Saengsuwan, T., Jantarang, S.: Diversity control approach to ant colony optimization for unit commitment problem. In: 2004 IEEE Region 10 Conference, TENCON 2004 (Volume C), vol. 3, pp. 488–491, November 2004

21. Phillips, K.W.: How diversity makes us smarter. Sci. Am. **311**(4) (2014)
22. Rusin, M., Zaitseva, E.: Hierarchical heterogeneous ant colony optimization. In: Proceedings of Federated Conference on Computer Science and Information Systems (2012)
23. Sekara, M., Michal-Kowalski, A.B., Indurkhya, B., Kisiel-Dorohinicki, M., Samson, D., Lenaerts, T.: Multi-pheromone ant colony optimization for socio-cognitive simulation purposes. Procedia Comput. Sci. **51**, 954–963 (2015). International Conference On Computational Science (ICCS), 2015 Computational Science at the Gates of Nature
24. Vrancx, P., Nowé, A., Steenhaut, K.: Multi-type ACO for light path protection. In: Tuyls, K., 't Hoen, P.J., Verbeeck, K., Sen, S. (eds.) LAMAS 2005. LNCS (LNAI), vol. 3898, pp. 207–215. Springer, Heidelberg (2006)

Multiwinner Voting in Genetic Algorithms for Solving Ill-Posed Global Optimization Problems

Piotr Faliszewski, Jakub Sawicki, Robert Schaefer$^{(\boxtimes)}$, and Maciej Smołka

AGH University of Science and Technology,
Al. Mickiewicza 30, 30-059 Kraków, Poland
{faliszew,schaefer,smolka}@agh.edu.pl, jsawicki@student.agh.edu.pl

Abstract. Genetic algorithms are a group of powerful tools for solving ill-posed global optimization problems in continuous domains. In case in which the insensitivity of the fitness function is the main obstacle, the most desired feature of a genetic algorithm is its ability to explore plateaus of the fitness function, surrounding its minimizers. In this paper we suggest a way of maintaining diversity of the population in the plateau regions, based on a new approach for the selection based on the theory of multiwinner elections among autonomous agents. The paper delivers a detailed description of the new selection algorithm, computational experiments that guide the choice of the proper multiwinner rule to use, and a preliminary experiment showing the proposed algorithm's effectiveness in exploring a fitness function's plateau.

Keywords: Ill-posed global optimization problems · New tournament-like selection · Fitness insensitivity

1 Introduction

Genetic algorithms (GAs) are a group of powerful tools for solving ill-posed global optimization problems (GOPs) in continuous domains

$$\arg\min_{x \in \mathcal{D}}\{f(x)\}, \ \mathcal{D} \subset \mathbb{R}^\ell, \ f : \mathcal{D} \to \mathbb{R}, \tag{1}$$

where $\mathcal{D}$ is a closed, bounded domain with a nonempty interior and sufficiently regular boundary (e.g., Lipschitz boundary [1]). The ill-conditioning of (1) is frequently due to the existence of many solutions (i.e., due to the multimodality of the objective function f over its domain $\mathcal{D}$) or/and its weak sensitivity in areas surrounding the global or local minimizers (for example, the chart of f may have an almost flat plateau at the level of local minimum $f(\hat{x})$, surrounding the minimizer $\hat{x}$). If the insensitivity is the main obstacle, the genetic algorithm should exhaustively explore plateaus of the fitness function.

The work presented in this paper has been partially supported by Polish NCN grant no. DEC-2015/17/B/ST6/01867 and by the AGH grant no. 11.11.230.124.

G. Squillero and P. Burelli (Eds.): EvoApplications 2016, Part I, LNCS 9597, pp. 409–424, 2016.
DOI: 10.1007/978-3-319-31204-0_27

The taxonomy of managing diversity classifies more than twenty groups of methods (see, e.g., the surveys of Črepinšek [2] and Gupta and Ghafir [3]). The most commonly used are, perhaps, *niching* and *sharing* (see, e.g., Schaefer's book [4] and the work of Goldberg and Richardson [5]). These methods generally lead to populations with sufficient differences in both location and fitness values among the individuals. On the other hand, methods used to increase the efficiency of Multi-Objective Evolutionary Optimization (MOEA) focus on location diversity [6]. There are only a few selection operators designed especially for diversity boosting (see, e.g., the works of Hutter [7] and Matsui [8]).

In this paper we suggest another approach to maintaining diversity of the population in the plateau regions only, based on a new approach for the selection operation. More specifically, we design the selection process for evolutionary algorithms based on the theory of multiwinner elections among autonomous agents. We refer to our approach as the Multiwinner Selection, or MWS for short.

2   Multiwinner Elections

An election is a pair $E = (C, V)$, where $C = \{c_1, \ldots, c_m\}$ is a set of candidates and $V = (v_1, \ldots, v_n)$ is a collection of voters. Each voter v_i has an associated preference order $\succ_i$ that ranks the candidates from the most desirable one to the least desirable one (from the point of view of this voter). For example, if $C = \{a, b, c\}$, then voter v who likes a best, then b, and then c, would have preference order $a \succ b \succ c$.

Given an election $E = (C, V)$, we write $\text{pos}_v(c)$ to denote the position of candidate $c \in C$ in the preference order of voter $v \in V$ (the candidate ranked first has position 1, the next one has position 2, and so on). The exact election that we mean will always be clear from the context.

A multiwinner voting rule is a function $\mathcal{R}$ that given an election $E = (C, V)$ and a positive integer k, $k \leq \|C\|$, outputs a size-k subset of C, the elected committee (ties among winning committees may occur, but we disregard them). So far, multiwinner rules received much less attention from the research community than the single-winner ones. Based on the discussion given by Elkind et al. [9], we consider seven rules inspired by scoring rules. A scoring rule is a function that given a position of a candidate on a voter's preference order returns this candidate's score. For example, k-Approval scoring rule, α_k, is defined so that $\alpha_k(i) = 1$ for $i \in \{1, \ldots, k\}$, and $\alpha_k(i) = 0$ for $i > k$ (α_1 is known as the plurality rule, i.e. the voter approves of a single candidate only). Borda scoring rule—in elections with m candidates—is defined as $\beta(i) = m - i$. Given an election $E = (C, V)$ and scoring function γ, the γ-score of candidate $c \in C$ is defined as $\sum_{v \in V} \gamma(\text{pos}_v(c))$.

Perhaps the easiest way to generalize a scoring rule to the multiwinner case is as follows: Given an election $E = (C, V)$, a scoring protocol α, and the desired number of winners k, simply pick k candidates with the highest scores. This way we define the three following rules:

Single Non-Transferrable Vote (SNTV). Under SNTV we pick the k candidates with the highest plurality scores (i.e., k candidates ranked first most frequently).

k-Borda. Under k-Borda we pick k candidates with the highest Borda scores.

Bloc. Under Bloc we pick the k candidates with the highest k-Approval scores.

The next four rules are based on a somewhat different idea. We first introduce some additional notation and then define the rules of Chamberlin and Courant [10], of Monroe [11], and their approximate versions due to Lu and Boutilier [12] and Skowron et al. [13].

Let $E = (C, V)$ be an election. We say that Φ, $\Phi \colon V \to C$ is a k-CC-assignment function if for each voter $v \in V$, $\Phi(v)$ returns one of at most k candidates (in other words, we require that $\|\Phi(V)\| \leq k$). Intuitively, a k-CC-assignment function matches up to k winners of the election to the voters. Given a k-CC-assignment function Φ and a scoring function α, both for the same election $E = (C, V)$, we say that the score of Φ is:

$$\mathrm{score}_\alpha(\Phi) = \sum_{v \in V} \alpha(\mathrm{pos}_v(\Phi(v))).$$

In other words, each voter v gives score only to the candidate c assigned to him or her.

We now define the Chamberlin–Courant rule (the CC rule). Given an election $E = (C, V)$ and a positive integer k, it picks a k-CC-assignment function Φ with the highest score with respect to the Borda scoring protocol, and returns the committee $\Phi(V)$. Intuitively speaking, the rule picks some k candidates and then assigns each voter to this one of them that this voter ranks highest. The rule picks these k winners in such a way that the sum of the Borda scores that voters give to "their" candidates is highest. In some sense, both the CC-rule and the k-Borda rule are generalizations of the single-winner Borda rule to the multiwinner case. The former, however, divides the electorate into k "districts" of likely-minded voters, and the latter picks k winners that form some sort of a global consensus. Interestingly, the "districts" created by the CC rule can be of very different sizes. In some applications this is undesirable and, thus, Monroe proposed a different variant of this rule.

We say that Φ is a k-Monroe-assignment function if it is a k-CC-assignment function that additionally satisfies the following condition: Let n be the number of voters in the election. For each candidate $c \in C$, it holds that either $\|\Phi^{-1}(c)\| = 0$ or $\lfloor \frac{n}{k} \rfloor \leq \|\Phi^{-1}(c)\| \leq \lceil \frac{n}{k} \rceil$ (in other words, the Monroe condition requires that either a given candidate is not a winner or is a winner and is matched to roughly the same number of voters as the other winners). Monroe rule is the same as the CC rule except that it chooses among k-Monroe-assignments.

While the CC rule and the Monroe rule are quite appealing, it is NP-hard to compute their winners [12,14]. (The situation becomes a bit better if one assumes one of the standard restrictions on the votes such as single-peakedness or single-crossingness; then the CC-rule becomes polynomial-time computable but the Monroe rule seems to remain intractable [15–17]). Fortunately, there are approximation algorithms for both rules.

Lu and Boutilier [12] introduced a simple greedy algorithm for the CC rule, based on the classic approximation result for submodular functions [18]. This algorithm proceeds as follows. We are given an election $E = (C, V)$ and a positive integer k, the number of winners that we are interested in. We construct the set of winners W. Initially W is empty and we add candidates to it one by one, by executing the following steps k times:

1. For each candidate $c \in C \setminus W$, we compute a k-CC-assignment function Φ_c that assigns each voter v to the candidate in $W \cup \{c\}$ that v ranks highest.
2. We compute for which candidate $c \in C \setminus W$ the score of Φ_c is highest (breaking ties in an arbitrary way).
3. We add this candidate c into W.

We refer to this algorithm as Greedy-CC. Lu and Boutilier have shown that Greedy-CC always picks a committee that under the CC rule would obtain at least fraction $1 - \frac{1}{e}$ of the score of the optimal solution.

For the case of the Monroe rule, there is an approximation algorithm due to Skowron et al. [13]. This algorithm also proceeds greedily, but it makes sure to satisfy the Monroe condition. As for the case of Greedy-CC, this algorithm builds the set of winners by adding candidates to it one by one, but it also maintains the set of available voters. Formally, we have the following algorithm.

Let $E = (C, V)$ be an election and let k be the number of winners that we seek. For the ease of exposition, we assume that k divides $\|V\|$ exactly. The algorithm first sets the set of current winners to be $W = \emptyset$ and the set of available voters to be $A = V$. Then it executes the following steps k times to find the set of k winners and to build a k-Monroe-assignment function Φ:

1. For each candidate $c \in C \setminus W$, compute set

$$A_c = \operatorname{argmax}_{A' \subseteq A, \|A'\| = \frac{\|V\|}{k}} \sum_{v \in A'} \beta(\operatorname{pos}_v(c))$$

   (break ties arbitrarily, if needed; intuitively, A_c is a set of $\frac{\|V\|}{k}$ voters from A that jointly rank c highest). For each c let score(c) be $\sum_{v \in A_c} \beta(\operatorname{pos}_v(c))$.
2. Pick candidate $c \in C \setminus W$ with the highest score (breaking ties arbitrarily, if needed). Add c to the set W, remove the voters from A_c from A, and for each voter v in A_c set $\Phi(v) = c$.

Finally, the algorithm returns the set W. We refer to the algorithm as Greedy-Monroe and treat it as a voting rule in its own right. Skowron et al. have shown that the score of the k-Monroe-assignment function Φ returned by Greedy-Monroe is at least $n(m - 1) \left(1 - \frac{k-1}{2(m-1)} - \frac{H_k}{k}\right)$, where n is the number of voters, m is the number of candidates, and H_k is the k'th harmonic number (i.e., $H_k = \sum_{i=1}^{k} \frac{1}{i}$). For the case where k is relatively large and $\frac{k}{m}$ is relatively small, this value is very close to the highest possible score under the Monroe rule, $n(m - 1)$, achieved by an assignment function that matches every voter with his or her most preferred candidate.

3   Selection Based on Multiwinner Voting

The input for the selection procedure consists of election group C, which is a subset of population $X_t = \langle x^{(1)}, \ldots, x^{(\mu)} \rangle$, the multiset of candidate solutions at the particular t-th epoch of evolutionary optimization process (each $x^{(i)}$ is a point in an Euclidean space). As long as it does not lead to ambiguity, we do not specify the dependency of each individual in X_t on the epoch number. For every point $x^{(i)} \in C$, we have its fitness value $f(x^{(i)})$, the smaller the fitness value the better (since in the multiwinner voting we maximize rather than minimize, we will have to apply appropriate transformations of these values). We are to pick k points from C that will be the parents in the following mixing phase. (Typically, a selection procedure is applied multiple times, each invocation producing a single individual. In our case we will still invoke the selection procedure several times, but each invocation will output a collection of individuals; we formalize this in Sect. 3.4).

Our idea is to consider C as a group of individuals who need to decide which k of them would survive to the next epoch. These individuals are driven by two, perhaps conflicting, desires.

1. Foremost, each individual would like to survive itself. If, however, the individual were not to survive, it would like some as similar as possible individual to survive. Intuitively, a similar individual would have similar genes that would be passed to the next generation.
2. The second desire is that the selected subpopulation is as fit as possible.

To model these desires, we introduce for each individual $x^{(i)}$ its utility function $u_i \colon C \to \mathbb{R}$ (we will provide some examples of utility functions very shortly). For each two individuals $x^{(i)}$ and $x^{(j)}$, the value $u_i(x^{(j)})$ expresses how much value $x^{(i)}$ attaches to $x^{(j)}$ being selected (the higher the value, the more $x^{(i)}$ would like $x^{(j)}$ to be chosen). Naturally, given the principles outlined above, for each two individuals $x^{(i)}, x^{(j)}$, we have that

$$u_i(x^{(i)}) \geq u_i(x^{(j)}).$$

That is, foremost each individual is selfish and its desire to survive is stronger than anything else.

Given the set of individuals and their utility functions, we define an election $E = (C, V)$ as follows:

1. The set of voters V is the same as the set of candidates (the individuals), that is, $V = C$.
2. For each individual $x^{(l)}$, we set its preference order so that if $u_l(x^i) > u_l(x^{(j)})$ then $x^{(l)}$ prefers $x^{(l)}$ to $x^{(j)}$. (We break the ties arbitrarily, should they occur.)

Now, given election E, we simply apply a multiwinner voting rule of choice to pick k winners.

3.1   Utility Functions

There are two crucial choices in the design of MWS. The choice of the multi-winner voting rule and the choice of the utility function. Here we outline several possibilities for the latter.

Perhaps the most natural idea is to use the following approach. For each two individuals $x^{(i)}$ and $x^{(j)}$, let $d(x^{(i)}, x^{(j)})$ be a distance between them (one could use any metric, but for simplicity we use the Euclidean distance). We also assume to have "reversal" function h such that for each two individuals $x^{(i)}$ and $x^{(j)}$ it holds that $f(x^{(i)}) \leq f(x^{(j)})$ if and only if $h(f(x^{(i)})) \geq h(f(x^{(j)}))$. Then we define the utility of individual $x^{(j)}$ from the point of view of individual $x^{(i)}$ to be:

$$u_i^p(x^{(j)}) = \frac{h(f(x^{(j)}))}{d(x^{(i)}, x^{(j)})}.$$

We refer to these functions as proportional utilities because they are directly proportional to the "reversed" fitness values and inversely proportional to the distances between the individuals (hence the symbol p in u_i^p).

It is easy to see that proportional utilities satisfy the basic desiderata outlined in the above section. Since they are inversely proportional to the distance between the individuals, each individual assigns the highest utility ($+\infty$) to itself and decreases its utility with increasing the distance (with increasing the dissimilarity) to the other individuals. Since the utilities are proportional to the reversed fitness values, the individuals assign value selecting individuals as fit as possible.

Naturally, it might be the case that the proportional utilities either put too much stress on the fitness values or too much stress on the distances between the agents. Thus, to temper this behavior, we might need to use the following variant of proportional utilities. Let γ and δ be two functions ($\gamma, \delta \colon \mathbb{R} \to \mathbb{R}$), where γ is increasing and δ is decreasing. We define (γ, δ)-proportional utility function to be:

$$u_i^{\gamma, \delta}(x^{(j)}) = \gamma(h(f(x^{(j)}))) \cdot \delta(d(x^{(i)}, x^{(j)}))$$

For example, by taking $\gamma(x) = x$ and $\delta(x) = \frac{1}{x}$ we obtain the proportional utility functions.

The choice of functions γ and δ is likely to have very strong impact on the quality of our selection procedure. Indeed, we believe that exploring various functions may be an interesting research project in its own right. In this work we will focus on one particularly appealing type of (γ, δ)-proportional utilities, where γ and δ are of the following form: For two positive numbers r and s, we take $\gamma_r(x) = x^r$ and $\delta_s(x) = x^{-s}$. Naturally, even in this case, choosing appropriate values of r and s is not obvious.

3.2   The Selection Procedure and the Choice of the Multiwinner Rule

The pseudocode of our selection procedure is given as Algorithm 1. This code is quite general and can use any (γ, δ)-utilities. It would also be straightforward

---

Algorithm 1. The Multiwinner Selection (MWS) procedure. The goal is to pick k individuals from the election group C, based on their locations and fitness values, using multiwinner voting rule $\mathcal{R}$.

Notation:
$C = \langle x^{(1)}, \ldots, x^{(n)} \rangle \subseteq X_t \leftarrow$ the election group
$f \leftarrow$ the fitness function
$h \leftarrow$ the "reversal" function
$d \leftarrow$ the metric over $\mathbb{R}^\ell$
$\gamma, \delta \leftarrow$ the functions defining (γ, δ)-proportional utilities
$k \leftarrow$ the number of individuals to pick
$\mathcal{R} \leftarrow$ the multiwinner voting rule

// prepare the election among the individuals
1   $V \leftarrow (v_1, \ldots, v_n)$
2   **for** $l \leftarrow 1$ **to** n **do**
3   **foreach** $i, j \in \{1, \ldots, n\}$, $i < j$ **do**
4       $u_l(x^{(i)}) \leftarrow \gamma(h(f(x^{(i)}))) \cdot \delta(d(x^{(l)}, x^{(i)}))$
5       $u_l(x^{(j)}) \leftarrow \gamma(h(f(x^{(j)}))) \cdot \delta(d(x^{(l)}, x^{(j)}))$
6   **if** $u_l(x^{(i)}) > u_l(x^{(j)})$ **then**
7   set v_l's preference order so that $x^{(i)} \succ_l x^{(j)}$
8   **else**
9   set v_l's preference order so that $x^{(j)} \succ_l x^{(i)}$
10   form election $E = (C, V)$

11   $W = \mathcal{R}(E, k)$;   // hold the virtual election among the individuals
12   output W

---

to adapt it to any other natural form of utilities. The code can also use any arbitrary multiwinner voting rule $\mathcal{R}$. However, it should be quite clear that the quality of the procedure will deeply depend on the choice of the rule. Indeed, if we used SNTV which simply counts how many times each candidate is ranked first, our procedure would—in essence—reduce to randomly selecting a group of k individuals. This is so because each individual ranks itself first and, thus, every individual would simply have one point.

Similarly, we believe that k-Borda and Bloc would not perform very well. The single-winner variant of k-Borda is designed to find a "consensus" winner, that is, a candidate that is in some sense acceptable to as many voters as possible. In effect, k-Borda finds a collection of such "consensus" candidates and they are likely to be very similar. Bloc rule might be a bit better because it is based on the k-Approval scoring protocol and, effectively, under Bloc the score of each candidate depends on local information only. This might, however, lead to picking individuals from large clusters only.

On the other hand, we believe that the CC rule and the Monroe rule would not suffer from the above-described problems. The reason is that under these rules, when a candidate is assigned to some voter, this voter cannot contribute to the score of any other candidate. As a result, members of a big cluster would only be able to promote *some* of their number into the winning set, preventing overwhelming of smaller clusters.

For the same reason, it seems that Greedy-CC and Greedy-Monroe should perform well, and should be faster than CC and Monroe (since computing CC and Monroe requires solving NP-hard problems). Indeed, using Monroe and CC for any non-trivial setting seems impossible. On the other hand, Greedy-CC and Greedy-Monroe are computed through simple, polynomial-time algorithms and, in effect, the computational cost of using multiwinner selection based on them is negligible (as compared to the cost of fitness evaluations for engineering applications for which our techniques are intended). We also note that using multiwinner selection requires the same number of fitness evaluations as, say, tournament selection.

3.3 A Simple Experiment

The main reason for developing our Multiwinner Selection procedure is to enhance the plateau exploration capabilities of genetic algorithms. In the preceding section we have argued theoretically that to achieve this effect we should, likely, use Greedy-CC or Greedy-Monroe rules. However, instead of relying on theoretical analysis only, we believe that it would be informative to perform a simple computational experiment that would give us some further insight into the behavior of MWS depending on the applied voting rule.

To this end, we have performed the following experiment. We have generated 500 points distributed uniformly on the two-dimensional square $[-3,3] \times [-3,3]$ and have applied our selection procedure to pick 50 points. We have assumed that each point has the same fitness value (thus the choice of function h is irrelevant) and we have used the unmodified 2-dimensional Euclidean distance (that is, we have used $\delta(x) = \frac{1}{x}$). This setting models the situation in which the genetic algorithm has hit the plateau and now the goal is to explore it. We would like to obtain as much diversity among the selected individuals as possible.[1]

The results of the experiment, presented in Fig. 1, are quite striking and fully support the theoretical discussion from the preceding section: k-Borda picks a centrally located cluster of individuals, Bloc picks candidates from areas where they are concentrated (due to the random choice of their positions), whereas Greedy-CC and Greedy-Monroe pick candidates that are, approximately, uniformly distributed among the individuals. In effect, we believe that k-Borda and Bloc would be poor choices for our application. Greedy-CC and Greedy-Monroe would, likely, perform comparatively well. However, Greedy-CC is faster to compute and since it does not have to respect the Monroe criterion, it is less likely to

---

[1] From the point of view of the elections theory, our setting is an example of two-dimensional Euclidean single-peaked preferences. Under two-dimensional Euclidean preferences, every voter and every candidate is a point in a two-dimensional Euclidean space and every voter (in our case, every individual) derives his or her preference orders by sorting the candidates (in our case, the individuals) with respect to their Euclidean distance from him or herself.

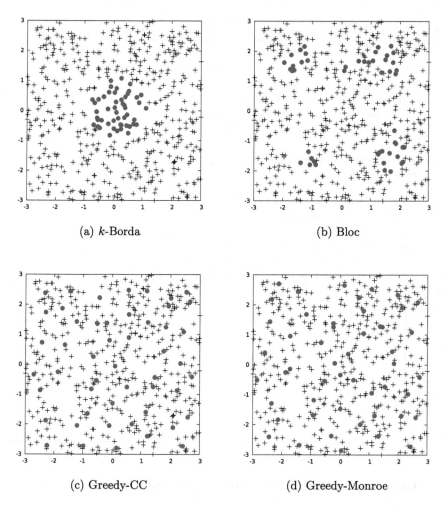

(a) *k*-Borda

(b) Bloc

(c) Greedy-CC

(d) Greedy-Monroe

Fig. 1. The result of using Multiwinner Selection to pick 50 out of 500 individuals, distributed uniformly on a $[-3, 3] \times [-3, 3]$ square, using *k*-Borda, Bloc, Greedy-CC, and Greedy-Monroe. All the individuals have the same fitness value, simulating the plateau scenario (in effect, the utilities, and the preference orders, are derived based on the distance between points only). For the case of each rule we use the same 500 points.

pick two very similar points (this would happen for the case of Greedy-Monroe if these two points were surrounded by a denser-than-usual cluster of points). Thus, from now on in MWS we will use Greedy-CC only.

We have shown a result of a single invocation of our experiment, but we have run it repeatedly and each time we obtained similar results. Since for now we are interested in qualitative results rather than quantitative ones, we believe the obtained evidence is sufficient to focus on Greedy-CC only.

Algorithm 2. Outline of a genetic algorithm using MWS.

Notation:
$X_t = \langle x_t^{(1)}, \ldots, x_t^{(\mu)} \rangle \in \mathcal{D}^\ell \leftarrow$ the population in the t-th epoch
$C \subseteq X_t \leftarrow$ the election group
$n \leftarrow$ the size of the election group $(n \leq \mu)$
$k = 2p \leftarrow$ the number of individuals to pick
MWS $\leftarrow$ Multiwinner Selection procedure

1 Sample the initial population X_0
2 Evaluate X_0
3 $t \leftarrow 0$
   // the main loop over the epochs
4 **while** *Stopping_Condition*(X_t) **do**
5   *Offspring* $\leftarrow \emptyset$
6   **for** $i \leftarrow 1$ **to** μ/p **do**
7   Choose the election group C of size n from X_t
           // Select k individuals from the election group
8          $\{c_1, \ldots, c_k\} \leftarrow$ MWS(C, k)
9   **for** $j \leftarrow 1$ **to** p **do**
10             $r \leftarrow 2j - 1$
11             $a \leftarrow cross(c_r, c_{r+1})$
12             $a \leftarrow mutate(a)$
13   *Offspring* $\leftarrow$ *Offspring* $\cup \{a\}$
14     $t \leftarrow t + 1$
15     $X_t \leftarrow$ *Offspring*
16   Evaluate X_t

17 output X_t

3.4 Genetic Algorithm Using Multiwinner Selection

Let μ be the number of individuals in the population in a single epoch. Further, we choose two numbers, n and p, where n is the size of the election group (i.e., the number of individuals over which we will carry a multiwinner election), and p, $p < \mu$, is a number that divides μ and such that we pick $k = 2p$ individuals from the election group.

A single iteration of our algorithm proceeds as follows. We pick the election group, that is, a set $C \subseteq X_t$ of individuals of size n, where X_t is the current population. (There are at least several ways of choosing X_t and we will discuss two possibilities shortly; for now let us simply think of it as some stochastic process that picks n individuals). Then we apply Multiwinner Selection procedure to pick k individuals out of these n. Finally, we iterate over these k individuals, pick consecutive pairs, apply the crossing operation to the individuals from the pair, obtaining a single individual, and finally we apply the mutation operator to this individual (we assume that the crossing operator and the mutation operator encapsulate the stochastic choices as to whether they should, all in all, be applied or not). In effect, we obtain a group of p individuals that we insert into the next epoch's population. We repeat this process μ/p times, to form a population of size μ.

Let us now move back to the issue of picking the election group. We suggest two ways in which this can be done:

1. We obtain the election group C by sampling without replacement n times from the current population X_t (using the uniform distribution). This approach is inherited from the standard tournament selection procedure.
2. The second way is composed of four steps. In the first step we select a single individual x_{seed} from X_t using some conventional selection procedure (e.g., the proportional one or the tournament one). In the second step we chose the normal, ℓ-dimensional distribution with a density function ρ, whose expectation is x_{seed}. The standard deviation σ of this distribution is a parameter of the procedure. Next, we create the probability distribution χ on the multiset $X_t \setminus \{x_{seed}\}$ by normalizing the vector $\{\rho(x^{(i)})\}$, $x^{(i)} \in X_t$, $i = 1, \dots \mu$, $x^{(i)} \neq x_{seed}$. Finally, we obtain the election group C by $(n-1)$-times sampling without replacement from $X_t \setminus \{x_{seed}\}$, according to the probability distribution χ. Finally, we add x_{seed} to the election group.

From now on, we will focus only on the first, far simpler, way of picking the election group. However, naturally, it has drawbacks. For example, the election group picked in this way might be too diverse in case of a relatively large search domain. We believe that the second procedure for picking the election group would resolve this problem. Measuring the extent to which this problem indeed occurs in practice is beyond the scope of this paper.

4   Experimental Evaluation

In this section we provide evidence that our Multiwinner Selection method indeed achieves its goal, i.e., it leads to the exploration of plateau areas (without necessary focusing on a single local or global optimum). To this end we consider the following experiment. Let $f(x, y)$ be a function, $f \colon \mathbb{R} \times \mathbb{R} \to \mathbb{R}$, defined as:

$$f(x,y) = \max\left(\frac{1}{2} - e^{-2(x^2+y^2)}, 0\right) + \frac{20 + x^2 - 10\cos(2\pi x) + y^2 - 10\cos(2\pi y)}{2000}$$

While $f(x, y)$ might look somewhat complicated at first, its definition is in fact very simple. The first term of $f(x, y)$, $\max(\frac{1}{2} - e^{-2(x^2+y^2)}, 0)$ is a simple Gaussian function reversed, translated up to $\frac{1}{2}$, and cut off at 0. In effect, this part of the function creates a well, with a plateau in the shape of a circle with the center at point $(0, 0)$ and radius $r = \sqrt{\frac{-\ln\frac{1}{2}}{2}} \approx 0.58$. The second summand is the Rastrigin function (downscaled by a factor of 2000) to introduce small perturbations. We plot the function in Fig. 2.

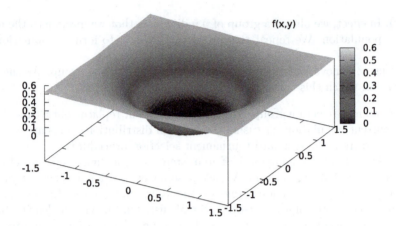

Fig. 2. The plot of the function $f(x, y)$. It is, in essence, a well with the center $(0, 0)$; the small perturbations due to the Rastrigin function are barely visible.

The idea of our experiment is to run a genetic algorithm whose goal is to find minimizers of function $f(x, y)$ (on the domain $[-2, 2] \times [-2, 2]$). Naturally, a standard algorithm would very quickly find the global minimum at point $(0, 0)$. However, what we are interested in is not finding the global minimum, but exploring the plateau area in the circle of radius $r \approx 0.58$, centered at $(0, 0)$. Arguably, all the points in this area are of very similar quality, and from our point of view it is important to cover as much of this area as possible.

To this end, we compare two algorithms. Our algorithm from the previous section, using the Multiwinner Selection procedure, and a simple standard genetic algorithm using a form of the tournament selection procedure. Both algorithms use the following basic parameters:

1. The population in each epoch contains $\mu = 100$ individuals.
2. The initial population is picked by drawing μ points from $[-2, 2] \times [2, 2]$ uniformly at random.
3. The probability of performing the crossover operation is 1%, whereas the probability of mutation is 100%. Mutation is executed by adding to the individual a value drawn randomly using the normal distribution with standard deviation 0.1.

Arguably, this setting of the parameters (especially the mutation rate and the standard deviation of the normal distribution used for mutation) is quite extreme. However, what we are modeling here is a situation where the algorithm already, roughly, identified the part of the domain with a plateau (thus we look at the domain $[-2, 2] \times [-2, 2]$) and now the goal is to fill in this plateau. For this task we want the population to be exploration-oriented and, thus, we want the mutation operator to have strong effect, and we want the crossover operation to have very small impact. Thus, We use the following parameters:

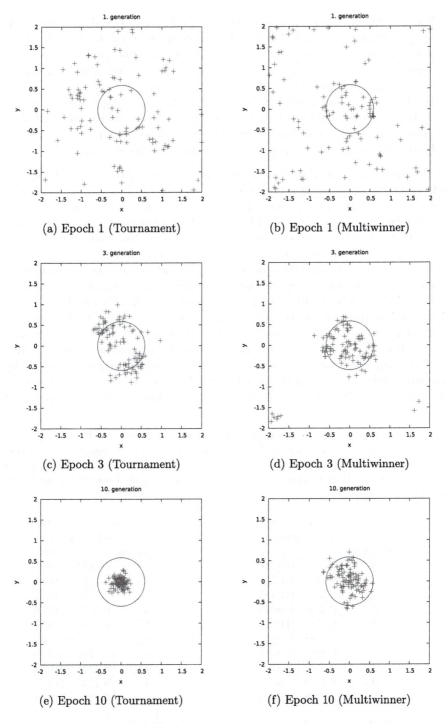

Fig. 3. Sample run of our algorithms with tournament and Multiwinner selection procedures (populations X_1, X_3, and X_{10}). The blue circle depicts the plateau region.

1. For the Multiwinner Selection, we use (γ, δ)-proportional utilities, with $\gamma(x) = x^6$ and $\delta(x) = \frac{1}{x}$. We use the "reversal" function $h(x) = \frac{1}{1+x}$. Further, we set $n = 30$ and $\mu = 5$ (that is, every selection process consists of holding $\frac{100}{5} = 20$ elections among randomly chosen 30 individuals, of whom, eventually, 5 are selected as the parental individuals).
2. For the tournament selection, we have used the following process: To pick each parental individual we pick 5 randomly selected individuals and then draw one of them with probability proportional to their fitness value (the better the fitness value, the better the chance of being selected).

We should mention that the above parameter setting is largely ad-hoc. We did not try to optimize the values and we have performed only a handful of preliminary experiments to asses the behavior of the system. This is in sync with the fact that what we present here is nothing more than a proof of concept that Multiwinner Selection is a tool worth developing in the context of extending the capabilities of plateau exploration for genetic algorithms.

We have run our algorithms on function $f(x, y)$ for 20 times each, in each case executing 10 epochs (i.e., starting with population X_0, and computing populations $X_1, \ldots, X_{10}$). In Fig. 3 we present populations X_1, X_3 and X_{10} for both algorithms, from representative runs. This figure shows that our intuition that Multiwinner Selection should lead to better exploration of the plateau region is correct. However, we also assessed the extent of this advantage quantitatively.

To this end, we have computed how much of the plateau region is covered by the individuals from each of the populations. The plateau region is a circle with the radius $r \approx 0.58$. Since mutation in our algorithm modifies the position of

Table 1. Comparison of the average fraction of the plateau covered by the algorithm using tournament selection and Multiwinner selection. For each algorithm we report the average fraction of the plateau covered by populations $X_1, \ldots, X_{10}$ and its standard deviation. (The average and the standard deviation is computed over the 20 test runs that we have executed.)

population	covered fraction of the plateau average	st. dev.	population	covered fraction of the plateau average	st. dev.
X_1	0.1124	0.0431	X_1	0.1152	0.0385
X_2	0.2165	0.0712	X_2	0.2511	0.0634
X_3	0.3521	0.0696	X_3	0.3613	0.0454
X_4	0.3846	0.0400	X_4	0.4011	0.0278
X_5	0.3586	0.0554	X_5	0.4202	0.0255
X_6	0.3118	0.0545	X_6	0.4139	0.0261
X_7	0.2765	0.0455	X_7	0.4251	0.0195
X_8	0.2588	0.0305	X_8	0.4276	0.0262
X_9	0.2474	0.0152	X_9	0.4207	0.0175
X_{10}	0.2428	0.0120	X_{10}	0.4234	0.0161

(a) Tournament Selection   (b) Multiwinner Selection

each individual according to a two-dimensional normal distribution with standard deviation 0.1, we have assumed that each individual covers an area of the plateau equal to the circle of radius 0.05, centered at the individual. Then, we have computed the fraction of the area of the plateau covered by at least one individual. We have done so for both algorithms, for every run of our algorithm, and for every epoch within the run. We present the results in Table 1.

The results confirm our intuition. The algorithm using Multiwinner selection achieves a better coverage (nearly twice as large as that using tournament selection) and does not converge to the local optimum. Instead, it maintains the diversity among the individuals. Since the standard deviation for the covered fraction of plateau in the tenth epoch is very small (for both algorithms), this is good indication that our results are meaningful.

We should mention that the results from Table 1 are very sensitive to our definition of what it means for an individual to "cover a given area of the plateau." For example, it depends on the radius "covered" by a single individual. However the main message, that Multiwinner Selection ensures noticeably better diversity of the individuals than tournament selection remains true.

5 Conclusions

We have put forward a new idea for a selection procedure for genetic algorithms. This selection procedure is based on the theory of multiwinner voting, and its goal is to maintain diversity of the population within plateau regions. We have considered several voting rules that can form the basis of the selection procedure and we have concluded that a greedy approximation algorithm for Chamberlin–Courant's rule is promising (however, the approximation algorithm for Monroe's rule is promising as well). Then we have tested our selection procedure within a very simple genetic algorithm. We have applied the algorithm to a toy example of a function with a plateau. We have shown that compared to the same algorithm with tournament selection, our algorithm achieves noticeably better diversity of its populations. This is seen both by a quantitative assessment and by inspecting the populations visually.

The research presented in this paper is very preliminary. It is very likely that there are better ways of employing our ideas, there are better parameter settings, etc. Nonetheless, we believe that our results are sufficient to support our main message: Selection based on the theory of multiwinner voting is an interesting idea that can lead to diversity among the individuals in genetic algorithms.

References

1. Zeidler, E.: Nonlinear Functional Analysis and its Application: II/A: Linear Monotone Operators. Springer, New York (2000)
2. Črepinšek, M., Liu, S.H., Mernik, M.: Exploration and exploitation in evolutionary algorithms: a survey. ACM Comput. Surv. **45**(3), 3–35 (2013)

3. Gupta, D., Ghafir, S.: An overview of methods maintaining diversity in genetic algorithms. Int. J. Emerg. Technol. Adv. Eng. **2**(5), 56–60 (2012)
4. Schaefer, R.: Foundation of Genetic Global Optimization (with Chap. 6 by Telega H.). Studies in Computational Intelligence Series, vol. 74. Springer, Heidelberg (2007)
5. Goldberg, D., Richardson, J.: Genetic algorithms with sharing for multimodal function optimization. genetic algorithms and their applications. In: Proceedings of 2nd International Conference on Genetic Algorithms, Lawrence Erlbaum Associates, pp. 41–49 (1987)
6. Bosman, P., Thierens, D.: The balance between proximity and diversity in multiobjective evolutionary algorithms. IEEE Trans. Evol. Comput. **7**(2), 174–188 (2003)
7. Hutter, M.: Fitness uniform selection to preserve genetic diversity. In: Proceedings of the 2002 Congres of Evolutionary Computation, pp. 783–788 (2002)
8. Matsui, K.: New selection method to improve the population diversity in genetic algorithms. In: Proceedings of IEEE International Conference on Systems, Man and Cybernetics, pp. 625–630 (1999)
9. Elkind, E., Faliszewski, P., Skowron, P., Slinko, A.: Properties of multiwinner voting rules. In: Proceedings of the 13th International Conference on Autonomous Agents and Multiagent Systems, pp. 53–60, May 2014
10. Chamberlin, B., Courant, P.: Representative deliberations and representative decisions: proportional representation and the Borda rule. Am. Polit. Sci. Rev. **77**(3), 718–733 (1983)
11. Monroe, B.: Fully proportional representation. Am. Polit. Sci. Rev. **89**(4), 925–940 (1995)
12. Lu, T., Boutilier, C.: Budgeted social choice: from consensus to personalized decision making. In: Proceedings of the 22nd International Joint Conference on Artificial Intelligence, pp. 280–286 (2011)
13. Skowron, P., Faliszewski, P., Slinko, A.: Fully proportional representation as resource allocation: approximability results. In: Proceedings of the 23rd International Joint Conference on Artificial Intelligence, pp. 353–359. AAAI Press (2013)
14. Procaccia, A., Rosenschein, J., Zohar, A.: On the complexity of achieving proportional representation. Soc. Choice Welfare **30**(3), 353–362 (2008)
15. Betzler, N., Slinko, A., Uhlmann, J.: On the computation of fully proportional representation. J. Artif. Intell. Res. **47**, 475–519 (2013)
16. Skowron, P., Yu, L., Faliszewski, P., Elkind, E.: The complexity of fully proportional representation for single-crossing electorates. In: Vöcking, B. (ed.) SAGT 2013. LNCS, vol. 8146, pp. 1–12. Springer, Heidelberg (2013)
17. Elkind, E., Faliszewski, P., Skowron, P.: A characterization of the single-peaked single-crossing domain. In: Proceedings of the 28th AAAI Conference on Artificial Intelligence, pp. 654–660 (2014)
18. Nemhauser, G., Wolsey, L., Fisher, M.: An analysis of approximations for maximizing submodular set functions. Math. Program. **14**(1), 265–294 (1978)

EvoENERGY

A Decentralized PSO with Decoder for Scheduling Distributed Electricity Generation

Jörg Bremer[✉] and Sebastian Lehnhoff

University of Oldenburg, 26129 Oldenburg, Germany
{joerg.bremer,sebastian.lehnhoff}@uni-oldenburg.de

Abstract. A steadily increasing pervasion of the distribution grid with rather small renewable energy resources imposes fluctuating and hardly predictable feed-in and thus calls for new predictive load planning strategies. On the other hand, combined with controllable, shiftable loads and electrical storages, these energy units set up a flexibility potential for fine-grained control. To tap the full potential, distributed control strategies are discussed for scheduling due to the expected large number of controlled entities. Decoder strategies for unit independent algorithm implementation and feasibility assurance had recently been applied to some first optimization approaches for scheduling in smart grid. We extended a distributed particle swarm to harnesses such decoder approach for model independent constraint-handling and achieved a higher accuracy compared with other approaches. A multi swarm is integrated after the island model into a decentralized agent-based solution and compared with an established decentralized approach for predictive scheduling within virtual power plants. We demonstrate the superiority of the particle swarm in terms of achieved solution accuracy and the competitiveness in terms of sent messages.

Keywords: Particle swarm optimization · Decoder · Constrained optimization · Distributed scheduling · Distributed generation

1 Introduction

A steadily increasing share of renewable energy resources within today's distribution grid imposes fluctuating and hardly predictable feed-in and demands new management strategies to allow the transition to a future smart grid. On the other hand, combined with controllable, shiftable loads and electrical storages, these energy units set up a new flexibility potential that may be used to full capacity when harnessing appropriate ICT-based control.

In European countries, especially in Germany where currently a financial security of guaranteed feed-in prices is given, the share of distributed energy resources (DER) is rapidly rising. Following the goal defined by the European Commission [1], a concept is needed for integration into electricity markets for both: active power provision and ancillary services [2,3] to reduce subsidy dependence. In order to get grip on the growing complexity caused by the large number

© Springer International Publishing Switzerland 2016
G. Squillero and P. Burelli (Eds.): EvoApplications 2016, Part I, LNCS 9597, pp. 427–442, 2016.
DOI: 10.1007/978-3-319-31204-0_28

of distributed generators, grouping strategies are discussed as a possible solution. Virtual power plants (VPP) are a well-known instrument for aggregating and controlling DER [4]. Connected and coordinated by communication means, such groups of DER can jointly operate and thus be seen as a single entity from a grid operator's perspective. VPP concepts for several purposes (commercial as well as technical) have been developed. A use case commonly emerging within VPP control is the need for predictive scheduling [3,5] the load of participating DER. Independently of the specific objective at hand, a schedule for each DER has to be found such that the aggregated schedule of the whole group fulfills some objective and each schedule that finally is assigned to a DER is operable without violating any technical constraint.

For large scale problems, distributed (usually agent based) approaches are currently discussed not least due to further advantages like ensured privacy issues or self-* traits. Some recent implementations are [5–7]. Distributed organization and self-organized control is also especially a characteristic of dynamic virtual power plants [8]. But, also many centralized approaches had already been developed [9]. Unfortunately, most of them do not take into account that an algorithm should work independently of the controlled DER as changes in the setting at runtime must not lead to changes in controller implementation. Thus, they are not appropriate for self-organizing scenarios.

Ensuring the feasibility of a VPP scheduling solution is a crucial task [10]. Feasibility depends on predictions about the initial operational state of the unit from which the schedule will be operated. These predictions determine operation boundaries for instance for attached storages and many more. At the same time feasibility depends on a set of unit specific technical constraints.

Recently, both problems led to the development of so called decoders as constraint-handling technique that ensures feasibility of solutions by harnessing a systematic way of constructing feasible solutions [11]. At the same time, such decoders abstract from the underlying DER model, its constraints or current setting and thus allow for an equal treatment of arbitrary DER by offering a standardized interface and a common representation to any algorithmic approach. Nevertheless, accuracy of the solution depends on the decoder as well as on the used heuristics that uses the decoder for generating solution candidates.

In this contribution we integrate the decoder approach that ensures feasibility of the assigned schedules with a decentralized particle swarm approach in order to enhance the so far achieved accuracy. The rest of the paper is organized as follows. After a study of the related work, we start with a description of the schedule interpretation and a brief recap of the used decoder. We describe the integration into PSO in general and the extension to a decentralized, agent based multi swarm approach. We test our method with simulations based on an establish co-generation model and compare with the heuristic COHDA [12].

2   Related Work

The operational management of an energy system comprises complex tasks ranging from technical monitoring aspects to organizational measures and business management coupled within an energy management system. Traditionally, energy management is implemented as centralized control. However, given the increasing share of DER as well as flexible loads in the distribution grid today, the evolution of the classical, rather static (from an architectural point of view) power system to a dynamic, continuously reconfiguring system of individual decision makers (e. g. as described in [13]), it is unlikely for such centralized control schemes to be able to cope with the rapidly growing problem size. Consequently, the seminal work of Wu et al. [14] identified the need for decentralized control.

The International Energy Agency (IEA) proposed the concept of a virtual utility (which was originally introduced in the late nineties) as one means to address the transition to decentralized control by a flexible and market-driven collaboration of independent entities [4,15]. As a consequence, the concept of virtual power plants (VPP) with different operational targets (commercial as well as technical) has been derived and studied extensively [16,17]. A list of realizations can for example be found in [9,18]. Usually, such VPP focus on the long-term aggregation of generators (and sometimes battery storages) only and are mostly still operated in a centralized manner. Recently, decentralized control schemes have become a research topic. For instance, [19,20] survey the use of agent-based control methods for power engineering applications. Exemplary applications can e. g. be found in [6,21,22].

In order to address the integration of the current market situation, [8] introduces the concept of a dynamic virtual power plant. In this approach VPPs gather dynamically together with respect to concrete electricity products at an energy market and will diverge right after delivery. Additionally, fully distributed control algorithms are used that – in conjunction with the dynamic composition of DER – do no longer allow for a static, a priori formulation of the optimization problem that has to be solved by the VPP for scheduling.

Each DER first and foremost has to serve the purpose it has been built for. But, usually this purpose may be achieved in different alternative ways. For example, it is the main task of a co-generation plant (CHP) to deliver enough heat for the varying heat demand in a household at every moment in time to ensure some comfortable temperature. Nevertheless, if heat usage can be decoupled from heat production by use of a thermal buffer store, different production profiles emerge for generating the heat. This leads, in turn, to different respective electric load profiles that may be offered as alternatives to a VPP controller. The set of all schedules that a DER may operate without violating any technical constraint (or soft constraint like comfort) is the sub-search-space with respect to this specific DER from which a scheduling algorithm may choose solution candidates. Geometrically seen, this set forms a sub-space $\mathcal{F} \subseteq \mathbb{R}^d$ in the space of all possible schedules. In [10] a model has been proposed to derive a description for this sub-space of feasible solutions that abstracts from any DER model and its specific constraint formulations. These surrogate models for the search spaces of different DER may be automatically combined to a dynamic optimization model by serving as a means that guides an arbitrary algorithm where to look for feasible solutions. Due to the abstract formulations, all DER may be treated the same by the algorithm. A detailed description of the approach may be found in [11]. This approach has so far already been used e. g. in [5,8,11,23].

Evolution strategies have been successfully applied to optimization problems with rugged, multi modal fitness landscapes, to non-linear problems, and to derivative free optimization. Evolutionary algorithms have shown excellent performance in global optimization especially when it comes to complex multi-modal, high dimensional, real valued problems [24,25]; they are originally designed for non-constrained problems but are suitable for integration with decoders for constraint-handling [26].

Especially particle swarm optimization [27] has been applied to a large selection of real world problems and shown good performance due to mimicking swarm behavior of individual social entities that benefit from each other while exploring different parts of large search spaces concurrently [28–30].

3   Algorithm

3.1   Scenario

For this paper we go with the example of predictive scheduling for active power planning in day-ahead scenarios (not necessarily 24 hours but for some given future period).

One of the crucial challenges in operating of a VPP arises from the complexity of the scheduling task due to the large amount of (small) energy units in the distribution grid [19]. In the following, we consider predictive scheduling, where the goal is to select exactly one schedule s_i for each energy unit U_i from a search space of feasible schedules with respect to a future planning horizon, such that a global objective function (e.g. a target power profile for the VPP) is optimized by the sum of individual contributions [31]. A basic formulation of the scheduling problem is given by

$$\delta \left(\sum_{i=1}^{m} s_i, \zeta \right) \rightarrow \min \tag{1}$$

such that

$$s_i \in \mathcal{F}^{(U_i)} \ \forall U_i \in \mathcal{U}. \tag{2}$$

In Eq. (1) δ denotes an (in general) arbitrary distance measure for evaluating the difference between the aggregated schedule of the group and the desired target schedule ζ. W.l.o.g., in this contribution we use the Euclidean distance $\|\cdot\|_2$. To each energy unit U_i exactly one schedule s_i has to be assigned. The desired target schedule is given by ζ. $\mathcal{F}^{(U_i)}$ denotes the individual set of feasible schedules that are operable for unit U_i without violating any (technical) constraint.

Solving this problem without unit independent constraint handling leads to specific implementations that are not suitable for handling changes in VPP composition or unit setup without having changes in the implementation of the scheduling algorithm [8].

Additionally, due to the expected complexity of the predictive scheduling entailed by the large number of distributed generation units (if applicable bundled with batteries and controllable consuming devices), decentralized solutions are discussed as a more scalable solution [3, 8]. In this paper, we discuss the use of particle swarm optimization, which may be integrated as a centralized approach or as a decentralized multi swarm approach within a multi agent environment. In our scenarios, we assume that a group of distributed energy resources already exists (whether a static group [18] or a dynamically re-organized one [8]) and wants to determine an operation schedule for each unit within the group according to some already known energy product that has to be delivered jointly by the group. In the decentralized case, we assume a multi-agent system with exactly one controlling agent per energy unit. Agents are connected by a communication grid that is not necessarily homomorphous with the electrical grid integration of the units, but as a minimum requirement forms a connected graph. Agents have access to the controlled units and may communicate with agents in direct neighborhood by sending and receiving messages.

Each individually configured energy unit has its own individual flexibilities and capabilities of altering electrical generation (or consumption) for a given, future planning horizon. Several, individual constraints have to be obeyed when it comes to the feasibility of an operation schedule for a specific energy unit. For handling the individual constraints and feasible regions $\mathcal{F}^{(U_i)}$ of arbitrary units without a need for a model integration or a domain specific algorithm implementation we used a decoder based constraint-handling technique proposed by [11].

3.2 Constraint-Handling by Decoder

We start with a brief recap of the used constraint-handling technique. We use a so called decoder that imposes a relationship between a decoder solution and a feasible solution and gives instructions on how to construct feasible solutions [24, 32, 33]. An in-depth discussion of the approach can for example be found in [11].

The basic idea is to start with a set of feasible example schedules derived from a simulation model of the respective energy unit and use this sample as a

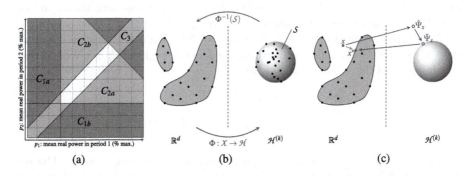

Fig. 1. General scheme of the used support vector decoder approach. Figure 1(a) on the left shows the geometrical interpretation of operable schedules and restricting technical constraints, Fig. 1(b) sketches the used surrogate based on a classical SVDD and 1(c) shows the decoder scheme on top of the surrogate model.

stencil for the region (the sub-space in the space of all schedules) that contains merely the feasible schedules.

In this sense, we regard a schedule of an energy unit as a vector $s = (s_0, \ldots, s_d) \in \mathcal{F}^{(U_i)} \subset \mathbb{R}^d$ with each element s_i denoting mean active power generated (or consumed) during the ith time interval. Figure 1(a) shows a simple 2-dimensional example. With a maximum of 100 % rated power (and a minimum of zero) as the only constraint, the space of feasible schedules would be $\mathcal{F} = [0, 1]^d$. That would be the whole area depicted in Fig. 1(a). Different constraints prohibit the use of different regions. Figure 1(a) shows the following examples for constraints on the operation of a single modulating CHP:

C_1 **Energy level restrictions:** Modulating CHP are usually able to vary their electricity generation between a minimum (p_{min}) and a maximum (p_{max}) load. Shutting down the device ($p = 0$) is sometimes an additional valid option. Therefore, such constraint prohibits the region $p_1 \in]0, p_{min}[$ (C_{1a}) and $p_2 \in]0, p_{min}[$ (C_{1b}).

C_2 **Limited acceleration:** Due to inertia effects, physical devices cannot instantaneously change behavior. Sometimes, additional reasons prohibit a too quick change of the energy generation level. Constraint C_2 prohibits all schedules with a difference $\|p_1 - p_2\| > d_{max}$ above a given threshold d_{max}. This constraint also gives an example for a not continuous region. It decomposes into C_{2a} and C_{2b}.

C_3 **Nearly charged storage:** Every co-generation plant needs to have the concurrently generated thermal energy used or buffered. If the buffer is nearly charged and no heat is used, the sum $p_1 + p_2$ might be limited by an upper bound as shown with constraint C_3.

Only the remaining region (white color) represents the scope of action of the energy resource (from which schedules are to be taken for consideration during optimization. The feasible part at the axes (zero power) has been enlarged for visualization purposes. In general it can be stated that the remaining part usually consist of a set of (separated) regions [34]. In addition, it has been shown in [35] that the scope of action is not always a convex polytope.

As has been shown in [36], it is advantageous from a machine learning point of view to use scaled schedules for learning the feasible region. Thus, we construct a training set $\mathcal{X}$ by a normalization with

$$\sigma : \mathbb{R}^d \rightarrow \mathcal{X} \subset [0, 1]^d$$

$$s \mapsto x = \sigma(s), \text{ with } x_i = \frac{s_i - p_{min}}{p_{max} - p_{min}}; \qquad (3)$$

p_{min} and p_{max} denoting minimum and maximum power respectively. The set $\mathcal{X}$, generated from a set of randomly sampled example schedules, is then used as a training set for a support vector approach [37] that derives a geometrical description of the sub-space that contains the given data; in our case: the set of feasible schedules [35]. Given a set of data samples, the inherent structure of the scope of action of the respective energy unit can be derived as follows:

After mapping the data to a high dimensional feature space $\mathcal{H}$ by means of an appropriate kernel, the smallest enclosing ball in this feature space is determined. When mapping back this ball to data space, the ball's pre-image forms a set of contours enclosing the given data sample (cf. Fig. 1(b)).

This approach is known as support vector data description (SVDD) [37]. Data description in general is concerned with the characterization of data sets and SVDD calculates a spherical envelope that contains the data seen in a training set [37]. Whether or not a so far unseen data point belongs to the same class derived from the training set is distinguished with the help of a distance function R (a result of the SVDD training process) that determines the distance of an arbitrary point's image (in $\mathcal{H}$). A subset of the training set (support vectors as another outcome of the training process) is mapped right onto the surface of the ball. Thus, the distance of a support vector determines the radius of the ball R_S and serves as a reference. As kernel we have chosen a Gaussian kernel as it is known to produce smooth decision boundaries [38].

At this point, the set of feasible schedules is represented as pre-image of a high-dimensional ball $\mathcal{S}$. Figure 1(c) shows the situation. This representation has some advantageous properties. Although the pre-image (the actual feasible region) might be some arbitrary shaped non-continuous blob in $\mathbb{R}^d$, the high-dimensional representation is still a ball and thus geometrically easier to handle. The relation is as follows: If a schedule is feasible, i.e. can be operated by the unit without violating any technical constraint, it lies inside the feasible region (grey area on the left hand side in Fig. 1(b)). Thus, the schedule is inside the pre-image (that represents the feasible region) of the ball and thus its image in the high-dimensional representation lies inside the ball. An infeasible schedule (e. g. x in Fig. 1(c)) lies outside the feasible region and thus its image $\hat{\Psi}_x$ lies outside the ball. But some relations are known: the center of the ball, the distance of the image from the center and the radius of the ball. Hence, we can move the image of an infeasible schedule along the difference vector towards the center until it touches the ball. Finally, we calculate the pre-image of the moved image $\tilde{\Psi}_x$ and get a schedule at the boundary of the feasible region: a repaired schedule x^* that is now feasible. Figure 1(c) shows the respective commutative diagram of sub-functions and geometric interpretation. We do not need a mathematical description of the original feasible region or of the constraints to do this. More sophisticated variants of transformation are e. g. given in [11]. Formally, the result is a mapping (the decoder γ)

$$\gamma : [0, 1]^d \rightarrow \mathcal{F}_{[0,1]} \subseteq [0, 1]^d$$
$$x \mapsto \gamma(x)$$

(4)

that transforms any given (maybe in-feasible) schedule into a feasible one. Thus, we are able to transform the given scheduling problem into an unconstrained formulation.

With these preliminaries in constraint handling we can now reformulate our optimization problem as

$$\delta \left(\sum_{i=1}^{m} \sigma_i^{-1} \circ \gamma(\boldsymbol{x}_i), \zeta \right) \rightarrow min, \tag{5}$$

where γ_i is the decoder function of unit i that produces feasible, scaled schedules from $\boldsymbol{x} \in [0,1]^d$ and σ_i^{-1} scales them unit specific entrywise to correct active power values (inverse to Eq. (3)) resulting in schedules that are operable by that unit. Please note, that this is a constraint free formulation. With this problem formulation, many standard algorithms for optimization can be easily adapted as there are no constraints (apart from a simple box constraint $\boldsymbol{x} \in [0,1]^d$) to be handled and no domain specific implementation (regarding the energy units and their operation schedules) has to be integrated. Equation (5) is used as a surrogate objective to find the solution to the constrained optimization problem Eq. (1).

Using a decoder fairly eases the implementation of a solver because no complex constraints have to be considered. On the other hand, such a decoder may introduce additional complexity into the optimization problem by the transformation. For this reason, we scrutinized the fitness landscapes of both problems (untransformed and transformed) to gain insight into the problem structure with means from standard fitness landscape analysis [39]. Indeed, our findings indicate a slightly growing complexity by an increased ruggedness with a growing number of local minima (removed reference). But, this situation can be easily countered by using a heuristics that copes well with rugged non-linear problems like particle swarm optimization.

3.3   Decentralized Particle Swarms

PSO is a stochastic, population based heuristic invented by Eberhard and Kennedy [27]. In a different way from evolutionary algorithms, PSO does not harness evolution operators but borrows movement concepts from schooling. In this way, particles represent individual solutions that move through solution space, attracted by currently best solutions. Each particle i represents a solution $\boldsymbol{x}_i \in \mathbb{R}^n$. Movement of the particles is simulated by an iterative update of the current positions in search space:

$$\boldsymbol{x}_i = \boldsymbol{x}_i + \boldsymbol{v}_i. \tag{6}$$

The position update is accomplished by adding an individual velocity vector $\boldsymbol{v}_i$ to each position. The velocity

$$\boldsymbol{v}_i = w\boldsymbol{v}_i + c_1 r_1(\hat{\boldsymbol{x}}_i - \boldsymbol{x}_i) + c_2 r_2(\boldsymbol{x}^* - \boldsymbol{x}_i) \tag{7}$$

is updated itself prior to the position updates. In Eq. (7) the so far best position seen by the ith particle is denoted by $\hat{\boldsymbol{x}}_i$ while $\boldsymbol{x}^*$ denotes the global best the whole swarm has seen so far. For a stochastic influence, r_1 and r_2 are random values regenerated for each update. Parameters w, c_1 and c_2 are user supplied for tuning the algorithm. Each term in (7) has a specific role [40]. The inertia term $w\boldsymbol{v}_i$ prevents a particle from changing direction too abruptly; user controlled by inertia coefficient w. The so called cognitive component $c_1 r_1(\hat{\boldsymbol{x}}_i - \boldsymbol{x}_i)$ provokes an orientation on already found good regions on the path of a particle. It serves as a sort of memory. The social component $c_2 r_2(\boldsymbol{x}^* - \boldsymbol{x}_i)$ lets a particle be attracted by globally good solutions.

The update process of position and velocity of each particle is iteratively repeated until some stopping criterion is met like maximum number iterations, a detected stalling of the procedure or similar.

Many modifications have been proposed to this basic version of PSO. A good overview on variants can e. g. be found in [28]. Traditional PSO are suitable for numerical optimization with real-parameter representation. Only few attempts have been made to adopt PSO for combinatorial problems [29,30]; with all approaches regarding discrete problems. In contrast to problems like [29] we do not have to deal with a discrete set of feasible schedules.

The main problem with Eq. (5) is the rapidly growing dimension $m \cdot d$ of the search space with number of units m and schedule dimension d for each unit. For this reason, we use a cooperative approach and distribute the PSO to all agents in the setting.

In order to solve Eq. (5) with an PSO approach, a solution and thus a particle position $\boldsymbol{x}$ is encoded as $\boldsymbol{x} \in [0,1]^{m \cdot d}$. Associated with $\boldsymbol{x}$ is the decoder solution (that is evaluated by the objective function)

$$\boldsymbol{x}' = (\sigma_1^{-1} \circ \gamma_i[(x_1, \ldots, x_d)], \sigma_2^{-1} \circ \gamma_2[(x_{d+1}, \ldots, x_{2d})], \ldots, \sigma_m^{-1} \circ \gamma_i[(x_{(m-1) \cdot d + 1}, \ldots, x_{m \cdot d})]). \quad (8)$$

Each decoder γ_i encapsulates the traits of the feasible region $\mathcal{F}^{(U_i)} \subseteq \mathbb{R}^d$ of a specific (unknown to the algorithm) DER.

It is known, that distributed, cooperative swarms better cope with high dimensional search spaces [41]. We also decided to use a multi-swarm approach, because it can be easily extended to a fully-fledged decentralized approach as follows. We use a multi swarm approach as for instance described in [41]. In this way, multiple instances of PSO are executed independently on the same problem. Usually, from time to time some interaction takes place, like the exchange of a share of k best particles that replace the k worst of the receiver swarm. This approach corresponds to the island model of EAs [42]. Normally, this execution is synchronized and clocked. We integrate this approach with an asynchronously executed agent model. Each agent controls his own swarm and evolves autonomously for a short while. Each agent can randomly decide on (within some given bounds) when to send the k best particles to a randomly chosen neighbor agent. Concurrently sent is the so far reached sender fitness. If an agent receives a particle set from some other agent, a replacement of the worst own particles takes only place if the own fitness is worse than the sender's one.

4   Simulation Results

4.1   Setup

Apart from the simple centralized case with a single swarm, we tested our app-roach with an agent based simulation study. Each energy unit is associated with a controlling agent who is responsible for conducting local optimization with a PSO sub-swarm and communication with other agents for particle exchange.

As a model for distributed energy resources we used a model for co-generation plants that has already served in several studies and projects for evaluation [5,11,34,35,43]. This model comprises a micro CHP with 4.7 kW of rated elec-trical power (12.6 kW thermal power) bundled with a thermal buffer store. Con-straints restrict power band, buffer charging, gradients, min on and off times, and satisfaction of thermal demand. The relationship between electrical (active) power and thermal power was modeled after engine test benches from [44]. Ther-mal demand is determined by simulating a detached house (including hot water drawing) according to given weather profiles. For each agent the model is individ-ually (randomly) configured with state of charge, weather condition, temperature range, allowed operation gradients, and similar. From these model instances, the respective training sets for building the decoders have been generated with the sampling approach from [45]. Prior to optimization each agent receives decoders from all other agents in order to calculate Eq. (5). Please note that only decoders are given away and that they do not disclose any internals about the foreign units. Unless specified differently, all result are averaged over 1000 runs with differently generated problem instances.

A decentralized detection of convergence or for finding the best solution out of all decentralized found solution is not a topic of this paper as there are readily available solution that can be used for these sub-problems [46,47].

4.2   Results

Figure 2(a) shows the result of a simulation run with a 5 CHP scenario. The scenario has been solved with a central multi-swarm approach with $n = 10$ particles each and with different numbers of particle exchange events c as well as with a single swarm with 50 particles for a fairer comparison with multi swarm solving. The MSPSO($n = 10$, $k = 3$, $c = 0$) series additionally shows the result for 5 individually operating swarms without any particle exchange. The time horizon was 96 intervals of 15 min. The target schedule has been selected in a way that ensures operability. Thus, the optimal solution can be reached exactly and has an error of zero. Depicted are the mean values of 1000 simulations runs each. Neither the multi-swarm without particle exchange nor the PSO with a respectively larger swarm converge to an acceptably good solution. Multi swarm versions with a rather agile particle exchange converge faster in the beginning but a rather low exchange rate succeeds in the long run. Figure 2(b) shows the same situation for a larger scenario with 10 CHP; depicting the effect more prominently.

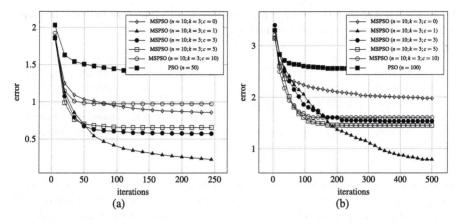

Fig. 2. Comparison of different swarm organization approaches for two different problem (and swarm) sizes (number of particles n). Compared are single swarm (PSO) and multi swarm (MSPSO) settings; where there latter has been parametrized with different amount of communication and particle exchange (number of exchanges of k particles between c swarms).

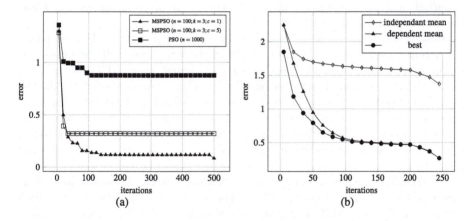

Fig. 3. Left: scenario with 10 CHP and swarm size of 100 particles for each multi-swarm and 1000 for the single swarm respectively; right: comparison of the mean convergence behaviour for coupled and non-coupled multi swarms.

Figure 3(a) shows the result for the 10 CHP scenario with larger swam sizes (100 particles and 1000 respectively for the single swarm). Again, it is not guaranteed that the optimum (an error of zero) is found in every run; especially not for isolated swarms. All swarms had been stopped after 250 and 500 iterations in these scenarios.

Figure 3(b) shows the development of the mean best values of the swarms during optimization. As can be seen, without any particle exchange the mean best value of the swarms stays worse than the case with particle exchange as

Table 1. Comparison of the decentralized multi swarm PSO and the fully decentralized heuristics COHDA for two different problem sizes.

# chp	algo	error	# messages	# evaluations
5	COHDA	0.0771±0.3475	**51.672 ± 15.379**	121.407 ± 41.023
	MSPSO$_{25}$	$1.459 \times 10^{-6} \pm 2.735 \times 10^{-6}$	593.397±4.583	9921.041±281.779
	MSPSO$_{100}$	$\mathbf{5.132 \times 10^{-7} \pm 8.601 \times 10^{-7}}$	593.683±4.505	39585.96±1157.81
50	COHDA	0.4095±0.4908	2521.22±832.16	**3968.28 ± 1786.14**
	MSPSO	$\mathbf{5.973 \times 10^{-4} \pm 2.575 \times 10^{-4}}$	**168.61 ± 56.74**	58749.40 ±12763.23

always a part of the swarms gets stuck in some local optimum (at least within the limited number of scrutinized iterations) due to the high dimensionality of the problem. With particle exchange the mean best value of all swarms converges towards the absolute best value of all swarms. Hence, all parallel operated swarms end up with an equally good solution.

Additionally, the multi swarm approach with particle exchange is clearly superior to the single swarm approach even with respectively larger number of particles. These experiments were mainly for getting an impression on the impact of particle migration and hence an idea for the design of the agent system. Thus, the limited number of iterations.

Results for an agent based simulation are shown in Table 1. We compared our decentralized multi swarm approach with the established fully decentralized Combinatorial Optimization Heuristic for Distributed Agents (COHDA).

COHDA was originally introduced in [12] and has been successfully applied to predictive scheduling within VPPs. The general idea behind COHDA is an asynchronous iterative approximate best-response behavior [48], where each agent reacts to updated information from other agents by adapting its own selected schedule with respect to local and global objectives, preferably in a way that the joint objective is served best.

COHDA has previously been successfully applied to the day-ahead load planning problem scrutinized here [8,12,31]. For both agent simulations we applied COHDA agents and MSPSO agents to the same set of CHP models for each simulation; connected by the same decoder for solution generation. Both multi agent systems used the same small word communication network [49] with the same local neighborhood. In all cases, coordination was done merely by sending and receiving messages from agents within the direct neighborhood.

Again, we started with a small problem size of 5 CHP and compared with two different swarm sizes. For the sake of simplicity we had each swarm stopped after 2000 iterations at latest or earlier if 5 consecutive iterations without any improvement had been detected. Other stopping criteria or a decentralized detection of convergence could be easily integrated. As can be seen, the MSPSO achieves much better results in the end. Also COHDA is as a heuristic not immune for getting stuck too early [12]. Obviously the MSPSO has better escape capabilities but at the cost of a comparably high number of objective function evaluations. Column 4 compares the number of sent messages which is also notably higher for the MSPSO if initiated at a constant rate. In case of high accuracy, MSPSO is the better choice.

For larger scenarios, i.e. 50 CHP with 96-dimensional schedules each, the effect is not that pronounced but still significant. Again at the cost of notably more objective evaluations. On the other hand, this result was achieved with notably fewer messages by decreasing the likelihood of sending particles to another agent over time (cf. Fig. 2(b)).

5   Conclusions

We presented a hybrid approach for ahead of time scheduling for active power production of distributed energy resources. Our approach takes into account individual operating constraints without a need for integrating specifically adapted models. Thus, the approach shows good generalizability for the integration of arbitrary types of energy units and almost always produces feasible solutions that are operable by the different units. For the first time, we integrated a decentralized PSO with an decoder for load planning and showed its good distributability property even for fully decentralized scenarios.

Compared with state of the art heuristics for agent based decentralized optimization we achieved an improved accuracy although at the cost of a larger number of objective evaluations. Additionally the amount of necessary communication might be minimized.

Depending on the use case and the targeted accuracy, the presented approach represents an alternative with the potential for further improvements regarding the mechanisms for solution migration or swarm evolution.

References

1. European Parliament & Council: Directive 2009/28/ec of 23 on the promotion of the use of energy from renewable sources and amending and subsequently repealing directives 2001/77/ec and 2003/30/ec, April 2009
2. Abarrategui, O., Marti, J., Gonzalez, A.: Constructing the active european power grid. In: Proceedings of WCPEE 2009, Cairo (2009)
3. Nieße, A., Lehnhoff, S., Tröschel, M., Uslar, M., Wissing, C., Appelrath, H.J., Sonnenschein, M.: Market-based self-organized provision of active power and ancillary services. In: Complexity in Engineering (COMPENG). IEEE, June 2012
4. Awerbuch, S., Preston, A.M. (eds.): The Virtual Utility: Accounting, Technology & Competitive Aspects of the Emerging Industry. Topics in Regulatory Economics and Policy, vol. 26. Kluwer Academic Publishers, Boston (1997)
5. Hinrichs, C., Bremer, J., Sonnenschein, M.: Distributed hybrid constraint handling in large scale virtual power plants. In: IEEE PES Conference on Innovative Smart Grid Technologies Europe (ISGT Europpe 2013). IEEE Power & Energy Society (2013)
6. Ramchurn, S.D., Vytelingum, P., Rogers, A., Jennings, N.R.: Agent-based homeostatic control for green energy in the smart grid. ACM Trans. Intell. Syst. Technol. 2(4), 35:1–35:28 (2011)

7. Kamphuis, R., Warmer, C., Hommelberg, M., Kok, K.: Massive coordination of dispersed generation using powermatcher based software agents. In: 19th International Conference on Electricity Distribution, May 2007

8. Nieße, A., Beer, S., Bremer, J., Hinrichs, C., Lünsdorf, O., Sonnenschein, M.: Conjoint dynamic aggrgation and scheduling for dynamic virtual power plants. In: Ganzha, M., Maciaszek, L.A., Paprzycki, M. (eds.) Federated Conference on Computer Science and Information Systems - FedCSIS 2014, Warsaw, Poland, September 2014

9. Coll-Mayor, D., Picos, R., Garciá-Moreno, E.: State of the art of the virtual utility: the smart distributed generation network. Int. J. Energy Res. **28**(1), 65–80 (2004)

10. Bremer, J., Rapp, B., Sonnenschein, M.: Encoding distributed search spaces for virtual power plants. In: IEEE Symposium Series on Computational Intelligence 2011 (SSCI 2011), Paris, France, April 2011

11. Bremer, J., Sonnenschein, M.: Constraint-handling for optimization with support vector surrogate models - a novel decoder approach. In: Filipe, J., Fred, A. (eds.) ICAART 2013 - Proceedings of the 5th International Conference on Agents and Artificial Intelligence, Barcelona, Spain, vol. 2, pp. 91–105. SciTePress (2013)

12. Hinrichs, C., Lehnhoff, S., Sonnenschein, M.: A decentralized heuristic for multiple-choice combinatorial optimization problems. In: Helber, S., et al. (eds.) Operations Research Proceedings 2012, pp. 297–302. Springer International Publishing, Switzerland (2014)

13. Ilić, M.D.: From hierarchical to open access electric power systems. Proc. IEEE **95**(5), 1060–1084 (2007)

14. Wu, F., Moslehi, K., Bose, A.: Power system control centers: past, present, and future. Proc. IEEE **93**(11), 1890–1908 (2005)

15. International Energy Agency: Distributed Generation in Liberalised Electricity Markets. OECD Publishing (2002)

16. Lukovic, S., Kaitovic, I., Mura, M., Bondi, U.: Virtual power plant as a bridge between distributed energy resources and smart grid. In: Hawaii International Conference on System Sciences, pp. 1–8 (2010)

17. Tröschel, M., Appelrath, H.-J.: Towards reactive scheduling for large-scale virtual power plants. In: Braubach, L., van der Hoek, W., Petta, P., Pokahr, A. (eds.) MATES 2009. LNCS, vol. 5774, pp. 141–152. Springer, Heidelberg (2009)

18. Nikonowicz, Ł.B., Milewski, J.: Virtual power plants - general review: structure, application and optimization. J. Power Technol. **92**(3), 135–149 (2012)

19. McArthur, S., Davidson, E., Catterson, V., Dimeas, A., Hatziargyriou, N., Ponci, F., Funabashi, T.: Multi-agent systems for power engineering applications - Part I: concepts, approaches, and technical challenges. IEEE Trans. Power Syst. **22**(4), 1743–1752 (2007)

20. Ramchurn, S.D., Vytelingum, P., Rogers, A., Jennings, N.R.: Putting the 'smarts' into the smart grid: a grand challenge for artificial intelligence. Commun. ACM **55**(4), 86–97 (2012)

21. Negenborn, R.R., Lukszo, Z., Hellendoorn, H. (eds.): Intelligent Infrastructures. Intelligent Systems, Control and Automation: Science and Engineering, vol. 42. Springer, The Netherlands (2010)

22. Anders, G., Siefert, F., Steghöfer, J.P., Seebach, H., Nafz, F., Reif, W.: Structuring and controlling distributed power sources by autonomous virtual power plants. In: IEEE Power and Energy Student Summit (PESS 2010). IEEE Power & Energy Society (2010)

23. Bremer, J., Sonnenschein, M.: Parallel tempering for constrained many criteria optimization in dynamic virtual power plants. In: 2014 IEEE Symposium on Computational Intelligence Applications in Smart Grid (CIASG), pp. 1–8, December 2014

24. Kramer, O.: A review of constraint-handling techniques for evolution strategies. Appl. Comp. Intell. Soft Comput. **2010**, 1–19 (2010)

25. Ulmer, H., Streichert, F., Zell, A.: Evolution strategies assisted by gaussian processes with improved pre-selection criterion. In: IEEE Congress on Evolutionary Computation, CEC 2003, pp. 692–699 (2003)

26. Koziel, S., Michalewicz, Z.: A decoder-based evolutionary algorithm for constrained parameter optimization problems. In: Eiben, A.E., Bäck, T., Schoenauer, M., Schwefel, H.-P. (eds.) PPSN 1998. LNCS, vol. 1498, pp. 231–240. Springer, Heidelberg (1998)

27. Kennedy, J., Eberhart, R.: Particle swarm optimization. In: 1995 Proceedings of IEEE International Conference on Neural Networks, vol. 4, pp. 1942–1948. IEEE, November 1995

28. Van Den Bergh, F.: An analysis of particle swarm optimizers. Ph.D. thesis, University of Pretoria, Pretoria, South Africa, South Africa AAI0804353 (2002)

29. Lapizco-Encinas, G., Kingsford, C., Reggia, J.: A cooperative combinatorial particle swarm optimization algorithm for side-chain packing. In: 2009 Swarm Intelligence Symposium, SIS 2009, pp. 22–29. IEEE, March 2009

30. Poli, R., Kennedy, J., Blackwell, T.: Particle swarm optimization. Swarm Intell. **1**(1), 33–57 (2007)

31. Sonnenschein, M., Hinrichs, C., Niee, A., Vogel, U.: Supporting renewable power supply through distributed coordination of energy resources. In: Hilty, L.M., Aebischer, B. (eds.) ICT Innovations for Sustainability. Advances in Intelligent Systems and Computing, vol. 310, pp. 387–404. Springer International Publishing, Switzerland (2015)

32. Koziel, S., Michalewicz, Z.: Evolutionary algorithms, homomorphous mappings, and constrained parameter optimization. Evol. Comput. **7**, 19–44 (1999)

33. Coello Coello, C.A.: Theoretical and numerical constraint-handling techniques used with evolutionary algorithms: a survey of the state of the art. Comput. Methods Appl. Mech. Eng. **191**(11–12), 1245–1287 (2002)

34. Neugebauer, J., Kramer, O., Sonnenschein, M.: Classification cascades of overlapping feature ensembles for energy time series data. In: Aung, Z., et al. (eds.) DARE 2015. LNCS, vol. 9518, pp. 76–93. Springer, Heidelberg (2015). doi:10.1007/978-3-319-27430-0_6

35. Bremer, J., Rapp, B., Sonnenschein, M.: Support vector based encoding of distributed energy resources' feasible load spaces. In: IEEE PES Conference on Innovative Smart Grid Technologies Europe, Chalmers Lindholmen, Gothenburg, Sweden (2010)

36. Juszczak, P., Tax, D., Duin, R.P.W.: Feature scaling in support vector data description. In: Deprettere, E., Belloum, A., Heijnsdijk, J., van der Stappen, F. (eds.) Proceedings of the 8th Annual Conference of the Advanced School for Computing and Imaging, ASCI 2002, pp. 95–102 (2002)

37. Tax, D.M.J., Duin, R.P.W.: Support vector data description. Mach. Learn. **54**(1), 45–66 (2004)

38. Ben-Hur, A., Siegelmann, H.T., Horn, D., Vapnik, V.: Support vector clustering. J. Mach. Learn. Res. **2**, 125–137 (2001)

39. Vassilev, V.K., Fogarty, T.C., Miller, J.F.: Information characteristics and the structure of landscapes. Evol. Comput. **8**(1), 31–60 (2000)

40. Abdul-Rahman, S., Bakar, A.A., Mohamed-Hussein, Z.-A.: An improved particle swarm optimization via velocity-based reinitialization for feature selection. SCDS 2015. CCIS, vol. 545, pp. 3–12. Springer, Heidelberg (2015). doi:10.1007/978-981-287-936-3_1

41. Vanneschi, L., Codecasa, D., Mauri, G.: A comparative study of four parallel and distributed PSO methods. New Gener. Comput. **29**(2), 129–161 (2011)

42. Fernández, F., Tomassini, M., Vanneschi, L.: An empirical study of multipopulation genetic programming. Genet. Program Evolvable Mach. **4**(1), 21–51 (2003)

43. Nieße, A., Sonnenschein, M.: A fully distributed continuous planning approach for decentralized energy units. In: Cunningham, D.W., Hofstedt, P., Meer, K., Schmitt, I. (eds.) Informatik 2015. GI-Edition - Lecture Notes in Informatics (LNI), vol. 246, pp. 151–165. Bonner Köllen Verlag, Bonn (2015)

44. Thomas, B.: Mini-Blockheizkraftwerke: Grundlagen, Gerätetechnik. Vogel Buchverlag, Betriebsdaten (2007)

45. Bremer, J., Sonnenschein, M.: Sampling the search space of energy resources for self-organized, agent-based planning of active power provision. In: Page, B., Fleischer, A.G., Göbel, J., Wohlgemuth, V. (eds.) Proceedings of the 27th International Conference on Environmental Informatics for Environmental Protection, Sustainable Development and Risk Management, EnviroInfo 2013, 2–4 September 2013, Hamburg, Germany, pp. 214–222. Berichte aus der Umweltinformatik, Shaker (2013)

46. Bahi, J., Contassot-Vivier, S., Couturier, R., Vernier, F.: A decentralized convergence detection algorithm for asynchronous parallel iterative algorithms. IEEE Trans. Parallel Distrib. Syst. **16**(1), 4–13 (2005)

47. Santoro, N.: Design and Analysis of Distributed Algorithms (Wiley Series on Parallel and Distributed Computing). Wiley-Interscience, New York (2006)

48. Littman, M.L., Stone, P.: Leading best-response strategies in repeated games. In: Seventeenth Annual International Joint Conference on Artificial Intelligence Workshop on Economic Agents, Models, and Mechanisms (2001)

49. Watts, D.J.: Networks, dynamics, and the small-world phenomenon. Am. J. Sociol. **105**, 493–527 (1999)

Comparison of Multi-objective Evolutionary Optimization in Smart Building Scenarios

Marlon Braun[1]([⊠]), Thomas Dengiz[1], Ingo Mauser[2], and Hartmut Schmeck[1,2]

[1] Karlsruhe Institute of Technology – Institute AIFB, 76128 Karlsruhe, Germany
{marlon.braun,schmeck}@kit.edu, thomas.dengiz@student.kit.edu
[2] FZI Research Center for Information Technology, 76131 Karlsruhe, Germany
mauser@fzi.de

Abstract. The optimization of operating times and operation modes of devices and systems that consume or generate electricity in buildings by building energy management systems promises to alleviate problems arising in today's electricity grids. Conflicting objectives may have to be pursued in this context, giving rise to a multi-objective optimization problem. This paper presents the optimization of appliances as well as heating and air-conditioning devices in two distinct settings of smart buildings, a residential and a commercial building, with respect to the minimization of energy costs, CO_2 emissions, discomfort, and technical wearout. We propose new encodings for appliances that are based on a combined categorization of devices respecting both, the optimization of operating times as well as operation modes, e.g., of hybrid devices. To identify an evolutionary algorithm that promises to lead to good optimization results of the devices in our real-world lab environments, we compare four state-of-the-art algorithms in realistic simulations: *ESPEA*, *NSGA-II*, *NSGA-III*, and *SPEA2*. The results show that ESPEA and NSGA-II significantly outperform the other two algorithms in our scenario.

Keywords: Energy management system · Smart building · Evolutionary algorithm · Multi-objective optimization

1 Introduction

In the face of a potential climate change that may be induced by man-made carbon dioxides, many countries started to change their power generation from fossil to renewable energy sources (RES). Coal-fired power plants are being replaced by plants emitting less carbon dioxide, such as wind turbines and photovoltaic (PV) systems that exploit solar radiation. Usually, their generation is intermittent and hardly controllable. This is already leading to high production peaks causing voltage problems and overloads in distribution grids as well as to periods with barely any generation by RES, meaning that almost all the needed power still has to be produced by conventional power generation [1].

© Springer International Publishing Switzerland 2016
G. Squillero and P. Burelli (Eds.): EvoApplications 2016, Part I, LNCS 9597, pp. 443–458, 2016.
DOI: 10.1007/978-3-319-31204-0_29

One way to deal with this problem of intermittent generation is to build additional energy storage systems that are able to balance the fluctuations out. Unfortunately, this is quite expensive, as for instance batteries, and often leads to public resistance, e. g., in case of pumped hydro storage plants. Another way would be a change in the central paradigm of power generation and consumption. At any time, the generation in the electricity grid has to be in balance with the consumption to keep the system stable. Currently, *generation follows consumption* in the energy system. This may change to *consumption follows production* by using measures of demand side management (DSM). DSM targets on the flexibilization of the traditional demand side of the power grid, which is nowadays also generating an increasing share of power, in particular by RES, and on increasing the self-consumption as well as self-sufficiency of local energy systems [2].

We consider the optimization of two different settings of smart buildings in a multi-objective context: one scenario consists of a smart residential building (SRB), the other scenario of a smart commercial building (SCB). In both scenarios, the operation times and operation modes of household appliances as well as heating, ventilation, and air-conditioning (HVAC) devices are optimized with respect to the minimization of energy costs, CO_2 emissions, and technical wearout as well as to the maximization of comfort, i.e., minimization of discomfort. To optimize the appliances and devices, we propose novel encodings for appliances that have advantages over existing encodings, which have been used so far in the community. These encodings are based on a combined categorization of the appliances and devices that respects both, the optimization of operation times as well as the optimization of operation modes, e.g., of hybrid appliances. We compare four multi-objective evolutionary algorithms (MO-EA) on these scenarios—*ESPEA* [3], *NSGA-II* [4], *NSGA-III* [5], and *SPEA2* [6]—that have either been applied successfully in past studies or appear to be suitable candidates for optimizing the scenarios found in smart buildings.

Section 2 provides an overview of similar approaches to multi-objective optimization in energy management systems. In Sect. 3, the general approach, the modeling of devices, and the encodings of the devices in evolutionary algorithms are outlined. Section 4 presents the scenarios and setups that have been chosen to evaluate different algorithms for multi-objective optimization. The simulation results are analyzed and discussed in Sect. 5. We summarize and conclude the paper in the final Sect. 6 and provide an outlook on further research.

2   Related Work

There are numerous building energy management systems (BEMS) that have been used in simulations and in real-world environments. Usually, they are based on some kind of categorization of the devices to generalize and ease their optimization by deriving appropriate device representations. In the following sub-sections, we first present different device categorizations, before giving an overview of approaches to optimization in BEMS and showing some typical device representations.

2.1   Categorization of Devices in Buildings

In the literature, there is a multitude of different device categorizations. For instance, Kok et al. [7] propose device categories that focus on the dimensions of *when is it or may it be operated*, e.g., stochastically or based on user interaction, and *what is the constraint of operation*, e.g., limiting buffer or storage.

In contrast, Ha et al. [8] propose a categorization that focuses on the services that are provided by the devices. The first dimension is similar to the previous categorization and targets on time and availability (*permanent services* versus *temporary* or *timed services*). The second dimension distinguishes whether the service is provided directly to the user or whether it is an ancillary service only (*end-user services* versus *support services*). The third dimension addresses the modifiability of the service by the BEMS (*modifiable* versus *non-modifiable*).

Nestle et al. [9] add the differentiation whether it is *load* or *generation*, whereas Soares et al. [10] emphasize the potentials of control and modifiability by the BEMS by distinguishing *uncontrollable*, *reparameterizable*, *interruptible*, and *shiftable loads*. Mauser et al. [11] separate modifiability into two dimensions. The first dimension, which is called *temporal degree of freedom*, defines whether a device is *deferrable* or *interruptible*. The second dimension is called *energy-related degree of freedom* and defines whether a device has *alternative profiles* or is *multivalent/hybrid*, i.e., supports the utilization of different energy carriers.

2.2   Heuristic and Multi-objective Optimization in Building Energy Management Systems

Energy management systems have to optimize the energy consumption and generation iteratively to adapt them continuously to changing conditions and states. This process of reoccurring optimization in BEMS shall be executed on computers that have a low energy consumption and thus small memory space and low computation power. Therefore, heuristics promise to achieve this task of optimization efficiently and in particular *evolutionary algorithms* have been used quite often, for instance in [11–15]. Other approaches also use other meta-heuristics, e.g., particle swarm optimization [16,17], to optimize the configuration [12] or the scheduling [11,13–17] of devices.

Usually, the optimization is done with respect to a single objective, which most of the time is the total costs. Nevertheless, there are also approaches taking other objectives into account that are not directly related to costs. For instance, De Oliveira et al. [18], who use a direct solver, consider economic costs and user comfort. Soares et al. [14] use a MO-EA, more precisely the genetic algorithm *NSGA-II* [4], to optimize the operation times of devices in a residential building with respect to energy costs and user satisfaction. Salinas et al. [19] develop their own MO-EA, which they call *Load Scheduling With an (ε-Approximate) Evolutionary Algorithm*. They optimize the device operation in multiple buildings with respect to the minimization of the total energy consumption costs and the maximization of usefulness, i.e., gross income of the overall community and comfort of individuals that living in it. They show that an approach that enforces diversity in the population leads to better results within shorter time.

2.3   Representation of Devices

Allerding et al. [13] use a matrix of binary values determining the starting times of the devices as the representation of the optimization problem. Therefore, at minute resolution, the next 24 h are discretized to 1440 slots and every column represents one minute. A similar approach has also been used by Zhao et al. [15], where the operation starting times of appliances and HVAC devices are optimized using a vector of binary strings that encodes the slot of the day, when the appliance with its fixed operation duration will be started.

Mauser et al. [20] propose a more distinct approach to the encoding of the operating times of the devices in the genotypes: the overall optimization problem is represented by a bit string. Every device uses a dedicated part of the bit string for the individual encoding. The starting times of appliances are encoded as relative times within the time window of optimization. Interruptible devices and devices with alternative load profiles, such as hybrid devices, are also supported by adding more bits to the bit string that determine the length of interruptions or enumerate the alternatives, respectively. The operating times of a micro combined heat and power plant (micro-CHP) are encoded as transitions of an automaton.

Soares et al. [14] encode the starting times of the appliances in a string of integers, i.e., the overall problem is represented by a vector of integer values. The optimization horizon is set to the next 36 h and discretized to time slots of one minute. This approach is similar to Salinas et al. [19], who use a representation that uses a vector of real values.

3   Approach and Modeling of Devices

This section outlines the general approach, the categories for the devices used in this paper, the encodings of the devices in evolutionary algorithms, and the objective functions used in the optimization.

3.1   General Approach

The general approach of this paper to the optimization of devices in smart buildings is as follows: Firstly, we define several device categories and assign devices that are typically found in buildings to these categories. Then, we use a dedicated generic encoding for each category that fits the optimization of such devices. Finally, we distinguish several objectives that are usually conflicting and thus leading to a multi-objective optimization problem.

3.2   Device Categories

The device categories used in this paper are based on the categorizations in [11,20] and can be found in Table 1. Firstly, we identify whether the device is optimizable. Secondly, we distinguish whether an appliance has load-flexibility,

Table 1. Proposed device categories.

Category			
Load-flexibility	Interruptibility	Description	Example
Non-optimizable		Device may only be controlled by the user	Hob, oven; PV system
None	(Yes)	Starting time of the device may be optimized within a defined period	Traditional dishwasher, tumble dryer, washing machine
None	Yes	Operation cycle of device may be split into one or more phases that are separated by pauses	Interruptible dishwasher, tumble dryer, washing machine; micro-CHP
Yes	None	Device has alternative load profiles for the same operation cycle	Lighting, heat pump, heating element; load-flexible appliances
Yes	Yes	*combination of the two above*	Air-conditioning, gas-fired boiler; interruptible load-flexible appliances

i.e., alternative load profiles, and whether the device's operation cycle may be deferred or may be even interrupted. Deferability is simply handled as a special form of interruptibility, which is limited to a single interruption at the very beginning of a possible operation cycle.

In contrast to [11,20], the micro-CHP, which is usually a gas-fired engine that generates electricity and hot water simultaneously, i.e., working as power generator and central heating, is modeled similarly to the appliances. The micro-CHP is a device having unlimited interruptibility but no load-flexibility, whereas the air-conditioning has not only unlimited interruptibility but also load-flexibility in form of different cooling power levels.

The flexibility of the devices is either limited to a temporal flexibility window that, in reality, would be set individually by a user per operation cycle or it is limited due to the capacity of an energy storage, such as a hot water storage tank.

3.3 Device Modeling and Encoding

Interruptible, load-flexible appliances form the basis of our modeling approach as all other device categories can be derived from them. For example, interruptible devices may be interpreted as interruptible, load-flexible devices that possess one operation mode only in every phase. Each device j possesses a user-provided earliest starting time, denoted by release time r^j, and a latest finishing time, called deadline d^j.

Interruptible devices possess multiple phases of operation. The length of an individual phase i is denoted by p_i^j. We model interruptibility by associating

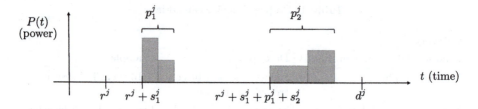

Fig. 1. Visualization of the encoding of interruptible devices.

every phase with a decision variable s_i^j. Each s_i^j is an integer that states the time, at which phase i is executed. In contrast to other approaches, every s_i^j uses its earliest time of execution instead of the system time as reference point. Hence, $s_i^j = 0$ implies that phase i is started directly after phase $i - 1$. Figure 1 illustrates the approach. Constraint-handling is simplified using this encoding, since it is sufficient to check whether the last phase terminates before d^j. Different operation modes are modeled by integer encoding as well. For each phase i, the variable a_i^j represents the operation mode that is chosen.

HVAC devices can be started an arbitrary number of times during the optimization horizon. We incorporate this feature by introducing an integer encoded variable n^j that states the number of possible interruptions and thus operation cycles. Minimum and maximum operation as well as minimum interruption time are implemented for HVAC devices limiting the number of feasible operation cycles. We apply a thermodynamical model to describe the changes in room and boiler temperatures that considers heat losses, energy conversion efficiencies, and body heat dissipation[1]. Finally, lighting is modeled as an interruptible device with multiple operation modes reflecting different light intensities.

Summarizing our approach, a single device is modeled using the decision variables listed below. Note that variables are omitted if they are not applicable for the given device and the sizes of $\mathbf{s}^j$ and $\mathbf{a}^j$ change dynamically depending on n^j:

$\mathbf{s}^j$: vector of starting times for each individual phase

$\mathbf{a}^j$: vector of operation modes for each phase

n^j : number of operation cycles

3.4   Objective Functions

Our model considers a fixed, discretized time horizon for optimization. For the sake of clarity, most functions in this section only compute values at any time t. Self-evidently, the objective values are obtained by summing up function values across all time slots. All objective functions in our model are minimized.

---

[1] A complete description of this model goes beyond the scope of this paper, but is available on request.

Energy costs are based on the electricity $P_b(t)$ bought from the utility for a time-variable price $p_b(t)$, the gas $G(t)$ consumed at a fixed price p_g per unit and the exceedance of a time-variable load limitation, which is penalized using a cost function $S(t, P_b(t))$. Excess energy generated by the PV system and the micro-CHP is sold to the utility for $C_f(t)$. Remuneration depends on the proportion of energy generated by the PV system and the micro-CHP, respectively.

$$C(t) = p_b(t) \cdot P_b(t) - C_f(t) + p_g \cdot G(t) + S(t, P_b(t)). \tag{1}$$

Carbon dioxide emissions are related to the electricity that is consumed from the grid and to local gas consumption. A time-variable signal $e(t)$ that depends on the current energy mix in the electricity grid determines the CO_2 caused by consuming electricity from the grid. The constant e_g describes the emissions per unit of gas consumed.

$$E(t) = e(t) \cdot P_b(t) + e_g \cdot G(t). \tag{2}$$

Thermal user discomfort occurs if the room temperature $T(t)$ falls below a predefined lower threshold T^l or exceeds an upper bound T^u. Otherwise, no discomfort occurs.

$$D_T(t) = \left(T^l - T(t)\right) \cdot 1_{T(t)<T^l} + (T(t) - T^u) \cdot 1_{T(t)>T^u}. \tag{3}$$

Lighting user discomfort is modeled in the same manner as thermal discomfort. We also consider a lower bound L^l and an upper illumination threshold L^u. In our model, the current illumination level $L(t)$ is the sum of daylight and artificial light.

$$D_L(t) = \left(L^l - L(t)\right) \cdot 1_{L(t)<L^l} + (L(t) - L^u) \cdot 1_{L(t)>L^u}. \tag{4}$$

Technical wearout is modeled as the number of total interruptions and thus start-ups across all HVAC devices. Let J^{HVAC} denote the set of all HVAC devices.

$$W = \sum_{j \in J^{HVAC}} n^j. \tag{5}$$

Smart residential buildings are optimized with respect to total energy costs, carbon dioxide emissions, thermal discomfort, and technical wearout, whereas smart commercial buildings consider all five objectives, i.e., including lighting-based user discomfort.

4   Scenarios and Simulation Setup

The simulation has been implemented in version 4.5 of the jMetal framework [21] and its code is publicly available on *Sourceforge*[2]. All data that was used in the study is contained within the repository.

---

[2] http://sourceforge.net/projects/jmetalbymarlonso/.

4.1   Scenarios

We consider the seven scenarios in our study that are listed in Table 2. These scenarios reflect different challenges that potentially affect the effectiveness of the optimization effort. All SRBs comprise a washing machine, a tumble dryer, a dishwasher, a micro-CHP, a condensing boiler, and a PV system. Household appliances possess different operation modes as depicted in Table 3. Temporal flexibilities are defined by release times and deadlines. The SCB possesses a micro-CHP, a condensing boiler, and a lighting system.

Power consumption profiles of the appliances have been obtained from measurements in our laboratory environments. Real, existing products, which are listed in Table 4, served as blueprints for the simulated HVAC devices used in this study. Technical data was up-scaled for the commercial building scenario and missing data was amended by our own considerations.

Floor spaces and ceiling heights were set to $130\,\mathrm{m}^2$ and $2\,\mathrm{m}$ (SRB) and $900\,\mathrm{m}^2$ and $3\,\mathrm{m}$ (SCB), respectively, taking German legislation for occupational safety into consideration. Outside temperatures have been extracted from the online weather portal *wetter.com* at an hourly resolution. Measurements were taken in Karlsruhe (in case of the SRB) and Freiburg (CB). Solar radiation data is depicted in Fig. 2.

The German standard load profiles of households H0 (SRB) and small enterprises G1 (CB) provided by the German Association of Energy and Water Industries (BDEW) served as estimates for the load of non-deferrable appliances. Time-variable electricity prices correspond to those employed in the project *iZeus* [22]. We used the same load limitation signal as employed by Allerding et al. [13]. The penalty for exceedance was set as paying twice the current price. A natural gas price of 9.16 Cents per kWh was chosen, which reflect current tariffs in Germany. Feed-in tariffs are based on the German Renewable Energy Act with 12.56 Cents per kWh for PV systems and 8.53 Cents per kWh for micro-CHPs. CO_2 emissions of power obtained from the grid is based on data from the Fraunhofer

Table 2. Overview of the scenarios examined.

Scenarios	Building type	Date	Optimization horizon
RW1 - RW3	Smart residential building	7 January 2015	00:00 – 23:59
RS1 - RS3	Smart residential building	4 June 2015	00:00 – 23:59
CB	Smart commercial building	13 March 2014	07:00 – 19:00

Table 3. Household appliances flexibility.

	RW1/RS1	RW2/RS2	RW3/RS3	Temporal flexibility
Washing machine	deferrable	interruptible	deferrable and load-flexible	08:00 – 17:00
Tumble dryer	deferrable	interruptible	interruptible and load-flexible	18:00 – 21:00
Dishwasher	deferrable	interruptible	interruptible and load-flexible	10:00 – 18:00

Table 4. Technical data of HVAC devices and lighting.

Appliance	Product	Manufacturer
Micro-CHP	ecoPOWER 1.0	*Vaillant*
Condensing boiler	Logamax plus GBH172	*Buderus*
Hot water storage tank	VITOCELL 100-E	*Viessmann*
Lighting	SP482P	*Philips*

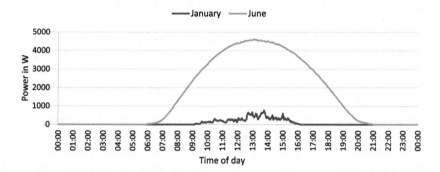

Fig. 2. Solar radiation obtained from measurements in our laboratory environments.

Institute of Solar Energy[3]. Hot water consumption profiles were obtained from Directive 2010/30/EU of the European Commission[4] and data for computing the visual comfort was taken from Reinhart et al. [23, p. 176]. Lower and upper bounds for temperature and lighting were set to 19.5°C and 20.5°C (January), 21.5°C and 22.5°C (June), and 20.5°C and 21.5°C (March), and 750 Lux and 1000 Lux, respectively.

4.2   Algorithms and Settings

We chose four different state-of-the-art evolutionary algorithms for optimizing the presented scenarios. All algorithm configurations were taken from their original publications if not stated otherwise.

NSGA-II's capability of solving problems in energy informatics has been demonstrated in several publications [14,24], making it an ideal candidate for this study. The algorithm uses non-dominated sorting to rank its population of solutions into different tiers of non-dominated fronts for deciding which solutions prevail to the next generation. The selection mechanism is amended by the crowding distance metric if a front cannot be fully accommodated in the next generation. Crowding distance is a density measure that quantifies the volume of the smallest cuboid than can be circumscribed around a solution without enclosing any other members of the current front.

---

[3] https://www.energy-charts.de/power_de.htm.

[4] http://eur-lex.europa.eu/legal-content/EN/TXT/?uri=celex:32013R0812.

The successor to NSGA-II is the NSGA-III algorithm that is specifically tailored to solve problems in three and higher dimensions. It uses reference points instead of crowding distance as diversity preserving mechanism. Objective values are normalized using extreme solutions and the currently best estimate of the utopia point. Reference points are projected on the hyperplane spanned by the translated extreme solutions. Solutions are associated with the reference point to which their Euclidean distance of the line between origin and reference point is minimal. Solutions are selected such that they are equally distributed over all reference points. Equidistant reference points from Deb and Jain [5] were chosen for this study.

SPEA2 has also been successfully applied in solving problems in energy informatics. The algorithm maintains a fixed-sized archive of solutions. The archive is trimmed using environmental selection, whereas fitness values are computed by combining the number of population members that a solution dominates and its distance to the k-nearest neighbor.

The recently developed steady-state ESPEA has exhibited excellent results on many popular artificial benchmark problems [3], which is why it was chosen for this study. It mimics the physical phenomenon of electrostatic potential energy by finding a Pareto front approximation that minimizes the overall sum of all pairwise inverse distances between individual solutions in the objective space. For this purpose, the algorithm maintains a variable-sized archive of non-dominated solutions. After the archive has reached its maximum size, a new solution can only replace an existing archive member if the replacement reduces the sum of pairwise inverse distances.

All algorithms were run using a population size of 100. Each run was terminated after 50,000 function evaluations. Binary crossover and bit flip mutation were applied as genetic operators. A crossover probability of 0.9 and a mutation probability of one by the number of decision variables were chosen. Every algorithm was run 100 times on each respective scenario.

5   Results and Discussion

Our analysis is divided into two parts. We first assess how the selected algorithms perform in finding a representative approximation of the Pareto front of the different problem scenarios. The second part concerns itself with how well the multi-objective approach is suited to achieve individual objectives by comparing extreme solutions. The temporal resolution was set to 60 s in the first and one second in the second part of the analysis.

5.1   Median Performance Evaluation

We use the *hypervolume* [25] and the *inverted generational distance* (IGD) [26] to evaluate performances, since the two indicators capture both diversity and convergence. Both indicators were computed for every final population of each algorithm and run. Hypervolume measures the space that is covered by the

Table 5. Hypervolume. Medians and inter-quartile ranges.

	ESPEA	NSGA-II	NSGA-III	SPEA2
RW1	$2.72e-01_{5.5e-02}$	$2.42e-01_{5.3e-02}$	$1.67e-01_{5.8e-02}$	$2.37e-01_{5.8e-02}$
RW2	$2.69e-01_{4.3e-02}$	$2.42e-01_{3.4e-02}$	$1.77e-01_{5.0e-02}$	$2.52e-01_{4.0e-02}$
RW3	$2.68e-01_{3.8e-02}$	$2.37e-01_{2.8e-02}$	$1.78e-01_{5.6e-02}$	$2.51e-01_{4.3e-02}$
RS1	$3.97e-01_{4.5e-02}$	$3.98e-01_{2.5e-02}$	$2.91e-01_{1.2e-01}$	$3.90e-01_{2.6e-02}$
RS2	$4.61e-01_{3.6e-02}$	$4.59e-01_{2.0e-02}$	$3.53e-01_{1.5e-01}$	$4.48e-01_{4.1e-02}$
RS3	$3.77e-01_{3.7e-02}$	$3.77e-01_{2.7e-02}$	$2.79e-01_{9.6e-02}$	$3.68e-01_{3.3e-02}$
CB	$2.45e-01_{1.8e-02}$	$2.30e-01_{1.5e-02}$	$4.39e-02_{3.8e-02}$	$2.04e-01_{3.1e-02}$

Pareto front approximation with respect to a given reference point, so a larger value is considered to be superior. The IGD indicator is computed by discretizing the Pareto front in a finite set of points and calculating the minimum distance of each said point to the Pareto front approximation. The average of all distances yields the IGD value, which makes smaller values preferable. Since the true Pareto fronts of all problem instances are unknown, reference fronts were created by combining the non-dominated solutions from all individual runs. Objective values were normalized before indicator values were calculated.

An *Anderson-Darling* test rejected the hypothesis of normal distribution for both indicator values across all scenarios, which is why we report medians and inter-quartile ranges. Table 5 shows medians and inter-quartile ranges (as subscripts) of the hypervolume indicator across all 100 runs. Best and second-best median performances are colored in dark and light gray, respectively. We observe that ESPEA and NSGA-II exhibit the overall largest hypervolumes. SPEA2 is only able to achieve two second-best performances and NSGA-III obtains no best or second-best performance. Performance differences were checked for significance using a *Kruskal-Wallis* test in combination with a post-hoc analysis. Significant differences between NSGA-II and ESPEA could be confirmed for scenarios RW2, RW3 and OB. Since the differences for scenario RS1 and RS3, in which NSGA-II outperforms ESPEA, were not found to be significant, we can draw the conclusion that ESPEA achieves the overall best performance with respect to the hypervolume indicator.

A similar pattern can be observed for the IGD indicator, whose results are displayed in Table 6. ESPEA and NSGA-II again nearly achieve all best and second-best performances. ESPEA even outperforms NSGA-II on all scenarios with the exception of the commercial building. Performance differences between both algorithms were found to be significant with the exception of the scenarios of the residential buildings in summer.

The values in Tables 5 and 6 support the conclusion that all algorithms are capable of finding good approximations of the Pareto fronts. This enables users to choose a schedule among multiple balanced options that serves their interests best. However, at the same time, we observe that significant performance differences exist among the algorithms analyzed in this study. Table 7 provides a summary of the significant performance differences that are observed in the data. The numbers support our previous analysis that ESPEA is the top performing algorithm closely followed by NSGA-II.

We believe that the performance differences observed can be explained by the algorithms' selection mechanisms. The smart building optimization problem as presented in this study is discrete. Reference point based methods such as NSGA-III, although tailored for high dimensional problems, may struggle to find a diverse approximation of the Pareto front. Points of the Pareto front might be unequally distributed, which could make an association with reference points that span the entire front an unfavorable density estimate. The objective space being discrete might also impair SPEA2's environmental selection mechanism as k-nearest distances are largely affected by the density of points along each objective. Crowding distance based selection mechanisms are known to perform poorly in higher dimensions [27]. NSGA-II's good results can be explained, however, by very small differences in objectives being less probable to occur. Marginal differences distort the crowding distance metric, as absolute differences to neighboring solutions are multiplied along each dimension. ESPEA's selection mechanism, on the other hand, is not negatively affected by discrete objective spaces, which is reflected by the results.

5.2   Comparison of Extreme Values

Increasing the temporal resolution to one second gives us a clear picture of the greatest improvements that can be achieved using optimization techniques. Tables 8 and 9 list the minimum values that were achieved for each objective and algorithm. Best and second-best performances are again colored in dark and light gray. Colorings are omitted if best and second-best performances are tied. Technical wearout is not deeper considered in this analysis, since optimal values are easily achieved and resulting objective vectors are highly dissimilar.

ESPEA and NSGA-II are again the best performing algorithms. At the same time however, the results of NSGA-III and SPEA2 greatly improve. This could be attributed to the higher temporal resolution, which is expected to increase the density of points in the objective space. We can also observe a clear difference

Table 6. IGD. Medians and inter-quartile ranges.

	ESPEA	NSGA-II	NSGA-III	SPEA2
RW1	$2.31e-03_{2.5e-03}$	$4.24e-03_{1.3e-03}$	$6.18e-03_{1.8e-03}$	$4.74e-03_{8.1e-04}$
RW2	$2.34e-03_{2.0e-03}$	$4.07e-03_{7.9e-04}$	$6.10e-03_{2.4e-03}$	$4.16e-03_{8.2e-04}$
RW3	$2.33e-03_{2.1e-03}$	$4.20e-03_{5.6e-04}$	$6.13e-03_{2.3e-03}$	$4.34e-03_{9.0e-04}$
RS1	$3.38e-03_{1.1e-03}$	$3.51e-03_{6.8e-04}$	$5.93e-03_{2.8e-03}$	$4.34e-03_{8.4e-04}$
RS2	$3.40e-03_{8.3e-04}$	$3.44e-03_{5.2e-04}$	$5.53e-03_{2.7e-03}$	$4.06e-03_{6.8e-04}$
RS3	$3.44e-03_{8.6e-04}$	$3.72e-03_{5.0e-04}$	$6.43e-03_{3.4e-03}$	$4.40e-03_{9.3e-04}$
CB	$2.03e-03_{2.5e-04}$	$1.81e-03_{1.1e-04}$	$4.57e-03_{9.9e-04}$	$1.96e-03_{1.6e-04}$

Table 7. Summary statistics for the hypervolume and IGD indicators. The number in each column states how many times the corresponding algorithm outperforms other algorithms with confidence across all scenarios and indicators at a five percent level.

	ESPEA	NSGA-II	NSGA-III	SPEA2
Performance	28	24	0	14

Table 8. Extreme solutions of RW and RS scenarios. Each cell entry is composed of the tuple costs, emissions, thermal discomfort, and wearout.

		Costs	CO_2 emissions	Thermal discomfort
RW1	ESPEA	(**1803**, 44165, 1.94, 3)	(1833, **44126**, 1.94, 3)	(2039, 49935, **0.0**, 3)
	NSGA-II	(**1818**, 44548, 1.85, 4)	(1838, **44418**, 1.91, 4)	(2044, 49982, **0.0**, 4)
	NSGA-III	(**1835**, 44560, 1.88, 5)	(1837, **44535**, 1.88, 5)	(2060, 50319, **0.0**, 4)
	SPEA2	(**1824**, 44757, 1.86, 4)	(1845, **44651**, 1.91, 4)	(2022, 50061, **0.0**, 4)
RW2	ESPEA	(**1800**, 44062, 1.94, 4)	(1825, **44045**, 1.94, 4)	(2091, 52415, **0.0**, 3)
	NSGA-II	(**1805**, 44231, 1.92, 4)	(1819, **44207**, 1.93, 4)	(2020, 49979, **0.0**, 4)
	NSGA-III	(**1818**, 44529, 1.91, 5)	(1824, **44525**, 1.91, 5)	(2030, 50167, **0.0**, 5)
	SPEA2	(**1815**, 44494, 1.87, 4)	(1833, **44429**, 1.89, 5)	(2024, 50095, **0.0**, 4)
RW3	ESPEA	(1809, 44392, 1.90, 3)	(1822, **44378**, 1.90, 3)	(2015, 49912, **0.0**, 3)
	NSGA-II	(**1797**, 44103, 1.93, 4)	(1818, **44086**, 1.93, 4)	(2012, 49841, **0.0**, 3)
	NSGA-III	(**1817**, 44624, 1.86, 4)	(1817, 44624, 1.86, 4)	(2046, 50645, **0.0**, 4)
	SPEA2	(**1801**, 44174, 1.93, 4)	(1835, **44032**, 1.94, 4)	(2023, 50095, **0.0**, 4)
RS1	ESPEA	(**-119**, 6224, 3.38, 3)	(-119, **6224**, 3.38, 3)	(374, 14650, **0.05**, 6)
	NSGA-II	(**-137**, 5713, 3.52, 3)	(-137, **5713**, 3.52, 3)	(363, 14106, **0.06**, 6)
	NSGA-III	(**-104**, 6551, 3.49, 4)	(-104, **6551**, 3.49, 4)	(310, 13935, **0.38**, 7)
	SPEA2	(**-119**, 6195, 3.57, 3)	(-119, **6195**, 3.57, 3)	(364, 14182, **0.06**, 5)
RS2	ESPEA	(**-183**, 4479, 4.03, 4)	(-183, **4479**, 4.03, 4)	(330, 13558, **0.08**, 6)
	NSGA-II	(**-122**, 6125, 3.50, 3)	(-122, **6125**, 3.50, 3)	(412, 15282, **0.10**, 8)
	NSGA-III	(**-150**, 5363, 3.82, 4)	(-150, **5363**, 3.82, 4)	(334, 13733, **0.09**, 6)
	SPEA2	(**-113**, 6354, 3.54, 3)	(-113, **6354**, 3.54, 3)	(356, 14148, **0.11**, 5)
RS3	ESPEA	(**-110**, 6445, 3.30, 3)	(-110, **6445**, 3.30, 3)	(318, 13706, **0.11**, 7)
	NSGA-II	(**-114**, 6373, 3.27, 3)	(-114, **6372**, 3.41, 3)	(333, 13868, **0.05**, 6)
	NSGA-III	(**-122**, 6095, 3.58, 3)	(-122, **6095**, 3.58, 3)	(306, 13571, **0.24**, 6)
	SPEA2	(**-118**, 6262, 3.47, 3)	(-118, **6262**, 3.47, 3)	(298, 13273, **0.21**, 6)

between the summer and winter scenarios. The PV system makes the residential household a net power supplier, as it provides more power than the household consumes. This circumstance leads the minimum cost and minimum emission solutions to coincide in summer, which demonstrates that private PV systems may reconcile those two conflicting objectives.

The comparison of summer and winter scenarios reveals that the additional degree of freedom provided by interruptible and load flexible appliances leads

Table 9. Extreme solutions of scenario CB. Each cell entry is composed of the tuple costs, emissions, thermal discomfort, visual discomfort, and wearout.

	Costs	CO_2 emissions
ESPEA	(**4800**, 127197, 3.45, 99.8, 3)	(5235, **99613**, 3.36, 99.0, 6)
NSGA-II	(**4848**, 128521, 3.40, 98.5, 2)	(5213, **112008**, 3.32, 98.5, 5)
NSGA-III	(**5288**, 140215, 3.14, 35.7, 4)	(5364, **138639**, 3.14, 20.0, 4)
SPEA2	(**4880**, 128856, 3.41, 96.1, 4)	(5371, **116884**, 3.30, 97.1, 8)

	Thermal discomfort	Visual discomfort
ESPEA	(7095, 189775, **0.3**, 99.0, 3)	(6338, 160802, 1.78, **0.00**, 2)
NSGA-II	(7546, 192995, **0.3**, 0.00, 3)	(7508, 200388, 0.31, **0.00**, 3)
NSGA-III	(7475, 199820, **0.3**, 6.60, 4)	(7475, 199820, 0.30, **6.60**, 4)
SPEA2	(7389, 196964, **0.3**, 28.8, 4)	(5473, 144760, 2.79, **27.5**, 4)

in most cases to an improvement in objective values. Consumption flexibility is therefore suitable to reduce costs and emissions, while raising comfort at the same time. Such an improvement, however, cannot be confirmed for all scenarios. We believe that deteriorating objective values that have been observed in some cases are caused by the additional complexity introduced by increasing the search space. We believe that focusing the search from the beginning on identifying tradeoff solutions that balance objective values, for example by applying methods from [28], might remedy this situation.

6   Conclusion and Outlook

We have presented a new approach of formulating a smart building multi-objective optimization problem. Our formulation introduces device encodings that allow for the generic modeling of interruptible and load-flexible appliances and make constraint satisfaction easier to achieve. Our computational study has revealed that current multi-objective evolutionary algorithms are able to compute representative approximations of the Pareto front and approximate extreme solutions alike. Among the four tested state-of-the-art algorithms, ESPEA and NSGA-II delivered the overall best results.

Future research may focus on developing specialized genetic operators and repair mechanisms tailored to our device encodings for improving MO-EA performances. Additionally, further focus can be placed on identifying tradeoff solutions that balance objective values.

An implementation of a multi-objective evolutionary algorithm based on the results in this paper will be transferred to the building energy management system in our laboratory environments and tested with real test users. This will help verifying the results obtained in the present analysis.

References

1. Gottwalt, S., Ketter, W., Block, C., Collins, J., Weinhardt, C.: Demand side management - a simulation of household behavior under variable prices. Energy Policy **39**(12), 8163–8174 (2011)
2. Palensky, P., Dietrich, D.: Demand side management: demand response, intelligent energy systems, and smart loads. IEEE Trans. Industr. Inf. **7**(3), 381–388 (2011)
3. Braun, M.A., Shukla, P.K., Schmeck, H.: Obtaining optimal pareto front approximations using scalarized preference information. In: Proceedings of the 2015 Annual Conference on Genetic and Evolutionary Computation, GECCO 2015, pp. 631–638. ACM, New York (2015)
4. Deb, K., Pratap, A., Agarwal, S., Meyarivan, T.: A fast and elitist multiobjective genetic algorithm: NSGA-II. IEEE Trans. Evol. Comput. **6**(2), 182–197 (2002)
5. Deb, K., Jain, H.: An evolutionary many-objective optimization algorithm using reference-point-based nondominated sorting approach, part I: solving problems with box constraints. IEEE Trans. Evol. Comput. **18**(4), 577–601 (2014)
6. Zitzler, E., Laumanns, M., Thiele, L.: SPEA2: improving the strength Pareto evolutionary algorithm. Technical report 103, Computer Engineering and Networks Laboratory (TIK), Swiss Federal Institute of Technology (ETH), Zurich, Switzerland (2001)
7. Kok, J.K., Warmer, C.J., Kamphuis, I.G.: PowerMatcher: multiagent control in the electricity infrastructure. In: Proceedings of the Fourth International Joint Conference on Autonomous Agents and Multiagent Systems, AAMAS 2005, pp. 75–82. ACM, New York (2005)
8. Ha, D.L., Joumaa, H., Ploix, S., Jacomino, M.: An optimal approach for electrical management problem in dwellings. Energy Build. **45**, 1–14 (2012)
9. Nestle, D., Bendel, C., Ringelstein, J.: Bidirectional energy management interface (BEMI)-integration of the low voltage level into grid communication and control. In: 19th International Conference on Electricity Distribution, pp. 21–24, Vienna (2007)
10. Soares, A., Gomes, Á., Antunes, C.H.: Integrated management of residential energy resources. In: EPJ Web of Conferences, vol. 33 (2012)
11. Mauser, I., Müller, J., Allerding, F., Schmeck, H.: Adaptive building energy management with multiple commodities and flexible evolutionary optimization. Renewable Energy **87**(Part 2), 911–921 (2016)
12. Ahmadi, P., Rosen, M.A., Dincer, I.: Multi-objective exergy-based optimization of a polygeneration energy system using an evolutionary algorithm. Energy **46**(1), 21–31 (2012)
13. Allerding, F., Premm, M., Shukla, P.K., Schmeck, H.: Electrical load management in smart homes using evolutionary algorithms. In: Hao, J.-K., Middendorf, M. (eds.) EvoCOP 2012. LNCS, vol. 7245, pp. 99–110. Springer, Heidelberg (2012)
14. Soares, A., Antunes, C.H., Oliveira, C., Gomes, Á.: A multi-objective genetic approach to domestic load scheduling in an energy management system. Energy **77**, 144–152 (2014)
15. Zhao, Z., Lee, W.C., Shin, Y., Song, K.B.: An optimal power scheduling method for demand response in home energy management system. IEEE Trans. Smart Grid **4**(3), 1391–1400 (2013)
16. Anvari-Moghaddam, A., Seifi, A., Niknam, T., Pahlavani, M.R.A.: Multi-objective operation management of a renewable MG (micro-grid) with back-up microturbine/fuel cell/battery hybrid power source. Energy **36**(11), 6490–6507 (2011)

17. Pedrasa, M., Spooner, T., MacGill, I.: Coordinated scheduling of residential distributed energy resources to optimize smart home energy services. IEEE Trans. Smart Grid **1**(2), 134–143 (2010)
18. De Oliveira, G., Jacomino, M., Ploix, S.: Optimal power control for smart homes. In: IFAC World Congréss, Milan, Italy, pp. 9579–9586 (2011)
19. Salinas, S., Li, M., Li, P.: Multi-objective optimal energy consumption scheduling in smart grids. IEEE Trans. Smart Grid **4**(1), 341–348 (2013)
20. Mauser, I., Dorscheid, M., Allerding, F., Schmeck, H.: Encodings for Evolutionary Algorithms in smart buildings with energy management systems. In: IEEE Congress on Evolutionary Computation (CEC), pp. 2361–2366. IEEE (2014)
21. Durillo, J.J., Nebro, A.J.: The jMetal framework for multi-objective optimization: design and architecture. Adv. Eng. Softw. **42**(10), 760–771 (2011)
22. Dallinger, D.: The contribution of vehicle-to-grid to balance fluctuating generation: comparing different battery ageing approaches. Technical report, Working Paper Sustainability and Innovation (2013)
23. Reinhart, C.F., Herkel, S.: The simulation of annual daylight illuminance distributions a state-of-the-art comparison of six RADIANCE-based methods. Energy Build. **32**(2), 167–187 (2000)
24. Graditi, G., Di Silvestre, M., Gallea, R., Sanseverino, E.R.: Heuristic-based shiftable loads optimal management in smart micro-grids. IEEE Trans. Industr. Inf. **11**(1), 271–280 (2015)
25. Zitzler, E., Thiele, L.: Multiobjective evolutionary algorithms: a comparative case study and the strength Pareto approach. IEEE Trans. Evol. Comput. **3**(4), 257–271 (1999)
26. Van Veldhuizen, D.A., Lamont, G.B.: Evolutionary computation and convergence to a Pareto front. In: Late Breaking Papers at the Genetic Programming 1998 Conference, pp. 221–228. Citeseer (1998)
27. Kukkonen, S., Deb, K.: Improved pruning of non-dominated solutions based on crowding distance for bi-objective optimization problems. In: Proceedings of the World Congress on Computational Intelligence, Vancouver, Canada, pp. 1179–1186. IEEE Press (2006)
28. Shukla, P.K., Braun, M.A., Schmeck, H.: Theory and algorithms for finding knees. In: Purshouse, R.C., Fleming, P.J., Fonseca, C.M., Greco, S., Shaw, J. (eds.) EMO 2013. LNCS, vol. 7811, pp. 156–170. Springer, Heidelberg (2013)

A Hybrid Genetic Algorithm for the Interaction of Electricity Retailers with Demand Response

Maria João Alves[1($\boxtimes$)], Carlos Henggeler Antunes[2], and Pedro Carrasqueira[3]

[1] Faculty of Economics, University of Coimbra/INESC Coimbra, Coimbra, Portugal
mjalves@fe.uc.pt
[2] Department of Electrical Engineering and Computers,
University of Coimbra/INESC Coimbra, Coimbra, Portugal
ch@deec.uc.pt
[3] INESC Coimbra, Coimbra, Portugal
pmcarrasqueira@net.sapo.pt

Abstract. In this paper a bilevel programming model is proposed for modeling the interaction between electricity retailers and consumers endowed with energy management systems capable of providing demand response to variable prices. The model intends to determine the optimal pricing scheme to be established by the retailer (upper level decision maker) and the optimal load schedule adopted by the consumer (lower level decision maker) under this price setting. The lower level optimization problem is formulated as a mixed-integer linear programming (MILP) problem. A hybrid approach consisting of a genetic algorithm and an exact MILP solver is proposed. The individuals of the population represent the retailer's choices (electricity prices). For each price setting, the exact optimal solution to the consumer's problem is obtained in a very efficient way using the MILP solver. An illustrative case is analyzed and discussed.

Keywords: Genetic algorithm · Bilevel problem · Mixed-integer linear programming · Demand response · Electricity retail market

1 Introduction

The retail electricity market has been mostly working as a one-way communication scheme. The retailer buys the energy in the wholesale market at variable prices, which depend on the purchase time and the demand profile. As the consumers are, in general, charged at a flat rate, they are indifferent to price oscillations and lack the stimulus to engage in distinct consumption patterns according to their flexibility of use of appliances. If consumers could see prices changing along the day, i.e. they were offered dynamic tariffs within some contracted bounds, they would expectedly adopt actions, namely by means of automated energy management systems, to schedule their loads to minimize the electricity bill without jeopardizing the quality of the energy services provided by the appliances, namely comfort requirements. Profiting from the flexibility consumers have in scheduling load operation would be of utmost importance to

© Springer International Publishing Switzerland 2016
G. Squillero and P. Burelli (Eds.): EvoApplications 2016, Part I, LNCS 9597, pp. 459–474, 2016.
DOI: 10.1007/978-3-319-31204-0_30

several players in the electricity industry chain. Consumers could see their electricity bill decreasing and retailers could make more judicious decisions regarding buying and selling electricity. Also network companies, both at distribution and transmission levels, would benefit because they could use demand-side resources to mitigate congestion and make a better management of the availability of distributed generation based on renewable sources.

In this setting, the retailer could reflect the acquisition conditions onto the consumers, e.g. by determining variable electricity prices, which is not feasible in the traditional one-side electricity market. Retailers and consumers have conflicting goals. Retailers want to maximize profits by selling electricity subject to the regulation framework, and consumers want to minimize costs subject to requirements of quality of the energy services associated with the operation of loads in appropriate time slots. In addition to conflicting goals, there is a hierarchical relation between retailers and consumers as the former determine prices and the latter react by scheduling their loads accordingly. This is a bilevel optimization problem.

Communication capabilities associated with the evolution of the electricity system to smart grids lay the foundations for bi-directional interaction between retailers and consumers. This enables that time-varying price information is sent by the retailer, which in turn receives the response of the consumer by adjusting the operation schedule of the loads with the aim to minimize the electricity bill.

Several models have been proposed in literature concerning demand side management (DSM) in the residential setting [1]. DSM has many beneficial effects, enabling a better usage of available generation capacity and network infrastructures, contributing to avoid or postpone new investments, decreasing peak load demand, reducing the carbon emission levels and improving the overall grid sustainability. DSM programs have re-emerged in the smart grid context allowing end-users reshaping their energy consumption pattern and taking advantage of dynamic tariffs.

Several approaches have been proposed using time-varying pricing strategies to decrease peak load. In the day-ahead hourly pricing strategy, the consumers receive the next 24 h prices a day or some hours before. Consumers should then react accordingly by scheduling their appliances to get a satisfactory trade-off between minimizing the electricity bill and maximizing or imposing constraints on their welfare regarding comfort requirements.

In this paper a bilevel programming model is proposed for modeling the interaction between the electricity retailer and consumers. Bilevel models for this purpose have been studied in the literature. Zugno et al. [2] considered a theoretical game to establish a Stackelberg relationship between retailers and consumers in a dynamic pricing framework, with a stochastic component. In Bu et al. [3] consumers aim at maximizing utility, which derives from the amount of energy consumed, the price and an individual factor associated with each consumer. The retailer aims at maximizing the profit considering that it can buy energy from two suppliers with distinct prices and degree of certainty. Yang et al. [4] incorporated DSM through an interaction game between retailers and consumers according to their utility functions. The consumer's objective function derives from energy cost and the utility of energy consumption, which depends on the difference between the actual consumption and a target. Zhang et al. [5] presented a bilevel model with multiple objective functions in the upper level, which aim at maximizing the profit of supply companies, and a single objective in the lower level, which aims at minimizing the consumer's electricity bill. The consumer can choose the supply company.

Hybrid approaches have been proposed by Meng and Zeng [6,7] to solve bilevel problems with one leader (retailer), who wants to maximize profit, and multiple consumers, who want to minimize their bills. Meng and Zeng [6] considers interruptible, non-interruptible and curtailable appliances, which lead to three lower level separate sub-optimization problems. In addition to consumers using demand optimization, [7] also considers costumers whose energy consumption patterns are not known to the retailer. Therefore, these patterns should be learned by the retailer with the purpose of retail price determination. Both approaches [6,7] use genetic algorithms to solve the profit maximization problem at the retailer's side and an LP solver to derive optimal solutions at the consumers' side.

The bilevel model proposed in this paper intends to determine the optimal pricing scheme to be established by the retailer and the optimal load schedule adopted by the consumer under this price setting. Consumers are able to deviate consumption of shiftable loads, i.e. cyclic loads as dishwashers, laundry etc., to lower price periods subject to time slot constraints for load operation, which can decrease the retailer's profits. The structure of the paper is as follows. In Sect. 2 the main concepts of bilevel models are outlined. In Sect. 3 new bilevel formulations for modeling the interaction between the retailer and consumer optimization problems are presented. In Sect. 4 an algorithmic approach combining a genetic algorithm (GA) with a mixed-integer linear programming (MILP) exact solver is described. Numerical results and the ensuing discussion are presented in Sect. 5. In Sect. 6 the main conclusions are drawn.

2   Bilevel Programming

In bilevel optimization problems the upper level decision maker (*leader*) controls decision variables x, while the lower level decision maker (*follower*) controls decision variables y. The two decision makers have their own objective functions, which are subject to interdependent constraints. The decision process is sequential as the leader makes his decisions first by setting the values of x. Then, the follower reacts by choosing the y values that optimize his objective function on the feasible solutions restricted by the fixed x. The goal of the leader is to optimize his objective function, but he must incorporate into the optimization process the reaction of the follower because it affects the leader's objective value.

The general bilevel programming problem can be stated as follows (BP):

$$\min_{x \in X} F(x, y)$$
$$s.t. \ \ G(x, y) \leq 0$$
$$y \in \arg\min_{y \in Y} \{f(x, y) : g(x, y) \leq 0\}$$

where $X \subset \mathbb{R}^{n_1}$ (n_1 being the number of upper level variables) and $Y \subset \mathbb{R}^{n_2}$ (n_2 being the number of lower level variables) are closed sets. $F(x,y)$ and $f(x,y)$ are the leader's and the follower's objective functions, respectively. Since the follower optimizes $f(x,y)$ after x has been selected, x is a constant vector whenever $f(x,y)$ is optimized. For fixed $x \in X$, the set $Y(x) = \{y \in Y : g(x,y) \leq 0\}$ is the *feasible set* of the follower. The set $\Psi(x) = \left\{ y \in Y : y \in \underset{y' \in Y(x)}{\arg\min} f(x,y') \right\}$ is called the follower's *rational reaction set* to a given x. The feasible set of (BP), also called the *induced region*, is $IR = \{(x,y) : x \in X, G(x,y) \leq 0, y \in \Psi(x)\}$. It is difficult to find global optimal solutions to bilevel optimization problems due to their inherent non-convexity. Even the linear bilevel problem is NP-hard [8].

3   Bilevel Formulations for the Interaction Between the Retailer's and Consumer's Optimization Problems

We consider a bilevel problem to model the interaction between the retailer and consumers. The retailer buys energy in the wholesale market and sells it to consumers. The retailer wants to maximize profit while consumers aim at minimizing the cost of their energy consumption. In this model a partially flexible consumer is considered, who can decide on the allocation of some shiftable loads based on a price schedule communicated by the retailer. Shiftable loads are typically cyclic loads, such as dishwashers or laundry machines, whose operation cycle can be shifted in time but not interrupted. The model considers a cluster (aggregation) of consumers with similar consumption and demand response profiles, thereafter referred to as the consumer.

The problem has a bilevel structure, where the retailer (*leader*) determines the prices x_i to be charged to the consumer (*follower*) in each predefined sub-period P_i ($i = 1, \cdots, I$) of the planning period T. Thus, the number of upper level variables is I, i.e., the number of sub-periods P_i. As proposed by Zugno et al. [2], in order to enforce market competitiveness of retailer prices, we introduce constraints on x_i imposing minimum $(\underline{x}_i)$ and maximum $(\overline{x}_i)$ values in each sub-period P_i and an average price (x^{AVG}) value during T.

Knowing the electricity prices, the consumer determines the time (z_j) each flexible load $j \in \{1, \ldots, J\}$ must start to minimize the cost of electricity and ensuring that the operation cycle of load j is within a specified *comfort time slot* $T_j = [T1_j, T2_j] \subseteq T$.

Data:

T = number of intervals (minutes, quarter-hour, half-hour or other period of time) of the planning period ($t = 1, \cdots, T$). Let $T = \{1, \cdots, T\}$.

J = number of shiftable loads to be managed by the consumer ($j = 1, \cdots, J$).

I = number of sub-periods of time $P_i \subset T$ in which different prices of electricity (time-of-use tariffs) are charged by the retailer to the consumer ($i = 1, \cdots, I$).

$P1_i$, $P2_i$: points in time that delimit each sub-period P_i, $i = 1, \cdots, I$, such that $P_i = [P1_i, P2_i]$ and $\bigcup_{i=1}^{I} P_i = T$. Let $\bar{P}_i$ denote the amplitude of P_i, i.e. $\bar{P}_i = P2_i - P1_i + 1$.

$\underline{x}_i$ = minimum price charged to the consumer in sub-period P_i.

$\bar{x}_i$ = maximum price charged to the consumer in sub-period P_i.

x^{AVG} = average price charged to the consumer in T.

π_t = energy price seen by the retailer in the spot market at time $t \in$ T (€/KWh $\times (m/60)$ where m is the number of minutes in one unit of time t).

C_t = contracted power by the consumer at time t of the planning period (KW).

b_t = non-controllable base load at time t of the planning period (KW), i.e. amount of load that cannot be scheduled by the consumer's energy management system.

d_j = duration of the operation cycle of shiftable load j.

$f_j(r)$ = power requested by load j at time r of its operation cycle ($r = 1, \cdots, d_j$) (KW).

$Tj = [T1_j, T2_j] \subseteq$ T: time slot in which load j is allowed to operate.

Upper level decision variables:

x_i = price charged by the retailer to the consumer during sub-period P_i (€/KWh $\times (m/60)$ where m has the same meaning as above), $i = 1, \cdots, I$.

Lower level decision variables:

z_j = starting time of the operation cycle of load j, $j = 1, \cdots, J$.

Auxiliary lower level variables:

u_{jt} = binary variable representing whether the operation cycle of load j is "on" or "off" at time t of the planning period, $j = 1, \cdots, J, t = 1, \cdots, T$.

p_{jt} = power requested to the grid by load j at time t of the planning period (KW), $j = 1, \cdots, J, t = 1, \cdots, T$.

Bilevel Model 1.

$$max\ F = \sum_{i=1}^{I} \sum_{t \in P_i} x_i(b_t + \sum_{j=1}^{J} p_{jt}) - \sum_{t=1}^{T} \pi_t(b_t + \sum_{j=1}^{J} p_{jt}) \tag{1}$$

s.t.

$$x_i \leq \bar{x}_i, i = 1, \cdots, I \tag{2}$$

$$x_i \geq \underline{x}_i, i = 1, \cdots, I \tag{3}$$

$$\frac{1}{T} \sum_{i=1}^{I} \bar{P}_i x_i = x^{AVG} \tag{4}$$

$$min\ f = \sum_{i=1}^{I} \sum_{t \in P_i} x_i(b_t + \sum_{j=1}^{J} p_{jt}) \tag{5}$$

s.t.

$$u_{jt} = \begin{cases} 1 \text{ if } z_j \leq t \leq z_j + d_j \\ 0 \quad \text{otherwise} \end{cases}, \quad j = 1, \cdots, J; t = 1, \cdots, T \tag{6}$$

$$p_{jt} = f_j(t - z_j + 1)u_{jt}, \quad j = 1, \cdots, J; t = 1, \cdots, T \tag{7}$$

$$\sum_{j=1}^{J} p_{jt} + b_t \leq C_t, \qquad\qquad t = 1, \cdots, T \qquad\qquad (8)$$

$$T1_j \leq z_j \leq T2_j - d_j + 1 \qquad\qquad j = 1, \cdots, J \qquad\qquad (9)$$

The objective function at the upper level (1) is the maximization of the retailer's profit (revenue from selling energy to consumer minus cost of purchasing the energy in the spot market). Constraints (2) to (4) define the limits for the energy prices charged to the consumer in each sub-period P_i and set an average price in T.

The formulation of the lower level problem in Model 1 is based on the DSM model proposed in [1]. The objective function (5) consists in the minimization of the consumer's total cost; (6) sets the value of the auxiliary binary variables u_{jt} as function of the variables z_j and time t; variables u_{jt} are, in turn, used in equations (7), which set the value of the power requested to the grid by each load j at each time t according to the load operation cycle; constraints (8) impose that the contracted power is not exceeded at any time and constraints (9) impose time limits for the operation of each load according to the time slots defined by the consumer.

Model 1 can be written in an equivalent manner by reformulating the lower level problem as a MILP problem. Thus, bilevel Model 2 is presented below, which is a mathematical programming model equivalent to bilevel Model 1. It does not consider variables z_j, $j = 1, \cdots, J$ (which specify the starting time t of the operation cycle of load j), but rather binary variables w_{jrt} that indicate whether the load j is "on" or "off" at time t of the planning period and it is at time r of its operation cycle. Explicit variables u_{jt} are no longer necessary because they can be expressed in terms of w_{jrt}: $u_{jt} = \sum_{r=1}^{d_j} w_{jrt}$. The upper level variables (x_i) are the same as in Model 1.

Data:

The data are the same as in Model 1 with a single difference: the functions $f_j(r)$ are replaced by series of discrete values, consisting of one f_{jr} value for each combination j, r. Thus:

f_{jr} = power requested by load j at time r of its operation cycle ($r = 1, \cdots, d_j$) (KW).

To avoid ambiguity between points in time of the operation cycle and points in time of the planning period, we refer to the "time r" of the operation cycle as "stage r".

Lower level decision variables:

w_{jrt} = binary variable representing whether load j is "on" or "off" at time t of the planning period and at stage r of its operation cycle.

In order not to unnecessarily increase the number of w_{jrt} variables, they are defined only for t in the time slot allowed for the operation for each load. Therefore, w_{jrt} are defined for $j = 1, \cdots, J$, $r = 1, \cdots, d_j$, $t = T1_j, \cdots, T2_j$.

Auxiliary lower level variables:

p_{jt} = power requested to the grid by load j at time t of the planning period (KW), $j = 1, \cdots, J$, $t = 1, \cdots, T$.

Bilevel Model 2.

$$max\ F = \sum_{i=1}^{I}\sum_{t \in P_i} x_i(b_t + \sum_{j=1}^{J} p_{jt}) - \sum_{t=1}^{T}\pi_t(b_t + \sum_{j=1}^{J} p_{jt}) \tag{10}$$

s.t.

$$x_i \leq \bar{x}_i\ , i = 1, \cdots, I \tag{11}$$

$$x_i \geq \underline{x}_i\ , i = 1, \cdots, I \tag{12}$$

$$\frac{1}{T}\sum_{i=1}^{I}\bar{P}_i x_i = x^{AVG} \tag{13}$$

$$min\ f = \sum_{i=1}^{I}\sum_{t \in P_i} x_i(b_t + \sum_{j=1}^{J} p_{jt}) \tag{14}$$

s.t.

$$p_{jt} = \sum_{r=1}^{d_j} f_{jr} w_{jrt}, \qquad j = 1, \cdots, J; t = T1_j, \cdots, T2_j \tag{15}$$

$$p_{jt} = 0, \qquad j = 1, \cdots, J; t < T1_j \vee t > T2_j \tag{16}$$

$$\sum_{j=1}^{J} p_{jt} + b_t \leq C_t, \qquad t = 1, \cdots, T \tag{17}$$

$$\sum_{r=1}^{d_j} w_{jrt} \leq 1, \qquad j = 1, \cdots, J; t = T1_j, \cdots, T2_j \tag{18}$$

$$w_{j(r+1)(t+1)} \geq w_{jrt}, \quad j = 1, \cdots, J; r = 1, \cdots, d_j - 1; t = T1_j, \cdots, T2_j - 1 \tag{19}$$

$$\sum_{t=T1_j}^{T2_j} w_{jrt} = 1, \qquad j = 1, \cdots, J; r = 1, \cdots, d_j \tag{20}$$

$$\sum_{t=T1_j}^{T2_j - d_j + 1} w_{j1t} \geq 1, \qquad j = 1, \cdots, J \tag{21}$$

$$w_{jrt} \in \{0, 1\}, \quad j = 1, \cdots, J; r = 1, \cdots, d_j; t = T1_j, \cdots, T2_j$$
$$p_{jt} \geq 0 \qquad j = 1, \cdots, J; t = 1, \cdots, T \tag{22}$$

The upper-level problem (10)–(13) and the lower-level objective function (14) are the same as in Model 1.

Constraints (15)–(16) correspond to (7) in Model 1 and aim at setting the auxiliary variables p_{jt}. Since these variables are defined for every $t = 1, \cdots, T$, these constraints comprise two groups: (15) which define p_{jt} for t within the time slot allowed for load j to operate (for which w_{jrt} variables have been defined) and (16) for t outside this time slot in which p_{jt} is always zero.

Constraints (17) are the same as (8) in Model 1.

Constraints (18) ensure that, at time t of the planning period, each load j is either "off" or is "on" at only one stage r of its operation cycle.

Constraints (19) ensure that, for each load j, if it is "on" at time t and at stage $r \leq d_j - 1$ of its operation cycle, then it must be also "on" at time $t + 1$ and at stage $r + 1$.

Constraints (20) ensure each load j is operating at stage r exactly once. Note that constraints (19) do not prevent that a load j starts at a time after $T2_j - d_j + 1$ and, as it cannot finishes until $T2_j$, it continues from $T1_j$. For instance, consider that a load j is at stage $r = 1$ at $t = T2_j - 1$, $r = 2$ at $t = T2_j$ and then skips to $r = 3$ at $t = 1$, $r = 4$ at $t = 2$, etc.; this operation scheme is not feasible in practice but it does not violate constraints (19). Thus, constraints (21) are imposed, which ensure that each load j starts its operation (stage $r = 1$) at most at time $T2_j - d_j + 1$ so that it can finish not later than $T2_j$, i.e. within its allowed comfort time slot. Constraints (19) together with (20) and (21) ensure that load j is operating exactly d_j consecutive time intervals, forcing w_{jrt} to be 0 when load j is "off".

4   A Hybrid Genetic Algorithm with MILP Solver

A hybrid approach consisting of a GA and an exact MILP solver is proposed to solve the bilevel programming problem formulated in Model 2 for the interaction between the electricity retailer and the consumer.

The GA applies to the upper level problem (10)–(13). Each individual of the population represents an electricity price setting $x' = (x'_1, x'_2, \cdots, x'_I)$. For each x' the lower level problem (14)–(22) with $x = x'$ is exactly solved. Let y' be the optimal solution obtained for this lower level instance (note that the lower level decision variables are w_{jrt} and p_{jt}). Each solution (x', y') to the bilevel problem is then evaluated by the upper level objective function F in (10). Hence, the fitness function is $F(x, y)$.

The lower level problem has been modeled using the AMPL language [9] and the GA has been coded in Delphi for Windows. For each individual x', the lower level MILP problem is exactly solved by the CPLEX solver called from the GA. The electricity prices (x_i) are the only parameters that change from one call to another one. The general description of the GA is presented below.

Genetic Algorithm:

1: Create the initial population Pop of N individuals $x' = (x'_1, x'_2, \cdots, x'_I)$ satisfying constraints (11)–(13), as described in Sect. 4.1.

2: For each individual x' in Pop, solve the lower level problem (14)–(22) with $x = x'$ using the MILP solver. Let y' be the optimal solution obtained.

3: Evaluate the fitness of each solution (x', y') to the bilevel problem by calculating $F(x', y')$ according to (10).

4: **while** the stopping condition is not met **do**

5:    **repeat**

6:       Select two parents x' and x'' from Pop and apply crossover to generate a child x^c.

7:      Apply mutation to x^c with probability P_m.

8:      Repair x^c to satisfy constraints (11)–(13) or discard it if it is not repairable.

9:    **until** N children have been generated, which form the set *Offspring* (see Sect. 4.2).

10:   For each x^c in *Offspring* solve the lower level problem (14)–(22) with $x = x^c$ using the MILP solver. Let y^c be the optimal solution obtained.

11:   Evaluate the fitness of each solution (x^c, y^c) by calculating $F(x^c, y^c)$ according to (10).

12:   Create *NextPop* by copying the best solution obtained thus far (which is either in *Pop* or in *Offspring*) and performing $N - 1$ binary tournaments without replacement between individuals of *Offspring* and *Pop*. Update the current population *Pop* with *NextPop*.

13: **end while**

14: **return** (x', y') of *Pop* with the highest fitness.

4.1   Initial Population

The initial population consists of N individuals $x' = (x'_1, x'_2, \cdots, x'_I)$ in which each x'_i is randomly generated in the range $[\underline{x}_i, \bar{x}_i]$. In order to ensure that x' also satisfies the average price constraint (13), the following repair procedure is applied.

    Repair x':

1: Compute $s = \sum_{i=1}^{I} \bar{P}_i x'_i$
   Let A be the set of indices i of x'_i that are allowed to be changed. Initially, $A = \{1, 2, \cdots, I\}$.

2: **if** $s \neq Tx^{AVG}$ **then**

3:    Let $\Delta = Tx^{AVG} - s$

4:    Let $P = \sum_{i \in A} \bar{P}_i$

5:    **for** each $x'_i, i \in A$ **do**

6:       $x'_i \leftarrow x'_i + \Delta/P$

7:    **end for**

8: **end if**

9: **for** $i = 1$ to I **do**

10:   **if** $x'_i < \underline{x}'_i$ **then**

11:      $x'_i \leftarrow \underline{x}'_i$ and $A \leftarrow A \setminus \{i\}$

12:   **else if** $x'_i > \bar{x}'_i$ **then**

13:      $x'_i \leftarrow \bar{x}'_i$ and $A \leftarrow A \setminus \{i\}$

14:   **end if**

15: **end for**

16: Compute $s = \sum_{i=1}^{I} \bar{P}_i x'_i$

17: **if** $s = Tx^{AVG}$ **then**

18:   Stop and return x'

19: **else if** $A = \oslash$ **then**

20:   Stop and discard x'

21: **else**
22:     **go to** 2:
23: **end if**

The process of randomly generating x' and repairing it using the above procedure is repeated until N individuals are generated. In our experiments the repair procedure converged in few iterations.

4.2   Reproduction Process

The reproduction process creates N offspring, each one generated from a different selection of two parents in the current population Pop. A binary tournament is applied in which two individuals from Pop are chosen at random and the best one is selected to be one parent. In the crossover process there is a 50 % of chance of this individual being the first or the second parent. The other parent is randomly selected from Pop. A one-point crossover operator is then applied to produce an offspring from the two selected parents. Hence, if the first parent is $x' = (x'_1, x'_2, \cdots, x'_I)$ and the second one is $x'' = (x''_1, x''_2, \cdots, x''_I)$, the offspring is $x^c = (x'_1, \cdots, x'_{i_1}, x''_{i_1+1}, \cdots, x''_I)$ where i_1 is the crossover point drawn at random between 2 and $I - 1$.

Mutation is then applied to x^c with a probability P_m of changing each gene of x^c. For a given x^c_i, the mutation consists of adding or subtracting a positive perturbation randomly generated in the range between 0 and $0.2(\bar{x}_i - \underline{x}_i)$. If x^c_i is out of the bounds imposed by constraints (11) and (12), then it is pushed to the closest bound and its index is excluded from the set A of variables that are allowed to be changed in the repair procedure. x^c is then repaired (using the procedure described in Sect. 4.1) to satisfy constraint (13) or is discarded if it is not repairable. This process is repeated until N offspring have been generated. They form the *Offspring* population, which will compete with Pop to determine the population for the next generation.

The individual with the best fitness obtained thus far always survives from one generation to the next (i.e. an elite set with one element is considered). This is the first individual inserted into the next population. The other $N - 1$ individuals are selected by binary tournament selection between an individual from the *Offspring* population and an individual from the current population Pop (the parents), both chosen at random. The individual with the highest fitness wins and is selected to integrate the next population. Any individual included in the new population is removed from its original population (*Offspring* or Pop), so the same individual cannot be selected twice.

5   Numerical Results and Discussion

An illustrative case is discussed in this section. Most data were obtained from actual audit information and some values were estimated. A 24 h planning period divided into intervals of 15 min is considered. Thus, 1 unit of time (t) is a quarter-hour, which leads to a planning period of $T = 96$ units of time, T={$1, \cdots, 96$}.

Since original data have been collected for 1 min periods, those values were aggregated (by considering average values) for the quarter-hour intervals.

Five shiftable loads were considered ($J = 5$): dishwasher, laundry machine, electric water heater (EWH), electric vehicle and clothes dryer. Figure 1 shows the operation cycles of these loads, i.e. f_{jr} values. The time slots $[T1_j, T2_j]$ allowed for the different loads j are displayed in Fig. 2.

Seven sub-periods of time $P_i \subset T$, $i = 1, \cdots, 7$, were considered for defining the electricity prices to be charged by the retailer to the consumer. The maximum and minimum prices ($\bar{x}_i$ and $\underline{x}_i$) in each sub-period are displayed in Fig. 3. The last time point $P2_i$ that delimits each sub-period P_i, $i = 1, \cdots, I$, is also presented in Fig. 3 below the curve of the minimum price. The first time point $P1_i$ is always given by $P2_{i-1} + 1$, being 1 for $i = 1$.

The energy prices the retailer has to pay (π_t) for the electricity bought in the spot market are displayed in Fig. 4. All prices in Figs. 3 and 4 are in €/KWh, so they were then converted to periods of quarter-hours (i.e. divided by 4) to feed the model. The average price $x^{AVG} = 0.116$ €/KWh was considered.

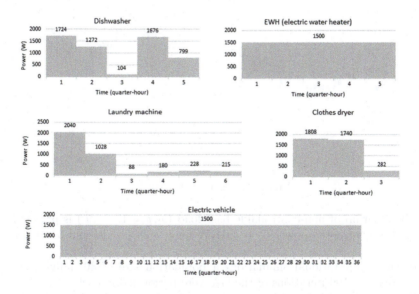

Fig. 1. Operation cycles of the loads to be managed.

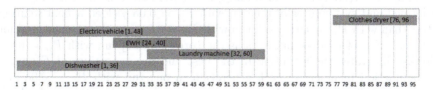

Fig. 2. Comfort time slots allowed for the operation of each load.

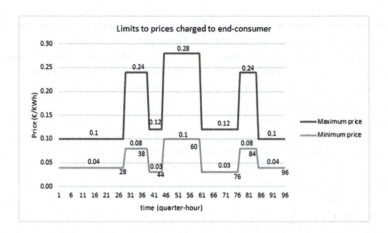

Fig. 3. Minimum and maximum electricity prices charged to the consumer.

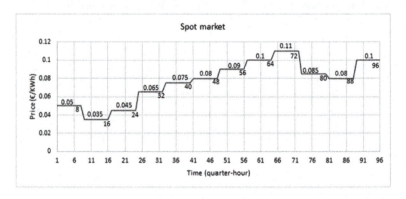

Fig. 4. Prices at spot market.

The diagram of non-controllable base load ($b_t, t = 1, \cdots, T$) is presented in Fig. 5. The contracted power C_t is 4.6 KW for $t = 28, \cdots, 84$ and 3 KW for the other $t \in$ T.

In the computational simulations a population size of 30 individuals was considered and 100 iterations of the GA were performed in each run.

We started by tuning the parameters and two values for the probability of mutation were tested: $P_m = 0.05$ and $P_m = 0.01$. Five runs for each P_m were performed and systematically the run with $P_m = 0.05$ yielded a final solution better than with $P_m = 0.01$. Each run with $P_m = 0.01$ considered equal seed for the generation of random numbers as the corresponding run with $P_m = 0.05$. Figure 6 illustrates the evolution of the best and average values of F among the population members of each generation in one run with $P_m = 0.05$ and the corresponding run with $P_m = 0.01$. In general, the algorithm with $P_m = 0.01$ converges quickly for solutions with higher fitness F but afterwards it has difficulty in improving them significantly. Thus, we have adopted $P_m = 0.05$.

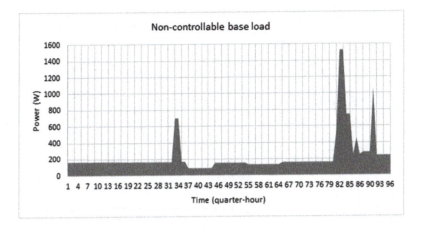

Fig. 5. Power requested to the grid by the base load.

Analyzing the solutions computed over the generations, we could observe that the schedule plans of the shiftable loads (optimal solutions to the lower level problem for each setting of the upper level variables) are few and there are not many differences among them. However, there are very significant differences in the retailer's profit (F) and the follower's cost (f) among the solutions due to the variation of the electricity prices (values of the upper level variables). Remind that the bilevel problem is the retailer's problem, so the optimal solution is the one that offers the retailer the highest profit knowing that the consumer will schedule loads at his best convenience considering the retailer's price setting.

For each upper level variable configuration given by the GA, the lower level is exactly solved. In the present case, this is a MILP problem with 2230 binary variables (w_{jrt}), 151 continuous variables (p_{jt}, excepting those that are necessarily 0 due to constraints (16)), 2233 inequality constraints and 206 equality constraints. Despite being a large problem it is solved very fast by CPLEX as the optimal solution was always found at the root of the branch-and-bound tree and only a few simplex iterations (less than 20) were needed to solve each instance. So, the exact resolution of the lower level problem has revealed to be a very interesting option for this model. The simulations were done in a computer with an Intel Core i7-2600K CPU@3.4 GHz and 8 GB RAM. On average, the time spent in performing one complete generation with a population of 30 individuals (i.e. steps 5: to 12: of the Genetic Algorithm, including solving 30 lower level problems) was 7 s.

Twenty independent runs with 100 iterations each (considering $P_m = 0.05$) were performed and the best solution in each run was recorded (solution with highest fitness, F^{best}). Let us call this set of solutions the *20-best set*. The maximum, minimum, mean and standard deviation of F in the *20-best set* are reported in Table 1. Among these best solutions there are only two distinct schedule plans for the shiftable loads. Plan 1 considers the following starting times for the loads: dishwasher - $t = 1$; laundry machine - $t = 41$; EWH - $t = 36$;

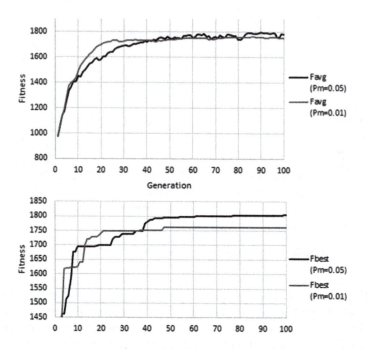

Fig. 6. Evolution of F^{avg} and F^{best} over 100 generations for $P_m = 0.05$ and $P_m = 0.01$.

Table 1. Statistics of F^{best} (€) in 20 independent runs.

maximum F^{best}	minimum F^{best}	mean F^{best}	standard deviation of F^{best}
1825.66	1770.68	1807.43	14.38

electric vehicle - $t = 5$; clothes dryer - $t = 85$. Plan 2 only differs from Plan 1 in the time operation of the laundry machine. The starting times in Plan 2 are the following: dishwasher - $t = 1$; laundry machine - $t = 48$; EWH - $t = 36$; electric vehicle - $t = 5$; clothes dryer - $t = 85$.

The solution with maximum F^{best} in the *20-best set* presents a retailer's profit (F) of 1825.66 and the consumer's cost (f) is 3368.24; the electricity prices in each of the 7 sub-periods of time are: $x_1 = 0.1$, $x_2 = 0.23998$, $x_3 = 0.119964$, $x_4 = 0.106572$, $x_5 = 0.0301812$, $x_6 = 0.235864$, $x_7 = 0.095124$ €/KWh. The schedule plan is Plan 2 whose load diagram is depicted in Fig. 7. The minimum F^{best} is 1770.68 and the consumer's cost in this solution is 3308.91. It corresponds to schedule Plan 1. All monetary values (F and f) are in € and refer to a period of 24 h and a cluster of 1000 consumers with similar consumption and demand response profiles.

We made further experiments with a higher number of iterations (1000). These experiments led to small improvements (0.37 % better than the best solution described above) in the upper level solution and similar plans for the lower level solutions but with a significantly heavier computation burden.

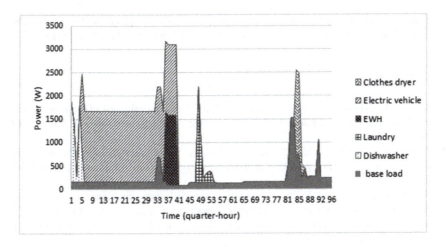

Fig. 7. Load diagram corresponding to the best solution.

6   Conclusions

In this paper new bilevel formulations for modeling the interaction between the retailer and consumers in the electricity market were presented. A bilevel problem is a programming problem where a (lower level) optimization problem is embedded as a constraint in another (upper level) optimization problem. In the present model, the electricity retailer is the upper level decision maker, which buys energy in the spot market and sells it to consumers. The retailer determines prices to be charged to the consumer, subject to the regulation framework, with the aim of maximizing its profit. The consumer (lower level decision maker) reacts to the prices by scheduling loads to minimize his electricity bill.

Bilevel programming problems are very difficult to solve due to their inherent non-convexity. In this paper we proposed a genetic algorithm combined with an exact MILP solver to tackle the problem.

An illustrative case was studied considering real data for the loads, which were obtained through audits. The exact resolution of the lower level problem for each upper level setting (electricity prices) has revealed to be a very interesting option, thus leading to a very efficient hybrid approach combining a GA with a MILP solver. Although the lower level problem is a high dimensional MILP problem, it is solved very fast.

As future work we intend to include other type of loads, in particular thermostatic controlled loads. We also plan to expand the model to consider multiple objective functions at the lower level to analyze consumer's cost vs. comfort trade-offs. For this purpose, time slot constraints for the load operation can be replaced by preferred time slots and the deviation from those slots (as measure of discomfort) is minimized.

Acknowledgments. This work has been supported by projects MITP-TB/CS/0026/2013, PO Centro 08/SI/2015-3266 IMMO and UID/MULTI/00308/2013.

References

1. Soares, A., Antunes, C.H., Oliveira, C., Gomes, A.: A multi-objective genetic approach to domestic load scheduling in an energy management system. Energy **77**, 144–152 (2014)
2. Zugno, M., Morales, J.M., Pinson, P., Madsen, H.: A bilevel model for electricity retailers' participation in a demand response market environment. Energy Econ. **36**, 182–197 (2013)
3. Bu, S., Yu, F.R., Liu, P.X.: A game-theoretical decision-making scheme for electricity retailers in the smart grid with demand-side management. In: 2011 IEEE International Conference on Smart Grid Communications (SmartGridComm), pp. 387–391. IEEE (2011)
4. Yang, P., Tang, G., Nehorai, A.: A game-theoretic approach for optimal time-of-use electricity pricing. IEEE Trans. Power Syst. **28**(2), 884–892 (2013)
5. Zhang, G., Zhang, G., Gao, Y., Lu, J.: Competitive strategic bidding optimization in electricity markets using bilevel programming and swarm technique. IEEE Trans. Industr. Electron. **58**(6), 2138–2146 (2011)
6. Meng, F.L., Zeng, X.J.: An optimal real-time pricing for demand-side management: a stackelberg game and genetic algorithm approach. In: 2014 International Joint Conference on Neural Networks (IJCNN), pp. 1703–1710. IEEE (2014)
7. Meng, F.L., Zeng, X.J.: Appliance level demand modeling and pricing optimization for demand response management in smart grid. In: 2015 International Joint Conference on Neural Networks (IJCNN), pp. 1–8. IEEE (2015)
8. Dempe, S.: Foundations of Bilevel Programming. Springer Science & Business Media, Berlin (2002)
9. Fourer, R., Gay, D., Kernighan, B.: AMPL- A Modeling Language for Mathematical Programming. Duxbury Press, Belmont (2002)

Stigmergy-Based Scheduling of Flexible Loads

Fredy H. Rios S.$^{(\boxtimes)}$, Lukas König, and Hartmut Schmeck

AIFB Karlsruhe Institute of Technology, Karlsruhe, Germany
fredy.rios@partner.kit.edu,{lukas.koenig,hartmut.schmeck}@kit.edu

Abstract. In this paper, we address the rescheduling of shiftable loads in a sub-section of the power grid (micro-grid) to maximize the utilization of renewable energy sources (RES). The objective is to achieve a schedule for all customers in the micro-grid such that the RES output utilization is maximized. Customers correspond to residential households provided with intelligent appliances with the ability to recalculate their operation times. We propose an approach based on stigmergy to efficiently find a close-to-optimal solution to the general problem. An empirical analysis of the internal functioning of the algorithm is performed. Furthermore, the performance of the algorithm is compared to a price-based approach.

Keywords: Demand side management · Load balancing · Multi agent coordination · Stigmergy

1 Introduction

The existing power grids are mostly designed to provide a permanent electricity supply from large centralized power plants, where flexibility is not a major concern [1]. However, power grids are expected to experience a dramatic change in the upcoming decades, as a consequence of the continuous increasing share of RES in future power systems. RES, specifically wind and solar power generation, are characterized by being weather dependent. This fact makes them fundamentally different to conventional generation, for which the systems were originally designed for [2]. RES energy output is intermittent, hard to predict and non-dispatchable, therefore it should be utilized as soon as it is generated. However, frequently RES generation is available on times of the day when customer consumption is low. Then, unless this power is stored, which might be highly expensive ([3]), this power cannot be utilized and is lost.

In this context, customer's load flexibility can have a crucial impact on increasing utilization of these RES. Residential households provided with intelligent appliances and electric vehicles (EV) are able to select intelligently their execution times, in compliance with customers-defined restrictions. For example, a customer could load his or her dishwasher at a certain hour of the day, programming a certain flexibility interval for the run time of the appliance. This way, the dishwasher can be scheduled flexibly within the interval to establish their operation at times where RES availability is larger, increasing the utilization of this resource [4,5]. This flexibility, however, raises a challenge for power

© Springer International Publishing Switzerland 2016
G. Squillero and P. Burelli (Eds.): EvoApplications 2016, Part I, LNCS 9597, pp. 475–490, 2016.
DOI: 10.1007/978-3-319-31204-0_31

grid operators. On one hand, a wrongly designed mechanism to schedule operation times for flexible appliances can generate avalanche effects and power grid
instability [1]. On the other hand, sub-sections of the power grid usually include
tens of thousands of households with many appliances, making the scheduling
process computationally expensive. Therefore, the efficient scheduling of these
loads becomes a relevant issue for future decentralized power supply systems.

In this paper we present a novel mechanism to calculate the execution times
of flexible appliances in a micro-grid in a 24 h simulation scenario. Our approach, called stigmergy-based load-control (SLC), is inspired by a fundamental
cooperation mechanism in nature, namely stigmergy. It is characterized by indirect communication between agents and probabilistic decisions by these agents
based on a shared signal or stimulus [6]. Furthermore, SLC is able to solve the
problem of scheduling flexible loads and increasing the utilization for different
RES outputs. Additional benefits of the approach include an improvement of
privacy and scalability, since no direct communication between participants is
required for the rescheduling process. The remainder of this paper is structured
as follows: In Sect. 2 we present the theoretical background of our approach.
In Sect. 3 we perform an empirical analysis of the behavior of the algorithm.
Furthermore, SLC is compared to a synchronized closed-loop pricing approach.
Finally, a conclusion and an outlook to future work is provided.

2   Theory and Implementation

2.1   Stigmergy in the Context of Optimization

The term stigmergy corresponds to a coordination mechanism which is characterized by indirect communication between individuals through alterations of
signals or stimuli embedded in a shared environment [6]. These alterations are
anonymous in the sense that they are not traceable to a specific individual.
Moreover, individuals perform their tasks induced by these stimuli, whose value
or concentration level is in permanent change. As a consequence, coordination
and cooperation appears spontaneously in stigmergic systems. For example, an
individual may be induced to a behavior by the value of the stimuli, and during
the performance, it reactively modifies the stimuli. Eventually, other individuals read the new value of the stimuli and modify (or enforce) their behavior
influenced by this new value, again modifying it, and so on. This feedback loop
enables individuals in stigmergic systems to trigger an auto-catalytic effect which
gives rise to coordination and cooperation [6].

The foraging of ants is a classical example of stigmergic systems in nature.
Here, ants find the shortest path between a food source and the colony's nest
by leaving pheromone deposition (stigmergic stimuli) on their way. Then, they
probabilistically select their route according to this pheromone concentration.
Initially, every path will have the same pheromone deposition, however, as the
ants travel, the shortest route will increase its pheromone deposition since the
density of traffic on it will be larger. As a consequence, the probability to visit this
specific path will increase, as the pheromone deposition on it will be significantly

larger in comparison to other alternatives. Eventually, a fairly short path is likely to be revealed in terms of having the highest pheromone concentration which, in turn, leads to most of the ants following it [6].

The most studied artificial stigmergic system is Ant Colony Optimization (ACO; [7]). ACO has been succesfully applied to many combinatorial optimization problems such as the Travelling Salesperson Problem (TSP), the Job-Shop Scheduling Problem (JSP) and the Quadratic Assignment Problem (QAP), proving the applicability of stigmergic systems in the context of optimization problems [7,8].

2.2 Problem Description

In SLC, we assume an isolated sub-section of the distribution grid, from now on, the micro-grid. This micro-grid is populated with residential households, which are provided with intelligent appliances and electric vehicles (EV). Appliances have specific load profiles, therefore, the aggregated load profiles of the appliances and their times of execution shape the load profile of the household. The load profiles of all households are aggregated and together they constitute the micro-grid load profile. Then, the objective is to calculate execution times for each device such that the aggregated load profiles of the households match an RES output as closely as possible, maximizing RES utilization. The information flow in SLC is observed in Fig. 1.

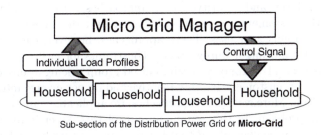

Fig. 1. Information flow in an idealized micro-grid managed by SLC.

In order to obtain these schedules, a micro-grid manager (MGM) performs an iterative optimization process. In the first iteration, the MGM directly transforms this output into a signal, which in our context corresponds to the stigmergic stimuli. This signal expresses on which hours of the day it is more desirable to increase power consumption and, implicitly, the usage of RES generation. This signal is given to every household in the simulation.

On a household level, the received signal is utilized in conjunction with the load profiles of the devices and the user defined flexibility intervals for each appliance, to calculate a new schedule. This schedule represents an updated load profile for the household in the next iteration. The load profiles of every household are aggregated by the MGM who builts an updated load profile for the micro-grid. This load profile is considered to update the signal utilized to influence the rescheduling of appliances. The updated signal is given to every household in the following iteration and the process repeats. This process incrementally approaches to close-to-optimal schedules with progressing iterations. The described iterative search for individual schedules in order to incrementally achieve global objectives corresponds to a form of stigmergy.

The challenge for the algorithm is to minimize the difference between the load profile of the micro-grid and the RES output. To achieve this, the optimal starting times for the devices in the micro-grid need to be calculated. This problem can be seen as an extension of the load management problem in smart homes [5]. A major difference is that in [5], the starting point is calculated for the appliances in a single residential household. On the contrary, we aim to obtain the execution times of the appliances for a large number of households. Therefore, the size of our problem is significantly larger in most instances. Another relevant difference is that we do not consider pricing signals, but rather gain information about the quality of a solution directly from the difference between the load profile of the micro-grid and the RES output.

Other approaches have also considered stigmergy to face this problem. In [9], a coordination mechanism is proposed to addressed the supply and demand matching of large number of distributed actors. Their approach considers achieving an equilibrium state through alterations on a definition of a shared environment. Nevertheless, in [9], the identity and a direct communication channel between actors is required to construct solutions. This can carry serious threats to customers privacy and security. Other approaches consider real-time pricing to guide the customers consumption behavior by reducing electricity prices when RES availability increases. These approaches, however, can create additional load peaks and might imply privacy constraints for customers [1,10]. To prevent this issues, [11][1] proposed a randomized group pricing approach. In this approach, however, customers are only reactive and do not engage into any form of communication. Moreover, these approaches are applicable almost exclusively to this specific problem. In this sense, the aim of this paper is to propose an approach which can be generalized to other optimization problems besides load scheduling.

2.3   Formalization

Stimulus Signal Determination. In our model, we assume a discrete time horizon indexed by $t \in [T]$, with $[T] = \{0, \ldots, T\}$. We consider an RES output vector $\mathbf{g} = (g_0, \ldots, g_T)$ where $g_t \geq 0$ for all $t \in [T]$. The signal in each iteration i is given by the vector $\mathbf{s}^i = \left(s_0^i, \ldots, s_T^i\right)$, with $s_t^i \in [0, 1]$. In the first iteration $(i = 0)$ the values in $\mathbf{s}^i$ are given by $s_t^0 = \frac{g_t}{\max \mathbf{g}}$.

In the following iterations, the MGM receives the load profile from the residential households of the previous iteration. These schedules are aggregated to

---

[1] It has to be noted that the approach utilized in Sect. 3.3, is also utilized in [11] for performance comparisons.

obtain the overall system load, which corresponds to vector $\mathbf{l}^i = \left(l_0^i, \ldots, l_T^i\right)$ with $l_t^i \geq 0$. This way, the stigmergic stimulus (the signal) is updated in each iteration $i > 0$. The values of the updated signal $\tilde{\mathbf{s}}^i$ are given by:

$$\tilde{s}_t^i = s_t^{i-1} + \alpha \cdot b_t^i, \ i > 0, \ t \in [T] \tag{1}$$

where $\mathbf{b}^i = \left(b_0^i, \ldots, b_T^i\right)$ is the vector that defines the adaptation of the signal. The value $\alpha \in [0, 1]$ specifies the weight of this vector for signal updating. The larger the value of α the larger the effect of $\mathbf{b}^i$ and vice versa. We can say that α regulates the level of exploration and exploitation in the search for an adequate global schedule. As previously mentioned, the values in $\mathbf{b}^i$ are calculated considering the micro-grid load and the RES output:

$$b_t^i = \frac{g_t - l_t^i}{\max\left\{\max \mathbf{g}, \max \mathbf{l}^i\right\}} \tag{2}$$

Since the signal that guides the rescheduling process of the households requires values in the interval $[0, 1]$, the updated control signal $\tilde{\mathbf{s}}^i$ has to be normalized:

$$\hat{s}_t^i = \frac{\tilde{s}_t^i + \left|\min \tilde{\mathbf{s}}^i\right|}{\max \tilde{\mathbf{s}}^i + \left|\min \tilde{\mathbf{s}}^i\right|} \tag{3}$$

To prevent step responses and oscillating behavior, a final filtering of the stimuli is performed. Hence, the broadcast signal is defined as:

$$s_t^i = \frac{s_t^{i-1} + \hat{s}_t^i}{2} \tag{4}$$

Residential Household Response. After receiving the signal, each household decides where to schedule its appliances in compliance with the user-defined flexibility intervals. For this, each individual household processes the signal and obtains a vector $\mathbf{r}^i = \left(r_0^i, \ldots, r_T^i\right)$. The values in $\mathbf{r}^i$ are interpreted as an indicator of where it is more desirable to schedule the execution times of its devices in iteration i.

In principle, the additional hierarchical level conformed by the households could be omitted by directly delivering the signal to the intelligent devices. However, the signal's detour over this abstract concept called household has a direct correlation with reality. Households can be categorized according to their device composition, allowing more advanced forms of interpretation of the signal (see Sect. 3.4). Nevertheless, in this paper we assume all households to utilize the same transformation function $\mathbf{r}^i = \mathbf{s}^i$.

In the following, the scheduling process is described for one exemplary appliance. Given an appliance a which runs once a day, we want to schedule this run. Let t_s^a be the time where the appliance is ready for operation and t_e^a the last possible starting time. In real world applications both values can either be defined by the user or automatically derived to meet appliance constraints. The flexibility interval for appliance a is defined as $F^a = [t_s^a, t_e^a]$, with $t_s^a, t_e^a \in T$ and $t_s^a \leq t_e^a$. Additionally, let δ_a be the duration of one execution of a. The load profile of this appliance is a static vector defined as: $\tau^a = (\tau_0, \ldots, \tau_\delta)$ where $\tau_j \geq 0$ for all $j \in [\delta_a]$.

The process of selecting a time of execution for the appliances is performed sequentially within each household. For this, the control signal $\mathbf{r}^i$ is utilized to build a vector which defines a *probability distribution* for the execution time of a: $\mathbf{p}_a^i = (p_0^i, \ldots, p_T^i)$. For the construction of $\mathbf{p}_a^i$, the consumption profile τ^a and the flexibility interval F^a are considered. Through this distribution a new starting time for a is probabilistically selected. Values in $\mathbf{p}_a^i$ are given by:

$$p_t^i = \begin{cases} \dfrac{\sum_{m=0}^{\delta} r_{t+m}^i \cdot \tau_m}{\sum_{k=t_s}^{t_e} \sum_{m=0}^{\delta} r_{k+m}^i \cdot \tau_m} & \text{if } t \in F_a^i, \\ 0 & \text{otherwise} \end{cases} \tag{5}$$

Vector $\mathbf{p}^i$ can be different for each appliance. Furthermore, $\mathbf{p}^i$ can change in each iteration for the same appliance, as the processed signal $\mathbf{r}^i$ changes. Devices such as electric-vehicles are only power and energy constrained, and require a specific energy level during their flexibility interval. In this case, the consumption on each individual time slot is rescheduled as an appliance with a load profile with one time slot of length ($\delta = 1$) by Eq. 5. As a consequence, the load profile of an EV does not need to have a fixed shape, and can change in the progressive iterations. After all devices in the household have been rescheduled through this process, a new schedule is obtained and, implicitly, a new load profile for the household as well.

At the end of each iteration i, each household sends its load profile to the MGM. The profiles are aggregated and a new micro-grid load profile $\mathbf{l}^i$ is obtained. With this information the MGM calculates an updated control signal for the next iteration and the process repeats. Through this process, we obtain a schedule for each appliance in the micro-grid at the end of each iteration. The performance of this global schedule is evaluated through the difference of the areas between the RES output and the micro-grid load profile. This way, at the end of each iteration, the new schedule is compared with the current best performing schedule. If the new schedule has a better performance, it replaces the old one as the new current best. By utilizing the distance between current load profile and desired load profile of the micro-grid as a mean stimulate the system to obtain a probabilistic response of individuals, SLC is able to search for solutions that increase RES utilization.

2.4   SLC and ACO

Even though the presented approach is inspired by the same coordination mechanism as ACO, some major differences exist. Firstly, the construction of solutions by the swarm is different. When looking at a common Traveling Salesperson (TSP) application of ACO, a colony of agents (ants) is selected to progressively create solutions which correspond to routes in a graph. In every iteration, each individual constructs a solution, guided by a heuristic value and an artificial pheromone. At the end of the iteration, the best solution is selected. Then the components of that solution are enforced with additional artificial pheromones in order to guide the search in the neighborhood of that good solution. In the next iteration, the agents utilize this new pheromone deposition to probabilistically select components and construct a new solution for the current iteration, and so on. Since at each iteration, there is only one *currently best solution*, low quality solutions are discarded [7]. On the contrary, in SLC each agent (household in our case) is in charge of one component of the solution (the household's load profile) and all components are needed to create a solution. Therefore, we cannot voluntarily neglect the results obtained by a single agent. Moreover, at each iteration, we generate only one solution, not a set of solutions to choose the best one from. Additionally, the concept of *population* is different in the presented approach. In SLC, the size of the swarm is given by the problem instance (a micro-grid with 50,000 households should have 50,000 agents in the simulation). In ACO the swarm size is a modifiable parameter of the optimization algorithm.

Another important distinction is the stimuli utilized. Ant systems usually utilize artificial pheromones to trigger the auto-catalytic effect that gives rise to coordination and cooperation in the solution construction process. These pheromones are local (they relate to individual arcs in a TSP context) but modifiable by any agent that utilizes the solution component. Regarding SLC, the stimulus corresponds to a signal, which is broadcast to every agent in the population. Then, agents react modifying their schedules as a response to the stimulus, which in turn, modifies the stimulus for the next iteration. These updated stimulus will be different for the whole population. However, not every agent will be able to modify the stimulus/signal, since, as a consequence of the user defined flexibility, agents are able to reschedule only over a subset of the time slots. Additionally, in SLC the stimulus reflects the level of fulfillment of the objective. For example, the larger the difference between the RES output and micro-grid load profile in a specific time slot, the larger will the value of the signal be, in order to guide agents to schedule load on that time slot. From this point of view, the signal can be seen as an indicator of the *distance* from achieving the objective in specific time slots. This does not occur in ACO, where pheromones express the quality of a specific component by means of how much it was utilized in previous iterations.

Regarding exploration and exploitation in the search, contrary to ACO, SLC does not utilize an explicit parameter to increase exploration on detriment of exploitation[2]. In our case, both aspects of the search are managed by the same parameter, α in Eq. 1, as it will be depicted in Sect. 3.

These features allow SLC to model and face problems such as load scheduling in a direct manner. ACO on the other hand, requires a graph representation any problem. This occurs because ACO is based on the foraging behavior of ants, rather than their cooperation and coordination mechanism.

---

[2] It has to be noticed though, that initial versions of Ant Systems did not have this mechanism either [7].

3   Evaluation and Discussion

In this section we describe the scenario in which SLC is evaluated. A 24 h day is discretized in 96 intervals of 15 min. An isolated micro-grid, composed of a variable number of buildings, is assumed. Although SLC can be used to schedule load for commercial buildings and apartments, in this paper homogeneous households are considered. These households are equipped with intelligent devices. Specifically washing machines, dishwashers, dryers and electric vehicles, with a penetration on the micro-grid's population of 100 %, 80 %, 50 % and 25 %, respectively. The flexibility intervals, which define the restriction on the execution times of the appliances, and the daily consumption for the appliances, are described in Table 1. The initial execution time of every device is uniformly distributed within the corresponding flexibility interval.

Table 1. Flexiblity intervals and daily consumption of appliances.

Washing machine		Dryer		Dishwasher		Electric vehicle	
Interval	Share	Interval	Share	Interval	Share	Interval	Share
00:00-06:30	20 %	00:00-08:00	20 %	00:00-06:30	20 %	00:00-13:45	30 %
06:30-12:00	32 %	08:00-13:30	35 %	06:30-12:00	30 %	00:00-07:15	30 %
12:00-20:00	46 %	13:30-20:00	40 %	12:00-17:30	40 %	15:00-00:00	20 %
20:00-00:00	2 %	20:00-00:00	5 %	17:30-00:00	10 %	16:15-00:00	20 %
Daily consumption [kWh]							
0.89		2.45		1.2		4.8	

The German Transmission System Operators provide wind and solar power (PV) generation data on in their balancing areas. This information has a 15-minutes time resolution. Wind data is obtained from the balancing zone of 50 Hz[3] and solar data from Transnet[4], both for 2014. Additionally, the RES output is scaled to the total available load on micro-grid.

3.1   Exemplary Run

Figure 2 shows an example run of SLC with 4, 000 households and $\alpha = 0.05$. The micro-grid load profile evolves through the iterations increasingly resembling the RES output. This shows how the algorithm gradually converges achieving close-to-optimal solutions. Figure 3 shows the behavior of the signal in the solution searching process. The curve corresponds to the stigmergic stimuli, and the values reference the requirement to shift load to specific time slots. Households are influenced by these values and modify the stimuli for future iterations by rescheduling their appliances, as explained in Sect. 2.3. In the first iteration, the

---

[3] http://www.50hertz.com/de/Kennzahlen.
[4] http://www.transnetbw.de/de/kennzahlen.

stimulus corresponds to the RES output scaled in $[0, 1]$. The shape of the stimuli changes between iterations, expressing the distance between the target behavior and the real behavior. This change increases the desirability to shift load to time slots where RES output has not been matched, and reduces the desirability on time slots where overload exists.

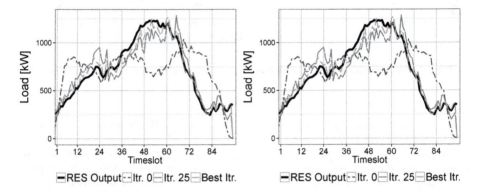

Fig. 2. Performance improvement throughout 100 iterations of a SLC run.

Fig. 3. Signal updating in different iterations of a SLC run.

In the following, the effect of different population sizes is studied on the performance of SLC. Moreover, the effect of parameter α from Eq. 1 is analyzed. This parameter has an impact on how much the distance between the objective (RES output) and the current solution (current micro-grid load profile) affects the adaptation of the signal, which guides the search. Finally, the performance of SLC is compared against a pricing-signal based approach.

3.2 Population Size Variations and Convergence of SLC

To analyze the scalability of SLC, we evaluated the impact of modifying α and the population size. For each α and population size combination, ten runs were performed.

A tendency was found for all α configurations, in which performance seems to improve as the population size increases. An example of this behavior can be observed in Fig. 4. After assessing that the data is not normally distributed, and that significant differences exist between medians of the groups, a statistical analysis was performed which confirms this observation. A subset of the analysis is observed in Tables 2 and 3. A likely explanation for this is that with a smaller population, there are less options to reschedule the appliances within their flexibility intervals and minimize the difference between the RES output and the micro-grid load profile. Additionally, since the RES output is scaled to the total amount of energy available (which depends on the number of devices), with a smaller population, the impact of rescheduling a single device is larger than with a large populations.

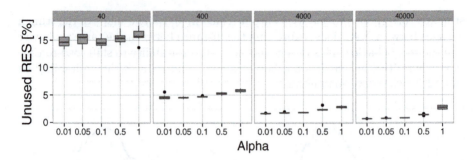

Fig. 4. Performance of SLC for different α values in different population sizes.

Table 2. Wilcoxon rank sum test for pairwise comparisons with Bonferroni correction. Data is grouped according to population size, with $\alpha = 0.05$. Values in the cells correspond to p-values.

	40	400	4,000
400	$6.5e-05$	–	–
4,000	$6.5e-05$	$6.5e-05$	–
40,000	$6.5e-05$	$6.5e-05$	$6.5e-05$

Table 3. Summary of statistical information for SLC. Data is grouped according to population size, with $\alpha = 0.05$. Values represent percentage of unused RES.

	40	400	4,000	40,000
Min.	13.32	4.27	1.64	0.69
1st Qu.	14.35	4.44	1.71	0.70
Median	15.51	4.52	1.72	0.71
Mean	15.37	4.52	1.74	0.73
3st Qu.	16.03	4.57	1.77	0.74
Max	17.34	1.703	1.93	0.79

In addition, although the differences in performance between $4,000$ and $40,000$ populations is significant, this difference is rather small, in comparison to the difference between 400 and $4,000$ households. Apparently, after a certain population size, the effect of individual devices is reduced to the point where only aggregated behavior has a relevant impact on the overall performance. This behavior is traditionally observed in natural stigmergic systems [6].

The purpose of parameter α in the solution construction process is to balance the exploration level of SLC. Intuition tells that larger values, close to 1.0, should increase the exploration ability of the SLC. However, the algorithm might not be able to direct the search to globally good schedules and might get stuck in a randomly oscillating search process. On the other hand, if values for the parameter are too small (close to zero) the algorithm might get trapped in a local optimum, reducing the search to a hill-climb in a non-optimal search space region. As discussed above, the effect of the population size follows the same tendency in every parameter configuration. Hence, a single population size, $4,000$ households, is selected for analyzing the effect of α in convergence and performance.

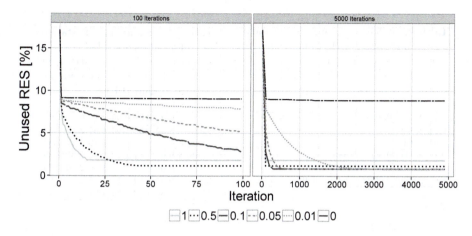

Fig. 5. Performance of single SLC runs throughout 100 and 5,000 iterations for different α values on a 4,000 households micro-grid.

Figure 5, *100 Iterations* shows how the performance changes when SLC schedules 10,200 intelligent devices with different α values throughout 100 iterations. As predicted, with $\alpha = 1.0$ our approach converges very fast during the first iterations, however, it does not seem to make major improvements after iteration 15. Something similar can be observed for $\alpha = 0.5$. For $\alpha = 0.1$ and $\alpha = 0.05$ a less steep, but in return more continuous improvement can be observed. On the other hand, for $\alpha = 0.01$, only a rather slow improvement can be observed. In this case, performance does not seem to largely improve and is comparable to the no-signal-adaption ($\alpha = 0.0$) performance. As predicted, high values for the parameter generally bring fast convergence.

Nevertheless, when SLC runs for 5,000 iterations, $\alpha = 0.01$ eventually achieves the best performance, even though the improvement is much slower (Fig. 5, *5,000 Iterations*). On the other hand, although $\alpha = 0.05$ and $\alpha = 0.1$ reach competitive solutions, they seem to be trapped in local optima throughout most of the run-time. The other parameter configurations also seem to get trapped in local optima early in the execution of the algorithm, with much worst performances.

The overall results of these experiments, in conjunction with Fig. 4, support the previous statement. The best performances are obtained with small values for α, specifically 0.01. This was further confirmed by a statistical analysis. A subset of the analysis is depicted in Tables 4 and 5 (Data was not normally distributed, and significant differences between medians of the groups were found). Therefore, it can be said that convergence is reached faster with higher α values, but the final outcome is better with lower values. Furthermore, when the signal is not updated ($\alpha = 0.0$), SLC is not able to improve the current solutions and therefore, stays trapped in local optima. Hence, it is shown that the adaptation of the signal has an impact in improving the quality of solutions.

Table 4. Wilcoxon rank sum test for pairwise comparisons with Bonferroni correction. Data is grouped according to α, with 4,000 households. Values in the cells correspond to p-values.

	0.0	0.01	0.05	0.1	0.5
0.01	0.0002	–	–	–	–
0.05	0.0002	0.0031	–	–	–
0.1	0.0002	0.0007	1.0000	–	–
0.5	0.0002	0.0002	0.0002	0.0002	–
1.0	0.0002	0.0002	0.0002	0.0002	0.0313

Table 5. Summary of statistical information for SLC. Data is grouped according to α, with 4,000 households. Values represent percentage of unused RES.

	0.0	0.01	0.05	0.1	0.5	1.0
Min.	8.39	1.54	1.64	1.65	2.10	2.42
1st Qu.	8.69	1.58	1.71	1.73	2.22	2.66
Median	9.02	1.61	1.72	1.80	2.31	2.86
Mean	9.06	1.61	1.74	1.80	2.36	2.80
3st Qu.	9.37	1.62	1.77	1.85	2.34	2.91
Max	9.86	1.70	1.93	1.97	3.15	3.14

These results imply that α does not regulate exploration and exploitation in the same way as it is known from traditional ACO algorithms ([8]). This is explained by the internal differences between both algorithms. Even when large α values allow faster convergence, they do not enable the algorithm to perform a continuous and effective search. This is not the case with smaller values of the parameter, where better results are unveiled continuously, and eventually runs with larger α values are outperformed. Nevertheless, it has to be pointed out that even when better results are reached with smaller values for α, the convergence speed might be too slow. Hence, an adaptive parameter control adaptation approach for α might be suitable to improve applicability in practical scenarios. This is further discussed in Subsect. 3.4.

3.3 Comparative Results

The performance of SLC was compared with a synchronized close-loop pricing approach (PS). In PS, a pricing signal is broadcast to a single household. This signal expresses the electricity price for each timeslot, in reference to RES availability. Hence, larger RES availability imply cheaper prices. The household selects the cheapest operation time for its devices. Then, it sends its updated profile to a centralized micro-grid price manager. This entity discounts the received load profile from the RES output. Afterwards, it updates the pricing signal according to this new RES availability. The updated pricing signal is broadcast to the next household and the process repeats. When all households have rescheduled their appliances, a single iteration is finished. In the next iteration, the price-based signal regains its original value. At the end of each iteration, the new schedule is compared with the current best schedule. The best performing one is selected as current best solution.

The process of synchronously selecting the cheapest timeslots, and adapting the price signal according to the updated RES output within the same iteration, allows the approach to obtain close-to-optimal solutions ([11]). For comparing the performance of SLC with PS, ten RES outputs have been randomly selected. A fixed population size has been chosen, and for SLC $\alpha = 0.01$ has been selected, since this value provided the best performance. Furthermore, $10,000$ iterations have been performed with both approaches. In addition, a non-load control (NLC) approach is considered, in which the consumption of residential households cannot be influenced. In this last case, residential load is distributed according to Table 1.

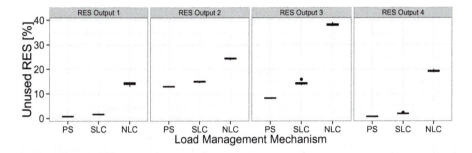

Fig. 6. Example performances of SLC, PS NLC with for four RES outputs and a $4,000$ households population.

Results show that PS outperformed SLC in every evaluated scenario. Furthermore, both approaches largely outperformed NLC. Four example scenarios, with different RES output each, are depicted in Fig. 6. In some scenarios, the differences in performance between PS and SLC does not seem to be large. A close examination of the RES outputs provided that the largest differences were observed when RES outputs had a pronounced load peak at midday, without generation in the morning nor afternoon. These profiles are representative of sunny days without wind power generation. In those scenarios, customers load is restricted to early and later hours, and will always be misplaced, regardless of the ability of the algorithm to relocate it. Additionally, the search space reduces in comparison to other scenarios, where the output is more spread.

Nevertheless, PS is only effective if there is synchronization between participants. If this requirement is not met, extreme load peaks are obtained ([1]). Furthermore, PS requires that the households receive the pricing signal in a specific sort. In isolated experiments, it was observed that if households with BEVs received the signal first, then the price-based approach emulates the output with gaps (Fig. 7). In this case, although the result is close to the optimum, PS is not able to exploit alternatives which imply a deviation from the RES output shape. This occurs when the most flexible device (BEV) is scheduled at the beginning of an iteration. Therefore, the micro-grid has no highly flexible loads available to fill gaps at the end of it. Hence, to achieve high quality solutions the price-based approach requires knowledge of the load composition of individual households, such that the most flexible devices are scheduled at last.

On the other hand, in SLC the probabilistic decision process, based on the current distance to the objective, allows the algorithm to outperform the price-

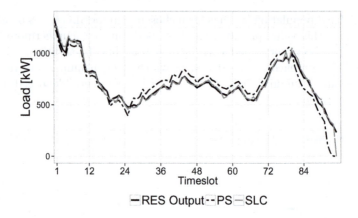

Fig. 7. Comparison of SLC and an unsorted PS on RES output 1.

based approach in this specific scenario. This is certainly a positive result, since in SLC, communication is asynchronous. Moreover, households are considered as *black boxes*, from which a load profile is obtained. Therefore, when an optimal solution is found, the MGM can reference the desired load profile of each household, without knowledge of its load composition nor their specific schedule.

These examples depict benefits and weaknesses of SLC. The approach is able to extend the search and diverge from the shape of the RES output to construct solutions which, in the long run, perform better (Fig. 7). Nevertheless, the probabilistic decision process that enables SLC to perform well when the RES output is balanced throughout the day, prevents the algorithm from exploiting large single load peaks. Hence, it appears that SLC performs better when the solution space becomes larger. With all this in consideration, the advantage of SLC is that it does not require synchronization, nor customers sorting for broadcasting the control signal. Moreover, since SLC does not require knowledge of the load composition and appliances flexibility, the risks of violating customers privacy are reduced.

3.4   Advanced Scenarios and Future Work

As discussed, larger values for parameter α enable SLC to achieve fast convergence in early stages. However, in the long run smaller α values perform better. This suggests that α could be modified in run time. From literature, two standard methods are suitable [12]. On one hand a *deterministic parameter control* approach can be applied. In this case, the value of α would be modified using a fixed function to go from large values in the early stages of the search, to ever lower values during the run. On the other hand an *adaptive parameter control* approach could be utilized. In this case, a feedback mechanism would provide information regarding the state of the search. Then, the MGM would decide how to modify α as the search stagnates. Both approaches are planned for future studies.

From the demand side perspective, to increase usage of single load peaks, the interpretation mechanism of the stimulus by the households could be modified. In this case, the values of the signal can be filtered such that the desirability to increase load in those timeslots remains high until the load is utilized. This matter, however, has to be studied with care, since modifying the interpretation mechanism might trigger chaotic global behavior.

Other enhancements consider residential generation through micro PV. This form of generation shares the same characteristics as large scale PV generation, with the exception that it is local and applies only to the building that generates it. From a household perspective the use of this locally generated RES should be maximized in the first place, and then globally generated RES should be utilized. An approach to this could be the inclusion of an additional term in the decision making process of households, which would be different for each household as it would depend on their geographical location.

Regarding privacy improvements, SLC could be complemented with mechanisms such as the SMART-ER protocol [10]. This protocol enables customers to hide their individual load profiles from the MGM, such that only the final aggregated micro-grid load profile is available. Considering that in SLC, the main identifiable feature of customers is their load profile, this mechanism could enhance customers privacy even further.

The described enhancements justify the utilization of a hierarchy, since residential generation is specific and unique to each household. Furthermore, it has to be considered that different optimization problem definitions might be required for these new optimization scenarios. Future scenarios for SLC include the adaptation of the algorithm to other combinatorial optimization problems, with the aim to extend the set of practical scenarios. In this sense, resource-constrained project scheduling problem, two dimensional bin packing problems and cutting stock problems are the most promising candidates.

4 Conclusion

In this paper we presented a novel approach to treat the problem of scheduling flexible loads, as given in households when residents are willing to shift energy-consuming workloads in time, in order to increase the utilization of a RES output. In this context, the RES output corresponds to a limited resource whose utilization must be maximized through the rescheduling of flexible loads, which must be in compliance with customer-defined restrictions. We proposed a stigmergy-based heuristic to generate a series of increasingly improved solution candidates which can be interrupted when a desired solution quality is achieved or due to external constraints. In various tested scenarios, SLC converged quickly to close-to-optimal solutions and it is therefore well-suited for practical applications. The algorithm is, in principle, applicable to more general scenarios within and outside of energy research, and even, as a kind of meta-heuristic, to other optimization problems. Therefore, we discussed the benefits of our approach over ACO, and we gave an analysis of an important parameter to the algorithm, the α value. We showed that α can be used to control properties that are related to those called *exploration* and *exploitation* in ACO, and we discussed the main similarities and differences to ACO in this regard.

Additionally, the algorithm was compared to a closed-loop pricing approach. Results showed that although SLC did not outperformed this alternative, differences were not large in various problem instances. Furthermore, it was shown that if the price-based approach did not sort their customers according to their flexiblity, SLC could outperform the alternative. This was identified as an important advantage of SLC, given by the fact that the algorithm is asynchronous and it does not require knowledge about individual customers load composition, protecting their privacy and identity.

Regarding future work, two main directions should be considered. The first corresponds to increasing the complexity of the proposed scenario, including residential generation. The second corresponds to the adaptation of the approach to other combinatorial optimization problems, in order to evaluate how our algorithm faces other scenarios.

References

1. Gottwalt, S., Ketter, W., Block, C., Collins, J., Weinhardt, C.: Demand side management - a simulation of household behavior under variable prices. Energy Policy **39**(12), 8163–8174 (2011)
2. Subramanian, A., Garcia, M., Dominguez-Garcia, A., Callaway, D., Poolla, K., Varaiya, P.: Real-time scheduling of deferrable electric loads. In: American Control Conference (ACC), pp. 3643–3650 (2012)
3. Zakeri, B., Syri, S.: Electrical energy storage systems: a comparative life cycle cost analysis. Renew. Sustain. Energy Rev. (2015). doi:10.1016/j.rser.2014.10.011
4. Multin, M., Allerding, F., Schmeck, H.: Integration of electric vehicles in smart homes - an ICT-based solution for V2G scenarios. In: 2012 IEEE PES Innovative Smart Grid Technologies (ISGT), pp. 1–8. IEEE, January 2012
5. Allerding, F., Premm, M., Shukla, P.K., Schmeck, H.: Electrical load management in smart homes using evolutionary algorithms. In: Proceedings of the Main European Events on Evolutionary Computation, pp. 99–110 (2012)
6. Theraulaz, G., Bonabeau, E.: A brief history of stigmergy. Artif. life **5**(2), 97–116 (1999)
7. Dorigo, M., Bonabeau, E., Theraulaz, G.: Ant algorithms and stigmergy. Future Gener. Compu. Syst. **16**(8), 851–871 (2000)
8. Stutzle, T., Lopez-Ibanez, M., Pellegrini, P., Maur, M., Montes de Oca, M., Birattari, M., Dorigo, M.: Parameter adaptation in ant colony optimization. In: Hamadi, Y., Monfroy, E., Saubion, F. (eds.) Autonomous Search, pp. 191–215. Springer, Heidelberg (2012)
9. Hinrichs, C., Vogel, U., Sonnenschein, M.: Approaching decentralized demand side management via self-organizing agents. In: Proceedings of 10th International Conference on Autonomous Agents and Multiagent Systems (2011)
10. Finster, S., Baumgart, I.: SMART-ER: peer-based privacy for smart metering. In: Proceedings of IEEE INFOCOM, pp. 652–657 (2014)
11. Gottwalt, S.: Managing flexible loads in residential areas. Ph.D. thesis, Karlsruher Institut für Technologie (2015)
12. Eiben, A., Hinterding, R., Michalewicz, Z.: Parameter control in evolutionary algorithms. IEEE Trans. Evol. Comput. **3**(2), 124–141 (1999)

Electrical Load Pattern Shape Clustering Using Ant Colony Optimization

Fernando Lezama[(✉)], Ansel Y. Rodríguez, Enrique Muñoz de Cote,
and Luis Enrique Sucar[(✉)]

Department of Computer Science, Instituto Nacional de Astrofísica,
Óptica Y Electrónica, Puebla 72840, Mexico
{f.lezama,ansel,jemc,esucar}@inaoep.mx

Abstract. Electrical Load Pattern Shape (LPS) clustering of customers is an important part of the tariff formulation process. Nevertheless, the patterns describing the energy consumption of a customer have some characteristics (e.g., a high number of features corresponding to time series reflecting the measurements of a typical day) that make their analysis different from other pattern recognition applications. In this paper, we propose a clustering algorithm based on ant colony optimization (ACO) to solve the LPS clustering problem. We use four well-known clustering metrics (i.e., CDI, SI, DEV and CONN), showing that the selection of a clustering quality metric plays an important role in the LPS clustering problem. Also, we compare our LPS-ACO algorithm with traditional algorithms, such as k-means and single-linkage, and a state-of-the-art Electrical Pattern Ant Colony Clustering (EPACC) algorithm designed for this task. Our results show that LPS-ACO performs remarkably well using any of the metrics presented here.

Keywords: Ant colony optimization · Electrical load patterns · Clustering · Clustering quality metrics

1 Introduction

With the inclusion of energy markets into the smart grid, customer clustering plays an important role in identifying client niches to assist the tariff formulation process. To this end, the curves known as load pattern shape (LPS) represent a customer's typical energy consumption during a day (24 h periods). LPSs are key to design effective customer tariffs and can be crucial to take decisions like promoting strategies for peak load reduction or to exploit the willingness of a client to accepts price-based demand conditions by demand programs [1].

The formation of customer classes based in LPS is a multi-stage approach involving the gathering of LPSs curves from customers, and applying *"clustering"* techniques based on pre-defined features extracted from the gathered information.

Broadly speaking, clustering is an unsupervised machine learning technique for partitioning an unlabeled data set into groups, in which the objects that

© Springer International Publishing Switzerland 2016
G. Squillero and P. Burelli (Eds.): EvoApplications 2016, Part I, LNCS 9597, pp. 491–506, 2016.
DOI: 10.1007/978-3-319-31204-0_32

belong to each group share some similarity between themselves, and are dissimilar to objects of other groups [2]. However, electrical LPSs have some characteristics that make their analysis different from other pattern recognition applications. For instance, the number of features (e.g., time series reflecting the electrical consumption of customer's typical day) is relatively high. Moreover, the power of such curves normalized in the [0,1] range and considering the same period of observation causes the overlapping of the patterns.

For the LPS clustering, some of the most studied algorithms include the classical k-means, fuzzy k-means and hierarchical clustering [3,4]. More recently, a clustering algorithm based on a centroid model using some concepts of the Ant Colony Optimization (ACO) algorithm was proposed for LPS clustering [1,5]. The algorithm accepts the number of cluster as input and incorporates a mechanism to guarantee the persistence of the centroids in the iteration process.

The ACO algorithm was first introduced by Dorigo et al. [6]. ACO is a swarm intelligence technique inspired by the behavior of real ants and targets discrete combinatorial optimization problems. The ACO algorithm is also a source of inspiration for the design of novel algorithms for the solution of optimization and distributed control problems.

In this paper, a modified ACO algorithm is applied to the LPS clustering problem, considering a specific number of clusters and an initial centroid model. Our proposed LPS-ACO algorithm implement two traces of pheromone that allow a more efficient information exchange between ants. We compare our LPS-ACO implementation against some classical algorithm, such as k-means and single-linkage, and also against the Electrical Pattern Ant Colony Clustering (EPACC) which is an heuristic algorithm that uses some ACO principles [1].

2   Related Work

Electrical LPS clustering has particular characteristics concerning other pattern recognition application. For this reason, new and modified clustering algorithms have been proposed to date. In [4], a framework for the consumer electricity characterization based on a combination of unsupervised and supervised learning techniques, such as k-means or fuzzy k-means, was proposed. Other works, as in [7], used a probabilistic neural network (PNN) as a method for allocating consumer's load profiles. An interesting comparison on the use of various unsupervised clustering algorithms (e.g., modified follow-the-leader, hierarchical clustering, K-means, fuzzy K-means) and the self-organizing maps (SOM) can be found in [3]. Also, LPSs clustering can be used in forecasting household-level electricity demand to assure balance in low-voltage networks [8].

Different approaches using evolutionary and bio-inspired algorithms have been proposed in the literature for the general problem of clustering [2,9,10]. More recently, In particular for the LPS problem an algorithm based on a user centroid model and assisted by ant colony principles was presented in [1,5]. The authors cleverly shape the information related to the distances of each LPS to the initial model selected by user. Nevertheless, their approach fails by ignoring useful information that ants could provide for the reinforcement of the pheromone trail, making its algorithm less flexible. Moreover, they compare their results against k-means only and under a criterion of just two quality metrics.

Against this background, here we propose an algorithm inspired by [1,5], but as opposed to them, our LPS-ACO implement two pheromone trails. One pheromone trail captures the information on the distance to the initial centroids defined by the user. The other pheromone trail uses the information related to the fitness (measured with a quality metric) of the solutions found. In this way, we allow information exchange between ants in a more efficient way, providing our algorithm with more flexibility to optimize different metrics.

3   Problem Formulation

A pattern is an object, physical or abstract, that has attributes called features that make it distinguishable from others [2]. In this paper, each customer is characterized as an LPS $p_i \in P$ containing $d \in \mathcal{R}$ points representing the energy consumption for a given period of observation. The initial profile data matrix $P_{n \times d}$ contains the n LPSs corresponding to each customer.

In general, given a set of objects P (customers LPSs in this case), a clustering algorithm tries to find a partition $C = \{c_1, c_2, ..., c_K\}$ of K classes such that the similarity between patterns of each class is maximum, and the patterns of different classes differ as much as possible. The partition should maintain three properties [2]: (1) Each cluster should have at least one pattern assigned; (2) Two different clusters should have no pattern in common; (3) Each pattern should be attached to a cluster.

The clustering problem can be formally defined as an optimization problem of finding the clustering partition C^* for which [9]:

$$f(C^*) = \min_{C \in \Omega} f(C) \tag{1}$$

where Ω is the set of feasible clusterings, C is a clustering of a given set of data P, and f is a statistical-mathematical function (i.e., the objective function) that quantifies the goodness of a partition by similarity or dissimilarity between patterns.

3.1   Clustering Quality Metrics

Clustering quality metrics are mathematical functions used to evaluate the accuracy of a clustering algorithm. Ideally, a quality metric should take care of two aspects: Cohesion (i.e., the patterns in one cluster should be as similar as possible) and separation (i.e., the clusters should be well separated).

Some of the most popular metrics used in the literature are the Dunn's index (DI) [11], the Davies-Bouldin index (DB) [12], the Chou-Su (CS) index [13], among others. Because of their optimizing nature (i.e., the maximum or minimum value indicates a proper partition), these metrics are used in association with evolutionary algorithms, such as GA, PSO or DE [2]. For the particular problem of LPS clustering, in [5], other quality metrics named the Clustering Dispersion Indicator (CDI) and the Scatter Index (SI), were used to guide the population and assess the quality of the solutions found.

Despite the effectiveness of such metrics for some applications, it is important to keep in mind that these metrics by themselves optimize some criterion but sometimes do not reflect good partitions from a broad perspective.

For that reason, in the next subsection, we explain the use of some other metrics used as objective functions (i.e., fitness function) in this paper.

3.2 Objective Functions

The selection of a proper metric as an objective function to guide the algorithm is a key aspect of our LPS-ACO algorithm. In this paper, we analyze two quality metrics (CDI and SI) defined in [5], and two complementary metrics, one based on compactness and another based on connectedness, defined in [9]. The analysis of these metrics gives a broad perspective of the quality of the solutions found.

Before presenting the mathematical definition of the analyzed metrics, some considerations have to be stated:

- A pattern is a physical or abstract structure of objects distinguish from others by a set of attributes called features [2]. In the LPS problem, $P_{n \times d}$ is a data matrix with n (i.e., the number of LPSs) d-dimensional patterns p_i. Each element in $p_i \in P$ corresponds to the jth real-value (i.e., feature $j = 1, 2, ..., d$) of the ith pattern.
- Clustering is a process of grouping patterns based on some similarity measures. The most used way to evaluate similarity is by using some distance measure. In this paper, we adopt the use of Euclidean distance as a similarity measure, since is widely used and is well-known [2].

CDI and SI Metrics. The CDI metric is defined under the principle that better clustering partitions have relatively low internal variation, and the centroids $(\delta_j \in \Delta)$ of each clusters $c_j \in C$ are as far as possible from each other. CDI is defined as:

$$\text{CDI}(C) = \frac{\sqrt{\frac{1}{K} \sum_{j=1}^{K} d^2(c_j)}}{d(\Delta)} \tag{2}$$

where $d(\bullet)$ is a function that computes the average distance between patterns in cluster $c_j \in C$, and Δ is the set of centroids belonging to each cluster.

The SI metric is calculated with reference to a pooled scatter $\bar{p} = \frac{1}{n} \sum_{i}^{n} p_i$, corresponding to the average of the data set P. SI is defined as:

$$\text{SI}(C) = \frac{\sum_{i=1}^{n} d^2(p_i, \bar{p})}{\sum_{j=1}^{K} d^2(\delta_j, \bar{p})} \tag{3}$$

where δ_j is the centroid of cluster c_j, n is the number of LPS and K is the number of clusters.

Lower values of CDI and SI indicate better clustering solutions. The main problem with these two metrics, in the particular problem of the LPS clustering, is that the optimal value of CDI or SI favors the creation of clusters under the principle of connectedness. That behavior can be explained observing Eq. (2). Since all the LPS are normalized in the [0,1] range (i.e., there is a little spatial separation between them), the distance of an LPS to its centroid is on average the same. On the other hand, if a cluster is made of just one LPS, the distance of that LPS to its centroid is 0 (i.e., the LPS $p_i \in c_j$ is its centroid). Due to this behavior, the metric will be optimized putting many LPSs as single clusters (i.e., the contribution of the distance to its centroids is 0), and letting a unique cluster with the rest of the patterns.

For that reason, we also analyze two complementary metrics, overall deviation (DEV) and Connectivity (CONN), as proposed in [9]. These two metrics are related to compactness and connectedness. In that way, we can analyze from a broad perspective which metric suits better for the specific problem of LPS clustering.

DEV and CONN Metrics. The DEV metric measures the compactness of a solution and is defined as:

$$\text{DEV}(C) = \sum_{c_j \in C} \sum_{p_i \in c_j} d(p_i, \delta_j) \tag{4}$$

where C is the set of clusters, δ_j is the centroid of cluster c_j and $d(\bullet)$ is a distance function (e.g., Euclidean distance). This metric computes the overall summed distances between patterns and their corresponding cluster center. Again, lower values of DEV indicates better clustering performance.

On the other hand, to compute the CONN metric, first it is necessary to compute a proximity matrix $Prox_{n \times L}$. $Prox_{n \times L}$ will contain the L closer patterns $\forall p_i \in P$. A distance function (e.g., Euclidean distance) is used once in the initialization process to determine the L closer patterns to each p_i. Once $Prox_{n \times L}$ is computed, CONN is defined as:

$$\text{CONN}(C) = \sum_{i=1}^{n} \left(\sum_{l=1}^{L} x_{i,nn_{il}} \right)$$

$$\text{where} \quad x_{i,nn_{il}} = \begin{cases} 1/l, \text{if} \ \not\exists \ c_j : p_i \in c_j \wedge p_{nn_{il}} \in c_j \\ 0 \quad \text{otherwise} \end{cases} \tag{5}$$

where nn_{ij} is the jth nearest neighbor of pattern p_i, n is the number of LPSs, and L is a parameter determining the number of neighbors that contribute to the connectivity measure. CONN metric evaluates the degree of neighboring data points placed in the same cluster by penalizing neighboring data belonging to different clusters. As DEV, lower values of CONN indicates better clustering quality.

4   Ant Colony Optimization and LPS-ACO

ACO algorithm exploits self-organized principles of collaboration between artificial agents (ants) to converge to a good solution. The artificial ants represent specific solutions to the problem and share information via pheromone trails to converge to an optimal solution.

In ACO, iterations involve three main phases: an initialization phase, a solution construction phase and an updating of the pheromone trail phase. The algorithm is run until a stop criterion is met.

4.1   Load Pattern Shape Ant Colony Optimization Algorithm

In the initialization phase, the LPS-ACO parameters are set. These parameters include the size of the colony (M), the threshold γ reflecting a number of iterations without a change in the solutions found and used as a stop criterion, among others specified in Sect. 5.1. After set the parameters, the proximity matrix $Prox_{n \times L}$ (defined in Sect. 3.2) with the L neighbors of each LPS p_i is computed.

The encoding of a solution is crucial in the LPS-ACO algorithm. Each ant is represented as a vector of size n (i.e., the number of LPSs) such as: $a_m = [p_{1 \to c_j}, p_{2 \to c_j}, ..., p_{n \to c_j}]$. The elements of the ant represent the allocation of the ith pattern p_i to a specific class $c_j \in C$.

The pheromone trails are also necessary to the success of the algorithm. In this approach, two pheromone trails that store the information learned from the ants are modeled as matrixes τ and η of dimension $n \times K$. Both pheromone trails reflect the desirability of assigning the LPS $p_i \in P$ to the class $c_j \in C$.

The first pheromone trail (also called the fitness-pheromone trail) $\tau_{i,j}$ is reinforced by the contribution of ants based on the quality of the solution found (measured by a metric). On the other hand, the second pheromone trail (also called the distance-pheromone trail), $\eta_{i,j}$, contains information related to the distance of LPSs to the corresponding centroids of the clusters.

In the solution construction phase, each ant a_m assigns the pattern p_i to the cluster c_j based on the next decision rule: if a random number in the range $[0, 1]$ is less than a probability θ, the ant a_m assigns the pattern p_i to the class c_j that has the major pheromone concentration. Otherwise, the assignation of p_i is determined thorough a stochastic decision policy based on the pheromone levels with probability:

$$prob^m_{p_i \to c_j} = \alpha * \tau_{i,j} + (1 - \alpha) * \eta_{i,j} \tag{6}$$

where α is a constant weight that controls the importance of one or another pheromone matrix.

After the construction of all the solutions, the pheromone trails are updated with particular rules as explained next.

First, the fitness-pheromone trail values are increased on the clusters that ants have selected during their solution construction phase. The fitness-pheromone trail update is implemented as:

$$\tau_{i,j} \leftarrow \tau_{i,j} + \sum_{m=1}^{M} \Delta\tau_{i,j}^{m} \qquad (7)$$

where $\Delta\tau_{i,j}^{m}$ is the amount of pheromone ant a_m deposits on the clusters it has selected. $\Delta\tau_{i,j}^{m}$ is defined as:

$$\Delta\tau_{i,j}^{m} = \begin{cases} \text{fit}_m, \text{if } p_i \in c_j \in a_m \\ 0 \quad \text{otherwise} \end{cases} \qquad (8)$$

where fit_m is the cost of the solution built by a_m. Moreover, we use the concept of elitist ant system (EAS) to provide additional reinforcement to the assignations that belong to a_{best}, which is the best solution found by the algorithm at each iteration. The additional reinforcement is achieved by adding $e_w * \text{fit}_{best}$ to the assignations done by a_{best}.

On the other hand, the distance-pheromone $\eta_{i,j}$ is updated with the information of the inverse distance from each LPS to the centroids δ_{best} corresponding to the solution found by a_{best}.

After the initial calculation and every update of both pheromone matrixes, using the concept of hyper-cube ACO [14], the values of both trails are normalized. This normalization not only avoids the continuous increase of pheromone trails during iterations but also eliminates the necessity of an evaporation rate constant. The normalization is done by dividing each row by its greater value.

A detailed pseudocode of the LPS-ACO is shown in Fig. 1.

5   Results and Discussion

We applied the LPS-ACO algorithm for the clustering of LPSs. The initial data set $P_{n\times d}$ has been obtained from real measures considering $n = 250$ customers[1]. Each LPS was normalized, in the $[0,1]$ range, and was obtained taking the average measurements of the five days of a regular week (without weekend days) in 30 minutes intervals (i.e., each LPS is an average of five days with $d = 48$ points). Figure 2 shows the set of LPS of each customer.

The numerical results section is divided into two parts. First we present the parameter tuning of the LPS-ACO algorithm. After that, the performance of the LPS-ACO algorithm using different metrics and comparing our results with some classical algorithms for clustering is presented.

---

[1] For simplicity, in this work only the first 250 LPSs were considered from the available data on-line: http://www.ucd.ie/issda/data/commissionforenergyregulationcer.

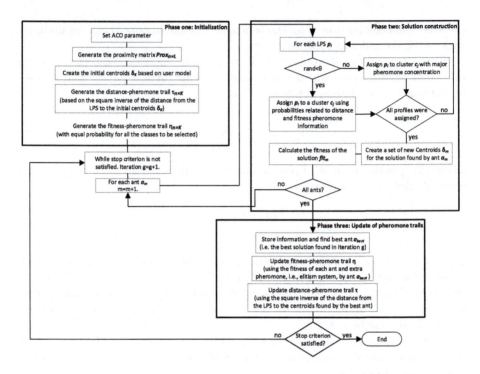

Fig. 1. LPS-ACO clustering algorithm.

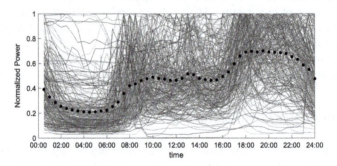

Fig. 2. 250 normalized load pattern shapes. The black dotted line correspond to the pooled scatter $\bar{p}$ used to compute Eq. (3).

5.1   Tuning of the Parameters

It is worth noting that the LPS-ACO algorithm requires the manipulation of few control parameters. These parameters include: θ controlling whether the assignation of p_i to c_j is done based on the major pheromone concentration or in a probabilistic choice; α controlling whether the ants follow the trace of fitness-pheromone, the trace of distance-pheromone, or a combination of both; υ used as stop criterion reflecting the maximum number of iterations without change in the value of the objective function; M is the size of the colony (i.e., the number of ants); and e_w which is an Elitism weight used in the EAS to provide extra pheromone by the best ant.

These parameters have an impact on the quality of the solutions. We performed a parametric analysis using a default value for each parameter. The default values are set to: $\theta = 0.95$, $\alpha = 0.5$, $\upsilon = 20$, $M = 20$ and $e_w = 1$. We use a sweeping technique to assess their impact on the solutions, one at a time. The reported values correspond to the average of 50 independent tests. The 95% confidence intervals (CI) and the standard deviation (Std) were also calculated. The experiments were done optimizing CDI, DEV and CONN metrics[2]. Lower values reflect better clustering performance. The best values found in each experiment are shown in bold.

The results of the first experiment are present in Table 1. This table shows the importance of selecting a proper value of the θ parameter. Notice that the best results with the three metrics were obtained when the θ value is set to 1 (shown in bold). The value of 1 means that the decision on the assignation of p_i to the class c_j is always done selecting the cluster with the major pheromone concentration and letting aside the probabilistic decision rule. Nevertheless, in Table 2, a more fine tuning is presented in the range of [0.9,1] with steps of 0.01. In this range of analyzed values, the best setting for θ to optimize the CDI and the DEV metrics were 0.97 and 0.95 respectively. On the other hand, if the user is interested in optimizing the CONN metric, the results suggest that the probabilistic choice is not necessary, and we can limit our algorithm to follow the trace of the major pheromone concentration.

For the second experiment, we analyze the impact of the α parameter used in Eq. (6). Table 3 shows an interesting behavior related to this parameter. For instance, if we are interested in optimizing CDI or CONN metrics, the results suggest that the α parameter should be set to 1. A value of 1 implies that we neglect the effect of the distance-pheromone matrix η in Eq. (6). In other words, the ants will take the decision based just on the fitness-pheromone matrix τ. On the other hand, if the user wants to optimize the DEV metric, the value of α should be set to 0, which means exactly the opposite (i.e., ants will only follow the trace of the distance-pheromone matrix η). Even when the best values for each metric were found activating a single pheromone trace (i.e., either η or τ), it is important to recall that the real optimal solution might be found in a trade-off of different metrics (e.g., the complementary metrics DEV and CONN) [9]. This suggest the possibility of an application of a multi-objective approach, let it as a further work and out of the scope of this research.

When using heuristic algorithms, it is well-known that setting a maximum number of iterations as a stop criterion is practically ineffective. A best practice is the use of a criterion that reflect the evolution and no variations of the solutions during the iteration process, and the use a maximum number of iterations just to prevent lack of convergence or an excessive computation time. For this reason,

---

[2] SI metric was not taken into account in the parameter tuning since its behavior is similar to the results obtained with CDI metric.

the third experiment is related to the γ parameter. This parameter is used as a stop criterion, reflecting a max number of consecutive iterations without a change in the best solution found. In this way, the stop criterion reflects the evolution of the solutions in the iterative process. Table 4 shows variations of the quality of the solutions when γ vary from 1 to 100. The best values are presented in bold. However, notice that the difference between the means of the tested values do not have a high variation (e.g., for the CONN metric, the best value with $\gamma = 100$ is almost equal to that with $\gamma = 70$). Moreover, increasing the value of γ will have an impact on the number of iterations needed before the algorithm stops. For that reason, a medium value (i.e., in the [20,50] range) is enough to find acceptable solutions without increase the execution time.

The results of the fourth experiment are related to the size of the colony M and are presented in Table 5. It can be noticed that the variations of the quality of the solutions in the range of [10,100] are small. However, more ants imply a greater number of constructions in the second phase of the algorithm, taking more time of execution if the ants construct their solutions sequentially rather than in parallel. Moreover, with the increase in the size of the colony, more objective functions evaluations are needed to determine the fitness of each ant slowing the execution of the algorithm. To reduce the number of objective function evaluations an M value in the range of [10,30] is recommended.

Finally, the fifth experiment is related to the e_w parameter used to control an additional contribution of pheromone by the best ant in the elitist ant system. Table 6 shows the values of the analyzed metrics varying e_w parameter. It can be noticed that this parameter has a small impact on the CDI metric. However, when the DEV metric is under optimization, small values of e_w are preferred for the algorithm (i.e. the best value was found when the e_w is set to 1). On the other hand, for the CONN metric an intermediate value of the parameter e_w (e.g., in the range of [20,50]) gives the best results.

Table 1. θ parameter tuning in the range [0, 1] with 0.1 steps

		0	0.1	0.2	0.3	0.4	0.5	0.6	0.7	0.8	0.9	1
CDI	Mean	1.7036	1.5728	1.5455	1.3494	1.1804	0.9788	0.8341	0.7268	0.5972	0.4792	**0.4378**
	± CI 95 %	0.1002	0.0699	0.0659	0.0420	0.0320	0.0243	0.0187	0.0196	0.0118	0.0070	**0.0000**
	Std	0.3525	0.2460	0.2318	0.1476	0.1127	0.0855	0.0657	0.0688	0.0416	0.0247	**0.0000**
DEV	Mean	331.06	326.24	317.33	307.75	294.42	281.26	265.57	249.89	235.97	224.85	**224.54**
	± CI 95 %	0.54	0.40	0.80	1.01	1.32	1.33	1.86	1.81	1.43	1.39	**0.00**
	Std	1.92	1.40	2.80	3.55	4.64	4.69	6.56	6.38	5.04	4.90	**0.00**
CONN	Mean	838.17	835.12	818.29	792.67	760.38	723.36	666.10	612.55	529.97	453.83	**366.08**
	± CI 95 %	1.09	2.60	3.97	4.68	5.69	5.97	6.33	7.60	8.12	5.61	**0.00**
	Std	3.84	9.14	13.96	16.48	20.03	21.00	22.27	26.74	28.57	19.72	**0.00**

As a summary, the recommended values for each metric and each parameter are presented in Table 7. These values are used in Sect. 5.2 when comparing our LPS-ACO against other clustering algorithms.

Table 2. θ parameter fine tuning in the range $[0.9, 1]$ with 0.01 steps

		0.9	0.91	0.92	0.93	0.94	**0.95**	0.96	**0.97**	0.98	0.99	1
CDI	Mean	0.4736	0.4676	0.4556	0.4522	0.4460	0.4419	0.4338	**0.4299**	0.4311	0.4309	0.4378
	± CI 95 %	0.0054	0.0087	0.0046	0.0064	0.0066	0.0088	0.0034	**0.0022**	0.0069	0.0067	0.0000
	Std	0.0190	0.0307	0.0163	0.0225	0.0231	0.0310	0.0118	**0.0078**	0.0244	0.0234	0.0000
DEV	Mean	223.74	223.24	223.05	222.23	220.88	220.71	220.91	220.89	222.05	223.15	224.54
	± CI 95 %	1.25	1.02	0.89	1.22	0.58	**0.80**	0.72	0.68	0.57	0.20	0.00
	Std	4.40	3.59	3.13	4.29	2.06	**2.80**	2.53	2.40	2.01	0.69	0.00
CONN	Mean	447.64	444.90	437.41	427.43	419.02	411.44	398.53	393.24	386.17	376.38	**366.08**
	± CI 95 %	5.86	6.42	5.39	5.43	4.59	4.22	3.32	3.66	2.92	2.09	**0.00**
	Std	20.61	22.59	18.97	19.10	16.14	14.86	11.68	12.86	10.29	7.35	**0.00**

Table 3. α parameter tuning

		0	0.1	0.2	0.3	0.4	0.5	0.6	0.7	0.8	0.9	1
CDI	Mean	0.4228	0.4175	0.4270	0.4436	0.4384	0.4407	0.4376	0.4367	0.4336	0.4296	**0.1759**
	± CI 95 %	0.0089	0.0047	0.0057	0.0040	0.0033	0.0056	0.0040	0.0034	0.0026	0.0026	**0.0065**
	Std	0.0314	0.0164	0.0201	0.0139	0.0116	0.0197	0.0139	0.0119	0.0092	0.0092	**0.0229**
DEV	Mean	**215.18**	215.79	217.02	217.73	219.70	221.38	223.57	225.84	232.29	241.35	311.54
	± CI 95 %	**0.77**	0.61	0.82	0.72	0.86	0.71	0.87	0.76	1.13	0.67	0.91
	Std	**2.69**	2.14	2.88	2.53	3.04	2.49	3.06	2.68	3.98	2.37	3.21
CONN	Mean	401.83	406.61	406.18	406.12	413.24	408.22	405.98	410.27	409.92	395.71	**129.12**
	± CI 95 %	3.98	4.17	4.48	4.57	3.69	3.86	4.57	4.51	4.34	5.35	**5.08**
	Std	14.00	14.67	15.77	16.08	12.99	13.59	16.07	15.87	15.27	18.81	**17.86**

Table 4. γ parameter tuning

		1	10	20	30	40	50	**60**	**70**	80	90	100
CDI	Mean	0.6429	0.4433	0.4391	0.4356	0.4307	0.4308	0.4305	**0.4284**	0.4270	0.4241	0.4263
	± CI 95 %	0.0299	0.0052	0.0043	0.0038	0.0032	0.0034	0.0030	**0.0031**	0.0046	0.0034	0.0039
	Std	0.1051	0.0183	0.0151	0.0135	0.0113	0.0118	0.0104	**0.0110**	0.0161	0.0118	0.0138
DEV	Mean	254.23	221.92	220.65	220.06	219.37	219.43	**218.91**	218.96	219.15	218.54	219.02
	± CI 95 %	1.30	0.91	0.76	0.73	0.81	0.71	**0.62**	0.54	0.76	0.68	0.67
	Std	4.57	3.20	2.67	2.56	2.86	2.50	**2.16**	1.89	2.67	2.41	2.36
CONN	Mean	408.69	407.85	411.79	407.36	410.61	406.85	407.93	405.39	410.50	408.08	**405.30**
	± CI 95 %	4.45	3.94	4.84	4.87	4.62	4.91	4.20	4.51	4.62	5.02	**4.50**
	Std	15.66	13.85	17.02	17.15	16.26	17.29	14.77	15.85	16.24	17.67	**15.84**

Table 5. M parameter tuning

		1	**10**	20	30	40	50	60	70	80	90	100
CDI	Mean	0.6467	0.4421	0.4413	0.4389	0.4394	0.4336	0.4322	0.4318	0.4355	0.4308	**0.4301**
	± CI 95 %	0.0256	0.0049	0.0069	0.0046	0.0063	0.0035	0.0031	0.0038	0.0118	0.0028	**0.0031**
	Std	0.0903	0.0172	0.0242	0.0162	0.0221	0.0122	0.0109	0.0133	0.0415	0.0098	**0.0109**
DEV	Mean	254.01	**219.87**	220.55	222.47	223.52	224.63	226.12	226.72	228.18	228.74	230.92
	± CI 95 %	0.93	**0.96**	0.78	0.62	0.58	0.71	0.61	0.72	0.61	0.68	0.51
	Std	3.27	**3.37**	2.75	2.17	2.04	2.49	2.14	2.53	2.14	2.38	1.81
CONN	Mean	412.11	409.20	405.81	412.30	409.56	407.45	403.38	405.00	404.02	405.35	**401.71**
	± CI 95 %	3.94	5.60	4.00	4.28	5.22	4.54	5.31	4.94	5.21	5.64	**5.46**
	Std	13.85	19.69	14.06	15.06	18.38	15.98	18.68	17.40	18.33	19.86	**19.20**

Table 6. e_w parameter tuning

		1	10	20	30	40	50	60	70	80	90	100
CDI	Mean	0.4391	**0.4241**	0.4306	0.4444	0.4312	0.4318	0.4389	0.4335	0.4323	0.4353	0.4256
	± CI 95 %	0.0050	**0.0065**	0.0089	0.0138	0.0098	0.0090	0.0143	0.0094	0.0107	0.0120	0.0084
	Std	0.0175	**0.0229**	0.0313	0.0486	0.0345	0.0315	0.0502	0.0330	0.0378	0.0421	0.0296
DEV	Mean	**220.31**	221.75	223.69	225.83	227.51	229.67	231.25	229.39	233.84	231.03	235.23
	± CI 95 %	**0.75**	0.69	1.32	1.26	1.29	1.67	1.75	1.59	1.98	1.70	1.79
	Std	**2.63**	2.42	4.66	4.44	4.55	5.89	6.14	5.59	6.98	5.98	6.29
CONN	Mean	410.78	404.16	**380.52**	382.48	385.19	383.30	386.55	385.80	390.34	386.74	391.53
	± CI 95 %	4.74	4.33	**3.50**	3.99	5.24	4.46	4.13	4.30	5.46	3.62	4.45
	Std	16.69	15.22	**12.33**	14.05	18.45	15.70	14.52	15.12	19.22	12.73	15.65

Table 7. Summary of the parameter tuning.

CDI:	$\theta = 0.97,$	$\alpha = 1,$	$\gamma = 70,$	$M = 100,$	$e_w = 10$
DEV:	$\theta = 0.95,$	$\alpha = 0,$	$\gamma = 60,$	$M = 10,$	$e_w = 11$
CONN:	$\theta = 1,$	$\alpha = 1,$	$\gamma = 50,$	$M = 100,$	$e_w = 20$

5.2   LPS-ACO Algorithm Application

To assess the performance of our LPS-ACO algorithm, we compare our results with two well-known classical clustering algorithms and two recently developed heuristic algorithms. The former two algorithms are the classical k-means [15] algorithm and the hierarchical agglomerative clustering single-linkage [16]. These two classical algorithms were selected because of their intrinsic nature of optimizing the DEV (in the case of k-means) and CONN (in the case of single-linkage) metrics.

The first heuristic algorithm was the Electrical Pattern Ant Colony Clustering (EPACC) [1][3]. This algorithm uses some principles of the ACO algorithm to guide the construction phase. Nevertheless, the main drawback of this algorithm is the lack of a pheromone trail that captures the information related to the metric under optimization. This makes the algorithm less flexible to move from one metric to another.

The second heuristic used for comparison was created by a slight modification of the EPACC algorithm. To explain this, it is important to recall that the original EPACC algorithm [1] uses a roulette wheel selection process for each assignation of p_i to c_i. The roulette wheel selection implies that a random number is generated in each assignation. In our EPACC modification, we use a principle called stochastic universal sampling (SUS) for the creation of the random number. Our algorithm called EPACC-SUS is identical to the original EPACC, with the only difference that a unique random number is created for all the assignations of p_i to c_j, rather that a particular random number for each p_i.

---

[3] The set of parameters used for the EPACC algorithm were those reported as the best set of parameters in [1].

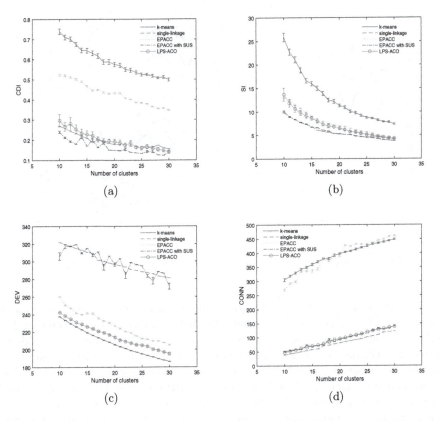

Fig. 3. Comparison of algorithms using different metrics. (a) CDI metric, (b) SI metric, (c) DEV metric, (d) CONN metric.

This simple modification results in an exceptional performance related to CDI and CONN metrics as shown in the results.

The algorithms were run 50 times varying the number of clusters K from 10 to 30. The results correspond to the mean value of those 50 test. The 95 % CI were also calculated and plotted for each point of the results.

Figure 3 shows the comparison of the tested algorithms under different scenarios. The first thing to remark is that k-means present a poor performance related to CDI, SI and CONN metrics (as shown in Fig. 3(a), (b), (d)). This poor performance could mislead the user to the conclusion that k-means is a bad algorithm for the LPS clustering problem. However, it can be observed in Fig. 3(c) that k-means present the best performance among all the algorithms. The opposite case occurs with the single-linkage and the EPACC-SUS algorithms. Both of them present the best performance related to CDI, SI and CONN metrics (as shown in Fig. 3(a), (b), (d)) but the worse performance related to DEV metric (shown in Fig. 3(c)).

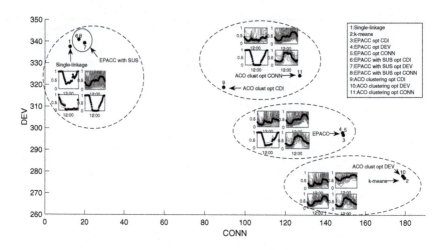

Fig. 4. Comparison of algorithms based on DEV and CONN metrics

This particular behavior when optimizing a single metric open a question on the validity of the chosen metric to measure the quality of the LPS clustering. For instance, observing just Fig. 3(a), (d), k-means and EPACC could be considered as bad algorithms for the LPS clustering problem. That conclusion completely changes by observing Fig. 3(c), in which those two algorithms present a good performance (in fact, k-means is the best algorithms when DEV metric is used to evaluate the performance of the algorithms) related to DEV metric. Overall, our LPS-ACO algorithm present a good performance for all the analyzed metrics without being the best in any case.

To go deep in the analysis of a proper LPS clustering, we developed an experiment using the same data set P but decreasing the number of clusters to $K = 4$ for an easy visualization of the patterns in each cluster. Again, each algorithm was run 50 times, and the mean values were stored. Moreover, EPACC, EPACC-SUS, and LPS-ACO algorithms were run three times each changing the optimized metric (i.e., CDI, DEV and CONN). SI metric was not taken into account since its behavior is similar to CDI.

Figure 4 shows the mean values of the 50 runs of each algorithm related to DEV and CONN metrics. Also, some graphical clusterings (corresponding to one of the solutions found) are shown in some areas of the plot to observe the kind of solutions generated by each algorithm. The first thing to recall is that single-linkage and k-means are on the opposite sides of the plots and represent the best solution for CONN and DEV metrics respectively (notice that this is a minimization problem). Moreover, it can be observed (on the top left of the figure) that single-linkage generates solutions that classified single patterns as a group and let one unique cluster with the rest of the patterns. This kind of classification will optimize CDI, SI or CONN metric, but the validity of such classification can be questioned when we analyze graphically the formed clusters. On the opposite side, we can visualize the kind of solutions provided by an algorithm like k-means on the bottom right of the figure. These solutions seem wrong from the perspective of metrics like CDI or CONN, but forms compact and well-defined clusters optimizing DEV metric. Nevertheless, such algorithms might fail in the identification of atypical patterns.

After those observations, it is logical to think that the correct solution (from the perspective of external knowledge) can be situated in the Pareto front between this two optimal solutions (represented by metrics DEV and CONN). For that reason, an algorithm that can generate optimal solutions considering the trade-off between these two metrics, such as our LPS-ACO algorithm, should be able to access to more proper clusterings solutions for the LPS problem.

6   Conclusion and Future Work

In this paper, we propose the application of Ant Colony Optimization (ACO) to the electrical LPS clustering problem. We analyze the performance of the algorithm searching for the optimization of different metrics proposed in the literature. As in most heuristics, ACO performance depends on good parameter settings. We present a careful analysis of the impact of ACO parameters in the quality of the solutions. Additionally, we compare our LPS-ACO algorithm with some classical clustering algorithms, such as k-means and single-linkage, and other heuristics, such as EPACC and EPACC-SUS. The results have shown that the selection of a proper metric to validate the performance of an algorithm plays a key role determining the effectiveness of a clustering algorithm. For instance, depending on the chosen metric for evaluating the algorithms, we can wrongly rank a clustering algorithm as good or bad if the selected metric is not suitable for the problem under analysis. For that reason, it is important to consider different metrics to measure the performance of an algorithm. Overall, the results obtained by our LPS-ACO algorithm are promising and promote further study to improve the ACO performance. As future work, we will examine the design of a proper metric that reflects good partitions in the particular domain of the electrical LPS clustering problem. We are also interested in the design of a multi-objective ACO approach that allows us to move in all the Pareto front or to determine the proper number of cluster automatically for the electrical LPS problem.

References

1. Chicco, G., Ionel, O.M., Porumb, R.: Electrical load pattern grouping based on centroid model with ant colony clustering. IEEE Trans. Power Syst. **28**(2), 1706–1715 (2013)
2. Das, S., Abraham, A., Konar, A.: Automatic clustering using an improved differential evolution algorithm. IEEE Trans. Syst. Man Cybern. Part A: Syst. Hum. **38**(1), 218–237 (2008)
3. Chicco, G., Napoli, R., Piglione, F.: Comparisons among clustering techniques for electricity customer classification. IEEE Trans. Power Syst. **21**(2), 933–940 (2006)

4. Figueiredo, V., Rodrigues, F., Vale, Z., Gouveia, J.: An electric energy consumer characterization framework based on data mining techniques. IEEE Trans. Power Syst. **20**(2), 596–602 (2005)
5. Chicco, G., Ionel, O.M., Porumb, R.: Formation of load pattern clusters exploiting ant colony clustering principles. In: IEEE EUROCON, pp. 1460–1467 (2013)
6. Dorigo, M., Maniezzo, V., Colorni, A.: Ant system: optimization by a colony of cooperating agents. IEEE Trans. Syst. Man Cybern. Part B: Cybern. **26**(1), 29–41 (1996)
7. Gerbec, D., Gasperic, S., Smon, I., Gubina, F.: Allocation of the load profiles to consumers using probabilistic neural networks. IEEE Trans. Power Syst. **20**(2), 548–555 (2005)
8. Chaouch, M.: Clustering-based improvement of nonparametric functional time series forecasting: application to intra-day household-level load curves. IEEE Trans. Smart Grid **5**(1), 411–419 (2014)
9. Handl, J., Knowles, J.: An evolutionary approach to multiobjective clustering. IEEE Trans. Evol. Comput. **11**(1), 56–76 (2007)
10. Li, M., Ming-ming, S.: An improved ant colony clustering algorithm based on dynamic neighborhood. In: IEEE International Conference on Intelligent Computing and Intelligent Systems, vol. 1, pp. 730–734 (2010)
11. Dunn, J.C.: Well separated clusters and optimal fuzzy partitions. J. Cybern. **4**, 95–104 (1974)
12. Davies, D.L., Bouldin, D.W.: A cluster separation measure. IEEE Trans. Pattern Anal. Mach. Intell. **1**(2), 224–227 (1979)
13. Chou, C.H., Su, M.C., Lai, E.: A new cluster validity measure and its application to image compression. Pattern Anal. Appl. **7**(2), 205–220 (2004)
14. Blum, C., Dorigo, M.: The hyper-cube framework for ant colony optimization. IEEE Trans. Syst. Man Cybern. Part B Cybern. **34**(2), 1161–1172 (2004)
15. Lloyd, S.: Least squares quantization in pcm. IEEE Trans. Inf. Theory **28**(2), 129–137 (1982)
16. Day, W.H., Edelsbrunner, H.: Efficient algorithms for agglomerative hierarchical clustering methods. J. Classif. **1**(1), 7–24 (1984)

Optimization of Operation and Control Strategies for Battery Energy Storage Systems by Evolutionary Algorithms

Jan Müller[1](✉), Matthias März[1], Ingo Mauser[2], and Hartmut Schmeck[1,2]

[1] Karlsruhe Institute of Technology – Institute AIFB, 76128 Karlsruhe, Germany
matthias.maerz@student.kit.edu, {jan.mueller,schmeck}@kit.edu
[2] FZI Research Center for Information Technology, 76131 Karlsruhe, Germany
mauser@fzi.de

Abstract. To support the utilization of renewable energies, an optimized operation of energy systems is important. Often, the use of battery energy storage systems is stated as one of the most important measures to support the integration of intermittent renewable energy sources into the energy system. Additionally, the complexity of the energy system with its many interdependent entities as well as the economic efficiency call for an elaborate dimensioning and control of these storage systems. In this paper, we present an approach that combines the forward-looking nature of optimization and prediction with the feedback control of closed-loop controllers. An evolutionary algorithm is used to determine the parameters for a closed-loop controller that controls the charging and discharging control strategy of a battery in a smart building. The simulation and evaluation of a smart residential building scenario demonstrates the ability to improve the operation and control of a battery energy storage system. The optimization of the control strategy allows for the optimization with respect to variable tariffs while being conducive for the integration of renewable energy sources into the energy system.

Keywords: Energy management system · Smart building · Evolutionary algorithm · Battery energy storage system · Closed-loop control

1 Introduction

One of today's biggest challenges is the fight against climate changes. One of the main measures is the change from fossil to renewable energy sources (RES) [1]. This includes an increased utilization of distributed generation (DG), e.g., wind power by wind turbines and solar radiation by photovoltaic (PV) systems. The intermittent generation by RES and the increasing power feed-in into distribution grids by DG is already leading to voltage and overload problems [2]. To compensate for the intermittency of solar radiation, battery energy storage systems (BESS) are often used in combination with DG by PV systems. BESS increase the self-consumption and self-sufficiency of local energy systems, i.e.,

© Springer International Publishing Switzerland 2016
G. Squillero and P. Burelli (Eds.): EvoApplications 2016, Part I, LNCS 9597, pp. 507–522, 2016.
DOI: 10.1007/978-3-319-31204-0_33

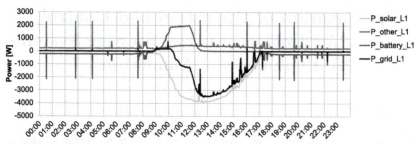

(a) Original control strategy that charges the battery with the complete available power until a maximum charging power of about 2000 W is reached (measured)

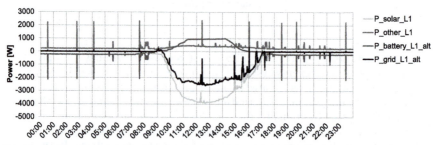

(b) Control strategy that charges the battery with only half of the available power until a maximum charging power of about 1000 W is reached (simulated)

Fig. 1. PV generation (P_solar_L1), other local consumption (P_other_L1), battery (dis-)charging (P_battery_L1), and resulting grid exchange (P_grid_L1) at the FZI Research Center for Information Technology on the 6th of March 2015

buildings that comprise DG. Hence, BESS decrease the feed-in of power into the electricity grid and may help easing the problems that arise from RES.

BESS have high acquisition costs and potential revenues are often low. As a consequence, the usage of distributed BESS in buildings is disputed: In [3] it is stated that "investing in storage is first profitable when large differences in the electricity price frequently occur", which is usually not yet the case. Thus, the economic dimensioning of their capacity is crucial because the costs of BESS are linked to the capacity. The dimensioning of BESS targets on a size that is economic all year round. Normally, the economic dimensioning leads to systems that are undersized for the high generation on days with heavy solar radiation. This reduces the positive effects of BESS on the grid dramatically because it does not reduce the maximum power feed-in into the grid, as it is depicted in Fig. 1 (in black). The figure visualizes the electricity generation by a combined PV and BESS which is located in Karlsruhe, Germany[1]. In times of high solar radiation, the BESS is fully charged by noon and the power generated by the PV system is then fed-back to the electricity grid. Economic dimensioning of BESS in the distribution grid, i.e., a similar ratio of capacity and consumption

---

[1] The overall research environment is more closely described in [4].

of all BESS, leads to a similar finish time of the charging process of each BESS. Thus, all PV systems will start feeding into the distribution grid at about the same time, which does not ease the problems we have today.

The optimization of the control strategies offers a chance to model the maximum feed-in to the electricity grid by charging the battery not with the maximum power possible but by a refined value. For example, the charging process of the battery can be stretched over a longer period as depicted in Fig. 1b. Limiting the charging power and thus stretching the charging period or shifting the time of charging may lead to the battery being not fully charged during the time of PV generation, although enough energy would actually have been available. As a consequence, the parameters have to be adjusted flexibly taking the predicted generation of the PV system into account. Additionally, the energy consumption by other devices in the building has to be respected, which may be achieved by an automated building energy management system (BEMS) that covers all devices as well as the PV system and the BESS.

The major contribution of this paper is the formulation and implementation of an approach optimizing the control strategy of a BESS using an *Evolutionary Algorithm* (EA) in an existing BEMS [5] that is deployed to real buildings. The approach combines the forward-looking nature of optimization and prediction with the feedback control of closed-loop controllers. The EA optimizes the parameters of the charging and discharging strategy of a BESS as well as the operating times of intelligent appliances in a smart building. It enables a joint optimization of a BESS together with all other devices while respecting the interdependencies between them. The EA has proven to optimize heterogeneous devices in various scenarios in a modular and customizable manner [5].

Section 2 provides an overview of related optimization approaches and evaluations regarding BESS. In Sect. 3, the general approach and the concept of combining scheduling, parameterization, and control strategy in the evolutionary optimization in a BEMS is outlined. Section 4 presents the scenario that has been chosen to demonstrate and evaluate the implementation of this novel approach, before discussing the results in Sect. 5. Finally, we conclude the paper in Sect. 6 and provide an outlook on further work.

2 Related Work

This section outlines several related approaches to BEMS and optimization in general as well as approaches to the simulation and optimization of BESS.

2.1 Building Energy Management Systems and Optimization

Optimization in energy systems is done on different levels of the energy system, with varying abstraction, and with respect to diverse objectives. Many publications focus on the optimization of specifications of technical systems, i.e., the dimensioning of the components [6,7]. Other publications focus on an abstracted global perspective, in particular energy markets [8] and long-term developments

[9]. In contrast, BEMS focus on the detailed level of individual buildings and consider individual devices. BEMS require adequate architectures that provide the means for the adaptability and modularity that is necessary to fit the system to a dedicated setup of devices and systems. Examples of such architectures include the *Energy Flexibility Platform and Interface* (EF-Pi) [10], TRIANA [11], and the *Open Gateway Energy Management* (OGEMA) [12]. All target on the exploitation of load flexibilities in the distribution grid by scheduling the operating times of appliances, devices, and in particular DG.

The optimization of device operation, i.e., scheduling, is usually done using a temporal resolution of 15 to 60 min [11,13], rarely in a resolution of five [14] or even one minute [15]. Effects that result from averaging over such time periods lead to so-called *averaging effects* that reduce load peaks in the sense of consumption as well as of generation [11,16]. This causes an overestimation of self-consumption and self-sufficiency rates and, thus, an underestimation of energy costs and maximum feed-in to the grid [16]. Commonly, the optimization problems are formulated as linear programming (LP) [17], mixed integer LP [13,18], or mixed integer non-LP [19] problems and solved using direct solvers. Modeling these kinds of optimization problems requires thousands of variables and constraints, even if the temporal resolution is fifteen minutes [13]. Optimizing such problems often leads to requirements with respect to computing power and memory requirements that are not reasonable for BEMS, which are running continuously and permanently. Additionally, formulating such problems in environments that have a variety of different devices and systems in different combinations and that may change over time is not practicable [5].

A wide range of optimization problems has been addressed quite successfully and efficiently using meta-heuristic approaches [20]. Main advantages include low memory requirements, short run times, and flexible and modular algorithms. Consequently, heuristics have been used in the optimization of energy systems, e.g., *evolutionary algorithms* and *particle swarm optimization* in the optimization of design parameters of used devices [6,7] or to solve scheduling problems [5,15,21]. The approach of [15] is similar to the approach presented hereafter. Nevertheless, it is limited to electrical loads, does not respect interdependencies between multiple devices and energy carriers, and does not cover the optimization of the actual control strategy, i.e., the (dis-)charging, of BESS.

2.2 Simulation of Battery Energy Storage Systems

To enable the simulation of batteries, different types of models have been developed. The structure of these models heavily depends on the application area and the battery type. Common models are electrochemical models, often used in battery design [22], electrical circuit models, which focus on the electrical properties of the battery and that are often used in electrical engineering [23], and analytical models like the *Kinetic Battery Model* [24], which describe the battery in a more abstracted way.

In BEMS, often more simple battery models are used. In [10], batteries and other energy storage systems, e.g., hot water storage tanks, are modeled similarly as buffers. Specific physical characteristics of batteries are not considered; Buffers simply produce or consume energy between a minimum and maximum fill level, oversimplifying the characteristics of BESS. The presented simulation uses an extended simple battery model described in Sect. 4.2.

2.3   Optimization of Battery Energy Storage Systems

There are various approaches to the integration of BESS into BEMS. Most of them use optimization methods to determine charging and discharging powers for concrete time steps under the constraint of a finite battery size. Normally, the optimization is done to find an optimal schedule for a given device setup for a defined optimization horizon. In [11,25,26], this is done using dynamic programming, whereas [27] uses mixed-integer linear programming. Even though these approaches are able to find good solutions for scheduling problems, they do not account for changes in the energy system during the execution of the schedule, as it is done in this paper [28].

The authors of [29] stress the importance of a control strategy that is able to cope with feed-in limits and other incentives, such as variable tariffs. They present an optimization of the configuration of a PV and BESS. The optimized system is then used to determine the peak power of a household. Nevertheless, their approach uses only a simplified model of households with a temporal resolution of ten minutes and is limited to two simple control strategies for the BESS. In [30], two different operational strategies for BESS are proposed that decrease the maximum feed-in into the grid based on predictions. The strategies are similar to those presented in this paper. Unlike this paper, the strategy remains fixed within a scenario, which disregards the possibility to adapt the control strategy to changing situations. Works described in [25,26] utilize dynamic programming to optimize the control of a BESS. Although an adaptive self-learning scheme is introduced, their approach is limited to optimize only three operational modes—charging, idle, and discharging—which provides only a limited subset of possible operational modes of a BESS.

Normally, Commercially available BESS use techniques from control theory to control the battery based on the current PV generation and local consumption, while neglecting predictions or variable tariffs. Sometimes the approaches are combined with BEMS optimizing all devices of the energy system but the BESS [31], whereas this paper optimized all devices and the BESS.

3   Approach

In this section, the approach to the optimization of the battery operation and control strategy is described. Since this approach is integrated into an existing BEMS, we introduce the BEMS briefly before describing and explaining the optimization of the battery more closely.

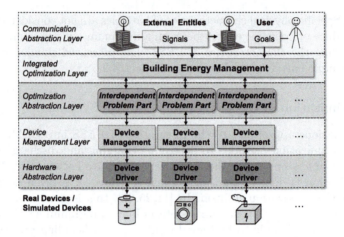

Fig. 2. Overview of organic smart home architecture

3.1 Building Energy Management System

To enable the optimization of the operation of a combined PV and battery system, a central control unit is needed, which gathers all the information necessary to ensure an optimal operation. Since an optimal operation of BESS in buildings depends on a variety of parameters, an integration into BEMS is beneficial.

Although the approach presented in this paper is in principle independent of a particular BEMS, it has been implemented using the *Organic Smart Home* (OSH)[2] framework. The OSH is a BEMS which is capable of optimizing the energy flows in a building while respecting interdependencies between devices. It uses a modular approach towards devices and their optimization. Its general architecture and components are more closely described in [5,32,33] and depicted in Fig. 2.

Devices and systems, e.g., the appliances and the BESS, are abstracted using device drivers. These drivers are connected to units that are dedicated to device management, such as operation strategies, and enable control actions without central optimization. Every device management unit analyzes and aggregates the data from the associated device, abstracts it and passes it to the global unit that is responsible for optimization—the integrated building energy management.

This abstraction is done using so-called *Interdependent Problem Parts* (IPP), which represent sub-problems of the optimization. They contain the current state, a physical model, and an optimization model of the associated device. The physical model is used to simulate and evaluate the future behavior and states of the device, whereas the optimization model provides an interface to the optimization algorithm. This optimization model converts the device representation in the optimization, i.e., solution candidates, into future behavior and, thus, energy consumption or generation that is reflected in load profiles.

---

[2] http://www.organicsmarthome.com.

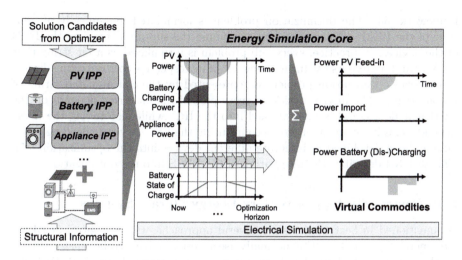

Fig. 3. Evaluation of *Interdependent Problem Parts* (IPP) in the *Energy Simulation Core* using solution candidates and structural information

In the Optimization Layer, an EA is used to find solution candidates which maximize the user goals. The *Energy Simulation Core* (ESC) simulates and evaluates the energy flows in the building based on the IPPs, the solution candidates from the optimizer, and structural information about the interconnections between the devices (see Fig. 3). Hence, it is able to respect the interdependencies between the devices and the devices' constraints, e.g., storage capacities. The simulation is done using discrete time steps over an optimization horizon that depends on the devices that are part of the current optimization process. This horizon always starts at the time of the current optimization run. The simulation of energy flows in the ESC allows to estimate the power flows over the optimization horizon, which is then used to calculate the fitness of the solution candidate. This is done by minimizing the total electricity costs according to the following formula of the electricity price $p_{\text{Consumption},t}$, the feed-in compensation $p_{\text{Feed-In},t}$, the electric energy obtained from the grid $E_{\text{Consumption},t}(X)$, and the electric energy feed-in to the grid $E_{\text{Feed-In},t}(X)$ for the current solution candidate X comprising the device control actions, i.e., the operation mode of the BESS and the delay of the appliances (see [5] for a more detailed description), at the time step t of N total steps:

$$f(X) = \sum_{t=1}^{N} (p_{\text{Consumption},t} \cdot E_{\text{Consumption},t}(X) + p_{\text{Feed-In},t} \cdot E_{\text{Feed-In},t}(X)). \quad (1)$$

3.2 Optimization

In the present system, new solution candidates are created using an EA based on an adapted version of the *generic Genetic Algorithm* from the *jMetal* framework [34]. The optimization problem is formulated dynamically at run-time of the BEMS. This is done by combining the sub-problems, i.e., the IPPs of the devices. In practice, every sub-problem is formulated as a bit string that encodes and represents the future behavior of its respective corresponding device. The overall optimization problem is represented by the concatenated string of all bit strings. The interpretation of every part of the bit string depends on the particular device and is programmed into the IPP, while also taking the inter-dependencies between them into account using the ESC. Several encodings have already been developed to integrate different devices into the optimization [35]. The Sect. 3.4 describes the integration of the BESS into the optimization.

3.3 Control Strategy of Battery Energy Storage Systems

As mentioned in Sect. 2, there are several approaches to the control of BESS. Commercially available BESS normally use a closed-loop controller that is similar to the one described in [31]: It determines the battery charge and discharge powers based on the generation of the PV system and the grid exchange power while ensuring a valid state of charge (SoC) of the battery. Defining that the power exchange from the building to the grid P_{Grid} is negative when power is fed into the grid and positive if power is consumed by the local energy system from the grid, the power for battery charging $P_{\text{Battery},t}$ at time step t depends on the power exchange to the grid in t-1 and is determined as:

$$P_{\text{Battery},t} = -P_{\text{Grid},t-1} + P_{\text{Battery},t-1} \tag{2}$$

This formula describes a simple closed-loop control of the battery charging, which aims at balancing the power exchange with the grid without optimization. In addition, $P_{\text{Battery},t}$ has to fulfill the following conditions that are based on the PV generation $P_{\text{PV},t}$, the SoC of the BESS E_{SoC}, and the mechanism limiting battery charging to the current PV power generation:

$$E_{\text{SoC},t} = E_{\text{SoC},t-1} + \Delta t \cdot P_{\text{Battery},t} \tag{3}$$

$$0 \leq E_{\text{SoC},t} \leq E_{\text{SoC}}^{\text{max}} \tag{4}$$

$$P_{\text{Battery},t} \leq P_{\text{PV},t} \tag{5}$$

$$|P_{\text{Battery},t}| \leq P_{\text{Battery}}^{\text{max}} \tag{6}$$

In this formula, $E_{\text{SoC},t}$ is the SoC at time step t, $E_{\text{SoC}}^{\text{max}}$ is the maximal SoC, and $P_{\text{Battery}}^{\text{max}}$ is the maximal charge and discharge power of the battery. This basic battery control system is also depicted in Fig. 4.

3.4 Optimization of Battery Energy Storage Systems

The approach presented in this paper aims at extending this closed-loop control by the introduction of new dynamic parameters: an upper limit ρ_t^{max} and a lower

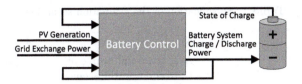

Fig. 4. Scheme of a basic battery control system diagram

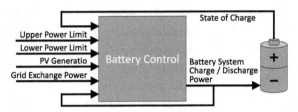

Fig. 5. Scheme of the proposed improved battery control system diagram

limit $\rho_t^{\min}$ of the charge and discharge power at time step t (see Fig. 5). After the introduction of the additional parameters, Eq. 6 changes to:

$$\rho_t^{\min} \cdot P_{\text{Battery}}^{\max} \leq |P_{\text{Battery},t}| \leq \rho_t^{\max} \cdot P_{\text{Battery}}^{\max} \tag{7}$$

$$0 \leq \rho_t^{\min} < \rho_t^{\max} \leq 1 \tag{8}$$

In our approach, the two new parameters $\rho_t^{\max}$ and $\rho_t^{\min}$ are optimized by the framework presented in Sect. 3.1 and its optimization presented in Sect. 3.2. The parameters are defined to be constant over a time of one hour. In the EA, the two parameters are gray-encoded as a bit string. In this paper, four bits have been chosen for each parameter to enable a granularity of 16 settings for the maximum and the minimum charging. Therefore, eight additional bits per hour in the optimization horizon are introduced in the optimization. For instance, when having an optimization horizon of 18 h this results in a total of 144 bits for the BESS in the optimization.

Based on the new parameters, the control strategy of the BESS reacts differently in every hour, because it is adapted by the BEMS to the current situation, which depends on the intermittent generation by the PV system, and the current as well as the future consumption by other devices. It is important to note that $\rho_t^{\max}$ limits the battery charge and discharge power while $\rho_t^{\min}$ forces the battery to charge or discharge. This can be utilized to comply with power limits provided by grid operators.

4 Scenario and Simulation Setup

To evaluate the performance of the approach, a scenario comprising household appliances, a PV system, and a BESS is analyzed (see Fig. 6 and Table 1). The total energy costs are optimized by the BEMS with respect to variable tariffs, the

Table 1. Specification of the evaluation scenarios

General	Number of persons:	4
Appliances	Avg. temporal degree of freedom in the scenario:	
(dishwasher, washing machine,	basic and improved control:	0 h
tumble dryer, hob, oven)	advanced optimization:	8 h
Photovoltaic system	Maximum power:	3.8 kW
	Annual production:	4500 kWh
Battery energy	Maximum power:	3.5 kW
storage system	Usable capacity:	3 kWh
	Charging efficiency:	93 %
Tariff and limit	Electricity from grid:	0.28 EUR/kWh
	Photovoltaic feed-in:	Time dependent

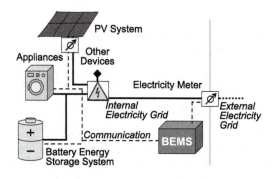

Fig. 6. Overview of the evaluation scenario

operation times of the appliances, and—most importantly—the control strategy of the BESS. To concentrate on the BESS, this paper does not consider the heating system of the building.

To evaluate the new approach, a scenario with the improved battery control (see Sect. 3.4) is compared to a scenario using the common basic battery control (see Sect. 3.3). In addition, an advanced scenario with intelligent household appliances is investigated to assess the integration of the proposed approach into an advanced BEMS with demand side management functionality of appliances. To assess the dependance on the solar radiation, simulations in January, April, July and October are carried out.

In the optimization, the genetic algorithm utilizes binary tournament selection, single-point-crossover with two offspring, and bit-flip-mutation with an elitist (μ, λ)-strategy and a rank based survivor selection. Parameters of the optimization have been calibrated manually and set to a crossover probability of 0.7, a mutation probability of 0.001, and 100 generations of 100 individuals [33]. Every scenario has been run 20 times with different random seeds.

4.1 Household Appliances and Baseload

To simulate the electrical load of the household, we use five major appliances: dishwasher, washing machine, tumble dryer, hob, and oven. The average number of yearly operation cycles as well as the starting times of the appliances are simulated using statistical data and real recorded load profiles as presented in [5]. In the advanced scenario with intelligent, deferrable appliances, the operating times of the appliances are optimized as described in [5].

In addition to the appliances, a load profile based on the German standard load profile *H0* is used to model further electricity consumption. The load profile has been scaled to match the total yearly electricity consumption of 4500 kWh of the household when added to the consumption of the simulated appliances.

4.2 Photovoltaic and Battery Energy Storage System

Electricity is produced by a PV system with a maximum power of 3.8 kW and an annual production of 4500 kWh. The production profile is taken from measurements in our laboratory in Karlsruhe, Germany. The profiles have a resolution of one minute, reflecting the change of generation during the time of the year as well as short-term intermittency.

The battery has a usable capacity of 3 kWh and a maximum technical charge and discharge power of 3.5 kW. In the present scenario, the battery is charged by the PV system only. The battery is modeled having a round-trip efficiency of 93 % and a typical constant voltage charging phase at the end of the charging process. The standing loss is omitted because it is rather small when considering a lithium-ion battery. More detailed physical characteristics of batteries are not included, which reflects the fact that commercially available BESS often include specialized systems managing the physical processes in the battery. We assume that these systems will provide an abstracted interface to the battery.

4.3 Electricity Tariffs

As mentioned in Sect. 1, BESS lead only to significant profitability when having variable electricity prices. Even though variable tariffs are not yet common, it can be assumed that future electricity tariffs for consumption as well as feed-in will reflect variable prices at the energy exchanges. In the present evaluation scenarios, a time-variable feed-in tariff is used to represent possible future electricity pricing. The feed-in tariff is modeled based on the electricity prices at the intra-day market at the *European Power Exchange* in September 2015 (see magenta line in Fig. 8). In addition, it is assumed that the state of the grid is contained in the feed-in tariff, i.e., the compensation is lower when the grid operator wants to avoid additional feed-in. For the sake of simplicity, power limits are not invcluded.

Table 2. Simulation results for a simulation duration of one week and 20 simulations per scenario

Month	Scenario	Electricity costs in cent			Improvement with respect to the basic control scenario in %
		Min	Max	Average	
January	basic control scenario	–	–	2856	–
	improved control scenario	2859	2872	2863	−2
	advanced scenario	2861	2870	2863	−2
April	basic control scenario	–	–	904	–
	improved control scenario	745	826	792	14
	advanced scenario	760	807	785	15
July	basic control scenario	–	–	−366	–
	improved control scenario	−500	−580	−540	47
	advanced scenario	−503	−613	−556	55
October	basic control scenario	–	–	1181	–
	improved control scenario	1088	1137	1120	5
	advanced scenario	1108	1154	1135	4

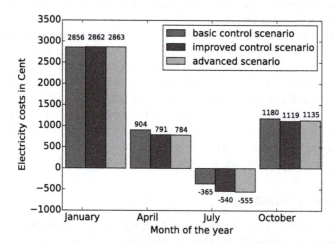

Fig. 7. Electricity costs in dependence on the scenario and the time of the year

5   Results and Discussion

The simulation results (see Fig. 7 and Table 2) show that the weekly electricity costs depend heavily on time of the year. Only in April and July, which are times of relatively high solar radiation, a significant improvement of the weekly electricity costs can be achieved by an improved battery control strategy. In

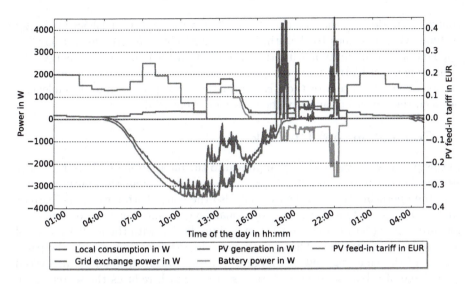

Fig. 8. Electricity generation and consumption in the simulation of one day in July in the improved control scenario(Color figure online)

July, the improved control scenario shows an average improvement of the weekly revenue by 47 % with respect to the basic control scenario.

Figure 8 depicts the power consumption, the power exchange with the electricity grid, the PV generation, and the PV feed-in tariff over the time of one day in one of the simulations. The peak in the power consumption at 13:00 represents the power used to charge the battery. Taking the PV feed-in tariff (magenta) into account, the battery is charged during a time of low PV feed-in compensation, showing the desired behavior. This allows for an active load shaping using selective battery charging and discharging that respects variable feed-in tariffs reflecting the state of the grid. At 19:00, a washing machine is started. When comparing the local consumption in blue with the battery discharge power in cyan, it is visible that the battery compensates the local consumption using an upper limit on the discharge power determined by the optimization. The compensation of the fluctuations is done by the open-loop control which adapts the battery discharge power appropriately. In comparison to predefined fixed discharge powers, the open-loop control avoids unnecessary discharge caused by unexpected load fluctuations.

The figure supports the results in Table 2, asserting that the improved battery control is able to find an optimized charging strategy while keeping the flexibility of the open-loop control and the number of necessary evaluations in the optimization low. The advanced control scenario shows an average improvement of 55 % with respect to the basic control scenario. This is caused by the additional flexibility that is introduced by the deferrable appliances. Compared to the improvement introduced by the improved battery control, the improvement achieved by the introduction of the deferrable appliances in the advanced control scenario is rather small. This indicates the ability of BESS in energy systems when it comes to the utilization of flexibility. Nevertheless, deferrable appliances in combination with the optimized BESS improve the energy systems. When considering more complicated energy systems with different energy carriers, the integration of the BESS in a BEMS respecting interdependencies between the devices promises to be even more beneficial.

6   Conclusion and Outlook

In this paper, we present a novel approach to the integration of a battery energy storage system into a building energy management system that optimizes not only the operating times of devices but also the control strategy of the battery energy storage system using an evolutionary algorithm. Thus, it combines the forward-looking nature of optimization and prediction with the feedback control of closed-loop controllers. The evolutionary algorithm optimizes only the parameters of the charging and discharging control strategy instead of the overall load shape of a battery energy storage system, which reduces the search space in comparison to optimizing the power of the battery energy storage system for every time step in the optimization directly.

Although battery energy storage systems offer the chance to reduce the residual load of a local energy system by their very nature as a storage system, the dimensioning of their capacity is crucial because costs are strongly linked to the capacity. Batteries that are dimensioned with respect to an economic use all year round are typically undersized for the heavy generation of photovoltaic systems on sunny days or wind turbines on stormy days. This reduces their positive effects on the grid dramatically. Optimized control strategies enable to reduce the maximum feed-in to the grid and adapt the overall load shape, taking variable tariffs into account that reflect the global balancing of generation and consumption or possibly also the local grid state.

Simulation results of a smart residential building scenario show that the optimization of the charging and discharging strategy of the battery system reduces the energy costs successfully in case of variable feed-in compensation tariffs. This helps to ease the integration of renewable energy sources into the distribution grid, where in particular intermittent feed-in by photovoltaic and wind power systems leads to a synchronized feed that will be reflected by variable feed-in compensation tariffs in the future.

Future work will extend the approach to the optimization of a parameter that defines the correlation between the maximum possible and the actually performed feed-in. The present simulation results will be verified with data and field test results from our test buildings, which are equipped with battery energy storage systems and optimized using the same building energy management system as this paper. To analyze the impact of the optimization of the control strategy on entire distribution grids, we will simulate multiple buildings concurrently in a bottom-up simulation of a distribution grid. Furthermore, we will extend the single objective to multi-objective optimization, to take additional objectives into account, such as carbon dioxide emissions and the battery health.

References

1. International Energy Agency (ed.): Tracking Clean Energy Progress 2015. International Energy Agency (2015)
2. von Appen, J., Stetz, T., Braun, M., Schmiegel, A.: Local voltage control strategies for PV storage systems in distribution grids. IEEE Trans. Smart Grid 5(2), 1002–1009 (2014)
3. Wirth, H., Schneider, K.: Recent Facts about Photovoltaics in Germany. Fraunhofer Institute for Solar Energy Systems (2015)
4. Becker, B., Kern, F., Lösch, M., Mauser, I., Schmeck, H.: Building energy management in the FZI house of living labs. In: Gottwalt, S., König, L., Schmeck, H., Steber, D., Bazan, P., German, R., Pruckner, M., Dethlefs, T., et al. (eds.) EI 2015. LNCS, vol. 9424, pp. 95–112. Springer, Heidelberg (2015). doi:10.1007/978-3-319-25876-8_9
5. Mauser, I., Müller, J., Allerding, F., Schmeck, H.: Adaptive building energy management with multiple commodities and flexible evolutionary optimization. Renew. Energ. 87(Part 2), 911–921 (2016)
6. Ahmadi, P., Rosen, M.A., Dincer, I.: Multi-objective exergy-based optimization of a polygeneration energy system using an evolutionary algorithm. Energy 46(1), 21–31 (2012)
7. Kavvadias, K., Maroulis, Z.: Multi-objective optimization of a trigeneration plant. Energ. Policy 38(2), 945–954 (2010)
8. Weidlich, A., Veit, D.: A critical survey of agent-based wholesale electricity market models. Energ. Econ. 30(4), 1728–1759 (2008)
9. Henning, H.M., Palzer, A.: A comprehensive model for the German electricity and heat sector in a future energy system with a dominant contribution from renewable energy technologies - Part I: Methodology. Renew. Sustain. Energ. Rev. 30, 1003–1018 (2014)
10. van der Waaij, B., Wijbrandi, W., Konsman, M.: White Paper Energy Flexibility Platform and Interface (EF-Pi) - How the Developments in the Energy Flexibility Market can be Accelerated by the Energy Flexibility Platform and Interface. Technical report, TNO, NRG, Alliander, and Technolution (2015)
11. Toersche, H., Hurink, J., Konsman, M.: Energy management with triana on FPAI. In: 2015 IEEE Eindhoven PowerTech, pp. 1–6 (2015)
12. Nestle, D., Ringelstein, J., Waldschmidt, H.: Open energy gateway architecture for customers in the distribution grid. IT - Inf. Technol. 52(2), 83–88 (2010)
13. Ha, D.L., Joumaa, H., Ploix, S., Jacomino, M.: An optimal approach for electrical management problem in dwellings. Energ. Build. 45, 1–14 (2012)
14. Chen, Z., Wu, L., Fu, Y.: Real-time price-based demand response management for residential appliances via stochastic optimization and robust optimization. IEEE Trans. Smart Grid 3(4), 1822–1831 (2012)
15. Soares, A., Antunes, C.H., Oliveira, C., Gomes, Á.: A multi-objective genetic approach to domestic load scheduling in an energy management system. Energy 77, 144–152 (2014)
16. Wright, A., Firth, S.: The nature of domestic electricity-loads and effects of time averaging on statistics and on-site generation calculations. Appl. Energ. 84(4), 389–403 (2007)
17. Mohsenian-Rad, A.H., Leon-Garcia, A.: Optimal residential load control with price prediction in real-time electricity pricing environments. IEEE Trans. Smart Grid 1(2), 120–133 (2010)

18. Sou, K.C., Weimer, J., Sandberg, H., Johansson, K.H.: Scheduling smart home appliances using mixed integer linear programming. In: 2011 50th IEEE Conference on Decision and Control and European Control Conference (CDC-ECC), pp. 5144–5149 (2011)
19. Shirazi, E., Jadid, S.: Optimal residential appliance scheduling under dynamic pricing scheme via HEMADAS. Energ. Build. **93**, 40–49 (2015)
20. Mesghouni, K., Hammadi, S., Borne, P.: Evolutionary algorithms for job-shop scheduling. Int. J. Appl. Math. Comput. Sci. **14**(1), 91–104 (2004)
21. Pedrasa, M., Spooner, T., MacGill, I.: Coordinated scheduling of residential distributed energy resources to optimize smart home energy services. IEEE Trans. Smart Grid **1**(2), 134–143 (2010)
22. Doyle, M.: Modeling of galvanostatic charge and discharge of the lithium/polymer/insertion cell. J. Electrochem. Soc. **140**(6), 1526 (1993)
23. Chen, M., Rincon-Mora, G.: Accurate electrical battery model capable of predicting runtime and i-v performance. IEEE Trans. Energ. Convers. **21**(2), 504–511 (2006)
24. Manwell, J.F., McGowan, J.G.: Lead acid battery storage model for hybrid energy systems. Sol. Energ. **50**(5), 399–405 (1993)
25. Boaro, M., Fuselli, D., Angelis, F., Liu, D., Wei, Q., Piazza, F.: Adaptive dynamic programming algorithm for renewable energy scheduling and battery management. Cogn. Comput. **5**(2), 264–277 (2013)
26. Huang, T., Liu, D.: Residential energy system control and management using adaptive dynamic programming. In: The 2011 International Joint Conference on Neural Networks (IJCNN), pp. 119–124 (2011)
27. Morais, H., Kdr, P., Faria, P., Vale, Z.A., Khodr, H.: Optimal scheduling of a renewable micro-grid in an isolated load area using mixed-integer linear programming. Renew. Energ. **35**(1), 151–156 (2010)
28. Lu, B., Shahidehpour, M.: Short-term scheduling of battery in a grid-connected pv/battery system. IEEE Trans. Power Syst. **20**(2), 1053–1061 (2005)
29. von Appen, J., Braun, M.: Grid integration of market-oriented PV storage systems. In: ETG Congress 2015: Die Energiewende - Blueprint for the new energy age, VDE in press (2015)
30. Zeh, A., Witzmann, R.: Operational strategies for battery storage systems in low-voltage distribution grids to limit the feed-in power of roof-mounted solar power systems. Energ. Procedia **46**, 114–123 (2014)
31. Castillo-Cagigal, M., Caamaño-Martín, E., Matallanas, E., Masa-Bote, D., Gutiérrez, A., Monasterio-Huelin, F., Jiménez-Leube, J.: PV self-consumption optimization with storage and active DSM for the residential sector. Sol. Energ. **85**(9), 2338–2348 (2011)
32. Allerding, F., Schmeck, H.: Organic smart home: architecture for energy management in intelligent buildings. In: Proceedings of the 2011 Workshop on Organic computing, pp. 67–76. ACM (2011)
33. Mauser, I., Feder, J., Müller, J., Schmeck, H.: Evolutionary optimization of smart buildings with interdependent devices. In: Mora, A.M., Squillero, G. (eds.) EvoApplications 2015, pp. 239–251. Springer, Heidelberg (2015)
34. Durillo, J., Nebro, A., Alba, E.: The jMetal framework for multi-objective optimization: Design and architecture. CEC **2010**, 4138–4325 (2010)
35. Mauser, I., Dorscheid, M., Allerding, F., Schmeck, H.: Encodings for evolutionary algorithms in smart buildings with energy management systems. In: IEEE Congress on Evolutionary Computation (CEC), pp. 2361–2366. IEEE (2014)

EvoGAMES

Orthogonally Evolved AI to Improve Difficulty Adjustment in Video Games

Arend Hintze[1]([✉]), Randal S. Olson[2], and Joel Lehman[3]

[1] Michigan State University, Michigan, USA
hintze@msu.edu
[2] University of Pennsylvania, Pennsylvania, USA
[3] IT University of Copenhagen, Copenhagen, Denmark

Abstract. Computer games are most engaging when their difficulty is well matched to the player's ability, thereby providing an experience in which the player is neither overwhelmed nor bored. In games where the player interacts with computer-controlled opponents, the difficulty of the game can be adjusted not only by changing the distribution of opponents or game resources, but also through modifying the skill of the opponents. Applying evolutionary algorithms to evolve the artificial intelligence that controls opponent agents is one established method for adjusting opponent difficulty. Less-evolved agents (i.e., agents subject to fewer generations of evolution) make for easier opponents, while highly-evolved agents are more challenging to overcome. In this publication we test a new approach for difficulty adjustment in games: orthogonally evolved AI, where the player receives support from collaborating agents that are co-evolved with opponent agents (where collaborators and opponents have orthogonal incentives). The advantage is that game difficulty can be adjusted more granularly by manipulating two independent axes: by having more or less adept collaborators, and by having more or less adept opponents. Furthermore, human interaction can modulate (and be informed by) the performance and behavior of collaborating agents. In this way, orthogonally evolved AI both facilitates smoother difficulty adjustment and enables new game experiences.

Keywords: Difficulty adjustment · Coevolution · Evolutionary computation · Markov Networks

1 Introduction

A challenge in designing computer games is to match a game's difficulty appropriately to the skill level of a human player. Most commonly, game developers design explicit levels of difficulty from which a user can select. Evolving artificial intelligence (AI), i.e., applying evolutionary algorithms to adapt agents, has often been applied to improve video games [1,2], particularly for adjusting their difficulty [3,4]. Such difficulty adjustment approaches generally fall into two categories. In one, the player is immersed in a world where the computer-controlled

© Springer International Publishing Switzerland 2016
G. Squillero and P. Burelli (Eds.): EvoApplications 2016, Part I, LNCS 9597, pp. 525–540, 2016.
DOI: 10.1007/978-3-319-31204-0_34

game agents evolve in *real-time* as the game is played. In the other, opponent AI is evolved *offline*, and options for difficulty are extracted by exploiting the evolutionary history of the opponent AI (among many others [5–16]).

This work extends from the latter category, but instead of evolving offline only AI for opponent agents, AI is evolved for two kinds of game agents that have orthogonal motives. One class of AI agent opposes the player (called opponent agents), while the other class helps the player (called collaborator agents). Such AIs are evolved through competitive co-evolution, i.e., one opponent population and one collaborator population compete with each other. The idea is that pitting such populations against each other can result in an arms race [17,18] in which both agents become more competent as evolution progresses.

One naïve application of such orthogonally-evolved opponent and player AIs is to discard the evolutionary history of the player AI (because a human player will fill that role in the game), and to use only the evolutionary history of the evolved opponent AI to derive a range of opponent difficulties that can be deployed within the game. However, an interesting idea is to instead use *both* evolutionary histories. In particular, if co-evolution is conducted as a competition between a population of player-friendly collaborators (with similar capabilities as the player) and a population of player-antagonistic opponents, then the final game can include both evolved collaborative and opponent agents of different adaptedness. Here we investigate the advantages that such an orthogonal evolutionary approach has for difficulty adjustment.

The motivation is that a richer set of player experiences can result from players interacting both with evolved opponents and evolved collaborators. That is, the distinct evolutionary histories of the orthogonally-evolved populations yield two independent axes for difficulty adjustment. In this way, the player can interact with opponent and collaborator AIs taken from separate evolutionary time points. The hypothesis is that such orthogonal evolution gives game designers more options to adjust game difficulty and provide diverse player experiences.

Experiments in this paper are conducted through a browser-based game in which a player competes with evolved opponents and is assisted by evolved collaborators. In particular, a scientific predator-prey simulation [19] is adapted such that the player controls one prey agent among a group of AI-controlled prey collaborators with the objective of avoiding being consumed by an AI-controlled predator agent. Play tests conducted through Amazon's Mechanical Turk collected data relating player survival time to the adaptedness of both the AI-controlled predator and prey agents. Supporting the hypothesis, the results demonstrate that independently adjusting the level of adaptation of opponents and collaborators creates unique difficulty levels for players. The conclusion is that orthogonally-evolved AI may be a promising method for game designers to adjust game difficulty more granularly and to provide a wider variety of experiences for players.

2   Background

The next sections review previous mechanisms to create adjustable difficulty in video games, and the Markov Network encoding applied in the experiments to represent controllers for game agents.

2.1   Difficulty Adjustment in Video Games

Difficulty adjustment is important for video games, because how a user experiences (and potentially enjoys) a game is impacted directly by the fit between their capabilities and those necessary to progress in the game [4]. The traditional approaches are to either have a fixed level of difficulty, or to allow a player to choose from a set of hand-designed difficulties. However, a universal difficulty level may fail to satisfy many players; and hand-designed difficulties may require significant design effort, provide only a coarse means of adjustment, and require the player to self-rate their capabilities before interacting with the game. Thus, many methods for automatic difficulty adjustment have been explored [1, 3, 4, 20].

One important facet of game difficulty is the adeptness of non-player character (NPC) agents in the game. For example, how far from optimality does the behavior of opponent agents stray? That is, the more optimal the opponent is, the more difficult the player challenge will be. This paper focuses on applying evolutionary algorithms (EA) to create agent controllers as a means to generate game scenarios that vary in difficulty. EAs are appropriate for automatic controller design because they provide a powerful and flexible mechanism to do so given only a measure of adeptness, and have therefor been used often in the past for such purposes [1, 16, 20].

Evolutionary approaches to difficulty adjustment can be categorized generally in two ways. In the first category, the computer-controlled agents evolve in real-time as the game is played. The idea is to dynamically alter the game AI based on player interaction, which can create unique and personalized player experiences [20–27]. However, such approaches require the game to be designed around AI adaptation, enabling new types of games but limiting their application to game AI in general. For example, in Creatures [21] the main game mechanic is to guide and teach a species of AI agents, while in NERO [23] a similarly-inspired mechanic is to train a battalion of agents to fight other ones. By their nature, such mechanics lead to unpredictable outcomes and can expose players to degenerate NPC behavior, which while compelling in their own right may also undermine a designer's ability to craft specific and predictable player experiences.

This study focuses on a second category in which AI for opponents is optimized offline, i.e., it remains unchanged during gameplay. The benefit of offline adaptation is that player experience can be more tightly controlled, enabling it potentially to be applied more broadly. One popular mechanism for such offline AI design is to use EAs to evolve agent controllers. In particular, if selection in an EA is oriented towards stronger AI behaviors, the difficulty of the game can then be adjusted by exploiting the evolutionary history of the opponent AI [5–16].

In this way, less evolved AI (e.g., from early generations of evolution) can serve as a player's opponent in early or easy levels, and more sophisticated AI (e.g., from later generations of evolution) can be featured in more difficult levels. However, most previous approaches focus singularly on the most optimal behavior evolved [8,9,16], and those that consider evolving interesting or diverse opponents [1,28] do not fully explore the possibilities enabled by competitive coevolution in this context. One such possibility (which is the central focus of this paper) is to leverage as a source of diverse difficulties the separate evolutionary trajectories of populations of agents with asymmetric abilities and conflicting motivations.

The next section reviews the encoding used to represent and evolve agent behaviors in the experiments.

2.2   Markov Networks

The experiment in this paper leverages a browser-game derived from the predator-prey simulation in [29]. Agents in the simulation (and thus the game) are controlled by Markov Networks (MNs), which are probabilistic controllers that makes decisions about how an agent interacts with its environment. Because a MN is responsible for the control decisions of its agent, it can be thought of as an *artificial brain* for the agent it controls. Although MNs are the particular artificial brain applied in the simulation, other methodologies for evolving controllers could also be used, such as neuroevolution or genetic programming. This section briefly describes MNs, but a more detailed description can be found in [30].

Agents in the game have sensors and actuators, as shown in Fig. 1. Every simulation time step, the MNs receive input via those sensors, perform a computation on inputs and any hidden states (i.e., their internal memory), then place the result of the computation into hidden or output states (e.g., actuators). When MNs are evolved with a GA, mutations affect (1) which states the MN pays attention to as input, (2) which states the MN outputs the result of its computation to, and (3) the internal logic that converts the input into the corresponding output.

When agents are embedded into a game simulation, sensory inputs from its retina are input into its MN every simulation step (labeled "retina" and "Markov Network", respectively in Fig. 1). The MN is then activated, which allows it to store the result of the computation into its hidden and output states for the next time step. MNs are networks of Markov Gates (MGs), which perform the computation for the MN. In Fig. 2, we see two example MGs, labeled "Gate 1" and "Gate 2." At time t, Gate 1 receives sensory input from states 0 and 2 and retrieves state information (i.e., memory) from state 4. At time $t + 1$, Gate 1 then stores its output in hidden state 4 and output state 6. Similarly, at time t Gate 2 receives sensory input from state 2 and retrieves state information in state 6, then places its output into states 6 and 7 at time step $t + 1$. When MGs place their output into the same state, the outputs are combined into a single

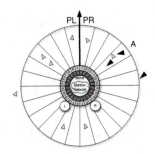

Fig. 1. An illustration of the agents in the model. Light grey triangles are prey agents and the dark grey triangles are predator agents. The agents have a 360° limited-distance retina (200 virtual meters) to observe their surroundings and detect the presence of other agents. The current heading of the agent is indicated by a bold arrow. Each agent has its own Markov Network, which decides where to move next based off of a combination of sensory input and memory. The left and right actuators (labeled "L" and "R") enable the agents to move forward, left, and right in discrete steps.

output using the OR logic function. Thus, the MN uses information from the environment and its memory to decide where to move in the next time step $t+1$.

In a MN, states are updated by MGs, which function similarly to digital logic gates, e.g., AND & OR. A digital logic gate, such as XOR, reads two binary states as input and outputs a single binary value according to the XOR logic. Similarly, MGs output binary values based on their input, but do so with a probabilistic logic table. Table 1 shows an example MG that could be used to control a prey agent that avoids nearby predator agents. For example, if a predator is to the right of the prey's heading (i.e., PL = 0 and PR = 1, corresponding to the second row of this table), then the outputs are move forward (MF) with a 20 % chance, turn right (TR) with a 5 % chance, turn left (TL) with a 65 % chance, and stay still (SS) with a 10 % chance. Thus, due to this probabilistic input-output mapping, the agent MNs are capable of producing stochastic agent behavior.

Table 1. An example MG that could be used to control a prey agent which avoids nearby predator agents. "PL" and "PR" correspond to the predator sensors just to the left and right of the agent's heading, respectively, as shown in Fig. 1. The columns labeled P(X) indicate the probability of the MG deciding on action X given the corresponding input pair. MF = Move Forward; TR = Turn Right; TL = Turn Left; SS = Stay Still.

PL	PR	P(MF)	P(TR)	P(TL)	P(SS)
0	0	0.7	0.05	0.05	0.2
0	1	0.2	0.05	0.65	0.1
1	0	0.2	0.65	0.05	0.1
1	1	0.05	0.8	0.1	0.05

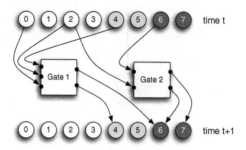

Fig. 2. An example Markov Network (MN) with four input states (white circles labeled 0-3), two hidden states (light grey circles labeled 4 and 5), two output states (dark grey circles labeled 6 and 7), and two Markov Gates (MGs, white squares labeled "Gate 1" and "Gate 2"). The MN receives input into the input states at time step t, then performs a computation with its MGs upon activation. Together, these MGs use information about the environment, information from memory, and information about the MN's previous action to decide where to move next.

A circular string of bytes is used to encode the genome, which contains all the information necessary to describe a MN. The genome is composed of *genes*, and each gene encodes a single MG. Therefore, a gene contains the information about which states the MG reads input from, which states the MG writes its output to, and the probability table defining the logic of the MG. The start of a gene is indicated by a *start codon*, which is represented by the sequence (42, 213) in the genome.

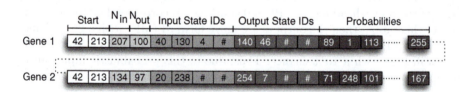

Fig. 3. Example circular byte strings encoding the two Markov Gates (MGs) in Fig. 2, denoted Gene 1 and Gene 2. The sequence (42, 213) represents the beginning of a new MG (white blocks). The next two bytes encode the number of input and output states used by the MG (light grey blocks), and the following eight bytes encode which states are used as input (medium grey blocks) and output (darker grey blocks). The remaining bytes in the string encode the probabilities of the MG's logic table (darkest grey blocks).

Figure 3 depicts an example genome. After the start codon, the next two bytes describe the number of inputs (N_{in}) and outputs (N_{out}) used in this MG, where each $N = 1 + (\text{byte} \mod N_{max})$. Here, $N_{max} = 4$. The following N_{max} bytes specify which states the MG reads from by mapping to a state ID number with the equation: $(\text{byte} \mod N_{states})$, where N_{states} is the total number of input, output, and hidden states. Similarly, the next N_{max} bytes encode which states the MG writes to with the same equation as N_{in}. If too many inputs or outputs are specified, the remaining sites in that section of the gene are ignored, designated by the # signs. The remaining $2^{N_{in}+N_{out}}$ bytes of the gene define the probabilities in the logic table.

All evolutionary changes such as point mutations, duplications, deletions, or crossover are performed on the byte string genome. During a point mutation, a random byte in the genome is replaced with a new byte drawn from a uniform random distribution. If a duplication event occurs, two random positions are chosen in the genome and all bytes between those points are duplicated into another part of the genome. Similarly, when a deletion event occurs, two random positions are chosen in the genome and all bytes between those points are deleted. Crossover for MNs was not implemented in this experiment to allow for a succinct reconstruction of the line of descent of the population, which was important in the original study [29].

3   Approach

In typical applications of video game difficulty adjustment through evolved AI, only one class of AI (typically the opponent) is evolved, and the evolutionary history of the evolved AI yields a variety of differentially-adapted AIs. Near the beginning of evolutionary training we expect AIs to be incapable or maladapted, while after many generations of selection the AI becomes increasingly competent at performing the task it was selected for. This range of behaviors forms a continuum from which one can tailor the difficulty of player game experiences. Here instead of evolving only a single population of opponent AI agents, we co-evolve both the opponent agent and collaborative agents that help the player; these distinct agent types can have different capabilities and will have orthogonal fitness functions (because their motivations are in conflict).

The advantage of orthogonally-evolved AI is that it can enable players to interact not only with collaborative and opponent AIs taken from the same generation of evolution, but also with agents taken from separate, arbitrary generations. For example, the player can play not only with opponents and collaborators that are both capable (i.e., opponents and collaborators taken from the end of an evolutionary run), but can also face a more difficult situation if a well-adapted opponent is combined with a weakly-adapted and largely incapable team of player-collaborative agents. Or conversely, to engineer an easier game experience, a well-adapted collaborating team can be combined with an incapable opponent taken from an early generation of its evolution. The idea is that combining opponents and collaborators from different points of evolution will result in increased possibilities for player game experiences, as illustrated by Fig. 4.

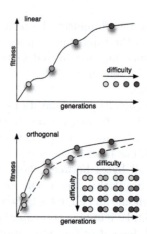

Fig. 4. Comparison of linear vs. orthogonal evolution of AI. The top figure (linear) shows a typical application of evolved AI. The difficulty (ability of the evolved AI) increases with generations of evolution. The bottom Figure (orthogonal) shows an example of orthogonal evolution where two populations are co-evolved with orthogonal incentives. Because AIs from both populations can be mixed arbitrarily, many more game situations can be constructed.

4   Experiment

The main hypothesis explored here is that the described method of using orthogonally evolved AIs helps to expand options for game difficulty. One means to test this hypothesis is for players to interact in a game setting with various mixtures of adapted and unadapted AIs for both opponents and collaborators. If the hypothesis is true, the expectation is that player performance will vary over all tested combinations. Conversely, if the hypothesis is false then there should be no additional significant differences in player performance from varying agent adapatedness across opponents and collaborators.

Testing this hypothesis requires a video game implementation with NPCs that can play alongside the player, which may not be possible or appropriate in every game. In this paper, the particular video game used for experimentation is derived from a simple predator-prey simulation from [29]. In the original simulation, predator and prey agents are controlled by evolved Markov networks inspired by computational abilities of biological brains. A single predator on a 2d surface is evolved to catch as many of the coexisting prey agents as possible. In contrast, the group of prey agents is collectively evolved to resist being caught (for detailed explanation see: [29]). In this way, the motivations of the predator and prey are orthogonal. Over generations, the predator evolves to more efficiently catch prey agents by learning to attack the outside of a swarm of prey agents more consistently. Prey agents in the simulation evolve to swarm together, because those that can not successfully swarm become isolated from the rest of the prey agents, and are more easily caught. The resulting evolved swarming behavior is explained by the selfish herd hypothesis [19,31], which the simulation was designed to investigate.

A game was created which implemented the same simulation rules, but substituted a human player for one of the swarming prey agents. The human's objective is to evade the predator as long as possible (Fig. 5). All of the other agents in the game are controlled by MNs; importantly, the non-player prey agents are controlled by MNs taken from a separate population (and potentially a separate point in that population's evolutionary history) from that of the predator agent.

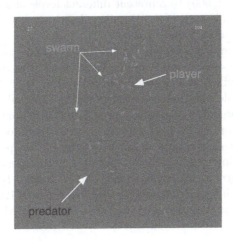

Fig. 5. Typical game situation. The player uses the keyboard's left and right keys to control the player agent (bright green) within a group of other collaborating agents (dim green). The predator agent (red) can kill the other agents if close enough to them. The player has 120 s (remaining time is shown at the top right) to evade the predator. Note that the number of remaining collaborating agents is shown at the top left. Figure is best viewed in color (Color figure online).

The simulation was adapted into a browser-based game, and human players were recruited from Amazon's Mechanical Turk to play. In the game, a group of swarming agents is antagonized by a predator agent. The player acts as one of the swarm agents and is tasked with avoiding the predator. All agents other than the player (i.e., the predator and the remaining prey agents) are controlled by previously evolved Markov networks (i.e., no evolution takes place while the game is played). The predators were evolved to catch prey while prey agents were evolved to flee predators and eventually swarm together to avoid capture. Predator and prey Markov networks can either come from a relatively *unevolved* stage (generation 900) or from a relatively *evolved* stage (generation 1,900). At the beginning of the game one of the four possible combinations of adaptation for the predator and prey Markov networks is randomly chosen. The game ends either if the predator catches the player or after 120 s pass.

4.1   Experimental Details

First, several evolutionary runs were performed using the EOS framework [29] with its default settings. Organisms on the line of descent [32] from the predator as well as from the prey populations were saved every 25 generations. A particular evolutionary run was then selected that showed a large gain in swarming and predation capability at end of the run compared to the beginning. The motivation was to ensure a significant recognizable difference between the capabilities of the AI over its evolutionary history. Agents were chosen from two time points (generations 900 and 1, 900) to represent different levels of adaptation (referred to here as *evolved* and *unevolved*). Detailed description of the simulation and evolutionary setup can be found in [29].

The game was implemented in Processing [33] and can be run in a web browser using ProcessingJS. 200 Amazon Mechanical Turk users were recruited to play the game[1], which was embedded in a website. At the start of the game, one of the four possible experimental conditions was randomly chosen: unevolved prey & unevolved predator, unevolved prey & evolved predator, evolved prey & unevolved predator, or evolved prey & evolved predator. Each player was required to play for either 120 s or until caught by the predator. The game difficulty implicitly increases with time because as the predator decimates the prey agents, the player is increasingly likely to be hunted by the predator. At the end of the game how long the player survived (at best 120 s) and how many other prey agents were still alive at that point, was recorded.

5   Results

The results of the experiment show that in the four tested combinations, average player survival time for each individual combination significantly differs from each of the others (Fig. 6). Intuitively, one might expect that the game's difficulty depends mostly upon the predator's ability to catch prey, because only the predator poses direct danger to the player. This intuition suggests that the two environments with the unevolved predator should be the easiest and the two environments containing the evolved predator should be the most challenging. Interestingly, however, the results instead show that the difficulty of the game depends more on the ability of the prey to swarm than on the ability of the predator to catch. The more evolved the prey is, the easier the game becomes, while the predator's ability is only of secondary importance. In this way, the results highlight that evolving opponents and collaborators with orthogonal objectives, like in this predator prey example, indeed allows for more combinations of difficulty (Fig. 4). Thus, choosing AIs for distinct roles from different evolutionary time points can facilitate a smoother (and potentially more complex) progression of game difficulties.

---

[1] Our study was exempt by the Office of Research Support at the University of Texas at Austin. Number 2013-09-0084. Due to the exemption, by not taking any personal data, and due to the anonymity of the subjects, we did not need written consent.

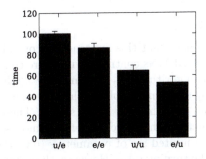

Fig. 6. Comparison of player performance. The average time in seconds players survived before being caught by the predator, for four different conditions, from left to right: (u/e) predator from an early point in evolution (generation 900) paired with evolved prey (generation 1,900), (e/e) both AIs from a late point in evolution (generation 1,900), (u/u) both AIs from an early point in evolution (generation 900), (e/u) predator from a late point in evolution (generation 1,900) paired with unevolved prey (generation 900)

Further, in this kind of interactive environment it is not only the swarming agents that influence player survivability. Conversely, the player actions can effect how well the swarm agents survive. When comparing the swarm agents' survival rate in the presence of the player, to a situation where the player's agent is controlled by the same AI as all the other prey agents, the result is that player interactions reduce the prey survivability only when the prey agents are controlled by the more evolved AI (Fig. 7). When prey agents are taken from an early generation the effect is more subtle. This result shows that not only can the difficulty the player experiences can be modulated by the degree of adaptation of prey agents, but that evolved prey agents are also more influenced by player actions if they are more evolved themselves.

Note that in order to assess the influence players have on the survivability of prey agents in the game, the following exponential decay function was used:

$$n = e^{a - \frac{a}{1 - (\frac{x}{a})^2}} \tag{1}$$

As an approximation for the number of organisms alive over time in the presence of the player, the number of organisms alive after 120 s or at the time point the player died was used to fit Eq. 1. To assess prey survivability without the presence of the player, for each of the four possible conditions the game was run without a player, and the player organism was controlled by the same AI as the other swarm agents. For these runs, the data was aggregated to estimate the average number of prey, and also was fit to Eq. 1.

For both data sets the residuals against each of the fitted functions were computed and a Mann Whitney U test was performed to show that the residuals of each others fit were significantly different from one another.

6   Discussion

First, it is important to note that this approach of controlling swarming agents in video games with evolved MNs contrasts with more conventional approaches. For example, the Boids algorithm [34] is commonly applied to control swarming agents, and works by uniformly applying three elementary forces (separation, alignment, and cohesion) to each agent in the swarm. These simple forces govern the entire swarm and dictate where each individual moves. Swarm behavior can be varied by adjusting a limited set of parameters (e.g., the radius of influence, the force applied, and the turning rate). However, the potential for novelty is limited because such parameters do not change the fundamental underlying forces. Furthermore, in general adapting the Boids model to particular capabilities of antagonisitic agents (like the predator) requires specific human insight. To overcome the limitations of simpler models (like the Boids algorithm) the EOS model is applied here, where agents are individually controlled by an evolved Markov network (which could in theory approximate the Boids algorithm through learning). Such Markov networks have to our knowledge not been applied to video games before, and our work demonstrates their feasibility. An interesting benefit of such networks is that in contrast to more computationally demanding AI algorithms, once evolved Markov networks are computationally tractable to embed even within javascript browser games. Each agent in a swarm is controlled by its own Markov network, allowing for novelty and variability between swarm agents. In some video games it is likely more interesting and visually appealing for the player to have heterogeneous swarms, which the use of evolved Markov networks enables.

More broadly, the results present an elaboration on previous approaches to difficulty adjustment in video games, which primarily focus on the evolution of opponent AI to create a single axis for difficulty adjustment: The more generations over which the opponent evolves, the more challenging it is to overcome. In contrast, coevolving the opponent agent with collaborative agents enables a wider spectrum of possibilities. Instead of exploiting only the evolutionary history of the opponent to adjust difficulty, the player can engage with different combinations of opponents and collaborators, which in the case studied here allows for smoother difficulty adjustment. In this way, the results demonstrate that coevolving separate populations of agents with orthogonal objectives is a viable method to improve difficulty adjustment.

One surprising result is that the difficulty in the explored game depends more on the capability of the collaborator (prey) than on the opponent (predator). While it is difficult to pinpoint the exact reasons for such behavior, it appears that when the player interacts with collaborators that can effectively swarm and evade the opponent, the player has an effective example to mimic, which shortens player learning time and thereby improves the player's performance. Another reason for the importance of the collaborative agents is that in the swarming example applied here, the evolved swarm of collaborators actively aggregates, and thereby protects the player as long as the player stays within the bounds of the swarm. This type of altruistic group behavior can improve the

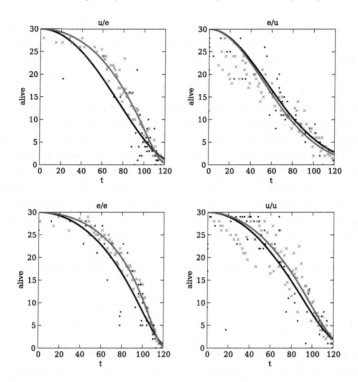

Fig. 7. Comparison of prey survivability. The figure compares the survival over time of prey agents, when interacting with a human-controlled prey, or when interacting only with computer-controlled agents. Each plot is titled by two letters that indicate the adaptedness of both the predator (first letter) and prey (second letter). The letter itself indicates whether agents come from an early point in evolution (u for "unevolved"; generation 900) or a late point in evolution (e for "evolved"; generation 1, 900). Thus the top left plot titled "u/e" reflects pairing a predator from generation 900 with prey from generation 1, 900). Black dots indicate the number of prey living when the player dies (when interacting with a human-controlled agent) and red Xs indicate the average number (over 5 sampled runs) of prey alive when there is no human interference. Curves are fit to the data points, and their color reflects the color of the data points from which they are derived. The distribution of the residuals between the red and black data points and their corresponding fit is significantly different between all four cases ($p < 0.001$). The conclusion is that player interaction influences the effectiveness of the other prey agents (Color figure online).

player's survivability, thereby providing an interesting example of an emergent game mechanism that is automatically discovered by orthogonal coevolution.

An additional idea explored in this paper is transplanting agents evolved originally in a scientific setting to create or enhance an entertaining video game. In particular, this paper creates a video game by exapting AI agents evolved in a simulation exploring biological hypotheses for the evolution swarming behavior [19]. A possible benefit is that through directly interacting with evolved AI in a game-like environment, a reader of a paper can potentially more easily understand the paper, as well as better judge the quality and sophistication of the results. In this way, video games based on scientific simulations can potentially assist wider understanding of scientific results by non-experts. Future work will investigate the plausibility of such ideas.

A limitation of the approach is that it may not always be appropriate or easy to formulate a game situation in terms of orthogonally evolved populations. However, computer games are increasingly multiplayer and increasingly incorporate massive game worlds, providing natural opportunities for game designers to augment games with collaborative agents in addition to more typical confrontational opponents. In particular, MMORPGs not only commonly contain companions and enrich their environments using NPCs, but such games (or perhaps real time strategy games) may also benefit from integrating evolutionary mechanisms *into* their gameplay, and thereby allow opponents and NPCs to evolve as a game progresses. As shown in the results, player action indeed influences prey performance, which supports the idea that agents can adapt to players within the game.

7   Conclusion

This paper introduced the concept of orthogonal coevolution, and tested its effectiveness in a browser-based game adapted from a scientific simulation. The results demonstrate that evolving opponents in conjunction with evolved companions can lead to smoother difficulty adjustment and allow players to experience more varied situations. The conclusion is that such orthogonal coevolution may be a promising approach for adjusting video game difficulty.

Acknowledgments. We would like to thank Chris Adami for insightful comments and discussion of the project.

References

1. Yannakakis, G.N.: AI in Computer Games (2006)
2. Browne, C.: Evolutionary Game Design. Springer, Heidelberg (2011)
3. Spronck, P., Sprinkhuizen-Kuyper, I., Postma, E.: Difficulty scaling of game AI. In: Intelligent Games (2004)
4. Hunicke, R., Chapman, V.: AI for dynamic difficulty adjustment in games. In: Challenges in Game Artificial Intelligence AAAI (2004)
5. Overholtzer, C.A., Levy, S.D.: Evolving AI opponents in a first-person-shooter video game. In: AAAI Proceedings of the 20th National Conference on Artificial Intelligence (2005)
6. Cole, N., Louis, S.J., Miles, C.: Using a genetic algorithm to tune first-person shooter bots. Trans. IRE Prof. Group Audio **1**, 131–139 (2004)
7. Tan, T.G., Anthony, P., Teo, J., Ong, J.H.: Neural network ensembles for video game AI using evolutionary multi-objective optimization. In: Transactions of the IRE Professional Group on Audio, pp. 605–610, December 2011

8. Yau, Y.J., Teo, J., Anthony, P.: Pareto evolution and co-evolution in cognitive game AI synthesis. In: Obayashi, S., Deb, K., Poloni, C., Hiroyasu, T., Murata, T. (eds.) EMO 2007. LNCS, vol. 4403, pp. 227–241. Springer, Heidelberg (2007)

9. Yau, Y.J., Teo, J., Anthony, P.: Pareto evolution and co-evolution in cognitive neural agents synthesis for Tic-Tac-Toe. In: IEEE Symposium on Computational Intelligence and Games, pp. 304–311. IEEE (2007)

10. Mayer, H.A., Maier, P.: Coevolution of neural go players in a cultural environment. Trans. IRE Prof. Group Audio **2**, 1012–1017 (2005)

11. Lubberts, A., Miikkulainen, R.: Co-evolving a go-playing neural network. In: Algorithms Upon Themselves (2001)

12. Chellapilla, K., Fogel, D.B.: Evolving an expert checkers playing program without using human expertise. IEEE Trans. Evol. Comput. **5**(4), 422–428 (2001)

13. Chellapilla, K., Fogel, D.B.: Evolution, neural networks, games, and intelligence. In: Proceedings of the IEEE, pp. 1471–1496 (1999)

14. Lim, C.U., Baumgarten, R., Colton, S.: Evolving behaviour trees for the commercial game DEFCON. In: Di Chio, C., Cagnoni, S., Cotta, C., Ebner, M., Esparcia-Alcazar, A.I., Goh, C.-K., Merelo, J.J., et al. (eds.) EvoApplicatons 2010. LNCS, vol. 6024, pp. 100–110. Springer, Heidelberg (2010)

15. Hagelbäck, J., Johansson, S.J.: Using multi-agent potential fields in real-time strategy games. In: AAMAS 2008: Proceedings of the 7th International Joint Conference on Autonomous Agents and Multiagent Systems, International Foundation for Autonomous Agents and Multiagent Systems, May 2008

16. Priesterjahn, S., Kramer, O., Weimer, A., Goebels, A.: Evolution of human-competitive agents in modern computer games. In: IEEE International Conference on Evolutionary Computation, pp. 777–784. IEEE, November 2005–2006

17. van Valen, L.: A new evolutionary law. Evol. Theor. **1**, 1–30 (1973)

18. Bell, G.: The Masterpiece of Nature: The Evolution and Genetics of Sexuality. CUP Archive (1982)

19. Olson, R.S., Knoester, D.B., Adami, C.: Critical interplay between density-dependent predation and evolution of the selfish herd. In: GECCO 2013: Proceeding of the 15th Annual Conference on Genetic and Evolutionary Computation Conference, ACM Request Permissions, July 2013

20. Yannakakis, G.N., Hallam, J.: Evolving opponents for interesting interactive computer games. In: From Animals to Animats (2004)

21. Grand, S., Cliff, D., Malhotra, A.: Creatures: artificial life autonomous software agents for home entertainment. In: AGENTS 1997: Proceedings of the 1st International Conference on Autonomous Agents, ACM, February 1997

22. Pollack, J., Blair, A.: Co-evolution in the successful learning of backgammon strategy. Mach. Learn. **32**, 225–240 (1998)

23. Stanley, K.O., Bryant, B.D., Miikkulainen, R.: Evolving neural network agents in the NERO video game. In: Proceedings of the IEEE (2005)

24. Hastings, E.J., Guha, R.K., Stanley, K.O.: Evolving content in the galactic arms race video game. In: IEEE Symposium on Computational Intelligence and Games (CIG), pp. 241–248. IEEE (2009)

25. DeLooze, L.L., Viner, W.R.: Fuzzy Q-learning in a nondeterministic environment: developing an intelligent Ms. Pac-Man agent. In: CIG 2009: Proceedings of the 5th International Conference on Computational Intelligence and Games. IEEE Press, September 2009

26. Handa, H.: Constitution of Ms. PacMan player with critical-situation learning mechanism. Int. J. Knowl. Eng. Soft Data Paradig. **2**(3), 237–250 (2010)

27. Tong, C.K., Hui, O.J., Teo, J., On, C.K.: The evolution of gamebots for 3D first person shooter (FPS). Transactions of the IRE Professional Group on Audio, pp. 21–26, September 2011
28. Agapitos, A., Togelius, J., Lucas, S.M., Schmidhuber, J., Konstantinidis, A.: Generating diverse opponents with multiobjective evolution. In: IEEE Symposium on Computational Intelligence and Games, CIG 2008, pp. 135–142. IEEE (2008)
29. Olson, R.S., Hintze, A., Dyer, F.C., Knoester, D.B., Adami, C.: Predator confusion is sufficient to evolve swarming behaviour. J. Roy. Soc. Interface **10**(85), 20130305 (2013)
30. Marstaller, L., Hintze, A., Adami, C.: The evolution of representation in simple cognitive networks. Neural Comput. **25**(8), 2079–2107 (2013)
31. Hamilton, W.D.W.: Geometry for the selfish herd. J. Theor. Biol. **31**(2), 295–311 (1971)
32. Lenski, R.E., Ofria, C., Pennock, R.T., Adami, C.: The evolutionary origin of complex features. Nature **423**(6), 139–144 (2003)
33. Fry, B., Reas, C.: Processing Library for Visual Arts and Design
34. Toner, J., Tu, Y.: Flocks, herds, and schools: a quantitative theory of flocking. Trans. IRE Prof. Group Audio (April 1998)

There Can Be only One: Evolving RTS Bots via Joust Selection

A. Fernández-Ares[(✉)], P. García-Sánchez, A.M. Mora, P.A. Castillo, and J.J. Merelo

Department of Computer Architecture and Technology,
University of Granada, Granada, Spain
antares.es@gmail.com

Abstract. This paper proposes an evolutionary algorithm for evolving game bots that eschews an explicit fitness function using instead a match between individuals called *joust* and implemented as a selection mechanism where only the winner survives. This algorithm has been designed as an optimization approach to generate the behavioural engine of bots for the RTS game Planet Wars using Genetic Programming and has two objectives: first, to deal with the noisy nature of the fitness function and second, to obtain more general bots than those evolved using a specific opponent. In addition, avoiding the explicit evaluation step reduce the number of combats to perform during the evolution and thus, the algorithm time consumption is decreased. Results show that the approach performs converges, is less sensitive to noise than other methods and it yields very competitive bots in the comparison against other bots available in the literature.

1 Introduction

Evolutionary algorithms (EAs) have been successfully applied to games for some time, despite the fact that the evaluation of strategies is *noisy* [19] in the sense that there is an inherent uncertainty in the *true* fitness or actual score of the bot, since it will depend on several stochastic factors: the game rules or status, the opponents' behaviour or the random initial conditions, which obviously have an influence on the score obtained by the agents. This problem also arises when the opponents follow non-deterministic Artificial Intelligence (AI) behavioural models, i.e. when they are Non-Playing Characters (NPC) or *bots*, since their behaviour considers stochastic factors which can influence the result of the game, and can vary from time to time.

Planet Wars, the RTS game introduced in the Google AI Challenge 2010[1] also presents this problem. It has been used by several authors for the study of computational intelligence in RTS games, such as generation of bots or map design [9,10,16,31]. As a summary, the objective of the player is to conquer

---

[1] http://planetwars.aichallenge.org/.

© Springer International Publishing Switzerland 2016
G. Squillero and P. Burelli (Eds.): EvoApplications 2016, Part I, LNCS 9597, pp. 541–557, 2016.
DOI: 10.1007/978-3-319-31204-0_35

enemy and neutral planets in a space-like simulator. Each player owns planets (resources) that produce ships (units) depending of a growth-rate. The player must send these ships to other planets (literally, crashing towards the planet) to conquer them. A player win if he/she is the owner of all the planets or the opponent forces have been completely defeated. As requirements, only one second is the limit to calculate next actions (this time window is called $turn^2$), and no memory about the previous turns can be used. Figure 1 shows a screen capture of the game.

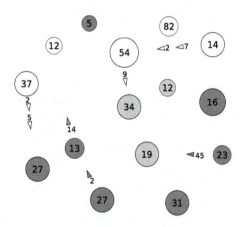

Fig. 1. Simulated screenshot of an early stage of a 1 vs 1 match in Planet Wars. White planets belong to the player, dark grey belong to the opponent, and light grey planets belong to no player. The triangles are fleets, and the numbers (in planets and triangles) represent the ships. The planet size models the growth rate of the amount of ships in it (the bigger, the higher).

This game presents the aforementioned problem in the fitness calculation phase [19]. Usually several matches are carried out for the same individual (maybe in different maps or against different opponents) and then its fitness value is computed as an average or sum of all the obtained results. This way, ideally, a more accurate (less noisy) measure of the individual's quality is obtained, but it is still not a completely reliable solution because it depends on some values (such as the number of victories or number of created ships) or on the rival's performance, which could be a previously created bot.

Even if we could obtain a statistically significant fitness evaluation, the way this fitness is obtained might include an additional bias due to opponent selection. This issue concerns the overfitting of the population with respect the selected rival/s, i.e. the individuals learn to play against it/them, and could behave poorly against another type of enemy [18,19].

The present paper proposes a co-evolutionary EA [22] for improving bot's AI in Planet Wars by means of an implicit fitness evaluation, based in the *survival*

---

^{2} Although we use this term, note that the game is always performed in real time.

of the individuals. To this end the selection process is transformed into a *tournament* (or *joust*, to distinguish it from the classical tournament in EAs) in which just the winners will survive and become parents of the new offspring. This way, the fitness computation is omitted and thus, the influence of noise is reduced. Moreover, it does not require the usual ad-hoc parameters, such as the number of battles or the score values, neither a previously existing bot to compare.

This model is closer than the canonical evolutionary algorithm to the *real natural selection process* which happens in nature [7], where just the fittest individuals survive. Thus, the approach can be described as a *competitive co-evolutionary algorithm* [14,24], where the (implicit here) fitness of an individual depends on competition with other individuals.

A Genetic Programming (GP) [15] approach has been implemented, due to the additional flexibility factor that this method offers with respect to a Genetic Algorithm (GA) [12], i.e. GP is able to create new sets of behavioral rules meanwhile while GA is devoted to optimize the parameters of previously designed rules. This technique has yielded excellent results in previous works related with agent generation in videogames [8,11].

Since the algorithm runs in Planet Wars, the joust is modeled as a battle in the game. The winner is moved to the next generation and also becomes a parent for the next set of players; the loser is removed from the pool. In addition, considering all the individuals in the population as opponents, and not using a specific one, makes the training (evolution) more general, and thus, the obtained individuals would, potentially, be able to face a wider amount of possible rivals.

Several experiments have been conducted, in order to measure the convergence and the noise impact, with comparisons with other bots available in the literature. We aim to solve the next research questions:

- Is the implicit fitness evaluation proposed a feasible way to evaluate individuals?
- Is this approach less sensible to noise than others?
- How does affects not using a previously defined opponent?
- How good is the behaviour of the generated bots?

2   Background and State of the Art

Evolving an AI engine to control the NPCs that play the game has become an usual and successful technique in the videogames scope. The first approach is to create a set of rules by a human expert, and then optimize the set of parameters which determine how this bot behaves. This kind of improvement has been previously performed in [6,9,10,13,20] by means of off-line (before the game) Evolutionary Algorithms. Furthermore, the use of GP [8,11] dispenses the human expert to define the set of rules, as these rules, along with their numerical parameters, are created and evolved automatically during the run. Thus, it is a more flexible approach for defining the behavioural engine of the bots, which can find rules that a human expert cannot imagine at all. For this reason GP has been used in this paper to generate the engines.

One of the shortcomings of previous works is that they depend on a baseline bot, taken as rival during the evolution, or an ad-hoc fitness function that involves some kind of parameterization (for example, number of matches or scoring the actions). To avoid this, co-Evolutionary Algorithms (CEAs) have been previously used in this scope, as it is a natural choice to use in problems where the behaviour of one agent is related to the behaviour of others [26].

The co-evolutionary scheme was initially used in puzzle and board games such as Backgammon [23], or Go [25]. The first work proposed a very simple hill-climbing algorithm to evolve a population of neural networks, playing among them as rivals, in a competitive co-evolutionary approach. The second presented a co-evolutionary learning approach which performed well once the EA was correctly tuned, moreover, this method yields better players to solve small Go boards since every individual is evaluated against a diverse population of rivals. In the same line, there are some other works in the card games area, such as [28], aimed to create Poker agents, considering a co-evolution process in which the players are part of the learning process. This meant a difficult process to get robust strategies, due to the variation in opponents, but the results shown to fit with some recommended strategies according to experts. The aim of this work is to conduct a study with a similar co-evolutionary approach, being competitive in the selection of individuals as parents for the next offspring, but cooperative since all the opponents are also part of the same learning process (same population).

In recent years, this type of EAs has been also applied to videogames, enclosed in the Computational Intelligence (CI) branch of AI. For instance Togelius et al. [29] studied the co-evolution effects of some populations in car racing controllers, comparing the performance of a single population against various, implementing both generational and a steady-state approaches. Avery and Michalewicz introduced in [2] a co-evolutionary algorithm for the game TEMPO, which used humans as rivals for the individuals in the evolutionary process. Cook et al. [5] presented a cooperative co-evolutionary approach for the automated design of levels in simple platform games. And recently Cardona et al. [3] studied the performance of a competitive algorithm for the simultaneous evolution of controllers to both Ms. PacMan and the Ghost Team which has to chase her.

Co-evolution has also been used in the RTS scope. Livingstone [17] compared several AI-modelling approaches for RTS games, and proposed a framework to create new models by means of co-evolutionary methods. He considered two levels of learning in a hierarchical AI model (inside an own-created RTS), evolving at the same time different partners in different strategic levels, so it was a cooperative approach. It is different to the one proposed here, since in the present work the co-evolution occurs at the same level for all the individuals. The work by Smith et al. [27] presents an analysis on how a co-evolutionary algorithm can be used for improving students' playing tactics in RTS games. Other authors proposed using co-evolution for evolving team tactics [1]. However, the problem is how tactics are constrained and parametrised and how the overall score is computed. Nogueira et al. [21] considered in a recent publication the use of a Hall of

Fame as a set of rivals (in the evaluation function) inside a co-evolutionary algorithm to create autonomous agents for the RTS game *RobotWars*. An updated version of this algorithm was also applied to Planet Wars game [4]. This approach is based in a self-learning algorithm similar to the one we are proposing, but focused in a subset of individuals (the elite) which might have a negative effect in the generalisation factor or the bots' knowledge. Moreover, they use an ad-hoc fitness function with specific parameters, taking into account several battles and extra score measurement. Also, using the evolution to a fixed set of players could not lead to strong players [26].

The approach presented in this work implements a survival-based co-evolutionary scheme, which omits an explicit fitness computation. Instead, the agents or bots (individuals) compete against the rest in the so-called *joust tournaments*. Thus, just the survivors will remain in the population and will reproduce to generate the next offspring. This tries to minimize the influence of a *noisy fitness function* [19] in the evolution of the individuals; i.e. a good fitness value could be assigned to a bad player by chance, and the other way round. Moreover the proposed scheme has two advantages with respect to previous works: not adding ad-hoc parameters (such as the number of victories), and not using a specific bot as rival during evolution, which would lead to a specialisation of the individuals to fight against it.

3   Survival Bots

This section describes the algorithm proposed in this work to generate competent bots (called *SurvivalBots*). A Genetic Programming [15] algorithm to generate the agent's behavior is combined with different selection and replacement policies, using an implicit fitness computation in a co-evolutionary way.

3.1   Bot Generation Using GP

To generate the bot's behavior the so-called *GPBot* algorithm (presented in [11]) has been used as a reference. However, the proposed method follows a different philosophy, based in the survival of the individuals, highly inspired by the crude natural evolutionary process. To this end new selection and replacement mechanisms have been adopted in the SurvivalBot approach.

In our approach, as in GPBot, a GP algorithm is used to evolve a set of rules which, in turn, models a Decision Tree. During the evolution, every individual in the population (a tree) must be evaluated. To do so, the tree is set as the behavioural engine of an agent, which is then placed in a map against a rival in a Planet Wars match.

Thus, during the match the tree will be used (by the bot) in order to select the best strategy at every moment, i.e. for every planet a target will be selected along with the number of ships to send from one another.

These Decision Trees are binary trees of expressions composed by two different *types of nodes*: *Decision nodes*, which include a logical expression composed by a variable, a less than operator ($<$), and a number between 0 and 1, equivalent to a "primitive" in the field of GP and *Action nodes*, a leave of the tree (therefore, a "terminal"), which is the name of a function, and a ratio between 0 and 1; the function indicates to which target planet the bot must send a percentage of the available amount of ships in the planet (from 0 to 1). As the bot applies the tree one time per planet it uses each time the information of the current planet.

The *decisions* are based in the values of different *variables* which are computed considering some parameters in the game. They were defined by a human expert in [11], and are shown next:

– *myShipsEnemyRatio* : Ratio between the player's ships and enemy's ships.
– *myShipsLandedFlyingRatio* : Ratio between the player's landed and flying ships.
– *myPlanetsEnemyRatio* : Ratio between the number of player's planets and the enemy's ones.
– *myPlanetsTotalRatio*: Ratio between the number of player's planet and total planets (neutrals and enemy included).
– *actualMyShipsRatio*: Ratio between the number of ships in the specific planet that evaluates the tree and player's total ships.
– *actualLandedFlyingRatio*: Ratio between the number of ships landed and flying from the specific planet that evaluates the tree and player's total ships.
– *Random* : This decision was not included in the original GPBot. It has been included in the list to add a stochastic component to the agent, with the aim of performing the noise study presented in this work (in Sect. 4). It is, essentially, a probability added to select one branch or the other.

Finally, the possible *actions* are:

– *Attack Nearest (Neutral—Enemy—NotMy) Planet*: The objective is the nearest planet.
– *Attack Weakest (Neutral—Enemy—NotMy) Planet*: The objective is the planet with less ships.
– *Attack Wealthiest (Neutral—Enemy—NotMy) Planet*: The objective is the planet with the highest growth rate.
– *Attack Beneficial (Neutral—Enemy—NotMy) Planet*: The objective is the more profitable planet, that is, the one with the highest value for growth rate divided by the amount of ships.
– *Attack Quickest (Neutral—Enemy—NotMy) Planet*: The objective is the easiest planet to be conquered: the lowest product between the distance from the planet being evaluated by the tree and the number of ships in the objective planet.
– *Attack (Neutral—Enemy—NotMy) Base*: The objective is the planet with more ships (that is, the base).
– *Attack Random Planet.*
– *Reinforce Nearest Planet*: Reinforce the nearest player's planet to the planet that is being evaluated by the tree.

- *Reinforce Base*: Reinforce the player's planet with the highest amount of ships.
- *Reinforce Wealthiest Planet*: Reinforce the player's planet with highest growth rate.
- *Reinforce Weakest Planet*: Reinforce the player's planet with less ships.
- *Do nothing*.

The bot's general behaviour is described in Algorithm 1.

Algorithm 1. Pseudocode of a GPBot. The same tree is used during all the agent's execution

---

/ At the beginning of the execution the agent receives the tree */*
tree ← readTree()
while game not finished **do**
  / starts the turn */*
  calculateGlobalPlanets() */* e.g. Base or Enemy Base */*
  calculateGlobalRatios() */* e.g. myPlanetsEnemyRatio */*
  for Each p in PlayerPlanets **do**
    calculateLocalPlanets(p) */*e.g. NearestNeutralPlanet to p*/*
    calculateLocalRatios(p) */* e.g. actualMyShipsRatio */*
    executeTree(p,tree) */* Choose and Send a percentage of ships to destination*/*
  end for
end while

3.2 Joust-Based Selection

The algorithm presented in [11] is combined with an *implicit fitness evaluation* for selection and replacement. This 'evaluation' is, in essence, a match between individuals, called *joust* (to distinguish it from the classical tournament in EAs), in a battle map of the game. Thus, the selection of the two mating parents is performed each one in a battle. The winner of the match is selected to mate and the loser will be definitely removed from the population (as it will be explained below).

    This selection mechanism tries to emphasize the survival of the fittest individuals, since just the best bots will be chosen as parents, and thus, will remain one more generation. Actually, in this algorithm, the concept of 'iteration' is used as a synonym of generation, since it is not a classical evolutionary process, as will be deeply explained in the next section.

    The use of such a selection/survival process tries to reduce the noise added by the fitness evaluation in the evolutionary process [19]. So, the individuals which are not able to win in a match are strongly penalised, and thus, removed from the population of the next iteration.

3.3   Replacement of Losers

The classical Steady-State EA approach [30] has been implemented as replacement policy. In it, the majority of the population remains the same in the following generation, and just a small subset of individuals are substituted (usually just the worst). This method aims to increase the exploitation factor in the EA, in order to increase the convergence, which is an interesting factor in a noisy search space as the scope of videogames is.

Thus, the proposed approach follows this idea and just performs two battles (or jousts) per generation, the aforementioned selection policy. The contenders are randomly selected from all the individuals in the population (ensuring that the same individuals are not chosen for both battles). The two winners of the battles will be the parents for the *crossover operation*, which generates two new individuals (offspring), which will be also mutated. In this paper, sub-tree crossover and 1-node mutation operators have been used, as they obtain good results in generation of bots using GP [8]. These individuals are inserted in the population after being created, substituting the bots that lose the jousts.

This approach presents a higher random component than the original, due to the lack of a fitness value which can value every individual with a simple number. The random selection of all the individuals also increases the chance of reducing the presence of noisy bots, i.e. those which are not good enough to remain in the population. This will be a key factor in the resolution of this problem, as will be proved in the experiments.

Algorithm 2 shows the combination of the GP approach, together with the implicit fitness evaluation, and the selection and replacement mechanisms.

4   Experiments and Results

Several experiments have been conducted in order to study different issues of the proposed approach, but having in mind that the main objective is not just the generation of competitive bots as usual. In this paper, the aim is firstly to demonstrate the validity of the proposed co-evolutionary algorithm with joust-based selection, i.e. we want to prove the correct convergence of the method, the low noise influence in the results, and finally, once these issues are demonstrated, the quality and characteristics of the obtained bots. Thus, the experiments are separated in three subsections, one per objective.

The set of parameters considered in our co-evolutionary GP (Co-GP) algorithm, SurvivalBot, is shown in Table 1. These parameters are the same as the authors used in [11], to obtain competitive bots. Since GPBot is the basis of the present proposal, we have considered it as a base for comparisons in the experiments. Thus, to do a fair comparison with that method and the results it yields, the termination criteria in SurvivalBot has been set to 8000 battles (therefore, 4000 generations/iterations), since GPBot considered 32 individuals * 5 combats per evaluation * 50 generations = 8000 evaluations/combats. The five maps are representative of different distributions and sizes of planets.

Algorithm 2. Pseudocode of the proposed SurvivalBot.

population ← initializePopulation()
while stop criterion not found **do**
    offspring,losers,selected ← {}
    /* Two random contenders for the joust */
    contenders ← selectContenders(population-selected)
    /* The contenders fight and the winner and loser are obtained */
    winner1,loser1 ← battle(contenders)
    /* Previously selected bots not participate again in the tournament */
    selected ← selected + winner1 + loser1
    /* Contenders of the second joust */
    contenders ← selectContenders(population-selected)
    /* The contenders fight and the winner and loser are obtained */
    winner2,loser2 ← battle(contenders)
    selected ← selected + winner2 + loser2
    /* The losers will be removed from the population */
    losers ← losers + loser1 + loser2
    /* Evolutionary process */
    son1,son2 ← crossover(parent1,parent2);
    son1,son2 ← mutation(son1,son2)
    offspring ← offspring + son1 + son2
    /* Replacement of the losers */
    population ← population - losers
    population ← population + offspring
end while

Table 1. Parameter set used in the experiments.

Parameter name	Value
Population size	32
Initialization	Random (trees of 3 levels)
Crossover type	Sub-tree crossover
Crossover rate	0.5
Maximum number of turns per battle	1000
Mutation	1-node mutation
Mutation step-size	0.25
Selection	2-tournament
Replacement	Steady-state
Stop criterion	4000 iterations
Maximum tree depth	7
Runs per configuration	30
Maps used in each evaluation	1 random chosen among maps #76 #69 #7 #11 #26

30 runs of the Co-GP have been performed to obtain statistically significant results.

4.1   Analysis of the Runs

The first set of experiments is devoted to analyze the convergence of the proposed method, since it is desirable that the method, even without a fitness function, performs similarly to other classical approaches. However, it is difficult to show the convergence of the populations using an implicit fitness evaluation, as the evaluated individuals (and therefore, the average fitness of the population) do not count with a numeric value to be plotted in time.

To solve this, we propose a scoring method that takes into account the number of victories, turns to win and turns resisted before being defeated by a rival. After the execution of our algorithm all the generated individuals during all the 30 runs have been confronted versus the best GPBot obtained in [11] (as it is the fairest opponent in terms of actions and parameters used). Then each of the individuals i has obtained a score using the next formulae:

$$Score_i = \alpha + \beta + \gamma \tag{1}$$

where

$$\alpha = v, \alpha \in [0, N] \tag{2}$$

$$\beta = N \times \frac{t_{win} + \frac{1}{N \times t_{MAX} + 1}}{\frac{t_{win}}{v+1} + 1},$$
$$\beta \in [0, N], \tag{3}$$
$$t_{win} \in [0, N \times t_{MAX}]$$

$$\gamma = \frac{t_{defeated}}{N \times t_{MAX} + 1},$$
$$\gamma \in [0, 1], \tag{4}$$
$$t_{defeated} \in [0, N \times t_{MAX}]$$

The terms used are: the number of battles (N) to test, the number of victories of the individual against GPBot (v), the total number of turns used to win GPBot (t_{win}), the total number of turns when the individual has been defeated by GPBot ($t_{defeated}$) and the maximum number of turns a battle lasts (t_{MAX}). This score aims to favour the victories against the turns to win and turns to be defeated, giving different ratios to each section. Therefore α has the highest ratio. The term β add extra score taking into account the number of turns when the individual wins (lower numbers to win implies better bots), following a exponential curve. Finally, the γ term adds score from the turns to be beaten (higher number is better, as it is difficult to be beaten). The 1 in all denominators is used to avoid dividing by 0.

Each individual has been tested three times in ten different maps (the 5 used during evolution and other new 5 ones from the Google set), therefore $N = 30$ (for all maps and 15 for each of the subsets of them), and the limit of turns is the default of the competition ($t_{MAX} = 1000$). As previously said, this score has two shortcomings that we are trying to avoid in this paper: it requires parametrization and an existing opponent. However, we will use this score as a way to measure and show the performance of our fitness-less approach.

Figure 2 presents the boxplots of the score of the whole current population (of all 30 original runs) in different stages of the evolution in different maps (those used with the evaluation and the new set). As it can be seen, the score grows during the evolution (something desirable), i.e. there exist some improvement of the population along the execution of the Co-GP algorithm. As expected, the figure shows a better performance of the individuals in the maps considered during the evolution (training maps).

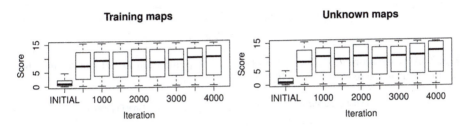

Fig. 2. Score for the confrontation for all SurvivalBots obtained during runs against the best GPBot.

Figure 2 shows a convergence trend, but the effect is clearly shown in Fig. 3, which plots the average score of all individuals, and the average score of all the best (one per run), obtained during the evolution of the SurvivalBots. This figure shows how there exist an increasing performance in the best (and also in average) individuals during the runs. Moreover, a lightly noisy factor is present, but the oscillations are not as striking as in previous approaches [19]. This effect will be better studied in the experiment in Sect. 4.2.

The study in complemented with two other graphs, first, Fig. 4 shows the percentage of individuals (normalized between 0 and 1) that wins a certain number of times (from 0 to 30) against GPBot in the initial and the final populations. As it can be seen there exists strong differences in the number of victories which are 0 (always lose) for more than the 40 % of individuals in the initial population with around 1 % of winners, and which is turned into a 10 % of 'completely losers' and around a 18 % of 'completely winners' against GPBot. Moreover, the increase in the number of victories for other values is also clear in the graphs. So we can conclude that an effective improvement has been done in the populations from the start to the end of the algorithm run.

Studying the *age* of the evolving bots can help to understand the dynamics of the evolution; this is shown in Fig. 5 for during one run. It is interesting how the age has an upper limit that does not change along run, meaning that a truly good bot is not generated at the very beginning and stays alive along all generations. This leads us to conclude that the population as a whole is effectively improved, since the good bots are beaten by their offspring a few generations down the road. The extreme values which live up to 50 generations happen due to the

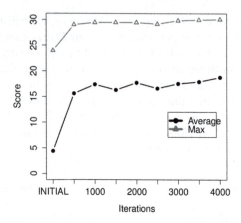

Fig. 3. Average score of the best and of the whole population for all runs of matches of SurvivalBots against the best GPBot.

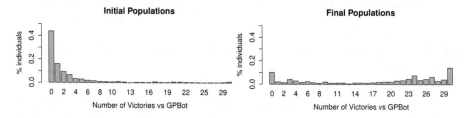

Fig. 4. Histogram of number of victories against GPBot of all the individuals in initial and final populations in 30 maps.

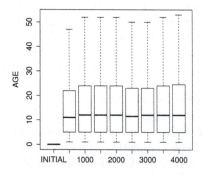

Fig. 5. Boxplots of the age (generations) of the population during one run.

random selection of contenders, which could make bot skip fights for several generations. Having checked that the algorithm works as intended we need to know if it effectively reduces uncertainty in scores, one of our main objecties. We will do this next.

4.2 Analysis of the Score Uncertainty

To conduct this study we will compare the 30 best bots obtained by GPBot and SurvivalBot across 30 runs; these bots have been fixed with a hard to defeat rival, such as the expert/specialized bot named ExpGeneBot [9]. The same 10 maps as in previous experiments have been considered, and 30 battles in each one have been done, computing the score in Eq. 1. Then a *noise factor* has been calculated for every bot, as the difference between the maximum and the minimum obtained scores in the 30 matches. This is because the noise in the scope of optimization in videogames is defined as the differences in performance that the same individual/bot could show in the same conditions (map and rival), due to the pseudo-stochasticity (there are some random events/actions) present in the opponent's behaviour and sometimes in the game itself.

Figure 6 shows the boxplots of the 30 bots of GPBot and SurvivalBot. According to the definition of our noise factor, a big distance between values means a higher uncertainty in the results; this figure shows that the results for Survival-Bot are better than those for GPBot, having a lower variance. Thus, we can conclude that the resulting bots for our method are more reliable in terms of behaviour and thus, show a more robust and predictable behavior. But of course the bots are designed to win battles. We will examine their performance in this area next.

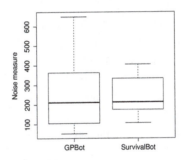

Fig. 6. *Noise factor*, that is, maximum - minimun scores for every map, of the best 30 bots obtained using GPBot and SurvivalBot approaches, evaluated in 10 different maps (5 previously trained and 5 not previously trained), 30 times/battles per map.

4.3 Analysis of the Generated Bots

Firstly, all the SurvivalBots obtained at the end of the runs have been tested against other bots available in the literature. To this end, jut one SurvivalBot per run must be chosen, so first the *best* individual of every run has been selected by confronting all the individuals of the last generation in an *all vs. all tournament*. The bot who has won more times is considered as the best. This method has been applied in order to avoid the usage of the score function (and therefore, the shortcomings we are trying to avoid).

Then, we have confronted the 30 best bots obtained in each configuration again with several bots available in the literature, in the 100 example maps provided by Google with the competition framework. These have been used to validate if the obtained SurvivalBots can be competitive in terms of quality in maps not used during evolution, and against unknown bots (as a difference to the other approaches). Table 2 presents the bots used as opponents.

Figure 7 shows the boxplots of the percentage of victories of the SurvivalBots against every rival. Note that it only shows the victories, not the draws. The most interesting result is that GPBot is clearly outperformed, even if the number of battles has been the same to train SurvivalBot than for GPBot, as we set. Therefore, our method can generate competitive bots without using existing ones in the training. The HoFBot, which was also obtained using co-evolution, has also been defeated more than 50 % times by most of the best SurvivalBots. However, highly trained bots (GeneBot and ExpGenebot), which applied 4 times more evaluations to be generated have been difficult to beat. It is interesting to mention that in [11] GPBot was able to beat these two bots in a higher value, but this happened because they were used to train GPBot, so it was focused only in beating them.

Table 2. Bots available in the literature used for measuring the quality of the SurvivalBots.

Bot name	&Reference	Simulations in training	Max. Turns
Bullybot	Google AI Web	None	None
Survivalbot	proposed here	8000	1000
Genebot	[19]	32000	1000
ExpGenebot	[9]	32000	1000
GPbot	[11]	8000	1000
HoFbot	[4]	180000	500

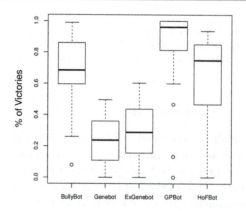

Fig. 7. Victories confronting the 30 best SurvivalBots against existing bots in the literature.

5   Conclusions and Future Work

This paper presents an implementation of a quite simple co-evolutionary app-roach for the generation of RTS bots: to omit the fitness-based selection mech-anism in the evolutionary process. Thus, it simulates the evolution in a more natural way: conducting real battles between individuals to select a survivor for the next generation. This method has been applied on the improvement of the behavioral parameters and rules of the bot's AI in the RTS game Planet Wars. Thus, the classic tournament selection mechanism has been modeled as a battle in the game (called *joust*), in which just the winner will remain (as a parent for the next offspring) in the population. The loser will be deleted.

According to the results in the experiments, the analysis performed, and the reached conclusions, this approach offers three main benefits: It yields more general-fighting bots, i.e. non-specialized in fighting against specific opponents, since it is a competitive co-evolutionary approach which does not need the use of a rival for the evaluation of an individual. It actually uses the same individuals in the population as opponents; it is less affected by the effect of the noise, since it is usually inserted in the loop by the fitness evaluation function. The high pressure included here by the loss of just a match (the individual will be removed from the population in that case), makes it very difficult that non-really-good individuals/bots survive and reduces the number of battles to be conducted in the evolution, in order to yield competitive bots, so it could reduce the computational time of the runs. The generated bots, named SurvivalBots, have been tested against some state-of-the-art rivals, getting excellent results even in the comparison with highly evolved and specialized bots.

As future work, we will firstly focus in obtaining more competitive bots. Thus, more possible actions and decisions will be added to the proposed Genetic Programming algorithm. In the same line, some tests will be done using an unlimited tree depth, which could lead to get much more complex (and effective) behavioural engines.

Acknowledgments. This work has been supported in part by projects EPHEMECH (TIN2014-56494-C4-3-P, Spanish Ministerio de Economía y Competitividad), PROY-PP2015-06 (Plan Propio 2015 UGR), PETRA (SPIP2014-01437, funded by Dirección General de Tráfico), CEI2015-MP-V17 (awarded by CEI BioTIC Granada), and PRY142/14 (funded by Fundación Pública Andaluza Centro de Estudios Andaluces en la IX Convocatoria de Proyectos de Investigación).

References

1. Avery, P., Louis, S.: Co-evolving team tactics for a real-time strategy game. In: Proceedings of the 2010 IEEE Congress on Evolutionary Computation (CEC 2010), pp. 1–8 (2010)
2. Avery, P.M., Michalewicz, Z.: Adapting to human game play. In: IEEE Symposium on Computational Intelligence and Games (CIG 2008), pp. 8–15 (2008)

3. Cardona, A., Togelius, J., Nelson, M.: Competitive coevolution in Ms. Pac-Man. In: Proceedings of the IEEE Congress on Evolutionary Computation (CEC 2013), pp. 1403–1410 (2013)
4. Collazo, M.N., Cotta, C., Leiva, A.J.F.: Virtual player design using self-learning via competitive coevolutionary algorithms. Nat. Comput. **13**(2), 131–144 (2014)
5. Cook, M., Colton, S., Gow, J.: Initial results from co-operative co-evolution for automated platformer design. In: Di Chio, C., Agapitos, A., Cagnoni, S., Cotta, C., de Vega, F.F., Di Caro, G.A., Drechsler, R., Ekárt, A., Esparcia-Alcázar, A.I., Farooq, M., Langdon, W.B., Merelo-Guervós, J.J., Preuss, M., Richter, H., Silva, S., Simões, A., Squillero, G., Tarantino, E., Tettamanzi, A.G.B., Togelius, J., Urquhart, N., Uyar, A.Ş., Yannakakis, G.N. (eds.) EvoApplications 2012. LNCS, vol. 7248, pp. 194–203. Springer, Heidelberg (2012)
6. Cotta, C., Fernández-Leiva, A.J., Sánchez, A.F., Lara-Cabrera, R.: Car setup optimization via evolutionary algorithms. In: Rojas, I., Joya, G., Cabestany, J. (eds.) IWANN 2013, Part II. LNCS, vol. 7903, pp. 346–354. Springer, Heidelberg (2013)
7. Darwin, C.: On the Origin of Species by Means of Natural Selection. Murray, London (1859)
8. Esparcia-Alcázar, A., Moravec, J.: Fitness approximation for bot evolution in genetic programming. Soft Comput. **17**(8), 1479–1487 (2013)
9. Fernández-Ares, A., García-Sánchez, P., Mora, A.M., Merelo, J.J.: Adaptive bots for real-time strategy games via map characterization. In: 2012 IEEE Conference on Computational Intelligence and Games, CIG 2012, pp. 417–721. IEEE (2012)
10. Fernández-Ares, A., Mora, A.M., Merelo, J.J., García-Sánchez, P., Fernandes, C.: Optimizing player behavior in a real-time strategy game using evolutionary algorithms. In: IEEE Congress on Evolutionary Computation, 2011 CEC 2011, pp. 2017–2024 June 2011
11. García-Sánchez, P., Fernández-Ares, A., Mora, A.M., Castillo, P.A., González, J., Guerv, J.J.M.: Tree Depth Influence in Genetic Programming for Generation of Competitive Agents for RTS Games. In: Esparcia-Alcázar, A.I., Mora, A.M. (eds.) EvoApplications 2014. LNCS, vol. 8602, pp. 411–421. Springer, Heidelberg (2014)
12. Goldberg, D.E.: Genetic Algorithms in Search, Optimization and Machine Learning. Addison Wesley, Boston (1989)
13. Jaśkowski, W., Krawiec, K., Wieloch, B.: Winning ant wars: evolving a human-competitive game strategy using fitnessless selection. In: O'Neill, M., Vanneschi, L., Gustafson, S., Esparcia Alcázar, A.I., Falco, I., Cioppa, A., Tarantino, E. (eds.) EuroGP 2008. LNCS, vol. 4971, pp. 13–24. Springer, Heidelberg (2008)
14. Kim, Y., Kim, J., Kim, Y.: A tournament-based competitive coevolutionary algorithm. Appl. Intell. **20**(3), 267–281 (2004)
15. Koza, J.R.: Genetic Programming: On the Programming of Computers By Means of Natural Selection. MIT Press, Cambridge (1992)
16. Lara-Cabrera, R., Cotta, C., Leiva, A.J.F.: On balance and dynamism in procedural content generation with self-adaptive evolutionary algorithms. Nat. Comput. **13**(2), 157–168 (2014)
17. Livingstone, D.: Coevolution in hierarchical ai for strategy games. In: IEEE Symposium on Computational Intelligence and Games (CIG 2005), IEEE (2005)
18. Merelo-Guervós, J.J.: Using a Wilcoxon-test based partial order for selectionin evolutionary algorithms with noisy fitness. Technical report, GeNeura group, university of Granada (2014). http://dx.doi.org/10.6084/m9.figshare.974598
19. Mora, A., Fernández-Ares, A., Guervós, J.M., García-Sánchez, P., Fernandes, C.: Effect of noisy fitness in real-time strategy games player behaviour optimisation using evolutionary algorithms. J. Comput. Sci. Technol. **27**(5), 1007–1023 (2012)

20. Mora, A.M., Fernández-Ares, A., Merelo-Guervós, J.-J., García-Sánchez, P.: Dealing with noisy fitness in the design of a RTS game bot. In: Di Chio, C., Agapitos, A., Cagnoni, S., Cotta, C., de Vega, F.F., Di Caro, G.A., Drechsler, R., Ekárt, A., Esparcia-Alcázar, A.I., Farooq, M., Langdon, W.B., Merelo-Guervós, J.J., Preuss, M., Richter, H., Silva, S., Simões, A., Squillero, G., Tarantino, E., Tettamanzi, A.G.B., Togelius, J., Urquhart, N., Uyar, A.Ş., Yannakakis, G.N. (eds.) EvoApplications 2012. LNCS, vol. 7248, pp. 234–244. Springer, Heidelberg (2012)
21. Nogueira, M., Cotta, C., Fernández-Leiva, A.J.: An analysis of hall-of-fame strategies in competitive coevolutionary algorithms for self-learning in RTS games. In: Nicosia, G., Pardalos, P. (eds.) LION 7. LNCS, vol. 7997, pp. 174–188. Springer, Heidelberg (2013)
22. Paredis, J.: Coevolutionary computation. Artif. Life **2**(4), 355–375 (1995)
23. Pollack, J.B., Blair, A.D.: Co-evolution in the successful learning of backgammon strategy. Mach. Learn. **32**, 225–240 (1998)
24. Rosin, C.D., Belew, R.K.: New methods for competitive coevolution. Evol. Comput. **5**(1), 1–29 (1997)
25. Runarsson, T.P., Lucas, S.M.: Co-evolution versus self-play temporal difference learning for acquiring position evaluation in smallboard go. IEEE Trans. Evol. Comput. **9**(6), 628–640 (2005)
26. Samothrakis, S., Lucas, S.M., Runarsson, T.P., Robles, D.: Coevolving game-playing agents: measuring performance and intransitivities. IEEE Trans. Evol. Comput. **17**(2), 213–226 (2013)
27. Smith, G., Avery, P., Houmanfar, R., Louis, S.: Using co-evolved rts opponents to teach spatial tactics. In: IEEE Symposium on Computational Intelligence and Games (CIG 2010), pp. 146–153 (2010)
28. Thompson, T., Levine, J., Wotherspoon, R.: Evolution of counter-strategies: Application of co-evolution to texas hold'em poker. In: IEEE Symposium on Computational Intelligence and Games (CIG 2008), pp. 16–22. IEEE (2008)
29. Togelius, J., Burrow, P., Lucas, S.M.: Multi-population competitive co-evolution of car racing controllers. In: Proceedings of the IEEE Congress on Evolutionary Computation (CEC), pp. 4043–4050 (2007)
30. Whitley, D., Kauth, J.: GENITOR: A different genetic algorithm. In: Proceedings of the 1988 Rocky Mountain Conference on Artificial Intelligence. pp. 118–130. Computer Science Department, Colorado State University (1988)
31. Ziółko, B., Kruk, M.: Automatic reasoning in the planet wars game. Annales UMCS, Informatica **12**(1), 39–45 (2012)

Constrained Level Generation Through Grammar-Based Evolutionary Algorithms

Jose M. Font[1(✉)], Roberto Izquierdo[2], Daniel Manrique[2], and Julian Togelius[3]

[1] U-tad, Centro Universitario de Tecnologíay Arte Digital,
C/Playa de Liencres 2-bis, 28290 Las Rozas, Madrid, Spain
jose.font@u-tad.com

[2] Departamento de Inteligencia Artificial, Escuela Técnica Superior de
Ingenieros Informáticos, Universidad Politécnica de Madrid, Campus de
Montegancedo S/n, 28660 Boadilla del Monte, Madrid, Spain
r.iamo@alumnos.upm.es, d.manrique@fi.upm.es

[3] Department Computer Science and Engineering, New York University,
2 Metrotech Center, Brooklyn, NY 11201, USA
julian@togelius.com

Abstract. This paper introduces an evolutionary method for generating levels for adventure games, combining speed, guaranteed solvability of levels and authorial control. For this purpose, a new graph-based two-phase level encoding scheme is developed. This method encodes the structure of the level as well as its contents into two abstraction layers: the higher level defines an abstract representation of the game level and the distribution of its content among different inter-connected game zones. The lower level describes the content of each game zone as a set of graphs containing rooms, doors, monsters, keys and treasure chests. Using this representation, game worlds are encoded as individuals in an evolutionary algorithm and evolved according to an evaluation function meant to approximate the entertainment provided by the game level. The algorithm is implemented into a design tool that can be used by game designers to specify several constraints of the worlds to be generated. This tool could be used to facilitate the design of game levels, for example to make professional-level content production possible for non-experts.

Keywords: Procedural content generation · Genetic programming · Evolutionary computation

1 Introduction

The problem of level generation is that of generating good level content for computer games, where "level" is the spatial content through which one or several player-controlled characters move. Depending on the game type, the level might be a dungeon [4], a map [14], a race track [22], etc. Level generation is an important problem for computer game development, as human authoring of this type

© Springer International Publishing Switzerland 2016
G. Squillero and P. Burelli (Eds.): EvoApplications 2016, Part I, LNCS 9597, pp. 558–573, 2016.
DOI: 10.1007/978-3-319-31204-0_36

of game content constitutes a very large part of the development costs of modern computer games. Procedural generation of levels can provide large cost savings for game developers [20], and make game development feasible in small teams on limited budgets, but can also make new types of games possible that rely on user-adaptive run time content generation [16]. Additionally, the level generation problem has similarities with many other problems in automated design and creativity (from circuit design to music generation) and in many cases methods developed for one of these problems can be adapted to work on related problems.

A common and successful approach to level generation is to cast it as an optimization problem, so that a space of levels is searched for levels that best satisfy certain criteria; this is called "search-based procedural content generation" [21]. The literature contains numerous studies using evolutionary algorithms for this problem, but also methods based on e.g. Answer Set Programming [18].

While successful methods have been found for many simple domains, the general problem of level generation is hard [19] and for many domains we cannot yet provide effective and efficient level generators. Part of this is because it is hard to automatically judge the quality of a level [8], and the best we can do is provide several imperfect heuristic evaluation functions that might be partly conflicting [11]; at the same time there are a number of hard constraints that need to be taken into account, most obviously that the level needs to be possible to complete. As a result, the optimization algorithm will need find a balance between multiple objectives and constraints. Also contributing the difficulty of the level design problem is that the search space is often vast and disconnected, because levels in many games feature a large number of elements, and many configurations of these are impossible. It is therefore important to devise a level representation that minimizes the dimensionality of the search space while making good levels reachable.

Other desirable qualities of level generators are controllability, the ability of a designer or algorithm to control important characteristics of the results of the generation process, and diversity in the output of the generator. While user control is often addressed through adding objectives that can be tweaked by the generator, diversity can be achieved through modifications to the optimization algorithm [14]. However, such added objectives and modifications often come at the price of a drop in efficiency. One idea for improving search efficiency while guaranteeing diversity is to divide up the generation process in several phases. For example, one can use an optimization algorithm to optimize "templates" for levels, that another algorithm then expands into multiple different levels. This has the benefits that the templates do not need to include all details so the search space can be lower-dimensional, and that each template can give rise to multiple levels which helps diversity. In [7] the authors evolved agent-based systems that could generate levels; however, the evolved level generators could not guarantee any properties of the final levels, and user control was limited and indirect.

Another approach to level generation is based on grammar expansion, where the general shape of a level is encoded as a grammar, and through variations in how the grammar is expanded variations of a level that still conform to structure decided by the grammar can be produced. The roots of producing content that could be used in games go back at least to L-systems, which were proposed plant generation by Prusinkiewicz and Lindenmayer [15]. Dormans brought grammar-based PCG to game level, devising a method for generating Zelda-like dungeons using grammar expansion, where both dungeon structure and quests were generated together [4]. Others have used grammar-based PCG for generating other kind of game levels, such as van Linden [10], or integrated grammar-based generation into mixed-initiative authoring tools [6].

In the above work, the grammars used for level generation were designed by humans. It is also possible to evolve grammars for content generation, as for example demonstrated by Ochoa who evolved L-systems for plant generation [13]; Shaker et al. used grammatical GP to evolve Super Mario Bros levels [17]. However, these approaches to evolving grammars for content generation do not take the ability of grammars to produce multiple levels from the same design constraints into account.

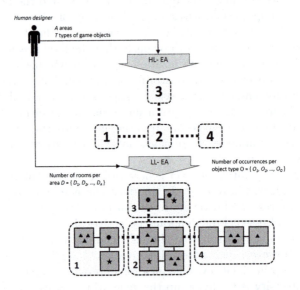

Fig. 1. Overview of the two-step Evolutionary Algorithm. HL-EA generates an acyclic graph of areas, each of those is later evolved by the LL-EA in order to be replaced by a cyclic graph of rooms.

In this paper, we introduce a two-stage method for creating levels that adhere to strong level design constraints, while allowing for considerable a diversity and designer control and a fast generation time. The core idea is to generate grammars that evolve levels through grammar-guided genetic programming. In a second step, these levels are expanded to yield complete maps that obey constraints while varying in a number of dimensions as permitted by the designer.

We demonstrate the method through a system that evolved dungeons for Zelda-like games, complete with lock and key puzzles that pose challenges for other methods.

2   The Evolutionary World Designer

The proposed system is an evolutionary tool that generates worlds for adventure games from scratch. This software displays an intuitive graphic interface that allows setting up and running an evolutionary algorithm (EA). The EA follows a two-step sequence, where two different evolutionary processes run sequentially to define, respectively, the high-level and low-level structures of a game world.

As it is depicted in Fig. 1, the high-level evolutionary algorithm (HL-EA) inputs the number of areas set by a human designer, and outputs a numbered acyclic graph of interconnected areas. For each area in this graph, a low-level evolutionary algorithm (LL-EA) runs. Each LL-EA inputs the number of rooms that the human designer set for its related area, as well as the different kinds and number of objects that it must contain. Therefore, each LL-EA outputs a cyclic graph of interconnected rooms, which contains a set of objects that populate the game world (i.e. monsters, chests, keys). The whole set of interconnected areas and their underlying sets of interconnected rooms composes a sufficient structure for codifying a world for an adventure game.

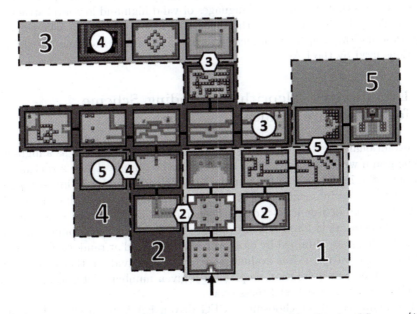

Fig. 2. Gnarled Root Dungeon, from The Legend Of Zelda: Oracle of Seasons [12]

Figure 2 shows the map called Gnarled Root Dungeon, from The Legend of Zelda: Oracle of Seasons [12], represented as a two-level structure. This map will be used as an explanatory example during the following sections, also showing that the proposed system is capable of coding actual adventure-game maps.

The first level is an acyclic graph whose nodes are the areas (dotted lines) named 1, 2, 3, 4, and 5. Each area contains a low-level cyclic graph whose nodes are rooms. Edges represent doors that connect pairs of rooms. Notice that edges do not imply direction, because doors can be traversed in both ways. Some doors connect a pair of areas as well, named with the number of the forthcoming area (in hexagons). Keys, named after the hexagon door they open (in circles), are required to open those doors, allowing the player to move to the next area. This stands as a lock and key game mechanic typical from adventure games. Lock and key mechanisms serve to connect missions and spaces, translating strong prerequisites in a mission into spatial constructions that enforce the relationships between tasks [1].

Though lock and key mechanisms can adopt several forms, for the purpose of this research they are implemented as actual locks and keys in the game map. The player starts in the first room of area 1 (pointed by an incoming arrow). Areas are named from 1 to 5, indicating the order in which the player must traverse them. This way, the player enters the map in area 1, needing to find the key to area 2 before he can move to that area. This process repeats until the adventure ends in the last room of area 5 (the one at the right), triggering some final event (e.g. facing the final boss).

The presented EAs use context-free grammars (CFG) in order to codify the syntactic rules that produce the languages of valid high and low-level structures (individuals) that comply with the depicted game worlds. The following sections describe the proposed codification system as well as the fitness evaluation functions used by the EAs.

3   Evolutionary Algorithm Encoding Scheme for Adventure Games

The evolutionary system represents a world for adventures games as a high-level acyclic graph whose nodes contain low-level cyclic graphs. These structures are represented as individuals in the populations of the different EAs involved in the searching process by means of context-free grammars (CFG). Thus, these EAs are grammar-guided genetic programs, engineered by Whigham crossover and mutation operators [3,23]. Each EA operates over a fixed size population, where new individuals are obtained by crossover, mutation and at random during every evolutionary iteration. After evaluating every new individual, the population is replaced by elitism. Evolution stops after a given number of iterations without an improvement on the best fitness score.

The main reason for choosing a CFG driven EA is to prevent the system from generating syntactically non-valid individuals (unfeasible solutions, worlds), without using any kind of repair operator. This operator would raise the overall computational cost by parsing, and possibly fixing, any individual generated during the initialization, crossover, and mutation steps.

The CFG to be used in each EA depends on the constraints set by the human designer on the systems interface. For this reason, a CFG is deterministically generated ad-hoc from the set of constraints to feed its correspondent EA. The overall process is as follows: A single CFG that becomes the input of the HL-EA is automatically generated from the features set by the human designer. Then, the HL-CFG generates the language of all possible high-level acyclic graphs that match the constraints defined by the designer. Once the optimal (satisfying) high-level graph has been evolved, a different LL-EA starts for each area in the high level graph. To do so, again, a LL-CFG is automatically generated for each LL-EA from the designers parameters to avoid the generation of unfeasible maps. The proposed CFG encoding schemes for cyclic and acyclic graphs is explained in the following subsections. It extends the work presented in [5] about coding acyclic graphs for evolving Bayesian Network architectures.

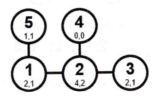

Fig. 3. High-level acyclic graph of the map Gnarled Root Dungeon.

3.1   High-Level Representation

Figure 3 shows the high-level acyclic graph that codes the areas, their connections (doors), and their amounts game objects in the Gnarled Root Dungeon:

- Each node is named after a single area. It contains the number of the area, as well as the kinds and amount of game objects distributed among the rooms of the area. In this example, there are only two kinds of game objects (treasure chests and monsters). Area 1 contains 2 chests and 1 monster. Area 4 contains neither chests nor monsters.
- The edges show the existing doors between pairs of areas. Notice that, though the final area (5) is directly connected to the first one (1), the player cannot access it until he finds the 5th key located in area 4. This order is strict, and this high-level graph representation does not accept cycles in order to preserve it. Adding a connection between 2 and 5 would create a cycle that allows the player to move directly from 2 to 5, before finding the required 5th key in 4. This would make areas 3 and 4 become useless parts of the map.

In order to encode graphs like this as individuals of a grammar-guided genetic programming system (sentences belonging to the CFG), the proposed encoding scheme includes the meaningful information of the graph as a sequence of natural numbers and separators. The sentence that encodes the high-level graph of the Gnarled Root Dungeon map is 2:1 ; 4:2 ; 2:2:1 ; 2:0:0 ; 1:1:1.

Figure 4 explains the meaning of every number in this sentence: it is composed by five sections, one per area (1, 2, 3, 4, 5), separated by;. The first number of each section points to the area that is connected to this one (input connection), excepting areas 1 and 2. Notice that the label input is missing. In both cases, information about their input connections is not provided because their connections are implicitly represented: area 1 will never have an input connection (because it is the first area in the map), and area 2 will always have an input connection from area 1. Due to the absence of cycles, areas cannot have more than one input connection, though they can have multiple output connections. Each of the remaining numbers indicates the amount of game objects placed in this area.

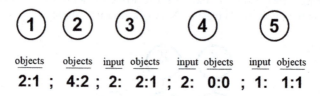

Fig. 4. Example of the proposed encoding scheme for high-level graphs.

The HL-CFG that has been automatically generated for this example is able to produce the language of all possible acyclic graphs containing 5 areas and two kinds of game objects. Figure 5 shows the generic template that generates the corresponding HL-CFG given A the number of areas, T the number of kinds of game objects, and D, a list with the number of rooms per area: $D = D_1,$ $\dots, D_A$., all set by the human designer. S is the axiom of the grammar, NTS, the set of non-terminal symbols, TS, the set of terminals, and $P_{A,T,D}$ the set of production rules that generate the language. The axiom S produces A sections for each area, separated by; where inputs and number of objects will be placed. Non-terminal symbols Z_i and C_{ij} produce, per section and respectively, its input connection and its amount of game objects for each kind. Notice that the first two sections produced by the axiom do not contain the symbol Z_i, because their connections are implicitly encoded. Z_i symbols always produce connections to preceding areas, and C_{ij} symbols always produce a number of game objects lower than or equal to the amount of rooms in the area.

3.2   Low-Level Representation

The low-level representation encodes the content of every area in the high-level representation. Thus, every node in the high-level graph leads to a different low-level graph. Figure 6 shows the low-level graph that represents the content in area 1, where nodes and edges encode rooms and (unlocked and bidirectional) doors, respectively. The information coming from node 1 in the high-level graph is the following: area 1 must have connections (doors) to areas 2 and 5, as well as two game objects of the first kind (monsters) and one object of the second

$$G_{A,T,D} = (S, NTS, TS, P_{A,T,D})$$
$$NTS = \{C_{11}, C_{12}, \dots, C_{1T}, C_{21}, C_{22}, \dots, C_{2T}, C_{31}, C_{32}, \dots, C_{3T}, \dots, C_{AT}, Z_3, \dots, Z_A\}$$
$$TS = \{:, ;, 0, 1, \dots, 9\}$$
$$P_{A,T,D} = \{$$
$$S ::= C_{11} : C_{12} : \dots : C_{1T};$$
$$C_{21} : C_{22} : \dots : C_{2T};$$
$$Z_3 : C_{31} : C_{32} : \dots : C_{3T};$$
$$\dots$$
$$Z_A : C_{A1} : C_{A2} : \dots : C_{AT}$$
$$Z_3 ::= 1|2$$
$$\dots$$
$$Z_A ::= 1|2| \dots |A-1$$
$$C_{11} ::= 0|1|D_1$$
$$\dots$$
$$C_{1T} ::= 0|1|D_1$$
$$\dots$$
$$C_{A1} ::= 0|1|D_A$$
$$\dots$$
$$C_{AT} ::= 0|1|D_A$$
$$\}$$

Fig. 5. Generic HL-CFG template for A number of areas, T kinds of game objects, and the list of rooms per area $D = D_1, \dots, D_A$.

kind (treasures). Area 1 has two additional implications: there must be an initial room, and one room must contain the second key: the key to area 2.

The graph in Fig. 6 depicts all this information, being room 1 the entry point to the map (marked with an asterisk), and rooms 2 and 6 those which lead to areas 2 and 5 (respectively). Two monsters can be found in rooms 5 and 6, as well as one treasure chest in room 5. Many subtypes of monsters and treasures can be represented, that is what the numbers in m1, m3, and t2 stand for. The key to area 2 is hold in room 3. In this level, numbers do not imply any kind of order, and cycles are allowed. Doing this increases the diversity of paths that players can follow during the gameplay. The proposed low-level encoding scheme employed in the LL-EA encodes this cyclic graph as the sentence: 01: 010 : 0011 : 00001 ; 1 : 2 : 6 : 4 ; 5 : 6 ; m1 : m3 ; 5 ; t2. This sentence contains several sections separated by;:

– The first section 01 : 010 : 0011 : 00001 is a binary codification of the input connections from room 3 to 6, separated by:. Room 1 will never have any input connections and room 2 only connects backwards to room 1, so this information is implicit. The existence of cycles imply that a given room R can have incoming connections from rooms 2 to R-1. In this order, the first subsection 01 means that room 3 does not connect to room 1 but it certainly does to room 2. The following subsections code this information for rooms 4, 5, and 6. The length of this section is directly proportional to the number of rooms in the area, information provided by the designer during the setup.

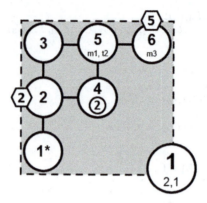

Fig. 6. Low-level cyclic graph for area 1 of the Gnarled Root Dungeon map.

- The second Section 1 : 2 : 6 : 4 implicitly indicates the name of the rooms that contain, in this order, the following items: the initial room (only present in area 1), the door to area 2, the door to area 4, and the key. The length of this section varies depending on the number of areas connected to this one. This information is coded in the high-level graph.
- The third Section 5 : 6 ; m1 : m3 codes first the locations for every monster in the area, and then the subtype of each of them. This means that room 5 contains a type 1 monster, and room 6 contains a type 3 monster.
- Analogously, the last Section 5; t2 codes the location and subtype for every treasure chest. The lengths of these last two sections depend on the amounts of monsters and enemies, respectively, defined by the designer and coded in the high-level graph.

For this example, a LL-CFG is automatically generated for creating the language of all possible cyclic graphs containing 6 rooms, two doors, one key, two monsters, and 1 treasure chest. Given the number of rooms, connections to other areas (doors), and a list with the different kinds of game objects existing in the area, the LL-CFG that generates the corresponding language is created using a generic template that is analogous to the one shown in Fig. 5.

4   The Evaluation Functions

The proposed evolutionary system uses different fitness functions for the HL-EA and the LL-EA. In both cases, evaluation comes from an analysis performed over the graph structures coded in the individuals, without needing to test the encoded maps in a real game. This way the system saves computational time as it evolves general-purpose maps, with optimal features that are independent of any game where they can be played in.

There are four features evaluated by the HL-EA fitness function for a given high-level graph, and other two features in case of LL-EA fitness function for low-level graphs. The goal of the HL-EA fitness function is to evaluate the general structure of the high-level graph and the expected game experience. The four evaluated features are the following:

- Branching: measures the average distance in nodes (calculated by an A* algorithm) between every pair of subsequent areas z_i and z_{i+1}, by applying the following equation:

$$\sum_{i=1}^{n-1} \frac{dist(z_i, z_{(i+1)})}{(i-1)} \tag{1}$$

  Maximizing this equation avoids the generation of corridor-like or star-like maps, where the proximity of subsequent areas does not encourage players to branch up their path to the exit. Figure 7, section B, shows an evolved high-level graph of five areas ($n = 5$) using the proposed high-level encoding scheme. In this case, the branching equation scores 9/4 out of 10/4, which nearly maximizes this feature. The real world example in Fig. 4 achieves a score of only 7/4 for the same number of zones.
- Recoil: applies a penalty every time an area has to be visited more than once. This balances the branching effect, fostering alternative paths, but penalizing recurrent areas in excess. The example shown in Fig. 7 has a reduced recoil level, focused mainly in area 1, which is expected to be visited four times. However, the game map lets the player peek the last levels before the player gains access to them, creating, this way, long-term objectives. The example in Fig. 4 compensates its lower branching with a lower recoil penalty, with section B being the most visited at three times.
- Progression: scores whether the occurrence of game objects increases as the player progresses, i.e. the number of game objects is greater in the later areas than in the first ones. This is calculated by applying the following equation:

$$\sum_{j=1}^{m} \left(\sum_{i=1}^{n-1} \frac{count_{j,i}}{rooms_i} - \frac{count_{j,i+1}}{rooms_{i+1}} \right) \tag{2}$$

  where $count_{j,i}$ is the amount of instances of type j game objects in the area i, and $rooms_i$ is the number of rooms in the area i. Maximizing this equation results is a better game experience because the player will encounter a higher density of game objects as he or she progresses, providing a motivation to keep on playing to discover new objects. Some corner cases can lead to odd progression scores, for example area 4 in Fig. 3 contains only one room, so it can only be completely filled or empty of content. This feature depends on the designer's input parameters.
- Designers input parameters: the designer manually sets the types and amount of game objects to be included in the map. A penalty applies for every asked object that is not present in the map. This parameter in the high-level fitness function establishes a limit in the maximization of the progression feature because the amount of game objects proposed by the designer may be insufficient to reach a high level of progression. Figure 7, section A, shows some of the objects that have been set by the designer: 5 monsters, 3 treasure chests and a ratio of 6 monsters for every 5 treasures.

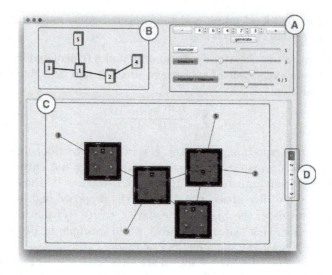

Fig. 7. Screenshot of the evolutionary world designer tool.

All these features are linearly combined to obtain the HL-fitness function as follows: $HLfitness(individual) = Branching + Progression - Recoil - Designer$.

There is a low-level graph per area in the high-level graph. Figure 7, section C, shows the low-level graph for the area 1, composed by four rooms. So, there exists one LL-EA fitness function for each low-level graph that evaluates the following two different features to get an enjoyable game:

- Critical path: given a low-level graph, this is the shortest path from the entrance to the exit passing through the room with the key, needed to go to the next area. The critical path is calculated by means of an A* algorithm. This feature applies a penalty to every room that, being outside of the critical path, does not contain any game objects. All of these rooms outside the critical path are optional, and they are worthy from an enjoyable point of view only if they include additional content, e.g. rewards or treasures after defeating a monster. Otherwise, these kind of rooms are not interesting to be visited. The critical path in the example case of Fig. 7, section C, starts with the room marked with an asterisk, then go to the room on the top right, collect the key to area 2 and take the exit to this area throughout the door marked with the number two in a circle. There are two rooms outside the critical path, located on the left and at the bottom, however these rooms contain two different kinds of game objects: treasure and monsters; so, no penalty is applied in this case. An example of the value of this scoring is that in the real world example of Fig. 4, the only room out of the critical path that doesn't contain a treasure or a monster is occupied by a one-time tutorial, and not just empty.

– Clustering: calculates the average distance in rooms (calculated by an A* algorithm) between every pair of game objects of the same kind. This average distance is calculated for each kind of game object j and then, all average distances are summed up:

$$\sum_{j=1}^{m} \left(\sum_{k=1}^{p-1} \frac{dist(n_{j,k}, n_{j,k+1})}{p-1} \right) \tag{3}$$

where p is the number of rooms that contain objects of the kind j, and $n_{j,k}$, $n_{j,k+1}$ are two diffent rooms that contain objects of the same kind j. Maximizing this equation encourages the dispersion of game objects among the existing rooms in the same area, and avoids clusters of the same kind of objects. The example shown in Fig. 7, section C, has an average distance of 2 for monsters and 1 for treasures, giving as a result a clustering level of 3. Taking into account the topology of the low-level graph and the number of different game objects present in the rooms, it is considered that the proposed evolutionary system has generated a good design for the low-level graph of area 1.

All these features are also linearly combined to obtain the following LL-fitness function for each low-level graph: $LLfitness(individual) = Clustering - Critical\_Path$.

5 Results

The proposed system is distributed as a software tool called GraphQuest [2]. This software displays an intuitive graphic interface that allows setting up and running the proposed evolutionary algorithm. Figure 7 displays an overview of the software interface with a sample evolved game level. Notice that the evolutionary process is not deterministic, so that given a fixed set of input features, many different evolved levels can be obtained. The following sections are displayed by the software interface:

– Section A contains the set-up menu that designers use to specify content constraints for the generated maps. The list of features that can be constrained are: number of areas, number of rooms per area, number of monsters, number of treasure chests, and the desired monster/treasure ratio per area. As explained in the previous sections, these features shape the HL-CFG and the LL-CFG, as well as part of the HL evaluation. The system is scalable, accepting any number of additional features required by the designer.
– Section B shows the high-level graph depicting a general overview of the generated map and its interconnected areas. Five different areas named from 1 to 5 compose the level in this case. Area 1 contains the entrance room and Area 5 contains the final room. The evolutionary process evolved this map to a high branching level with a reduced recoil level, focused mainly in Area 1.

– Section C depicts the low-level graph for every area as a set of interconnected
rooms. Each node displays a background image resembling a room in a dungeon, with as many doors as the number of rooms connected to it. Numbers in
circles represent the areas to which it is possible to go from a room when the
corresponding key has been collected. Different monsters and treasure chests
are displayed according to the content of every room: monsters (skeletons)
and a treasure (jewel) on the left, a treasure chest in the middle, monsters
(skeletons) and a treasure at the bottom, and, finally, a treasure and the key
to area 2 on the top right. The start room is the one with an input connection marked with an asterisk. Clicking on each button in D changes the area
displayed in C. In this example, there are five buttons to display each of the
five areas of the game level.

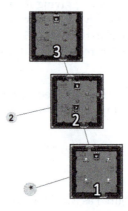

Fig. 8. Detail of an evolved low-level graph.

Figure 8 shows the low-level graph for the first area from a different evolved map.
In this case, the low-level EA was constrained to three rooms including three
treasure chests and a five to one monster/treasure ratio. Room 1 is the entrance
to the area. Room 2 contains the key and the door to Area 2, so the critical
path is the following: enter Room 1, kill the monsters and get the treasure; then
go to Room 2, kill the monsters, get the treasure and the key; then exit Area
1. Though Room 3 stays out of the critical path, it offers an optional challenge
to the player comprised by six monsters protecting another treasure chest. This
is achieved by analyzing the critical path as part of the low-level evaluation
step. It is also important to notice that the treasure chests in rooms 1 and 2
belong to different kinds. Nevertheless, the chest in Room 3 is the same kind as
that in Room 1. Evolution prevents object distribution from creating clusters of
identical objects in adjacent rooms.

Test runs have been carried out for both HL-EA and LL-EA, in order to
measure the average convergence speed in both cases. These tests were run for a
map with 5 areas containing 5, 6, 7, 5, and 3 rooms each, 24 treasure chests, and

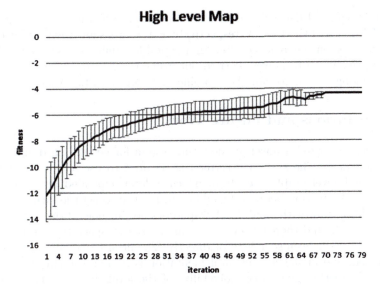

Fig. 9. Mean and standard deviation of the evolution of the best fitness score for high-level maps in 50 runs.

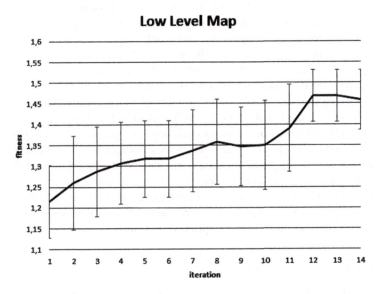

Fig. 10. Mean and standard deviation of the evolution of the best fitness score for low-level maps in 50 runs.

a 3/4 treasure/monster ratio. A population of 30 individuals has been used, considering that it converges after 6 iterations without an improvement on the best fitness score. At each iteration, 18 new individuals were obtained by crossover, 20 were generated by mutation, and 20 were randomly generated. Each experiment was run 50 times. Figures 9 and 10 show the mean best fitness scores obtained by the HL-EA and the LL-EA, respectively. Error bars show the standard deviations. Convergence is reached after an acceptable number of iterations: 79 and 14, respectively. All tests were carried out using a standard personal computer, taking no longer than 10 s for the combined HL-EA and LL-EA to finish.

6   Conclusions and Future Work

This paper presented a novel method and system for generating dungeon-type levels for games. The system described is the first to use a two-step method where a high-level graph is evolved, which is later expanded into a low-level graph. This distinction between two different levels of generation carries several benefits. Importantly, the high-level graph does not include all the details of the low-level graph, and therefore defines a smaller search space than if a low-level graph would have been searched for directly. The fitness evaluation can also be streamlined, and in the current implementation is computationally lightweight. The system further guarantees solvability of the levels by construction rather than by generate-and-test, and allows the designer a relatively high level of control.

Computational experiments show that the method reliably and quickly generates what to the authors of the current paper are apparently good levels. However, in future work we plan to carry out extensive testing with human player and designers, to verify that the generated levels are perceived as well-designed, and investigate to what extent the existing control options answer to game designers' need to parametrize their level generators. At this point, it would be also desirable that the combination weights of fitness evaluation components for both, HL-EA and LL-EA, were left to the designer's choice.

It would also be desirable to extend the current generator so that it can be embedded in a mixed initiative game design tool, such as Sentient Sketchbook [9]. Here, the designer would be able to make changes to the design at any time, and the system would give feedback about the changes from different angles (using different "computational critics"). Suggestions for changes could be generated using evolution from the current high-level description; these could be accepted or ignored by the user in a form of optional interactive evolution.

References

1. Adams, E., Dormans, J.: Game Mechanics: Advanced Game Design. New Riders, San Francisco (2012)
2. Izquierdo, R.: GraphQuest (2015). http://robertoia.github.io/GraphQuest/
3. Couchet, J., Manrique, D., Porras, L.: Grammar-guided neural architecture evolution. In: Mira, J., Álvarez, J.R. (eds.) IWINAC 2007. LNCS, vol. 4527, pp. 437–446. Springer, Heidelberg (2007)
4. Dormans, J.: Adventures in level design: generating missions and spaces for action adventure games. In: Proceedings of the 2010 Workshop on Procedural Content Generation in Games (2010)

5. Font, J.M., Manrique, D., Pascua, E.: Grammar-guided evolutionary construction of bayesian networks. In: Ferrández, J.M., Álvarez Sánchez, J.R., de la Paz, F., Toledo, F.J. (eds.) IWINAC 2011, Part I. LNCS, vol. 6686, pp. 60–69. Springer, Heidelberg (2011)
6. Karavolos, D., Anders, B., Bidarra, R.: Mixed-initiative design of game levels: integrating mission and space into level generation. In: Proceedings of the 10th International Conference on the Foundations of Digital Games (2015)
7. Kerssemakers, M., Tuxen, J., Togelius, J., Yannakakis, G.N.: A procedural procedural level generator generator. In: IEEE Conference on Computational Intelligence and Games, CIG 2012, Granada, pp. 335–341 (2012)
8. Koster, R.: Theory of Fun for Game Design. O'Reilly Media Inc, California (2013)
9. Liapis, A., Yannakakis, G.N., Togelius, J.: Sentient sketchbook: computer-aided game level authoring. In: Proceedings of ACM Conference on Foundations of Digital Games (2013)
10. van der Linden, R., Lopes, R., Bidarra, R.: Designing procedurally generated levels. In: Ninth Artificial Intelligence and Interactive Digital Entertainment Conference (2013)
11. Myerson, R.: Game Theory: Analysis of Conflict. Harvard University Press, Cambridge (1991)
12. Nintendo: The Legend of Zelda: Oracle of Seasons (2001)
13. Ochoa, G.: On genetic algorithms and Lindenmayer systems. In: Eiben, A.E., Bäck, T., Schoenauer, M., Schwefel, H.-P. (eds.) PPSN 1998. LNCS, vol. 1498, pp. 335–344. Springer, Heidelberg (1998)
14. Preuss, M., Liapis, A., Togelius, J.: Searching for good and diverse game levels. In: 2014 IEEE Conference on Computational Intelligence and Games, CIG 2014, Dortmund, pp. 1–8 (2014)
15. Prusinkiewicz, P., Lindenmayer, A.: The Algorithmic Beauty of Plants. Springer Science and Business Media, Chicago (1990)
16. Shaker, N., Togelius, J., Nelson, M.J.: Procedural Content Generation in Games: A Textbook and an Overview of Current Research. Springer, New York (2015)
17. Shaker, N., Nicolau, M., Yannakakis, G.N., Togelius, J., O' Neill, M.: Evolving levels for super mario bros using grammatical evolution. In: IEEE Conference Computational Intelligence and Games (CIG), pp. 304–311 (2012)
18. Smith, A.M., Mateas, M.: Answer set programming for procedural content generation: a design space approach. In: IEEE Transactions onComputational Intelligence and AI in Games, vol. 3, pp. 187-200(2011)
19. Sorenson, N., Pasquier, P.: Towards a generic framework for automated video game level creation. In: Di Chio, C., et al. (eds.) EvoApplicatons 2010, Part I. LNCS, vol. 6024, pp. 131–140. Springer, Heidelberg (2010)
20. Sorenson, N., Pasquier, P., DiPaola, S.: A generic approach to challenge modeling for the procedural creation of video game levels. IEEE Trans. Comput. Intell. AI Games 3, 229–244 (2011)
21. Togelius, J., Yannakakis, G.N., Stanley, K.O., Browne, C.: Search-based procedural content generation: a taxonomy and survey. IEEE Trans. Comput. Intell. AI Games 1, 172–186 (2011)
22. Togelius, J., De Nardi, R., Lucas, S.M.: Towards automatic personalised content creation for racing games. In: IEEE Symposium on Computational Intelligence and Games, CIG 2007, Honolulu, pp. 252–259 (2007)
23. Whigham, P.A.: Grammatically-based genetic programming. In: Proceedings of the Workshop on Genetic Programming: From Theory to Real-World Applications, pp. 33–41 (1995)

Evolving Chess-like Games Using Relative Algorithm Performance Profiles

Jakub Kowalski$^{(\boxtimes)}$ and Marek Szykuła

Institute of Computer Science, University of Wrocław, Wrocław, Poland
{jko,msz}@cs.uni.wroc.pl

Abstract. We deal with the problem of automatic generation of complete rules of an arbitrary game. This requires a generic and accurate evaluating function that is used to score games. Recently, the idea that game quality can be measured using differences in performance of various game-playing algorithms of different strengths has been proposed; this is called Relative Algorithm Performance Profiles.

We formalize this method into a generally application algorithm estimating game quality, according to some set of model games with properties that we want to reproduce. We applied our method to evolve chess-like boardgames. The results show that we can obtain playable and balanced games of high quality.

Keywords: Procedural content generation · Evolutionary algorithms · Relative algorithm performance profiles · Simplified board games · General game playing

1 Introduction

The idea of Procedural Content Generation has been found widely applicable to creating various parts of computer games. It can be used for generating every single part of the game, from textures and items, through levels and music, to AI opponents [1]. However, the most challenging task here is to generate a complete game [2–4]. The core of that task concerns game rules, which will specify the environment, the player role, and the plot. So far, several such attempts were made, creating games belonging to some restricted classes of possible rules. The most notorious one is Browne's Ludi system [5], which created two commercially successful boardgames. Also it is worth mentioning that METAGAME, one of the first general game playing systems producing symmetric chess-like games, also was equipped with a generator of rules [6].

Of course attempts of game generation do not end on boardgames, and some results concerning different genres have been published. Let us just mention here card games [7], strategy games [8], and Pac-Man-like grid-world games [9].

J. Kowalski—Supported in part by the National Science Centre, Poland under project number 2014/13/N/ST6/01817. M. Szykuła—Supported in part by the National Science Centre, Poland under project number 2013/09/N/ST6/01194.

© Springer International Publishing Switzerland 2016
G. Squillero and P. Burelli (Eds.): EvoApplications 2016, Part I, LNCS 9597, pp. 574–589, 2016.
DOI: 10.1007/978-3-319-31204-0_37

Arcade-style video games become a domain of special interest. The ANGELINA is an ongoing project generating complete games including rules, levels, and game characteristics [10]. Alternatively, a Video Game Description Language (VGDL) [11] game generator has been recently proposed in [12].

Although not so little research in the direction of generating complete games was made, the main question – how to judge the quality of arbitrary games and how to distinguish the good ones from the bad ones – still remains a key problem. As it is usually not very difficult to generate syntactically valid game rules belonging to some of the existing general game description languages, there are numerous syntactically valid and playable games which are not interesting from the human point of view.

In this paper we investigate the method of evaluation based on the assumption that games should be primarily sensitive to the skill of players. We present an extension of the method called Relative Algorithm Performance Profiles (RAPP) [13], which makes use of score differences between strong and weak AI players to evaluate games. We applied our method in the domain of the simplified boardgames [14], a modern language to describe chess-like boardgames, which is more concise and less restricted than METAGAME.

Instead of computing the quality of a game basing just on the assumption that we should be interested in games for which good algorithms play significantly better than bad ones, we present more methodical and sophisticated approach. First, for a given set of player strategies, we train our evaluating function using a set of *model* games, i.e. games which have desired properties and should be considered interesting. Then, to reduce computational effort, we use a generic tactic to choose the subset of strategies that will best reflect relations observed for the model. This allows us to value the game according to the level of similarity between relations of the algorithms for the evaluated game and for the model.

1.1   General Game Playing

The aim of *General Game Playing* (GGP) is to develop a system that can play a variety of games with previously unknown rules and without any human intervention. Using games as a test-bet, the goal of GGP is to construct universal algorithms performing well in different situations and environments. As such, GGP was identified as a new Grand Challenge of Artificial Intelligence and from 2005 the annual International General Game Playing Competition (IGGPC) is taking place to foster and monitor progress in this research area [15]. The General Description Language (GDL) [16] created for that purpose can describe any n-player, turn-based, finite, deterministic game with full information. The expressiveness of the language was even enhanced by some extensions: GDL-II [17] which adds nondeterminism and imperfect information, and rtGDL [18] which adds asynchronous real-time events.

A somewhat concurrent approach is represented by the General Video Game AI (GVG-AI) Competition [19]. This is a brand new, yet very rapidly growing area of GGP research. The special Video Game Description Language (VGDL) was designed to describe limited class of Atari-like 2D arcade video games [11].

However, the players are not provided with the game rules which they have to understand; instead, they have a simulation engine which they can use to learn how the game behaves.

Although widespread research in GGP domain was initiated and motivated by IGGPC, it should be pointed out that the idea is dated much earlier. The first approach, concerning a class of fairy-chess boardgames has been done by Pitrat in 1968 [20], and the second contribution, aforementioned METAGAME, contains several works of Pell from 1992 [6].

1.2   Relative Algorithm Performance Profiles

The concept of player performance profiles has been used as a method for improving the level of game playing programs by comparing them with various opponent's strategies [21,22]. The idea behind the RAPP is to use a comparison of playing agents not to evaluate their strategies, but to evaluate a game. This is an indirect approach, which focuses not on the fact how the game is built, but rather how it behaves. That makes RAPP a promising method for application in the GGP domains, where game descriptions are complicated and very sensitive (what actually applies to all GGP languages including GDL and VGDL).

The initial study concerning RAPP focuses on verifying if the concept is applicable in the domain of VGDL games [13]. The authors show that in human-designed games the differentiation between scores obtained by strong and weak algorithms is greater than in randomly generated or mutated games. This leads to the conclusion that good games should magnify the differences between the results of distinct algorithms.

More insightful research has been presented in [12], where RAPP has been used in a fitness function evaluating VGDL games. The function compares the scores of the following two algorithms: *DeepSearch* presented in details in the paper, and *DoNothing* which always returns the *null* action. A number of games were generated; many of them evaluated with near-perfect fitness, and some of them had interesting properties and features. Yet, for creating high quality games, comparable to human-designed ones, the necessity of refining the fitness function to identify more aspects of the game has been stated.

1.3   Simplified Boardgames

Simplified boardgames is a class of games introduced in [14] and slightly extended in [23] (see [24] for an alternative extension). The simplified boardgames language describes turn-based, two player, zero-sum games on a rectangular board with piece movements being a subset of a regular language. The player can win by moving a certain piece to a fixed set of squares, by capturing a fixed amount of the opponent's pieces of a certain type, or by bringing the opponent into the position where he has no legal moves. Every game has assigned a turnlimit, whose exceedance causes a draw.

The language can describe many of the Fairy Chess variants, including games with asymmetry and moves that can capture own pieces. However, the regularity of the description, besides being easily processable and concise, poses some important limitations. Actions like castling, en-passant, or promotions are impossible to express, as all the moves depending on the move history or (according to the statement within the original language description) moves depending on the absolute location of the piece. However, the latter restriction can be bypassed, so it is possible to describe e.g. chess pawn initial two-square advance.

The set of legal moves rules for each piece is the set of words described by a regular expression over an alphabet Σ containing triplets $(\Delta x, \Delta y, on)$ where Δx and Δy are relative column/row distances, and $on \in \{e, p, w\}$ describes the content of the destination square: e indicates an empty square, p a square occupied by an opponent piece, and w a square occupied by an own piece.

Consider a rule $w \in \Sigma^*$, such that $w = a_1 a_2 \ldots a_k$, each $a_i = (\Delta x_i, \Delta y_i, on_i)$, and suppose that a piece stands on a square $\langle x, y \rangle$. Then, the rule w is applicable if and only if, for every i such that $1 \leq i \leq k$, the content condition on_i is fulfilled by the content of the square $\langle x + \sum_{j=1}^{i} \Delta x_j, y + \sum_{j=1}^{i} \Delta y_j \rangle$. If move rule w is applicable in the current game position, then the move transferring a piece from $\langle x, y \rangle$ to $\langle x + \sum_{i=1}^{k} \Delta x_k, y + \sum_{k=1}^{k} \Delta y_i \rangle$ is legal.

2   Method

RAPP is an approach to evaluate the quality of games by measuring results of different controllers playing them. To use this method, one needs a set of algorithms serving as controllers, and a set of approved games that will be used as the model set. Then, a game is evaluated by running the algorithms on it, and comparing their results with those obtained by playing the model set of games.

2.1   Games in the Example Set

As a set of exemplary, well-founded games, we have chosen ten human-written variants of fairy-chess (including chess itself). Most of these games use the orthodox chess pieces, and have a similar starting state, e.g. the first line of pawns, one king, or pawns promotion.

Our set of example games contains: *Gardner, Action Man's Chess, Petty Chess, Half Chess, Demi-chess, Los Alamos Chess, Cannons and Crabs, Small-Deacon Chess, Shatranj*, and *Chess*. The detailed rules of each game can be seen in [25]. In some cases, we had to omit some of the special game rules (eg. castling, en-passant, promotion) to fit the game within the simplified boardgames framework. The concept of chess check is replaced by the goal of capturing the king. We also changed promotion of the weakest piece into the winning condition. To each game we assigned a turn limit, whose exceeding causes a draw.

2.2   Evaluating Algorithms

As the evaluating algorithms we have used a min-max search with a constant depth and different heuristic functions evaluating game states. In total, there are 16 distinct player profiles, which differ by strategic aspects they cover. Because such a big number of profiles is impractical to evaluate a large number of games, by analyzing their behavior on the example games, we narrowed this set to a subset of algorithms that produce most characteristic results.

All the heuristic functions are sums of weighted game features. The primary set of features contains material and positional features, which are general approaches to evaluate a state in a chess-like game [26]. For a given type of piece, the material feature is the difference between the numbers of pieces of the two players. The value of a positional feature for a given piece and square is 1 if such piece occupy the square, and 0 otherwise. Thus, the weight of this feature determines the willingness of the player to put a piece of that type on the square.

Implemented profiles use two strategies of assigning weights to material and positional features. The first is **Constant**, and it assigns to every material feature the square root of the number of squares on the board, and zero to all positional features. The alternative strategy is **Weighted**, which bases on the mobility of each piece as presented in [23]. For every available move of a piece, a probability that this move will be legal is estimated. The weight of the positional features is the sum of the probabilities of all moves that are legal from the given square divided by the number of squares on the board. The weight of the material features is the sum of all positional features for the given piece.

This primary set of algorithms can be improved by using the following more subtle heuristics:

Mobility. This computes for each square the square root of the number of legal moves ending on this square, and adds to the score of the game state the difference between the sum of these values between players. The aim of this heuristic is to promote expansion of pieces and covering a large area of the board.

Control. This is a similar, yet in some sense opposite, strategy to the previous one. For every square it computes which player wins a maximal sequence of captures, i.e. who has more capturing moves, assuming that the square is occupied by an opponent's piece. The score of the game state is modified by the difference between the numbers of squares controlled by the players. This strategy assists protection of own pieces and points out holes in opponent's defense, posing a threat to unprotected pieces.

Goal. This modifies the values of pieces and squares that are crucial to win, according to the method presented in [23]. The weight of pieces occurring in the game's terminal conditions are increased depending on their numbers in the initial state. The larger it is, the less important is an individual piece. Moreover, for the pieces whose aim is to reach some squares, the weights of positional features are increased for the nearby squares (using the number of moves required to move from a square to another as the distance measure). This heuristic, in contrast to Mobility and Control, is computed at the beginning of the game and does not tune weights during the gameplay.

We have used Constant or Weighted material heuristics with all combinations of the three positional ones. From this point, we will use shortened names to identify these combinations; for example *CGM* stands for Constant+Goal+Mobility, while *WC* denotes Weighted+Control.

2.3 Results

We have tested the performance of all sixteen algorithms by playing against each other the games from the example set. For every game, each pair of the heuristics played 100 times using three-ply deep min-max search, which gives 12,000 plays to create a single game profile. The gathered data is the matrix of the average scores between every pair of the heuristics. A win of a play was counted as 1 point, a draw as 0.5, and a loss as 0.

To evaluate how good each heuristic is, we took the average of its results from playing against all other heuristics. We illustrate the performance of the algorithms taking its average score for the games from the example set. This is presented by the bars of *example games* in Fig. 1.

From this point, we can see a clear tendency in the cases of some algorithms. Heuristic *CG* is undoubtedly the worst, with *C* being the second worst. All heuristics based on weighted material and position values performed above the average score of 0.5. *WGC* is counted as the best heuristic, yet the differences between the top three are very small (*WGC* − 0.66360, *WC* − 0.64927, *WCM* − 0.64780). However, the standard deviation shows that the set of example games is not consistent, and some games have very different profiles than the others.

3 Selection of the Model

The results from the previous experiment show us that some games from the example set behave substantially different in the set of example algorithms. For example, the performance of *CM* and *WGM* in *Petty chess* (0.653 and 0.745, respectively) is vitally better than in all other games, while the performance of *CGM* for that game is the lowest (0.12). Hence, our first task is to narrow the set of example games into a smaller set, which will contain games with similar behavior. Then it will be used as the point of reference – the model for generating game rules.

Due to the computational cost, we also needed to narrow the set of algorithms used as evaluation profiles. Choosing the set of representative heuristics for a given model set of games is the second task covered in this section.

3.1 Selecting Model Games

For two games $\mathcal{G}$ and $\mathcal{H}$, given their profile matrices $P_\mathcal{G}$, $P_\mathcal{H}$ obtained by running n heuristics, we can calculate their level of resemblance in a standard way as the pairwise distance between these matrices:

$$\text{dist}(\mathcal{G}, \mathcal{H}) = \frac{\sum_{i=1}^{n-1} \sum_{j=i+1}^{n} (P_\mathcal{G}[i,j] - P_\mathcal{H}[i,j])^2}{n(n-1)/2}. \tag{1}$$

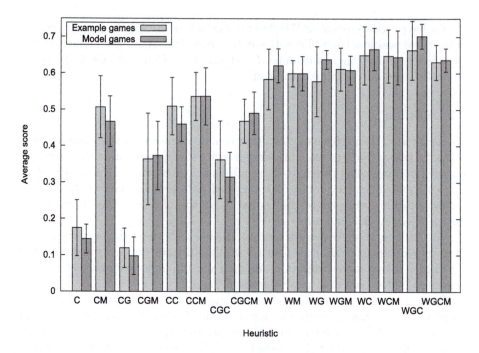

Fig. 1. The average score and the standard deviation of the sixteen algorithms on the sets of example and model games.

We can extend this definition to sets of games in the way that the distance of a set is the sum of the distances between every pair of games from this set.

We computed distances of all subsets of the example games, and decided to divide the example set into two subsets. The larger set in the best partition is *Action Man's Chess, Cannons And Crabs, Chess, Los Alamos, Shatranj* and *Small-Deacon Chess*. This subset was evaluated as tenth among the sets containing six games. On the other hand, the set of the remaining games is one of the worst four element sets.

From the obtained results, we draw a conclusion that the larger subset represents a type of games which we are interested in. These games are significantly different, yet share enough similarities to be relatively close in terms of performance of the algorithms. The set also contains *Chess*, which is arguably one of most popular board games with very desired strategic properties. Thus, we decided to use this set as our *model set* of games.

The difference in performances of particular algorithms between playing the model set and the full example set of games is presented in Fig. 1. There are significant variations in scores of some algorithms and, as expected, in most cases the standard deviation is smaller.

3.2   Selecting Evaluation Algorithms

Having the model set, we can evaluate how similarly a given game behave to the model, by comparing the relative performances of the heuristics. The more

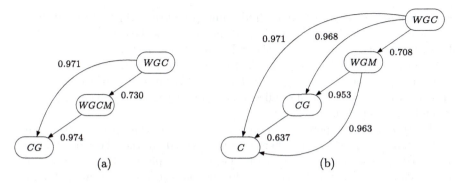

Fig. 2. Representation graphs of the relative performances in the models with (a) 3 algorithms and (b) 4 algorithms. The arrows are directed from the stronger toward the weaker algorithm, and their labels contain the average score.

heuristics we will use, the more accurate evaluation will be. The drawback is that the required number of tests grows quadratically in the number of heuristics. For this reason, we needed to find a method of selecting a small number of algorithms, which stand as the best representatives for the model games.

We have decided to selected the algorithms that are most distant to each other in terms of their relative performances. The reasoning behind this idea is twofold. First, the selected heuristics are substantially different, which allows further evaluation to incorporate more game features. Second, we take into account only the most essential differences in performance, which are easier to capture during evaluation using a smaller number of test plays. If the difference in the performance of two algorithms is small, it could be easily misjudged, depending on the number of test plays.

Let P_{model} be the matrix containing the average relative performances on the model games of some k algorithms. We can define its spread as follows:

$$\text{spread}(P_{\text{model}}) = \sum_{i=1}^{k-1} \sum_{j=i+1}^{k} |P_{\text{model}}[i,j] - 0.5|^2. \tag{2}$$

Then, for the given k, we can compute the matrices for all k-subsets of the example algorithms and calculate their spreads. We performed this computation for $k \in \{3, 4, 5\}$, and found the subsets with the largest spreads. For $k = 3$ this is CG, WGC, and $WGCM$; for $k = 4$ this is C, CG, WGM, and WGC; for $k = 5$ this is C, CG, WGM, WGC, and $WGCM$. In further experiments, we will use two models, using the best 3-subset and the best 4-subset, respectively. A visualization of the relative performance of the algorithms in these models is shown in Fig. 2.

3.3 Fitness Function

According to our RAPP method assumptions, the fitness of a game should be based on the distance between the matrix the algorithm performances for that game and the average performance matrix of the model games. This, however, is not enough to ensure that desired game properties will be fulfilled. There are several features that may be important for a good boardgame, e.g. balance between players, complexity of the rules, branching factor, game length, and usage of the pieces [27]. Our goal was to keep the fitness formula as simple and general as possible, so we decided to choose two features we perceive as the most crucial in the case of the simplified boardgames: balance between players and game length being not too short.

Let $score_w$ be the percent of points scored by the white player during p test plays, and $score_b$ be the number of points scored by the black player. Then $B = |score_w - score_b|/p$ is the balance between the players. Let say that a play is *too short* if it ends in 10 turns. If during the test s games were qualified as *too short*, then $Q = \frac{s}{p}$ is the measure of the game's *quickness*. At last, let D be the modified formula (1) of distance between the game's matrix of the algorithm performances and the model matrix, where the absolute value of the difference is used instead of the power of two. Then, our fitness function is calculated as follows:

$$f = \begin{cases} (1-D)(1-B)(1-Q) & \text{if the game is } playable, \\ -1 & \text{otherwise.} \end{cases} \qquad (3)$$

The game is considered as *playable* if during the test all plays fit within the given timelimit. This function roughly matches with our intuition about that when a given game should be considered better than the others. We made some experiments using the unmodified formula (1), yet because it usually returns very small values, the overall fitness was high, and the differences between good games and slightly-less-good games were very small. For this reason we decided to stricter results to increase the impact of D in the formula, and make the differences in fitness more visible.

4   Evolving New Games

Our goal is to generate games which have similar properties, in the sense of efficiency of strategies, to the games in the model. We use RAPP to achieve that, and evolutionary search over a constrained subset of the simplified boardgames.

4.1   Generating Games

For the sake of efficiency and reduction of the search space, we restricted our generating mechanism to produce only chess-like games. By that, we mean the games fulfilling the following additional conditions:

- The initial position is symmetric and contains two rows of pieces for both players. The front row contains only the pieces called *pawns* or empty squares. The back row contains, among other pieces, one piece of the *king* type.
- The terminal conditions are symmetric. A player wins by capturing the enemy king, or by reaching the opponent's back row with a pawn.

To generate a new game, we start by selecting randomly parameters *width* and *height* of the board, the number of *non-winning* types of figures (in addition to always present *king* and *pawn*). For the purpose of our experiments, *width* and *height* are taken from $\{6, 7, 8\}$, and the number of figures from $\{3, 4, 5\}$. Parameter *turnlimit* is computed by the formula $3 \times width \times height + r$, where r is a uniformly random number taken from $\{0, \ldots, 19\}$.

Generating the initial position is straightforward: In the front row every square can be either empty with probability 0.1 or occupied by a pawn. The king is placed uniformly at random in a column of the back row, and the remaining squares are either empty with probability 0.1 or filled by a non-winning piece chosen uniformly at random.

The remaining part, generating regular expressions for descriptions of piece rules, is more challenging. First, we generate a lot of *raw* move patterns; they will be used as building blocks for piece rules. A single raw move pattern is a list of tuples $(\Delta x, \Delta y, on, star)$, where Δx and Δy are signed integers that define the horizontal and vertical shifts of the moves, *on* is from $\{e, p, w, ep, ew, pw\}$ and defines allowed contents on the destination square (e – empty, p – enemy's piece, w – own piece), and *star* is a logical value indicating whether this tuple can be repeated with Kleene star. There is also a set of modifiers extending the pattern: FB mirrors the pattern across the horizontal axis, and ROT rotates the pattern through 90°, 180°, and 270° about the relative origin $(0, 0)$. For example, the orthodox rook can be described as $(0, 1, e, true)(0, 1, ep, false)[ROT])$.

Generating raw move patterns is based on a number of parameters. The process begins with picking at random a list of $(\Delta x, \Delta y)$ pairs. Exemplary probability values for each possible pair are listed in Table 1(a). New elements are added to the list as long as another probability test is passed. Next, we extend the list by adding *on* values. The exemplary probability table used for that purpose is shown in Table 1(b). We use a separate set of probabilities to determine the content of the destination square in the last part of a pattern. In particular, we do not allow self capturing, while it is possible to step over own figures during the movement. The last step is to add modifiers. As long as the probability test of the *modifier_prob* parameter (e.g. 0.1) is passed, one modifier is added to the pattern (see Table 1(c) for exemplary values). When $STAR$ modifier is drawn, it is applied to a random tuple in the list by changing the *star* value to true. Also, if $\Delta x > 0$ occurs in the pattern, we mirror it across the vertical axis to avoid side asymmetry. This completes generating raw move patterns.

Next, we generate each piece by choosing raw patterns, whose sum will be used as the movement language. This process is guided by the following two parameters: the chance of adding a new raw pattern to the set, and the maximal mobility of the piece. Because the values of these parameters are different for king and pawn, and non-wining pieces, we can enforce limited mobility for kings and pawns, to make resulting games more chess-like.

A weak point of this approach is that it heavily relies on human designed values. We have used four different settings, including the one presented here. The main differences in the settings we used were in probabilities of long-range jumps,

Table 1. Exemplary probability parameters used in generating games for picking at random: (a) relative coordinates, (b) square content, (c) pattern modifiers.

$\Delta y \backslash \Delta x$	0	+1	+2	+3
+3	0.6%	0.6%	0.6%	0.6%
+2	7%	7%	3.5%	0.6%
+1	20%	14%	5%	0.6%
0	0%	16%	7%	0.6%
-1	7%	7%	1.4%	0.6%

(a)

on	e	p	w	ep	ew	pw
normal	42%	14%	14%	7%	7%	14%
last	33%	33%	0%	33%	0%	0%

(b)

modifier	FB	ROT	STAR
probability	50%	25%	25%

(c)

probabilities of backward moves (also with additional possibility of $\Delta y = -2$), chances of applying modifiers, and probability of enlarge a raw move pattern.

4.2   Genetic Operators

Let n be an even population size. Given fitness values, we select $n/2$ pairs of parents using the roulette-wheel method. Although it never happened in practice, in this method if all games from the population have fitness less than or equal to zero, the uniform selection is used instead. Every pair of parents produce two offspring using the uniform crossover, which independently swaps some squares of the initial position (except the squares containing kings) and some piece rules.

There are two types of mutation, both independently modifying an offspring with some small probability. The *piece mutation* regenerates the rule of a random piece. The *position mutation* changes the content of a random square. If the square belongs to the first row, a pawn is replaced by the empty square and vice versa. If the square is occupied by the king, it is swapped with some other second row square. In the remaining cases, the square content is replaced by a random non-winning piece or the square is left empty (with uniform probability).

The next generation is created by choosing the best n games from the population of parents and children. The evolution process stops when the maximum number of generations is done.

5   Experiments

We tested our method using the extracted set of six model games and the two variants using 3 and 4 algorithms. We generated 200 games (using four sets of parameters), and evaluated each of them by 3 or 4 algorithms. In all our tests, we have used 50 min-max plays (25 per side) with depth 3 to compare every pair of algorithms and obtain the values to the profile matrix of the game. In the same way, we also evaluated all the example games.

The last test used evolution. For both variants with 3 and 4 algorithms we made 12 runs of evolution with populations of size 10, and both mutation rates equal to 5 %. Additionally, we made 12 runs for the variant with 4 algorithms, and increased population size to 16 and mutation rates to 10 %. In all these cases, the number of generations was 20.

5.1   Overview

We say that a game is *promising* if its balance factor B of the fitness function is less than 0.2 and its Q factor is less than 0.05. Such games may not have a great score taking into account the set of model games, but their properties assure that they should be at least human playable.

For all sets of games (example, generated, and evolved), we used two measures of comparison: the maximal fitness within the set and the average of the fitness of *promising* games. We are convinced that the last one is very important, as it shows the potential of the set. It may happen that the top game is not the best one from the human point of view; then the other candidates should be as good as possible in terms of accordance with the model. At last, we show the percent of *promising* games within the sets. These results are presented in Table 2.

Table 2. Comparison of game evaluation between the sets of example games, randomly generated games, and evolved games. The last column contains the results of the test with increased population size and mutation rate.

Variant	3 algs.			4 algs.			4 algs., population 16		
	Max.	Avg.	Promising	Max.	Avg.	Promising	Max.	Avg.	Promising
Evolved	0.971	0.911	100%	0.959	0.916	100%	0.978	0.922	100%
Generated	0.907	0.671	29%	0.942	0.704	34%			
Example	0.858	0.811	90%	0.959	0.859	100%			

As we can see, in all variants the best individual score was obtained by the evolutionary approach. Moreover, the average score of all final populations is also the highest. On the other hand, a pure generation can create a game even better than the best one from the example set (as in the variant with 3 algorithms), yet it is not so likely. The average score is low, as it is the fraction of promising games. Thus, the pure generation is too general and too random to be seen as a proper method just by itself, and there is a necessity of combining it with some score-improving method as evolution process. The score of the example games is not so high as one might think; this is due to the fact that the model reflects the average relations observed between the games, so every individual is likely to be distinct to that average. However, the games from the model set were usually rated better than the other example games that are outside the model.

There are two important observations. One is that, a larger number of algorithms indeed yields a better ordering of games and evaluations closer to expected, as the score is obtained using a more detailed model. The second is that, the evaluation method is very play-sensitive, i.e. the scores of one game

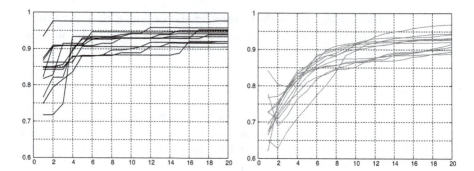

Fig. 3. Visualization of our evolutionary runs in the variant with 4 algorithms and population of size 16. The left and right charts show respectively the maximum and average fitness values of playable games in every iteration.

obtained by consecutive tests can vary. Although the general tendency, whether a game is good, average, or poor, usually remains clear, the detailed results may be significantly different. This can influence the ordering of games and so the evolution process. However, to achieve better stability, it should be sufficient to increase the number of plays used for game evaluation.

5.2   Evolution Results

Figure 3 presents the course of the evolution in the variant with 4 algorithms and population of size 16. This look similar in the other variants, but in the variant with 3 algorithms, the density of final scores distribution is visibly lower.

It can be seen that the process of evolution can increase the quality of the population to a decent level even when the initial population is poor. What is to be expected, the improvement of the best individual takes place stepwise, as it is not so easy to obtain a better game in every generation. On the other hand, the improvement of the average population is more smooth. Also, usually, except the first few generations, all games remaining in the populations are *promising*, so dysfunctional individuals are quickly discarded.

An example of an evolved game is presented in Fig. 4. This game appeared in the 16-th generation and obtained score 0.9538. Its rules are not too complicated and human-readable, and it requires a non-trivial strategy since the beginning. Note that an opening using ♘ blocks the pawn advance of the opponent. Thus, to be able to move out the other pieces, and simultaneously prevent opponent's expansion, the player has to move his own pawns in a clever way.

Although pawns seem to be powerful and advance rapidly, it is impossible to move them to the opponent's backrank, so capturing the king remains the only reachable goal. It is common between best-scored games that pawn usefulness is very limited, e.g. they cannot advance but only move sideways. It seems that pawn movements are so crucial and hard to control that it is easier to gain better game flow stability by reducing their impact on the game than by finding a set of moves that is not too strong yet still substantially useful.

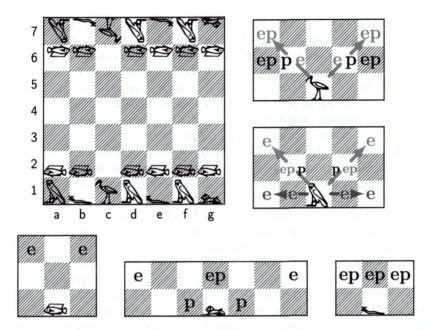

Fig. 4. Initial position and piece rules for the evolved game. Destination square content (from $\{e, p, w\}$) designates legal moves. Moves with consecutive requirements are distinguished by colors and arrows. The pawn symbol is ⬟, and the king symbol is ⬟. The turnlimit is 160.

Also we have found that the rules which seem pretty understandable in their regular expression form are often very hard to be translated on the board and remembered during a play. It seems that by generating we can easily produce movement rules which are reasonable, yet over-complicated from the human point of view. This observation addresses the problem of differences between generating game rules for humans and for AI players (e.g. to be used during GGP competitions).

6   Conclusion

We proposed an extension of the RAPP method evaluating the quality of a game using the performances of game-playing algorithms. The original approach was based on the assumption that in well-designed games, strong algorithms should significantly outperform weak ones. In this paper, we presented a more sophisticated solution, which uses more detailed information concerning the mutual relations of the given algorithms, and thus, it is able to catch and evaluate more subtle aspects of games. However, to make this working, one needs a set of model games, which have desired strategic properties, and use it to actually train the evaluation function.

The previous application of RAPP concerns VGDL games [12], which are one-player puzzles focusing on maximizing the player's score. We have shown its application in the domain of two-player, zero-sum games belonging to the class of simplified boardgames. We also tested it with different numbers of algorithms used for performance comparison in the evaluation function. In contrast with the previous study using only 2 algorithms, we provided an evaluation function in a more general form, which can be applied with any number of strategies, and allowed us to perform tests with 3 and 4 algorithms.

Finally, we applied our method to an evolutionary system, which was able to produce in some cases even "more model" games than these actually used in the model set. Not all of the top rated games could be evaluated as well-designed by human players. The reason for that is mostly abnormality and large complexity of movement rules. However, this issue can be solved in two ways: by improving move generation techniques and their parameterization, or by extending the fitness function to take into account features like complexity of the rules or pieces usefulness.

A nice property of the RAPP method is its generality. The presented methods of extracting the set of model games and model algorithms are domain independent, and can be used for any types of games. Unfortunately, the task of finding sets of example games and algorithms requires human expert knowledge on the subject. Making these tasks knowledge-free is a potential field of further study. In the case of example algorithms, the simplest (yet probably of low quality) approach is to use an MCTS algorithm, where the strength of a player is determined by its timelimit.

We confirm the conclusions from [12], that the best use for a RAPP-based game generation is being a fine sieve, with a human intervention at the last stage, where a man can choose the best game from the given results. Also, we confirm that the game evaluation, as it requires a significant number of simulations, is very time-consuming. However, given very promising results, it is a choice of preferring quality over quantity. The expected practical usage of such generated games, concerning e.g. providing interesting and novel games for the GVG-AI competition [19], requires obtaining just over a dozen games per year.

References

1. Shaker, N., Togelius, J., Nelson, M.J.: Procedural Content Generation in Games: A Textbook and an Overview of Current Research. Springer, Heidelberg (2015)
2. Nelson, M.J., Mateas, M.: Towards automated game design. In: AI* IA 2007: Artificial Intelligence and Human-Oriented Computing, pp. 626–637 (2007)
3. Togelius, J., Nelson, M.J., Liapis, A.: Characteristics of generatable games. In: Proceedings of the FDG Workshop on Procedural Content Generation (2014)
4. Zook, A., Riedl, M.O.: Automatic game design via mechanic generation. In: AAAI, pp. 530–537 (2014)
5. Browne, C., Maire, F.: Evolutionary game design. IEEE Trans. Comput. Intell. AI Games **2**(1), 1–16 (2010)
6. Pell, B.: METAGAME in symmetric chess-like games. In: Programming in Artificial Intelligence: The Third Computer Olympiad (1992)

7. Font, J.M., Mahlmann, T., Manrique, D., Togelius, J.: A card game description language. In: Esparcia-Alcázar, A.I. (ed.) EvoApplications 2013. LNCS, vol. 7835, pp. 254–263. Springer, Heidelberg (2013)
8. Mahlmann, T., Togelius, J., Yannakakis, G.: Modelling and evaluation of complex scenarios with the strategy game description language. In: CIG, pp. 174–181 (2011)
9. Togelius, J., Schmidhuber, J.: An experiment in automatic game design. In: CIG, pp. 111–118 (2008)
10. Cook, M., Colton, S.: Multi-faceted evolution of simple arcade games. In: CIG, pp. 289–296 (2011)
11. Schaul, T.: A video game description language for model-based or interactive learning. In: CIG, pp. 1–8 (2013)
12. Nielsen, T., Barros, G., Togelius, J., Nelson, M.: Towards generating arcade game rules with VGDL. In: CIG, pp. 185–192 (2015)
13. Nielsen, T., Barros, G., Togelius, J., Nelson, M.: General video game evaluation using relative algorithm performance profiles. In: Mora, A.M., Squillero, G. (eds.) Applications of Evolutionary Computation. Lecture Notes in Computer Science, vol. 9028, pp. 369–380. Springer, Switzerland (2015)
14. Björnsson, Y.: Learning rules of simplified boardgames by observing. In: ECAI, FAIA, vol. 242, pp. 175–180. IOS Press (2012)
15. Genesereth, M., Love, N., Pell, B.: General game playing: overview of the AAAI competition. AI Mag. **26**, 62–72 (2005)
16. Love, N., Hinrichs, T., Haley, D., Schkufza, E., Genesereth, M.: General game playing: game description language specification. Technical report, Stanford Logic Group (2008)
17. Thielscher, M.: A general game description language for incomplete information games. In: AAAI, pp. 994–999 (2010)
18. Kowalski, J., Kisielewicz, A.: Game description language for real-time games. In: GIGA, pp. 23–30 (2015)
19. Perez, D., Samothrakis, S., Togelius, J., Schaul, T., Lucas, S., Couëtoux, A., Lee, J., Lim, C., Thompson, T.: The 2014 general video game playing competition. In: CIG (2015)
20. Pitrat, J.: Realization of a general game-playing program. In: IFIP Congress, pp. 1570–1574 (1968)
21. Jaśkowski, W., Liskowski, P., Szubert, M., Krawiec, K.: Improving coevolution by random sampling. In: GECCO, pp. 1141–1148 (2013)
22. Szubert, M., Jaśkowski, W., Liskowski, P., Krawiec, K.: The role of behavioral diversity and difficulty of opponents in coevolving game-playing agents. In: Mora, A.M., Squillero, G. (eds.) Applications of Evolutionary Computation. Lecture Notes in Computer Science, vol. 9028, pp. 394–405. Springer, Switzerland (2015)
23. Kowalski, J., Kisielewicz, A.: Testing general game players against a simplified boardgames player using temporal-difference learning. In: CEC, pp. 1466–1473. IEEE (2015)
24. Gregory, P., Björnsson, Y., Schiffel, S.: The GRL system: learning board game rules with piece-move interactions. In: GIGA, pp. 55–62 (2015)
25. The Chess Variant Pages. http://www.chessvariants.org/
26. Droste, S., Fürnkranz, J.: Learning the piece values for three chess variants. Int. Comput. Games Assoc. J. **31**(4), 209–233 (2008)
27. Hom, V., Marks, J.: Automatic design of balanced board games. In: AIIDE, pp. 25–30 (2007)

Online Evolution for Multi-action Adversarial Games

Niels Justesen[1], Tobias Mahlmann[2(✉)], and Julian Togelius[3]

[1] IT University of Copenhagen, Copenhagen, Denmark
njustesen@gmail.com
[2] Lund University, Lund, Sweden
tobias.mahlmann@lucs.lu.se
[3] New York University, New York, USA
julian@togelius.com

Abstract. We present *Online Evolution*, a novel method for playing turn-based multi-action adversarial games. Such games, which include most strategy games, have extremely high branching factors due to each turn having multiple actions. In Online Evolution, an evolutionary algorithm is used to evolve the combination of atomic actions that make up a single move, with a state evaluation function used for fitness. We implement Online Evolution for the turn-based multi-action game *Hero Academy* and compare it with a standard Monte Carlo Tree Search implementation as well as two types of greedy algorithms. Online Evolution is shown to outperform these methods by a large margin. This shows that evolutionary planning on the level of a single move can be very effective for this sort of problems.

1 Introduction

Game-playing can fruitfully be seen as search: the search in the space of game states for desirable states which are reachable from the present state. Thus, many successful game-playing programs rely on a search algorithm together with a heuristic function that scores the desirability (usually related to the probability of winning given that state). In particular many adversarial two-player games with low branching factors, such as Checkers and Chess, can be played very well by the Minimax algorithm [15] together with a state evaluation function. Other games have higher branching factors, which greatly reduces the efficacy of Minimax search, or make the development of informative heuristic functions very hard as many game states are deceptive. A classic example is *Go*, where computer players for a long time performed poorly. For such games, Monte Carlo Tree Search (MCTS) [4] tends to work much better; MCTS handles higher branching factors well by building an unbalanced tree, and performs state estimations by Monte Carlo simulations until the end of the game. The advent of the MCTS algorithm caused a qualitative improvement in the performance of Go-playing programs [2].

© Springer International Publishing Switzerland 2016
G. Squillero and P. Burelli (Eds.): EvoApplications 2016, Part I, LNCS 9597, pp. 590–603, 2016.
DOI: 10.1007/978-3-319-31204-0_38

Many games, including all one-player games and many one-and-a-half-player games (where the player character faces non-player characters), are not adversarial [6]. These include many puzzles and video games. For such games, the game-playing problem is similar to a classic planning problem, and methods based on best-first search become applicable and in many cases effective. For example, a version of A* plays *Super Mario Bros* very well given reasonably linear levels [20]. But MCTS is also useful for many non-adversarial games, in particular with high branching factors, hidden information and/or non-deterministic outcomes.

First-Play Urgency (FPU) is one of many enhancements to MCTS for games with large branching factor [7]. FPU encourages early exploitation by assigning a fixed score to unvisited nodes. Rapid Action Value Estimation (RAVE) is another popular enhancement that has been shown to improve MCTS in Go [9]. Script-based approaches such as Portfolio Greedy Search [5] and Script-based UCT [11] deals with the large branching factor of real-time strategy games by exploring a search space of scripted behaviors instead of actions.

Recently, a method for playing non-adversarial games called *rolling horizon evolution* was introduced [17]. The basic idea is to use an evolutionary algorithm to evolve a sequence of actions to perform and during the execution of these actions a new action sequence is evolved. This process is continued until the game is over. This use of evolution differs sharply from how evolutionary algorithms are commonly used in game-playing and robotics, to evolve a controller that later selects actions [3,21,22]. The fitness function is the desirability of the final state in the sequence, as estimated by either a heuristic function or Monte Carlo playouts. This approach was shown to perform well on both the Physical Travelling Salesman Problem [16] and many games in the General Video Game Playing benchmark [13]. However, rolling horizon evolution cannot be straightforwardly applied to adversarial games, as it does not take the opponent's actions into account; in a sense, it only considers the best case.

In this paper, we consider a class of problems which has been relatively less studied, and for which none of the above described methods perform well. This is the problem of multi-action turn-based adversarial games, where each player each turn takes multiple separate actions, for example by moving multiple units or pieces. Games in this class include strategy games played either on tabletops or using computers, such as *Civilization, Warhammer 40k* or *Total War*; the class includes games more similar to classic board games, such as *Arimaa*, and arguably many real-world problems involving the coordinated action of multiple units. The problem with this class of games is the branching factor. Whereas the average branching factor hovers around 30 for Chess and 300 for Go, a game where you move six units every turn and each unit can do one out of ten actions has a branching factor of a million. Of course, neither MiniMax nor MCTS work very well with such a number; the trees become very shallow. The way such games are often played in practice is by making strongly simplifying. For example, if you assume independence between units your branching factor is only 60, but this assumption is typically wrong.

Rolling horizon evolution does not work on the class of games we consider either for the reason that they are adversarial. However, evolution can still be useful here, in the context of selecting *which actions to take during a single move*. The key observation here is that we are only looking to know which turn to take next, but finding the right combination of actions to compose that turn is a formidable search problem in itself. The method we propose here, which we call *online evolution*, evolves the actions in a single turn and uses an estimation of the state at the end of the turn (right before the opponent takes their turn) as a fitness function. It can be seen as a single iteration of rolling horizon evolution with a very short horizon (one turn).

In this paper, we apply online evolution to the game *Hero Academy*. It is contrasted with several other approaches, including MCTS, random search and greedy search, and shown to perform very well.

2   Methods

This section presents our testbed game, our methods for reducing the search space and evaluating game states, and search algorithms we test, including MCTS and Online Evolution.

2.1   Testbed Game: Hero Academy

Our testbed, a custom-made version[1] of *Hero Academy*[2], is a two-player turn-based tactics game inspired by chess and is very similar to the battles in the *Heroes of Might &Magic* series. Figure 1 shows a typical game state. Players have a pool of combat units and spells at their disposal to deploy and use on a grid-shaped battle field. Tactical variety is achieved by different unit classes that fulfil different combat roles (fighter, wizard, etc.) and the mechanic of "action points". Each turn, the active player starts with five action points, which can be freely distributed among units on the map, deploy new units, or cast spells. Especially noteworthy is that a player may chose to distribute more than one action point per unit, i.e. let a unit act twice or more times per turn. A turn is completed once all five action points are used. The game itself has no turn limit while our experiments did implement a limit of 100 turns per player. The first player to eliminate the enemy's units or base *crystals* wins the game. For more details on the implementation, rules, and tactics on the game, we kindly ask the reader to refer to the Master thesis referenced as [10].

The action point mechanic makes *Hero Academy* very challenging for decision making algorithms due to the number of possible future game states which is significantly higher than in other games. Many different action sequences may however, lead to the same end turn game state as units can be moved freely in any order. In the following, we present and discuss different methods in regard to this problem.

---

[1] https://github.com/njustesen/hero-aicademy.

[2] http://www.robotentertainment.com/games/heroacademy/.

Fig. 1. A typical game state in Hero Academy. The screenshot is from our own implementation of the game.

2.2   Action Pruning &Sorting

Our implemented methods used action pruning to reduce the enormous search space of a turn by removing (pruning) redundant swap actions and sub-optimal spell actions from the set of available actions in a state. Two swap actions are redundant if they swap the same kind of item and one can be removed as they produce the same outcome. A spell action is sub-optimal if another spell action covers the same or more enemy units. In this way spells that do not target any enemy units will also be pruned because it is always possible to target the opponent's crystals.

For some search methods, it makes sense to investigate the most promising moves first and thus a method for sorting actions is needed. A simple way would be to evaluate the resulting game state of each action, but this is usually a slow method. The method we implemented rates an action by how much damage it deals or how much health it heals. If an enemy unit is removed from the game, it is given a large bonus. In the same way, healing actions are awarded a bonus if they are saving a knocked out unit. In this way, critical attack and healing actions are rated high and movement actions are rated low.

2.3   State Evaluation

Several of our algorithms require an evaluation of how "good" a certain state for a player is. For this case, we used a heuristic to evaluate the board in a given state. This heuristic is based on the difference between the values of both players' units, assuming it as the main indicator for which player is winning. This includes the units on the game board and those which are still at the players' disposal. Furthermore, the value of a unit u is calculated using a linear combination as follows:

$$v(u) = u_{hp} + \underbrace{u_{maxhp} \times up(u)}_{\text{standing bonus}} + \overbrace{eq(u) \times up(u)}^{\text{equipment bonus}} \quad (1)$$
$$+ \underbrace{sq(u) \times (up(u) - 1)}_{\text{square bonus}}$$

whereas u_{hp} is the number of health points u has, $sq(u)$ adds a bonus based on the type of square u stands on, and $eq(u)$ adds a bonus based on the unit's equipment. For brevity, we will not discuss these in detail, but instead list the exact modifiers in Table 1. Lastly, the modifying term $up(u)$ is defined as:

$$up(u) = \begin{cases} 0, & \text{if } u_{hp} = 0 \\ 2, & \text{otherwise} \end{cases} \quad (2)$$

This will make standing units more valuable than knocked out units.

Table 1. For completeness, we list the modifiers used by our game state evaluation heuristic.

	Dragonscale	Runemetal	Helmet	Scroll
Archer	30	40	20	50
Cleric	30	20	20	30
Knight	30	-50	20	-40
Ninja	30	20	10	40
Wizard	20	40	20	50

(a) Bonus added to units with items.

	Assault	Deploy	Defence	Power
Archer	40	-75	80	120
Cleric	10	-75	20	40
Knight	120	-75	30	30
Ninja	50	-75	60	70
Wizard	40	-75	70	100

(b) Bonus added to units with items.

2.4   Tree Search

Game-tree based methods have gained much popularity and have been applied with success to a variety of games. In short, a game tree is a acyclic directed graph with one source node (the current game state is the root) and several leaf nodes. Its nodes depict hypothetical future game states and its edges define the players' actions that would lead to these states. A node has therefore as many edges leading from it, as the number of actions available for the *active player* in that game state. Additionally, each edge is assigned a value, and the edge leading from the actual gamestate (the root node of the tree) with the highest value is considered the best current move. In adversarial games, it is common that players take turns and hence the active player alternates between plies of the tree. The well-known Minimax algorithm makes use of this. However, in Hero Academy players take several actions before their turn ends. One possibility would be to encode multiple actions as one multi-action, e.g. as an array of actions, and assign it to one edge. Due to the number of possible permutations, this would raise the number of child nodes for a given game state immensely. Therefore, we decided to model each action as its own node, trading tree breadth for depth.

As the number of possible actions is variable, depending on the current game state, determining the exact branching factor is hardly possible. To get an estimate, we manually counted the number of possible actions in a recorded game to be 60 on average. We therefore estimate the average branching factor per turn to be $60^5 = 7.78 \times 10^8$ as each player has five actions. If we further assume through observation that the average game length is 40 turns and both players take a turn each round, we can calculate the average game-tree complexity to $((60^5)^2)^{40} = 1.82 \times 10^{711}$. As a comparison: Shannon calculated the game-tree complexity of Chess to be 10^{120} [19].

In the following, we will present three game-tree based methods, which were used as a baseline for our online evolution method.

Greedy Search Among Actions. The *Greedy Action* method is the most basic method developed. It makes a one-ply search among all possible actions, and selects the action that leads to the most promising game state based on our heuristic. It also uses action pruning described earlier. The Greedy Action search is invoked five times to complete a turn.

Greedy Search Among Turns. *Greedy Turn* performs a five-ply depth-first search corresponding to a full turn. Both action pruning and action sorting are applied at each node. The heuristic described earlier rates all states at leaf nodes and then chooses the action sequence that leads to the highest-rated state. A transposition table is used so that already visited game states will not be visited again. This method is very similar to a Minimax search that is depth-limited to only search in the first five ply. Except for some early and late game situations *Greedy Turn* is not able to make an exhaustive search of the space of actions, even with a time budget of a minute.

Monte Carlo Tree Search. Monte Carlo Tree Search has successfully been implemented for games with large branching factors such as the strategy game Civilization II [1] and it thus seems to be an important algorithm to test in *Hero Academy*. Like the two greedy search variants, the Monte Carlo Tree Search algorithm was implemented with an action based approach, i.e. one ply in the tree represents an action, not a turn. Hence the search has to reach the depth of five to reach the beginning of the opponent's turn. In each exploration phase, one child is added to the node chosen in the selection phase, and a node will not be selected unless all of its siblings have been added in previous iterations. Additionally, we had to modify the standard backpropagation to handle two players with multiple actions. We solved this with an extension of the *Backup-Negamax* [2] algorithm (see Algorithm 1). This backpropagation algorithm uses a list of edges corresponding to the traversal during the selection phase, a Δ value corresponding to the result of the simulation phase and a boolean $p1$ that is *true* if player one is the max player and *false* otherwise.

Algorithm 1. Alteration of the BackupNegamax [2] algorithm for multi-action games.

```
 1: procedure MULTINEGAMAX(Edge[] T, Double Δ, Boolean p1)
 2:     for all Edge e in T do
 3:         e.visits++
 4:         if e.to ≠ null then
 5:             e.to.visits ++
 6:         if e.from = root then
 7:             e.from.visits ++
 8:         if e.p1 = p1 then
 9:             e.value += Δ
10:         else
11:             e.value −= Δ
```

The ϵ-greedy approach was used in the rollouts that combine random play with the highest rated action (rated by our actions sorting method). The MCTS agent was given a budget of b milliseconds. As agents in Hero Academy have to select not one but five actions, we experimented with two approaches: the first approach was to request one action from the agent five times each with a time budget of $\frac{b}{5}$. The second approach was to request five actions from the agent with a time budget of b. The second approach proved to be superior as it gives the search algorithms more flexibility.

2.5   Online Evolution

Evolutionary algorithms have been used in various ways to evolve controllers for many games. This is done by what is called *Offline Learning* where a controller first goes through a training phase in which it learns to play the game. In this section we will present an evolutionary algorithm that, inspired by the rolling horizon evolution, evolves strategies while it plays the game. We call this algorithm *Online Evolution*. The online evolution was implemented to play *Hero Academy* and aims to evolve the best possible action sequence each turn. Each individual in a population thus represent a sequence of five actions. A brute force search, like the Greedy Turn search, is not able to explore the entire space of action sequences within a reasonable time frame and may miss many interesting choices. An evolutionary algorithm on the other hand can explore the search space in a very different way and we will show that it works very well for this game.

An overview of the online evolution algorithm will now be given and is also presented in pseudocode (see Algorithm 2). The online evolution first creates a

Algorithm 2. Online Evolution (Procedures *Procreate* (Crossover and Mutation), *Clone* and *Eval* are omitted)

```
 1: procedure ONLINEEVOLUTION(State s)
 2:     Genome[] pop = ∅                              ▷ Population
 3:     Init(pop, s)
 4:     while time left do
 5:         for each Genome g in pop do
 6:             clone = Clone(s)
 7:             clone.update(g.actions)
 8:             if g.visits = 0 then
 9:                 g.value = Eval(clone)
10:             g.visits++
11:         pop.sort()                      ▷ Descending order after value
12:         pop = first half of pop                    ▷ 50 % Elitism
13:         pop = Procreate(pop)               ▷ Mutation &Crossover
14:     return pop[0].actions                  ▷ Best action sequence
15:
16: procedure INIT(Genome[] pop, State s)
17:     for x = 1 to POP_SIZE do
18:         State clone = clone(s)
19:         Genome g = new Genome()
20:         g.actions = RandomActions(clone)
21:         g.visits = 0
22:         pop.add(g)
23:
24: procedure RANDOMACTIONS(State s)
25:     Action[] actions = ∅
26:     Boolean p1 = s.p1                          ▷ Who's turn is it?
27:     while s is not terminal AND s.p1 = p1 do
28:         Action a = random available action in s
29:         s.update(a)
30:         actions.push(a)
31:     return actions
```

population of random individuals. These are created by repeatedly selecting a random action in a forward model of the game until no more action points are left. In our case we were able to use the game implementation itself as a forward model.

In each generation all individuals are rated using a fitness function which is based on the hand-written heuristic described in the previous section, where after the worst individuals are removed from the population. The remaining individuals are then each paired with another random individual to breed an offspring through uniform crossover. An example of the crossover mechanism for two action sequences in Hero Academy can be seen on Fig. 2. The offspring will the represent an action sequence that is a random combination of its two parents'. Crossover can however in its simplest form easily produce illegal action sequences for Hero Academy. E.g. moving a unit from a certain position obviously requires that there is a unit on that square, which might not be true due to an earlier action in the sequence. Illegal action sequences could be allowed but we believe the population would be swarmed with illegal sequences doing so. Instead actions are only selected from a parent if it is legal and otherwise the action will be selected from the other parent. If both actions are illegal it will try the same approach on the next action in the parents sequences and if they are illegal as well a completely random available action is finally selected.

Some offspring will also be mutated to introduce new actions in the gene pool. Mutation simply changes one random action to another legal action. Legal en respect to the previous actions only. In some cases this will still result in an illegal action sequence. If this happens the following part of the sequence is changed to random but legal actions as well.

Attempts were made to use rollouts as the heuristic for the online evolution to incorporate information about possible counter moves. In this variation the fitness function is altered to perform one rollout with a depth limit of five actions i.e. one turn. The goal of introducing rollouts is to rate an action sequence by the outcome of the best possible counter-move. Individuals in the population that

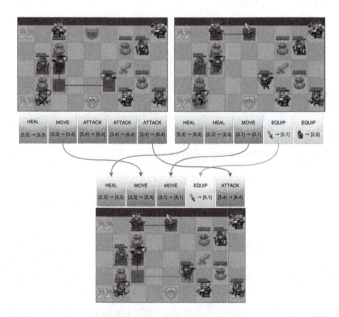

Fig. 2. An example of the uniform crossover used by the online evolution in *Hero Academy*. Two parent solutions are shown in the top and the resulting solution after crossover in the bottom. Each gene (action) are randomly picked from one of the parents. Colours on genes represent the type of action they represent. Healing actions are green, move actions are blue, attack actions are red and equip actions are yellow (Color figure online).

survive several generations will also be tested several times and in this case only the lowest found value is used. A good action sequence can thus survive many generations until a good counter-move is found. To avoid that such a solution re-enters the population the worst known value for each action sequence is stored in a table. Despite our efforts of using stochastic rollouts as a fitness function no significant improvement was observed compared to a static evaluation. The experiments of this variation are thus not included in this paper.

3   Experiments and Results

In this sections we will describe our experiments and present the results of playing each of the described methods against each other.

3.1   Experimental Setup

Experiments were made using the testbed described earlier. Each method was played against each other method 100 times, 50 times as the starting player and 50 times as the second player. The map seen on Fig. 1 was used and all methods played as the Council team. The testbed was configured to be without randomness and hidden information to focus further on the challenge of performing multiple actions. Each method was not allowed to use more than one processor and had a time budget of six seconds each turn. The winning percentages of each matchup will be presented where draws counts as half a win for each player. The rules of Hero Academy does not include draws, but we enforced this when no winner was found in 100 rounds. The experiments were carried out on a Intel Core i7-3517U CPU with 4×1.90 GHz cores and 8 GB of ram.

3.2   Configuration

The following configurations were used for our MCTS implementation. The traditional UCT tree policy $\overline{X}_j + 2C_p\sqrt{\frac{2\ln n}{n_j}}$ was used with the exploration constant $C_p = \frac{1}{\sqrt{2}}$. The default policy is ϵ-greedy, where ϵ=0.5. Rollouts were depth-limited to one turn, using the heuristic state evaluator described above. Action pruning and sorting are used as described above. A transposition table was used with the descent-path only backpropagation strategy and thus values and visit counts are stored in edges. n_j in the tree policy is thus in fact extracted from the child edges instead of the nodes.

Our experiments clearly show that short rollouts are preferred over long rollouts and that rollouts of just one turn gives the best results. Also by adding some domain knowledge to the rollouts with the ϵ-greedy policy the performance is improved. ϵ-greedy picks a greedy action equivalent to the highest rated action by the action sorting method with a probability of ϵ and otherwise a random action is picked.

Online evolution used a population size of 100, survival rate 0.5, mutation probability 0.1 and uniform crossover. The heuristic state evaluator described earlier is also used by the online evolution.

Table 2. Win percentages of the agents listed in the left-most column in 100 games against agents listed in the top row. Any win percentage of 62 % or more is calculated to be significant with a significance level of 0.05 using the Wilcoxon Signed-Rank Test.

	Random	Greedy Action	Greedy Turn	MCTS	Online Evolution
Greedy Action	100 %	-	36 %	51.5 %	10 %
Greedy Turn	100 %	64.0 %	-	88.0 %	19.5 %
MCTS	100 %	48.5 %	22.0 %	-	2 %
Online Evolution	100 %	90.0 %	80.5 %	98 %	-

3.3   Performance Comparison

Our results, shown in Table 2, show a clear performance ordering between the methods. Online evolution was the best performing method with a minimum winning percentage of 80.5 % against the best of the other methods. *GreedyTurn* performs second best. In third place, MCTS plays on the same level as *Greedy-Action*, which indicates that it is able to identify the action that gives the best immediate reward while it is unable to search sufficiently through the space of possible action sequences. All methods convincingly beat random search.

3.4   Search Characteristic Comparison

To further understand how the methods explores the search space, let us investigate some of the statistics gathered during the experiments, in particular the number of different action sequences each method is able to evaluate within the given time budget. Since many action sequences produce the same outcome, we have recorded the number of unique outcomes evaluated by each method. The *GreedyTurn* search was on average able to evaluate 579,912 unique outcomes during a turn. Online Evolution evaluated on average 9,344 unique outcomes, and MCTS only 201. Each node at the fifth ply of the MCTS tree corresponds to one unique outcome and the search only manages to expand the tree to a limited number of nodes at this depth. When looking into more statistics from MCTS, we can see that the average depth of leaf nodes in the final trees is 4.86 plies, while the deepest leaf node of each tree reached an average depth of 6.38 plies. This means that the search tree just barely enters the opponents' turn even though it manages to run an average of 258,488 iterations per turn. The Online Evolution ran an average of 3,693 generations each turn but seems to get stuck at a local optima very quickly as the number of unique outcomes evaluated is low. This suggests that it would play almost equally good with a much lower time budget, but also that the algorithm could be improved.

4   Discussion

The results strongly suggest that online evolution searches the space of plans more efficiently than any of the other methods. This should perhaps not be too surprising, since MCTS was never intended to deal with this type of problem, where the "turn-level branching factor" is so high that it all possible turns cannot even be enumerated during the time allocated. MCTS have also failed to work well in Arimaa which has only four actions each turn [12]. In other words, the superior performance of evolutionary computation on this problem might be due more to that very little research has been done on problems of this type. Given the similarities of Hero Academy to other strategy games, and to that these games model real-life strategic decision making, this is somewhat surprising. More research is clearly needed.

One immediately promising avenue for further research is to try using evolutionary algorithms with diversity maintenance methods (such as niching [14]), given that many strategies in the method used here seems to have been explored multiple times. Tabu-search could also be effective [8]. Exploration of a larger number of strategies is likely to lead to better performance.

Finally, it would be very interesting to try and take the opponents' move(s) into account as well. Obviously, a full Minimax search will not be possible, given that the first player's turn cannot even be explored exhaustively, but it might still be possible to explore this through competitive coevolution [18]. The idea here is that one population contains the first player's turn, and another population the second player's turn; the fitness of the second population's individuals is the inverse of that of the first population's individuals. There is a major unsolved problem here in that the outcome of the first turn decides the starting conditions for the second turn so that most individuals in the second population would be incompatible with most individuals in the first population, but it may be possible to define a repair function that addresses this.

5   Conclusion

This paper describes online evolution, a new method for playing adversarial games with very large branching factors. This is common in strategy games, and presumably in the real-world scenarios they model. The core idea is to use an evolutionary algorithm to search for the next turn, where the turn is composed of a sequence of actions. We compared this algorithm with several other algorithms on the game Hero Academy; the comparison set includes a standard version of Monte Carlo Tree Search. MCTS is the state of the art for many games with high branching factor. Our results show that online evolution convincingly outperforms all other methods on this problem. Further analysis shows that it does this despite considering fewer unique turns than the other algorithms. It should be noted that other variants of the MCTS algorithm are likely to perform better on problems of this type, just as other variants of Online Evolution might; we are not claiming that evolution outperforms all types of tree search. Future work will go into investigating how well this performance holds up in related games, and how to improve the evolutionary search. We will also compare our approach with more sophisticated versions of MCTS, as outlined in the introduction.

References

1. Branavan, S., Silver, D., Barzilay, R.: Non-linear monte-carlo search in civilization ii. In: AAAI Press/International Joint Conferences on Artificial Intelligence (2011)
2. Browne, C.B., Powley, E., Whitehouse, D., Lucas, S.M., Cowling, P., Rohlfshagen, P., Tavener, S., Perez, D., Samothrakis, S., Colton, S., et al.: A survey of monte carlo tree search methods. IEEE Trans. Comput. Intell. AI Games 4(1), 1–43 (2012)
3. Cardamone, L., Loiacono, D., Lanzi, P.L.: Evolving competitive car controllers for racing games with neuroevolution. In: Proceedings of the 11th Annual conference on Genetic and evolutionary computation, pp. 1179–1186. ACM (2009)
4. Chaslot, G., Bakkes, S., Szita, I., Spronck, P.: Monte-carlo tree search: a new framework for game ai. In: AIIDE (2008)
5. Churchill, D., Buro, M.: Portfolio greedy search and simulation for large-scale combat in starcraft. In: 2013 IEEE Conference on Computational Intelligence in Games (CIG), pp. 1–8. IEEE (2013)
6. Elias, G.S., Garfield, R., Gutschera, K.R.: Characteristics of Games. MIT Press, Cambridge (2012)
7. Gelly, S., Wang, Y.: Exploration exploitation in go: uct for monte-carlo go. In: NIPS: Neural Information Processing Systems Conference On-line trading of Exploration and Exploitation Workshop (2006)
8. Glover, F., Laguna, M.: Tabu Search*. Springer, New York (2013)
9. Helmbold, D.P., Parker-Wood, A.: All-moves-as-first heuristics in monte-carlo go. In: IC-AI, pp. 605–610 (2009)
10. Justesen, N.: Artificial intelligence for hero academy. Master's thesis, IT University of Copenhagen (2015)
11. Justesen, N., Tillman, B., Togelius, J., Risi, S.: Script-and cluster-based uct for starcraft. In: 2014 IEEE Conference on Computational Intelligence and Games (CIG), pp. 1–8. IEEE (2014)
12. Kozelek, T.: Methods of mcts and the game arimaa. Charles University, Prague, Faculty of Mathematics and Physics (2009)
13. Levine, J., Congdon, C.B., Ebner, M., Kendall, G., Lucas, S.M., Miikkulainen, R., Schaul, T., Thompson, T., Lucas, S.M., Mateas, M., et al.: General video game playing. Artif. Comput. Intell. Games 6, 77–83 (2013)
14. Mahfoud, S.W.: Niching methods for genetic algorithms. Urbana 51(95001), 62–94 (1995)
15. Neumann, J.V.: Zur Theorie der Gesellschaftsspiele. Math. Ann. 100(1), 295–320 (1928)
16. Perez, D., Rohlfshagen, P., Lucas, S.M.: Monte-Carlo tree search for the physical travelling salesman problem. In: Di Chio, C., et al. (eds.) EvoApplications 2012. LNCS, vol. 7248, pp. 255–264. Springer, Heidelberg (2012)
17. Perez, D., Samothrakis, S., Lucas, S., Rohlfshagen, P.: Rolling horizon evolution versus tree search for navigation in single-player real-time games. In: Proceedings of the 15th Annual Conference on Genetic and Evolutionary Computation, pp. 351–358. ACM (2013)
18. Rosin, C.D., Belew, R.K.: New methods for competitive coevolution. Evol. Comput. 5(1), 1–29 (1997)
19. Shannon, C.E.: XXII. programming a computer for playing chess. Lond. Edinb. Dublin Philos. Mag. J. Sci. 41(314), 256–275 (1950)

20. Togelius, J., Karakovskiy, S., Baumgarten, R.: The 2009 mario ai competition. In: 2010 IEEE Congress on Evolutionary Computation (CEC), pp. 1–8. IEEE (2010)
21. Togelius, J., Karakovskiy, S., Koutník, J., Schmidhuber, J.: Super mario evolution. In: IEEE Symposium on Computational Intelligence and Games, 2009, CIG 2009, pp. 156–161. IEEE (2009)
22. Zhou, A., Qu, B.Y., Li, H., Zhao, S.Z., Suganthan, P.N., Zhang, Q.: Multiobjective evolutionary algorithms: a survey of the state of the art. Swarm Evol. Comput. 1(1), 32–49 (2011)

The Story of Their Lives: Massive Procedural Generation of Heroes' Journeys Using Evolved Agent-Based Models and Logical Reasoning

Rubén H. García-Ortega[1]($\boxtimes$), Pablo García-Sánchez[1], Juan J. Merelo[1], Aránzazu San-Ginés[2], and Ángel Fernández-Cabezas[1,2]

[1] Department of Computer Architecture and Technology,
University of Granada, Granada, Spain
rhgarcia@ugr.es

[2] Department of Philosophy I, University of Granada, Granada, Spain

Abstract. The procedural generation of massive subplots and backstories in secondary characters that inhabit Open World videogames usually lead to stereotyped characters that act as a mere backdrop for the virtual world; however, many game designers claim that the stories can be very relevant for the player's experience. For this reason we are looking for a methodology that improves the variability of the characters' personality while enhancing the quality of their backstories following artistic or literary guidelines. In previous works, we used multi agent systems in order to obtain stochastic, but regulated, inter-relations that became backstories; later, we have used genetic algorithms to promote the appearance of high level behaviors inside them. Our current work continues the previous research line and propose a three layered system (Evolutionary computation - Agent-Based Model - Logical Reasoner) that is applied to the promotion of the monomyth, commonly known as the hero's journey, a social pattern that constantly appears in literature, films, and videogames. As far as we know, there is no previous attempt to model the monomyth as a logical theory, and no attempt to use the sub-solutions for narrating purposes. Moreover, this paper shows for the first time this multi-paradigm three-layered methodology to generate massive backstories. Different metrics have been tested in the experimental phase, from those that sum all the monomyth-related tropes to those that promote distribution of archetypes in the characters. Results confirm that the system can make the monomyth emerge and that the metric has to take into account facilitator predicates in order to guide the evolutionary process.

Keywords: Procedural generation · Monomyth · Agent-based model · Emergent behavior

1 Introduction

Non-player characters (NPC for further references) usually interact with the main player in order to challenge them, offer information to complete their goals

© Springer International Publishing Switzerland 2016
G. Squillero and P. Burelli (Eds.): EvoApplications 2016, Part I, LNCS 9597, pp. 604–619, 2016.
DOI: 10.1007/978-3-319-31204-0_39

or provide life-likeness to a virtual world. Following Szymanezyk et al. in [1], crowds of NPCs enhance the game-play experience of open-world videogames. This kind of games are objective-oriented and open-landscaped videogames, following the classification by [2], where a player can explore freely a virtual world.

Open worlds are usually created using Procedural Content Generation (PCG) techniques. Recently, PCG is boosting its popularity, mainly due to two reasons: the cost reduction that the automatic generation implies to the developers, and the re-playability that it offers to the players. As a recent example, *No man's sky*, an indie adventure videogame where planets, fauna and flora are created procedurally, won three Game Critics Awards, including Best Original Game and the Special Commendation for Innovation, in the past Electronic Entertainment Expo 2014, a reference annual trade show for the video game industry[1].

Szymanezyk et al. in [1] remark that virtual humans composing a crowd are often modeled only in terms of individuals and that research in crowd behavior identifies that a large majority of persons in real crowds do not act in individualistic terms. Characters need to explore a social network in order to augment their believability and, as simulated crowds, need to consider group aspects, hence they propose to create a network-type data structure with the help of ongoing sociological work. The different inter-relations between the NPCs generate events and the goal of our research is to use those events to generate quality backstories for the NPCs, that are an important part of the game narrative, as defined by Bateman and Adams in [3].

Backstories are the stories leading up to the events of the game, and they give the player the information they need to immerse themselves in the fiction. Since backstories are relevant for the immersion of the game, in this research, we create them massively using social interactions between the NPCs.

Archetypes are recurring thematic and linguistic patterns in folklore and literature [4]. In videogames, archetypes are classically related to the nature of the characters (for example, thief, warrior or wizard), which reminds of the concept of stereotype, the unfair belief that all people or things with a particular characteristic are the same. An archetype is a role that a character plays in a given moment or period, regardless of the nature of their characteristics. A character could play many different archetypes in its life, even at the same time. But, what could be the minimum set of archetypes to design if we want to create interesting backstories? We found a good answer in the monomyth.

The monomyth, commonly known as the hero's journey, is a pattern where different archetypes behave in a specific way and conform to ancient and modern myths from cultures all over the world: it is the story of someone who is considered a hero, that has to deal with his/her shadow, who is waiting at the end of a journey, where different characters appear like mentors, allies or obstacles. The monomyth was studied by Joseph Campbell in 'The Hero with a Thousand Faces' [5], and later by Vogler in 'The writer's journey' [6]. The monomyth has been typically applied to literature and traditional media, but it actually manifests also in modern videogames like Mass Effect or Skyrim [7] among others.

---

[1] http://www.gamecriticsawards.com/winners.html.

In videogames, the *monomyth* is used to design the main plots that the player can empathize with: in his work [8], Bartle identifies and examines the application of key elements of the monomyth to videogames, and discovers that players play virtual worlds as a mean for self-discovery, by subconsciously, following a predetermined path: the monomyth. Our work uses the monomyth as a frame for backstories in videogames, since it provides a basic but omnipresent set of high level behaviors that can be found in the daily life but also in the biggest adventures. We use the monomyth as metric of interest in order to promote the archetypes present in it.

Our research uses a virtual world inhabited by autonomous agents, but the idea of using Agent-Based Models (ABM) of the world to generate stories is not new, as remarked in previous works [9,10], especially in the area of the interactive drama. In 2002, the Virtual Storyteller [11] used agents that made up dialog using techniques from improv theater, a plot guide and a narrator. Our technique uses the same approach, but there is no plot guide. Instead, a Genetic Algorithm (GA) sets the mood of the backstories created by finding 'archetypes'. Mei et al. created in 2005 a system called Thespian [12], where the agents' personalities, their goals, are fitted so that they are motivated to perform according to the scripts. They use look-ahead search in a decision-theoretic framework to determine the best way to achieve their goals and they are prepared to respond to the user interaction in a consistent way. Like Thespian, our work is also focused in the final script, not in modeling the agent's personality, but in our target application, open-world videogames, scripts are auto-generated in order to be re-playable. In 2008, Peinado et al. studied in [13] the Belief-Desire-Intention (BDI), a cognitive model that reinforces narrative causality insofar as motivations and where beliefs are causal links that enrich characters. BDI is a theory that is starting to be considered a promising tool for modeling sophisticated characters, but it is not suitable for our research since we need to use more basic agents, easy to model and parametrize: as discussed by Sanchez and Lucas in [14], the analysis of relatively simple simulations using ABMs can, nonetheless, be quite complex, and in our case we use them in massively inhabited virtual worlds hard to analyze and evolve.

In previous works we used Finite State Machines (FSMs) to model the agent behavior, but FSMs and its variants have limitations in developing game Artificial Intelligence (AI). For this reason in our present research we have used Behavior Trees (BT) instead, following Lim's arguments in [15]: BTs simplify the design of behavior by allowing the re-usability of tasks without increasing the complexity of the nodes and transitions. Moreover, BTs are the most successful method to model AIs in videogames, and many game engines allow the possibility to create them, like Unity3D or Unreal Engine. A behavior tree consists of different kinds of tasks that are the nodes in a hierarchical structure, following the description by Trembley [16]: conditions, that check properties of the environment, actions, that alter the state of the environment, and compositions, whose result is calculated from the children tasks.

Our previous work proved that a hybrid *Evolutionary Computation - Agent Based Model* (EC-ABM) methodology can be used to achieve the emergence of archetypes, according to the structure studied by Cioffi et al. in [17]. Here we add a new layer in charge of evaluating the backstories: the Logical Reasoning (LR). We use an ABM for the execution of the virtual world, whose execution is parametrised by a set of integer and float values, a Logical Reasoner that evaluates the events generated by each simulation and a Genetic Algorithm (GA) to find the set of parameters with highest fitness. The LR uses a mixed imperative-declarative paradigm, following Denti's approach in [18] to make high level deductions from the simulation's events, that are also expressed as logical predicates, using a logical theory that is modeled from the monomyth. The three layered architecture proposed can be observed in Fig. 1a.

2   A Backstory Generator that Promotes the Monomyth

In this section we will present a system that uses the three-layered system EC-ABM-LR to generate and evaluate backstories in the virtual world. Firstly, Sect. 2.1 will present the high level architecture of the system and the tasks assigned to each logical module. Later, in Sect. 2.2 we give the details of the ABM layer including the virtual world, the agent's model, the parametrisation of the simulation and the output format. Then, in Sect. 2.3, we propose an implementation of the logical model of the monomyth, and a method to evaluate the quality of the backstories using logical reasoning on that model. Finally, in Sect. 2.4, we argue the usage of a GA to promote the emergence of the monomyth.

2.1   The System Architecture

The proposed system is described in the Dataflow Diagram in Fig. 1b. Different modules have been defined: the Optimizer, the Simulator, the Reasoner, the Narrator and the Customizer. The Optimizer, formerly the EC layer, uses a GA whose individuals are simulation configurations and returns the best individual found. The fitness function of each individual consists in several simulations, performed by the Simulator, i.e. the ABM layer, that uses a multi-agent system in order to generate a sequence of events. These events are evaluated by the LR, the third layer, that uses predicate logic to assign a numerical value to the events following guidelines extracted from the monomyth. The Simulator runs the virtual environment with each configuration in different trials and returns the average (the fitness) to the Optimizer. On the other hand, the Narrator takes a given solution and transforms it to an agnostic narration, that is a selection of relevant predicates without final names for the narration elements. The Customizer is in charge of assigning names and structures to the agnostic narration in order to apply a specific literary setting, for example, medieval, futuristic, steampunk or Shakespearian.

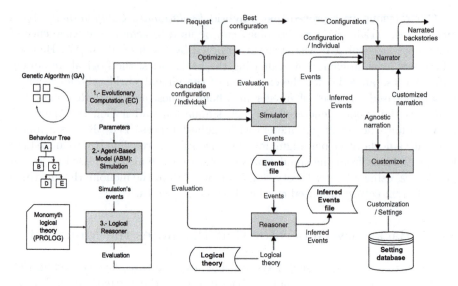

Fig. 1. high-level architecture of the system.

2.2 ABM Layer: An Agent-Based Model that Generates Simple Events

The virtual environment that we propose in this work is called ColorForms, and has been designed as a board-like game that focuses on the four independent, ontic dimensions that conform the ludo-narrative design space described by Aarseth in his work [19]: world, objects, agents and events. In our virtual environment, pieces will compete and collaborate in order to reach their desired color. The narrative, as Aylett studied in [20], emerges through small interactions between the characters. The main elements are described as follows:

World: Our backstories need spatio-temporal dimensions, namely, a place where they occur and a time that sequences and orders the events. In ColorForms, we propose to use a very simple map: the checkerboard, that consists of 64 squares (8 × 8). The time is measured in virtual days.

Agents: our agents are pieces that can occupy one cell each. There are three shapes for the pieces (circle, square and triangle) and a zero-sum game relation between them, rock-paper-scissors alike, in order to establish a method to decide the winner and the loser in individual direct conflicts while keeping a global balance. Agents change along the timeline: every piece has a background color and a foreground color that are constantly modified. The background color varies randomly in a slowly way for every piece and represents the continuous internal changes in the character needs. The foreground color can be changed by the piece or by other pieces under certain circumstances and is used to improve the happiness: the piece is happier when the two colors are equal.

Objects: we use one type of object the characters must compete or collaborate for: the color spot. In our environment, the agents can stain with the spot in order to become colored. The color spots appear or disappear randomly.

Events: the characters interact between them and the environment, although the number of actions is very limited: move to a free adjacent cell, push another agent to a depending on the shape, interchange the background color and stain with a color spot.

As the we remarked in the introduction and following [1], characters need to explore a social network in order to augment their believability. The social component is defined in ColorForms by an affinity matrix: each piece has an affinity with every other piece in the board, that will depend on their shape, background and foreground colors. This affinity is dynamic and will influence the decisions that piece makes; for instance, the happiness of the piece will be influenced by the ones with high affinity with it.

In our proposal, events are represented as logical predicates. Formerly, Kim defined the events in his theory of the structured complexes [21] using the canonical notation $[x, P, t]$, where x is the substance, P is the property it exemplifies and t is the time. The existence condition implies that $Event[x, P, t]$ exists just in case that substance x has the property P at the time t. In our work, the events that compound the backstories are defined as meaningful logical predicates with notation $Predicate(t, x_0, \ldots, x_n)$ where t is the time, and each x is an element of the world (a character, a cell, a moment, the property of an element or a value). Each predicate has a name, a signature (or arguments) and an interpretation, that is a description of the event in natural language. We contemplate two types of predicates: those that have been generated by the agents in the virtual environment (world's facts, present in Fig. 3) and those that are inferred using the Reasoner module, related to the monomyth (world's deductions, present in Fig. 4). White nodes in the Figure represent the predicates that can be generated by the ABM layer, while the gray ones are inferred by the Reasoner. The light gray nodes (called *helpers*) help to promote the dark-gray predicates, or *facilitators*. These are the ones that are present in the monomyth pattern, and they are interesting by themselves. Therefore, the *monomyth* node implies the appearance of all the *facilitators*. This node represents the 'pure' literary monomyth.

The virtual environment created for this work, the pieces (characters) are emotion-driven and can perform different strategies to interact with each other in order to generate events that will conform later backstories. The behavior tree used, described in Fig. 2, has condition nodes and action nodes; when a condition is fulfilled it generates an event from the Fig. 3c and when an action is executed correctly it generates an event from Fig. 3d. Our implementation of behavior tree executes the tree in pre-order until a leaf node is evaluated as 'success'.

The Simulator takes an array of 52 values as input, maps them to the environment internal variables and performs the execution of the virtual world by running sequentially the parametrised agents' behavior tree. Finally, when a specific number of virtual days is reached, the simulation finishes and the generated

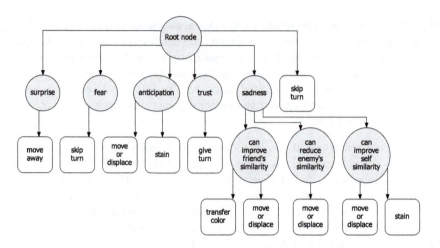

Fig. 2. Proposed agents' behavior tree: circular nodes represent conditions and rectangular nodes represent actions.)

events are sent to the Reasoner for further analysis. The structure of the configuration array and the meaning of each element are described in Fig. 5. As we will see in further sections, the configuration array of the Reasoner will be the chromosome in the Optimizer, in other words, the individual of the GA.

2.3   LR Layer: Logical Reasoning to Construct the Monomyth

As remarked in Sect. 1, the monomyth is a pattern that conforms the quintessence of heroism in literature and is composed of different archetypes. Vogler in [6] encountered that archetypes are not 'rigid character roles' but 'flexible character functions' performed to achieve certain effects in the story, hence the same character can play different archetypes along a story. 'There are, of course, many more archetypes; as many as there are human qualities to dramatize in stories', but we will focus in the eight archetypes that conform the monomyth: the **Hero**, the **Mentor**, the **Threshold Guardians**, the **Herald**, the **Shapeshifter**, the **Shadow**, the **Ally** and the **Trickster** (see Table 1 for detailed definitions).

In our approach, we try to define the archetypes in terms of the world's events described in Sect. 2.2. The Reasoner uses a logical theory in PROLOG, that is a powerful approach since it gives us the possibility to compound predicates and use time-frameworks. The terms in our theory are shown in Fig. 4 in an ordered way following a dependency graph. The monomythic archetypes are open to interpretations and can be instantiated in many different ways inside a specific story, as many as different films, books, comics or videogames with a hero we could think about. For this reason, we have implemented our own interpretation of the Monomyth archetypes in ColorForms, that are described in the third column of Table 1.

a) Events for the board configuration

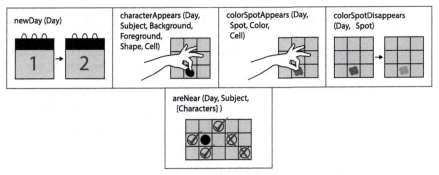

b) Daily pieces' events

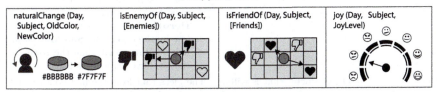

c) Conditions in the Behavior Tree (emotions and checks)

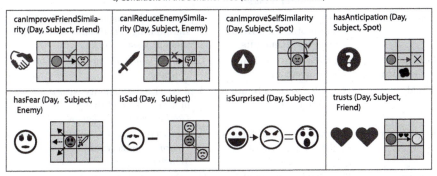

d) Actions in the Behavior Tree (Strategies)

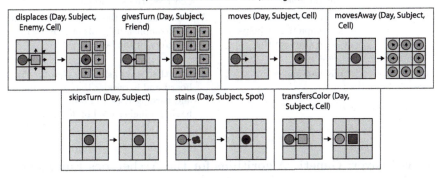

Fig. 3. Predicates generated by the Simulator (world's events).

Table 1. Comparison of theoretical archetypes in the Monomyth and their interpretations in ColorForms.

Trope	Brief definition	Interpretation in ColorForms
Conflict	(Not present in the definition of monomyth)	A sequence of events that succeed in a certain moment where two characters take part (the winner and the loser). A conflict might appear when a piece scares or displaces other, or when it takes a color spot that the other desires. If two pieces are friends and one of them is displaced by a third one, the two friends have a *Conflict* with it.
Journey	(not present in the definition of monomyth)	The time interval between *Conflicts* with the same pieces. The first *Conflict* must be lost by one of them (the **Hero**) and the second one has to be lost by the other (the **Shadow**), who must be 'more evil' than the *Hero*, that is, have helped less friends.
Hero	The character that is willing to sacrifice his own needs on behalf of others.	The piece that loses the first conflict and wins the second one in a *Conflict*.
Mentor	The character that teaches and protects the *Hero* and gives them gifts.	A piece that has high affinity with the *Hero* during the *Journey* and helps them at least once.
Threshold Guardians	The forces that the hero must overcome.	Losers of *Conflicts* against the *Hero* in a *Journey*
Herald	The caller to the adventure for the hero.	A piece with high affinity with the *Hero* in a *Journey* and that lost the first *Conflict* with the *Shadow*.
Shapeshifter	The character that changes along the journey.	A piece that plays *Ally* and *Threshold Guardian* archetypes in the same *Journey*.
Shadow	The villain or antagonist; the dark side.	The character that wins the first conflict and loses the second one in a Journey.
Ally	The character that accompanies or helps the *Hero* through the *Journey*.	A piece that gives the *Hero* an extra turn to accomplish its goals or stays aside the *Hero* in the board.
Trickster	A mischief-maker.	A piece that stays aside the *Hero* in the board and is trickier than it (joking and turning tricks, in other words, having a higher level of Joy).

2.4   EC Layer: Genetic Algorithms for the Archetypes Emergence

The objective of our work is to generate the archetypes present in the monomyth. The usage of a GA is proposed, since we have to deal with a big amount of

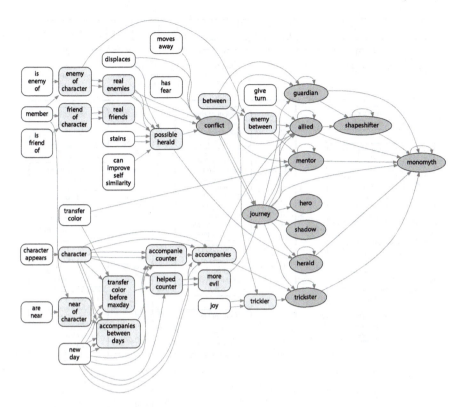

Fig. 4. Predicate dependencies in the proposed logical theory of the monomyth: white nodes represent predicates that can be generated by the ABM layer and gray predicates (helpers) are inferred by the Reasoner. Dark-gray predicates (facilitators) conform the monomyth.

parameters. Also, because the result of the simulations is uncertain, due the stochastic nature of the virtual world and the unpredictable inter-relationships of the agents. This technique has been proved to work successfully in generative storytelling by Nairat in his work [22], and later by [9,10].

The Optimizer module uses a GA to evolve configurations of the Simulator in order to find the one with the best fitness, that is calculated in base of the predicates found by the Reasoner. The use of a GA is adequate since it is not possible to deduce the effects of the parameter values when the agents interact: the fitness is the result of high level interactions or specific sequences of them (phenotype) that cannot be predicted. This uncertainty is increased by the stochastic nature of the simulation, since different executions of the Simulator provide distinct results that improves the re-playability of the videogame.

An individual for our GA is a chromosome of 52 genes, where the first 5 are natural numbers in different ranges and the following 47 are real values in the range [0–1]. The structure of the chromosome is detailed in Fig. 5.

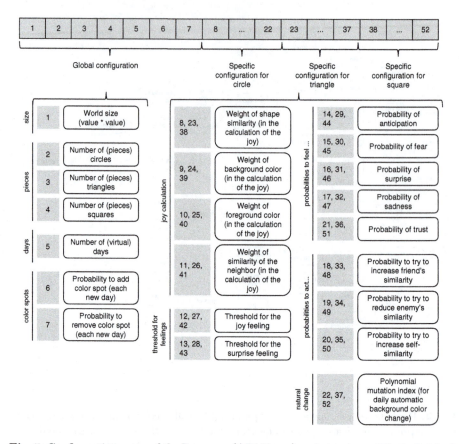

Fig. 5. Configuration array of the Reasoner (ABM layer) and structure of the individual of the Optimizer (EC layer). Gray rectangles stand for the index of the parameters.

The fitness used takes into account the number of archetypes that have been deduced by the Reasoner since we are interested in the promotion of the monomyth, as described in Sect. 2.3.

Let A_i an archetype in the set of all T desired predicates A:

$$A = \{Conflict, Journey, Shadow, Hero, Mentor, Herald, Guardian,$$
$$Allied, Trickster, Shapeshifter, Monomyth\} \qquad (1)$$

and $|A_i|$ the number of occurrences of that predicate generated by the Reasoner at the end of the simulation, we define the next three metrics to evaluate the fitness F:

– Amount of Monomyth Occurrences Found (M): counts the number of occurrences of the predicate 'monomyth', hence it stands for the most basic metric we can implement.

$$F_M = |A_{Monomyth}| \qquad (2)$$

– Amount of All Occurrences Found (A): counts the number of occurrences of every monomyth-implication predicate in the simulation, defined as 'facilitators' in Sect. 2.3, and also shown as dark-gray nodes in Fig. 4.

$$F_A = \sum_{i=0}^{T} |A_i| \qquad (3)$$

– Amount of All Occurrences Distributed by Predicates (AD): counts the number of occurrences, but promotes the appearance of new archetypes. That is, the occurrence of two different predicates gets a bigger fitness than two occurrences of the same predicate.

$$F_{AD} = \sum_{i=0}^{T} \log_{10}(|A_i| + 1) \qquad (4)$$

Although our goal is to make the monomyth emerge, and this can be achieved with fitness metric M (Eq. 2), we have added the extra metrics defined in Eqs. 3 and 4 to observe the effects of guiding the evolution by promoting the appearance of the facilitators.

More complicated metrics could be chosen, for example, using human interaction as proposed by Takagi in [23], but the chosen ones can be considered the simplest ones to detect the emergence of the monomyth.

3   Experiments and Results

The objective of this paper is to demonstrate that it is possible to create a multi-paradigm three-layered system (AC-LR-ABM) that is able to generate massive backstories following the monomyth guidelines. As secondary objectives, we try to test if the chosen metrics are valid to allow the emergence in the evolutionary process of the GA.

We use a generational model, with selection by binary tournament and no elitism. Our mutation operator is polynomial and our crossover operator is blx-α, both good elections when our genes are coded as numerical values. All these decisions are supported by Eiben and Smith in their book [24]. The number of individuals in the populations is 30, the termination condition is to reach 40 generations and the maximum time to deduce the predicates from the world's events is 300 seconds. To reduce uncertainty in the evaluation of an individual, we use explicit averaging [25], using the average of 15 trials. The range for the number of pieces is [0–15] for each shape, hence the final solution can have [0–45] pieces. The size of the world can vary from 8×8 (64 cells) to 16×16 (256). The range of virtual days for each simulation is [2–128]. These values have been selected empirically after preliminary tests, and are summarized in Table 2.

The experiments are executed with the three fitness functions previously defined in Sect. 2.4: amount of Monomyth Occurrences Found (M), Amount of Monomyth Occurrences Found (M) and Amount of All Occurrences Distributed by Predicates (AD). The results obtained can be seen in Fig. 6.

Table 2. Parameters of the GA used in the experiments.

Parameter	Value
GA parameters	
Type	generational
Population size	30
Selection	binary tournament
Mutation operator	polynomial (theta = 20)
Crossover operator	blx-α ($\alpha = 0.5$)
Termination condition	30 generations
Fitness	M / A / AD
Gene ranges	
World size gene	8–16
Number of circles	0–15
Number of triangles	0–15
Number of squares	0–15
Number of virtual days	2–128
ABM parameters	
Maximum reasoning time	300 s
Number of trials in fitness	15

The predicate monomyth depends directly and indirectly on a huge amount of predicates (world's events, helpers and facilitators, as the dependency graph in Fig. 4 reflects) hence we cannot search it directly, as seen in the graphs called 'Fitness M' and 'Monomyth in fitness M'. However, if we guide the evolutionary process by taking into account also the facilitator predicates, as A and D do, we can make the whole monomyth emerge with our system.

The fitness AD gets better solutions than fitness A, since we promote the appearance of new the predicates (and all of them are necessary for the monomyth). In graph 'Monomyth fitness A' we can observe that monomyth emerges punctually but the GA loses it in favor of new appearances of the *Ally*, as the graph 'facilitators in fitness A' shows; nevertheless, it does not happen in AD, as shown in 'Monomyth in fitness AD', where the number of monomyths fluctuates but keep a tendency.

Regarding the facilitators found, their rate of appearance is aligned with the real applications of the monomyth: *Allies*, *Threshold Guardians* and *Tricksters* have much more occurrences than the rest. The *Hero* and the *Shadow* are the leading roles and they must appear only the necessary. Finally *Shapeshifters*, *Mentors* and *Heralds* are uncommon characters that depends on predicates that are hard to satisfy. These appearance rates also confirms that our logical model for the monomyth is aligned with the one described by Vogler. However, the fitness AD promotes less difference in the number of appearances than fitness A.

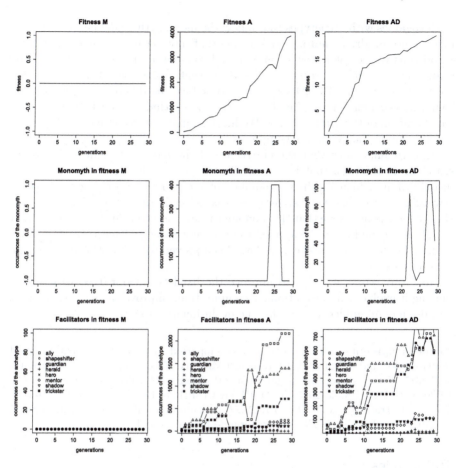

Fig. 6. Graph showing the results of running ColorForms with the three different fitness functions; the first column belongs to M (promotion of the monomyth), the second column to A (promotion of the facilitators) and the third column to AD (promotion of the distribution of facilitators). The first row shows the evolution of the best individual, the second one the number of occurrences of the monomyth found and the third one facilitators.

As we are not interested in the number of specific archetypes but in the number of monomyths found, we favor the use of AD.

The source code and the experiments are distributed under GPLv3 license, and are freely available at the official project's page[2].

4   Conclusions

The present paper uses agent-based modeling, logical reasoning and genetic algorithms to promote the appearance of massive backstories that can later be used

---

[2] https://github.com/raiben/made.

in videogames with archetypes that follow the monomyth guidelines. For this, we model emotion-driven autonomous agents that interact in a virtual world called ColorForms where color pieces compete and collaborate in order to change their colors. The virtual environment generates events using logical predicates that are later used by a PROLOG reasoner to deduce the different archetypes that conforms the monomyth. Finally, a genetic algorithm is used to optimize the configuration of the virtual world. We have compared three different fitnesses in order to guide the evolutionary process.

Experiments show that the emergence of the monomyth can be helped by searching facilitators, that is, the predicates that are found in the monomyth, instead of just focusing in the number of monomyth occurrences. Moreover, rewarding the diversity of archetypes usually leads to better solutions. Thus, these experiments prove that the monomyth can emerge in this kind of individual based models and it can be applied to massively generate backstories populated with archetypes; these archetypes have been proved to be emotionally engaging for the videogame player.

In future works we plan to improve the fixed Behaviour Tree structure by making it evolve using evolutionary computation, improve the narration engine and use human-guided metrics for the evolutionary process.

Acknowledgments. This work has been supported in part by projects TIN2014-56494-C4-3-P and TIN2012-32039 (Spanish Ministry of Economy and Competitivity), V17-2015 of the Microprojects program 2015 from CEI BioTIC Granada, PROY-PP2015-06 (Plan Propio 2015 UGR), PETRA (SPIP2014-01437, funded by Dirección General de Tráfico), and MSTR (PRY142/14, Fundación Pública Andaluza Centro de Estudios Andaluces en la IX Convocatoria de Proyectos de Investigación).

References

1. Szymanezyk, O., Dickinson, P., Duckett, T.: From individual characters to large crowds: augmenting the believability of open-world games through exploring social emotion in pedestrian groups. In: Proceedings Think Design Play: DiGRA Conference, ARA Digital Media Private Limited (2011)
2. Aarseth, E.: From hunt the wumpus to everquest: introduction to quest theory. In: Kishino, F., Kitamura, Y., Kato, H., Nagata, N. (eds.) ICEC 2005. LNCS, vol. 3711, pp. 496–506. Springer, Heidelberg (2005)
3. Bateman, C.M., Adams, E.: Game Writing: Narrative Skills for Videogames. Charles River Media, Boston (2007)
4. Garry, J., El-Shamy, H.: Archetypes and Motifs in Folklore and Literature. ME Sharpe, Armonk (2005)
5. Campbell, J.: The Hero with a Thousand Faces, vol. 17. New World Library, Novato (2008)
6. Vogler, C.: The Writer's Journey. Michael Wiese Productions, Studio City (2007)
7. Knopf, E., et al.: The Rationalist's Spirituality: Campbell's Monomyth in Single-Player Role-Playing Videogames Skyrim & Mass Effect. PhD thesis, University of Sydney (2013)

8. Bartle, R.A.: Massively multihero: Why people play virtual worlds. In: Proceedings of CGAIDE 2004, 5: th International Conference on Computer Games, Artificial Intelligence, Design and Education, Wolverhampton: The University of Wolverhampton, School of Computing and Information Technology (2004)
9. Garcia-Ortega, R.H., Garcia-Sanchez, P., Mora, A.M., Merelo, J.: My life as a sim: evolving unique and engaging life stories using virtual worlds. In: ALIFE 14: The Fourteenth Conference on the Synthesis and Simulation of Living Systems, vol. 14, pp. 580–587 (2014)
10. García-Ortega, R.H., García-Sánchez, P., Merelo, J.J., Arenas, M.I.G., Castillo, P.A.: How the world was MADE: parametrization of evolved agent-based models for backstory generation. In: Mora, A.M., Squillero, G. (eds.) EvoApplications 2015. LNCS9, vol. 9028, pp. 443–454. Springer, Heidelberg (2015)
11. Theune, M., Faas, S., Nijholt, A., Heylen, D.: The virtual storyteller. ACM SigGroup Bull. **23**(2), 20–21 (2002)
12. Si, M., Marsella, S.C., Pynadath, D.V.: Thespian: Using multi-agent fitting to craft interactive drama. In: Proceedings of the Fourth International Joint Conference on Autonomous Agents and Multiagent Systems, pp. 21–28. ACM (2005)
13. Peinado, F., Cavazza, M., Pizzi, D.: Revisiting character-based affective storytelling under a narrative BDI framework. In: Spierling, U., Szilas, N. (eds.) ICIDS 2008. LNCS, vol. 5334, pp. 83–88. Springer, Heidelberg (2008)
14. Sanchez, S.M., Lucas, T.W.: Exploring the world of agent-based simulations: simple models, complex analyses: exploring the world of agent-based simulations: simple models, complex analyses. In: Proceedings of the 34th Conference on Winter Simulation: Exploring New Frontiers, Winter Simulation Conference, pp. 116–126 (2002)
15. Lim, C.U.: An AI Player for Defcon: An Evolutionary Approach using Behaviour Trees. Imperial College, London (2009)
16. Tremblay, J.: Understanding and evaluating behaviour trees. McGill University, Modelling, Simulation and Design Lab, Technical report (2012)
17. Cioffi-Revilla, C., De Jong, K., Bassett, J.K.: Evolutionary computation and agent-based modeling: biologically-inspired approaches for understanding complex social systems. Comput. Math. Organ. Theory **18**(3), 356–373 (2012)
18. Denti, E., Omicini, A., Ricci, A.: tuProlog: a light-weight prolog for internet applications and infrastructures. In: Ramakrishnan, I.V. (ed.) PADL 2001. LNCS, vol. 1990, p. 184. Springer, Heidelberg (2001)
19. Aarseth, E.: A narrative theory of games. In: Proceedings of the International Conference on the Foundations of Digital Games, pp. 129–133. ACM (2012)
20. Aylett, R.: Narrative in virtual environments-towards emergent narrative. In: Working Notes of the Narrative Intelligence Symposium (1999)
21. Kim, J.: Events as Property Exemplifications. Springer, New York (1976)
22. Nairat, M., Dahlstedt, P., Nordahl, M.G.: Character evolution approach to generative storytelling. In: 2011 IEEE Congress on Evolutionary Computation (CEC), pp. 1258–1263. IEEE (2011)
23. Takagi, H.: Interactive evolutionary computation: fusion of the capabilities of ec optimization and human evaluation. Proc. IEEE **89**(9), 1275–1296 (2001)
24. Eiben, A.E., Smith, J.E.: What is an evolutionary algorithm? In: Eiben, A.E., Smith, J.E. (eds.) Introduction to Evolutionary Computing, pp. 25–48. Springer, New York (2003)
25. Jin, Y., Branke, J.: Evolutionary optimization in uncertain environments - a survey. IEEE Trans. Evol. Comput. **9**(3), 303–317 (2005). cited By (since 1996)576

Dangerousness Metric for Gene Regulated Car Driving

Sylvain Cussat-Blanc$^{(\boxtimes)}$, Jean Disset, and Stéphane Sanchez

University of Toulouse - IRIT - CNRS - UMR5505,
21 Allée de Brienne, 31042 Toulouse, France
{cussat,disset,sanchez}@irit.fr

Abstract. In this paper, we show how a dangerousness metric can be used to modify the input of a gene regulatory network when plugged to a virtual car. In the context of the 2015 Simulated Car Racing Championship organized during GECCO 2015, we have developed a new cartography methodology able to inform the controller of the car about the incoming complexity of the track: turns (slipperiness, angle, etc.) and bumps. We show how this dangerousness metric improves the results of our controller and outperforms other approaches on the tracks used in the competition.

Keywords: Gene regulatory network · Virtual car racing · Dangerousness measure

1 Introduction

The Simulated Car Racing competition (SCR[1]) aims to design a controller in order to race competitors on various unknown tracks. This competition is based on The Open-source Racing Car Simulator (TORCS[2]). Both scripted and evolutionary approaches can be used to control the virtual car. Within the frame of the 2015 competition that was held during GECCO2015, we have improved the approach based on gene regulation evolved with a genetic algorithm firstly presented at the previous edition in 2013 [21]. The SCR competition involves virtual car driver competing on 12 tracks, each of them having different properties (shape, track adherence, etc.). The virtual drivers can be hand-written scripts, optimized scripts or learning agents. For each track, the competitors first run alone for 5 laps (warm-up) during which they can learn the track. Then, the competitors run for a qualifying session during which they have to race against the clock in 5 laps. This qualifying session determines the starting order of the final round: the race. During this stage, all the competitors are running at the same time for 5 laps. The competitors are evaluated on this race based on the Formula One point system. The championship is composed of 12 of these races and the winner is the driver with the maximum number of points.

Amongst the existing controllers for virtual car racing games that use an evolutionary process to optimize the driver behavior, we can broadly consider two

---

[1] http://cs.adelaide.edu.au/~optlog/SCR2015/index.html.
[2] http://torcs.sourceforge.net/.

G. Squillero and P. Burelli (Eds.): EvoApplications 2016, Part I, LNCS 9597, pp. 620–635, 2016.
DOI: 10.1007/978-3-319-31204-0_40

kind of controllers. The first kind are indirect controllers where the inputs are not directly linked to the outputs of the virtual racing car. The inputs are computed and transmitted to driving policies based on hand-coded rules or heuristics that manage steering and throttle controls. Usually, these controllers optimize their driving policies with an evolutionary algorithm [4,18,19]. These kind of controllers are efficient and they have been at the top of the SRC competition since 2009. The second kind of controllers are direct controllers where sensors are directly mapped to the car effectors. These direct controllers actually learn how to drive the car using its sensors and actuators. They can be based on evolved artificial neural networks [2,5,22–24] or genetic programming [1]. These methods usually produced either very fast controllers but specialized on one specific track or slower more generic drivers. Because evolving a direct controller from scratch that can drive on-track and manage all car controls and race events is difficult, these controllers are sometimes mixed with hand-coded policies that modify the controller outputs to handle crash recovery or opponents, or that manage specific controls such as gear handling.

The controller we develop is a direct controller. Instead of designing complex hand-coded heuristics, we prefer to evolve the controller to drive the car by using a standard genetic algorithm. In this work, the controller is based on a Gene Regulatory Network (GRN). In order to optimize this GRN, we have used an incremental evolution (as in [4,24]) based on different fitnesses that gradually refine the controller's behavior. The improvement we detail in this paper is related to the first stage (warm-up) but it impacts the behavior of the driver on both other stages. It consists in modifying the GRN inputs depending on an on-the-fly dangerousness measure of the track. Based on that, the GRN modify its behavior in order to slow down or speed up based on the current situation of the car (position on the track, sliding of the car, etc.).

The paper is organized as follows. Section 2 presents gene regulatory networks in general, describing the existing computational models and the problems they are currently handling. This section also introduces the computational model we have used in this work. Section 3 summarizes how the GRN is connected to the car sensors and actuators and how the GRN is incrementally trained with a genetic algorithm to produce a basic driver. Then, Sect. 4 details our learning approach of the tracks, the dangerousness measure we produce and how it influences the inputs of the gene regulatory network. Section 5 proposes an experimental validation of our approach by evaluating the GRNDriver with and without the dangerousness measure against the drivers involved in the 2015 edition of the SCR competition.

2   Gene Regulatory Network

Gene Regulatory Networks (GRN) are biological structures that control the internal behavior of living cells. They regulate gene expression by enhancing and inhibiting the transcription of certain parts of the DNA. For this purpose, the cells use protein sensors dispatched on their membranes; these provide crucial information to guide the cells through their cycle. Many modern computational models of these networks exist. They are used both to simulate real gene regulatory networks [3,11,20] and to control agents [10,12,14,17].

When used for simulation purpose, a GRN is usually encoded within a bit string, as DNA is encoded within a nucleotide string. As in real DNA, a gene sequence starts with a particular sequence, called the *promoter* in biology [16]. In the real DNA, this sequence is represented with a set of four protein: *TATA* where *T* represents the thymine and *A* the Adenine. In [20], Torsten Reil is one of the first to propose a biologically plausible model of gene regulatory networks. The model is based on a sequence of bits in which the promoter is composed of the four bits *1010*. The gene is coded directly after this promoter whereas the regulatory elements are coded before the promoter. To visualize the properties of these networks, he uses graph visualization to observe the concentration variation of the different proteins of the system. He points out three different kinds of behavior from randomly generated gene regulatory networks: stable, chaotic and cyclic. He also observes that these networks are capable of recovering from random alterations of the genome, producing the same pattern when they are randomly mutated. In 2003, Wolfgang Banzhaf formulates a new gene regulatory network heavily inspired from biology [3]. He uses a genome composed of multiple 32-bit integers encoded as a bit string. Each gene starts with a promoter coded by any integer ending with the sequence "XYZ01010101". This sequence occurs with a 2^{-8} probability (0.39 %). The gene following this promoter is then coded in five 32-bits integers (160 bit) and the regulatory elements are coded upstream to the promoter by two integers, one for the enhancing and one for the inhibiting kinetics. Banzhaf's model confirms the hypothesis pointed out by Reil's one; the same properties emerges from his model.

From these seminal models, many computational models have been initially used to control the cells of artificial developmental models [10,11,14]. They simulate the very first stage of the embryogenesis of living organisms and more particularly the cell differentiation mechanisms. One of the initial problem of this field of research is the French Flag problem [26] in which a virtual organism has to produce a rectangle that contains three strips of different colors (blue, white and red). This simulates the capacity of differentiation in a spatial environment of the cells. Many models addressed this benchmark with cells controlled by a gene regulatory network [6,14,15]. More recently, gene regulatory networks have proven their capacity to regulate complex behaviors in various situations: they have been used to control virtual agents [9,13,17] or real swarm or modular robots [8,12].

2.1 Our Model

The gene regulatory network used to control a virtual car in this work is based on Banzhaf's model. It has already been successfully used in other applications. It is capable of developing modular robot morphologies [8], controlling cells designed to optimize a wind farm layout [25] and controlling reinforcement learning parameters in [7]. This model has been designed for computational purpose only and not to simulate a biological network.

This model is composed of a set of abstract proteins. A protein a is composed of three tags:

- the *protein tag* id_a that identifies the protein,
- the *enhancer tag* enh_a that defines the enhancing matching factor between two proteins, and
- the *inhibitor tag* inh_a that defines the inhibiting matching factor between two proteins.

These tags are coded with an integer in $[0, p]$ where the upper bound p can be tuned to control the precision of the network. In addition to these tags, a protein is also defined by its concentration that will vary over time with particular dynamics described later. A protein can be of three different types:

- *input*, a protein whose concentration is provided by the environment, which regulates other proteins but is not regulated,
- *output*, a protein with a concentration used as output of the network, which is regulated but does not regulate other proteins, and
- *regulatory*, an internal protein that regulates and is regulated by others proteins.

With this structure, the dynamics of the GRN are computed by using the protein tags. They determine the productivity rate of pairwise interaction between two proteins. For this, the affinity of a protein a for another protein b is given by the enhancing factor u_{ab}^+ and the inhibiting factor u_{ab}^- calculated as follows:

$$u_{ab}^+ = p - |enh_a - id_b| \quad ; \quad u_{ab}^- = p - |inh_a - id_b| \tag{1}$$

The proteins are then compared pairwise according to their enhancing and inhibiting factors. For a protein a, the total enhancement g_a and inhibition h_a are given by:

$$g_a = \frac{1}{N} \sum_b^N c_b e^{\beta u_{ab}^+ - u_{max}^+} \quad ; \quad h_i = \frac{1}{N} \sum_b^N c_b e^{\beta u_{ab}^- - u_{max}^-} \tag{2}$$

where N is the number of proteins in the network, c_b is the concentration of the protein b, u_{max}^+ is the maximum observed enhancing factor, u_{max}^- is the maximum observed inhibiting factor and β is a control parameter which will be detailed hereafter. At each timestep, the concentration of a protein a changes with the following differential equation:

$$\frac{dc_a}{dt} = \frac{\delta(g_a - h_a)}{\Phi}$$

where Φ is a normalization factor to ensure that the total sum of the output and regulatory protein concentrations is equal to 1. β and δ are two constants that influence the reaction rates of the network. β affects the importance of the matching factors and δ is used to modify the production level of the proteins in the differential equation. In summary, the lower both values are, the smoother the regulation is; the higher the values are, the more sudden the regulation is.

3   Using a GRN to Drive a Virtual Car

3.1   Linking the GRN to the Car Sensors and Actuators

The GRN can be seen as any kind of computational controller: it computes inputs provided by the problem it is applied to and it returns values to solve the problem. To use the gene regulatory network to control a virtual car, our main wish is to keep the connection between the GRN and the car sensors and actuators as simple as possible. In our opinion, the approach should be able to handle the reactivity necessary to drive a car, the possible noise of the sensors and unexpected situations. The car simulator provides 18 track sensors spaced $10°$ apart and many other sensors such as car fuel, race position, motor speed, distance to opponents, etc. However, in our opinion, all of the sensors are not required to drive the car. Reducing the number of inputs directly reduces the complexity of the GRN optimization. Therefore, we have selected the following subset of sensors provided by the TORCS simulator:

- 9 track sensors that provide the distance to the track border in 9 different directions,
- longitudinal speed and transversal speed of the car.

Figure 1 represents the sensors used by the GRN to drive the car. Before being computed by the GRN, each sensor value is normalized to $[0, 1]$ with the following formula:

$$norm(v(s)) = \frac{v(s) - min_s}{max_s - min_s} \qquad (3)$$

where $v(s)$ is the value of sensor s to normalize, min_s is the minimum value of the sensor and max_s is the maximum value of the sensor.

Once the GRN input protein concentrations are updated, the GRN's dynamics are run one time in order to propagate the concentration modification to the whole network. The concentrations of the output proteins are then used to regulate the car actuators. Four output proteins are necessary: two proteins o_l and o_r for steering (left and right), one protein o_a for the accelerator and one o_b for the brake. The final values provided to the car simulator are computed as follow:

$$steer = \begin{cases} 0 & \text{if } c(o_l) = c(o_r) = 0 \\ \frac{c(o_l) - c(o_r)}{c(o_l) + c(o_r)} & \text{otherwise} \end{cases} \qquad (4)$$

$$accel = max(0, ab) \qquad (5)$$

$$brake = min(-ab, 0) \qquad (6)$$

$$ab = \begin{cases} 0 & \text{if } c(o_a) = c(o_b) = 0 \\ \frac{c(o_a) - c(o_b)}{c(o_a) + c(o_b)} & \text{otherwise} \end{cases}$$

where $steer$ is the final steering value of the car in $[-1, 1]$, $accel$ is the final acceleration value in $[0, 1]$, $brake$ is the final brake value in $[0, 1]$, $c(o_*)$ is the

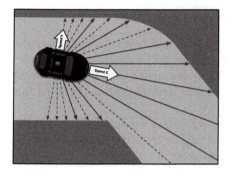

Fig. 1. Sensors of the car connected to the GRN. The red plain arrows are used track sensors whereas the gray dashed ones are the track sensors also available in the simulator but not used by the GRN. The plain arrows *Speed X* and *Speed Y* are respectively the longitudinal and the transversal car speeds (Color figure online).

Fig. 2. The GRN uses 9 track sensors and the longitudinal and transversal speeds to compute the steering, the acceleration and the brake of the car.

concentration of the output protein o_*. Figure 2 shows the connection of the GRN to the virtual car. When the both output proteins corresponding to the direction are equal to zero, the final steering is equal to zero, meaning that the car keeps its direction. Likewise, when both the accelerator and the brake proteins are equal to zero, the car is neither accelerating nor braking: the car will slowing reduce its speed with the engine brake.

Whereas other approaches use a noise reduction filter in addition to the standard anti-locking braking system (ABS) and the traction control systems (TCS), the GRN approach does not need any noise filter: it is naturally noise-resistant. The ABS and TCS are switched on because they provide a large support in the braking and acceleration zones. The impact of noise on the GRN reaction is detailed in [21].

3.2 GRN Genome

Before it can drive, the regulatory network needs to be optimized. In this work, we use a standard genetic algorithm to optimize the GRN's protein tags, enhancing tags and inhibiting tags. The GRN can be easily encoded in a genome. The genome contains two independent chromosomes. The first one is defined as a variable length chromosome of indivisible proteins. Each protein is encoded with three integers between 0 and p that correspond to the three tags. In this particular work, p is set at 32 and the genome proteins are organized with the input proteins first, followed by the output proteins and then regulatory proteins. The inputs and outputs presented in the previous section will be always be linked to the same protein.

This chromosome requires particular crossover and mutation operators:

- a *crossover* can only occur between two proteins and never between two tags of the same protein. This ensures the integrity of both subnetworks when the GRN is subdivided into two networks. When assembling another GRN, local connections are kept with this operator and only new connections between the two networks are created.
- three *mutations* can be equiprobably used: add a new random regulatory protein, remove one protein randomly selected in the set of regulatory proteins, or mutate a tag within a randomly selected protein.

A second chromosome is used to evolve the dynamics variables β and δ. This chromosome consists of two double-precision floating point values and uses the standard mutation and crossover methods. These variables are evolved in the interval $[0.5, 2]$. Values under 0.5 produce unreactive networks whereas values over 2 produce very unstable networks. These values are chosen empirically through a series of test cases.

3.3   Incremental Evolution

In order to optimize the GRN to drive a car, we use an incremental evolution in three stages[3]. This section summarizes these three stages. More details can be found in [21].

The first stage consists of training the GRN to drive as far as possible, with a minimum speed, on one track. We use CGSpeedway, which is simple with long turns and straight lines. Each tested GRNDriver is rewarded when going farther and faster with an innovative ticket system. The ticket represents the maximum time the GRNDriver must take to cover a sector of a certain distance. Furthermore, the ticket's value is reduced each time the GRNDriver validates a sector: the farther the GRN goes on the track, the faster it must drive. The fitness is given by the distance covered by the GRNDriver without getting out of the track or getting out of time on a sector. This first stage builds a basic network with which the GRNDriver can drive endlessly on this particular track. It can also drive on most of the tracks but hard turns, never encountered in CGSpeedway, are still problematic.

In order to generalize the network to any possible track, we evolved the previous GRN a second time with the same evolutionary process but on three different tracks. The tracks used are CGSpeedway (in order not to lose the driving capabilities of the previous GRN), Alpine and Street. The fitness function is the sum of the fitnesses of the first evolution stage successively applied to the three tracks. At the end of this evolution, the best GRN is able to drive on every

---

[3] During these stages, the same parameters have been used to tune the genetic algorithm. Only the fitness function is modified. The genetic algorithm parameters are: Population size: 500; Mutation rate: 15 %; Crossover rate: 75 %; GRN Size: [4, 20] regulatory proteins plus inputs and outputs.

possible track. It drives very safely, going at a suitable speed to go through every kind of turn and braking when it detects a turn.

The final stage of evolution consists in removing all possible imperfection of the best GRN obtained previously. We observed a few oscillatory behavior, very common with GRNs, that could be a problem in a racing car championship. To minimize the oscillatory behaviors, we evolve the best GRN one last time. This time we add to the fitness function another test case that penalizes the continuous oscillations of the car on straight lines and long turns or fast multiple steering changes from full right to full left. As with the ticket system used in the previous fitness functions, we simply stop the evaluation if we detect oscillatory behaviors.

4   Influencing the GRN with a Dangerousness Measure

In order to make the previously evolved GRN fast enough to win a car racing competition, we have designed a dangerousness measure that allows the GRN to anticipate the incoming complexity of the track. This measure needs a full cartography (turns, slipperiness, etc.) of the track and modify the longitudinal speed input of the GRN. This sections details these two parts.

4.1   Track Cartography

The cartography consists in recording a maximum of information about a track during one lap. Therefore, we have scripted a driver that strictly follows the middle of the track at a limited speed of 90 km/h. To do so, the behavior is decomposed as follows:

– *Steering wheel.* The angle of the wheel is given by the angle of the car with the track: if the car longitudinal axis is not aligned with the track tangent, the car driver must turn the wheel in the corresponding direction. This wheel angle direction is corrected with the track position in order to avoid any possible derivation of the car: if the car is shifted to the left (respectively right) hand side of the track, the wheel angle is augmented (respectively reduced) in order to recenter the car.
– *Gas and brake pedals.* The gas/brake pedals are regulated so that the car speed stay as close as possible to the target speed. To do so, the following formulas are used to compute the gas pedal value g and the brake pedal value b:

$$g = max(0, th) \tag{7}$$
$$b = min(0, -th)$$
$$\text{with } th = \frac{2}{1 + exp(s_c - s_t)} - 1$$

where s_c is the car current speed, s_t is the target speed.

This simple cartographic driver has been made so that the car can pass through all possible curves fast enough to produce some transversal speed (slides). This is important in order to evaluate in each turn the quality of the track and therefore the speed limits of the car. While driving, the script identifies track sectors according to the wheel angle and Z-axis speed. Sectors can be of following types:

– left if the left front sensor is greater than the right front sensor and actual steering is greater than 0.025 rad (steering left),
– right if the right front sensor is greater than the left front sensor and actual steering is smaller than -0.025 rad (steering right),
– jump if the Z-axis speed is smaller than -12.5 km/h,
– straight otherwise.

At each time step of the cartography, a new sector is created when the sector type is different to the sector type at the previous time step. This allows subdivision of the tracks to as a series of sectors. In each sector, the following values are stored:

– the sector length l_s,
– sum of transversal speeds s_y,
– sum of Z-axis speeds s_z,
– sum of wheel angles w_a.

With these data, two dangerousness measures are calculated for each sectors:

– the turn dangerousness, which expresses the turn complexity, given by $\frac{|w_a - s_y|}{l_s}$,
– the jump dangerousness, which expresses the jumping risk of the car, given by $\frac{50|s_z|}{l_s}$.

These measures are used by the GRN to regulate its speed before and during the turn. Before being used by the GRN, the track cartography is first filtered in order to remove micro sectors generated by sensor noise. To do so, all straight sectors which length is smaller than 50 meters are removed. Their data are reallocated to the neighbor sectors and their dangerousness measures are recalculated.

The turn dangerousness measure is refined during the 4 last laps of the warmup and the qualifying session in order to eliminate all possible risk of accident. When the GRNDriver gets off the track, the dangerousness value of the sector is increased so that the same mistake is avoided during the next laps. Figure 3 shows both the turn dangerousness measure obtained right after the cartography stage (upper figures) and the refined cartographies at the end of the qualifying session (lower figures). We can observe that the refining is necessary in some turns, under evaluated by the initial cartography (first two tracks: Wildno and

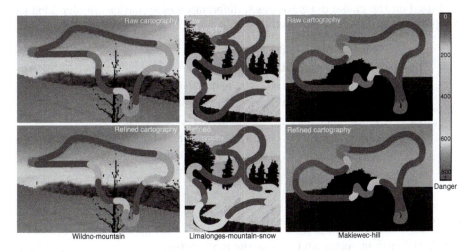

Fig. 3. Examples of cartographies obtained, before refining (upper figures) and after (lower ones).

Limalonges). In particular, this is necessary after long straights to build a braking zone which cannot appear with the low speed of the driver during cartography.

4.2   Influence on the GRN Inputs

In order to regulate the GRNDriver speed, we have decided to distort the car longitudinal speed sensor. Whereas most other approaches manipulated directly the output of the controller, we have decided to modify its input so that the GRNDriver keeps full control of the situation based on its full perceptions. For example, even if the dangerousness measure is low, the GRN can decide not to accelerate because the current car state is unstable (the car could be sliding, on a bad track position, etc.). To distort the longitudinal speed sensor, a speed regulator coefficient is applied to the direct sensor value. The aim of the coefficient is to let the GRN think the car is going slower than it really does when the dangerousness is low and faster when the dangerousness is high. This encourages the GRN either to speed up or slow down the car. To do so, the GRNDriver uses the incoming sectors (including the one it is currently in) up to the sector at a distance of $0.002 * s_x^2$, where s_x is the current longitudinal speed. This allows the GRNDriver to regulate the dangerousness estimation according to the car speed. With these sectors, the turn dangerousness d_t and the jump dangerousness d_j measures are summed in order to provide the incoming global dangerousness d_g.

Based on the global dangerousness, different cases scenarios are identified to compute the speed regulator coefficient C_s:

$$C_s = \begin{cases} 0.5 & \text{if } s_i \text{ is straight and } d_g < t_n \\ 0.66 & \text{if } s_i \text{ is not straight and } d_g < t_l \\ 0.66 & \text{if } s_i \text{ is not straight, } d_g < t_h \text{ and } dist(s_i) > 0.37 * l_s \\ 2 & \text{if } s_i \text{ is not straight, } d_g < t_m, s_x > 135\text{km/h and } dist(s_{i+1}) < s_x^2/700 \\ 2 & \text{if } s_{i+1} \text{ is jump,} s_x > 170 \text{ and } dist(s_{i+1}) < s_x^2/800 \\ 5 & \text{if } s_i \text{ is straight, } d_g > t_h, s_x > 120\text{km/h and } dist(s_{i+1}) < s_x^2/600 \\ 5 & \text{if } s_i \text{ is straight, } d_g < t_h, s_x > 135\text{km/h and } dist(s_{i+1}) < s_x^2/850 \\ 0.9 & \text{otherwise} \end{cases}$$

$$(8)$$

where

- s_i is the current sector and s_{i+1} is the next sector,
- d_g is the current global dangerousness,
- $t_n = 150$ is the no dangerousness threshold, $t_l = 250$ is the low dangerousness threshold, $t_m = 400$ is the medium dangerousness threshold and $t_h = 800$ is the high dangerousness threshold,
- $dist(s_i)$ (resp. $dist(s_{i+1})$ is the distance to the beginning of the current (resp. next) sector,
- s_x is the car current longitudinal speed.

In these formulas, the condition $dist(s_{i+1}) < s_x^2/y$, with $y = 600, 700, 800$ or 850, is used to evaluate the braking distance necessary to speed the car down to the target speed. All the parameters involved in this formula have been empirically chosen through test. A broader study could improve the approach and its results.

This speed regulator coefficient C_s is then simple multiplied to the car longitudinal speed s_x to provide the car speed input $c(i_{Sx})$ of the GRN:

$$c(i_{Sx}) = norm(s_x * C_s) \qquad (9)$$

This input is sufficient to modify because it is strongly linked to the acceleration output protein: other proteins are too (such as the front track sensor) but are harder to modify due to their implications in the driving. Modifying a track sensor might generate bad behavior when the GRNDriver is sliding in a turn for example. The modification of the car speed input seams to be the more direct and efficient way to impact the GRNDriver speed behavior.

5   Comparative Study

In order to evaluate the dangerousness measure approach, we have compared its benefits on the GRNDriver to other approaches submitted to the competition[4]

---

[4] The source code and a short description of these approaches are available on the competition website: http://cs.adelaide.edu.au/~optlog/SCR2015/.

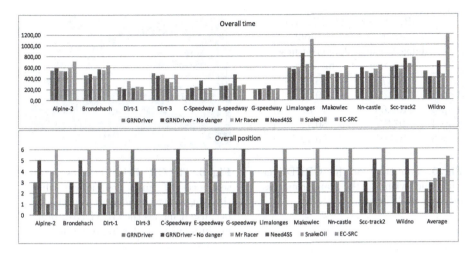

Fig. 4. The upper plot presents the overall time to cover the 5 laps in qualifying mode. The lower one presents the position of the drivers according to the overall time.

on the 12 tracks used during the competition. With this aim in mind, we have compared the GRNDriver with and without the speed regulator coefficient as well as the other approaches using the first two stage of the competition rules:

1. the drivers are run in warm-up mode during 5 laps in order to learn the track.
2. they are then run in qualifying mode for 5 more laps.

In this comparative study, only the results of the qualifying session are presented. Because the focus of the paper is the dangerousness measure we have introduced this year, the results of the race are not presented in this paper. However, they are also available on competition website. All the runs have been made with noisy sensors.

Figure 4 presents the overall time taken by the competitors to cover 5 laps and their position. Times (upper chart) are very close between competitors and it is hard to evaluate which approach is better than the other. However, when looking to position (lower chart), we can observe that the GRNDriver is most of the time first with the dangerousness measure on and second when switched off. This can be viewed with the last bars labelled "Average" which represent the positions of the drivers averaged over the tracks. The GRNDriver with the dangerousness measure finishes on average 2.25, the GRNDriver without dangerousness measure finishes on average 2.83 and the next closer competitor, Mr. Racer, finishes 3.33.

When only comparing the GRNDriver with and without the dangerousness metric, we can observe that the dangerousness measures improve the results of the GRNDriver on 8 out of 12 tracks. The Wildno track seams to be problematic for this approach, the GRNDriver with dangerousness measure having lower performances than expected. This bad performance implies a poor ranking on this particular track. Figure 5 provides the time of each lap on the Wildno track

	GRNDriver with dangerousness measure	GRNDriver without dangerousness measure
Lap 1	117.75	88.24
Lap 2	96.25	79.86
Lap 3	102.63	79.82
Lap 4	97.85	80.12
Lap 5	103.04	80.12

Fig. 5. Lap time of the GRNDriver with and without the dangerousness measure on Wildno.

Fig. 6. The GRNDriver crashes in two sequences of turns (red stars) when the dangerousness measure is activated because it makes the GRNDriver braking in a curve (Color figure online).

for the GRNDriver with and without dangerousness measure. The table shows a big instability of the driver over the laps. The GRNDriver with dangerousness measure actually gets out off the track often, which is unproductive. On this particular track the dangerousness measure seams to be too optimistic and makes the car hard to control for the GRN. More precisely, Fig. 6 shows where the GRNDriver with dangerousness measure crashes on the track: it is always in the middle of a sequence of turns. The dangerousness metric forces the GRN to brake in the turn which leads to a spin. However, the GRNDriver without dangerousness measures finishes the qualifying session in first position (see Fig. 4).

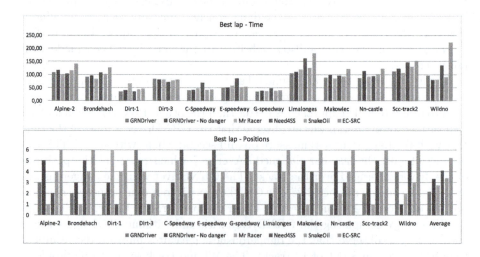

Fig. 7. The upper plot presents the best time of the 5-laps qualifying session. The lower one presents the position of the drivers according to the best lap.

Wildno is a complex mountain track with a complex track to manage: a safe approach of this track looks to be more productive than an aggressive one.

Figure 7 shows the best time of the same 5-laps qualifying session. The benefits of the dangerousness measure are here undeniable, the GRNDriver with this feature making 10 times out of 12 better best time than without it. This is an important measure because the best lap corresponds to a lap at the very end of the learning curve of the dangerousness measure. It corresponds to the data that will be used during the race session against other approaches: being faster on one lap is decisive at this point. When looking at the average ranking of the GRN-Driver with and without the dangerousness measure, we observe that enabling it allows the GRNDriver to overtake Mr Racer, which is the closest opponent to the GRNDriver. This shows, in race condition regarding to the cartography of the track, the significance of this feature.

6   Conclusion

In this paper, we have presented an improved version of our learning procedure of a track used to influence the inputs of a gene regulatory network that drives a simulated racing car. We have designed a dangerousness metric that evaluates the dangerousness of the track and helps the GRN to regulate the car speed in function of the turn difficulties (slipperiness, angle, etc.) and the possible jumps due to track bumps. We show the quality of this metric by comparing the GRNDriver with and without its influence on the 12 tracks used in the Simulated Car Racing Championship. In particular, we show that it greatly improves the best lap time of the GRNDriver, which is crucial when opposed to other drivers during a race session. At the end, the GRNDriver won the 2015 edition of the SCRC both because of the generalization capacity of the GRN (able to drive fast on any kind of track with no re-optimization) and because of the improvement brought by the dangerousness method.

To improve this work, multiple options have to be investigated. Our goal is to design a driver with as much automatic learning as possible. First, the use of the GRN as a racing driver requires the design of a track learning method to speed up the wise GRNs we generally obtain by evolution. We would like to teach the GRN to go faster by the use of a hierarchical architecture: a second GRN, pre-optimized on multiple tracks and reoptimized during the warm-up stage, could modify the inputs and/or the outputs of the driving GRN according to the current car state. The specialization capacity of the GRN observed in the first evolutionary step could be helpful during this warm-up stage.

This GRNDriver must also be improved in order to correctly handle opponents. For now, the perception of the GRN is modified by a hand-written script in order to overtake or avoid an opponent detected to close to the car. This approach is innovative in comparison to most other approaches because they usually directly impact the car actuators. Modifying the inputs instead of the output keeps the controller as the center piece of the algorithm. However, we want the GRN to learn to handle this move by itself because most overruns are currently due to this script. Having all the information the car can detect and letting the GRN decide the best move could reduce this issue.

The application of such an approach to a real car driving is still problematic because of the difficulty to prove the security of this approach: in order to drive a real car, it is necessary to strictly prove the algorithm. It is currently mathematically complex to make it on a gene regulatory network because of the complexity of the generated network. However, this approach could be exploited as a controller in a car racing game: multiple GRNs could evolve in parallel with the player, making non-scripted controllers with a large diversity. It could improve the interest of the game by producing different strategies to which the player would have to face.

References

1. Agapitos, A., Togelius, J., Lucas, S.M.: Evolving controllers for simulated car racing using object oriented genetic programming. In: Proceedings of the 9th Annual Conference on Genetic and Evolutionary Computation, pp. 1543–1550. ACM (2007)
2. Athanasiadis, C., Galanopoulos, D., Tefas, A.: Progressive neural network training for the open racing car simulator. In: IEEE Conference on Computational Intelligence and Games (CIG), pp. 116–123. IEEE (2012)
3. Banzhaf, W.: Artificial regulatory networks and genetic programming. In: Riolo, R.L., Worzel, B. (eds.) Genetic Programming Theory and Practice. Genetic Programming Series, vol. 6, pp. 43–62. Springer, Heidelberg (2003)
4. Butz, M.V., Lönneker, T.D.: Optimized sensory-motor couplings plus strategy extensions for the torcs car racing challenge. In: Proceedings of the 5th International Conference on Computational Intelligence and Games, CIG 2009, pp. 317–324. IEEE Press, Piscataway, NJ, USA (2009)
5. Cardamone, L., Loiacono, D., Lanzi, P.L.: Learning to drive in the open racing car simulator using online neuroevolution. IEEE Trans. Comput. Intell. AI Games **2**(3), 176–190 (2010)
6. Cussat-Blanc, S., Bredeche, N., Luga, H., Duthen, Y., Schoenauer, M.: Artificial gene regulatory networks, spatial computation: a case study. In: Proceedings of the European Conference on Artificial Life (ECAL 2011). MIT Press, Cambridge, MA (2011)
7. Cussat-Blanc, S., Harrington, K.: Genetically-regulated neuromodulation facilitates multi-task reinforcement learning. In: Proceedings of the on Genetic and Evolutionary Computation Conference, pp. 551–558. ACM (2015)
8. Cussat-Blanc, S., Pollack, J.: Cracking the egg: virtual embryogenesis of real robots. Artif. Life **20**(3), 361–383 (2014)
9. Cussat-Blanc, S., Sanchez, S., Duthen, Y.: Simultaneous cooperative and conflicting behaviors handled by a gene regulatory network. In: IEEE Congress on Evolutionary Computation (CEC), pp. 1–8. IEEE (2012)
10. Doursat, R.: Organically grown architectures: creating decentralized, autonomous systems by embryomorphic engineering. In: Doursat, R. (ed.) Organic Computing. Understanding Complex Systems, pp. 167–200. Springer, Heidelberg (2008)
11. Eggenberger Hotz, P.: Combining developmental processes and their physics in an artificial evolutionary system to evolve shapes. In: On Growth, Form and Computers, p. 302 (2003)

12. Guo, H., Meng, Y., Jin, Y.: A cellular mechanism for multi-robot construction via evolutionary multi-objective optimization of a gene regulatory network. BioSystems **98**(3), 193–203 (2009)

13. Joachimczak, M., Wróbel, B.: Evolving gene regulatory networks for real time control of foraging behaviours. In: Proceedings of the 12th International Conference on Artificial Life (2010)

14. Joachimczak, M., Wróbel, B.: Evolution of the morphology and patterning of artificial embryos: scaling the tricolour problem to the third dimension. In: Karsai, I., Szathmáry, E., Kampis, G. (eds.) ECAL 2009, Part I. LNCS, vol. 5777, pp. 35–43. Springer, Heidelberg (2011)

15. Knabe, J., Schilstra, M., Nehaniv, C.: Evolution and morphogenesis of differentiated multicellular organisms: autonomously generated diffusion gradients for positional information. Artif. Life XI **11**, 321 (2008)

16. Lifton, R., Goldberg, M., Karp, R., Hogness, D.: The organization of the histone genes in drosophila melanogaster: functional and evolutionary implications. In: Cold Spring Harbor Symposia on Quantitative BIology, vol. 42, pp. 1047–1051. Cold Spring Harbor Laboratory Press (1978)

17. Nicolau, M., Schoenauer, M., Banzhaf, W.: Evolving genes to balance a pole. In: Esparcia-Alcázar, A.I., Ekárt, A., Silva, S., Dignum, S., Uyar, A.Ş. (eds.) EuroGP 2010. LNCS, vol. 6021, pp. 196–207. Springer, Heidelberg (2010)

18. Onieva, E., Pelta, D.A., Alonso, J., Milanés, V., Pérez, J.: A modular parametric architecture for the torcs racing engine. In: Proceedings of the 5th International Conference on Computational Intelligence and Games, CIG 2009, pp. 256–262. IEEE Press, Piscataway, NJ, USA (2009)

19. Quadflieg, J., Preuss, M., Rudolph, G.: Driving faster than a human player. In: Cagnoni, S., Cotta, C., Ebner, M., Ekárt, A., Esparcia-Alcázar, A.I., Merelo, J.J., Neri, F., Preuss, M., Richter, H., Togelius, J., Yannakakis, G.N., Chio, C. (eds.) EvoApplications 2011, Part I. LNCS, vol. 6624, pp. 143–152. Springer, Heidelberg (2011)

20. Reil, T.: Dynamics of gene expression in an artificial genome — implications for biological and artificial ontogeny. In: Floreano, D., Mondada, F. (eds.) ECAL 1999. LNCS, vol. 1674, pp. 457–466. Springer, Heidelberg (1999)

21. Sanchez, S., Cussat-Blanc, S.: Gene regulated car driving: using a gene regulatory network to drive a virtual car. Genet. Program. Evolvable Mach. **15**(4), 477–511 (2014)

22. Stanley, K., Sherony, R., Kohl, N., Miikkulainen, R.: Neuroevolution of an automobile crash warning system. In: Proceedings of the Genetic and Evolutionary Computation Conference (GECCO) (2005)

23. Stanley, K.O., Miikkulainen, R.: Evolving neural networks through augmenting topologies. Evol. Comput. **10**, 99–127 (2002)

24. Togelius, J., Lucas, S.M.: Evolving robust and specialized car racing skills. In: IEEE Congress on Evolutionary Computation, CEC 2006, pp. 1187–1194. IEEE (2006)

25. Wilson, D., Awa, E., Cussat-Blanc, S., Veeramachaneni, K., O'Reilly, U.-M.: On learning to generate wind farm layouts. In: Proceeding of the Fifteenth Annual Conference on Genetic and Evolutionary Computation Conference, pp. 767–774. ACM (2013)

26. Wolpert, L.: Positional information and the spatial pattern of cellular differentiation. J. Theor. Biol. **25**(1), 1 (1969)

Using Isovists to Evolve Terrains with Gameplay Elements

Andrew Pech[✉], Chiou-Peng Lam, Philip Hingston, and Martin Masek

School of Computer and Security Science, Edith Cowan University, Perth, Australia
apech@our.ecu.edu.au, {c.lam,m.masek,p.hingston}@ecu.edu.au

Abstract. The virtual terrain for a video game generally needs to exhibit a collection of gameplay elements, such as some areas suitable for hiding and others for large scale battles. A key problem in automating terrain design is the lack of a quantitative definition of terrain gameplay elements. In this paper, we address the problem by proposing a representation for gameplay elements based on a combination of space-based isovist measures from the field of architecture and graph-connectivity metrics. We then propose a genetic algorithm-based approach that evolves a set of modifications to an existing terrain so as to exhibit the gameplay element characteristics. The potential for this approach in the design of computer game environments is examined by generating terrain containing instances of the "hidden area" game element type. Results from four preliminary tests are described to show the potential of this research.

Keywords: Procedural content · Genetic algorithm · Game terrain · Isovist

1 Introduction

Environment design is a major component of video game production and is typically conducted manually by experienced designers. Demand for larger game levels with greater amounts of visual detail have increased the time required to design and construct the environment of these game levels. With virtual terrains being one of the largest components of a game level in several game genres, techniques that automatically generate game-ready terrain would be valuable to game developers.

There are a number of techniques capable of generating virtual terrain with a primary focus on generating aesthetic qualities, defined as elements which contribute to a terrain's appearance. On the other hand, gameplay elements are defined as those that have a functionally-related gameplay purpose, with Hullet and Whitehead [1] identifying, amongst others: hidden areas, choke points, vantage points, and strongholds. Generation of such areas, apart from some examples with a narrow scope, is largely unaddressed, therefore the terrains generated by most existing techniques require post-editing to make them suitable for gameplay.

A barrier to the automated incorporation of terrain areas with the properties of specific gameplay elements is the lack of an objective, measurable definition for each of them. However, in the field of architecture, where environment design is also a focus, the objective measurement of spatial configuration of an environment is not new.

© Springer International Publishing Switzerland 2016
G. Squillero and P. Burelli (Eds.): EvoApplications 2016, Part I, LNCS 9597, pp. 636–652, 2016.
DOI: 10.1007/978-3-319-31204-0_41

In particular, isovist-based analysis seems to hold promise in providing objective measures for characterising gameplay elements. An isovist of a point in the environment is the volume of space visible from that point. Benedikt [2] first investigated the use of isovist-based measures in architecture, defining an isovist in terms of a set of radial vectors from the view point to the boundary of the visible volume. Measures of the isovists were largely used to provide a graphical indication to designers on the properties of an environment in the form of isovist fields, where a designer could see how aspects, such as isovist volume, change as one travels through the environment. This can be used in applications, such as finding an optimal patrol path for a guard. Benedikt also discussed how different spatial concepts translate to isovist measures, for example equating the privacy of a location to the visible volume of its isovist and the skewness of the set of isovist radial vectors. In terms of virtual world design, isovist-based measures were used by van Bilsen and Poleman [3] to give designers a tool from which to assess safety of areas for use in a serious game for training supervisors in an industrial environment.

The focus of this paper is a novel approach to the generation of virtual terrains that contain desired gameplay elements, quantified by isovist-based, as well as graph connectivity based measures. This approach takes an initial terrain, which may be generated with specific aesthetic qualities, and then adds the desired gameplay elements. This is accomplished by using a genetic algorithm (GA) to evolve individuals encoding a set of modifications to an initial terrain. The fitness function evaluates areas on a modified terrain against desired gameplay elements based on the set of isovist and graph-based measures to determine the fitness of the individual. The contributions of our approach are 2 folds: (1) quantifies properties of gameplay elements based on a set of generic measures of space, rather than specific knowledge of the game genre and (2) evolutionary method for incorporating specific gameplay elements in automatic terrain generation. The rest of this paper is structured as follows: In Sect. 2, previous work related to terrain generation is reviewed, followed by a detailed description of our approach in Sect. 3. Section 4 contains details of experiments and results, followed by a conclusion and some possible future research directions in Sect. 5.

2   Related Work

Existing work in automated terrain generation can be divided into approaches that consider purely aesthetics and those that attempt to include gameplay related considerations. This section reviews these approaches, focusing on limitations and opportunities for improvement in current techniques.

Many aesthetic based terrain generation methods include the use of coherent noise, such as Perlin noise [4], or other fractal techniques including midpoint displacement [5] and the diamond square algorithm [6], due to their ability to form hilly or mountainous terrain. For example, terrain generation software, such as World Machine [7] and Terragen [8], use these fractal techniques to generate a base terrain for users to edit. More complex techniques include the work of Belhadj and Audibert [9], who use "particles" to form networks of ridgelines and rivers and then apply midpoint displacement to fill in the terrain details.

By altering the particles' behaviour, the user is able to control how the ridge and river networks are formed. Another approach introduced by Doran and Parberry [10] uses various types of software agents to form islands with desired aesthetic elements including mountains, hills, rivers, and beaches. The user can control the generation of the terrain by specifying which agent types to use and by altering parameters associated with the agents' behaviour.

Other techniques generate terrain based on more explicit user input and usually require a near complete design to generate a terrain. Hnaidi [11] introduced such an approach which required the user to sketch a terrain using control curves, which would subsequently be used to generate a complete height-map. This approach requires the user to specify where each incline and decline is to be placed on a terrain, as well as specifying noise values to add terrain detail for each curve. The control curves typically form ridges and rivers, but, with correct user input, can form roads and flat areas suitable for a video game. Since this approach generates a terrain from a manually crafted design, it may be more suitable as a developer aid rather than an automated process. Another tool, introduced by Peytavie et al. [12], requires manual input from the user to make alterations to a terrain and then uses physically inspired simulation and other techniques to automatically generate terrain detail.

A tool to aid in the generation of virtual worlds, including terrain, was introduced by Smelik et al. [13], enforcing designer specified constraints. This approach used semantic constraints to modify features of the environment and enforce properties such as line-of-sight and concealment. An example of this is a semantic constraint being used to lower the terrain where it occludes line-of-sight to a target location.

Other techniques use GAs to evolve terrains either through an interactive evolutionary approach (Raffe et al. [14]), or by blending existing height-map samples together based on user input (Ong et al. [15]; Li et al. [16]). Raffe et al.'s approach represented terrain as a 2D grid of tiles, where each tile was a small height-map. The GA consisted of a small population of terrains generated from random configurations of tiles. Rather than using a fitness function or selection scheme, this approach presented the user with the population of terrains and allowed the user to select which terrains would go through the genetic operators to form the next generation. This technique is suitable as a design aid rather than an automated approach as user input is required to drive the evolution towards terrains with desired gameplay elements.

Ong et al.'s and Li et al.'s approaches were very similar and used a two phase GA, which took a sketch of terrain boundaries as input and then filled the boundaries with existing height-map samples. This approach required the user to create a sketch of terrain boundaries, where each boundary was assigned a specific terrain type such as hills or mountains. The first phase of the approach used a GA to displace the boundaries to form more natural transitions between different terrain types. The second phase used a GA to fill each boundary with existing height-map samples which had been designated the same terrain type. These two approaches suffer from similar drawbacks to the previous methods described, relying heavily on a designer, in this case to map desired gameplay elements to terrain types, and to explicitly lay out a level.

Some work has been done towards generating terrain to meet specific gameplay requirements, typically tailored to a particular genre of video game. Togelius et al. [17]

used a multi-objective GA to evolve Real-Time Strategy (RTS) maps, including place-ment of player bases, resources, and the terrain. In this approach maps are evaluated based on placement of resources, height advantages, and map symmetry. These are all gameplay elements, but the evaluation measures are specific to the type of game, with only the height advantage fitness evaluation having any impact on the terrain. Terrain generation was not the primary focus of this approach, producing simple terrains with low detail. Similar work was also performed in [18] with map symmetry, area control, and exploration potential being the measures incorporated.

Another approach which considered gameplay was presented by Olsen [19] who used an erosion algorithm to generate aesthetic terrain, while using Voronoi diagrams to control placement of hills and paths to increase the likelihood of creating flat areas of terrain. This approach was designed to generate terrains with a strong focus on the RTS game genre, where the flat sections of terrain are used to build structures, a common element in games of this genre. Although flat sections of terrain have uses in many game genres, most require more advanced terrain modifications to make the most of their game mechanics. Frade *et al.* [20] also attempted to generate terrain based on a flatness measure, evolving terrains to maximise the size of a connected flat area up to some threshold. Their technique was however limited in terms of aesthetics and range of terrain maps generated.

As can be seen from the literature, the generation of gameplay specific terrain elements is still an open research problem. Certain game genres, such as first person shooters (FPS) use many game elements, with Hullet and Whitehead [1] identifying, amongst others: hidden areas, choke points, vantage points, and strongholds. Though broad definitions of these area types are provided in [1] in terms of their affordances, consequences and relationships, means of quantifying these gameplay elements are largely undefined. This quantification would be a significant step towards such gameplay elements being generated automatically without manual alterations by the game designer, and our approach offers a solution.

3   The Approach

To quantifying game environment gameplay elements we use a collection of space-based metrics. These metrics seek to capture information about the area of open space for a particular location on a terrain, and its accessibility from other areas on that terrain. Significantly this leads to the approach being game-genre independent. As long as the game genre uses a terrain, or similar game environment, a particular terrain-based gameplay element can be quantified by gathering the metrics from examples of this element, rather than needing pre-existing knowledge of the genre. In the following sections, we detail the metrics that have been adopted, with an initial study of the "hidden area" game element [1] for first person shooters. This is followed by details of the GA that has been used to evolve a set of modifications for a terrain incorporating desired gameplay elements.

3.1   Measures and Characteristics of Gameplay Elements

The first step of our approach involved using isovist-based measures to characterise specific gameplay elements in a quantifiable way. We interpret a game terrain as a collection of distinct areas within the terrain and make the assumption that each area maps to a particular gameplay element which corresponds to one of those identified in [1]. The structure of the terrain defines how the player can traverse between areas, and in this way the terrain can be thought of as a graph-like structure, with each area as a node and edges connecting areas. Thus the spatial measure of a particular terrain area can consist of a set of isovist-based measures, based on the volume of space around a distinct area, and a set of graph-based measures, based on the accessibility of adjacent areas. Together, this collection of measures can be used to characterise an area. In terms of graph-based measures, in this paper we focus only on node degree. Node degree for a particular node is defined as the number of nodes that node is connected to. As such, this corresponds to the number of terrain areas that are accessible from a particular area. The isovist-based measures are now discussed.

To calculate isovist-based measures we represent an isovist as 32 radial vectors originating from the centre of a node's area, offset by eye height, and sweeping across 360° with even spacing. This is consistent with existing literature, including [2, 3], for constraining the calculated isovist to a horizontal plane at eye height. The magnitude of each vector is equal to the distance of the nearest occluding obstacle from the origin along the vector's direction, capped at the maximum view distance. Figure 1 illustrates a top down view of an example isovist with eight vectors. In this example, the magnitude of six of the eight vectors is equal to the maximum view distance, while the remaining two vectors terminate at an occluding obstacle resulting in a lower magnitude. The dark blue line shows the perimeter of the isovist's area. A wide variety of isovist-based measures exist, with the initial set defined in [2]. Common measures based on the radial vector representation, as used by van Bilsen and Poleman [3] among others, are listed in Table 1. The exception in the table is Node Size, which we have defined as the surface area of flat terrain surrounding the centre of a particular area and is obtained directly from the terrain.

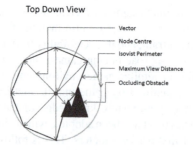

Fig. 1. A top down view of an example isovist with eight radial vectors.

Table 1. A list of isovist-based measures which can be used to identify different area types.

Node measures	Description
Area	Area that can be seen (is not occluded) from a given location
Average distance	Average distance from a given location to its isovist's perimeter
Minimum distance	Minimum distance from a given location to its isovist's perimeter
Maximum distance	Maximum distance from a given location to its isovist's perimeter
Perimeter	Perimeter of an isovist's area
Variance	Variation in distance from a given location to its isovist's perimeter
Skewness	Skewness of the distance from a given location to its isovist's perimeter
Standard deviation	Standard deviation of distance from a given location to isovist perimeter
Node size	Surface area of flat terrain around a given location

The isovist measures can be used to quantitatively determine several characteristics of an area. For example, if a location with a small isovist area implies that it is surrounded by occluding obstacles and thus well concealed. The variance, standard deviation, and skewness measures can be used to determine the dispersion of occluding obstacles and where the observation point sits in relation to them.

For a particular gameplay element, only some of these measures may be relevant. In these cases, only relevant measures, along with their desired characteristics, should be chosen. This paper described our preliminary investigation where we initially focus on a single gameplay element, the "hidden area" [1]. This is defined in [1] as one that is *off the main player path*, intended to reward exploration behaviour of the player, with the environment-related affordance of "the *ease of finding* and accessing the hidden area" [1]. In order to objectively represent these qualities, we have chosen three of the metrics from Table 1: the node degree, the node size, and the isovist area. The criterion of being *off the main player path* is enforced by setting the ideal node degree to 1, which means that particular area is not a means to access other areas. In terms of the *ease of finding* criterion, the ideal node size and isovist area are set to small values. Future work will focus on identifying relationships between a wider range of metrics and various gameplay elements. The minimum and maximum values of the three measures used in our experiments to categorise an area as hidden are listed in Table 2, where cs is a user defined constant which represents the diameter of a game character (this value was set to 25 in our experiments).

Table 2. List of the measures used to identify hidden areas.

Area type name	Measures	Min	Max
Hidden area	Node degree	1	1
	Node size	$2cs$	$8cs$
	Isovist area	$(1.5cs)^2.\pi$	$(6cs)^2.\pi$

Once relevant measures, along with their minimum and maximum values, have been chosen, an Area Type Similarity Measure (*atsm*) can be calculated to determine how similar an area is to a desired gameplay element. The *atsm* is a value in the range of [0, 1] where a value of one indicates the area meets all of the criteria of its desired element, and as the *atsm* nears zero, the further the area's measures are from the desired values. One method to calculate *atsm* is shown in Fig. 2, where n is the number of area measures employed for the area type, v_i is the current area measure, l_i is the minimum desired value for the area measure, h_i is the maximum desired value for the area measure, and m_i is the maximum possible value for the area measure.

---

atsm = 0
for each measure i to n
    if $l_i \leq v_i \leq h_i$ then *atsm* += 1.0 / n
    else if $v_i > h_i$ then *atsm* += $(1.0 - (v_i - h_i) / m_i) / n$
    else if $v_i < l_i$ then *atsm* += $(1.0 - (l_i - v_i) / m_i) / n$
end loop

Fig. 2. Pseudo code for calculating the *atsm* value.

3.2   Evolution of Terrain Based on Gameplay Element Metrics

We now present details of how our representation of gameplay element metrics can be used to drive a GA to evolve a terrain that contains certain gameplay elements. Figure 3 shows a flow chart of this process. The input consists of an initial terrain height map, chosen for its aesthetics, and the properties of desired gameplay elements. An initial population of individuals, each representing a set of modifications to the initial terrain height map, is then evolved. This is done through a repeated process of fitness

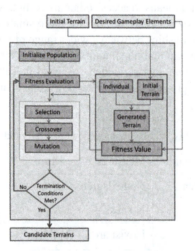

Fig. 3. A flow chart of the approach.

evaluation and the genetic operators selection, crossover and mutation until a termination condition is met. Fitness evaluation for an individual consists of generating a terrain based on the initial terrain and modifications encoded in the individual. A fitness value is then produced based on properties of the modified terrain, including similarity to particular gameplay elements. The rest of this section describes the details of the GA-based approach. The section titled "Population Representation" details the representation of individuals in the population and the terrain modifications they encode. This is followed with a discussion of fitness evaluation in the section titled "Fitness Evaluation". Finally the section titled "Genetic Operators" details the genetic operators used.

Population Representation. Each individual in the population encodes a graph consisting of a set of areas and connecting paths for modifying a terrain, an example shown in Fig. 4(a). The graph is encoded as two fixed length chromosomes, shown in Fig. 4(b), with indices used to map the graph to the chromosome. One chromosome stores information about the graph's nodes, where each gene represents a single node, and another chromosome stores information about the graph's edges, where each gene represents a single edge. Each gene in the node and edge chromosome is composed of a set of attributes which govern the effect that corresponding node or edge has on the terrain. Table 3 lists the attributes stored in a single gene of an individual's node chromosome, and Table 4 lists the attributes stored in a single gene of an edge chromosome.

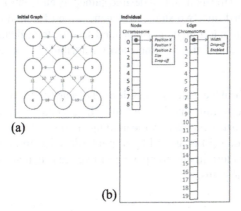

(a)

(b)

Fig. 4. (a) Illustration of an initial graph showing the layout of nodes and edges using indices. (b) The corresponding node and edge chromosome with indices corresponding to the graph.

Table 3. Node attributes stored in the node chromosome.

Attribute name	Attribute description
Position X, Y, Z	Geographic location of the node's centre along the X, Y, and Z axis
Size	Diameter of the node
Drop-off	Distance from the node's perimeter for blending the node's *Position Y* (height) with the initial terrain heights

Table 4. Terrain edge attributes stored in the edge chromosome.

Attribute name	Attribute description
Width	Edge width as a percentage of the smallest diameter node connected to the edge
Drop-off	Linear blend factor between initial terrain height and edge height
Enabled	Binary value indicating whether this edge is included in the modified terrain

The *Position* attributes of nodes govern the placement of the area in the terrain. In the initial population, areas are set to form a regularly spaced grid on the terrain. The node's *Size* attribute governs the diameter of the area where each cell of the terrain is set to the height of the *Position Y* attribute. The *Drop-off* attribute represents the distance from the area's perimeter used to transition between the area height and the height of the initial terrain.

Nodes adjacent to each other on the grid are 8-connected with edges. The edge *Width* attribute affects how wide their paths are, as a percentage of the diameter of the smallest node they connect. The path *Drop-off* attribute, similarly to the *Drop-off* in nodes, affects how the path is blended with the initial terrain. The edge's *Enabled* attribute determines if the path is to be included in the terrain.

Fitness Evaluation. The purpose of the fitness evaluation is to assign each individual a fitness value based on the inclusion of desired gameplay elements into the initial terrain. This process is divided into two steps, first, generation of a modified terrain using an individual and the initial terrain height-map, and second, the calculation of the individual's fitness by evaluating the modified terrain attributes. These steps will now be detailed.

Modifications to Initial Terrain Generation. To generate a modified terrain using an individual, the initial terrain that is input into the approach is altered to contain the network of areas and paths encoded by the individual. Figure 5 demonstrates an example of how an initial terrain (a) is altered by a graph (decoded from an individual) (b) to generate the modified terrain (c), composing of five areas that are all connected to one another by traversable paths.

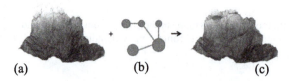

(a)   (b)   (c)

Fig. 5. Illustration of a modified terrain by superimposing an individual onto the initial terrain. The brown areas on the modified terrain highlight the areas and paths that were added (Color figure online).

The height map of the initial terrain is altered by changing its set of height values to blend in the nodes and edges associated with the graph encoded in the individual's chromosome. For each node, the height of the modified terrain within the radius of the node (*Size*/2) is set to the node's Y position attribute. Outside of this radius for a distance up to the *drop-off radius*, the height at a point in the modified terrain is determined using linear interpolation between the node height and the height of the initial terrain. The linear interpolation factor is based on the distance from the node perimeter, so that at the perimeter the final height is the node height and at a distance of *drop-off* away from the perimeter the final height is the initial terrain height.

Each edge from the graph whose *enabled* attribute is set to true is used to form a straight path between the two areas encoded by the nodes on either end of the edge. Height values of the initial terrain along the path are altered, using interpolation, between the heights of the two nodes. Height values between the edge's perimeter and its *Drop-off* radius are set using interpolation between the initial terrain height and the height of the path using distance from the path for the interpolation factor.

Once the terrain is generated, its attributes are extracted and a fitness value assigned to its associated individual. This process is described in the following section.

Fitness Calculation. During evaluation, each individual is assigned a fitness value equal to the average of three separate measures, gameplay elements (*ge*), node overlap (*no*), and edge intersection (*ei*) as shown in Eq. (1). The first measure (*ge*) is a rating of how well the areas within the modified terrain match desired gameplay elements, which is where the isovist-based and graph-based measures are employed. The last two measures (*no* and *ei*) are to help generate terrains with minimal visual deformities. The rest of this section describes these measures and how they are calculated in detail.

$$fitness = \frac{ge + no + ei}{3} \tag{1}$$

The gameplay element component of fitness, *ge*, requires calculating an *atsm* value (See Sect. 3.1) for each area of the terrain (node of the graph) to determine the areas similarity to a desired gameplay element. The resulting *atsm* values are then scaled based on desired constraints between areas. Constraint scaling is performed by assigning each unique pair of nodes a Node Pair Fitness (*npf*), calculated in Eq. (2) as a product of the average *atsm* value of the nodes and a constraint multiplier, *cm*. The constraint multiplier is a value in the range of [0, 1], where a value of one means the node pair constraint is completely satisfied. The closer the value approaches zero, the further the node pair is from satisfying the constraint.

$$npf = \frac{atsm_1 + atsm_2}{2} * cm \tag{2}$$

Once an *npf* value has been determined for each node pair, the game element (*ge*) measure can be calculated, specifics are documented in Sect. 4.

The node overlap measure, *no*, penalises the fitness value for any nodes in the graph that encode overlapping terrain areas. This measure is a value between 0 and 1, where 1 means no areas are overlapping, and a value of 0 indicates a maximum amount of overlap. This measure is calculated as the sum of the distances between the centres of each pair of nodes (*dst_i*), minus the radii of the nodes (*ns0_i*/2 and *ns1_i*/2), and their

Drop-off radii ($nd0_i$ and $nd1_i$), over the maximum possible overlap (the sum of combined radii, including *Drop-Off* radii, of each node pair). This is shown in Eq. (3), where n is the number of unique node pairs in the graph.

$$no = 1 - \frac{\sum_{i=1}^{n} \left| \min \left(dst_i - \left(\frac{ns0_i}{2} + nd0_i + \frac{ns1_i}{2} + nd1_i \right), 0 \right) \right|}{\sum_{i=1}^{n} \frac{ns0_i}{2} + nd0_i + \frac{ns1_i}{2} + nd1_i} \tag{3}$$

The edge intersection measure, ei, penalises the fitness value for any paths on the terrain that intersect. This measure is also a value between 0 (maximum number of paths are intersecting) and 1 (no paths are intersecting). This measure is calculated in Eq. (4) by dividing the number of overlapping path pairs, where n is the number of unique path pairs, by the maximum path pair overlaps and subtracting the value from 1. In this paper, we set this maximum value to the $(N-1)^2$, where N is the grid width, since this would be the number of edge pairs that intersect in an initial configuration if all edges are enabled. It is possible for more paths to intersect than the calculated maximum and in these cases ei is clamped to 0. The function c returns a value of 1 if the passed in edge pair are intersecting and a 0 otherwise.

$$ei = 1 - \frac{\min \left(\sum_{i=1}^{n} c(i), (N-1)^2 \right)}{(N-1)^2} \tag{4}$$

Genetic Operators. After all individuals in the population are evaluated, selected individuals undergo mutation and crossover to create offspring which form a new population. Our approach uses an elitist scheme where the first step towards forming the new population is copying the best individual from the old population into the new one. The following process is then repeated until all remaining slots in the new population have been filled. First an individual is selected using binary tournament selection and then a check is made against the crossover probability to determine if crossover should be applied. If crossover is used, a second individual is selected using binary tournament. Single point crossover is then applied to each of the chromosomes of the two selected individuals. Although each gene consists of a number of attributes, the traditional approach of splitting chromosomes for crossover at gene boundaries was followed. The mutation operator is applied to either the original selected individual (if crossover was not conducted) or the two new individuals (if crossover was conducted), by iterating over every attribute within each gene and with a set mutation probability replacing the value of the attribute with a new value randomly selected from a set range. The resulting individual (or individuals) are then placed into the new population.

4   Experiments and Results

As this investigation is focused on incorporating a single gameplay element, the "hidden area", we conducted four experiments, listed in Table 5, to evolve terrains that contain two hidden areas with some form of distance constraint between them. The constraints used in these experiments include minimising and maximising the Euclidean distance between the hidden areas, and obtaining desired values for the shortest path distance between the hidden areas. The shortest path distance is the distance between two nodes following the network of paths, calculated using the A* algorithm [21].

Table 5. Constraints between the two hidden areas.

Experiment	Constraint between hidden areas
1	Minimum Euclidean distance
2	Maximum Euclidean distance
3	Shortest path distance between 250 & 500
4	Shortest path distance between 5000 & 5500

To implement the constraints for the experiments, the *cm* parameter in the node pair constraints calculation (Eq. (2)) was set as follows. For maximising Euclidean distance between nodes, cmg_1 was used (Eq. (5)), where the Euclidean distance between nodes centres, $d(n_1, n_2)$, was normalised by the maximum possible distance across the terrain, which has the three dimensions length (*tl*), width (*tw*), and height (*th*). For minimising Euclidean distance between the node pair, cmg_2 was used (Eq. (6)), which is 1 minus the normalised Euclidean distance between the nodes perimeters (including the drop off radius), so as to avoid nodes overlapping. Here *ns* is the *Size* parameter of a node (diameter) and *nd* the *drop off radius* of a node.

$$cmg_1 = \frac{d(n_1, n_0)}{\sqrt{tl^2 + tw^2 + th^2}} \tag{5}$$

$$cmg_2 = 1 - \frac{d(n_1, n_0) - \left(\frac{ns_0}{2} + nd_0 + \frac{ns_1}{2} + nd_1\right)}{\sqrt{tl^2 + tw^2 + th^2}} \tag{6}$$

For experiments where a shortest path distance constraint was applied, *cma* in Fig. 6 was used as the constraint multiplier. Here, *v* is the shortest path distance between the nodes, in our experiments calculated using the A* algorithm [21]. The parameter *l* is the minimum desired distance, *h* is the maximum desired distance, and *m* is the maximum possible distance. Based on these constraint multipliers, in each of these four experiments the node pair fitness *npf* is calculated for all node pairs, and the maximum value is used as the *ge* measure for the fitness calculation in Eq. (1).

```
if l ≤ v ≤ h then cma = 1.0
else if v > h then cma = 1.0 – (v – h) / m
else if v < l then cma = 1.0 – (l – v) / m
```

Fig. 6. Pseudo code determining the shortest path distance constraint multiplier between nodes.

Table 6 lists the GA parameter settings used for each experiment. Note that the mutation probabilities are set to 1 over the length of the chromosome (l_n and l_e for the node and edge chromosomes respectively) multiplied by the number of attributes stored in each gene, resulting in an expected 1 attribute mutation per chromosome.

Table 6. List of parameter values used in each experiment.

Parameter name	Parameter value
Population size	100
Selection type	Binary tournament
Node mutation probability	$1/(l_n*5)$
Edge mutation probability	$1/(l_e*3)$
Crossover probability	0.6
Number of generations	100

The node chromosome is initialized to contain 9 nodes, where each node attribute is set based on a random value within a specified range, listed in Table 7. For the *Position X*, and *Position Z* attributes, these are initialized to form a grid on the terrain, to prevent node overlap and edge intersection in the initial population. From a regularly spaced grid, the *Position X* and *Position Z* attributes are offset by *Position X Offset* and *Position Z offset* respectively. The edge chromosome is initialized to contain 20 edges, where each edge attributes is set to a random value within a specified range. The ranges used for the edge chromosome are listed in Table 8. The initial terrain was a 2048 × 2048 height-map. The rest of this section describes the results from each of the four experiments that were conducted.

Table 7. List of minimum and maximum values used to initialize the node chromosome.

Attribute	Minimum	Maximum
Position X Offset	−300	300
Position Y	0	2048
Position Z Offset	−300	300
Size	0	500
Drop-off	50	500

Table 8. List of minimum and maximum values used to initialize the edge chromosome.

Attribute	Minimum	Maximum
Width	0.1	1
Drop-off	50	500
Enabled	0	1

4.1 Two Hidden Areas: Minimum and Maximum Euclidean Distance

Experiment 1 attempted to place the two hidden areas a minimal distance apart whilst avoiding overlap. Experiment 2 attempted to maximize the distance between the two hidden areas. Figure 7(a) shows the fittest terrain from the experiment 1 after 100 generations, where the two hidden areas are circled in red. A grayscale height-map (b) is also displayed and highlights both hidden areas' isovist in red, and outlines their drop-off perimeters (in green). Both areas are surrounded by elevated terrain and meet the 'hidden area' requirements as specified in Table 2. It can also be seen that the two hidden areas are placed very close to one another, as indicated by their drop-off perimeters just touching, without overlapping one another.

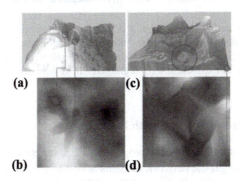

(a)   (c)

(b)   (d)

Fig. 7. The fittest terrains from the final population of the experiment 1 (a) and 2 (c), hidden areas circled in red. The height-map of each terrain is show in (b) and (d), ach hidden area's isovist highlighted in red and drop-off perimeters (for experiment 1) circled in green (Color figure online).

For experiment 2, Fig. 7(c) shows the terrain associated with the fittest chromosome after 100 generations, where both the hidden areas are circled. Again, the grayscale height-map (d) shows the hidden areas' isovists highlighted in red. The hidden areas are at opposite corners of the terrain, giving a near maximum distance between them. Another generated area can be seen at the front of the terrain, circled in blue, and was not classified as a hidden area as its size and isovist area were too large and it had a node degree greater than one, which is fitting as it is quite an obvious area on the terrain. The convergence graph for this experiment is presented in Fig. 8 and shows the GA converging by the 100th generation. In this paper, convergence was defined as the highest fitness minus the average fitness being below a threshold of 0.01 for five generations.

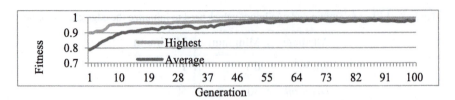

Fig. 8. Convergence graph for experiment 2 (Two hidden areas: maximum Euclidean distance) (Color figure online).

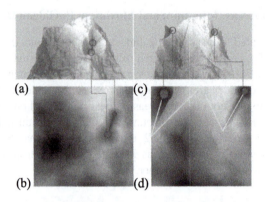

Fig. 9. The fittest terrains from experiment 3 (a) and 4 (c), attempting to separate two hidden areas by a shortest path distance between 250–500 and 5000–5500 respectively. Hidden areas are circled in red. Greyscale height-map of each terrain is shown in (b) and (d) with isovists highlighted in red and shortest path for experiment 4 (d) highlighted in yellow (Color figure online).

4.2   Two Hidden Areas: Shortest Path Distance

Experiments 3 and 4 attempted to find two hidden areas separated by a set shortest path distance, traveled through the network of paths added to the terrain. Experiment 3 set this shortest path distance to be between 250 and 500 units, while experiment 4 used an ideal a shortest path distance of between 5000 and 5500 units. Figure 9(a) shows the fittest terrain from experiment 3 after 100 generations, with the hidden areas circled in red. Figure 9(b) shows a grayscale heightmap of the terrain, with isovist areas' highlighted in red. The red line joining the hidden areas indicates that each area is visible from the other. Both of the hidden areas are surrounded by elevated terrain, and are connected by a path with a length of 375 units. So not only do the hidden area attributes fall within the desired ranges, but the A* distance constraint is also met.

Figure 9(c) shows the fittest terrain from experiment 4. Both hidden areas are circled with their isovist areas highlighted in red in Fig. 9(d). Figure 9(d) also indicates the shortest path between the two hidden areas in yellow. The shortest path distance between the two hidden areas, at 5380 units, falls within the desired range.

5   Conclusions and Future Work

With the advancement of technology, the time required to create more highly detailed video game content is becoming too expensive. Therefore techniques that automatically generate this content could be valuable for future game development. One of the largest components of many of today's video games is the virtual terrain, which makes its design and creation an important process. While there are existing techniques that automatically generate virtual terrain, the majority of them lack any game design consideration, with the ones that do typically being genre game-specific.

This paper introduced an automatic procedural approach for incorporating gameplay elements in virtual terrain generation. Using "hidden areas" as a specific example, we explored the idea of characterizing gameplay elements using a combination of graph connectivity metrics and a set of isovist-based metrics, established metrics that have been used by environment designers in the related field of architecture. We then proposed a GA-based approach where the representation was then used in the fitness function of a GA, to evolve modifications to an initial terrain to make it meet criteria based on the inclusion of desired gameplay elements.

The approach was demonstrated by using it to generate terrain which contains hidden areas and enforces distance constraints between them. The results demonstrate the potential for isovist-based measures as measures of fitness for driving evolution of computer game environments. Future work will include a more comprehensive exploration of area types, constraints and measures.

References

1. Hullett, K., Whitehead, J.: Design patterns in FPS levels. In: Proceedings of the 5th International Conference on the Foundations of Digital Games, pp. 78–85 (2010)
2. Benedikt, M.L.: To take hold of space: Isovist fields. Environ. Plann. B Plann. Des. **6**(1), 47–65 (1979)
3. van Bilsen, A., Poleman, R.: 3D visibility analysis in virtual worlds: the case of supervisor. In: Proceedings of the Construction Applications of Virtual Reallity, pp. 5, 6, 267–278 (2009)
4. Perlin, K., Hoffert, E.: Hypertexture. SIGGRAPH Comput. Graph. **23**(3), 253–262 (1989)
5. Lau, W.-C., Erramilli, A., Wang, J.L., Willinger, W.: Self-similar traffic generation: the random midpoint displacement algorithm and its properties. In: 1995 IEEE International Conference on Communications (ICC 1995), pp. 466–472 (1995)
6. Fournier, A., Fussell, D., Carpenter, L.: Computer rendering of stochastic models. Commun. ACM **25**(6), 371–384 (1982)
7. World Machine: World Machine (2015). http://www.world-machine.com/
8. Planetside: Planetside Software (2015). http://planetside.co.uk/
9. Belhadj, F., Audibert, P.: Modelling landscapes with ridges and rivers: bottom up approach. In: GRAPHITE 2005: Proceedings of the 3rd International Conference on Computer Graphics and Interactive Techniques, pp. 447–450 (2005)
10. Doran, J., Parberry, I.: Controlled procedural terrain generation using software agents. IEEE Trans. Comput. Intell. AI Games **2**(2), 111–119 (2010)
11. Hnaidi, H., Guérin, E., Akkouche, S., Peytavie, A., Galin, E.: Feature based terrain generation using diffusion equation. Comput. Graph. Forum **29**(7), 2179–2186 (2010)

12. Peytavie, A., Galin, E., Merillou, S. Grosjean, J.: Arches: a framework for modelling complex terrain. In: Proceedings of the Eurographics 2009 (2009)
13. Smelik, R., Galka, K., de Kraker, K.L., Kuijper, F., Bidarra, R.: Semantic constraints for procedural generation of virtual worlds. In: Proceedings of the 2nd International Workshop on Procedural Content Generation in Games. ACM (2011)
14. Raffe, W., Zambetta, F., Li, X.: Evolving patch-based terrains for use in video games. In: Proceedings of the 13th Annual Conference on Genetic and Evolutionary Computation, pp. 363–370 (2011)
15. Ong, T., Saunders, R., Keyser, J., Leggett, J.: Terrain generation using genetic algorithms. In: Proceedings of the Genetic and Evolutionary Computation Conference, pp. 1463–1470 (2005)
16. Li, Q., Wang, G., Zhou, F., Tang, X., Yang, K.: Example-based realistic terrain generation. In: Pan, Z., Cheok, D.A.D., Haller, M., Lau, R., Saito, H., Liang, R. (eds.) ICAT 2006. LNCS, vol. 4282, pp. 811–818. Springer, Heidelberg (2006)
17. Togelius, J., Preuss, M., Yannakakis, G.N.: Towards multiobjective procedural map generation. In: Proceedings of the 2010 Workshop on Procedural Content Generation in Games, pp. 1–8 (2010)
18. Liapis, A., Yannakakis, G.N., Togelius, J.: Towards a generic method of evaluating game levels. In: Proceedings of the Artificial Intelligence for Interactive Digital Entertainment Conference (2013)
19. Olsen, J.: Realtime procedural terrain generation. Technical report, University of Southern Denmark (2004)
20. Frade, M., de Vega, F.F., Cotta, C.: Evolution of artificial terrains for video games based on accessibility. In: Di Chio, C., Cagnoni, S., Cotta, C., Ebner, M., Ekárt, A., Esparcia-Alcazar, A.I., Goh, C.-K., Merelo, J.J., Neri, F., Preuß, M., Togelius, J., Yannakakis, G.N. (eds.) EvoApplicatons 2010, Part I. LNCS, vol. 6024, pp. 90–99. Springer, Heidelberg (2010)
21. Hart, P.E., Nilsson, L.J., Raphael, B.: A formal basis for the heuristic determination of minimum cost paths. IEEE Trans. Syst. Sci. Cybern. 4(2), 100–107 (1968)

A Spatially-Structured PCG Method for Content Diversity in a Physics-Based Simulation Game

Raúl Lara-Cabrera[1], Alejandro Gutierrez-Alcoba[2],
and Antonio J. Fernández-Leiva[1(✉)]

[1] Departmento de Lenguajes y Ciencias de la Computación,
Universidad de Málaga, Málaga, Spain
{raul,afdez}@lcc.uma.es
[2] Departamento de Arquitectura de Computadores,
Universidad de Málaga, Málaga, Spain
agutierrez@ac.uma.es

Abstract. This paper presents a spatially-structured evolutionary algorithm (EA) to procedurally generate game maps of different levels of difficulty to be solved, in *Gravityvolve!*, a physics-based simulation videogame that we have implemented and which is inspired by the n-body problem, a classical problem in the field of physics and mathematics. The proposal consists of a steady-state EA whose population is partitioned into three groups according to the difficulty of the generated content (hard, medium or easy) which can be easily adapted to handle the automatic creation of content of diverse nature in other games. In addition, we present three fitness functions, based on multiple criteria (i.e., intersections, gravitational acceleration and simulations), that were used experimentally to conduct the search process for creating a database of maps with different difficulty in *Gravityvolve!*.

Keywords: Content creation · Evolutionary algorithms · Physics-based game · Human evaluation

1 Introduction and Motivation

The economic costs of producing a video game are very high: the development is a slow process that requires a large team of heterogeneous professionals who, in addition, are required to be highly qualified and specialized. Therefore, any improvement that is able to optimize both the time and resources required to create a video game is always welcome.

According to a recent analysis published in [1], the field of computational intelligence in video games is a vibrant, active field, which attracts new researchers each year and generates new publications. There has been a steady growth in the number of authors, which was accentuated mid-decade 2000–2010. Moreover, the number of publications per year from the community has been increasing since 2005, thus supporting the continued growth of the community.

© Springer International Publishing Switzerland 2016
G. Squillero and P. Burelli (Eds.): EvoApplications 2016, Part I, LNCS 9597, pp. 653–668, 2016.
DOI: 10.1007/978-3-319-31204-0_42

One of the most promising areas in this field is Procedural Content Generation (PCG) which consists of generating game content through algorithms instead of creating it by hand, and refers to each component that makes up a video game except from the behaviour of the non-playable characters (NPCs) [2,3]. Some examples of content susceptible to generation are maps, terrains, weapons, items, music or even the game's rules [4].

There are many advantages of producing video game content algorithmically using PCG techniques. In the first place, it allows us to substantially reduce the memory consumption of the game, although nowadays it should be seen as a secondary improvement, it was the main reason for the research and development of such techniques. Another important reason is the high cost of generating some game content manually, as we have already mentioned. Additionally, the game content can be automatically adapted to given criteria, such as the player ability, in such a way that the game offers the player a continuous challenge. If the algorithm is able to generate the content at the same time as the player is playing the game then we are able to create infinite games which offer a different game experience each time the player starts a new game.

It is well known that games can be catalogued according to a set of different genres whose frontiers are usually fuzzy and intersect with the space of other game genre, and it is not difficult to find games simultaneously catalogued, as belonging to distinct genres. In the last few years, the so-called *physics-based simulation games* (PbSGs) have emerged as one of the most exciting classes of games in the video games universe as developers are required to simulate real life physics with the aim of providing more realism and, as a consequence, to create more believable games. This opens new lines of research up as stated in [5] "Physics-based gaming can give your game development repertoire a huge boost, enabling *sandbox-style* game mechanics and *emergent gameplay*." In general, PbSGs (e.g., Angry Birds or Crayon Physics, to name a couple) are easy to play and provide simple game mechanics, but they introduce a number of challenging problems like, unfortunately (or fortunately), the movements of the rigid bodies have to be perfectly simulated which is not an easy task as these are subjected to real Physics Laws and to the interaction with the environment (represented as the game world). As a consequence of all the possible interactions, between all the game objects and the game scenario itself, the number of new possible states (i.e., movements) is huge (in fact, usually infinite) and even unknown a priori. Therefore, the only way to proceed from one state of the game to the following one is to simulate the moves realistically, and this is mandatory as one needs to measure the quality of the state transition (i.e., movements).

In spite of the research interest of PbSGs, generating content for physics-based simulation games is an area that has been explored timidly and, as far as we know, only [6] uses Grammatical Evolution to automatically generate levels for a clone of *Cut the Rope*, a commercial physics-based puzzle game.

Not to mention that PCG algorithms must ensure that the generated content meets some criteria in a way similar to if it had been generated by hand, but this goal is not always easy to satisfy and, in addition, it is difficult to find good mechanisms to evaluate whether the generated artefacts meets the criteria in reality (i.e., according to the player's game experience).

Moreover, it is not enough to simply automatically generate a great number of elements as one might be more interested in creating components that are both diverse and of high quality [7].

In this context, this paper presents the following contributions: first, it introduces *Gravityvolve!*, a physics-based simulation video game that we have implemented, inspired by the n-body problem, a classical problem in the field of physics and mathematics, which can be used by the CI/AI community for research purposes [8]. Moreover, this paper proposes a method, that can be generalized to other (not-necessarily physics-based) games, to procedurally design maps of diverse solving complexity (i.e., of distinct levels of difficult); the proposal consists of an evolutionary algorithm (EA) which is spatially-structured in a number of sub-populations that are co-evolved separately according to different properties required by the individuals of each sub-population (which should supposedly guarantee diversity). In addition, it presents a preliminary experimental study performed to check the suitability of the method.

This paper is structured as follows: Sect. 2 provides a discussion of related work and Sect. 3 describes the game *Gravityvolve!*, its rules, objective and the physical laws that guide the gameplay. The map generation algorithm and the fitness functions that measure the difficulty level of the maps are defined in Sect. 4. Finally, Sect. 5 discusses the conclusions and future work.

2   Background

In addition to aforementioned related work, this section provides a general overview on Procedural Content Generation (PCG) and Physics-based Simulation Games (PbSGs), and it mentions a number of papers that are directly related with these issues. That being said, the list of papers is far from exhaustive as a review of these fields is not a goal of this paper.

We can make several distinctions regarding the procedures to follow when it comes to the automatic content generation for video games. Following the taxonomy proposed in [9], the content generation should be made online during the gameplay (which provides us with the aforementioned advantages) or offline during the development phase of the game. In the same way, PCG techniques might generate all the content using random seeds (purely stochastic), vectors of parameters (deterministic) or a combination of both.

According to the necessity of the procedurally generated content for the player's progression within the game, we should distinguish between essential and optional content; the former must meet more restrictive criteria than the latter.

Depending on the objectives we want to accomplish, the generation might be done in a constructive manner, ensuring that the content is always valid; or following a generate-and-test scheme, so the content is verified after its creation and if it does not pass the test then the algorithm discards and recreates it.

A widely used and well-known class of PCG algorithms is the so-called Search Based Procedural Content Generation (SBPCG) [10], which is based on looking for the desired content in the complete landscape of solutions. These algorithms follow a generate-and-test scheme and assign real values to each solution in order to measure its quality, instead of accepting or rejecting them. Although evolutionary algorithms are a common choice when developing SBPCG techniques, they are not unique and we are able to use other kind of algorithms in this context (like, for example, planning methods [11]).

Presently, procedural content generation is a vibrant field of research with a large number of papers related to these techniques (the reader can find an analysis about the diversity of the content that may be generated in a procedural manner in [3]). A recurring objective is to generate levels/maps for a platform game. For instance, Shaker et al. [12], proposed a system capable of adapting several parameters, which define the behavior of a level generator for a Super Mario Bros. clone, to the playing style of a certain player. Similarly, Pedersen et al. [13], researched the relationship between the parameters of a PCG algorithm and the game experience and feelings (frustration, fun, ...) that the generated levels provoked to the player.

Another type of content that is susceptible to evolution is a game's map/scenario. For example, Julian Togelius et al., designed a SBPCG multiobjective evolutionary algorithm whose objective was create maps for realtime strategy [14,15] and racing [16] games. In a similar way, Ferreira and Toledo [17] presented a SBPCG approach for generating levels for the physics-based videogame Angry Birds. Lara et al. [18] presented a search-based procedural content generation method in the context of the real-time strategy game[1] *Planet Wars* (i.e., the Google AI Challenge 2010) whose objective was to generate maps that resulted in an interesting gameplay, focusing on properties of balance and dynamism. Furthermore, the authors expanded their PCG method by considering both new geometrical properties and topological measures that were not affected by rotation, scaling and translation with the aim of avoiding the generation of symmetrical maps that are conceptually equivalent with respect to the gameplay [19]. The topological measures were obtained from the sphere-of-influence graph induced by each map [20]. In turn, Frade et al., proposed a fitness function to guide the generation of accessible terrains with application to the video game "Chapas" [21,22]. Hom and Marks [23] went further and they procedurally generated rules for a two-player board game with a certain requirement: maximize the balance between both players.

Moreover, there are several examples of PCG for optional content, such as the weapons that a player is able to use. Hastings et al. [24,25], proposed a SBPCG algorithm for the game "Galactic Arms Race". In this case, the fitness of the generated weapons was computed based on the amount of time the players used them, hence measuring the player satisfaction without requiring the explicit attention of the players. Collins [26] introduced to procedural music in video games, exploring several approaches to procedural composition that had been used in the past. Font et al. [27] presented initial research towards a system that is able to create the rules for different card games. The authors of [28] developed

---

[1] http://planetwars.aichallenge.org/.

a prototype of a tool that automatically produces design pattern specifications for missions and quests for the role-playing game *Neverwinter Nights*.

The reader wishing to know more about the current state of PCG applied to games is referred to [4] for more information. There exist, however, specific references about the application of PCG methods to particular areas of game AI such as, for instance, procedural methods to generate dungeon game levels [29] maze-like levels [30], or music generation [31], just to name a few.

As for Physics-based Simulation Games (PbSGs), it is easy to find evidence of their success in the commercial world as, for instance, *Angry Birds*, *Tower of Goo*, *Crayon Physics*, or *jelly Car*, to name but a few. As mentioned, PbSGs provide realism in the simulation of the game and, therefore, increase the immersion of the player which surely positively influences her satisfaction and favors their involvement with the game. In fact, physics can be found even in the early phases in the history of video games; so, Super Mario Bros already exhibited, in 1985, elementary concepts of physics in the form of jumps, forward/backward movements, and object throwing, executed by the main character. However, in modern video games, physics is generally referred to as rigid body physics simulation subject to real physics laws (e.g., Newton's Three Laws of Motion or Newton's Law of Universal Gravitation, just to name a couple associated to the classical Physics). PbSGs games are very interesting not only from the player's perspective but also from a research point of view; so, there have been interesting papers recently published on the Physical Travelling Salesman Problem (PTSP), a real-time game that consists of a ship that must visit a number of waypoints scattered around a 2-D maze full of obstacles [32]. This problem can be viewed as a PbSG as all actions applied to the ship are forces that influence its position, orientation and velocity at each step of the game. Precisely, [33,34] employ algorithms based on Monte Carlo tree search [35] to handle the problem.

Furthermore, "A slower speed of light" [36] is a game developed by the MIT Game Lab to help students understand and visualize the effects of special relativity by artificially lowering the speed of light to walking pace. The game, which is based on a first-person relativity visualization engine that has been released as *OpenRelativity*, is a prototype in which players navigate through a 3D space while picking up orbs that reduce the speed of light.

3   The Game: Gravityvolve!

There is a well-known problem inside the field of physics and mathematics, the so-called *n-body problem* the origin of which lies in Newton's Principia and classically consists of[2] "predicting the individual motions of a group of celestial objects interacting with each other gravitationally". This problem requires the existence of n rigid bodies and basically consists of determining the positions and velocities of these n particles in each instant of the time in accordance with Newton's Laws of Motion and of Universal Gravitatión, starting with an initial position and velocity for each particle and letting the gravitational forces act on

---

[2] Wikipedia. Accessed on 17th of January, 2016.

the set of particles. For $n = 2$ the problem represents, in certain form, the most fundamental kind of interaction between two bodies, and the problem has no analytical solution (for $n \geq 3$); moreover, generally it can be only simulated using numerical integration methods [37]. The n-body problem considers n particles with specific masses $m_1, \ldots, m_n$ moving in a three dimensional space under the influence of mutual gravitational attraction.

Gravityvolve! is a game implemented by the authors of this paper [8], that has been inspired by the *n*-body problem, which includes some additional features that transforms an interactive simulation into a playable environment. In the following paragraphs we discuss some articular features of our game:

In the first place, the simulation is constrained to a 2D environment, because using a 3D environment would result on a senseless increase of the complexity and we want to preserve the simplicity of traditional PbSG such as *Angry Birds*. Players have a simpler visualization of the game using a bi-dimensional environment: they are able to watch the full playing area all the time, with an aerial view of the plane where the particles are located. This way, the particles' trajectories are more natural and understandable in 2D than in 3D.

Anyhow, dealing with the problem in this way does not represent a decrease in the generality of 3D games, especially those games whose game terrain is a surface that can be defined as the graph of a certain function $f : \mathbb{R} \times \mathbb{R} \to \mathbb{R}$. Thus, although the map has a three-dimensional appearance, the position of its components may be defined with two coordinates and its topology matches the plane $\mathbb{R} \times \mathbb{R}$.

Secondly, there is only a particle (i.e., a ball) affected by the gravitational force of the remaining particles during the simulation. This particle is the only one the player is able to interact with, by changing its velocity vector and guiding it over the screen, while the remaining particles remain static on their initial position. Once again, this restriction is done for the sake of simplicity, hence reducing the amount of information the player has to process and avoiding highly chaotic and unpredictable behaviors.

Figure 1 shows two screenshots of the game running the version with 5 planets. There are $n = 5$ particles (i.e., planets) with an associated mass distributed over the screen; each particle is represented by a planet with a radius that is as long as its mass. These particles (i.e., planets) remain static throughout the game. There exist two other bodies that are positioned on the surface of two distinct planets: the *ball*, represented by a small red circle, and the *hole*, represented by a red circumference. In each step of the game, the user has to interact with the game to set the magnitude and direction of the velocity vector associated with the ball. A green line segment (as displayed in Fig. 1a) represents precisely these values to be fixed by the user (the orientation of this line segment indicates the direction in which the ball will be thrown and its length the magnitude of the force with which the ball is thrown). To help the player, a purple line shows the prediction of the ball's trajectory according to the user interaction; Fig. 1b shows the movement of the ball after being affected by the user in another 5-planet map. The objective of the game is to drive the ball from its point of

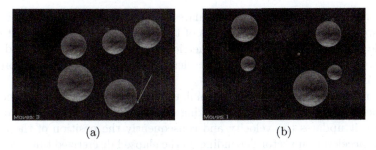

(a)   (b)

Fig. 1. Example of two different phases/maps of the game in a 5-planet version. (a) The length and orientation of the green line segment indicate the magnitude and direction of the velocity associated with the ball whereas the purple line is a hint (to the player) that provides a prediction about the trajectory of the ball according to the 'force' applied to the ball (b) The red ball has been thrown and is moving toward the objective according to both its velocity and the forces of gravitational system (Color figure online).

origin to where the hole is placed, in the minimum number of moves. In a way, *Gravityvolve!* is equivalent to a Golf Game in the sense that a ball has to reach the objective of a hole in a minimum number of hits, whereas the tuning of the values associated with the magnitude and direction of the ball in *Gravityvolve!* corresponds to hitting the ball in golf.

Given the aforementioned objective and game features, the task of defining a map for this game consists of defining the mass and position for each planet in the map, as well as the initial positions of both the ball (i.e., the movable particle) and the hole. An additional effort consists of obtaining maps of varying difficulty which clearly depends on the arrangement of the planets and the original placement of both the ball and the hole.

3.1   Game Physics

The gameplay in *Gravityvolve!* is based on Newton's Law of Universal Gravitation, as well as Newton's Three Laws of motion. In particular, the first law states that any two bodies in the universe attract each other with a force that is directly proportional to the product of their masses m_1, m_2 and inversely proportional to the square of the distance r between them. Note that as our objective was to create a simulation with our own magnitudes, the gravitational constant became irrelevant so we got by without it ensuring that the simulation keeps meeting Kepler's laws. That way, the acceleration that one body causes to another is the interesting magnitude, which is, in addition, an easily computable variable: the second of Newton's law states that the relationship between the acceleration a suffered by a body with a mass m_1 then the received force is $F = m_1 a$.

Considering this force F the one produced by another body with a mass m_2, then the first body is affected by an acceleration $a = G \frac{m_2}{r^2}$.

The aforementioned equation calculates the modulus of the acceleration vector, however, we need the coordinates of this vector in order to apply it to the movable particle. Using basic trigonometric equations: denoting by p_1 and p_2 the position vectors of both bodies, the acceleration that the second body provokes on the first one is $a = \frac{m_2}{r^3}(p_2 - p_1)$.

A simple implementation of the simulation that uses the aforementioned vector of acceleration consists of discretizing the time in such a way that, on each iteration, it updates the velocity and consequently the position of the particle with the acceleration vector depending on the elapsed discretized time. Although the Runge-Kutta integration method has more precision, we used the Newton integration method which is precise enough for this game and its rules. Magnitudes such as time discretization, the mean of the masses and the game screen dimension are strictly related with each other them while establishing the precision and velocity of the simulation.

4   The Procedural Map Generator

The map generator is based on a steady-state evolutionary algorithm with a structured population divided into three sets (i.e., subpopulations) with the same number of elements. The first set contains those best adapted individuals according to the fitness function, which measures the difficulty level of the map. According to this, each sub-population groups the individuals with similar difficulty levels, so hard maps (i.e., those with their fitness value ranged between the theoretical maximum fitness and a 66 % of this value) should rely on the first group as medium and easy maps should rely on second and third sub-populations, respectively.

Regarding the generation of the initial population, the algorithm generates random individuals and assigns them to their corresponding subpopulation until all of them are complete, hence ensuring a high population diversity at the beginning of the evolutionary process.

Upon each generation, the algorithm selects two random individuals from the population, which are then mutated and recombined using the operators described in Sect. 4.1. We decided to use this random selection mechanism to increase the diversity of the offspring. Then, the algorithm computes the fitness of the new individuals and inserts them into the population applying the following replacement policy: as the population is structured into three groups, the subpopulation where the new individual may be inserted depends on its fitness value and the theoretical maximum fitness value. Then, depending on the selected subpopulation, that is, Hard, Medium or Easy, the fitness of the new individual is compared to the fitness of the best, central or worst individual, respectively (see Fig. 2).

Each individual of the algorithm represents a map, and every map is defined by the planets included in it and the position of both the ball and the hole. Each planet is made up of three genes: its x and y coordinates and its mass m (which, in addition, corresponds to its radius). Regarding the ball and the hole, they are

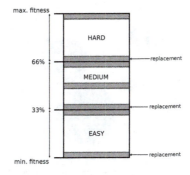

Fig. 2. Structure of the algorithm's population according to the fitness of their elements and both theoretical maximum and minimum fitness value. New individuals replace the worst individual of the corresponding subpopulation if their fitness is better. Gray coloured individuals are those selected for the experimental phase.

made of two genes each: the planet over which they are placed and the angle in radians that specify the position over the planet surface. Figure 3 shows an example of a map and its corresponding encoding using our evolutionary PCG method.

4.1 Operators

Regarding the mutation operator, on a map with n planets, the number of genes is $3n + 4$, one for each planet's X and Y coordinate and its mass, and two for both the ball and the hole, which corresponds to the planets on which ball and hole are placed as well as the angle in radians (see Fig. 3). Note that planets' genes and those that encode the properties of the ball and the hole follows a different mutation process. With respect to the planets' genes, their mutation probability is $p = \frac{1}{3n}$, so the number of mutated genes from each map follows a binomial distribution $X \sim B(3n, \frac{1}{3n})$ so the mean quantity of mutated genes turns out to be 1.

When a gene is selected for mutation, its coordinates are modified adding a random value Δ_c that follows a normal distribution with $\mu = 0$ and $s^2 = 50$ to each coordinate. A similar modification is made to the planet's mass in such a way so that the planet may increase or decrease its radius (mass), the mutation step follows the same distribution as the mutation step of the coordinates.

On the other hand, for the mutation of both the ball and the hole, a new random planet and angle are assigned to them, with a mutation probability of 0.15.

After the mutation step, the algorithm checks the validity of the map and, if the map is no longer valid (i.e., a planet has moved out of bounds or there are overlapping planets), the algorithm reverts the mutation and repeats the process until the map remains valid after applying the mutation changes.

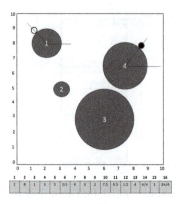

Fig. 3. An example of a 4-planet *Gravityvolve!* map in a scenario of size 10×10, and its corresponding encoding. Planets are numbered for clarity; the first 12 genes represent the information of the planets—i.e., 3 genes per planet to encode its position in the 2D grid and its radius (i.e., mass). Genes 13th and 14th (resp., genes 15th and 16th) provide information for the ball (resp. the hole) in the form of the planet number over which it is placed and angle in radians that indicates its position in the planet's surface taking the circumference relative to the surface of the planet as reference.

The crossover operator, whose definition follows, is inspired by the one point recombination, but geometrically. In the first step the operator computes a random line that splits the map area and then, using two random points, defines the equation of the line. Secondly, the operator builds two sets of planets, one for those that stand above the line and another for those under the line. A map is considered to be a member of one set or the other if its center is above or below the line. This cut should be accepted as valid if there is at least one planet on either side of the line and the number of planets on each side is the same. The crossover operator works over two parents and, after a valid cut is computed for each parent, their slices are swept (left/up slice from parent one combined with right/down slice from parent two and vice versa) so new individuals get one slice from each parent.

It might seems that considering a cut as valid only if there is the same number of planets on both slices is a very restrictive condition, because as the number of planets raises, the probability of making a valid cut decreases: given n planets, this probability is roughly $\frac{1}{n}$ (provided that we are omitting the unlikely case where there are no planets on a slice). Hence the mean number of attempts to achieve a valid cut is n. However, this problem can be avoided introducing a minor change into the algorithm: a cut will be valid if there is a difference of k ($k \ll n$) between the number of planets on both slices and, in this case, positioning the remaining planets at random positions or moving them into the parent with the lowest number of planets.

Figure 4 shows a valid example of a crossover. The red line is the line used to divide the map into two slices. For the sake of clarity, planets from each map are colored so it is easy to identify them before and after the crossover.

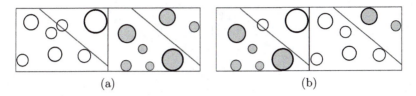

(a)   (b)

Fig. 4. Example of a crossover operation. The upper maps are the parents (a) and the lower are the children (b). Parent maps are divided into two slices following the same cut line and then swapped in order to obtain the children. Planets have been colored to distinguish which parent the planets on the child side have been taken from (Color figure online).

4.2   Measuring a Map's Difficulty Level

We propose three different fitness functions representing level properties that try to measure the difficulty level of a map. Contrary to the aforementioned methods, these functions are strictly related to the game and may be difficult to adapt and use for other games.

The first fitness function is based on the idea that the distance between the particle and the hole as well as the number of planets between them is a good estimator of how hard it is to place the particle over the hole. This method computes the equation of the line between the particle's center and the planet where the hole is located in order to find which planets intersect with this line and how far the hole is from the particle. Given the equation of the line $y = mx + n$ and the circumference of a certain planet $(x - a)^2 + (y - b)^2 = R^2$, the intersect points between that planet and the line are:

$$
\begin{cases}
x_1 = \pm \dfrac{\sqrt{R^2m^2 + R^2 - a^2m^2 + 2abm - 2amn - b^2 + 2bn - n^2}}{m^2 + 1} + \dfrac{a + bm - mn}{m^2 + 1} \\
\\
y_1 = mx_1 + n
\end{cases}
$$

The fitness value computed by this function is the sum of all these line segments defined by the aforementioned intersect points. We observe that, if there were no restrictions on the generated levels such as no two planets should overlap and there must exist a minimum distance between the planets so the ball is able to move between them without getting stuck, maps with a high fitness would be a level with all the planets aligned with a screen diagonal without any gap between them and the particle and hole placed on opposite corners of the screen, which defines, in turn, the map that evaluates to the maximum possible fitness value (see Fig. 5). However, the restrictions applied to the generated maps increase the variability of the procedurally generated maps.

There is a another way of measuring the difficulty level of the game's objective: minimize the ball's force of attraction. An easy way to achieve this could be minimizing the mass of the planet that hosts the hole and increasing the

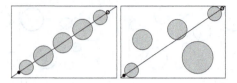

Fig. 5. Maps with maximum fitness values computed by the intersections (left) and simulations (right) methods. The black circle and circumference represents the ball and the hole, respectively. On the left side, the planets are aligned with a screen diagonal and the particle and hole are placed on opposite corners of the screen. On the right side, the ball is located on the corner after a simulated shoot and the hole is on the opposite corner, so the distance between them is the highest.

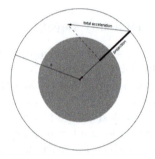

Fig. 6. Projection of the induced acceleration for an auxiliary particle by other planets over the vector between this particle and the center of the planet

respective masses of the other planets. For instance, the fitness function might be the defined as $\frac{m}{M}$, given the mass of the target planet m and the sum of the mass of the remaining planets M. Due to the simplicity of the solution, this could be implemented without using evolutionary algorithms but the resulting maps would be quite similar to each other and the position of the planets would not affect the generation process which in turn would lead to a loss in diversity in the solutions.

For these reasons, we consider a more complex fitness function, based on the force of attraction, that takes into account the mass and position of each planet. The function takes the mean acceleration $\bar{a}$ produced by all the planets over the planet with the hole (the target planet) as the fitness value for a certain map. The function puts $n = 50$ auxiliary particles uniformly distributed around the target planet within a certain distance d and then, computes the acceleration α induced for each auxiliary by the other p planets, which is projected over the vector between the auxiliary particle and the planet center (see Fig. 6):

$$\bar{a} = \frac{\sum_{i=1}^{n} \mathrm{proj}(\alpha_i)}{n}$$

Table 1. Each row (except the first row) represents the maps with a certain difficulty level from our testbed; column 1 indicates the fitness function employed to generate these maps whereas column 2 provides the group to which the map belongs according to the PCG method whose search was lead by the specified fitness function; columns 3 to 5 display the votes given by the users to each specific map according to their game experience, and column 6 shows the number of votes received for each map.

Fitness type	Level	Hard	Medium	Easy	Total
Intersections	Hard	10	45	54	109
	Medium	5	19	136	160
	Easy	12	56	44	112
Attraction	Hard	15	34	60	109
	Medium	26	40	106	172
	Easy	7	31	67	105
Simulations	Hard	17	44	49	110
	Medium	11	49	104	164
	Easy	2	21	89	112

In this case, the maximum fitness value, which defined the subpopulations, was the highest fitness value computed for the random individuals that were in the initial population.

Thirdly, we define a simulation-based fitness function that makes several shootings with different angles and forces, more precisely, the function simulates 10 throws towards the target planet with random velocities. So, the fitness of the map is the minimum distance between the ball after the shoot and the hole, hence as the fitness decreases, the difficulty level of the map also decreases. The reader should observe that the maximum theoretical value for this fitness function is reached when the ball, after the shoot, is located on a corner of the screen and the hole is on the opposite corner (see Fig. 5).

5   Conclusions

This paper has presented a physics-based video game inspired by the so-called *n-body problem*, a well-known problem in the field of physics and mathematics. The maps for the games have been procedurally generated using a steady-state evolutionary algorithm whose population was structured in three groups or difficulty levels; our spatially-structured proposal is directed to obtain content of diverse nature (in the case of this paper of varying difficulty) and can be easily generalized to other games. In addition, three different fitness functions have being proposed to drive the search process inside our evolutionary algorithm. These functions compute how difficult a map is depending on multiple criteria: the number of planets in straight line between the particle and the target, the gravitational acceleration induced on the target planet by the rest of the map's elements and the distance between the target and the particle after shooting it using different angles and forces (i.e., simulations).

In [8], we provide a web link where the reader can play a number of maps (with 5 planets), of diverse difficulty, that were generated by our proposed PCG method (and by considering all the fitness functions defined here). We have performed a preliminary study on the system's capabilities to generate maps of diverse difficulty: the web platform has collected the participation of 58 unique users and an average number of votes per map of 55, because not all participants played and evaluated each of the 21 maps (see Table 1).

Note that the generated maps contains only 5 planets and, thus the maps are not difficult to play. However, some players (a 10 %) considered some of our maps as hard. This is a promising result although, in general, players perceived that the maps were of medium or easy difficulty. This is a preliminary experiment and we plan to generate maps with varying number of planets (e.g., between 5 and 15) and conduct a deeper analysis of the results.

We are aware that some of the concepts described in this paper might seem very specific of the game, however, the schema of our spatially-structured method can be adjusted to other games, and demonstrating this is also part of a future work.

Acknowledgments. This work has been partially funded by Spanish MICINN under project UMA-EPHEMECH (http://blog.epheme.ch) (TIN2014-56494-C4-1-P), by Junta de Andalucía under project P10-TIC-6083 (DNEMESIS (http://dnemesis.lcc. uma.es/wordpress/)) and Universidad de Málaga. Campus de Excelencia Internacional Andalucía Tech.

References

1. Lara-Cabrera, R., Cotta, C., Fernández-Leiva, A.: An analysis of the structure and evolution of the scientific collaboration network of computer intelligence in games. Phys. A Stat. Mech. Appl. **395**, 523–536 (2014)
2. Togelius, J., Champandard, A.J., Lanzi, P.L., Mateas, M., Paiva, A., Preuss, M., Stanley, K.O.: Procedural content generation: goals, challenges and actionable steps. In: Lucas, S.M., Mateas, M., Preuss, M., Spronck, P., Togelius, J. (eds.) Artificial and Computational Intelligence in Games. Dagstuhl Follow-Ups, vol. 6, pp. 61–75. Schloss Dagstuhl - Leibniz-Zentrum fuer Informatik (2013)
3. Hendrikx, M., Meijer, S., Van Der Velden, J., Iosup, A.: Procedural content generation for games: a survey. ACM Trans. Multimed. Comput. Commun. Appl. (TOMCCApP) **9**(1), 1–22 (2013)
4. Shaker, N., Togelius, J., Nelson, M.J.: Procedural Content Generation in Games: A Textbook and an Overview of Current Research. Springer, New York (2014)
5. Rivello, S.A.: Developing physics-based games with adobe flash professional EDGE, article 7, April 2010. Accessed 15 Jan 2015
6. Shaker, M., Sarhan, M.H., Naameh, O.A., Shaker, N., Togelius, J.: Automatic generation and analysis of physics-based puzzle games. In: 2013 IEEE Conference on Computational Inteligence in Games (CIG), Niagara Falls, ON, Canada, 11–13 August 2013, pp. 1–8. IEEE (2013)

7. Preuss, M., Liapis, A., Togelius, J.: Searching for good and diverse game levels. In: 2014 IEEE Conference on Computational Intelligence and Games (CIG), pp. 1–8, August 2014

8. Lara-Cabrera, R., Fernández-Leiva, A.J.: Gravityvolve!

9. Togelius, J., Yannakakis, G.N., Stanley, K.O., Browne, C.: Search-based procedural content generation: a taxonomy and survey. IEEE Trans. Comput. Intell. AI Games **3**(3), 172–186 (2011)

10. Togelius, J., Yannakakis, G.N., Stanley, K.O., Browne, C.: Search-based procedural content generation: a taxonomy and survey. IEEE Trans. Comput. Intellig. AI Games **3**(3), 172–186 (2011)

11. Yannakakis, G., Togelius, J.: A panorama of artificial and computational intelligence in games. IEEE Trans. Comput. Intell. AI Games **PP**(99), 1 (2014)

12. Shaker, N., Yannakakis, G.N., Togelius, J.: Towards automatic personalized content generation for platform games. In: AIIDE (2010)

13. Pedersen, C., Togelius, J., Yannakakis, G.N.: Modeling player experience in super mario bros. In: IEEE Symposium on Computational Intelligence and Games, 2009, CIG 2009, pp. 132–139. IEEE (2009)

14. Togelius, J., Preuss, M., Yannakakis, G.N.: Towards multiobjective procedural map generation. In: Proceedings of the 2010 Workshop on Procedural Content Generation in Games, vol. 3. ACM (2010)

15. Togelius, J., Preuss, M., Beume, N., Wessing, S., Hagelback, J., Yannakakis, G.N.: Multiobjective exploration of the Starcraft map space. In: 2010 IEEE Symposium on Computational Intelligence and Games (CIG), pp. 265–272. IEEE (2010)

16. Togelius, J., De Nardi, R., Lucas, S.M.: Towards automatic personalised content creation for racing games. In: IEEE Symposium on Computational Intelligence and Games, 2007, CIG 2007, pp. 252–259. IEEE (2007)

17. Ferreira, L., Toledo, C.: A search-based approach for generating angry birds levels. In: 2014 IEEE Conference on Computational Intelligence and Games (CIG), pp. 1–8 (2014)

18. Lara-Cabrera, R., Cotta, C., Fernández-Leiva, A.: On balance and dynamism in procedural content generation with self-adaptive evolutionary algorithms. Nat. Comput. **13**(2), 157–168 (2014)

19. Lara-Cabrera, R., Cotta, C.: Geometrical vs. topological measures for the evolution of aesthetic maps in a RTS game. Entertainment Comput. **5**(4), 251–258 (2014)

20. Toussaint, G.T.: A graph-theoretic primal sketch. In: Toussaint, G.T. (ed.) Computational Morphology, pp. 229–260. Elsevier, Amsterdam (1988)

21. Frade, M., de Vega, F.F., Cotta, C.: Evolution of artificial terrains for video games based on obstacles edge length. In: 2010 IEEE Congress on Evolutionary Computation (CEC), pp. 1–8. IEEE (2010)

22. Frade, M., de Vega, F., Cotta, C.: Automatic evolution of programs for procedural generation of terrains for video games. Soft. Comput. **16**(11), 1893–1914 (2012)

23. Hom, V., Marks, J.: Automatic design of balanced board games. In: Proceedings of the AAAI Conference on Artificial Intelligence and Interactive Digital Entertainment (AIIDE), pp. 25–30 (2007)

24. Hastings, E.J., Guha, R.K., Stanley, K.O.: Automatic content generation in the Galactic Arms Race video game. IEEE Trans. Comput. Intell. AI Games **1**(4), 245–263 (2009)

25. Hastings, E.J., Guha, R.K., Stanley, K.O.: Evolving content in the Galactic Arms Race video game. In: IEEE Symposium on Computational Intelligence and Games, 2009, CIG 2009, pp. 241–248. IEEE (2009)

26. Collins, K.: An introduction to procedural music in video games. Contemp. Music Rev. **28**(1), 5–15 (2009)
27. Font, J.M., Mahlmann, T., Manrique, D., Togelius, J.: A card game description language. In: Esparcia-Alcázar, A.I. (ed.) EvoApplications 2013. LNCS, vol. 7835, pp. 254–263. Springer, Heidelberg (2013)
28. Onuczko, C., Szafron, D., Schaeffer, J., Cutumisu, M., Siegel, J., Waugh, K., Schumacher, A.: Automatic story generation for computer role-playing games. In: AIIDE, pp. 147–148 (2006)
29. van der Linden, R., Lopes, R., Bidarra, R.: Procedural generation of dungeons. IEEE Trans. Comput. Intell. AI Games **6**(1), 78–89 (2014)
30. Ashlock, D., Lee, C., McGuinness, C.: Search-based procedural generation of maze-like levels. IEEE Trans. Comput. Intell. AI Games **3**(3), 260–273 (2011)
31. Plans, D., Morelli, D.: Experience-driven procedural music generation for games. IEEE Trans. Comput. Intell. AI Games **4**(3), 192–198 (2012)
32. The physical travelling salesman problem. Accessed 10 December 2015
33. Perez, D., Powley, E.J., Whitehouse, D., Rohlfshagen, P., Samothrakis, S., Cowling, P.I., Lucas, S.M.: Solving the physical traveling salesman problem: tree search and macro actions. IEEE Trans. Comput. Intellig. AI Games **6**(1), 31–45 (2014)
34. Powley, E., Whitehouse, D., Cowling, P.: Monte carlo tree search with macro-actions and heuristic route planning for the physical travelling salesman problem. In: IEEE Conference on Computational Intelligence and Games (CIG), September 2012, pp. 234–241 (2012)
35. Browne, C., Powley, E., Whitehouse, D., Lucas, S., Cowling, P., Rohlfshagen, P., Tavener, S., Perez, D., Samothrakis, S., Colton, S.: A survey of monte carlo tree search methods. IEEE Trans. Comput. Intell. AI Games **4**(1), 1–43 (2012)
36. Kortemeyer, G., Tan, P., Schirra, S.: A slower speed of light: developing intuition about special relativity with games. In: International Conference on the Foundations of Digital Games, Chania, Crete, Greece, 14–17 May 2013, pp. 400–402 (2013)
37. Aarseth, S.J., Aarseth, S.J.: Gravitational N-Body Simulations: Tools and Algorithms. Cambridge University Press, Cambridge (2003)

Design and Evaluation of an Extended Learning Classifier-Based StarCraft Micro AI

Stefan Rudolph[✉], Sebastian von Mammen, Johannes Jungbluth,
and Jörg Hähner

Organic Computing Group, Faculty of Applied Computer Science,
University of Augsburg, Eichleitnerstr. 30, 86159 Augsburg, Germany
{stefan.rudolph,sebastian.von.mammen,
joerg.haehner}@informatik.uni-augsburg.de

Abstract. Due to the manifold challenges that arise when developing an artificial intelligence that can compete with human players, the popular realtime-strategy game *Starcraft: Broodwar* (BW) has received attention from the computational intelligence research community. It is an ideal testbed for methods for *self-adaption at runtime* designed to work in complex technical systems. In this work, we utilize the broadlys-used Extended Classifier System (XCS) as a basis to develop different models of BW micro AIs: the Defender, the Attacker, the Explorer and the Strategist. We evaluate theses AIs with a focus on their adaptive and co-evolutionary behaviors. To this end, we stage and analyze the outcomes of a tournament among the proposed AIs and we also test them against a non-adaptive player to provide a proper baseline for comparison and learning evolution. Of the proposed AIs, we found the Explorer to be the best performing design, but, also that the Strategist shows an interesting behavioral evolution.

1 Introduction

Starcraft and its expansion pack *Starcraft: Broodwar*[1] (BW, sometimes also referred to as only *Starcraft* or *Broodwar*), combined, are one of the most famous instances of real-time strategy (RTS) games. They were released in 1998 for PCs and since then nearly 10 million copies have been sold. Founded on this number and on a huge number of players attracted to the game until today, it is seen as one of the most successful RTS game to date. RTS games can be characterized by three main tasks that the player has to fulfill: (i) collecting resources, (ii) creating buildings/units and (iii) controlling the units.

BW takes place in a science fiction setting, where three species compete for dominance in the galaxy. This are *Terrans*, a human-like species, *Protoss*, a species that is very advanced in technology and has psionic abilities, and *Zerg*, an insect swarm inspired species. The game has been extensively used for competitions, i.e., tournaments and leagues. These competitions usually consist of several 1-on-1 matches.

---

[1] *Starcraft* and *Starcraft: Broodwar* are trademarks of Blizzard Entertainment.

© Springer International Publishing Switzerland 2016
G. Squillero and P. Burelli (Eds.): EvoApplications 2016, Part I, LNCS 9597, pp. 669–681, 2016.
DOI: 10.1007/978-3-319-31204-0_43

BW represents exactly the kind of training ground needed for testing and honing online learning methods and their capacity to function in complex real-world scenarios. BW challenges the learner through its great complexity, the arising dynamics, and the fact that the fitness landscapes targeted by the learner are self-referential [1]. In BW, we face a set of entities (units and buildings) that interact with an environment (map and units of other players) in non-trivial ways. Furthermore, the environment is only partially observable and brings different types of uncertainty with it. Compared to other games that have been used as scientific testbeds, such as Chess, Go or Poker, it creates a much bigger challenge. Another reason to chose BW as an application to test and hone online learning methods fit for real-world scenarios is the availability of an easy to use C++ library[2] that provides an interface to the game and therefore allows the development of artificial players as well as automated test runs of them.

Learning classifier systems, in particular variants of the extended learning classifier system (XCS), have been successfully deployed in various online learning tasks in real-world scenarios. In this work, we present an XCS-based model design for the artificial intelligence assuming the role of a player in *Starcraft: Broodwar*. The remainder of this paper is structured as follows. In Sect. 2, we touch upon various related works in the context of RTS and corresponding machine learning approaches. We also introduce XCS as the learning system our approach is based on. In Sect. 3 we detail our model and the specific *Starcraft: Broodwar* scenario it was developed for. Section 4 presents and discusses the results of our co-evolutionary learning experiments. Afterward, we conclude with a short summary and an outlook on potential future work.

2   Related Work

In this section, we first touch upon the numerous approaches to development and deploying artificial intelligence techniques and machine learning in the Starcraft domain. Second, we present the Extended Classifier System (XCS) as the machine learning system used as the learning method for BW AIs in this work.

2.1   AI Approaches in Starcraft

A recent survey covering bot architectures, i.e. the algorithmic architectures for automated players, is given in [2]. It identifies *learning and adaptation* as an open question in RTS game AI, which is addressed in this work. Numerous works in the field target prediction and handling uncertainty. In [3], for instance, a method is introduced to predict openings in RTS games. As another example, [4] presents an approach for estimating game states. In contrast, this work focuses on learning, but, the presented methods could be combined with the approach given here. Another direction of research is the exploration of methods for the engineering of bots. To this end, [5] proposes to follow the paradigm of agent-oriented

---

[2] https://github.com/bwapi/bwapi.

programming, and [6] presents a method for automated testing of bots. Some works concentrate on providing data sets of BW games, e.g., [7,8]. There are also works about making and executing plans, such as [9] that proposes a method for the opening strategy optimization, or [10], where a method for the navigation of units is presented. Another category of works are the ones that innovate on mechanisms of strategy selection, e.g., [11], or of choosing tactical decisions [12]. Recently, in [13], a framework for the generation of a complete strategy from scratch using a evolutionary approach has been proposed. Finally, there are several works on the control of units, e.g., [14], where a Bayesian network is utilized for the unit control, or [15], where Reinforcement Learning methods are applied to learn *kiting*, a hit and run technique for a special unit type. In this work, we propose the use of Learning Classifier Systems for providing an AI that both evolves new behaviours through evolutionary computation and hones and refines established ones through reinforcement learning. We provide four according AI designs which exhibit different focusses of the learning system's deployment.

2.2   Extended Learning Classifier Systems

A *Learning Classifier System* (LCS) has originally been proposed in [16] by Holland. Later, he reworked the idea and proposes what today is considered a standard LCS in [17]. The most common extension of his work is the *Extended Classifier System* (XCS) of Wilson. It has been originally introduced in [18].

Since we adopted this variant for this work, the essence of the XCS is presented now. The basic architecture of an XCS is depicted in Fig. 1. It represents a very elaborate learning system tailored towards real-world applications. Accordingly, in Fig. 1, we see that the XCS gets a situation description of the environment through detectors. The situation is in the basic version of the XCS encoded as a bitstring. The population consists of classifiers, which hold several values:

- The **condition** is a string of 0s, 1s and *don't cares* (often represented by X). The purpose of the condition is to determine, if the classifier matches the situation given by the detector. A match is given, if for every 0 in the situation there is a 0 or an X at same position in the condition.
- The **action** is also encoded as a bit string. The set of available actions is typically provided by the designer of the system and depends on the application.
- The **prediction** is a value that approximates the expected reward, given the action of this classifier is executed in the situations described by the condition. It is constantly adapted by taking new observations into account.
- The **prediction error** is a value that reflects how much the prediction deviated from the actual reward.
- The **fitness** expresses the accuracy of the prediction of the classifier.

The match set holds all classifiers that match the current situation. If it appears that this set is empty a covering procedure is started that generates classifier that match the given condition and propose a random action. Afterwards, the

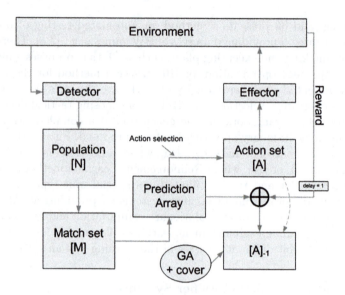

Fig. 1. The basic architecture of an extended learning classifier system or XCS.

XCS advances to the next step. Most often, the match set holds classifiers that suggest different actions. The purpose of the prediction array is to decide which action is applied. To this end, it uses a fitness-weighted average of the predictions for each action that is mentioned in the match set. The classifiers proposing the chosen action are transferred to the action set and the action is applied through the effector.

In the next step, a reward is provided by the environment that values the current state. The prediction, error and fitness values of all the classifiers in the previous action set are adjusted according to a given update rule, based on the reward. In addition, a genetic algorithm is applied to the action set in order to create more appropriate rules. It selects parent classifiers for generating new ones based on their fitness values and can apply different crossover and mutation operators, which is up to the designer of the system.

3   Approach

In this section, we first explain the general scenario the AIs we have developed had to train and prove themselves in. Based on this knowledge, it is easier to follow the motivation for their individual designs which follows next.

3.1   Competition Scenario

We let our AIs compete and train in so-called *micro* matches, which implies that each AI was only able to control one group of units. In the given scenario, we did not consider the collection of resources and the production of units and buildings. A game is won, if all the enemy units or all the enemy buildings are destroyed. The match comes to a draw, if neither of the two competing AIs wins within a period of five minutes of simulated time. Each player starts out with the following heterogeneous set of predefined units which is a subset of the available zerg units.

Zergling. Each player has control over 64 Zerglings at the start of the match. They are light units dealing little damage and they can only suffer little damage before they are destroyed. Zerglings are melee units, which means they can only attack, if they are close to enemy units.

Hydralisk. Hydralisks can attack over distance but are not robust units, i.e., they should try to keep distance to the enemy units since they can be destroyed fast if they are attacked. Each player has control over 12 Hydralisks at the beginning of a match.

Ultralisk. Each player only has two Ultralisks at their disposition. They are heavy units that are very robust, i.e. they can sustain a high number of hitpoints. Like Zerglings, Ultralisks are melee-only units.

Scourge. Scourges are airborne units. Primarily, they attack other airborne units. At the loss of the Scourge unit itself, it can crash into other units to explode and damage the enemy. The player is provided with four Scourges at the beginning of a match.

Zerg Queen. The Queen has no direct means of attack. Yet, it can slow down other units in a small quadratic area for 25 to 40 s, depending on the game speed. The enemies' movement velocity is halved, their rate of attack is reduced by between 10 to 33 %, depending on the affected unit type. In addition, the queen can hurl parasites at enemy units at a larger distances. The infested units' views extend directly add to the reconnaissance of the Queen's player.

The tournament map as well as the initial spatial arrangement of the given units is shown in Fig. 2. It has been established by the SCMAI[3] tournament. In Fig. 2a, we see the used map. The starting points of the players are marked with (1). At the positions marked with (2), there are buildings that can attack units that are within their range. Two of these buildings belong to each player. If a player destroys one of the opponent's buildings, additional Zerglings appear in the center of the map as a reinforcement. This is to encourage the players to engage and not just protect there own buildings.

3.2 AI Components

In analogy to the components of an extended learning classifier system (Sect. 2.2), we considered the following basic building blocks for creating an effective Starcraft AI. (1) Behavioral rules to classify and react to a given situation, (2) a reinforcement component to adjust the rules' attributes in order to

---

[3] Starcraft Micro AI Tournament.

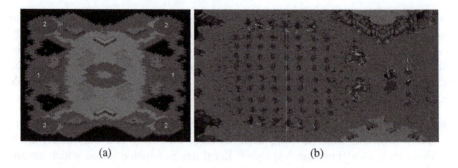

        (a)   (b)

Fig. 2. (a) The map used for tournaments in our co-evolutionary experiment setup. (b) The initial spatial arrangement of the units made available to the AIs.

increase the achieved reward, (3) a covering mechanism to generate rules and fit newly encountered situations, and (4) a genetic algorithm for evolving the existing set of behavioral rules. In addition, we considered the ability to progress in battle formation based on individual *boid* steering urges [19], see Fig. 3. Different

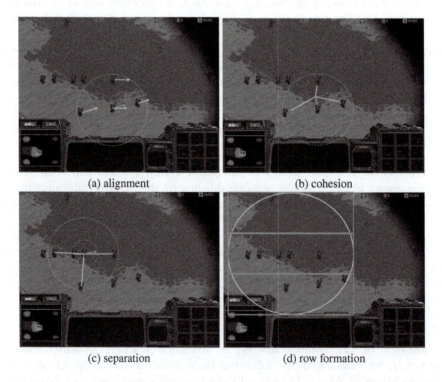

      (a) alignment   (b) cohesion

      (c) separation   (d) row formation

Fig. 3. The augmented screenshots (a) to (c) depict the steering urges as defined by Reynolds' flocking algorithm [19]. Based on these urges, formations emerge such as the one in (d).

from dynamically chosen but otherwise fixed formations, inferring the individual accelerations based on the units' neighborhoods results in emergent, adaptive formations [20]. In particular, the units sense their neighbors and (a) align their heading and speed with them, (b) tend towards their geometrical center, and (c) separate, if individual units get too close. As a result, battle formations such as the row formation in Fig. 3(d) emerge.

3.3   XCS-based AIs

We combined the AI components outlined above in four different ways, to trigger interesting competition scenarios and to trace and analyze the components' effectiveness. In particular, we implemented and co-evolved one defensive AI, one aggressive one, one that focuses on exploration and one where XCS takes global strategic decisions. Screen shots of their respective activities are seen in Fig. 4.

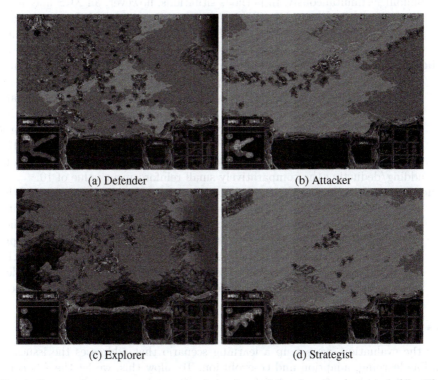

(a) Defender   (b) Attacker

(c) Explorer   (d) Strategist

Fig. 4. Screen shots of representative behaviors of the four implemented AIs. The Defender assembles his troops to defend the buildings. The Attacker storms towards the enemy's buildings to attack. The units of the Explorer swarm in different directions from the base. The Strategist decided on attacking enemy units.

The Defender. Right at the beginning of the match, all units move to the upper of two buildings on the map and stay there for its defense until the end. Using the full functionality of an XCS, the Hydralisks' as well as the Queen's behaviors are learned. For the Hydralisks, the condition part of the classifier rules considers the distance to the next visible enemy. Six actions are offered: (1) Approach and attack the closest ground or (2) airborne enemy, (3) move to a predefined point, (4) support a friendly unit, (5) protect the hatchery, or (6) burrow. For the Queen the proximity to the next enemy unit can trigger escape or *ensnare* airborne units or to hurl *parasites* at ground units. The XCS' reinforcement component positively rewards any attacks, whereas the other actions are only remunerated, if the player is attacked itself or if the units have built up a great distance to their buildings.

The Attacker. This is an offense-oriented AI. It will attack the next visible enemy. If none are in sight, the next enemy building is attacked. The units move in flocks, often they break into two clusters to attack both the enemy's buildings simultaneously. In perilous situations, however, an XCS may reinforce the attack or instigate retreat. Any kind of attack (the XCS also decides the Queen's mode of attack) is rewarded, suffering damage results in negative reinforcement.

The Explorer. The units are divided into two clusters that head into randomly chosen directions to explore the environment. When an enemy is sighted, an XCS determines to attack or to escape. Successful attacks directly translate into positive rewards, whereas loss of health points implies negative reinforcement. First strike is additionally greatly rewarded, whereas suffering a surprise attack results in an equally great loss—adding or subtracting 100 reward points, respectively. Similarly, winning and losing a match results in adding/deducting the comparatively small reinforcement value of 10.

The Strategist. Here, XCS is used to determine the overall strategy of the player. Based on the remaining time, the available and the opposing units, XCS determines whether to (1) attack enemy units or (2) buildings, whether to (3) defend one's buildings or (4) to idle. The respective strategies imply according convoy movements, if necessary. Independently of the strategy, the next sighted enemy is always attacked. The exhibited behavior is rewarded with the number of remaining units and buildings at the end of each match.

3.4   Learning Scenario

For the evaluation, we set up a learning scenario that addresses the issues of on-line learning, adaption and co-evolution. To allow this, we let the AIs compete with and learn from each other in several matches in a row. In particular, we first let the AIs train for 100 matches in a row with each of the other three AIs. In a second round, the previous adaptation is put to the test in the course of another 50 matches against each enemy AI. As all the AIs are designed to improve themselves by means of the combined reinforcement and evolutionary learning components of XCS, the tournament allows the AIs to co-evolve. Furthermore, for a better comparability of the approaches, we conducted additional

Table 1. Each of the four AIs trains with and competes against all the other ones. This table depicts the consecutive changes in the number of frames the simulation ran for, the health points successfully maintained by the AIs, and the frequency of winning situations.

Defender

enemy	matches	frames	hp left	winner
Explorer	100	+	+	0
Attacker	100	-	-	+
Strategist	100	+	-	-
Strategist	50	+	+	-
Explorer	50	+	+	+
Attacker	50	-	0	+

Strategist

enemy	matches	frames	hp left	winner
Attacker	100	-	+	-
Explorer	100	+	+	0
Defender	100	+	+	0
Explorer	50	-	0	-
Attacker	50	-	+	0
Defender	50	+	0	0

Explorer

enemy	matches	frames	hp left	winner
Defender	100	+	+	-
Strategist	100	+	-	-
Attacker	100	-	0	+
Attacker	50	-	0	-
Strategist	50	-	0	+
Defender	50	+	+	+

Attacker

enemy	matches	frames	hp left	winner
Strategist	100	-	+	+
Defender	100	-	-	+
Explorer	100	-	0	-
Explorer	50	+	+	-
Strategist	50	-	+	+
Defender	50	-	-	0

experiments that include a non-learning AI. Competing against a non-learning AI ensures that any observed improvements do not emerge from co-evolutionary dynamics but are the result of the individual learners themselves. In addition, the non-learning AI also provides a clear baseline against which all the other AIs can be measured against. The non-learning AI mainly defends its position by splitting the given units in two groups which defend the two buildings. It has no intention of winning the game by moving on to attack the enemy.

4   Evaluation and Discussion

Considering 450 matches, the Explorer with 269 won matches clearly represents the best designed AI. The Attacker is second best with 105 wins, followed by the Defender (30 wins) and the Strategist (7 wins). These performances are both the product of the AIs basic strategies but also of the sequence of their co-evolutionary learning experiences. Therefore, it is important to analyze the relative progress each of the AIs has achieved. An according, high-level summary is depicted in Table 1. It shows the consecutive changes regarding the number of frames a simulation runs, the number of health points left of a player and the number of experienced winning situations. A decrease in frames signals an increase in clarity regarding the winner, as draws become less likely and as quicker solutions take over. An increase of left-over health points of an AI may be considered an improvement. However, an AI may also learn to sacrifice more health points in order to win a match in the end. Therefore, an increase in wins statistically indicates an improved, i.e. learned, behavior.

However, as pointed out above, in an attempt to objectively compare the learning successes of each AI, we let all four of them train and compare against a simple, non-learning AI for another 200 matches. The results in terms of averaged fitness evolution as well as in terms of averaged prediction error can be seen in Fig. 5. Although they do not seem to improve much, the Defender and the Attacker AIs have rather high average fitness values to begin with. Their averaged prediction error does not change over the course of the evolutionary experiment either. The average fitness of the Strategist AI, however, rises continuously and converges quickly, despite the fact that its average prediction error rises sporadically as well. The most likely explanation for this discrepancy is that the prediction errors rise so uniformly across the whole population of classifiers that a greater error value would not impact the selection and thereby the whole interaction process.

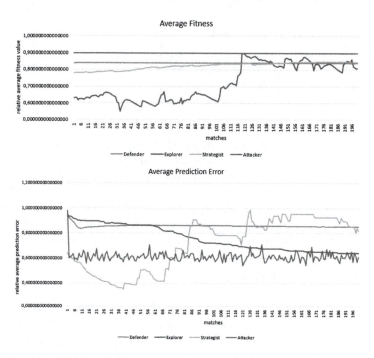

Fig. 5. The four XCS-based AIs' training progress when learning to compete against a simple non-learning AI.

4.1   Co-evolution Qualitatively

The Attacker AI reinforces its aggressive behavior, especially if it does not suffer from inflicted damages. As a consequence, it learns to ruthlessly exploit the Defender AI's feeble assaults by being even more fierce. When exposed to the other AIs, the Attacker AI quickly adapts to be slightly less aggressive. Similarly, the Strategist AI reinforces behaviors that minimize damage. As a result, when facing the Defender AI, the Strategist AI is not motivated to learn well-directedly, i.e., it does not tend to a more aggressive or defensive behavior. Instead, any behavior leads to success. When facing more aggressive opponents such as the Attacker or the Explorer, the Strategist AI receives smaller rewards, almost independently of the ingenuity of a selected strategy. The Defender AI adapts its Hydralisks to shy away from enemies as they get destroyed too quickly, otherwise. Towards simple AIs, such as the simple non-learning AI mentioned above, the Defender AI increases aggressive behaviors. The Defender's Queen is mostly on the run, too, as otherwise the distance to the enemy's units becomes too small. The Explorer as the overall best designed AI presented in this work is discussed more in depth in the next section.

4.2   Decentralized XCS-Concept

After the description of the evaluation and results, we want to provide further details about the Explorer, the most successful AI in the tournament. It utilizes a decentralized XCS-concept, i.e., it adopts multiple XCS, that act in common. It uses one for each unit type (which are presented in Sect. 3.1). Each XCS decides whether the units of the respective type will engage or retreat when faced with an enemy. For the XCS the following configuration is used. Regarding the genetic algorithm, we empirically found the following parameters to be effective. Crossover is applied with a probability of 1 % and mutation with a probability of 1.5 % for each bit in a classifier. The parents are chosen by means of tournament selection with a tournament size of 5. The reinforcement component is configured with a learning rate of $\beta = 0.2$, i.e., the adaption of the prediction value is rather careful, and a discount factor of $\gamma = 0.71$ is used, i.e., future rewards are valued rather high. The reward is the difference between the damage the units have dealt and damage they have took, as it has been proposed in [15]. Additionally, there are some rewards that are considered in special situation. There is an extra reward of 10 if the game has been won and negative reward of -10 if the game is a draw or lost. The action selection is ϵ-greedy with $\epsilon = 0.02$, i.e., the action is selected randomly with a probability of 2 %, otherwise the best action is selected. Additionally, a battle formation procedure based on Reynolds' flocking algorithm is utilized, where the parameters in the algorithm are optimized by a genetic algorithm. Through the optimization, the player evolves a very tight formation, where the Queen is positioned in the center.

Based on this architecture, the Explorer exhibits the following behavior. If there are no enemy units in sight, it creates two separate swarms from the set of available units. These two groups go in an individual, randomly chosen direction in order to explore the map. Utilizing a formation for movement can lead to a tactical advantage, if the opponent's forces are met. The Strategist develops different behavior against the different opponents. If facing the Defender, it will attack immediately, since it appears that this enemy will not be harmful for the strategist. Against the other two AIs, the Strategist is more reluctant since these more offensive AIs tend to deal more damage then the defender.

5   Summary and Future Work

Concluding the work, a motivation for *Starcraft: Broodwar* as an ideal test-bed for *self-adaption at runtime* is given in Sect. 1, followed by an overview over the state-of-the-art in the scientific developments in the BW domain and a short description of the XCS in Sect. 2. Next, in Sect. 3, four XCS-based AIs – the Defender, the Attacker, the Explorer and the Strategist – are presented. Furthermore, in Sect. 4, the evaluation scenario in a tournament scenario and against a non-learning AI with a special focus on the co-evolutionary behavior is presented and discussed.

Overall, we see two main results: The first is that the Explorer shows the best performance of all proposed XCS-based AIs. The second one is that, even though some players performed much worse than the Explorer, the *self-adaption at runtime* worked out for each player. This can be concluded since every player - after a period of adaption - shied away from the more aggressive AIs and, in turn, became more offensive against the more passive AIs.

For future work, we propose two directions. First, we want to refine the XCS-based approach for micro-management in BW. In particular, we want to provide all the units' degrees of freedom available to individual, decentralized XCS learners and also feed them with pre-processed data that indicates general trends in the evolution of the game by means of correlation factors (e.g. [21]) or ascertainment of structural emergence (e.g. [22]). Second, we want to develop an AI that considers a broader managerial scope including micro-management and strategic group activities. To this end, we deem a multi-layered AI architecture taking on different responsibilities through the consideration of different time-scales and levels of abstraction a first important step [1].

References

1. Müller-Schloer, C., Schmeck, H.: Organic computing - quo vadis? In: Müller-Schloer, C., Schmeck, H., Ungerer, T. (eds.) Organic Computing - A Paradigm Shift for Complex Systems, pp. 615–625. Birkhäuser, Verlag (2011)
2. Ontañón, S., Synnaeve, G., Uriarte, A., Richoux, F., Churchill, D., Preuss, M.: A survey of real-time strategy game AI research and competition in StarCraft. IEEE Trans. Comput. Intell. AI Games **5**(4), 293–311 (2013)
3. Synnaeve, G., Bessière, P.: A Bayesian model for opening prediction in RTS games with application to StarCraft. In: Cho, S.B., Lucas, S.M., Hingston, P. (eds.) CIG, pp. 281–288. IEEE (2011)
4. Weber, B.G., Mateas, M., Jhala, A.: Applying goal-driven autonomy to StarCraft. In: Youngblood, G.M., Bulitko, V. (eds.) AIIDE. The AAAI Press (2010)
5. Shoham, Y.: Agent-oriented programming. Artif. Intell. **60**(1), 51–92 (1993)
6. Blackadar, M., Denzinger, J.: Behavior learning-based testing of StarCraft competition entries. In: Bulitko, V., Riedl, M.O. (eds.) AIIDE. The AAAI Press (2011)
7. Robertson, G., Watson, I.: An improved dataset and extraction process for Star-Craft AI. In: The Twenty-Seventh International Flairs Conference (2014)
8. Weber, B.G., Ontañón, S.: Using automated replay annotation for case-based planning in games. In: ICCBR Workshop on CBR for Computer Games (ICCBR-Games) (2010)

9. Churchill, D., Buro, M.: Build order optimization in StarCraft. In: Bulitko, V., Riedl, M.O. (eds.) AIIDE. The AAAI Press (2011)

10. Hagelback, J.: Potential-field based navigation in StarCraft. In: IEEE Conference on Computational Intelligence and Games (CIG), Sep 2012, pp. 388–393 (2012)

11. Yi, S.: Adaptive strategy decision mechanism for StarCraft AI. In: Han, M.-W., Lee, J. (eds.) EKC 2010. SPPHY, vol. 138, pp. 47–57. Springer, Heidelberg (2011)

12. Synnaeve, G., Bessière, P.: A Bayesian tactician. In: Proceedings of the Computer Games Workshop at the European Conference of Artificial Intelligence 2012, pp. 114–125 (2012)

13. Garcia-Sanchez, P., Tonda, A., Mora, A., Squillero, G., Merelo, J.: Towards automatic StarCraft strategy generation using genetic programming. In: 2015 IEEE Conference on Computational Intelligence and Games (CIG), pp. 284–291, August 2015

14. Parra, R., Garrido, L.: Bayesian networks for micromanagement decision imitation in the RTS game StarCraft. In: Batyrshin, I., Mendoza, M.G. (eds.) MICAI 2012, Part II. LNCS, vol. 7630, pp. 433–443. Springer, Heidelberg (2013)

15. Wender, S., Watson, I.D.: Applying reinforcement learning to small scale combat in the real-time strategy game StarCraft: Broodwar. In: CIG, pp. 402–408. IEEE (2012)

16. Holland, J.H.: Adaptation*. In: Rosen, R., Snell, F.M. (eds.) Progress in Theoretical Biology, pp. 263–293. Academic Press, New York (1976)

17. Holland, J.H., Reitman, J.S.: Cognitive systems based on adaptive algorithms. SIGART Bull. **63**, 49–49 (1977)

18. Wilson, S.W.: Classifier fitness based on accuracy. Evol. Comput. **3**(2), 149–175 (1995)

19. Reynolds, C.W.: Flocks, herds, and schools: a distributed behavioral model. Comput. Graph. **21**(4), 25–34 (1987)

20. Lin, C.S., Ting, C.K.: Emergent tactical formation using genetic algorithm in real-time strategy games. In: Proceedings of the 2011 International Conference on Technologies and Applications of Artificial Intelligence, TAAI 2011, Computer Society, pp. 325–330. IEEE, Washington (2011)

21. Rudolph, S., Tomforde, S., Sick, B., Hähner, J.: A mutual influence detection algorithm for systems with local performance measurement. In: Proceedings of the 9th IEEE International Conference on Self-adapting and Self-organising Systems (SASO15), held September 21st to September 25th in Boston, USA, pp. 144–150 (2015)

22. Fisch, D., Jänicke, M., Sick, B., Müller-Schloer, C.: Quantitative emergence - a refined approach based on divergence measures. In: 4th IEEE International Conference on Self-Adaptive and Self-Organizing Systems (SASO), Sep 2010, pp. 94–103 (2010)

EvoIASP

A Wrapper Feature Selection Approach to Classification with Missing Data

Cao Truong Tran[✉], Mengjie Zhang, Peter Andreae, and Bing Xue

School of Engineering and Computer Science, Victoria University of Wellington,
PO Box 600, Wellington 6140, New Zealand
{cao.truong.tran,mengjie.zhang,peter.andreae,bing.xue}@ecs.vuw.ac.nz

Abstract. Many industrial and real-world datasets suffer from an unavoidable problem of missing values. The problem of missing data has been addressed extensively in the statistical analysis literature, and also, but to a lesser extent in the classification literature. The ability to deal with missing data is an essential requirement for classification because inadequate treatment of missing data may lead to large errors on classification. Feature selection has been successfully used to improve classification, but it has been applied mainly to complete data. This paper develops a wrapper feature selection approach to classification with missing data and investigates the impact of this approach. Empirical results on 10 datasets with missing values using C4.5 for an evaluation and particle swarm optimisation as a search technique in feature selection show that a wrapper feature selection for missing data not only can help to improve accuracy of the classifier, but also can help to reduce the complexity of the learned classification model.

Keywords: Missing data · Feature selection · Classification · C4.5 · Particle swarm optimisation

1 Introduction

Classification is one of the most important tasks in machine learning and data mining [14]. The input space plays a crucial role in most classification algorithms. Many classification algorithms such as decision trees and rule-based classifiers are not able to achieve adequate predictive performance when the input contains many features that are not necessary for predicting the desired output. Feature selection which finds a sufficient feature subset from original features is one approach to the problem [18].

Missing values are a common problem in many datasets [21,27]. For example, 45 % of the datasets in the UCI repository [1], which is one of the most popular data repository for benchmarking machine learning tasks, contain missing values [11]. Missing data causes a number of serious problems [2]. One of the most serious problems is non-applicability of data analysis methods because the majority of existing data analysis methods require complete data. Therefore,

© Springer International Publishing Switzerland 2016
G. Squillero and P. Burelli (Eds.): EvoApplications 2016, Part I, LNCS 9597, pp. 685–700, 2016.
DOI: 10.1007/978-3-319-31204-0_44

these data analysis methods cannot work directly with original data containing missing values. Furthermore, missing data may lead to biased results because of differences between missing and complete data.

In statistical analysis field, the problem of missing data has been tackled extensively [12, 21, 26, 27] and also, but with less effort, in the classification literature. There are two main approaches to classification with missing data. One approach is to use imputation methods that fill missing values by plausible values before using classifiers. The other approach is to use classifiers that are able to classify missing data. Although the two approaches are able to handle missing data, they often result in large errors on classification [10]. Therefore, further approaches to improving classification accuracy of missing data should be investigated.

Feature selection is the process of finding a subset of the original features that is sufficient to solve the classification problem. Feature selection has been widely used to improve classification for complete data [7, 18, 19]. In feature selection, two main ways of evaluating feature subsets are the wrapper approach and the filter approach (nonwrapper) [18]. The wrapper approach uses the performance of a classifier to evaluate feature subsets. In contrast, instead of using a particular classifier, the filter approach uses a measure such as information gain (IG) and information gain ratio (IGR) [23] to evaluate the feature subset. In [9], a filter approach to feature selection for missing data was proposed, and the experimental results showed that the filter feature selection method for missing data can increase the precision of the prediction models. However, a wrapper approach to feature selection for missing data has not been investigated. Therefore, whether a wrapper approach to feature selection can improve classification with missing data is still an open issue.

1.1 Research Goals

The overall goal of this paper is to develop a wrapper approach to feature selection on classification with missing data and investigate the impact of this approach. To achieve this goal, three different ways are used to classify missing data. Firstly, missing data is classified by using a classifier that is able to classify directly missing data. Secondly, missing values are filled with plausible values by using imputation methods before using a classifier. Thirdly, a wrapper feature selection method is used to select a feature subset from missing data before using classifiers. Results from the three processes are compared to answer the following questions:

1. Whether feature selection for missing data can improve classification accuracy and achieve dimensionality reduction compared to without using feature selection; and
2. Whether feature selection for missing data can improve classification accuracy and achieve dimensionality reduction compared to using imputation methods.

1.2   Organisation

The rest of the paper is organised as follows. Section 2 discusses related work. Section 3 outlines the method and experiment design. Section 4 presents empirical results and analysis. Section 5 draws conclusions and presents future work.

2   Related Work

This section discusses related work including classification with missing data, imputation methods, feature selection, C4.5 for classification with missing data and Particle Swarm Optimisation-based feature selection.

2.1   Classification with Missing Data

There are three major approaches to classification with missing data including deletion approach, imputation approach and machine learning approach [11].

Deletion Approach eliminates all instances containing missing values before using classifiers. This approach provides complete data that can be classified by any classifiers, but instances containing missing values are not included in the classification process [11].

Imputation Approach uses imputation methods that fill missing values with plausible values before using classifiers. By using imputation methods, missing data is transferred to complete data that can be then classified by any classifiers. Moreover, most imputation methods help improve classification accuracy when compared to classification without using imputation methods. Therefore, using imputation methods is a major approach to classification with missing data [10].

Machine Learning Approach builds classifiers that are able to classify directly missing data without using imputation methods. For example, C4.5 [25], CART [8] and CN2 [5] can deal with missing values in any feature for both training set and test set.

2.2   Imputation Methods

The purpose of imputation methods is to fill missing values with plausible values. By using imputation methods, missing data is transformed into complete data that can be then analysed by any data analysis methods. Therefore, using imputation methods is a popular approach to handling missing data [21, 26, 27]. This section presents three popular imputation methods which are used in this paper.

Mean Imputation fills missing values in each feature with the average of complete values in the same feature. This method maintains the mean of each feature, but it under-represents the variability in the data because all missing values in each feature are filled with the same value [11].

KNN-Based Imputation finds the K most similar instances of each instance containing missing values, and then fills missing values of the instance with the average of the values in the K most similar instances. KNN-based imputation is often better than mean imputation [3]. However, this method is often computationally intensive due to having to search through all instances to find the K most similar instances of each instance containing missing values [11].

Expectation Maximization-Based Imputation uses the Expectation Maximization(EM) algorithm to estimate a maximum likelihood variance-covariance matrix and vector of means that are then used to impute missing values [21,27]. This method is an iterative procedure that includes two main steps at each iteration: an E-step and an M-step. The E-step is used to estimate the means, variances and covariances from complete values and the current best guess of missing values. The M-step is used to estimate new regression equations for each attribute predicted by all others, after that the new regression equations are then used to update the best guess for missing values during the E-step of next iteration. EM-based imputation has been proven to be one of the most powerful imputation methods [12].

2.3   Feature Selection

Feature selection is the process of finding a subset of the original features that is sufficient to solve the classification problem. Feature selection can remove redundant features; hence, it helps to improve classification accuracy. Furthermore, feature selection results in dimensionality reduction, so it makes the learning and execution processes faster. Moreover, models constructed using a smaller set of selected features are often easier to interpret [19].

The two main components of a feature selection method are a search technique and an evaluation criterion. The search procedure is used to generate candidate feature subsets that are then examined by the evaluation procedure to determine their goodness. The quality of the final selected features depends strongly on both the search technique and the evaluation criterion [7].

Many search techniques have been applied to feature selection including conventional methods and evolutionary techniques. For example, sequential forward selection and sequential backward selection are two traditional search techniques used in feature selection [15]. Recently, evolutionary computation techniques such as Genetic Algorithms and Particle Swarm Optimisation (PSO) have been applied to feature selection [4,20,24,28,30].

The two main ways of evaluating selected features are the wrapper approach and the filter approach (nonwrapper) [7]. In the wrapper approach, the performance of a classifier is used to evaluate the subset and hence guide the search. Because every evaluation requires training a classifier and then testing its performance, the search process using a wrapper approach is typically computationally intensive. In the filter approach, instead of using a particular classifier in the evaluation function, the selected features are evaluated by a measure such as information gain and information gain ratio [23]. Because no classification algorithm is involved in the evaluation of selected features, the search process of the filter approach is expected to be more efficient and the results are expected to be more general. However, wrapper approaches often achieve better classification performance than filter approaches [18].

Feature selection has been mainly applied to complete data. A filter approach to feature selection for regression with missing data was proposed in [9], where mutual information was modified to evaluate feature subsets containing missing values. The experimental results showed that the filter approach to feature selection for missing data help improve the performance of the prediction models. However, a wrapper approach to feature selection for missing data has not been investigated.

2.4   C4.5 for Classification with Missing Data

In a wrapper feature selection algorithm for missing data, a classifier that is able to classify missing data is required to evaluate feature subsets. In this paper, the C4.5 algorithm that can classify directly missing data is used to evaluate feature subsets [25].

C4.5 can handle missing values in any feature for both training set and test set. C4.5 uses a probabilistic approach to handling missing values in both the training set and test set, but the way of handling missing values in the training stage is different from the testing stage. In the training stage, each value of each feature is assigned a weight: if a feature value is known, then the weight is assigned one; otherwise, the weight of any other values for that feature is the frequency of that values. In the testing stage, if a test case is unknown, from the current node, it finds all the available branches and decides the class label by using the most probable value [25].

2.5   PSO-Based Feature Selection

Particle swarm optimisation (PSO) is a swarm intelligence algorithm proposed by Kennedy and Eberhart in 1995 [16,17]. PSO is inspired by the movement of organisms such as a bird flocking or fish schooling. In order to optimize a problem, PSO builds a population of candidate solutions encoded as particles in the search space, and moves these particles around in the search space based on information of the particles' position and velocity. The movement of each particle is guided not only by its local best known position but also by the global best known position in the search space. When improved positions are discovered, these will be used to guide the movements of the swarm. This is expected to move the swarm toward the best solution. PSO does not require making assumptions about the problem being optimized and has ability to search very large spaces of candidate solutions. Therefore, PSO is able to be used for optimization problems that are partially noisy, irregular, change over time, etc. However, as like the majority of evolutionary computation algorithms, PSO does not ensure an optimal solution is ever found.

PSO has recently been applied to feature selection problems [29]. In PSO-based feature selection, PSO is used as a search technique to find feature subsets.

If n is the total number of original features in the dataset, then the dimensionality of the search space is n. Each particle in the swarm is often a vector of n real numbers. The value of particle i in the dth dimension, x_{id}, is usually in interval $[0, 1]$. To determine whether a feature will be selected or not, a threshold $0 < \theta < 1$ is required to compare with the real numbers in the position vector. If $x_{id} > \theta$, then the feature {d} will be selected; otherwise, feature {d} will be not selected.

Many PSO based feature selection algorithms have been proposed for both wrapper approaches and filter approaches. PSO has been proven to have the potential to address feature selection problems [4, 20, 28, 30]. However, the performance of PSO for feature selection on missing data has not been investigated.

3   Method and Experiment Design

This section shows detailed experiment design including the method, datasets, C4.5 algorithm, imputation methods and PSO parameter settings for feature selection.

3.1   The Method

The main objective of this study is to empirically evaluate the impact of a wrapper feature selection method on classification with missing data. To achieve this, three experimental setups are designed, as shown in Figs. 1, 2 and 3, respectively. The Fig. 1 shows classification with missing data by using a classifier that is able to classify missing data. The Fig. 2 shows classification with missing data by using an imputation method before applying a classifier. The Fig. 3 shows classification with missing data by using a feature selection algorithm before applying a classifier that is able to classify missing data.

In the three experimental setups, in case of complete data, firstly, missing values are introduced into complete data to generate missing data. Next, missing data is divided into training missing data and testing missing data. In the first setup, as shown in Fig. 1, the training missing data is directly put into a classifier to build a classification model that is then used to classify testing missing data. In the second setup, as shown in Fig. 2, both training missing data and testing missing data are put into an imputation method to generate imputed training data and imputed testing data, and then, the imputed training data is put into a classifier to build a classification model that is then used to classify the imputed testing data. In the third setup, as shown in Fig. 3, training missing data is used by a feature selection procedure to choose a suitable feature subset that is then used to build a data transformation. The data transformation is then used to transform the training missing data and the testing missing data into transformed training missing data and transformed testing missing data, respectively. The transformed training missing data is then put into a classifier to build a classification model that is then used to classify the transformed testing missing data.

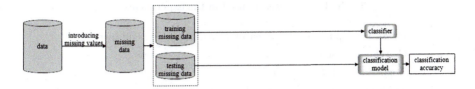

Fig. 1. Classification with missing data by using a classifier able to classify missing data

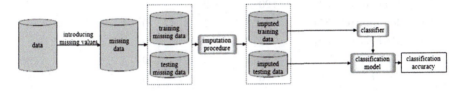

Fig. 2. Classification with missing data by using an imputation method before applying a classifier

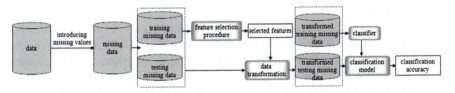

Fig. 3. Classification with missing data by using a feature selection method before using a classifier able to classify missing data.

3.2 Datasets

The experiments used 10 benchmark datasets selected from the UCI machine learning repository [1]. Table 1 summarises the main characteristics of each dataset including the number of instances, the number of features, the number of classes and the percentage of instances in the datasets which have at least one missing value.

The first five datasets have missing values in a "natural"way. There is not any information related to the randomness of missing values in the datasets, so we assume they are distributed in a missing at random (MAR) way [21,22].

To test the performance of the wrapper feature selection method for datasets with different levels of missing values, the missing completely at random mechanism (MCAR) [21] was used to introduce missing values into the last five complete datasets. Six levels of missing values: 5 %, 10 %, 20 %, 30 %, 40 % and 50 % were used to put into the datasets. For each dataset and each level of missing values, perform 30 times: choose randomly 50 % features of the dataset, and then put the level of missing values into the chosen features. Therefore, for each level of missing values on one dataset, 30 artificial missing datasets were generated. Hence, from one complete dataset, 180 (30×6) artificial missing datasets were

Table 1. The datasets used in the experiments.

Dataset	#Instances	#Features	#Classes	Missing Inst (%)
Cleveland	303	13	5	1.98
Hepatitis	155	19	2	48.39
Marketing	8993	13	9	23.54
Ozone	2536	73	2	27.12
Wisconsin	699	9	2	2.29
Climate	540	20	2	0
Ionosphere	351	34	2	0
Parkinsons	197	23	2	0
Robot	463	90	5	0
Sonar	208	60	2	0

generated and a total of 900 (180 × 5) artificial missing datasets were used in the experiments.

Since none of the datasets in the experiments comes with a specific test set and the number of examples in some datasets is relatively small, a ten-fold cross-validation approach was used to evaluate the performance of induced classification models. With the first five datasets containing natural missing values, a ten-fold cross-validation approach was performed 30 times for each dataset. With the last five datasets, for each level of missing values on one dataset, ten-fold cross-validation was performed on the 30 missing datasets. As a result, for each dataset in the first five datasets and each level of missing values on one dataset in the last five datasets, 300 couples of training set and testing set were generated.

3.3   Imputation Algorithms

The experiments used three imputation methods including mean imputation, KNN-based imputation, EM-based imputation. Mean imputation and KNN-based imputation were in-house implementations. For KNN-based imputation, the number of neighbors was set to 10. The experiments used WEKA's [13] implementation for EM-based imputation by setting their parameters as the default values.

3.4   Classification Algorithm

The experiments used C4.5 that has ability to classify missing data. C4.5 was used to classify data and evaluate feature subsets in feature selection. The experiments used WEKA's [13] implementation for C4.5 by setting its parameters as the default values.

3.5 PSO Settings

The experiments used PSO as a search technique for feature selection. The parameters in the PSO based feature selection algorithm were selected according to common settings proposed by Clerc and Kennedy [6]. The detailed settings were shown as follows: $\omega = 0.729844$, $c_1 = c_2 = 1.49618$, population size was 50, and the maximum iteration was 100. The fully connected topology is used. The threshold θ was set 0.6 as suggested by [29] to determine whether a feature is selected or not. For each dataset in the first five datasets and each level of missing values on one dataset in the last five datasets had 300 couples of training set and test set, so PSO repeated 300 times on each dataset.

4 Results and Analysis

Tables 2 and 3 present the average of classification accuracy along with standard deviation of the first five datasets and the last five datasets with six levels of missing values, respectively, by using C4.5 in different ways. With the first five datasets containing natural missing values, the average of classification accuracy were calculated on accuracy of 30 times performing ten-fold cross-validation on each dataset. With the last five datasets, for each dataset and each missing level, the averages of classification accuracy were calculated on accuracy of 30 generated missing datasets with the missing level.

Tables 4 and 5 present the average of size of decision trees (the number of nodes in the trees) generated by using C4.5 in different ways of the first five datasets and the last five datasets with six levels of missing values, respectively.

In the four tables, C4.5 column indicates results from the first experimental setup in Fig. 1; C4.5MI, C4.5KNNI and C4.5EMI columns indicate results from the second experimental setup in Fig. 2 by using mean imputation, KNN-based imputation and EM-based imputation, respectively; C4.5FS column indicates results from the third experimental setup in Fig. 3. In order to compare the classification performance of C4.5FS with other methods, t-tests at 95 % confidence level have been conducted to compare the classification performance achieved by C4.5FS with all other methods. "T" columns in Tables 2 and 3 indicate significant tests of the columns before them against C4.5FS, where "+" means C4.5FS

Table 2. Classification accuracy comparison of C4.5FS with C4.5, C4.5MI, C4.5KNNI and C4.5EMI on datasets containing natural missing values. The T columns indicate significant tests of the columns before them against C4.5FS.

Dataset	C4.5FS	C4.5	T	C4.5MI	T	C4.5KNNI	T	C4.5EMI	T
Cleveland	**56.64** ± 1.37	55.07 ± 1.88	+	53.89 ± 1.31	+	54.08 ± 1.53	+	53.77 ± 1.71	+
Hepatitis	**79.62** ± 1.93	78.59 ± 1.87	+	76.84 ± 2.56	+	77.24 ± 2.98	+	77.71 ± 2.08	+
Marketing	**32.85** ± 0.46	30.80 ± 0.41	+	29.99 ± 0.38	+	30.0 ± 0.39	+	29.98 ± 0.40	+
Ozone	**96.55** ± 0.31	96.28 ± 0.26	+	95.94 ± 0.32	+	95.93 ± 0.38	+	95.95 ± 0.31	+
Wisconsin	94.62 ± 0.55	94.73 ± 0.46	=	94.47 ± 0.47	=	**94.96** ± 0.48	−	**94.87** ± 0.48	−

Table 3. Classification accuracy comparison of C4.5FS with C4.5, C4.5MI, C4.5KNNI and C4.5EMI using several missing rates. The T columns indicate significant tests of the columns before them against C4.5FS.

Dataset	Missing rate (%)	C4.5FS	C4.5	T	C4.5MI	T	C4.5KNNI	T	C4.5EMI	T
Climate	5	**91.34** ± 0.66	90.07 ± 0.93	+	89.74 ± 0.98	+	89.86 ± 0.87	+	90.20 ± 0.98	+
	10	**91.28** ± 0.80	90.39 ± 0.95	+	89.42 ± 0.97	+	89.86 ± 1.13	+	89.63 ± 1.30	+
	20	**91.23** ± 0.55	90.39 ± 1.10	+	89.20 ± 1.19	+	89.46 ± 1.12	+	89.35 ± 1.21	+
	30	**91.38** ± 0.63	90.94 ± 1.09	+	89.30 ± 1.20	+	89.30 ± 1.15	+	89.29 ± 0.95	+
	40	**91.41** ± 0.52	91.22 ± 0.75	=	89.10 ± 1.07	+	88.95 ± 1.05	+	89.20 ± 0.92	+
	50	**91.48** ± 0.13	91.08 ± 0.91	+	89.30 ± 1.30	+	89.23 ± 1.92	+	89.43 * 1.20	+
Ionosphere	5	**90.87** ± 1.17	90.25 ± 1.58	+	89.68 ± 0.88	+	89.05 ± 1.06	+	89.29 ± 1.07	+
	10	**90.25** ± 1.67	89.36 ± 1.70	+	89.08 ± 1.50	+	88.47 ± 1.52	+	89.49 ± 1.35	=
	20	**90.30** ± 1.39	89.50 ± 1.39	+	89.03 ± 1.66	+	87.54 ± 2.46	+	89.12 ± 1.55	+
	30	89.46 ± 1.62	**89.33** ± 1.67	=	88.64 ± 1.62	+	88.18 ± 2.23	+	88.10 ± 1.70	+
	40	88.60 ± 1.99	**88.59** ± 2.44	=	87.44 ± 2.65	+	87.43 ± 2.70	+	**88.40** ± 2.19	=
	50	89.03 ± 2.24	**88.54** ± 2.64	=	86.36 ± 2.60	+	86.36 ± 2.60	+	87.56 ± 2.30	+
Parkinsons	5	**87.23** ± 2.24	85.84 ± 1.98	+	84.51 ± 2.70	+	84.09 ± 2.31	+	84.32 ± 2.39	+
	10	**86.89** ± 1.72	85.05 ± 2.34	+	84.36 ± 2.41	+	84.10 ± 2.62	+	84.52 ± 2.43	+
	20	**86.51** ± 2.10	85.47 ± 2.04	+	84.11 ± 2.64	+	84.09 ± 2.50	+	83.86 ± 2.26	+
	30	**86.79** ± 1.87	84.96 ± 2.50	+	83.90 ± 1.91	+	83.28 ± 2.35	+	83.16 ± 2.03	+
	40	**86.66** ± 2.11	85.42 ± 2.11	+	83.07 ± 2.54	+	83.13 ± 2.70	+	83.77 ± 2.93	+
	50	**86.68** ± 2.08	85.03 ± 2.09	+	83.19 ± 3.41	+	83.06 ± 3.41	+	83.15 ± 2.00	+
Robot	5	**36.21** ± 1.91	32.72 ± 2.16	+	31.97 ± 1.91	+	31.82 ± 2.11	+	32.53 ± 1.93	+
	10	**35.12** ± 2.11	33.10 ± 2.11	+	32.09 ± 1.63	+	32.36 ± 1.81	+	32.24 ± 1.95	+
	20	**35.87** ± 1.75	32.54 ± 1.96	+	33.54 ± 2.01	+	33.54 ± 2.02	+	33.39 ± 2.08	+
	30	**35.44** ± 1.92	33.67 ± 2.08	+	34.14 ± 2.19	+	34.14 ± 2.19	+	33.65 ± 1.92	+
	40	**36.69** ± 2.61	35.18 ± 2.01	+	34.60 ± 2.04	+	34.60 ± 2.04	+	**35.93** ± 1.90	=
	50	**38.39** ± 2.13	36.60 ± 1.63	+	33.82 ± 2.28	+	33.82 ± 2.28	+	35.66 ± 2.49	+
Sonar	5	**74.97** ± 3.04	72.96 ± 2.63	+	72.65 ± 3.00	+	**74.15** ± 2.72	=	72.68 ± 2.77	+
	10	**74.11** ± 3.20	72.60 ± 3.15	+	72.20 ± 2.78	+	**72.79** ± 2.93	=	72.19 ± 2.66	+
	20	73.94 ± 3.48	73.94 ± 3.34	+	71.58 ± 2.82	+	71.44 ± 2.77	+	**72.56** ± 2.76	=
	30	72.23 ± 3.24	**72.74** ± 2.43	=	70.94 ± 2.71	=	70.94 ± 2.71	=	**71.22** ± 3.19	=
	40	73.20 ± 4.17	72.49 ± 3.85	=	69.31 ± 3.82	+	69.31 ± 3.82	+	**71.70** ± 2.72	=
	50	73.71 ± 3.58	**72.85** ± 3.01	=	68.25 ± 3.25	+	68.25 ± 3.25	+	70.23 ± 3.55	+

was significantly more accurate, "=" means not significantly different, and "−" means significantly less accurate.

4.1   Classification Performance

It is clear from Table 2 that with the first five datasets containing natural missing values, C4.5FS achieves significantly better classification performance than other methods on the first four datasets, similar classification performance to C4.5 and C4.5MI on Wisconsin dataset and significantly worse classification performance to C4.5KNNI and C4.5EMI on Wisconsin dataset.

Figure 4 summarises the results from Table 3. It is clear from Fig. 4 that with artificial missing datasets, C4.5FS often achieves significantly better or at least similar classification accuracy to the other methods. C4.5FS has more times

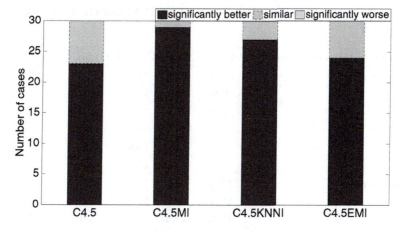

Fig. 4. Comparison of C4.5FS with C4.5, C4.5MI, C4.5KNNI and C4.5EMI

Table 4. Tree size of C4.5FS, C4.5, C4.5MI, C4.5KNNI and C4.5EMI on datasets containing natural missing values

Dataset	C4.5FS	C4.5	C4.5MI	C4.5KNNI	C4.5EMI
Cleveland	**32.7**	79.0	81.8	81.6	81.9
Hepatitis	**10.3**	17.3	19.8	21.3	18.6
Marketing	**304.8**	1367.1	1720.3	1665.2	1717.4
Ozone	**13.6**	24.8	29.4	30.7	30.1
Wisconsin	**15.8**	22.8	24.0	22.3	22.4

achieving significantly better than C4.5MI and followed by C4.5KNNI, C4.5EMI and C4.5.

In summary, with both natural and artificial missing datasets, in most cases, feature selection for missing data can help improve classification accuracy of C4.5.

4.2   Size of the Learned Models

According to Table 4, with the first five datasets containing natural missing values, in all cases, C4.5FS generates smaller decision trees than other methods. For example, in Marketing dataset, sizes of decision trees generated by C4.5FS are nearly one fifth the sizes of decision trees generated by C4.5 and more than one fifth of sizes of decision trees generated by using imputation methods before using C4.5.

Figure 5 shows minimum, average and maximum of ratio of tree size of C4.5, C4.5MI, C4.5KNNI and C4.5EMI with C4.5FS from Table 5. The minimum of ratio of tree sizes of the other methods with C4.5FS show that C4.5FS generates smaller trees than other methods. On average, C4.5 generates about 30 % bigger

Table 5. Tree size of C4.5FS, C4.5, C4.5MI, C4.5KNNI and C4.5EMI with several missing rates

Dataset	Missing amount (%)	C4.5FS	C4.5	C4.5MI	C4.5KNNI	C4.5EMI
Climate	5	**10.4**	23.8	25.4	25.7	24.8
	10	**9.9**	21.1	27.4	26.5	25.9
	20	**7.2**	14.5	26.5	24.8	23.9
	30	**5.6**	10.2	28.0	25.3	22.7
	40	**4.8**	8.7	28.6	27.7	22.3
	50	**3.8**	6.5	25.7	25.0	22.0
Ionosphere	5	**17.5**	24.3	26.0	26.0	25.9
	10	**17.4**	24.2	25.4	25.6	25.3
	20	**17.3**	23.4	25.8	25.9	25.1
	30	**17.7**	22.8	26.8	26.5	25.6
	40	**18.5**	22.5	28.3	28.2	25.5
	50	**18.2**	20.7	27.7	27.7	25.6
Parkinsons	5	**15.0**	17.8	19.0	18.7	18.8
	10	**15.5**	17.9	18.7	19.2	18.5
	20	**15.4**	17.8	19.9	19.5	18.7
	30	**15.0**	17.0	19.6	19.5	18.8
	40	**13.9**	15.8	19.7	19.6	18.3
	50	**13.3**	14.6	19.3	19.2	18.7
Robot	5	**63.6**	71.3	118.4	106.6	100.9
	10	**69.4**	76.0	133.9	133.9	120.7
	20	**73.1**	86.4	131.7	131.7	129.2
	30	**74.6**	85.4	129.2	129.2	126.8
	40	**70.8**	79.4	128.4	128.4	125.0
	50	**63.7**	73.1	129.2	129.2	121.3
Sonar	5	**25.1**	27.7	28.0	27.5	27.9
	10	**25.4**	28.1	28.7	28.3	27.9
	20	**24.8**	28.4	29.5	29.5	27.6
	30	**23.2**	27.5	30.3	30.3	28.2
	40	**22.4**	26.7	30.6	30.7	28.7
	50	**22.7**	26.3	31.8	32.0	29.2

than those generated by C4.5FS, and the other three methods generate trees over twice bigger than those of C4.5FS. Especially, the maximum of ratio of tree sizes of the other methods with C4.5FS shows that sizes of decision trees generated by using imputation methods before using classifier in some cases are dramatically bigger than C4.5FS. The main reason is likely that imputation methods often

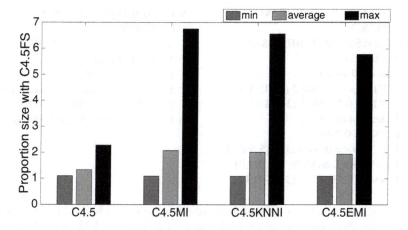

Fig. 5. Ratio tree size of C4.5, C4.5MI, C4.5KNNI and C4.5EMI with C4.5FS

generate further values for missing features; therefore, if the missing features are chosen to build decision trees, the further values make decision trees bigger.

In summary, with both natural and artificial missing data, in all cases, feature selection for missing data can help reduce complexity of the classification model generated by C4.5, especially compared to using imputation methods before using C4.5.

4.3 Analysis

To give a better picture of how C4.5FS can achieve better classification and smaller trees than the other methods, we looked carefully at the trees generated by C4.5 and C4.5FS on Climate dataset which has 20 features {V1,..,V20}. Climate dataset was chosen since the trees generated on the Climate dataset are not too big to analyse. Figures 6 and 7 show two typical pattern trees we observed.

Figure 6 shows trees generated by C4.5 and C4.5FS on Climate dataset with 20 % missing values in 9 features {V2, V3, V4, V12, V13, V14, V15, V17, V19}. After applying feature selection on the dataset, only seven features {V1, V4, V11, V15, V16, V17, V20} were chosen. The C4.5FS tree achieved slightly higher classification accuracy compared to the C4.5 tree, with 90.95 % and 89.91 %, respectively. Both of them had the same features in the top part of the trees. However, in the bottom part, the C4.5 tree had additional features which were not present in the C4.5FS tree because these features had been removed in the feature selection procedure. As a result, the C4.5FS achieved both better classification accuracy and a smaller tree than the C4.5.

Figure 7 shows trees generated by C4.5 and C4.5FS on Climate dataset with 20 % missing values in 10 features {V1, V4, V5, V7, V9, V10, V13, V14, V16, V19}. After applying feature selection on the dataset, only seven features {V7,

```
V4 <= 0.581307: 2 (290.98/7.18)              V4 <= 0.581307: 2 (290.98/7.18)
V4 > 0.581307                                V4 > 0.581307
|  V15 <= 0.515342: 2 (107.95/8.08)          |  V15 <= 0.515342: 2 (107.95/8.08)
|  V15 > 0.515342                            |  V15 > 0.515342
|  |  V16 <= 0.449815                         |  |  V16 <= 0.449815
|  |  |  V17 <= 0.24368: 2 (11.15/1.3)        |  |  |  V17 <= 0.24368: 2 (11.15/1.3)
|  |  |  V17 > 0.24368: 1 (30.78/11.19)       |  |  |  V17 > 0.24368: 1 (30.78/11.19)
|  |  V16 > 0.449815                          |  |  V16 > 0.449815: 2 (46.15/5.85)
|  |  |  V9 <= 0.279653
|  |  |  |  V19 <= 0.644323: 1 (5.66/1.21)
|  |  |  |  V19 > 0.644323: 2 (7.7/1.0)
|  |  |  V9 > 0.279653: 2 (32.79/0.4)
```

Fig. 6. Left tree generated by C4.5 and right tree generated by C4.5FS on Climate dataset with 20 % missing values in features {V2, V3, V4, V12, V13, V14, V15, V17, V19}

```
V3 <= 315: 2 (283.0/3.0)                     V16 <= 0.449517
V3 > 315                                     |  V15 <= 0.933518: 2 (196.78/21.27)
|  V15 <= 0.515342: 2 (109.0/8.0)            |  V15 > 0.933518
|  V15 > 0.515342                            |  |  V13 <= 0.245317: 2 (4.83/0.86)
|  |  V16 <= 0.450556                         |  |  V13 > 0.245317
|  |  |  V1 <= 1                              |  |  |  V8 <= 0.725071
|  |  |  |  V16 <= 0.144613: 2 (5.03/0.55)    |  |  |  |  V18 <= 0.65047: 1 (4.18/0.75)
|  |  |  |  V16 > 0.144613: 1 (12.51/2.29)    |  |  |  |  V18 > 0.65047: 2 (3.22)
|  |  |  V1 > 1: 2 (29.96/11.73)              |  |  |  V8 > 0.725071: 1 (4.71)
|  |  V16 > 0.450556                         V16 > 0.449517: 2 (273.27/11.73)
|  |  |  V11 <= 0.239041
|  |  |  |  V16 <= 0.802212: 1 (5.4/1.2)
|  |  |  |  V16 > 0.802212: 2 (3.6/0.8)
|  |  |  V11 > 0.239041: 2 (38.5/3.5)
```

Fig. 7. Left tree generated by C4.5 and right tree generated by C4.5FS on Climate dataset with 20 % missing values in features {V1, V4, V5, V7, V9, V10, V13, V14, V16, V19}

V8, V9, V13, V15, V16, V18} were chosen. In C4.5, when computing the information gain of a feature containing missing values, it computes the gain on the complete values and discounts it by the ratio of complete instances to all instances [25]. In other words, missing values discount information gain of missing features. Therefore, C4.5 biases towards choosing complete features to build decision trees, but the bias of choosing complete features to build decision trees is not always good. For example, on Fig. 7, while the first node of C4.5 tree is a complete feature V3, the first node of C4.5FS tree is a missing feature V16. However, the C4.5FS tree achieved both better classification accuracy (91.3 % vs 90.1 %) and smaller tree than the C4.5 tree. A possible reason could be that by removing less suitable features such as V3, feature selection helps to counteract the C4.5's bias towards choosing complete features to build decision trees.

In summary, feature selection is able to choose relevant features and remove irrelevant features. Therefore, feature selection helps to build better classifier.

5   Conclusions and Future Work

This paper presents research which has attempted to find the effect of a wrapper feature selection approach to classification with missing data. To undertake the research, three different experimental setups were designed: classification with missing data by using a classifier that is able to classify missing data, classification with missing data by using an imputation method before applying a classifier, and classification with missing data by using feature selection before using a classifier that is able to classify with missing data. The results from the three setups were compared on 10 datasets (five dataset containing natural missing values and five datasets with six levels of artificial missing values), using C4.5 for an evaluation and PSO as a search technique for feature selection. The empirical results showed that a wrapper feature selection approach to classification with missing data can help to improve classification performance of C4.5 and reduce the complexity of the learned classifier.

The experiment in this paper used C4.5 as a classifier because it can handle missing data. There are some other classifiers that are able to classify missing data such as CART [8] and CN2 [5]. Future work could repeat this investigation with CART and CN2. This paper used a wrapper-based approach to feature selection on classification problems with missing data. In [9], a filter-based approach to feature selection was applied to regression problems with missing data. The future work could explore the effectiveness of a filter-based approach to feature selection on classification problems with missing data.

References

1. Asuncion, A., Newman, D.: UCI machine learning repository (2007)
2. Barnard, J., Meng, X.-L.: Applications of multiple imputation in medical studies: from AIDS to NHANES. Stat. Methods Med. Res. **8**, 17–36 (1999)
3. Batista, G.E., Monard, M.C.: A study of K-nearest neighbour as an imputation method. In: HIS, vol. 87, pp. 251–260 (2002)
4. Chuang, L.-Y., Chang, H.-W., Tu, C.-J., Yang, C.-H.: Improved binary PSO for feature selection using gene expression data. Comput. Biol. Chem. **32**, 29–38 (2008)
5. Clark, P., Niblett, T.: The CN2 induction algorithm. Mach. Learn. **3**, 261–283 (1989)
6. Clerc, M., Kennedy, J.: The particle swarm-explosion, stability, and convergence in a multidimensional complex space. IEEE Trans. Evol. Comput. **6**, 58–73 (2002)
7. Dash, M., Liu, H.: Feature selection for classification. Intell. Data Anal. **1**, 131–156 (1997)
8. De'ath, G., Fabricius, K.E.: Classification and regression trees: a powerful yet simple technique for ecological data analysis. Ecology **81**, 3178–3192 (2000)

9. Doquire, G., Verleysen, M.: Feature selection with missing data using mutual information estimators. Neurocomputing **90**, 3–11 (2012)
10. Farhangfar, A., Kurgan, L., Dy, J.: Impact of imputation of missing values on classification error for discrete data. Pattern Recogn. **41**, 3692–3705 (2008)
11. García-Laencina, P.J., Sancho-Gómez, J.-L., Figueiras-Vidal, A.R.: Pattern classification with missing data: a review. Neural Comput. Appl. **19**, 263–282 (2010)
12. Graham, J.W.: Missing data analysis: making it work in the real world. Annu. Rev. Psychol. **60**, 549–576 (2009)
13. Hall, M., Frank, E., Holmes, G., Pfahringer, B., Reutemann, P., Witten, I.H.: The WEKA data mining software: an update. ACM SIGKDD Explor. Newsl. **11**, 10–18 (2009)
14. Han, J., Kamber, M., Pei, J.: Data Mining, Southeast Asia Edition: Concepts and Techniques. Morgan Kaufmann, Burlington (2006)
15. Jain, A., Zongker, D.: Feature selection: evaluation, application, and small sample performance. IEEE Trans. Pattern Anal. Mach. Intell. **19**, 153–158 (1997)
16. Kennedy, J.: Particle swarm optimization. In: Encyclopedia of Machine Learning, pp. 760–766 (2010)
17. Kennedy, J., Kennedy, J.F., Eberhart, R.C.: Swarm Intelligence. Morgan Kaufmann, San Francisco (2001)
18. Kohavi, R., John, G.H.: Wrappers for feature subset selection. Artif. Intell. **97**, 273–324 (1997)
19. Koller, D., Sahami, M.: Toward optimal feature selection (1996)
20. Lin, S.-W., Ying, K.-C., Chen, S.-C., Lee, Z.-J.: Particle swarm optimization for parameter determination and feature selection of support vector machines. Expert Syst. Appl. **35**, 1817–1824 (2008)
21. Little, R.J., Rubin, D.B.: Statistical Analysis with Missing Data. Wiley, New York (2014)
22. Luengo, J., García, S., Herrera, F.: A study on the use of imputation methods for experimentation with radial basis function network classifiers handling missing attribute values: the good synergy between rbfns and eventcovering method. Neural Netw. **23**, 406–418 (2010)
23. MacKay, D.J.: Information theory, inference, and learning algorithms, vol. 7. Citeseer (2003)
24. Oh, I.-S., Lee, J.-S., Moon, B.-R.: Hybrid genetic algorithms for feature selection. IEEE Trans. Pattern Anal. Mach. Intell. **26**, 1424–1437 (2004)
25. Quinlan, J.R.: C4.5: Programs for Machine Learning. Elsevier, New York (2014)
26. Schafer, J.L.: Analysis of Incomplete Multivariate Data. CRC Press, Boca Raton (1997)
27. Schafer, J.L., Graham, J.W.: Missing data: our view of the state of the art. Psycholog. Meth. **7**, 147 (2002)
28. Wang, X., Yang, J., Teng, X., Xia, W., Jensen, R.: Feature selection based on rough sets and particle swarm optimization. Pattern Recogn. Lett. **4**, 459–471 (2007)
29. Xue, B.: Particle Swarm Optimisation for Feature Selection in Classification. Victoria University of Wellington (2014)
30. Xue, B., Zhang, M., Browne, W.N.: Particle swarm optimization for feature selection in classification: a multi-objective approach. IEEE Trans. Cybern. **43**, 1656–1671 (2013)

Bare-Bone Particle Swarm Optimisation for Simultaneously Discretising and Selecting Features for High-Dimensional Classification

Binh Tran[(✉)], Bing Xue, and Mengjie Zhang

School of Engineering and Computer Science, Victoria University of Wellington,
PO Box 600, Wellington 6140, New Zealand
{binh.tran,bing.xue,mengjie.zhang}@ecs.vuw.ac.nz

Abstract. Feature selection and discretisation have shown their effectiveness for data preprocessing especially for high-dimensional data with many irrelevant features. While feature selection selects only relevant features, feature discretisation finds a discrete representation of data that contains enough information but ignoring some minor fluctuation. These techniques are usually applied in two stages, discretisation and then selection since many feature selection methods work only on discrete features. Most commonly used discretisation methods are univariate in which each feature is discretised independently; therefore, the feature selection stage may not work efficiently since information showing feature interaction is not considered in the discretisation process. In this study, we propose a new method called PSO-DFS using bare-bone particle swarm optimisation (BBPSO) for discretisation and feature selection in a single stage. The results on ten high-dimensional datasets show that PSO-DFS obtains a substantial dimensionality reduction for all datasets. The classification performance is significantly improved or at least maintained on nine out of ten datasets by using the transformed "small" data obtained from PSO-DFS. Compared to applying the two-stage approach which uses PSO for feature selection on the discretised data, PSO-DFS achieves better performance on six datasets, and similar performance on three datasets with a much smaller number of features selected.

Keywords: Particle swarm optimisation · Feature discretisation · Feature selection · Classification · High-dimensional data

1 Introduction

Feature selection is an important technique in data preprocessing, especially for datasets with thousands to tens of thousands of features. High-dimensional datasets, such as text, image and gene expression data, are automatically collected by machines. Therefore, they usually contain a significant number of irrelevant and redundant features, which negatively affects not only the learning process but also the system memory. To deal with this problem, feature selection has been applied to select only informative features. Results of many studies

© Springer International Publishing Switzerland 2016
G. Squillero and P. Burelli (Eds.): EvoApplications 2016, Part I, LNCS 9597, pp. 701–718, 2016.
DOI: 10.1007/978-3-319-31204-0_45

have shown the effectiveness of applying feature selection [1,2] in general as well as on high-dimensional data [3,4]. Existing feature selection methods can be generally classified into filter and wrapper approaches [5]. Filter approaches evaluate features based on their intrinsic characteristics. On the other hand, wrapper approaches use a learning algorithm to measure the classification performance of the selected features. Although filters are said to be faster than wrappers, they usually obtain lower classification accuracy than wrappers.

Besides feature selection, feature discretisation is also important in preprocessing data especially for high-dimensional data because of the following important reasons. Firstly, many commonly used machine learning techniques can only be applied or work efficiently on discrete data. High-dimensional datasets usually include continuous features that are automatically collected at the interval or ratio level such as images and gene expression data. Therefore, continuous or real-value features are required to be partitioned or discretised into a number of sub-ranges. Each sub-range is considered as a category. This process is called discretisation. Secondly, discretisation techniques aim at finding a discrete representation of each feature so that it contains enough information for the learning task while eliminating the minor fluctuations that may be noisy in the original data [6]. The report in [7] showed that feature discretisation helps learning be more accurate and faster. Finally, discrete features are more compact than continuous ones. Discretisation can reduce a significant amount of memory required to store data, enabling learning algorithms to work efficiently. As a result, feature selection and feature discretisation have been used as popular preprocessing techniques to obtain more compact datasets that are better representatives of the learning task. The use of these techniques has shown to improve classification performance, memory requirements and computation time [8].

Although many discretisation methods have been proposed, most of them are univariate, which means that only one continuous feature is discretised at a time. For the sake of efficiency, these methods work with an assumption that each feature independently influences the task. However, this assumption is not valid in problems in which feature interdependency occurs [9]. In these cases, multivariate discretisation is needed to consider multiple features at a time. However, it would certainly increase the time complexity for discretisation. Therefore, it is necessary to have an efficient search technique that can simultaneously discretise multiple features in a reasonable time.

A common practice using feature selection and discretisation in data preprocessing is to apply discretisation before the selection process. However, when using data discretised by a univariate discretisation method, feature selection methods may miss relevant features since information showing feature interaction may be destroyed in the discretisation process. Therefore, combining these two processes into a single stage may obtain better representation for the learning task.

Particle swarm optimisation (PSO), proposed by Kennedy and Eberhart [10], is a meta-heuristic algorithm inspired by social behaviours found in birds flocking. Each particle flies from one position to another in the problem search space with its own velocity. PSO has been applied to different tasks including feature selection using different versions such as continuous PSO [11,12] and binary PSO [13–15]. However, the performance of these PSO versions strongly depend on the control parameters for particles' velocity such as inertia weight, acceleration coefficients and velocity clamping [16]. Although many adaptive and dynamic methods have been proposed to overcome this problem, it is still challenging since setting these parameters depends on individual applications and needs to be adjusted for different problems. In 2003, Kennedy [17] suggested updating particles' position using a simple Gaussian sampling around the mean of the individual best position and its neighbours' best position. This "bare-bone" PSO (BBPSO) eliminates the use of velocity and all of the parameters mentioned above. The performance of this new method on function optimisation problems has shown to be superior than the canonical one. This method also attained promising results in [18] which proposed binary BBPSO for feature selection. While PSO has been widely used for feature selection, it has not been applied to feature discretisation and selection at the same time.

Goals

The aim of this study is to propose a new approach to the use of BBPSO for simultaneously discretising features and selecting relevant features in a single stage for high-dimensional continuous data. For presentation convenience, we call the new method PSO-DFS. PSO-DFS will be examined and compared with using all original feature set and the corresponding two-stage approach (feature discretisation and then feature selection) on ten public available high-dimensional datasets of varying difficulty. More specifically, we would like to investigate the following research objectives:

1. How to discretise multiple features simultaneously in BBPSO so that the discriminating power of the feature set is improved;
2. How to perform discretise and select features in a single stage;
3. Whether the features generated by PSO-DFS can produce better classification performance than using all features;
4. Whether PSO-DFS can outperform the corresponding two-stage approach in terms of classification accuracy, the number of features and the computation time.

2 Background and Related Work

2.1 Feature Discretisation

In order to discretise a continuous feature into a discrete one, a discretisation method determines the number of intervals and the corresponding cut-points for each interval. A large number of discretisation methods can be found in the literature. They can be categorised based on different axes [19–21] such as *direct* versus *incremental*, *supervised* versus *unsupervised*. While *direct* methods determine these intervals based on a user-defined parameter, *incremental* methods apply some criteria to further split or merge intervals forming *splitting* or *merging* methods, respectively. Methods which use class labels in the discretisation process are *supervised*; otherwise, they are *unsupervised*. Discretisation methods can also be categorised into *global* or *local* based on whether the entire instance space or a subset of instances is used in each discretisation step. While in *dynamic* methods, the discretisation process is done while the learner is building the model, *static* methods separate these two processes. Discretisation methods are also categorized into *univariate* where each feature is discretised independently and *multivariate* where multiple features are discretised at the same time so that feature interaction is also considered in the discretisation process [19].

Two simple unsupervised discretisation methods are equal-width and equal-frequency binning methods, which require a user-defined number of intervals m. While the former method partitions features into m intervals with the same width, the latter partitions features into m intervals that have the same number of instances. Although these binning methods are simple and easy to implement, they are sensitive to a given m which is usually unknown. They may not give good results on non-uniform distribution features and features with outliers, which are extreme values that strongly affect the ranges [22].

To overcome the shortcoming of unsupervised methods, supervised discretisation takes into account the interdependence between the discrete values and their class labels. Different ways of using class labels to find cut-points with higher class coherence have been proposed. A simple example is 1R [23] where cut-points have to lie between sorted instances of different classes and each bin has at least six instances except the right most bin. Many ways of evaluating a cut-point are proposed based on information theory [24,25], statistical measures [26,27], classification error rates [28,29], etc. Readers are referred to more comprehensive review in [19–21,30].

2.2   Minimum Description Length

Fayyad and Irani's minimum description length (MDL) [25] is a univariate incremental splitting discretisation method which uses the minimum description length principle (MDLP) as the stopping criterion. The algorithm starts with one interval containing all values of the feature and recursively partitions this interval until the criterion is met.

In each discretisation step, a cut-point is chosen to partition the corresponding interval into two sub-intervals. The algorithm considers all candidate cut-points which lie between instances of different classes. The best cut-point is the one with the highest information gain. Given S as the set of instances, T is a candidate cut-point of Feature A, S_1 and S_2 are the resulting subsets after partitioning S by T, information gain of T is calculated based on Eq. (1). A cut-point is only accepted if its information gain satisfies the MDLP criterion as shown in Eq. (2), where details about this equation can be seen from [25].

$$Gain(T, A; S) = E(S) - \frac{|S_1|}{|S|} E(S_1) - \frac{|S_2|}{|S|} E(S_2) \tag{1}$$

$$Gain(T, A; S) > \frac{log_2(|S| - 1)}{|S|} + \frac{\delta(T, A; S)}{|S|} \tag{2}$$

where

$$\delta(T, A; S) = log_2(3^k - 2) - [k_S E(S) - k_{S_1} E(S_1) - k_{S_2} E(S_2)] \tag{3}$$

and $|S|$ is the number of instances in the given set S, $E(S)$ is the entropy of S, and k_S is the number of classes appeared in S.

If the best cut-point of an interval is accepted by the MDLP criterion, then a recursive discretisation step is applied to each new sub-interval; otherwise, the discretisation process stops.

2.3 Feature Selection via Discretisation

Chi2 [31] is one of the first methods proposing selecting features via discretisation. It is an improvement of the ChiMerge [32] method. ChiMerge is a bottom up method which starts with each interval having one distinct value of the feature. In each iteration, it merges the pair of adjacent intervals with the lowest χ^2 test result. The merging process is continued until all pairs of intervals have χ^2 values exceeding the parameter determined by a predefined significant level. Chi2 has two phases. Phase 1 is a general version of ChiMerge. Instead of using a predefined significant level, Chi2 automatically determines this value from the data by gradually decreasing the significant level from 0.5. Consistency is used as a stopping criterion. After this first phase, each feature has a different significant level. Starting with the significant level determined in phase 1, phase 2 is a fining process used to further merge features in a round robin fashion until the inconsistency of the data above a given limit. At the end, if all intervals of a feature were merged into one, that feature was discarded.

PEAR [33] is also a supervised and univariate discretisation method. It performs simultaneously feature discretisation and selection. In this method, cut-points are chosen for a feature if they lie between instances of different classes and produces intervals in which the majority class has at least a predefined number of instances called minperint. Furthermore, two consecutive intervals should not have the same majority class occurrence and the ratio between these occurrences needs to be higher than a predefined ratio called mintofuse. Then, features are ranked based on the number of cut-points. Features with small numbers of cut points are considered as relevant features and therefore selected. Result on a medical image dataset using 17 % best features from a total of 140 showed that it maintained the precision of the full feature set. Its result was better than Relief [34] with the same number of features. However, domain knowledge or a significant number of trials need to be done to choose appropriate values for the parameters.

2.4   Particle Swarm Optimisation

PSO [10] is a population-based algorithm proposed by Kennedy and Eberhart in 1995. In this section, we will describe the standard continuous PSO and its variance, the bare-bone PSO [17].

Continuous PSO. PSO maintains a swarm of particles. These particles "fly" from one position to another in the search space based on the information shared by each other to find better solutions. The solution is represented in the position which is a vector of D real numbers, where D is the dimensionality of the problem. Each particle also has another vector of the same size called velocity showing the speed and direction that the particle should move in each dimension. At each iteration, velocity and position of a particle are adjusted based on the two best positions, one is the best position it has explored so far called *pbest* and the other best is shared from its neighbours called *gbest*. Equations (4) and (5) are used to update these vectors.

$$v_{id}^{t+1} = w * v_{id}^t + c_1 * r_{1i} * (p_{id}^t - x_{id}^t) + c_2 * r_{2i} * (p_{gd}^t - x_{id}^t) \tag{4}$$

$$x_{id}^{t+1} = x_{id}^t + v_{id}^{t+1} \tag{5}$$

where v_{id}^t and x_{id}^t are velocity and position of particle i in dimension d at time t, respectively. p_{id} and p_{gd} are *pbest* and *gbest* positions in dimension d. c_1 and c_2 are acceleration constants, and r_1 and r_2 are random values uniformly distributed in $[0, 1]$. These constants are important to control the behaviour of the particle [16]. It determines the type of trajectory the particle travels. w is the inertia weight controlling the impact of the last velocity to the current velocity.

Bare-Bone PSO. From Eqs. (4) and (5), we can see that PSO operates by sampling points on the search space. It uses discovered knowledge to guide the search. The position updating reflects how particles select the next point to explore in the search space. To investigate the trajectory of the particle swarm, Kennedy [17] plotted all the points that were visited in one million iterations of a standard PSO where *pbest* and *gbest* were set as constants. The obtained histogram is a tidy bell curve centered midway between these two best positions. This observation suggests that the difference between these two best points, *pbest* and *gbest*, is an important parameter for scaling the amplitude of particle's trajectory. The step size for particles' movement should be a function of consensus between these two points. Therefore, Kennedy [17] proposed bare-bone PSO (BBPSO) which uses a Gaussian sampling based on *pbest* and *gbest* to update particle positions as follows:

$$x_{id}^{t+1} = \begin{cases} \mathcal{N}\left(\frac{p_{id}^t + g_{id}^t}{2}, |p_{id}^t - g_{id}^t|\right), & rand() < 0.5 \\ p_{id}^t, & \text{otherwise} \end{cases} \tag{6}$$

In Eq. (6), $\mathcal{N}(\mu, \sigma)$ is a random number generator using Gaussian distribution with the mean μ centered between *pbest* and *gbest* and the standard deviation σ equal to the absolute difference between them. A probability *rand*() is used to retain the previous best position *pbest* in order to speed up convergence.

BBPSO eliminates the velocity component of the canonical PSO algorithm. The advantages of this strategy is not only that PSO does not need to optimize another D-dimensional vector (the velocity), but also there is no lag between when an adaptation is needed and when it occurs.

Both PSO and BBPSO have been used for feature selection, which can be seen from [1,18,35] (not detailed here due to the page limit). However, they have not been used for feature discretisation or simultaneously feature discretisation and feature selection, which could potentially improved the performance.

3   The Proposed Approach

This section describes the proposed approach for simultaneously feature discretisation and feature selection. To achieve these two tasks, a key component needs to be designed is the particle representation which represents a candidate solution for feature discretisation and selection, and also requires a new updating mechanism.

PSO Representation. As a binary discretisation method, our method evolves one cut-point for each feature. Therefore, a candidate solution, i.e. a particle's position, is encoded as a vector of length D, which is the number of original features in the dataset. Each element in the vector represents a cut-point for the corresponding feature. Therefore, each particle's position is a vector of real numbers (i.e. cut-points) whose values need to be in the range $[Min..Max]$ of the corresponding feature values. Figure 1 shows an example of a particle's position. During the updating process of the algorithm, if the updated value of a dimension corresponding to feature F is greater than Max_F, then it is set to Max_F. Similarly, it is set to Min_F if the updated value is smaller than Min_F.

To achieve *feature discretisation*, a feature's original continuous values are compared with the corresponding cut-point value in the particle's position vector, then these continuous values are converted/discretised to either 0 or 1

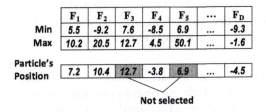

Fig. 1. Particle representation of PSO-DFS.

depending whether their continuous values are larger than the cut-point value or not. To achieve *feature selection*, we consider a feature to be relevant to the target concept if its discrete version has a better discriminating power indicated by an improvement in the classification performance of the learning algorithm. Therefore, if a feature is discretised to a single interval (i.e. all the original continuous values are converted to the same discrete value), which means that it is useless in differentiating instances of different classes, it can be considered irrelevant and should be discarded. In our method, if the cut-point of a feature equals to its minimum or maximum value, it is discretised into one interval. For example, in Fig. 1, Features F_3 and F_5 are not selected because the cut-point of F_3 equals to its maximum value and that of F_5 equals to its minimum value.

To *update the position*, we proposed to use the updating mechanism of the BBPSO instead of standard PSO that is usually used in PSO-based methods for feature selection. In most existing PSO-based feature selection methods, the PSO representation is a vector of real numbers whose values are all in the same range [0, 1] representing the probability to select features. A feature is selected if its probability is greater than a predefined threshold. Therefore, two evolved probabilities, one is slightly greater than the threshold and the other is significantly greater than the threshold, have the same effect on the solution, which may limit the performance of PSO for feature selection. In the new representation for feature discretisation and selection, a different evolved value leads to a different discrete feature. Therefore, the search needs to be fine tuned so that an appropriate value of cut-point can be found. Since BBPSO use a Gaussian random generator to explore new positions between the *pbest* and the *gbest*, it is likely to obtain this behaviour. The new position is sampled around the mean of *pbest* and *gbest* with a standard deviation equal to the absolute difference between them. Therefore, when the difference between these two bests is large, the variance enables particles to explore new regions in the space. When they are closer, the new position is limited to a smaller region around this mean.

PSO Initialisation. To speed up the evolutionary process, each particle's initial position is a random feature subset with the size restricted to 50 for two-class problems and 150 for multi-class problems. These values are taken as suggested in previous studies [36]. For each randomly selected feature, its cut-point is initialised using the best binary cut-point calculated based on MDLP (see Sect. 2.2) [25]. For those features that are not selected in the initial candidate solutions, their corresponding dimensions will be initialised to the corresponding maximum values. In the initialisation procedure, information gain of the best cut-point is used as a probability to choose features.

Fitness Function. The optimisation process is guided by the classification accuracy. To evaluate a particle, its evolved cut-points are used to discretise features in the training set. Features that are discretised into a single interval, which means all instances have the same feature value, will be discarded. Classification accuracy of the transformed training set is used to evaluate the performance of the particle. By evaluating the cut-points of multiple features together, the joint contribution of all cut-points is taken into account. Since many of these datasets are unbalanced data, fitness values are calculated based on the balanced classification accuracy [37] as follows:

$$fitness = \frac{1}{n} \sum_{i=1}^{n} \frac{TP_i}{|S_i|} \tag{7}$$

where n is the number of classes of the problem, TP_i is the number of correctly identified instances in class i and $|S_i|$ is the total number of instances in class i. Since there is no bias to any specific class, the weight here is set equally to $1/n$.

The Overall Approach. Figure 2 shows an overview of the proposed PSO approach to discretisation and feature selection in one stage. Figure 3 shows the two-stage approach (named PSO-FS) in which features are first discretised and then selected. In both systems, the input dataset is first divided into a training set and a test set. The training set is used to find the discretisation scheme as well as select relevant features. Based on the output scheme, the training and the test sets are transformed as inputs to the classification algorithm to evaluate the performance of both methods. The pseudo-code of PSO-DFS is presented in Algorithm 1.

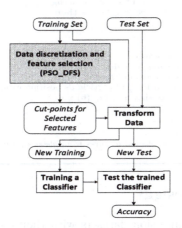

Fig. 2. Overview of the PSO-DFS system.

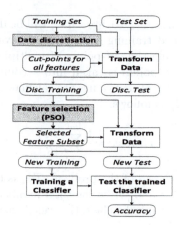

Fig. 3. Overview of the PSO-FS system.

Algorithm 1. The pseudo code of PSO-DFS

Input   : Training set with continuous data
Output: Cut-points for the selected features
begin
    Calculate the best binary cut-point for each feature based on the MDLP principle [25];
    Initialise particles;
    while *Maximum iterations or stopping criterion is not met* **do**
        for $i = 1$ *to Population Size* **do**
            $P_i \leftarrow$ Position of particle i;
            $Tr_i \leftarrow$ Transform training set based on P_i;
            $F_i \leftarrow$ Evaluate the accuracy of Tr_i using Eq. (7);
            if F_i *is better than pbest's fitness* **then**
                | Update *pbest* ;
            end
        end
        Update *gbest* of the swarm;
        for $i = 1$ *to Population Size* **do**
            for $j = 1$ *to Dimensionality* **do**
                | Update dimension j of particle i's position using BBPSO updating
                | mechanism as shown in Eq. (6) ;
            end
        end
    end
    Return cut-points for the selected features from *gbest*'s position;
end

---

4   Experiment Design

Datasets. Ten gene expression datasets with thousands of features are used to examine the performance of the proposed method on high-dimensional data. These datasets are publicly available on http://www.gems-system.org. Details about these datasets are shown in Table 1.

Baseline Methods. To test the effectiveness of PSO-DFS in data discretisation and feature selection, we compared the classification performance using the PSO-DFS transformed data and the original dataset. To see if combining both feature

Table 1. Datasets

Dataset	#Features	#Instances	#Classes	%Smallest class	%Largest class
SRBCT	2,308	83	4	13	35
DLBCL	5,469	77	2	25	75
9Tumor	5,726	60	9	3	15
Leukemia 1	5,327	72	3	13	53
Brain Tumor 1	5,920	90	5	4	67
Leukemia 2	11,225	72	3	28	39
Brain Tumor 2	10,367	50	4	14	30
Prostate	10,509	102	2	49	51
Lung Cancer	12,600	203	5	3	68
11Tumor	12,533	174	11	4	16

discretisation and feature selection in a single stage achieves better results than applying them in two stages, we also compared PSO-DFS with using PSO for feature selection on the discretised data i.e. the two-stage method (PSO-FS). In PSO-FS, data is first discretised by MDLP [25], then, PSO runs on the discrete data to find the best feature subset.

Parameter Settings and Termination Criteria. Table 2 shows the parameter settings used in the experiments for both PSO-FS and PSO-DFS. Because the numbers of features in the datasets are quite different, ranging from about two thousand to twelve thousand, the search spaces of these problems are very different. As a result, we set the population size proportional to the number of original features, i.e. $pop\_size = \#features/20$. However, due to the limitation of computer memory, this number is restricted to 300. The stopping criterion is either PSO reached a maximum iteration of 70 or the *gbest* does not improved after 10 iterations.

Table 2. Parameters for PSO

Parameters	Settings
Population size	#features/20 (restriction to 300)
Maximum iterations	70
Stopping criterion	*gbest* not improved for 10 iterations
Communication topology	Fully connected

Experiment Configuration. Since PSO-DFS is proposed as a wrapper method, classification performance of a specific classification algorithm will be used as a measure to evaluate the particles. In this study, we use k-nearest neighbour (KNN) with $k = 1$ as it is simple, non-parametric and used in previous papers on these datasets [4, 38].

Due to the small numbers of instances in these datasets, two loops of cross validation (CV) are used to avoid feature selection bias as suggested in [5]. The outer loop uses a stratified 10-fold CV on the whole dataset. One fold is kept as the unseen data to evaluate the performance of each method, and the remaining 9 folds are used to form the training set for feature discretisation and feature selection. In the fitness function, an inner loop of 10-fold CV on the training set is used to evaluate the evolved solution of each particle during the evolutionary process.

Since PSO is a stochastic method, 30 independent runs are executed with different random seeds for each method. Therefore, a totally 300 runs (30 runs × 10 fold CV) are executed for each method on each dataset. Experiment runs on PC with Intel Core i7-4770 CPU @ 3.4 GHz, running Ubuntu 4.6 and Java 1.7 with a total memory of 8 GB. The results of 30 runs from each method are compared using Wilcoxon test, a pair-wise statistical significance test, with the significance level 0.05.

5   Results and Discussions

Table 3 shows the experimental results of PSO-FS and PSO-DFS. "Full" means KNN using the original full set of continuous features. "Size" shows the average number of features selected by each method over the 30 runs. The best, the average and the standard deviation of the test accuracies achieved by each method on each dataset are displayed in the fourth and the fifth columns. The test accuracy is calculated using Eq. (7). The smallest subset size and the best classification accuracy in each dataset are bold. The last column displays the statistical Wilcoxon significance test results of the corresponding method over the proposed method. "+" or "−" means the result is significantly better or worse than the proposed method and "=" means they are similar in the Wilcoxon tests. In other words, the more "−", the better the proposed method.

5.1   PSO-DFS Versus Full

According to Table 3, the average number of features selected by PSO-DFS is always the smallest and significantly smaller than the total number of features. Less than 1 % of total number of features is selected in Prostate and DLBCL, 1 % to 3 % in other seven datasets and 6 % in SRBCT.

With the discretised and selected features, the classification performance is significantly improved over using all continuous features. An increase of more than 7 % in accuracy is achieved on five out of ten datasets and the highest improvement is 22 % on 9Tumor. PSO-DFS obtains a similar accuracy as using original features on Brain1 and Leuk2 and 2 % lower accuracy on Prostate with much smaller feature subsets (on average 54.93 features from 10,509 features).

Table 3. KNN average results over the 30 independent runs.

Dataset	Method	Size	Best	Mean ± Std	Sig. Test
SRBCT	Full	2,308		87.08	−
	PSO-FS	150.00	97.50	91.31 ± 2.71	−
	PSO-DFS	**137.25**	100.00	**96.89 ± 1.64**	
DLBCL	Full	5,469		83.00	−
	PSO-FS	101.84	96.67	80.03 ± 6.13	−
	PSO-DFS	**42.75**	94.17	**85.18 ± 5.46**	
9Tumor	Full	5,726		36.67	−
	PSO-FS	954.99	55.00	45.95 ± 4.93	−
	PSO-DFS	**138.54**	65.00	**58.22 ± 3.12**	
Leuk1	Full	5,327		79.72	−
	PSO-FS	150.00	92.22	81.60 ± 4.72	−
	PSO-DFS	**135.92**	95.56	**93.37 ± 1.83**	
Brain1	Full	5,920		72.08	=
	PSO-FS	317.34	78.75	71.00 ± 3.06	−
	PSO-DFS	**150.73**	79.17	**72.79 ± 3.48**	
Leuk2	Full	11,225		89.44	=
	PSO-FS	150.00	93.89	86.11 ± 3.97	−
	PSO-DFS	**139.94**	94.44	**89.93 ± 2.79**	
Brain2	Full	10,367		62.50	−
	PSO-FS	417.92	82.08	69.11 ± 5.89	=
	PSO-DFS	**152.78**	83.75	**70.76 ± 5.30**	
Prostate	Full	10,509		**85.33**	+
	PSO-FS	777.39	90.33	85.20 ± 2.35	=
	PSO-DFS	**54.93**	90.33	83.74 ± 3.55	
Lung	Full	12,600		78.05	−
	PSO-FS	686.23	85.73	**81.72 ± 2.08**	=
	PSO-DFS	**150.80**	85.58	80.60 ± 2.42	
11Tumor	Full	12,533		71.42	−
	PSO-FS	1,638.84	86.07	**82.62 ± 1.70**	+
	PSO-DFS	**149.93**	83.68	79.29 ± 2.11	

However, the best accuracies of PSO-DFS on these three datasets are still 7 %, 5 % and 5 % higher than using all features, respectively.

The results indicate that PSO-DFS can simultaneously discretise and select relevant features so that the discriminating power of the feature set is either significantly improved or maintained with a much smaller number of features.

5.2   PSO-DFS Versus PSO-FS

It can be observed in Table 3 that PSO-DFS always selects a much smaller number of features than PSO-FS. With the transformed features, PSO-DFS outperforms PSO-FS on six datasets with the highest improvement of about 12 % in accuracy on 9Tumor and Leuk1. While PSO-FS degrades the performance of KNN on DLBCL, PSO-DFS still attains a better performance than using the original feature set on this dataset with only half of the number of features. In general, PSO-DFS obtains the highest average accuracy in seven datasets.

The biggest difference between these two methods can be seen in the 9Tumor dataset. While PSO-FS selects 955 features to achieve an improvement of 9 % in classification accuracy, PSO-DFS improves 22 % accuracy with only 139 features. A similar pattern can be seen in the first six datasets. In Brain2, Prostate and Lung, both methods obtain a similar accuracy. However, PSO-DFS selects less than half of the number of features selected by PSO-FS. In Prostate and Lung, PSO-DFS selects only 54 and 150 features on average compared to 777 and 686 features selected by PSO-FS. In 11Tumor, PSO-DFS attains 3 % lower accuracy than PSO-FS with only 149 features while PSO-FS selects more than 1,600 features.

5.3   Further Analysis

To further analyse the performance of PSO-FS and PSO-DFS, we look at the training accuracy or fitness values of the returned solutions in the ten datasets. Table 4 shows the average fitness of the best solutions and its standard deviation obtained by both methods in the 30 runs.

Table 4. Fitness of the best solutions obtained by PSOFS and PSO-DFS.

Dataset	Method	Fitness	Dataset	Method	Fitness
SRBCT	PSO-FS	100.00 ± 0.00	Leuk2	PSO-FS	100.00 ± 0.00
	PSO-DFS	100.00 ± 0.00		PSO-DFS	100.00 ± 0.00
DLBCL	PSO-FS	100.00 ± 0.00	Brain2	PSO-FS	99.73 ± 0.12
	PSO-DFS	100.00 ± 0.00		PSO-DFS	98.38 ± 0.27
9Tumor	PSO-FS	97.49 ± 0.23	Prostate	PSO-FS	98.89 ± 0.10
	PSO-DFS	95.03 ± 0.22		PSO-DFS	98.56 ± 0.14
Leuk1	PSO-FS	100.00 ± 0.00	Lung	PSO-FS	97.77 ± 0.05
	PSO-DFS	100.00 ± 0.00		PSO-DFS	97.10 ± 0.14
Brain1	PSO-FS	100.00 ± 0.00	11Tumor	PSO-FS	99.80 ± 0.08
	PSO-DFS	99.33 ± 0.29		PSO-DFS	96.21 ± 0.19

It can be seen from the obtained fitness that both methods have converged to the optimal solutions in all runs on 4 datasets, namely SRBCT, DLBCL, Leuk1, and Leuk2. However, as seen in Table 3, PSO-DFS achieves significantly better test accuracy on these four datasets than PSO-FS. This indicates that the solutions evolved by PSO-DFS generalise better to unseen data than those of PSO-FS.

In the other six datasets, neither of the two methods achieves the optimal solutions for the training data. Since both methods stop running if *gbest* fitness does not improve after 10 iterations, they may need a better stopping criteria or a more effective mechanism to jump out of these local optima. In these problems, PSO-DFS obtains a slightly lower fitness with a much smaller feature subset than PSO-FS. The big difference between PSO-DFS and PSO-FS feature set sizes in these cases indicates that PSO-DFS might need to select more features to achieve a better performance.

Comparing the test accuracy in Table 3 and the training accuracy (or fitness) in Table 4, we can see that there is a big gap between these accuracies in most cases with the biggest difference in 9Tumor. This indicates that overfitting has occurred with different levels of effect in different datasets. The reason of this phenomenon is that the features' distribution in these datasets is very skew. Therefore, the training and test sets may have different distributions. The model learned from training data may not be generalised to the test data. This effect is worse in datasets with a small number of instances. Therefore, with only 50 and 60 instances, Brain2 and 9Tumor are the most affected cases. In addition, the class imbalance issue in these datasets also makes them challenging problems. With 9 classes, 9Tumor has worse results than Brain2 which has 4 classes.

In general, the results show that the proposed approach of combining data discretisation and feature selection in one stage performs better than separate these two steps in different stages. PSO-DFS can create a more compact and better discriminating representation for data than PSO-FS. This confirms our hypothesis that individually discretising features in the first stage of PSO-FS may lose important information including feature interaction. Since PSO-DFS evaluates the cut-points of all features simultaneously, such information is taken into account.

5.4   Computation Time

The average time in minutes to complete one run for PSO-FS and PSO-DFS is shown in Table 5. Using wrapper approach, both methods use the classification accuracy of KNN classifier running with 10-fold CV on the training data to guide the search. It is noticed that while PSO-FS only needs to transform the training set based on the selected features, PSO-DFS has to do both discretisation and selection in every particle's evaluation process. Therefore, its running time was expected to be higher than PSO-FS. However, the observation from Table 5 reflects an opposite trend. Compared to PSO-FS, PSO-DFS has a lower running time on eight datasets. PSO-DFS spends less than half of the running time used by PSO-FS on most datasets with only about one tenth on Lung and 11Tumor. A detailed inspection of the evolutionary process revealed that this is because PSO-DFS selected a significant smaller number of features than PSO-FS.

Table 5. Computation time (in minutes) of PSOFS and PSO-DFS.

Dataset	Method	Time	Dataset	Method	Time
SRBCT	PSO-FS	1.14	Leuk2	PSO-FS	7.85
	PSO-DFS	1.60		PSO-DFS	4.95
DLBCL	PSO-FS	3.69	Brain2	PSO-FS	6.95
	PSO-DFS	2.04		PSO-DFS	2.95
9Tumor	PSO-FS	16.74	Prostate	PSO-FS	30.23
	PSO-DFS	5.09		PSO-DFS	4.29
Leuk1	PSO-FS	3.01	Lung	PSO-FS	129.20
	PSO-DFS	3.61		PSO-DFS	17.01
Brain1	PSO-FS	10.14	11Tumor	PSO-FS	192.11
	PSO-DFS	5.09		PSO-DFS	13.00

6   Conclusions and Future Work

This paper proposes a new PSO-based method for feature discretisation and feature selection in a single stage. To achieve feature discretisation, a new PSO encoding scheme is proposed to evolve cut-points for multiple features simultaneously. Feature selection is accomplished by removing features that are discretised into only one interval. PSO-DFS is tested and compared with the two-stage approach, PSO-FS, in which features are individually discretised and then selected.

Experimental results on ten high-dimensional datasets show that PSO-DFS can effectively discretise multiple features to significantly improve or maintain the classification performance on most cases. Through discretisation, a much smaller number of relevant features is selected at the same time. Comparison between PSO-FS and PSO-DFS shows that conducting feature discretisation and feature selection in a single stage is more effective than applying these techniques in two different stages.

Although PSO-DFS obtained better solutions than PSO-FS for preprocessing data, it may still get stuck in local optima. Overcoming this problem enables PSO-DFS to achieve even better solutions. Multiple interval discretisation is another promising direction to investigate. Our future work will focus on these directions.

References

1. Chandrashekar, G., Sahin, F.: A survey on feature selection methods. Comput. Electr. Eng. **40**, 16–28 (2014)
2. Xue, B., Cervante, L., Shang, L., Browne, W., Zhang, M.: A multi-objective particle swarm optimisation for filter-based feature selection in classification problems. Connection Sci. **24**, 91–116 (2012)
3. Ferreira, A.J., Figueiredo, M.A.: Efficient feature selection filters for high-dimensional data. Pattern Recogn. Lett. **33**, 1794–1804 (2012)

4. Tran, B., Xue, B., Zhang, M.: Improved PSO for feature selection on high-dimensional datasets. In: Dick, G., et al. (eds.) SEAL 2014. LNCS, vol. 8886, pp. 503–515. Springer, Heidelberg (2014)
5. Kohavi, R., John, G.H.: Wrappers for feature subset selection. Artif. Intell. **97**, 273–324 (1997)
6. Ding, C., Peng, H.: Minimum redundancy feature selection from microarray gene expression data. J. Bioinf. Comput. Biol. **3**, 185–205 (2005)
7. Dougherty, J., Kohavi, R., Sahami, M., et al.: Supervised and unsupervised discretization of continuous features. In: Machine Learning: Proceedings of the Twelfth International Conference, vol. 12, pp. 194–202 (1995)
8. Ferreira, A.J., Figueiredo, M.A.: An unsupervised approach to feature discretization and selection. Pattern Recognit. **45**, 3048–3060 (2012)
9. Chao, S., Li, Y.: Multivariate interdependent discretization for continuous attribute. In: Third International Conference on Information Technology and Applications, vol. 1, pp. 167–172. IEEE (2005)
10. Eberhart, R., Kennedy, J.: A new optimizer using particle swarm theory. In: Proceedings of the Sixth International Symposium on Micro Machine and Human Science, pp. 39–43 (1995)
11. Xue, B., Zhang, M., Browne, W.: Particle swarm optimization for feature selection in classification: a multi-objective approach. IEEE Trans. Cybern. **43**, 1656–1671 (2013)
12. Xue, B., Zhang, M., Browne, W.N.: Particle swarm optimisation for feature selection in classification: novel initialisation and updating mechanisms. Appl. Soft Comput. **18**, 261–276 (2014)
13. Cervante, L., Xue, B., Zhang, M., Shang, L.: Binary particle swarm optimisation for feature selection: a filter based approach. In: IEEE Congress on Evolutionary Computation (CEC 2012), pp. 881–888 (2012)
14. Mohamad, M., Omatu, S., Deris, S., Yoshioka, M.: A modified binary particle swarm optimization for selecting the small subset of informative genes from gene expression data. Inf. Technol. Biomed. **15**, 813–822 (2011)
15. Zhou, W., Dickerson, J.A.: A novel class dependent feature selection method for cancer biomarker discovery. Comput. Biol. Med. **47**, 66–75 (2014)
16. Van den Bergh, F., Engelbrecht, A.P.: A study of particle swarm optimization particle trajectories. Inf. Sci. **176**, 937–971 (2006)
17. Kennedy, J.: Bare bones particle swarms. In: Proceedings of IEEE Swarm Intelligence Symposium (SIS 2003), pp. 80–87. IEEE (2003)
18. Zhang, Y., Gong, D., Hu, Y., Zhang, W.: Feature selection algorithm based on bare bones particle swarm optimization. Neurocomputing **148**, 150–157 (2015)
19. Garcia, S., Luengo, J., Sáez, J.A., Lopez, V., Herrera, F.: A survey of discretization techniques: taxonomy and empirical analysis in supervised learning. IEEE Trans. Knowl. Data Eng. **25**, 734–750 (2013)
20. Liu, H., Hussain, F., Tan, C.L., Dash, M.: Discretization: an enabling technique. Data Min. Knowl. Disc. **6**, 393–423 (2002)
21. Kotsiantis, S., Kanellopoulos, D.: Discretization techniques: a recent survey. GESTS Int. Trans. Comput. Sci. Eng. **32**, 47–58 (2006)
22. Catlett, J.: On changing continuous attributes into ordered discrete attributes. In: Kodratoff, Y. (ed.) EWSL 1991. LNCS, vol. 482, pp. 164–178. Springer, Heidelberg (1991)
23. Holte, R.C.: Very simple classification rules perform well on most commonly used datasets. Mach. Learn. **11**, 63–90 (1993)

24. Grzymala-Busse, J.W.: Discretization based on entropy and multiple scanning. Entropy **15**, 1486–1502 (2013)
25. Irani, K.B.: Multi-interval discretization of continuous-valued attributes for classification learning. Machine Learning (1993)
26. Cano, A., Nguyen, D.T., Ventura, S., Cios, K.J.: ur-CAIM: improved CAIM discretization for unbalanced and balanced data. Soft Comput. **20**, 173–188 (2014)
27. Yang, P., Li, J.S., Huang, Y.X.: Hdd: a hypercube division-based algorithm for discretisation. Int. J. Syst. Sci. **42**, 557–566 (2011)
28. Flores, J.L., Inza, I., Larrañaga, P.: Wrapper discretization by means of estimation of distribution algorithms. Intell. Data Anal. **11**, 525–545 (2007)
29. Ramirez-Gallego, S., Garcia, S., Benitez, J.M., Herrera, F.: Multivariate discretization based on evolutionary cut points selection for classification. IEEE Trans. Cybern. (2015)
30. Mahanta, P., Ahmed, H.A., Kalita, J.K., Bhattacharyya, D.K.: Discretization in gene expression data analysis: a selected survey. In: Proceedings of the Second International Conference on Computational Science, Engineering and Information Technology, pp. 69–75. ACM (2012)
31. Liu, H., Setiono, R.: Chi2: feature selection and discretization of numeric attributes. In: Proceedings of the Seventh International Conference on Tools with Artificial Intelligence, TAI 1995, p. 88. IEEE Computer Society (1995)
32. Kerber, R.: Chimerge: discretization of numeric attributes. In: Proceedings of the Tenth National Conference on Artificial Intelligence, pp. 123–128. AAAI Press (1992)
33. Sheela, J.L., Shanthi, D.V.: An approach for discretization and feature selection of continuous-valued attributes in medical images for classification learning. Int. J. Comput. Theory Eng. **1**, 154–158 (2009)
34. Kira, K., Rendell, L.A.: The feature selection problem: traditional methods and a new algorithm. In: Proceedings of the Tenth National Conference on Artificial Intelligence, pp. 129–134. AAAI Press (1992)
35. Tran, B., Xue, B., Zhang, M.: Overview of particle swarm optimisation for feature selection in classification. In: Dick, G., et al. (eds.) SEAL 2014. LNCS, vol. 8886, pp. 605–617. Springer, Heidelberg (2014)
36. Zhu, Z., Ong, Y.S., Dash, M.: Markov blanket-embedded genetic algorithm for gene selection. Pattern Recogn. **40**, 3236–3248 (2007)
37. Patterson, G., Zhang, M.: Fitness functions in genetic programming for classification with unbalanced data. In: Orgun, M.A., Thornton, J. (eds.) AI 2007. LNCS (LNAI), vol. 4830, pp. 769–775. Springer, Heidelberg (2007)
38. Chuang, L.Y., Chang, H.W., Tu, C.J., Yang, C.H.: Improved binary PSO for feature selection using gene expression data. Comput. Biol. Chem. **32**, 29–38 (2008)

Mutual Information Estimation for Filter Based Feature Selection Using Particle Swarm Optimization

Hoai Bach Nguyen[⊠], Bing Xue, and Peter Andreae

School of Engineering and Computer Science,
Victoria University of Wellington, Wellington, New Zealand
{hoai.bach.nguyen,bing.xue,peter.andreae}@ecs.vuw.ac.nz

Abstract. Feature selection is a pre-processing step in classification, which selects a small set of important features to improve the classification performance and efficiency. Mutual information is very popular in feature selection because it is able to detect non-linear relationship between features. However the existing mutual information approaches only consider two-way interaction between features. In addition, in most methods, mutual information is calculated by a counting approach, which may lead to an inaccurate results. This paper proposes a filter feature selection algorithm based on particle swarm optimization (PSO) named PSOMIE, which employs a novel fitness function using nearest neighbor mutual information estimation (NNE) to measure the quality of a feature set. PSOMIE is compared with using all features and two traditional feature selection approaches. The experiment results show that the mutual information estimation successfully guides PSO to search for a small number of features while maintaining or improving the classification performance over using all features and the traditional feature selection methods. In addition, PSOMIE provides a strong consistency between training and test results, which may be used to avoid overfitting problem.

Keywords: Feature selection · Mutual information estimation · Particle swarm optimization

1 Introduction

A feature refers to a property of an object. In classification problems, each instance in a dataset is a set of values, which are assigned to the instance's features. These values will be used by a classification algorithm to determine which category or class the instance belongs to. A set of instances is used to train the classification algorithm, which is called a training set. However, in many classification problems, a large number of features are used to describe the instances. Due to "the curse of dimensionality", the larger a set of features is, the more difficult the training is and the longer the training time may take. In addition,

© Springer International Publishing Switzerland 2016
G. Squillero and P. Burelli (Eds.): EvoApplications 2016, Part I, LNCS 9597, pp. 719–736, 2016.
DOI: 10.1007/978-3-319-31204-0_46

not all features provide useful information. Some features have no or little relevance to the class labels, which blur useful information from other features [1]. Such features may lead to classification performance reduction. Also, some features may provide the same information as other features, and therefore do not improve the classification performance but result in a longer training time. In order to reduce the number of features, two feature reduction approaches, including feature selection and feature construction, are proposed. Feature construction constructs a small number of new high-level features while feature selection [2] reduces the size of the feature set by removing irrelevant and redundant features, which hopefully maintains or even increases the classification performance compared with using all features. This paper focuses mainly on feature selection in classification.

Feature selection is a difficult task due to the complex interaction between features. For example, a weakly relevant feature, which may not individually provide useful information to determine the class label, can significantly improve the classification performance when used with other features. Furthermore, an individually relevant feature may become redundant when working with others. Another reason, which makes feature selection become a challenge task, is the large search space, where the search space's size grows exponentially with respect to the number of features. Suppose there are n original features, then the total number of possible subsets is 2^n. Therefore, the exhaustive search is too slow to perform over the large search space in most situations. In order to reduce the searching time, some greedy algorithms such as sequential forward selection [3] and sequential backward selection [4] are developed. However, these methods usually do not guarantee to find optimal solutions due to getting stuck at local optima. Evolutionary computation (EC) algorithms such as genetic programming (GP) [5], genetic algorithms (GAs) [6] or particle swarm optimization (PSO) [7] are considered global optimization methods, which are suitable for a problem with large search space like feature selection. Therefore, EC has been widely applied to solve feature selection problems in recent years. PSO is chosen as the search technique for this work because it has a natural representation for feature selection, in which each original feature is represented by an entry of a particle's position. In addition, PSO is also simple and converges more quickly than other EC algorithms. In [8], it has been shown that, to achieve the same effectiveness, PSO is more efficient than GAs.

According to the evaluation criterion, existing feature selection methods can fall into two categories: wrapper and filter approaches [9,10]. In a wrapper approach, a learning algorithm is used to calculate the fitness value of the selected features. Meanwhile, a filter approach is done in an independent way of learning algorithms. Therefore, wrapper methods usually can achieve better classification accuracy than filter ones. However, wrappers may produce a feature subset with poor generality, which is only good for the wrapped classification algorithm. In addition, in comparison with wrappers, filter methods are usually less expensive in terms of the computation complexity. Nowadays, there are many filter measures for feature selection problems, for example fisher score [11], consistency measure [12], correlation measure [13] and mutual information [14]. Among these measures, mutual information measure gains more attraction. The reason is that mutual information is fast and able to analyze the complex interaction between multiple features or between the class label and a set of features while most other filter measures like correlation coefficients mainly evaluate a pair of features or the class label and an individual feature. However, most existing mutual information based feature selection approaches consider two-way interactions between features and simply calculate probability distributions by counting instances, which results in an inaccurate mutual information. A solution for the above problem is using mutual information estimation [15], which is able to compute the mutual information between multiple features by an accurate estimation approach. However, mutual information estimation has never been used with any EC algorithm to solve feature selection problems. Therefore, this work will propose a new feature selection approach, which bases on PSO algorithm and mutual information estimation.

1.1   Goals

The overall goal of this paper is to propose a PSO based filter feature selection approach to evolve a small set of features, which achieves similar or better classification performance than using all features. To achieve this goal, a new fitness function is proposed, which is inspired by the nearest neighbor estimation for mutual information [16]. Specifically, we will investigate:

- whether the proposed feature selection approach (named PSOMIE) can select a small number of features and maintain or even improve the classification performance over using all features.
- whether PSOMIE can maintain or improve the classification accuracy than two traditional feature selection approaches, filter sequential forward and backward feature selection [3,4].

2   Background

2.1   Particle Swarm Optimization (PSO)

In 1995, Kennedy and Eberhart [17] proposed an EC technique, named PSO. Like other swarm intelligence algorithms, PSO maintains a set of particles, in which each particle represents a candidate solution for an optimization problem. The behaviour of the swarm in PSO originates from social behaviours such as bird flocking and fish schooling. In particular, each particle is guided by its own best experience, called *pbest* and its neighbors best position so far, called *gbest*, to explore the search space. The current position of a particle i is encoded as a vector $x_i = (x_{i1}, x_{i2}, \ldots, x_{iD})$, where D is the dimensionality of the search space. Particle i moves in the search space by using a velocity, which is defined by a vector $v_i = (v_{i1}, v_{i2}, \ldots, v_{iD})$. In PSO, each velocity component is limited by a predefined maximum velocity, called v_{max}, and $v_{id} \in [-v_{max}, v_{max}]$. The position and velocity of particle i are updated according to the following equations:

$$v_{id}^{t+1} = w * v_{id}^t + c_1 * r_{i1} * (p_{id} - x_{id}^t) + c_2 * r_{i2} * (p_{gd} - x_{id}^t) \tag{1}$$

$$x_{id}^{t+1} = x_{id}^t + v_{id}^{t+1} \tag{2}$$

where t denotes the tth iteration in the search process, d is the dth dimension in the search space, w is inertia weight, c_1 and c_2 are acceleration constants, r_{i1} and r_{i2} are random values uniformly distributed in $[0, 1]$, p_{id} and p_{gd} represent the position entry of *pbest* and *gbest* in the dth dimension, respectively.

2.2 Mutual Information

Basic Concepts: Entropy and mutual information are two well-known concepts in information theory [18], which are used to measure the information provided by random variables. Let X be a discrete variable, then its uncertainty can be measured by entropy $H(X)$ defined as:

$$H(X) = - \sum_{x \in X} P(X = x) * \log_2 P(X = x) \tag{3}$$

Joint entropy is used to measure the uncertainty of a joint variable, which consists of two random variables X and Y. Joint entropy $H(X, Y)$ is defined as:

$$H(X, Y) = - \sum_{x \in X, y \in Y} p(x, y) * \log_2 p(x, y) \tag{4}$$

where $p(x, y) = P(X = x, Y = y)$

When a variable is known and the other is unknown, the remaining uncertainty is measured by the conditional entropy as below

$$H(X|Y) = - \sum_{x \in X, y \in Y} p(x, y) * \log_2 p(x|y) \tag{5}$$

where $p(x|y) = P(X = x|Y = y)$.

Mutual information is a measure of shared information between two random variables. Mutual information between two random variables X and Y can be defined as

$$
\begin{aligned}
MI(X; Y) &= H(X) + H(Y) - H(X, Y) \\
&= H(X) - H(X|Y) \\
&= H(Y) - H(Y|X) \\
&= - \sum_{x \in X, y \in Y} p(x, y) * \log_2 \frac{p(x, y)}{p(x)p(y)}
\end{aligned}
\tag{6}
$$

Mutual information is very popular in feature selection problems because it is able to detect non-linear relationship between features. According to Eq. (6), if two variables X and Y have a strong relationship, their mutual information $MI(X;Y)$ will be large. In contrast, if X and Y are totally independent then $MI(X;Y) = 0$. Mutual information is also extended to measure the common information between more than two random variables, which is called multi-information, defined as

$$MI(X_1; X_2; \ldots; X_n) = \sum_{i=1}^{n} H(X_i) - H(X_1, X_2, \ldots, X_n) \tag{7}$$

Limitations of Current Work on Mutual Information for Feature Selection: Most of feature selection approaches aim to select a set of features which are most relevant to the class labels and do not contain any redundant features. These goals can be achieved by using mutual information criteria. In particular, the most relevant set of features will share the most information with the class label. A non-redundant set of features will have the least mutual information between the features. These conditions are expressed in the Eq. 8.

$$F_{mi} = Red - Rel \tag{8}$$

where

$$Rel = MI(S; C)$$
$$Red = MI(s_1; s_2; \ldots; s_m)$$

where S is a set of features and its size $|S| = m$, $s_i \in S$ and C is the class label. Rel and Red measure the relevance of feature set S and the redundancy between features in S.

A feature selection approach often aims to find a set of features which minimizes the fitness F_{mi} shown in Eq. (8). However the existing mutual information criteria for feature selection just consider the relationship between a single feature with the class label or between a pair of features. In other words, these approaches only consider two-way interactions between features. This does not ensure an optimal feature set being evolved because the interactions between features are more complex than two-way interactions.

Another limitation of the current mutual information based feature selection approaches is how the mutual information is calculated. To induce the mutual information, the joint probability of multi-variables needs to be known. In most current approaches, the probability distribution is achieved by counting the number of instances in the training set. However, counting approaches can only be applied to discrete variables not continuous variables. Furthermore, even for discrete variables, when the number of variables is large, the value distribution in the training set will be sparse, which leads to an inaccurate probability and mutual information. This is one of the reasons why only two-way interactions between single features are normally considered. To overcome these limitations, mutual information estimator has been developed. Currently, there are many mutual information estimators, such as basic histogram [19], kernel estimator [20] or nearest neighbors-based estimators [16]. Among them, the nearest neighbors-based estimators (NNE) has only one parameter and achieves the most accurate and consistent results with an independent hypothesis [21]. Therefore this work will use NNE incorporated with PSO to achieve feature selection.

2.3   Existing Feature Selection Approaches

Traditional Feature Selection Methods: A heuristic search, named Sequential Forward Selection (SFS) is proposed by Whitney [3], which starts with an empty set of features. At each step, a single feature, which gives the best fitness value with current selected features, will be added permanently to the feature subset. This process will stop when there is no single feature which is able to improve the current fitness. Another heuristic search is proposed by Maril and Green [4], which is called Sequential Backward Selection (SBS). The search starts with a full set of features. At each step, a single feature, whose removal results in the best score, is permanently removed from current feature set. SBS terminates when removing any feature from current feature set does not lead to any fitness improvement. Although SBS and SFS achieve better performance than feature ranking methods, they still suffer from the "nesting" problem, in which once a feature is added (or removed) from the feature set, it cannot be removed (or added) later. More works can be seen from [9, 22].

EC Approaches (Non-PSO) for Feature Selection: EC algorithms have been applied to feature selection problems, such as GAs [23], GP [24,25]. Sousa, et al. [26] proposed two ensemble GA-based feature selection approaches, where a set of classifiers are used together to evolve better solutions than a single classifier. The first algorithm is a simple filter approach, which uses Pearson correlation measure as the main criterion. In the second algorithm, the filter and wrapper measures are combined into a single fitness function. However, due to the complexity and time consuming process, only a proportion of population are evaluated by wrapper evaluation. This proportion is defined dynamically based on the similarity between two ranked lists by filter and wrapper measures. The experiments show that the proposed algorithms achieve better classification performance than the original GAs.

Two GP-based approaches are proposed by Bhowan, et al. [27] to evolve a set of features, which is used directly in the Watson system, an intelligent open-domain question answering system. The first approach extracts all features, which are used in the evolved best-of-run GP tree. The second approach considers all evolved trees. Particularly, from the set of GP trees, the top T features with the most frequency are chosen as extracting features. The experiment results show that, the set of features selecting from the best GP tree can only work well when the number of selected features is small. Meanwhile, selecting top T features from the whole set of trees produces good results on both small and large feature sets.

PSO-based Feature Selection Methods: Xue et al. [28] propose three new initialisation mechanisms, which mimic the sequential feature selection approach. While the small initialisation use about 10 % of original features to initialize the particles, particles in the large initialisation are constructed based on 50 % of original features. These two initialisation mechanisms are combined in the mixed initialisation, which use the small initialisation for most of particles and the large initialisation for the rest. The experimental results show that the new initialisation and updating mechanisms led to smaller feature subsets with better classification performance than the standard PSO. Two PSO based filter feature selection algorithms are proposed in [29], where mutual information and entropy are used in the fitness function to evaluate the relevance and redundancy of the selected feature subset. The experiments show that the proposed methods significantly reduce the number of features whilst achieve similar or better classification than using all features. Butler-Yeoman et al. [30] proposed a hybrid filter-wrapper approach named FastPSO, which mainly uses two-way mutual information as a fitness measure. In addition, a wrapper evaluation is used to determine whether or not a *pbest* need to be updated. The experiment results show that FastPSO outperformed not only using all features but also achieved better classification performance than PSO with mutual information as a fitness measure. A comprehensive EC-based feature selection survey can be seen in [31].

However most existing mutual information based feature selection approaches only consider two-way interaction between features and use the counting approach to calculate the mutual information. Therefore, the investigation of using mutual information estimation with an EC technique is still an open issue and the work conducted in this paper is the first effort in this area.

3   Proposed Feature Selection Approach

The key ideas of our proposed approach are to use mutual information as the measure of solution quality, calculated using NNE and to use PSO to search for an optimal set of features. This approach not only deals with numeric datasets but also considers the interaction between multiple features. The details of NNE and how we use NNE and PSO to solve feature selection problems are shown in the following sections.

3.1   Nearest Neighbors-Based Mutual Information Estimation

In order to estimate the mutual information between variable sets, it is usually necessary to estimate the underlying probability densities, which is a hard task. In addition, because the underlying probability densities will then be used together to induce mutual information, an inaccurate probability distribution is likely to result in a more inaccurate estimation of mutual information. To overcome this problem, Kraskov's Nearest Neighbors-based mutual information

Estimation (NNE) [16] directly estimates the mutual information by using nearest neighbors statistics instead of estimating probability densities. The main idea of NNE is that if the neighbors of an instance in X space are similar to the neighbors of that instance in Y space, then there must be a strong relationship between X and Y, i.e. the mutual information between X and Y is high. This is true when X and Y are single variables or sets of variables. Therefore, this estimation can be applied to multi-variate mutual information.

The mutual information is calculated via an estimation of entropy. The NNE based entropy estimation of a single variable X, which is $\hat{H}(X)$ is given by Eq. (9).

$$\hat{H}(X) = -\psi(k) + \psi(N) + \log c_d + \frac{d}{N} * \sum_{i=1}^{N} \log \epsilon_X(i) \qquad (9)$$

where ψ is the digamma function, N is the total number of instances in the training set, k is the number of nearest neighbors, d is the dimensionality of variable X, c_d is the volume of the d-dimensional unit ball, $\epsilon_X(i)$ is twice the distance from the ith instance to its kth nearest neighbor.

Given this entropy estimation, a multi-variate mutual information estimation $(\hat{MI})$ of a feature set $S = \{X_1, X_2, \ldots, X_m\}$ can be derived from Eq. (7), where the multi-variate mutual information is defined by Eq. (10).

$$\hat{MI}(X_1; X_2; \ldots; X_m) = \psi(k) - \frac{m-1}{k} + (m-1) * \psi(N) - \frac{1}{N} * \sum_{i=1}^{N} \sum_{j=1}^{m} n_{ij} \qquad (10)$$

where m is the number of single variables (features) in the variable (feature) set, n_{ij} is the number of neighbors whose distance from the ith instance in the space specified by X_j is not greater than $0.5 * \epsilon(i) = 0.5 * \max(\epsilon_{X_1}(i), \ldots, \epsilon_{X_m}(i))$.

An example of computing the n_{ij} of the NNE is given in Fig. 1, where there are only 2 variables ($m = 2$) X_1 and X_2, the number of neighbors is set to $k = 3$ and the ith instance is marked by a red point. Firstly, the 3rd nearest neighbor of the ith instance is found, which is marked by a blue point. The distances between the ith instance and its 3rd nearest neighbor in each dimensions X_1 and X_2 are calculated, respectively, $\Delta_{X_1}(i)$ and $\Delta_{X_2}(i)$. In this case, $\Delta_{X_1}(i) > \Delta_{X_2}(i)$ so $\epsilon(i) = \epsilon_{X_1}(i) = 2 * \Delta_{X_1}(i)$. After that, for each dimension X_1 and X_2, the total number of neighbors whose distance in that dimension from the ith instance is not greater than $0.5 * \epsilon(i)$ are counted. In this case, $n_{i1} = 7$ and $n_{i2} = 6$. Given the n_{ij}, $\hat{MI}$ can be calculated using Eq. (10).

3.2   Mutual Information Estimation for Feature Selection

By using NNE, mutual information can be used to evaluate the relevance between a set of features and the class label, and the redundancy within a set of features even when the data is too sparse to give good estimates of the probability of density. The aims are to improve the classification performance via maximising

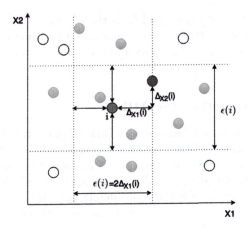

Fig. 1. Example of $\epsilon(i), n_{i1}, n_{i2}$ using the k-nearest neighbors distances, where $k=3$ and for the ith instance.

the relevance between the set of features and the class label, and to reduce the number of selected features by minimising the redundancy within the set of features. To achieve the above objectives in a PSO search, a new fitness function for PSO is proposed, which is shown by the Eq. (11).

$$Fitness = (1 - \alpha) * Red - \alpha * Rel \tag{11}$$

where

$$Red = \hat{M}I(X_1; X_2; \ldots; X_m)$$
$$Rel = \hat{M}I(S; C)$$

where $S = \{X_1, X_2, \ldots, X_m\}$ is the set of selected features, C is the class labels and α is a weight determining the contribution of Red and Rel in the fitness measure. Notice that in Red, the fitness function considers the multi-way interaction between features in S and in Rel, the whole set of features is considered a single variable using NNE. It is easy to calculate redundancy measure by directly applying the mutual information estimation given in Eq. (10). However, the relevance measure is harder to compute because all features in S are numeric variables while the class label is usually a categorical variable. In order to solve this problem, it is necessary to decompose $\hat{M}I(S; C)$, which is achieved by the Eq. (12).

$$\hat{M}I(S; C) = \hat{H}(S) - \hat{H}(S|C)$$
$$= \hat{H}(S) - \sum_{l=1}^{L} P(C = C_l) * \hat{H}(S|C_l) \tag{12}$$

where L is the number of classes and C_l is the lth class.

In the above formula, $\hat{H}(S)$ is easily calculated by using Eq. (9). In order to calculate $\hat{H}(S|C_l)$, Eq. (9) is also applied but only for the instances which belong to class C_l.

3.3 The New Algorithm: PSOMIE

The algorithm based on our approach is called PSOMIE ("PSO with Mutual Information Estimation"). The representation of a particle in PSOMIE is a vector of n real numbers, where n is the total number of features. Each position entry x_{id} falls in the range $[0, 1]$ and corresponds to the dth feature in the original feature set. A threshold θ is used to determine whether or not a feature is selected: if $x_{id} > \theta$ then the dth feature is selected, otherwise the dth feature is not selected.

4   Experiment Design

4.1   Datasets

Ten datasets (Table 1) chosen from the UCI machine learning repository [32] are used in the experiments. These datasets have different numbers of features, classes and instances. Since currently mutual information estimations are mainly used for continuous variables, the data in the selected datasets are continuous values.

For each dataset, all instances are randomly divided into a training set and a test set, which contain 70 % and 30 % of the instances respectively. The algorithm firstly runs on the training set to evolve a subset of features. After that the classification performance of the selected features will be calculated on both training and test set by K-nearest neighbors (KNN) classification algorithm where $K = 5$.

Table 1. Datasets.

Dataset	#features	#classes	#instances
Wine	13	3	178
Australian	14	2	690
Image Segmentation	19	7	210
Wall Robot	24	4	5456
Ionosphere	34	2	351
Lung	56	2	32
Musk1	166	2	476
LSVT	310	2	126
Isolet5	617	5	1559
Multiple Features	649	10	2000

4.2   Parameter Settings

In the experiments, the parameters of PSO are set as follows [33]: $w = 0.7298, c_1 = c_2 = 1.49618, v_{max} = 6.0$, population size is 30, the maximum number of iterations is 100. The fully connected topology is used. There are three α values being tested in this work, which are $\alpha = 0.9, \alpha = 0.95, \alpha = 1.0$. For each α value, PSOMIE is ran 40 independent times on each dataset. The threshold θ in the continuous PSO is set to 0.6. The number of neighbours used in nearest neighbour mutual estimation k is set to 4 [16]. A statistical significance test, Wilcoxon signed-rank test, is performed to compare between PSOMIE and using all features as well as two traditional feature selection approaches (SBS [4], SFS [3]). The significance level of the Wilcoxon test was set as 0.05.

5   Experiment Results

Experiment results are shown in Tables 2 and 3. Table 2 shows the comparison between PSOMIE and using all features, while Table 3 shows the comparison between PSOMIE and the two traditional methods. In these tables, "Full" means that all the original features are used for classification, "Ave-Size" stands for the average number of selected features over the 40 runs. "Ave" and "Std" represents the average and standard deviation of the training or test accuracies over the 40 runs. "Test" represents the significant test comparing between PSOMIE and other approaches. "+", "="or "−" mean that PSOMIE is respectively significantly better, similar or significantly worse than using all features or the traditional methods.

5.1   Comparison with All Features

According to the results shown in Table 2, in terms of test results, in almost all datasets, the number of features selected by PSOMIE is at least 70 % lower than the total number of original features. In seven datasets, PSOMIE always achieves significantly better classification accuracy than using all features. For example, in the Ionosphere dataset, by selecting around 4 features from 34 original features, PSOMIE significantly increases the classification accuracy about 6 % over using all features. In addition, in all datasets, the best solutions evolved by PSOMIE always outperform the set of original features in both classification performance as well as the number of selected features.

The results suggest that PSO with NNE for mutual information can significantly reduce the dimensionality of datasets, while maintains or even improves the classification accuracy over using all features.

5.2   Results of PSOMIE with Different α

As can be seen from Table 2, in almost all datasets, with at least one of α value, despite of selecting a small number of features, PSOMIE is still able to

Table 2. Results of PSOMIE with different α values

Dataset	Method	Ave-size	Training set			Test set		
			Best	Ave ± Std	Test	Best	Ave ± Std	Test
Wine	Full	13.0	83.87			82.72		
	α =0.9	3.68	96.77	93.95±3.95	+	96.30	89.46±3.79	+
	α =0.95	4.5	98.39	89.36±6.47	+	97.53	86.96±5.85	+
	α =1.0	6.12	96.77	85.61±4.18	+	97.53	83.7±4.86	=
Australian	Full	14	77.85			67.15		
	α =0.9	4.98	88.82	79.92±2.4	+	80.68	71.28±3.32	+
	α =0.95	4.92	83.64	79.71±1.77	+	75.36	70.13±2.37	+
	α =1.0	5.48	83.02	79.73±1.82	+	75.36	70.87±2.0	+
Image Segmentation	Full	19	94.75			93.65		
	α =0.9	2.86	96.89	95.23±1.0	+	98.19	95.06±1.65	+
	α =0.95	3.5	97.28	96.05±0.56	+	98.64	94.89±2.14	=
	α =1.0	7.54	97.86	97.41±0.33	+	98.64	98.11±0.37	+
Wall Robot	Full	24	94.87			92.33		
	α =0.9	2.06	97.6	95.25±1.33	=	96.18	92.85±2.12	=
	α =0.95	2.84	97.97	97.33±0.99	+	96.67	95.76±1.36	+
	α =1.0	3.76	98.28	97.93±0.22	+	97.43	96.6±0.42	+
Ionosphere	Full	34	89.02			81.9		
	α =0.9	3.8	93.09	90.04±1.73	+	95.24	87.83±3.4	+
	α =0.95	4.3	93.9	91.27±1.4	+	94.29	87.9±2.51	+
	α =1.0	13.12	92.68	89.93±0.98	+	89.52	83.9±2.39	+
Lung	Full	56	86.36			90.0		
	α =0.9	15.96	90.91	82.91±7.53	=	100.0	95.0±8.06	+
	α =0.95	16.54	90.91	83.18±5.77	−	100.0	93.2±9.47	+
	α =1.0	17.64	90.91	81.55±6.19	−	100.0	93.6±12.77	+
Musk1	Full	166	90.99			81.12		
	α =0.9	18.52	90.69	87.27±1.66	−	84.62	77.69±3.56	−
	α =0.95	18.98	90.69	87.79±1.89	−	88.11	78.29±3.54	−
	α =1.0	60.68	94.59	91.2±1.36	=	89.51	84.15±2.46	+
LSVT	Full	310	78.41			52.63		
	α =0.9	55.74	81.82	76.14±2.18	−	76.32	59.16±5.4	+
	α =0.95	54.56	82.95	76.36±2.22	−	65.79	58.32±3.74	+
	α =1.0	111.26	84.09	76.93±1.97	−	65.79	54.1±3.54	+
Isolet5	Full	617	99.16			98.37		
	α =0.9	107.28	99.15	98.87±0.1	−	98.5	98.09±0.2	−
	α =0.95	108.52	99.06	98.94±0.0	−	98.47	98.19±0.14	−
	α =1.0	210.14	99.28	99.14±0.0	=	98.8	98.45±0.14	+
Multiple Features	Full	649	99.1			99.0		
	α =0.9	121.66	99.37	98.87±0.4	−	99.1	98.43±0.51	−
	α =0.95	120.0	99.39	98.88±0.39	−	99.13	98.45±0.57	−
	α =1.0	256.52	99.51	99.13±0.26	=	99.37	98.89±0.33	−

achieve better classification accuracy than using all features. The only exception is the largest dataset, Multiple Features, which is not too surprising since filter approaches often can not scale well [34] and the classification accuracy is already very high (99.0 %) when using all features. In Multiple Features, although the feature set selected by PSOMIE is not better than all features according to the significant test, PSOMIE can select around only 18.5 % of the features and still achieve around 98.45 % classification accuracy, which is only 0.55 % lower than using all features. In addition, the best accuracy is better than using all features.

In these experiments, α is given 3 values, which corresponds to the contribution of relevant measure to the fitness function. When α is larger than 0.5, relevant measure will contribute more than redundant measure, which means that the searching process will focus more on improving classification accuracy than reducing the number of features. As can be seen from Table 2, on most datasets, when α is increased, PSOMIE tends to select more features and achieves better classification performance. In seven of the ten datasets, with $\alpha = 0.95$, PSOMIE only selects 16.7 % of the original feature set and still achieves better classification than using all features. On the other hand, when α is assigned to 1.0, although the feature sets evolved by PSOMIE are larger than when $\alpha = 0.95$, these sets achieve better classification accuracy than using all features in nine of the ten cases.

The results suggest that different weights for the two fitness components, i.e. the relevant and redundant measures, have significant effect on the searching process. In other words, by setting a proper value for α, PSOMIE is able to evolve a set of small number of features and achieve better classification accuracy than using all features.

5.3   Analysis Between Training and Test Results

In classification tasks, overfitting is a very common problem. After a long training time, the training accuracy still increases while the test accuracy becomes worse. This is because the classifier may remember all specific properties of training instances. As can be seen from Table 2, in almost all datasets, the results of training and test sets are very consistent. In particular, a small increase/decrease of the classification accuracy on the training set usually corresponds to a small increase/decrease of the accuracy on the test set. This consistency suggests that mutual information estimation is able to extract a consistent pattern, which is effective in both training and test (unseen data) set.

The results show that, by using NNE for mutual information in PSO, the overfitting problem in feature selection might be avoided. With this property, the testing accuracy can be improved by advancing the accuracy on the training set, which does not happen in many feature selection algorithms.

5.4   Comparison with SFS and SBS

The comparison between PSOMIE and two traditional methods, SFS [3] and SBS [4] are shown in Table 3. These traditional methods use standard mutual

Table 3. Comparison between PSOMIE and SFS, SBS

Dataset	Method	Ave-size	Ave	Std	Test	Time (ms)
Wine	PSOMIE	4.5	86.96	5.85		528
	SFS	5.0	81.48		+	1
	SBS	5.0	81.48		+	5
Australian	PSOMIE	4.9	70.13	2.37		1883
	SFS	4.0	67.63		+	1
	SBS	5.0	69.57		+	4
Image Segmentation	PSOMIE	3.5	94.89	2.14		685
	SFS	8.0	89.57		+	1
	SBS	9.0	89.57		+	5
Wall Robot	PSOMIE	2.8	95.76	2.12		142605
	SFS	4.0	94.77		+	1
	SBS	4.0	95.20		=	3
Ionosphere	PSOMIE	4.3	87.90	2.51		2427
	SFS	5.0	78.10		+	1
	SBS	5.0	85.71		+	4
Lung	PSOMIE	16.5	93.20	9.47		952
	SFS	18.0	80.00		+	8
	SBS	18.0	80.00		+	5
Musk1	PSOMIE	19.0	78.29	3.54		61789
	SFS	2.0	62.94		+	8
	SBS	3.0	62.24		+	213
LSVT	PSOMIE	54.56	58.32	3.74		19313
	SFS	3.0	76.32		−	9
	SBS	3.0	60.53		−	1599
Isolet5	PSOMIE	108.5	98.19	0.14		2981564
	SFS	13	95.83		+	43
	SBS	28	95.33		+	26419
Multiple Features	PSOMIE	120.0	98.45	0.57		5063384
	SFS	15	96.93		+	56
	SBS	91	99.17		−	28466

information with the counting approach to calculate the redundancy and relevance, which are combined into a fitness function as shown in Eq. (11). However, in the traditional methods, the counting approach is applied to calculate the mutual information. To ensure a fair comparison, the α value in the fitness function shown in Eq. (11) is set to 0.95 in both PSOMIE and traditional methods. Since SFS and SBS are deterministic methods, they produce a single solution on each dataset.

As can be seen from this table, in the first six datasets where the number of features is relatively small, although PSOMIE always selects a smaller number of features than both SFS and SBS, it still achieves better classification performance. For example, in Image Segmentation, the subset evolved by PSOMIE is three times smaller than the traditional methods, but the PSOMIE's subset is about 5 % better than the sequential selection in terms of classification performance. In the other four large datasets, PSOMIE selects more features than the traditional methods and obtains higher classification accuracy in three of the four cases. The results suggest that PSOMIE is able to adapt with different datasets to balance between the number of selected features and the classification performance.

However, in terms of efficiency, PSOMIE is much more expensive than sequential approaches, but the performance of SBS and SFS can not be improved by giving longer time since they are deterministic methods and stop when the fitness value is not further improved. For PSOMIE, although the Ionosphere dataset has a larger number of features than Wall Robot, the selection time of Wall Robot is still longer. Similarly, despite of having a smaller number of features than the LSVT dataset, the Musk1 dataset needs longer training time. This is caused by the large number of instances. Particularly Musk1 and Wall Robot datasets have more instances than Ionosphere and LSVT datasets, respectively. Because for each instance, PSOMIE needs to calculate a distance between the current instance to another instances, the total cost for finding distance for all instances is $O(N^2)$ where N is the total number of instances in the training set. After that, for each instance, there is another process which counts all neighbors which fall in a range around that instance in each dimension. Therefore, in order to estimate one mutual information, the worst cost will be $n * O(N^3)$ where n is the total number of features. So the computation cost of mutual information estimation increases with respect to the number of instances in the training set.

6   Conclusions and Future Work

The goal of this paper is to investigate a PSO based filter feature selection approach, which uses nearest neighbor mutual information estimation to evolve a set of small number of features while maintaining or improving the classification performance over using all features. The experiment results show that PSOMIE substantially reduces the dimensionality of the datasets and achieves the similar or better classification performance than using all features. PSOMIE also produces a strong consistency between training and test accuracies. In addition, by using mutual information estimation, PSOMIE can better balance between a smaller number of features in comparison with sequential feature selection approaches.

However, besides the strengths, PSOMIE still has several limitations, which will be addressed in our future work. Firstly, PSOMIE could not be applied to categorical datasets because NNE requires a numeric dataset. In order to solve this problem, a good distance measure needs to be developed, which can perform well on both numeric and categorical datasets. In addition, PSOMIE will have an expensive computation cost if there are a large number of instances in the training set. Therefore, developing an instance selection algorithm, which reduces the number of instances while maintaining most of important information, will lower the computation cost. In addition, scalability is a common issue in filter feature selection. So developing novel feature selection methods to solve feature selection tasks with thousands of features is also needed in future work. This work is investigates the use of mutual information estimation in standard PSO for feature selection. In the future, other advanced PSO algorithms, such as constriction factor version of PSO [35], and other methods will be investigated to improve the performance of filter feature selection on different problems.

References

1. Guyon, I., Elisseeff, A.: An introduction to variable and feature selection. J. Mach. Learn. Res. **3**, 1157–1182 (2003)
2. Liu, H., Motoda, H., Setiono, R., Zhao, Z.: Feature selection: an ever evolving frontier in data mining. FSDM **10**, 4–13 (2010)
3. Whitney, A.W.: A direct method of nonparametric measurement selection. IEEE Trans. Comput. **100**(9), 1100–1103 (1971)
4. Marill, T., Green, D.M.: On the effectiveness of receptors in recognition systems. IEEE Trans. Inf. Theory **9**(1), 11–17 (1963)
5. Nag, K., Pal, N.R.: A multiobjective genetic programming-based ensemble for simultaneous feature selection and classification. IEEE Trans. Cybern. **46**(2), 499–510 (2015)
6. Lin, F., Liang, D., Yeh, C.C., Huang, J.C.: Novel feature selection methods to financial distress prediction. Expert Syst. Appl. **41**(5), 2472–2483 (2014)
7. Chuang, L.Y., Chang, H.W., Tu, C.J., Yang, C.H.: Improved binary PSO for feature selection using gene expression data. Comput. Biol. Chem. **32**(1), 29–38 (2008)
8. Hassan, R., Cohanim, B., De Weck, O., Venter, G.: A comparison of particle swarm optimization and the genetic algorithm. In: Proceedings of the 1st AIAA Multidisciplinary Design Optimization Specialist Conference, pp. 1–13 (2005)
9. Dash, M., Liu, H.: Feature selection for classification. Intell. Data Anal. **1**(3), 131–156 (1997)
10. Kohavi, R., John, G.H.: Wrappers for feature subset selection. Artif. Intell. **97**(1), 273–324 (1997)
11. Duda, R.O., Hart, P.E., Stork, D.G.: Pattern Classification. Wiley, New York (2012)
12. Dash, M., Liu, H., Motoda, H.: Consistency based feature selection. In: Terano, T., Liu, H., Chen, A.L.P. (eds.) PAKDD 2000. LNCS (LNAI), vol. 1805, pp. 98–109. Springer, Heidelberg (2000)
13. Hall, M.: Correlation-based feature selection for discrete and numeric class machinelearning. In: Proceedings of 7th International Conference on Machine Learning, Stanford University (2000)
14. Kononenko, I.: On biases in estimating multi-valued attributes. In: IJCAI. vol. 95, pp. 1034–1040. Citeseer (1995)
15. Walters-Williams, J., Li, Y.: Estimation of mutual information: a survey. In: Wen, P., Li, Y., Polkowski, L., Yao, Y., Tsumoto, S., Wang, G. (eds.) RSKT 2009. LNCS, vol. 5589, pp. 389–396. Springer, Heidelberg (2009)

16. Kraskov, A., Stögbauer, H., Grassberger, P.: Estimating mutual information. Phys. Rev. E **69**(6), 066138 (2004)
17. Kennedy, J., Eberhart, R., et al.: Particle swarm optimization. In: Proceedings of IEEE International Conference on Neural Networks, vol. 4, pp. 1942–1948. Perth, Australia (1995)
18. Jaynes, E.T.: Information theory and statistical mechanics. Phys. Rev. **106**(4), 620 (1957)
19. Sturges, H.A.: The choice of a class interval. J. Am. Stat. Assoc. **21**(153), 65–66 (1926)
20. Parzen, E.: On estimation of a probability density function and mode. Ann. Math. Stat. **33**(3), 1065–1076 (1962)
21. Doquire, G., Verleysen, M.: A performance evaluation of mutual information estimators for multivariate feature selection. In: Carmona, P.L., Salvado Sánchez, J., Fred, A.L.N. (eds.) ICPRAM 2012. AISC, vol. 204, pp. 51–63. Springer, Heidelberg (2013)
22. Stearns, S.D.: On selecting features for pattern classifiers. In: Proceedings of the 3rd International Conference on Pattern Recognition (ICPR 1976), pp. 71–75. Coronado, CA (1976)
23. Zhu, Z., Ong, Y.S., Dash, M.: Wrapper-filter feature selection algorithm using a memetic framework. IEEE Trans. Syst. Man Cybern. B Cybern. **37**(1), 70–76 (2007)
24. Neshatian, K., Zhang, M.: Genetic programming for feature subset ranking in binary classification problems. In: Vanneschi, L., Gustafson, S., Moraglio, A., Falco, I., Ebner, M. (eds.) EuroGP 2009. LNCS, vol. 5481, pp. 121–132. Springer, Heidelberg (2009)
25. Hunt, R., Neshatian, K., Zhang, M.: A genetic programming approach to hyperheuristic feature selection. In: Bui, L.T., Ong, Y.S., Hoai, N.X., Ishibuchi, H., Suganthan, P.N. (eds.) SEAL 2012. LNCS, vol. 7673, pp. 320–330. Springer, Heidelberg (2012)
26. Sousa, P., Cortez, P., Vaz, R., Rocha, M., Rio, M.: Email spam detection: a symbiotic feature selection approach fostered by evolutionary computation. Int. J. Inf. Technol. Decis. Making **12**(04), 863–884 (2013)
27. Bhowan, U., McCloskey, D.: Genetic programming for feature selection and question-answer ranking in IBM watson. In: Machado, P., Heywood, M.I., McDermott, J., Castelli, M., García-Sánchez, P., Burelli, P., Risi, S., Sim, K. (eds.) EuroGP 2015. LNCS, vol. 9025, pp. 153–166. Springer, Heidelberg (2015)
28. Xue, B., Zhang, M., Browne, W.N.: Particle swarm optimisation for feature selection in classification: novel initialisation and updating mechanisms. Appl. Soft. Comput. **18**, 261–276 (2014)
29. Cervante, L., Xue, B., Zhang, M., Shang, L.: Binary particle swarm optimisation for feature selection: a filter based approach. In: 2012 IEEE Congress on Evolutionary Computation (CEC), pp. 1–8. IEEE (2012)
30. Butler-Yeoman, T., Xue, B., Zhang, M.: Particle swarm optimisation for feature selection: a hybrid filter-wrapper approach. In: 2015 IEEE Congress on Evolutionary Computation (CEC), pp. 2428–2435. IEEE (2015)
31. Xue, B., Zhang, M., Browne, W., Yao, X.: A survey on evolutionary computation approaches to feature selection. IEEE Trans. Evol. Comput. published online on 30 November 2015. doi:10.1109/TEVC.2015.2504420
32. Asuncion, A., Newman, D.: UCI machine learning repository (2007)
33. Van Den Bergh, F.: An analysis of particle swarm optimizers. PhD thesis, University of Pretoria (2006)

34. Zhai, Y., Ong, Y.S., Tsang, I.W.: The emerging big dimensionality. IEEE Comput. Intell. Mag. **9**(3), 14–26 (2014)
35. Eberhart, R.C., Shi, Y.: Comparing inertia weights and constriction factors in particle swarm optimization. In: Proceedings of the 2000 Congress on Evolutionary Computation, vol. 1, pp. 84–88. IEEE (2000)

Speaker Verification on Unbalanced Data with Genetic Programming

Róisín Loughran[1]($\boxtimes$), Alexandros Agapitos[1], Ahmed Kattan[2],
Anthony Brabazon[1], and Michael O'Neill[1]

[1] Natural Computing Research and Applications Group,
University College Dublin, Dublin, Ireland
{roisin.loughran,alexandros.agapitos}@ucd.ie
[2] Computer Science Department, Um Al-Qura University, Mecca, Saudi Arabia

Abstract. Automatic Speaker Verification (ASV) is a highly unbalanced binary classification problem, in which any given speaker must be verified against everyone else. We apply Genetic programming (GP) to this problem with the aim of both prediction and inference. We examine the generalisation of evolved programs using a variety of fitness functions and data sampling techniques found in the literature. A significant difference between train and test performance, which can indicate overfitting, is found in the evolutionary runs of all to-be-verified speakers. Nevertheless, in all speakers, the best test performance attained is always superior than just merely predicting the majority class. We examine which features are used in good-generalising individuals. The findings can inform future applications of GP or other machine learning techniques to ASV about the suitability of feature-extraction techniques.

Keywords: Speaker verification · Unbalanced data · Genetic programming · Feature selection

1 Introduction

Automatic Speaker Verification (ASV) is the process of accurately verifying that a speaker is who they claim to be. This is feasible because each individual's voice is audibly unique due to physical attributes such as length of vocal tract, size of larynx etc. along with habitual characteristics such as accent and inflection. ASV has important applications in the fields of phone banking, shopping and security systems.

ASV is inherently a highly unbalanced binary-classification problem since it requires to accurately recognise one speaker from everyone else; the first class A contains examples from to-be-verified speaker, whereas the second class B contains examples from the rest of the speakers (i.e. impostors). Class A is the *minority class* as it is often represented with a smaller number of training examples, while class B is the *majority class*. This imbalance in class distribution is a significant problem; it introduces a learning bias and often results in classification models that are not accurate in the cases of the to-be-verified speaker.

© Springer International Publishing Switzerland 2016
G. Squillero and P. Burelli (Eds.): EvoApplications 2016, Part I, LNCS 9597, pp. 737–753, 2016.
DOI: 10.1007/978-3-319-31204-0_47

In general, in class imbalance problems, the smaller the ratio of minority class examples to majority class examples, then the stronger this bias becomes and the harder it is for a classifier to generalise [4].

With the notable exception of [9], ASV is an application area that has received little attention from the GP community. This paper reports a preliminary empirical study that approaches the problem of ASV from the class imbalance perspective. A number of different fitness functions and training-data sampling techniques found in the literature are examined for their efficiency to evolve good-generalising programs. The simulations herein are performed on the TIMIT corpora [13], a regularly-used dataset for speaker recognition and verification.

Section 2 provides background information on ASV, and on methods for tackling class imbalance issues in pattern classification algorithms in general and in GP in particular. Section 3 details the scope of the experiments. Section 4 introduces the TIMIT corpora, presents methods for feature extraction, details the GP systems under comparison and the setup of the experiments. Section 5 analyses the empirical results, while Sect. 6 concludes and discusses future work.

2   Background

The aim of this section is to first provide an overview of traditional classification models for ASV. It then briefly describes the two main categories of methods for tackling class imbalance problems. The final part reviews GP work based on unbalance datasets.

2.1   Automatic Speaker Verification

Early speaker verification models were based on Vector Quantisations [6]. Another prominent classification method for ASV to emerge in the early 1990s was based on Gaussian Mixture models [31]. Over the next three decades other classification techniques were applied to the problem of speaker recognition and verification, such as Support Vector Machines (SVM) [7], Artificial Neural Networks [32], ensemble learning [25], and Genetic Programming [9].

A number of recent studies have focussed on methods to counteract inter-speaker and inter-session variability by examining channel compensation between recordings. Such studies have used feature mapping to transform obtained features into a channel-independent feature-space. Methods such as Joint Factor Analysis [20], i-vectors [21] and PDLA [10] were used for this purpose. This focus on channel effects is in part driven by the NIST Speaker Recognition Evaluation[1] which evaluates novel speaker recognition systems on a corpora of phone recordings.

---

[1] http://www.nist.gov/itl/iad/mig/ivec.cfm.

2.2   Tackling Class Imbalance

There exist a number of methods for learning good-generalising classifiers for class imbalance datasets. The taxonomy of these methods mainly consists of two major categories; those of *training data sampling* and *cost-sensitive training*. The work of [4] provides an excellent overview of these methods, with references from both statistical machine learning and GP. For the sake of completeness we very briefly introduce the two dominant classes in the following sections.

Training-Data Sampling. Balancing of training examples can be achieved either by *over-sampling* the minority class or *under-sampling* the majority class [2]. *Synthetic over-sampling* and *editing* have been often shown to be superior to the sampling techniques described above. Synthetic oversampling of the minority class creates additional examples by interpolating between several similar examples [3], while editing removes noisy or atypical examples from the majority class [23].

Cost-Sensitive Training. In a classification problem, we are given a training set of N examples $\{(x_i, y_i)\}_{i=1}^{N}$, where $x \in \mathbb{R}^d$ is a d-dimensional vector of explanatory variables and $y \in C = \{1, \ldots, c\}$ is a discrete response variable, with joint distribution $P(x, y)$. We seek a function $f(x)$ for predicting y given the values of x. The loss function $L(y, f(x))$ for penalising errors in prediction can be represented by a $K \times K$ cost matrix L, where $K = card(C)$. L will be zero on the diagonal and non-negative elsewhere, where $L(k, l)$ is the price paid for misclassifying an observation belonging to class C_k as C_l. Most often, in cases of balanced datasets, a *zero-one* loss function $L(y, f(x)) = I(y \neq f(x))^2$ is used, where all misclassifications are charged one unit. In the case of unbalanced datasets, the cost matrix can be adjusted to increase the cost of misclassifying the examples of the minority class.

2.3   GP on Unbalanced Datasets

Genetic Programming has been applied to unbalanced datasets in a number of studies. Work using the data sampling techniques of Random Sampling Selection (RSS) and Dynamic Subset Selection (DSS) is reported in [8,14]. In [8] a two-level sampling approach is first used to sample blocks of training examples using RSS and then select examples from within those blocks using DSS. In [14] DSS is used to bias the selection of training examples towards hard-to-classify examples, while RSS was used to bias towards the selection of minority class training examples.

Cost adjustment strategies usually focus on adapting the fitness function to reward programs which have good accuracy on both classes with better fitness, while penalising those with poor accuracy on one class with low fitness. The use

---

[2] $I(\cdot)$ is the indicator function.

of different misclassification costs to incorrect class predictions is reported in [18]. In the work of [12] an adaptive fitness function increases misclassification costs for difficult-to-classify examples. In [33] RSS and DSS are used in conjunction with three novel fitness functions with an application to a network intrusion detection problem. The work of [29] used both rebalancing of data and cost-sensitive fitness functions in comparing GP with other data-mining approaches to predict the rate of student failure in school. The work of [4] used six data sets with different class imbalance ratios and applied GP with a number of different fitness functions. A multi-objective GP approach for evolving accurate and diverse ensembles of GP classifiers that perform well on both minority and majority classes was proposed in [5]. A weighted average composed of error rate, mean squared error and a novel measure of class separability similar to Area Under Curve is used in [34]. In the work of [11], data sub-sampling is used in combination with the average of the geometric mean between minority and majority class accuracies and the Wilcoxon-Mann-Whitney statistic.

3   Scope of Research

To the best of our knowledge, the application of GP to speaker verification has only been reported in the work of [9]. One of the principal applications of ASV systems is remotely confirming the identity of a person for reasons of security such as telephone banking. The literature review conducted in [9] showed that while good results have been reported using a variety of different systems of statistical machine learning on noiseless input signals, most systems suffer heavily if the signal is transmitted over a noisy transmission path (i.e. a telephone network). In order to create a "noisy" environment, several datasets were derived from the original TIMIT corpora using filters that included both additive and convolutive noise. GP experiments were then set to evolve classifiers based on extracted features impaired by noise. Twenty-five speakers to-be-verified and forty-five "impostors" were selected from the TIMIT corpora. For each of the to-be-verified speakers the training set consists of fifteen seconds of to-be-verified speech and forty-five seconds of impostor speech (one second of randomly selected speech from each of the forty-five impostor individuals). This results in a minority class to majority class imbalance ratio of 1:3. A pool of hand-engineered features were extracted from the raw signal and populated the terminal set. The fitness function used was dynamically biased to concentrate on the most difficult-to-classify examples. Finally, an island model was employed to improve population diversity. Results showed that generated programs can be evolved to be resilient to noisy transition paths, which was mainly attributed to the speaker-dependent and environment-specific feature selection inherent in GP.

In this paper we investigate a different facet of the GP application to ASV. The research scope is two-fold. First, we study the generalisation of GP-evolved programs on ASV datasets that exhibit a high class imbalance ratio. Specifically, the experiments designed are based on datasets with a class imbalance ratio of 1:9. We used the original, noiseless TIMIT corpora. A number of different methods for cost-sensitive training and data sampling are compared in terms of their effectiveness to assist with the evolution of good-generalising programs.

The second aim of this paper is to inform future applications of GP and other machine learning algorithms to ASV in terms of the usefulness of different features for constructing classifiers. For this purpose, we extract 275 features from the raw signal to create program input, and rely on the inherent ability of GP to perform feature selection. We analyse the terminal-nodes of highly-performing programs, and calculate statistics on the frequency of usage of different features.

4 Methods

4.1 Speaker Corpus

The speech recordings used in this study are taken from the TIMIT corpora [13]. This was chosen due its very regular use in the speaker recognition and verification literature. The corpora consists of 630 speakers, 192 female and 438 male, from 8 American dialects each reading 10 phonetically rich sentences. Each sentence was recorded on a high quality microphone at a sampling rate of 16 kHz.

4.2 Training and Test Data

For these experiments we chose a random 10 speakers, 4 female and 6 male, from the corpus and developed a classifier for each speaker. For each experiment the audio from the given speaker is the to-be-verified minority class and the audio from the nine other speakers constitute the majority class. In this manner we create a 1:9 class imbalance ratio for each experiment.

Each speaker offers 10 utterances of approximately 3 s each. To increase the number of speech utterances, we split each sentence into three equal parts of approximately 1 s. Early analysis showed that the third part of each sentence was of lower timbral quality than the preceding sections. Thus only the first two thirds of each sentence were included in the learning dataset of 200 examples. In the experiments, a *training set* of size 120 examples is used to evolve programs, and a *test set* of 80 examples is used to assess generalisation.

The features calculated on this data are detailed in Sect. 4.4. Rather than reducing these features using the statistical mean or variance of the windowed signal, we employed Principal Component Analysis (PCA) on a number of the high-dimensional features. PCA was used on these results to record the maximum variance within each feature while reducing the dimensionality of the data. In total this resulted in 275 features calculated on 200 data samples.

4.3 GP Systems

A number of systems tailored to unbalanced classification problems from the literature were chosen for this study. These are detailed below.

ST. This system is trained using the original unbalanced dataset. We employ a version of the MSE-based loss function that has been shown [4] to improve upon the performance of fitness functions based on classification accuracy[3] or the weighted average of true positive and true negative rates. Given N training examples $\{(x_i, t_i)\}_{i=1}^{N}$ containing the examples of both majority and minority classes, L_{MSE} is defined as:

$$L_{MSE} = \frac{1}{N} \sum_{i=1}^{N} (\Phi(f(x_i)) - t_i)^2 \tag{1}$$

where

$$\Phi(x) = \frac{2}{1 + e^{-x}} - 1 \tag{2}$$

and $f(x_i)$, t_i is the program output and target value for the ith training case respectively. The sigmoid function in Eq. 2 scales $f(x)$ within the range of $\{-1, \ldots, 1\}$. Similarly to [4], the target value for the majority class is set to -0.5, while the target value for the minority class is set 0.5. Classification is based on a zero-threshold approach; positive program output is mapped to the minority class label, while negative output is mapped to the majority class label.

AVE. This system is trained using the original unbalanced dataset. The loss function uses a weighted-average classification accuracy of the minority and majority classes [4]. Minority accuracy corresponds to the true positive rate, whereas majority accuracy corresponds to true negative rate. The weighting coefficient between the two is $0 < w < 1$. When w is set to 0.5, the accuracy of both classes contributes equally to the loss function. In case of $w > 0.5$ the accuracy of the minority class contributes more to the loss function, lowering the contribution of the majority class accuracy. The loss function L_{AVE} is defined as:

$$L_{AVE} = 1.0 - \left(w \times \frac{TP}{TP + FN} + (1 - w) \times \frac{TN}{TN + FP} \right) \tag{3}$$

where TP, TN, FN, FP is the count of true positives, true negatives, false negatives and false positives respectively.

US. Since static under-sampling of the majority class examples can introduce unwanted sampling bias and discard potentially useful training examples, we resort to a dynamic version of under-sampling. At every generation, a new set of examples is drawn random-uniformly from the set of training examples of the majority class. Under-sampling ensures that the number of examples drawn from the majority class is the same as the number of examples for the minority class. The loss function used is given in Eq. 1.

---

[3] The number of examples correctly classified as a fraction of the total number of training examples.

RS. A type of random sampling technique in which programs are evaluated on a *single* example drawn uniform-randomly from the entire training dataset in each generation was shown to improve the generalisation of programs as compared to the use of the complete training set [15]. An obvious extension of this method to datasets with class imbalance is to populate the training set with *two* randomly-drawn examples (different in each generation), one each from the minority and majority class. The loss function used is given in Eq. 1.

4.4 Feature Extraction

High-level features describing fundamental frequency or rhythm are difficult to measure accurately, possible to mimic and susceptible to emotions. Thus lower level spectral, cepstral, spectro-temporal and statistical features are more common for speaker verification. A survey of the literature indicates that the following short-term spectral features are the recommended features to include [22].

Mel-frequency Cepstral Coefficients. MFCCs have become the standard measure of speech analysis for some time [30]. They consist of a set of coefficients that can represent the spectral quality within a sound according to a scale based on human hearing. Obtaining the MFCCs consists of windowing the sound, calculating amplitude spectrum of cepstral feature vector for each frame and then converting this to the perceptually derived *mel-scale* [26]. The dimensionally of the MFCCs were reduced in this study using PCA. The first four PCs of the first 12 MFCCs along with their derivatives are included in these experiments.

Linear Prediction Coefficients. Linear prediction calculates a given signal based on a linear combination of the previous inputs and outputs [28]. As a spectrum estimation it offers good interpretation in both the time and frequency domains. In the time domain, LP predicts according to

$$s[\tilde{n}] = \sum_{k=1}^{p} a_k s[n-k] \tag{4}$$

Where $s[\tilde{n}]$ is the predicted signal, s[n] is the observed signal and a_k are the predictor coefficients. The prediction error or residual is defined as the difference between the predicted signal and the observed signal:

$$e[n] = s[n] - s[\tilde{n}] \tag{5}$$

The linear predictive coefficients (LPCs), a[k], are determined by minimising this residual. This analysis leads to the *Yule-walker equations* that can be efficiently solved using *Levinson-Durbin recursion* [19]. Given the LPC coefficients $a[k]$, $k = 1, \ldots, p$, the linear predictive cepstral coefficients (LPCCs) are computed using the recursions:

$$c[n] = \begin{cases} a[n] + \sum_{k=1}^{n-1} \frac{k}{n} c[k] a[n-k] & \text{if } 1 \le n \le p \\ \sum_{k=n-p}^{n-1} \frac{k}{n} c[k] a[n-k] & \text{if } n > p. \end{cases} \tag{6}$$

Table 1. Function/Terminal sets and run parameters

	PRIMITIVE LANGUAGE	
Function set	$+, -, *, /$ (x/y returns x if $|y| < 10^{-5}$), sin, cos, e^x,	
	log (log(x) returns x if $x \leq 0$), $sqrt$ (sqrt(x) returns x if $x < 0$)	
Terminal set	275 features	
	40 uniform-randomly drawn constants in the range of $[-1.0, 1.0]$	
	GP PARAMETERS	
Evolutionary algorithm	elitist (1% of population size), generational	
Population size	1,000	
Tournament size	4	
No. of generations	51 for ST	
	51 for AVE	
	251 for US	
	3,001 for RS	
Population initialisation	ramped half-and-half (depths of 2 to 4)	
Max. tree depth	8	

The first 21 LPCs and 10 LPCCs (apart from the zeroth order) were included in our dataset. An equivalent measure to these that has become popular in speaker analysis is line spectral frequencies (LSFs) [19]. These can be useful in practice as they result in low spectral distortion and are deemed to be more sensitive and efficient than other equivalent representations.

Perceptual Linear Prediction. One downfall of the LP method is that it approximates the spectrum of speech equally well at all frequencies. In contrast, after 800 Hz, the human ear becomes less sensitive and spectral resolution decreases with frequency. This is compensated for by using Perceptual Linear Prediction (PLP) [16]. The RelAtive SpecTrAl (RASTA) [17] method was developed to make PLP more robust to linear spectral distortions by replacing the short-term spectrum by a spectral estimate. This suppresses any slow varying component making the spectral estimate of that channel less sensitive to slow variations.

Other Features. A number of descriptive spectral features were also included in our dataset. These included the Spectral Centroid, Inharmonicity, Number of Spectral Peaks, Zero Crossing Rate, Spectral Rolloff, Brightness, Spectral Regularity and the Spectral Spread, Skewness and Kurtosis. Many of these were calculated using the MIRToolbox [24], a Matlab toolbox dedicated to the extraction of musically-related features from audio recordings.

4.5   Primitive Language, Variation Operators, GP Parameters

The primitive language and the evolutionary run parameters are given in Table 1. Over 50 generations, the ST and AVE methods perform 6,000 fitness evaluations. To ensure a fair comparison the number of generations used in the US and RS methods are adjusted accordingly. Preliminary experiments revealed a tendency of all systems to overfit, thus the maximum tree-depth is set to 8 to restrict the complexity of the evolved programs.

The search strategy that we employed relies heavily on mutation-based variation operators. The operation of `pointMutation(x)` traverses the tree in a depth-first manner, and depending on the probability x it substitutes a tree-node by another random tree-node of the same arity. The operation of `subtreeMutation()` selects a node uniform-randomly and replaces the subtree rooted at that node with a newly generated subtree. The tree-generation procedure is *grow* or *full*, each applied with equal probability. To improve on the exploratory effect of the mutation operator, other than picking the tree-node to be replaced from the whole expression-tree, we devised an additional node-selection method. In this method a depth-level is picked uniform-randomly from the range of all possible depth-levels present in the expression-tree, and subsequently a node is picked uniform-randomly from the set of nodes that lie in the chosen depth-level. The decision between the two node-selection methods is governed by a probability set to 0.5 for both methods. Finally, our implementation of recombination operator is the standard subtree crossover defined for expression-tree representations. The probability of selecting an inner-node as a crossover point is set to 0.9, while the probability of selecting a leaf-node is set to 0.1.

In generating offspring, a probability is associated with applying either mutation of crossover, set to 0.7 in favour of mutation. If mutation is chosen, `pointMutation(0.1)` is applied with a probability of 0.1, `pointMutation(0.2)` is applied with a probability of 0.1, `pointMutation(2 / tree_size)` is applied with a probability of 0.2, and `subtreeMutation()` is applied with a probability of 0.6.

5   Results

We created 50 splits of the 200 learning examples into training and test sets. In each split, 120 examples are drawn uniform-randomly for the training set, while the remaining 80 examples populate the test set. Stratification ensures that the class imbalance ratio is maintained in both sets. Using each split, we performed 50 independent evolutionary runs using each GP system. Many practitioners use an equal weighing in the AVE system by setting $w = 0.5$ [4]. In this work the effectiveness of AVE is evaluated using a set of values for w, that of $W = \{0.5, 0.6, 0.7, 0.8\}$. In the experiments we performed no model selection, thus the fittest individual (on the training dataset) of the last generation is designated as the output of a run.

5.1   Generalisation Performance

Table 2 presents statistics of training and test *classification accuracy* for the different systems on all 10 speakers. In each case, we report the median, interquartile range, and maximum based on 50 independent runs. Note that in a classification setup, in which the true positive rate corresponds to the minority class accuracy, a classifier that always outputs the majority class label attains a classification accuracy of 0.9 (true positive rate of 0 %). Our first observation concerns the significant difference between training and test performance in all datasets.

Table 2. Performance summary. Interquartile range in parentheses.

	TRAINING CLASSIFICATION ACCURACY							
	AVE($w = 0.5$)		ST		RS		US	
Speaker id	Median	Max	Median	Max	Median	Max	Median	Max
FGRW0	0.97 (0.04)	1.00	0.98 (0.09)	1.00	0.96 (0.03)	0.99	0.99 (0.03)	1.00
FJEN0	0.97 (0.05)	1.00	0.95 (0.09)	1.00	0.97 (0.02)	0.98	0.98 (0.03)	1.00
FPJF0	0.99 (0.03)	1.00	0.93 (0.08)	1.00	0.97 (0.03)	1.00	0.98 (0.03)	1.00
FSAH0	0.97 (0.05)	1.00	0.90 (0.06)	0.98	0.97 (0.03)	0.99	0.98 (0.03)	1.00
MEFG0	1.00 (0.02)	1.00	0.98 (0.07)	1.00	0.98 (0.02)	1.00	0.99 (0.01)	1.00
MJDC0	0.97 (0.04)	1.00	0.97 (0.07)	1.00	0.96 (0.01)	0.98	0.99 (0.03)	1.00
MKDD0	0.98 (0.04)	1.00	0.97 (0.09)	1.00	0.97 (0.03)	1.00	0.99 (0.02)	1.00
MMGC0	0.95 (0.04)	1.00	0.92 (0.07)	0.99	0.96 (0.03)	0.98	0.97 (0.03)	1.00
MPGR1	0.97 (0.03)	1.00	0.97 (0.06)	1.00	0.96 (0.03)	0.98	0.98 (0.04)	1.00
MTRT0	0.96 (0.03)	1.00	0.90 (0.05)	0.98	0.96 (0.03)	0.99	0.97 (0.03)	1.00

	TEST CLASSIFICATION ACCURACY							
	AVE($w = 0.5$)		ST		RS		US	
Speaker id	Median	Max	Median	Max	Median	Max	Median	Max
FGRW0	0.90 (0.06)	0.97	0.91 (0.03)	0.96	0.90 (0.04)	0.97	0.93 (0.04)	0.97
FJEN0	0.85 (0.06)	0.91	0.91 (0.01)	0.95	0.89 (0.03)	0.96	0.88 (0.02)	0.96
FPJF0	0.86 (0.05)	0.95	0.90 (0.04)	0.95	0.88 (0.04)	0.94	0.89 (0.10)	0.96
FSAH0	0.84 (0.05)	0.91	0.90 (0.01)	0.95	0.89 (0.04)	0.96	0.90 (0.05)	0.95
MEFG0	0.93 (0.05)	0.99	0.90 (0.05)	0.99	0.94 (0.03)	0.97	0.91 (0.05)	0.96
MJDC0	0.90 (0.08)	0.96	0.90 (0.04)	0.96	0.90 (0.04)	0.94	0.88 (0.02)	0.94
MKDD0	0.93 (0.07)	1.00	0.90 (0.04)	0.95	0.91 (0.05)	0.97	0.94 (0.04)	1.00
MMGC0	0.81 (0.14)	0.94	0.91 (0.01)	0.94	0.86 (0.04)	0.93	0.86 (0.06)	0.94
MPGR1	0.79 (0.10)	0.94	0.92 (0.03)	0.93	0.88 (0.06)	0.94	0.85 (0.05)	0.93
MTRT0	0.89 (0.07)	0.94	0.90 (0.01)	0.94	0.88 (0.05)	0.94	0.90 (0.05)	0.95

This is indicative of overfitting, a typical problem in unregularised GP [1]. There are a number of reasons why overfitting is occurring in these preliminary experiments. First and foremost, this is attributed to the limited number of examples for the to-be-verified speakers in each dataset. A second reason is the absence of both model selection and regularisation from the learning process. In light of the above, we attempted to limit the syntactic complexity of the evolved programs by setting the maximum tree-depth allowed during search to 8, however this was not adequate for preventing overfitting.

The generalisation performance of different systems is presented in the second part of Table 2. Table 4 presents the p-values of a two-sided Wilcoxon rank sum test, which tests the *null* hypothesis that two data samples have equal medians, against the alternative that they don't. We set the significance level α to 0.05. Median test accuracy of ST is shown to be statistically superior to rest of the systems AVE, RS, US for speakers FJEN0, MMGC0, MPGR1. In addition, ST median is shown to be statistically superior against that of (a) AVE for speakers FPJF0, FSAH0; (b) RS for speakers FGRW0, FPJF0, MTRT0; and (c) US for speakers MJDC0. This result is consistent with the findings in [4], which showed that the MSE-based loss function of Eq. 1 routinely outperformed loss functions based on classification accuracy or the weighted average between true positive and true negative rates (Eq. 3). The results also suggest that ST, which uses the original unbalanced datasets, is often statistically superior or no different to the data-sampling methods of RS and US. Specifically ST is statistically better in 6/10 speakers, and statistically worse in 1/10 speakers against RS. Also, ST

Table 3. Test classification accuracy for AVE. Interquartile range in parentheses.

Speaker id	AVE($w = 0.5$) Median	Max	AVE($w = 0.6$) Median	Max	AVE($w = 0.7$) Median	Max	AVE($w = 0.8$) Median	Max
FGRW0	0.90 (0.06)	0.97	0.88 (0.11)	0.99	0.84 (0.10)	0.97	0.86 (0.14)	0.96
FJEN0	0.85 (0.06)	0.91	0.85 (0.06)	0.93	0.82 (0.10)	0.90	0.82 (0.10)	0.93
FPJF0	0.86 (0.05)	0.95	0.89 (0.12)	0.95	0.86 (0.09)	0.95	0.86 (0.06)	0.94
FSAH0	0.84 (0.05)	0.91	0.86 (0.06)	0.94	0.82 (0.10)	0.96	0.84 (0.06)	0.91
MEFG0	0.93 (0.05)	0.99	0.93 (0.06)	1.00	0.91 (0.06)	0.95	0.94 (0.05)	0.97
MJDC0	0.90 (0.08)	0.96	0.86 (0.06)	0.96	0.78 (0.12)	0.90	0.84 (0.08)	0.95
MKDD0	0.93 (0.07)	1.00	0.88 (0.12)	0.97	0.91 (0.10)	0.99	0.88 (0.12)	0.99
MMGC0	0.81 (0.14)	0.94	0.81 (0.08)	0.90	0.81 (0.09)	0.90	0.84 (0.07)	0.94
MPGR1	0.79 (0.10)	0.94	0.79 (0.06)	0.93	0.85 (0.09)	0.93	0.82 (0.10)	0.94
MTRT0	0.89 (0.07)	0.94	0.85 (0.10)	0.94	0.85 (0.07)	0.96	0.84 (0.12)	0.95

is statistically better in 4/10 speakers, and statistically worse in 1/10 speakers against US.

Overall, the median of ST is equal to 90 % in 6/10 speakers, and greater than 90 % in 4/10 speakers. This suggests that in 50 runs, the median generalisation performance of ST is consistently equal or better to the performance of a classifier that always outputs the majority class label. The median test accuracy is higher than or equal to 90 % in 4/10 speakers for AVE; 4/10 speakers for RS; and 5/10 speakers for US. Nevertheless, among 50 runs, the maximum test classification accuracy that is achieved through evolution is always higher than the one yielded from the classifier that always outputs the majority class label, for all speakers.

Table 3 presents the test accuracy statistics for the different values of w in the loss function (Eq. 3) of the Ave system. A two-sided Wilcoxon rank sum test is performed to test the difference in the median values. The table with

Table 4. p-values of Winlcoxon rank-sum test. AVE uses $w = 0.5$.

FGRW0	ST	RS	US	FJEN0	ST	RS	US	FPJF0	ST	RS	US
AVE	0.27	0.50	0.03	AVE	0.00	0.00	0.00	AVE	0.02	0.32	0.20
ST		0.03	0.08	ST		0.04	0.00	ST		0.02	0.92
RS			0.00	RS			0.09	RS			0.23

FSAH0	ST	RS	US	MEFG0	ST	RS	US	MJDC0	ST	RS	US
AVE	0.00	0.00	0.00	AVE	0.00	0.42	0.01	AVE	0.17	0.79	0.05
ST		0.17	0.47	ST		0.00	0.65	ST		0.03	0.00
RS			0.16	RS			0.00	RS			0.00

MKDD0	ST	RS	US	MMGC0	ST	RS	US	MPGR1	ST	RS	US
AVE	0.35	0.71	0.00	AVE	0.00	0.00	0.00	AVE	0.00	0.00	0.00
ST		0.12	0.00	ST		0.00	0.00	ST		0.01	0.00
RS			0.00	RS			0.87	RS			0.00

MTRT0	ST	RS	US
AVE	0.21	0.28	0.19
ST		0.00	0.80
RS			0.00

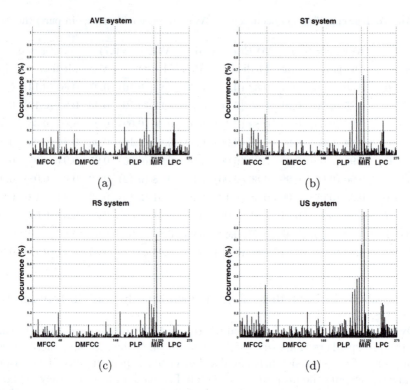

Fig. 1. Mean of the occurrence of all features in the top performing programs from independent runs for the AVE, ST, RS and US systems described in Sect. 4.3. Features are grouped into the mel-frequency cepstral coefficients (MFCC), their derivatives (DMFCC), perceptual linear prediction (PLP), other spectral features (MIR) and the linear prediction coefficients (LPC).

the p-values is omitted due to space limitations. We found that no value of w, where $w \neq 0.5$ shows a significantly better test classification accuracy compared to equal weighing. This finding is in accordance with the result reported in [4].

5.2   Feature Selection

We can determine the most beneficial features for the given problem by examining which features are most often chosen by high performing trees. A similar method has been used for feature selection in musical instrument analysis [27]. In examining these successful features, we only considered those programs from 50 independent GP runs that attained a classification accuracy greater than 90 %. A plot of the mean percentage of times each feature is chosen by a successful classifier is shown in Fig. 1. A more detailed account of the top 20 chosen features for each system is given in Table 5. This table names each of the top chosen features, reporting the mean percentage of times this features was chosen from the list of all 275 features and the standard error of this selection.

Table 5. Top 20 features for each system. Values are reported as the mean percentage of times each feature was chosen by a successful classifier (accuracy greater than 94 %) with the standard error in parenthesis.

Ave		ST		RS		US	
Feature	Mean(%)	Feature	Mean(%)	Feature	Mean(%)	Feature	Mean(%)
Inharm	0.89 (0.38)	Inharm	0.66 (0.2)	Inharm	0.85 (0.16)	Inharm	1.03 (0.23)
plp9_1	0.39 (0.18)	plp7_1	0.53 (0.1)	plp7_1	0.3 (0.11)	plp9_1	0.76 (0.15)
plp6_1	0.34 (0.27)	plp9_1	0.44 (0.11)	plp8_1	0.27 (0.06)	plp8_1	0.49 (0.18)
lpcc4	0.27 (0.1)	plp8_1	0.43 (0.11)	plp9_1	0.24 (0.06)	plp7_1	0.47 (0.09)
plpRast5_2	0.23 (0.13)	mfcc12_1	0.34 (0.08)	plpRast3_2	0.21 (0.01)	mfcc12_1	0.43 (0.1)
lpcc3	0.2 (0.15)	lpcc7	0.28 (0.09)	mfcc12_1	0.19 (0.03)	plp6_1	0.4 (0.16)
mfcc12_1	0.19 (0.04)	plp5_1	0.28 (0.1)	plp5_1	0.19 (0.05)	plp5_1	0.37 (0.1)
plp5_1	0.19 (0.1)	mfcc6_1	0.22 (0.1)	plp8_3	0.16 (0.05)	lpcc5	0.28 (0.14)
lpcc2	0.18 (0.17)	mfcc7_1	0.2 (0.08)	mfcc3_1	0.14 (0.04)	lpcc7	0.27 (0.16)
lpcc5	0.18 (0.1)	lpcc8	0.19 (0.09)	lpcc7	0.14 (0.03)	lpcc3	0.26 (0.1)
Dmfcc7_2	0.17 (0.07)	plp4_1	0.19 (0.07)	plp3_1	0.14 (0.09)	mfcc9_1	0.21 (0.06)
plp7_1	0.17 (0.06)	lpcc5	0.18 (0.09)	DDmfcc9_1	0.13 (0.02)	DDmfcc6_2	0.21 (0.06)
mfcc9_1	0.14 (0.07)	mfcc9_1	0.18 (0.06)	Centroid	0.12 (0.02)	ZeroCross	0.2 (0.09)
mfcc5_3	0.14 (0.09)	mfcc2_1	0.17 (0.08)	mfcc10_3	0.1 (0.00)	NoPeaks	0.19 (0.1)
plp3_1	0.13 (0.12)	mfcc8_1	0.12 (0.06)	Dmfcc5_2	0.1 (0.00)	mfcc4_1	0.18 (0.05)
plp4_1	0.12 (0.08)	plp9_2	0.12 (0.05)	DDmfcc1_2	0.1 (0.00)	mfcc6_1	0.17 (0.07)
DDmfcc6_1	0.12 (0.11)	mfcc10_2	0.12 (0.08)	lpcc3	0.1 (0.04)	lpcc9	0.17 (0.06)
plp8_1	0.11 (0.08)	DDmfcc4_1	0.12 (0.05)	lsf19	0.09 (0.09)	plp4_1	0.16 (0.04)
plpRast6_2	0.1 (0.05)	Dmfcc6_1	0.12 (0.07)	mfcc7_1	0.09 (0.04)	mfcc2_1	0.16 (0.11)
ZeroCross	0.1 (0.1)	lpcc3	0.11 (0.06)	mfcc6_1	0.08 (0.05)	DDmfcc11_1	0.15 (0.05)

From the plots in Fig. 1 it is clear that certain features are chosen more consistently by high performing classifiers than others. In each system investigated there is a strong peak at feature number 218. We can see from Table 5 that this corresponds to Inharmonicity. If a sound is perfectly internally 'harmonious' each of the upper partials will be integer multiples of the fundamental frequency. Inharmonicity is a measure of how much the spectral content of a sound differs from this ideal relationship. Although it has been generally used as a musical descriptor, its prominent and consistent selection in high performing classifiers in these experiments indicate that it may be a very strong indicator for voice verification also. Other individual spectral features are not strongly represented although the Zero Crossing Rate, Spectral Centroid and the Number of Spectral Peaks did appear in the top 20 features chosen by at least one system.

From Table 5 we can see that higher order PLPs were the next most selected feature. Within these only the first PC was chosen, indicating that the variance in the principle dimension for these features contains the most useful information. Surprisingly, the RASTA variations were not selected as frequently implying that the original implementation of the PLPs are more important for this problem. This may be because we used the high quality audio signal from the TIMIT database without adding noise. The RASTA method was developed to compensate for noisy channels, but as our signals are not noisy they are not found to be more beneficial than the standard PLP implementation.

The LPCCs were prominent among the highly selected features. LPCC3 was within the top 20 for each system and LPCCs 7 and 5 also featured in three of the four systems. Interestingly, the LPCs did not feature as strongly as their cepstral counterparts, indicating that in linear prediction for these problems the cepstral domain may be influential than the spectral domain. MFCCs have for a long time been one of the most widely used features in speech analysis. It may be surprising then to see that they did not appear as prominently as other features already discussed. In saying that, the first PC of a number of higher MFCCs did emerge as consistently chosen by successful classifiers. The derivatives of the MFCCs were among the least successful features.

6   Conclusions and Future Work

In this work we applied GP to evolve speaker verification programs on highly unbalanced training datasets. We found that using a number of independent runs, it is possible to evolve good-generalising programs, however good generalisation is not consistent in terms of median performance across all runs for the majority of systems. The MSE-based loss function that measures the discrepancy between program output and target value attained a median generalisation performance that is at least as good as the "majority classifier" for all speakers. This outperformed the loss function based on the weighted accuracy between minority and majority classes for most speakers. In addition, the MSE-based loss function performed better when used on the original unbalanced dataset than when used in combination with down-sampling in nearly all speakers. Finally, the use of non-equal weighted misclassification costs for the minority and majority classes did not significantly improve generalisation compared to an equal weighting.

In future work we plan to improve on the overfitting problem encountered during training. Restricting the size of the evolved solutions did not provide an effective remedy. We are experimenting with restricting the complexity of programs in the ST system through the use of a first-order Tikhonov-based regulariser that penalises functions that change rapidly. The Tikhonov function is minimised through multi-objective optimisation. Another possible solution against overfitting is the use of validation-based model selection for designating the output of the evolutionary process. Holding-out a validation set is however prohibitive using the limited training resources currently available. The use of analytical model selection methods that estimate the optimism of training error in terms of program complexity and training sample size is a possible line of attack. Having highlighted the limitations of standard GP fitness functions and data-sampling methods, we plan to experiment with additional fitness functions that were shown to be very effective in unbalanced datasets [4]. The use of the Area Under Curve as a fitness function is expected to drastically improve performance. We will compare these new experiments against current state of the art machine learning methods.

The feature selection reported is averaged across all successful speakers for each method. Our next experiment will evolve classifiers with GP using only the subset of top 20 features selected in this study. This will determine not only what features to use, but the best way to combine them for robust ASV. Furthermore, we will consider the selection of features by individual speakers to investigate the dependency of feature extraction on individual voice characteristics.

Acknowledgments. This work was carried out as a collaboration of projects funded by Science Foundation Ireland under grant Grant Numbers 08/SRC/FM1389 and 13/IA/1850.

References

1. Agapitos, A., Brabazon, A., O'Neill, M.: Controlling overfitting in symbolic regression based on a bias/variance error decomposition. In: Coello, C.A.C., Cutello, V., Deb, K., Forrest, S., Nicosia, G., Pavone, M. (eds.) PPSN 2012, Part I. LNCS, vol. 7491, pp. 438–447. Springer, Heidelberg (2012)
2. Batista, G., Prati, R.C., Monard, M.C.: A study of the behavior of several methods for balancing machine learning training data. SIGKDD Explor. Newsl. **6**(1), 20–29 (2004)
3. Batista, G.E.A.P.A., Prati, R.C., Monard, M.C.: Balancing strategies and class overlapping. In: Famili, A.F., Kok, J.N., Peña, J.M., Siebes, A., Feelders, A. (eds.) IDA 2005. LNCS, vol. 3646, pp. 24–35. Springer, Heidelberg (2005)
4. Bhowan, U., Johnston, M., Zhang, M.: Developing new fitness functions in genetic programming for classification with unbalanced data. IEEE Trans. Syst. Man Cybern. B Cybern. **42**(2), 406–421 (2012)
5. Bhowan, U., Johnston, M., Zhang, M., Yao, X.: Evolving diverse ensembles using genetic programming for classification with unbalanced data. IEEE Trans. Evol. Comput. **17**(3), 368–386 (2013)
6. Burton, D.: Text-dependent speaker verification using vector quantization source coding. IEEE Trans. Acoust. Speech Signal Process. **35**(2), 133–143 (1987)
7. Campbell, W.M., Sturim, D.E., Reynolds, D.A.: Support vector machines using gmm supervectors for speaker verification. IEEE Signal Process. Lett. **13**(5), 308–311 (2006)
8. Curry, R., Lichodzijewski, P., Heywood, M.I.: Scaling genetic programming to large datasets using hierarchical dynamic subset selection. IEEE Trans. Syst. Man Cybern. B Cybern. **37**(4), 1065–1073 (2007)
9. Day, P., Nandi, A.K.: Robust text-independent speaker verification using genetic programming. IEEE Trans. Audio Speech Lang. Process. **15**(1), 285–295 (2007)
10. Dehak, N., Kenny, P.J., Dehak, R., Dumouchel, P., Ouellet, P.: Front-end factor analysis for speaker verification. IEEE Trans. Audio Speech Lang. Process. **19**(4), 788–798 (2011)
11. Doucette, J., Heywood, M.I.: GP classification under imbalanced data sets: active sub-sampling and auc approximation. In: O'Neill, M., Vanneschi, L., Gustafson, S., Esparcia Alcázar, A.I., De Falco, I., Della Cioppa, A., Tarantino, E. (eds.) EuroGP 2008. LNCS, vol. 4971, pp. 266–277. Springer, Heidelberg (2008)
12. Eggermont, J., Eiben, A.E., van Hemert, J.: Adapting the fitness function in gp for data mining. In: Langdon, W.B., Fogarty, T.C., Nordin, P., Poli, R. (eds.) EuroGP 1999. LNCS, vol. 1598, pp. 193–202. Springer, Heidelberg (1999)

13. Garofolo, J.S., Lamel, L.F., Fisher, W.M., Fiscus, J.G., Pallett, D.S.: Darpa timit acoustic-phonetic continous speech corpus cd-rom. nist speech disc 1–1.1. NASA STI/Recon technical report n **93**, 27403 (1993)

14. Gathercole, C., Ross, P.: Dynamic training subset selection for supervised learning in genetic programming. PPSN III. LNCS, vol. 866, pp. 312–321. Springer, Jerusalem (1994)

15. Gonçalves, I., Silva, S., Melo, J.B., Carreiras, J.M.B.: Random sampling technique for overfitting control in genetic programming. In: Moraglio, A., Silva, S., Krawiec, K., Machado, P., Cotta, C. (eds.) EuroGP 2012. LNCS, vol. 7244, pp. 218–229. Springer, Heidelberg (2012)

16. Hermansky, H.: Perceptual linear predictive (plp) analysis of speech. J. Acoust. Soc. Am. **87**, 1738 (1990)

17. Hermansky, H., Morgan, N., Bayya, A., Kohn, P.: Rasta-plp speech analysis technique. In: IEEE International Conference on Acoustics, Speech, and Signal Processing, ICASSP 1992, vol. 1, pp. 121–124. IEEE (1992)

18. Holmes, J.H.: Differential negative reinforcement improves classifier system learning rate in two-class problems with unequal base rates. In: 3rd Annual Conference on Genetic Programming, pp. 635–642. ICSC Academic Press (1998)

19. Huang, X., Acero, A., Hon, H.W., et al.: Spoken Language Processing, vol. 15. Prentice Hall PTR, New Jersey (2001)

20. Kenny, P., Boulianne, G., Ouellet, P., Dumouchel, P.: Factor analysis simplified. In: Proceedings of ICASSP, vol. 1, pp. 637–640. Citeseer (2005)

21. Kenny, P., Boulianne, G., Ouellet, P., Dumouchel, P.: Joint factor analysis versus eigenchannels in speaker recognition. IEEE Trans. Audio Speech Lang. Process. **15**(4), 1435–1447 (2007)

22. Kinnunen, T., Li, H.: An overview of text-independent speaker recognition: from features to supervectors. Speech Commun. **52**(1), 12–40 (2010)

23. Kubat, M., Matwin, S.: Addressing the curse of imbalanced training sets: one-sided selection. In: Fisher, D.H. (ed.) Proceedings of the Fourteenth International Conference on Machine Learning (ICML 1997), Nashville, Tennessee, USA, July 8–12, 1997, pp. 179–186. Morgan Kaufmann (1997)

24. Lartillot, O., Toiviainen, P.: A matlab toolbox for musical feature extraction from audio. In: International Conference on Digital Audio Effects, pp. 237–244 (2007)

25. Liares, L.R., Garca-Mateo, C., Alba-Castro, J.L.: On combining classifiers for speaker authentication. Pattern Recogn. **36**(2), 347–359 (2003)

26. Logan, B., et al.: Mel frequency cepstral coefficient for music modelling. In: ISMIR (2000)

27. Loughran, R., Walker, J., O'Neill, M., McDermott, J.: Genetic programming for musical sound analysis. In: Machado, P., Romero, J., Carballal, A. (eds.) EvoMUSART 2012. LNCS, vol. 7247, pp. 176–186. Springer, Heidelberg (2012)

28. Makhoul, J.: Linear prediction: a tutorial review. Proc. IEEE **63**(4), 561–580 (1975)

29. Márquez-Vera, C., Cano, A., Romero, C., Ventura, S.: Predicting student failure at school using genetic programming and different data mining approaches with high dimensional and imbalanced data. Appl. Intell. **38**(3), 315–330 (2013)

30. O'Shaughnessy, D.: Speech communication: human and machine. Digital Signal Processing. Addison-Wesley, Reading (1987)

31. Reynolds, D.A., Quatieri, T.F., Dunn, R.B.: Speaker verification using adapted gaussian mixture models. Digital sig. process **10**(1), 19–41 (2000)

32. Sivaram, G.S., Thomas, S., Hermansky, H.: Mixture of auto-associative neural networks for speaker verification. In: INTERSPEECH, pp. 2381–2384 (2011)

33. Song, D., Heywood, M.I., Zincir-Heywood, A.N.: Training genetic programming on half a million patterns: an example from anomaly detection. IEEE Trans. Evol. Comput. **9**(3), 225–239 (2005)
34. Winkler, S.M., Affenzeller, M., Wagner, S.: Advanced genetic programming based machine learning. J. Math. Model. Algorithms **6**(3), 455–480 (2007)

Binary Tomography Reconstruction by Particle Aggregation

Mohammad Majid al-Rifaie$^{(\boxtimes)}$ and Tim Blackwell

Department of Computing Goldsmiths, University of London,
New Cross, London SE14 6NW, UK
m.majid@gold.ac.uk

Abstract. This paper presents a novel reconstruction algorithm for binary tomography based on the movement of particles. Particle Aggregate Reconstruction Technique (PART) supposes that pixel values are particles, and that the particles can diffuse through the image, sticking together in regions of uniform pixel value known as aggregates. The algorithm is tested on four phantoms of varying sizes and numbers of forward projections and compared to a random search algorithm and to SART, a standard algebraic reconstruction method. PART, in this small study, is shown to be capable of zero error reconstruction and compares favourably with SART and random search.

Keywords: Binary tomography · Discrete tomography · Particle aggregation · Underdetermined linear systems

1   Introduction

Tomographic reconstruction is the process of inferring the internal structure of an object from a set of projected images. The projected images are records of the quantity of penetrating radiation that has passed through, or has been emitted from the interior of, the object in question. There are many applications, ranging from medical imaging (CT, SPECT, PET and MRI) [3,4,16] to oceanography (seismic tomography) [14] and quantum tomography (quantum state tomography) [5].

Although an exact reconstruction is possible by use of the inverse Radon transform, in practice the discrete nature of the imaging, and the finite number of available projections, mean that approximate, discrete, techniques must be employed. The continuous density distribution of the object is modelled as a grid of pixels and the projections are acquired in bins becasue cameras consist of arrays of detectors of finite size [4].

Even after discrete modelling, the remaining mathematical problem may be ill-defined due to underdetermination: the number of independent relationships amongst the unknown quantities is fewer than their number. As a result, the solution of the inverse problem is not unique, and indeed very many solutions might exist.

This incompleteness of data arises from cost, time and geometrical concerns. For instance, the importance of cost reduction in industrial applications results

© Springer International Publishing Switzerland 2016
G. Squillero and P. Burelli (Eds.): EvoApplications 2016, Part I, LNCS 9597, pp. 754–769, 2016.
DOI: 10.1007/978-3-319-31204-0_48

in shortened scan duration and fewer projected images; similarly, in electron tomography, the damage caused to the sample by the electron beam reduces the number of collectable projections [13].

The classical filtered back projection [7,11] technique is a relatively quick and effective reconstruction procedure. However, increasing computation power means that algebraic reconstruction techniques (algebraic-RT or ART) are gaining eminence. This is due to ART's potential for greater accuracy, albeit at increased time of execution.

The first ART algorithm was a rediscovery [6] of the Kaczmarz method for solving linear equations [10]. An improved Kaczmarz method for image reconstruction, SART, (simultaneous-ART) was proposed by Andersen and Kac [1]. SART remains popular to this day and has been the subject of mathematical analysis (for example, [9]).

Prior knowledge can inform algorithms and speed up computation. For example, if it is known that the object is composed of just a few regions of homogenous density, discrete tomography can be employed. The aim is to reconstruct an image that is composed of just a few greyscale values. And, as an extreme instance of discrete tomography, if just two greyscale values are assumed, corresponding to the interior and exterior of the object, the problem is to find a binary reconstruction [8].

The aim of this paper is to investigate a new binary reconstruction technique based on the aggregation of particles. The idea is to suppose that pixel values 0 and 1 represent particles that may be absent or present in a particular cell (a pixel), and for isolated particles to move freely until they meet, and thereupon stick to, clusters of other particles. The underlying assumption is that the preferred solutions to the inverse problem will be those solutions that are more homogeneous.

The paper continues with an overview of tomography and of reconstruction. Then, the aggregation algorithm, Particle aggregate-RT (PART) is specified; a section detailing a sequence of experiments that tests and compares PART to SART and to a random search on a number of phantoms (i.e. pre-prepared exact images) follows. The results are then reported and evaluated. The paper ends with a summary of the main findings and suggestions for future research.

2   Tomography and Algebraic Reconstruction

There are two important imaging modalities, parallel beam and fan beam tomography. In either modality, an array of detectors is rotated to lie at a number of (usually) equally spaced angles in $[0, \pi)$. Figure 1 shows the two modalities and the pixellated representation of the object. Ideally, if the detectors have perfect collimators, each detector will record the amount of radiation received in a finite width beam.

However, an approximate model of the physical measurement must be built in order to formalise the mathematical reconstruction problem. This approximation is called the forward model. Beams are typically modelled by parallel rays

Left: parallel beam geometry; right: fan beam geometry

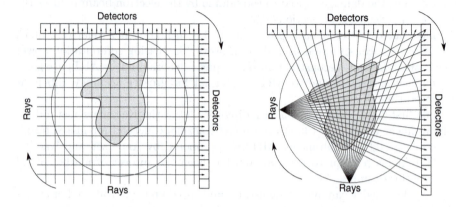

Fig. 1. Tomography geometry

(Fig. 1 left). Each ray is incident on the centre of each detector or projection bin. The imaging process is approximated by a projection matrix $A \in \mathbb{R}^{m \times n}_{\geq 0}$ where m is the total number of rays collected (equal to the number of rays at each projection angle multiplied by the number of projection angles) and n is the number of pixels in the reconstructed image. If $b \in \mathbb{R}^m$ is a vector of detector values, the continuous/discrete reconstruction problem can be stated as:

$$\text{find } x \begin{cases} \in \mathbb{R}^n \\ \in \{0, 1, \ldots, k-1\}^n, k > 1 \end{cases} \quad \text{such that } Ax = b. \tag{1}$$

The binary problem is $k = 2$ i.e. with $x \in \{0, 1\}^n$.

The methods used to compute the entries A_{ij} of the projection matrix vary. A simple procedure is to count the number of pixels that each ray passes through. The more refined *line model* uses the length of the intersection between the ray and the pixel.

Since the equation $Ax = b$ is, in general, underdetermined, it cannot be inverted. Instead an approximate solution y must be obtained (for example, by filtered back projection, or SART). This trial solution is forward projected according to the measurement model:

$$Ay = c$$

with an associated l_p error

$$\epsilon(y) = ||b - c||_p$$

where the l_p, $p \geq 1$, norm is defined

$$||v||_p \equiv \left(\sum |x|^p \right)^{\frac{1}{p}} .$$

An iterative scheme will produce a sequence of candidate solutions, $y^{(k)}, k = 1, 2, \ldots$, of decreasing error.

3   Reconstruction by Particle Aggregation

In many applications, the reconstructed image is expected to consist of patches of various sizes of uniform pixel value, since many physical objects of interest consist of uniform structures. Non-uniform regions with randomly varying pixel value would be construed as noisy and unphysical. Relevant reconstructed images are therefore those with low entropy.

This observation suggests the following assumption: given a number of distinct candidate reconstructions, $\{y : Ay = c\}$, with identical error $\epsilon(y)$, the preferred reconstruction is the one with the lowest entropy (or one of the reconstructions of lowest entropy, in the case of non-uniqueness). It would clearly be beneficial to equip a reconstruction algorithm with this assumption, under those conditions where the assumption might be expected to hold.

The principle idea underlying the aggregation technique proposed in this paper and motivated by the low-entropy assumption, is to suppose that pixel values are mobile particles, moving from pixel to pixel. The low-entropy assumption is implemented by requiring that particles stick together in clusters or aggregates uniform pixel value.

A model of aggregation for any random deposition process that is dominated by diffusive transport, for example electodeposition and mineral growth, was proposed in 1981 by Witten and Sander [18]. Their model, known as Diffusion Limited Aggregation or DLA, is remarkably simple: a particle is released from a random point on a boundary and subsequently follows a random walk until it strikes a stationary particle at some location within the enclosing boundary. The walking particle sticks to the stationary particle and another particle is released. Surprisingly complex dendrite-like clusters with fractal structure are formed by repeated application of this simple rule.

The reconstruction problem is converted into particle aggregation with the following correspondence:

- image x → configuration of particles,
- pixels → cells
- pixel values $1/0$ → presence/absence of a single particle,
- image → a grid of cells.

Furthermore, an objective function

- error function → objective function

converts the growth model into an *optimisation* problem: only those aggregates that lower the objective function are permitted to form.

A direct implementation of DLA as a reconstructive process would be very expensive since a randomly walking particle might pass by many isolated cells before arriving at a boundary cell; diffusion can be accelerated by causing a particle to jump from cell a to a vacant cell b, picked uniformly at random from all vacant cells. Although a jump has been made, the particle might not necessarily 'stick'.

Suppose a particle has jumped from a to b and that b is a boundary cell of a particle cluster. (Note that the boundary might lie within the cluster i.e. bounding a hole). We might suppose that whether the particle sticks or not to the cluster is conditional on the number of occupied neighbours of the boundary cell b relative to neighbour count for cell a, *and* on the fitness of the new configuration. There are a number of ways to deal with a particle that has jumped to a vacant cell but does not stick. For example, it could simply return to a. These modifications should ensure that particle diffusion builds aggregates which lower overall entropy and image error.

With these considerations in mind, the Particle Aggregate Reconstruction Technique (PART) can be specified[1]. Algorithm 1 specifies an application of PART to a single particle. Here, y is the reconstructed image, SELECT returns pixels $a, b \in y, a \neq b$, such that a is occupied and b is empty. n is the number of occupied cells in the neighbourhood (Moore or von Neumann) of a particular cell and $\epsilon(a \to b)$ is the error of the new image with the pixel a set to zero and pixel b set to 1. u is a sample drawn from $U(0, 1)$ (the uniform distribution on $[0, 1]$).

Particle Aggregation RT   Algorithm 1

1: $\{a, b\} = \text{SELECT}(y)$
2: **if** $n(a) = n(b)$ OR $u \sim U(0, 1) < p_1$ **then**
3:      **if** $\epsilon(a \to b) \leq \epsilon(y)$ OR $u \sim U(0, 1) < p_2$ **then**
4:          move particle from a to b
5:      **end if**
6: **end if**

The algorithm has two parameters p_1 and p_2. p_1 governs the influence of the local neighbourhood constraint: the requirement to move to a neighbourhood of higher local particle density. $p_1 = 1$ corresponds to a random search and the neighbourhood constraint is ignored. A move $a \to b$ will always be attempted even if the neighbourhood function n is lowered.

In contrast, p_2 governs the influence of the global constraint on the particle configuration as a whole. If $p_2 = 0$, a move $a \to b$ will always be rejected if it does not lower or equal the current error. The algorithm is greedy. If $p_2 > 0$, the algorithm is not greedy and a configuration with higher error will be accepted with probability p_2. Movement away from a local minima of ϵ can occur. In principle, p_2 might depend on the change in error (and on a steadily reducing temperature parameter as in simulated annealing).

---

[1] The source code for PART algorithm can be downloaded from http://doc.gold.ac.uk/~map01mm/PART/.

Algorithm 1 specifies a trial update of a single particle. Each application incurs a cost of a single function evaluation ($\epsilon(y)$). The algorithm is iterated until zero error or until a set number of function evaluations (FEs) has been achieved.

As stated by Reynolds [15], the three simple rules of interaction in flocks or swarms are collision avoidance, velocity matching and flock centring. The aggregating particles can be considered as individuals in a swarm. The dynamic rules of particles swarms are of the form:

1. **If** too close or colliding to neighbouring particles, move away
2. **Else if** too far from neighbours, move closer.

where rule 1 opposes crowding and rule 2 brings the particles together in a swarm. The single occupancy condition implements the anti-crowding rule, and the (conditional) move to a neighbourhood of higher particle density, as measured by the neighbourhood function n implements rule 2. The error function $\epsilon(y)$ imposes a global constraint on the swarm as a whole.

4   Experiments and Results

This section presents a series of experiments to investigate the performance of PART in binary image reconstruction. Three experiments were designed. The first, and preliminary, experiment, aims at finding a suitable value for the local constraint parameter p_1 for a single phantom of one size only. The second experiment investigates the convergence properties of PART and random search, which can be seen as a limiting case of PART. The final experiment provides a comparison between random search, the commonly used reconstruction algorithm, Simultaneous Algebraic Reconstruction Technique (SART), and PART with p_1 set to the empirical value determined in the first experiment.

4.1   Methodology

Forward Model. The acquisition geometry used for the experiments is parallel beam topology and the experiments use simulated objects (i.e. virtual phantoms). In all cases, the elements of the projection matrix were calculated from the line model.

Phantoms. Phantoms 1 and 2 (see Fig. 2) are commonly used in binary tomography [17] and the third and fourth phantoms resemble the Jaszczak phantoms used to calibrate the SPECT and PET scanning machines. The size of all the phantoms is 512×512. To carry out the experiments in images with different sizes, the phantoms or reference images have been scaled to create images of varying sizes (namely, 32×32, 64×64 and 128×128).

PART. PART is used with Moore neighbourhood. There are a number of alternatives for line 1 of Algorithm 1, the selection step in PART. The purpose of this step is to find an occupied cell, a, and a vacant cell, b. The following experiments use random selection: a and b are selected uniformly at random from the sets of all occupied/unoccupied cells. A list implementation would have been efficient, but since the numbers of occupied/unoccupied cells is roughly similar, uniform sampling over the entire grid y was used due to the ease of implementation and small time overhead. Algorithm 2 specifies SELECT; $U(y)$ is a uniform random selection of a single cell from the grid y. The value of the global constraint parameter p_2 was fixed, in all experiments, to zero.

Select Algorithm 2

1: **procedure** SELECT(y)
2:     $a \sim U(y)$
3:     **while** a is vacant **do**
4:         $a \sim U(y)$
5:     **end while**
6:     $b \sim U(y)$
7:     **while** b is occupied **do**
8:         $b \sim U(y)$
9:     **end while**
10:     **return** $\{a, b\}$
11: **end procedure**

Random Search (RS). For the purposes of these experiments, random search is defined as the PART algorithm with the neighbourhood parameter p_1 set to 1 with the consequence that a particle will always attempt a move to an unoccupied cell b even if the neighbour count of b, $n(b)$, is less than $n(a)$.

SART. The implementation of SART used here was based on Andersen and Kac's algorithm, [1]. The projection angles were selected uniformly at random ([2]). The value of the relaxation parameter λ was set to 1.9 in accordance with the recommendation of [12].

SART needs to be modified for binary reconstruction since in the unaltered form SART produces a continuum of pixel values. The reconstructed image also needs to be normalised in order to make error comparisons. The following modifications were made: any negative pixel values occurring after updating at any angle were set to zero; the final image y after updating all projection angles was normalised so that the total pixel value count of the phantom image and the reconstructed image were equal; y was thresholded at the average pixel value so that values below the average were set to zero, values above or equal tot he average were set to 1.

Measure. The principle performance measure is the l_1 norm, $\epsilon(y) = (\sum |b - c|)$ where, for the phantom image x, $Ax = b$ and for the reconstructed image, $Ay = c$. A secondary measure, used for comparison amongst iterated algorithms, is the number of evaluations of ϵ (function evaluations or FEs) needed to attain a given error.

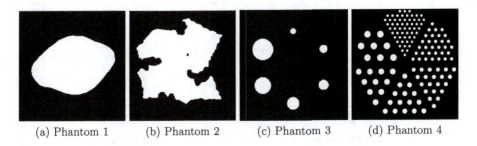

(a) Phantom 1   (b) Phantom 2   (c) Phantom 3   (d) Phantom 4

Fig. 2. Phantom images used in the experiments

4.2   Experiment I: Neighbourhood Constraint Parameter, p_1

The aim of this preliminary experiment is to find a suitable value for the parameter p_1 which determines the probability of ignoring the neighbourhood constraints. Phantom 3 is used in this experiment (see Fig. 2c), with dimension 32×32 and 32 angles of projections. The graphs in Fig. 3 show the result of a typical PART run. The termination condition is $50,000$ FEs.

As shown in Fig. 3(b), $p_1 = \{0.1, 0.2, 0.3\}$ produces zero error, and when the number of FEs needed to achieve this error is taken into account (Fig. 3(a)), the most suitable value is $p_1 = 0.1$.

4.3   Experiment II: PART and RS convergence

Figure 4 reports on a typical run of PART at $p_1 = 0.1, p_2 = 0$ and RS ($p_1 = 1, p_2 = 0$) on phantoms of size 64×64 with 32 angles ($64 \times 64 \times 32$). The termination condition is zero error or $50,000$ FEs.

As shown in the graphs in Fig. 4, PART outperforms RS before the maximum allowed iterations for phantoms 1-3. RS appears to be superior for phantom 4 for large numbers of FE's (beyond 30000). This is possibly due to the presence of isolated pixels with value 1 in the phantoms as a result of scaling down the image to 64×64; PART will struggle to place a particle on an isolated ($n(b) = 0$) cell.

The PART convergence curves begin to level off at about 10000 particle updates, i.e. at about 8% of the number of measurements ($64^2 \times 32$).

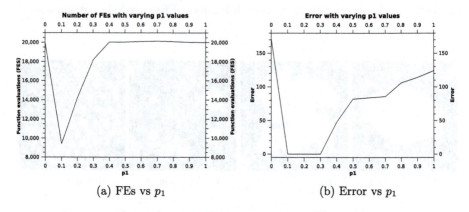

(a) FEs vs p_1 (b) Error vs p_1

Fig. 3. Error and function evaluations (FEs) with varying p_1 values

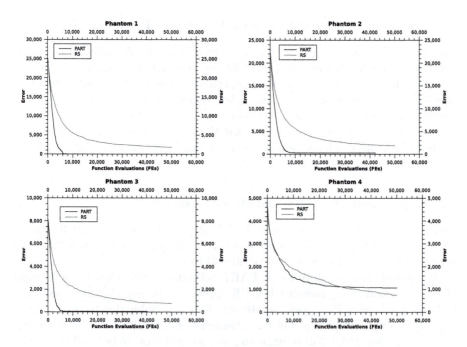

Fig. 4. Convergence plots of PART and RS algorithms

4.4   Experiment III: PART, RS and SART Comparisons

The aim of this experiment is to compare PART, RS and SART reconstructions of the four phantoms for various sizes and numbers of projection angles. 30 runs were conducted for each phantom and for the each algorithm in order to acquire adequate statistics. The termination condition for each run is zero error or 50,000 FEs. (For the purposes of this study, the number of FEs does not vary with the number of measurements.)

Figures 5 and 6 (and Table 2) illustrate the performance of the three algorithm in two separate sets of experiments, using phantoms of size 32×32 with 16 projections ($32 \times 32 \times 16$) and 64×64 with 32 projections ($64 \times 64 \times 32$) respectively. As shown in the figures, increasing the size of the phantoms, PART distances itself from RS and SART in terms of the error value and outperforms both algorithms (with the exception of phantom 4 for the reason stated before).

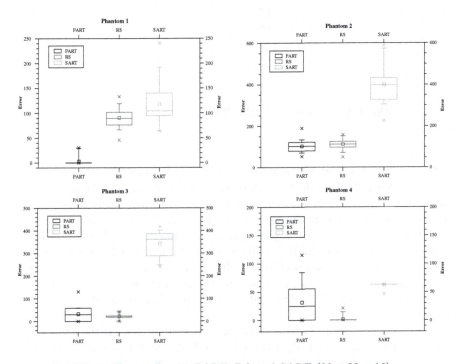

Fig. 5. Error values in PART, RA and SART ($32 \times 32 \times 16$)

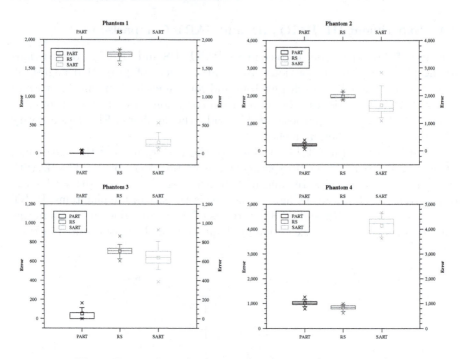

Fig. 6. Error values in PART, RA and SART ($64 \times 64 \times 32$)

In order to conduct the statistical analysis and identify the presence of any significant difference in the performance of the algorithms, Wilcoxon 1×1 non-parametric statistical test is deployed. Table 1 confirms the findings reported and shows that in the experiments for $64 \times 64 \times 32$, PART significantly outperforms RS in all cases (except in phantom 4); additionally it is shown that PART significantly outperforms SART in all cases in both $32 \times 32 \times 16$ and $64 \times 64 \times 32$ experiments. While RS outperforms SART in all cases in $32 \times 32 \times 16$ experiments, with scaling the image to 64×64 the results are reversed. Phantom 4 remains the only case where RS significantly outperforms both PART and SART algorithms.

Based on Wilcoxon 1×1 Non-Parametric Statistical Test, if the *error* difference between each pair of algorithms is significant at the 5 % level, the pairs are marked. X–o shows that the left algorithm is significantly outperforming its counterpart algorithm; and o–X shows that the right algorithm is significantly better than the one on the left. The figures, n – m, in the last row present a count of the number of X's and o's in the respective columns.

In the previous sections of the paper, the success of RS over PART and SART was cautiously attributed to the presence of isolated pixels of value 1 throughout phantom 4 due to scaling down the original phantom.

Table 1. Statistical analysis of the performance of the algorithms

(a) $32 \times 32 \times 16$

	PART − RS	RS − SART	PART − SART
Phantom 1	X − o	X − o	X − o
Phantom 2	−	X − o	X − o
Phantom 3	−	X − o	X − o
Phantom 4	o − X	X − o	X − o
Σ	1 − 1	4 − 0	4 − 0

(b) $64 \times 64 \times 32$

	PART − RS	RS − SART	PART − SART
Phantom 1	X − o	o − X	X − o
Phantom 2	X − o	o − X	X − o
Phantom 3	X − o	o − X	X − o
Phantom 4	o − X	X − o	X − o
Σ	3 − 1	1 − 3	4 − 0

To put this hypothesis to test, another experiment is designed which uses a larger phantom size of 128×128 with 16 number of angles in order to evaluate the performance of the algorithms. The results of this experiment is shown in Fig. 7 and Table 3. The results show, that while RS still outperforms SART, PART which is a simple neighbourhood aware algorithm, exhibits the best performance over all the 30 trials with statistically significant difference (see Table 4).

5   Conclusions

This paper has proposed a novel method of binary tomographic reconstruction. PART – particle aggregate reconstruction technique – is based on the idea that an image can be interpreted as a grid of cells populated by particles. Pixel values represent cell occupancy; particles are mobile and diffuse throughout the grid by random jumps, preferably landing adjacent to regions of increased particle density. The algorithm has two free parameters, a neighbourhood constraint probability, p_1, that controls the preference for jumping to increased density locations, and a global constraint p_2 that determines how likely the particle is likely to remain at a target cell if the reconstructed image error were to increase by the jump in question.

A series of experiments on four phantoms ranging in size from 32×32 to 128×128 suggest that, for $p_2 = 0$ (i.e. no uphill motion in the fitness landscape is allowed), and in the context of the specific trials conducted in this study:

– p_1 should be set to 0.1. This means that 10 % of jumps are to locations of decreased neighbourhood particle density.

Table 2. Comparing PART, RS and SART in 32 × 32 × 16 and 64 × 64 × 32

(a) 32 × 32 × 16

		Min	Max	Median	Mean	StDev
	PART	0.00	30.56	0.00	3.02	9.23
Phantom 1	**RS**	45.95	133.03	89.52	90.75	18.77
	SART	64.29	240.69	104.42	118.00	44.02
	PART	53.17	188.90	101.89	102.87	27.79
Phantom 2	**RS**	50.83	158.18	112.56	112.07	24.47
	SART	225.69	589.04	398.36	399.45	88.04
	PART	0.00	129.82	29.06	32.15	31.06
Phantom 3	**RS**	0.00	42.60	22.03	21.79	11.58
	SART	241.41	417.06	361.53	343.15	54.09
	PART	0.00	115.11	24.43	31.55	33.02
Phantom 4	**RS**	0.00	21.07	0.00	1.65	5.13
	SART	46.63	62.98	62.98	62.44	2.98

(b) 64 × 64 × 32

		Min	Max	Median	Mean	StDev
	PART	0.00	59.59	0.00	13.90	25.63
Phantom 1	**RS**	1571.18	1831.95	1745.67	1740.33	68.06
	SART	63.41	539.67	160.31	194.65	102.15
	PART	61.88	386.65	212.50	215.76	63.26
Phantom 2	**RS**	1841.01	2164.48	1943.93	1972.55	85.57
	SART	1096.49	2831.89	1537.65	1659.28	386.71
	PART	0.00	163.67	59.62	52.45	42.64
Phantom 3	**RS**	604.34	864.29	709.08	708.47	53.00
	SART	384.86	929.52	641.62	640.78	106.97
	PART	794.81	1262.07	1017.73	1021.08	93.57
Phantom 4	**RS**	623.06	995.50	854.67	849.62	88.21
	SART	3626.21	4663.74	4222.92	4142.60	325.57

- PART converges rapidly when compared to random search for phantoms with all nonzero pixel values occuring in connected regions. And in the case that there are isolated nonzero values pixels, PART will find better reconstructions at fewer iterations.
- PART performs (statistically) significantly well when compared to random search and a standard algebraic reconstruction technique for 32×32 and 64×64 phantoms, except for the case of isolated nonzero pixel values;

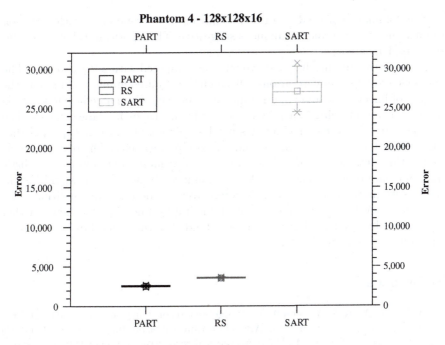

Fig. 7. Error values in reconstructing phantom 4 using PART, RA and SART in 128 × 128 × 16

Table 3. Reconstructing phantom 4 using PART, RS and SART in 128 × 128 × 16

		Min	Max	Median	Mean	StDev
Phantom 4	**PART**	2398.79	2719.48	2555.81	2564.35	84.00
	RS	3469.15	3677.52	3589.69	3584.50	58.11
	SART	24387.04	30642.35	27009.73	27110.61	1867.59

Table 4. Statistical analysis of the performance of the algorithms on phantom 4 in 128 × 128 × 16

	128 × 128 × 16		
	PART − RS	**RS − SART**	**PART − SART**
Phantom 4	X − o	o − X	X − o
Σ	1 − 0	0 − 1	1 − 0

– however for a larger (128 × 128) phantom with proportionally fewer angles of projection, PART wins out over random search and SART

The dominance of PART – or as its limiting case – random particle diffusion over a standard algebraic technique is suggestive. The algorithm is intuitive, and easily implemented.

However the findings listed above must be tempered with a few provisos. The technique requires many iterations (although the required number is less than the number of linear equations) and it remains to be seen if a more efficient selection mechanism for jumping can provide tolerable run times for large images. The $(p_1, p_2) \in [0, 1]^2$ parameter plane has not been explored and it is expected that some fitness landscape climbing would improve performance. The test phantoms used in this study are of limited size and diversity; larger and physically realistic phantoms and object images should be examined. Binary-PART would benefit from a systematic study of the parameters over a wide range of conditions.

This paper has considered binary tomography and the question whether aggregation by particle diffusion can be extended to the general discrete case, is the topic of an ongoing research.

References

1. Andersen, A., Kak, A.: Simultaneous algebraic reconstruction technique (SART): a superior implementation of the ART algorithm. Ultrason. Imaging **6**, 81–94 (1984)
2. Batenburg, K., Sijbers, J.: Dart: A practical reconstruction algorithm for discrete tomography. IEEE Trans. Image Process. **20**(9), 2542–2553 (2011)
3. Block, K.T., Uecker, M., Frahm, J.: Undersampled radial MRI with multiple coils. Iterative image reconstruction using a total variation constraint. Magn. Reson. Med. **57**(6), 1086–1098 (2007)
4. Bruyant, P.P.: Analytic and iterative reconstruction algorithms in spect. J. Nucl. Med. **43**(10), 1343–1358 (2002)
5. D'Ariano, G., Presti, P.L.: Quantum tomography for measuring experimentally the matrix elements of an arbitrary quantum operation. Phys. Rev. Lett. **86**(19), 4195 (2001)
6. Gordon, R., Bender, R., Herman, G.T.: Algebraic reconstruction techniques (art) for three-dimensional electron microscopy and x-ray photography. J. Theor. Biol. **29**(3), 471–481 (1970). http://www.sciencedirect.com/science/article/pii/0022519370901098
7. Herman, G.T.: Fundamentals of Computerized Tomography: Image Reconstruction From Projections. Springer Science & Business Media, Heidelberg (2009)
8. Herman, G.T., Kuba, A.: Advances in Discrete Tomography and Its Applications. Springer Science & Business Media, Heidelberg (2008)
9. Jiang, M., Wang, G.: Convergence of the simultaneous algebraic reconstruction technique (sart). In: Conference Record of the Thirty-Fifth Asilomar Conference on Signals, Systems and Computers, 2001, vol. 1, pp. 360–364, November 2001
10. Kaczmarz, S.: Angenaherte auflosung von systemen linearer gleichungen. Bull. Int. Acad. Polon. Sci. Lett. A **35**, 335–357 (1937)
11. Kak, A.C., Slaney, M.: Principles of computerized tomographic imaging. In: Society for Industrial and Applied Mathematics (2001)
12. Kazemini, E., Nedialkov, N.: An empirical study of algebraic reconstruction techniques. In: J., T., N., J.(eds.) Computational vision and Medical Image Processing IV, pp. 93–98. Taylor and Francis (2014)

13. Midgley, P.A., Dunin-Borkowski, R.E.: Electron tomography and holography in materials science. Nat. Mater. **8**(4), 271–280 (2009)

14. Nolet, G.: Seismic wave propagation and seismic tomography. In: Nolet, G. (ed.) Seismic Tomography: With Applications in Global Seismology and Exploration Geophysics. Seismology and Exploration Geophysics, vol. 5, pp. 1–23. Springer, Heidelberg (1987)

15. Reynolds, C.W.: Flocks, herds and schools: A distributed behavioral model. In: ACM Siggraph Computer Graphics. vol. 21, pp. 25–34. ACM (1987)

16. Ter-Pogossian, M.M.: Positron emission tomography (PET). In: Reba, R.C., Goodenough, D.J., Davidson, H.F. (eds.) Diagnostic Imaging in Medicine: Series E: Applied Sciences. NATO ASI Series, vol. 61, pp. 273–277. Springer, Heidelberg (1983)

17. Van Dalen, B.: Stability results for uniquely determined sets from two directions in discrete tomography. Discrete Math. **309**(12), 3905–3916 (2009)

18. Witten, T.A., Sander, L.M.: Diffusion-limited aggregation, a kinetic critical phenomenon. Phys. Rev. Lett. **47**, 1400–1403 (1981). http://link.aps.org//10.1103/PhysRevLett.47.1400

Population Based Ant Colony Optimization for Reconstructing ECG Signals

Yih-Chun Cheng[2(✉)], Tom Hartmann[1], Pei-Yun Tsai[2],
and Martin Middendorf[1]

[1] Parallel Computing and Complex Systems Group, Institute of Computer Science,
University Leipzig, Leipzig, Germany
{thartmann,middendorf}@informatik.uni-leipzig.de
[2] Department of Electrical Engineering, National Central University,
Taoyuan City, Taiwan
{101521083,pytsai}@cc.ncu.edu.tw

Abstract. A population based ant optimization algorithm (PACO) for reconstructing electrocardiogram (ECG) signals is proposed in this paper. In particular, the PACO algorithm is used to find a subset of nonzero positions of a sparse wavelet domain ECG signal vector which is used for the reconstruction of a signal. The proposed PACO algorithm uses a time window for fixing certain decisions of the ants during the run of the algorithm. The optimization behaviour of the PACO is compared with two random search heuristics and several algorithms from the literature for ECG signal reconstruction. Experimental results are presented for ECG signals from the MIT-BIT Arrhythmia database. The results show that the proposed PACO reconstructs ECG signals very successfully.

Keywords: Population based ACO · ECG signals · Signal reconstruction · Subset selection problem

1 Introduction

Due to population aging in many countries and rising level of cardiovascular diseases, the demand for patient-centralized healthcare systems or service is increasing. Wireless body sensor networks (WBSNs) provide a media for doctors to ubiquitously monitor patients health conditions and react in real-time when patients need urgent care. A major challenge in health monitoring systems is to avoid too much power consumption, so that a small battery can be offered for portable sensor devices. Thus, efficiently compressing long-term biomedical data is required. The advent of the Compressed Sensing (CS) technique [1] can keep signal quality while reducing data quantity. Therefore, the CS encoder should follow a simple, energy efficient, and real-time signal compression approach on wearable monitor sensor devices. In the CS decoder part, sparse signals in transformed domains are reconstructed.

© Springer International Publishing Switzerland 2016
G. Squillero and P. Burelli (Eds.): EvoApplications 2016, Part I, LNCS 9597, pp. 770–785, 2016.
DOI: 10.1007/978-3-319-31204-0_49

Using Electrocardiogram (ECG) signals for monitoring has often been discussed in the literature [2–4]. By using the state-of-the-art discrete wavelet transform (DWT) embedded ECG compression, the wavelet domain ECG signal can be approximated by a sparse signal. Despite the principle simplicity of the CS encoder, the specific design of the CS decoder to recover a sparse signal from a small number of measurements is a critical issue. To improve the signal reconstruction quality, a great amount of research efforts on solving related optimization problems have been done, e.g., [5–8]. For small problem instances exact algorithms can be applied. However, for larger problem instances and the application to ECG signals the main type of algorithms that are used for the reconstruction are greedy algorithms. The most often used greedy algorithms are orthogonal matching pursuit (OMP) [3], orthogonal multi-matching pursuit (OMMP)[9], compressive sampling matching pursuit (CoSaMP) [4], iterative hard thresholding (IHT) [4], and the recently proposed variable orthogonal multi-matching pursuit (vOMMP) [10]. Other CS reconstruction algorithms are [7,8], BP convex optimization [6], and regularized orthogonal least square methods [5]. For recent overviews see [2,11]. According to these overviews OMP is the preferred method for CS signal reconstruction with respect to reconstruction accuracy and computational time. The mentioned greedy algorithms iteratively recover signals based on previously reconstructed information. This causes the potential problem, that a signal recovering that was poor at some iteration leads also to poorly reconstructed signals in the following iterations. Therefore, it is interesting to look for other type of heuristics that can be used for CS signal reconstruction.

In this paper we propose to use the population based ant colony optimization (PACO) [12] metaheuristic for CS-based ECG signal reconstruction. PACO is an iterative method where the potentially useful information of previous iterations is stored as a population of good solutions. These solutions are used by artificial ants to create new solutions in the following iterations. The solution construction principle is similar to how solutions are constructed in Ant Colony Optimization (ACO) [13].

The PACO metaheuristic has been applied successfully to several combinatorial optimization problems (an overview is given in [14]). For example, in the recent paper [15] several evolutionary computation methods for the Traveling Salesperson problem (TSP) have been benchmarked and it was concluded that PACO is the method of choice. An additional advantage of PACO is that it is a relatively simple metaheuristic. All this makes PACO attractive as a candidate for the reconstruction of ECG signals.

In the proposed PACO algorithm the goal of the artificial ants is to correctly select the positions of K nonzero terms for a sparse wavelet domain ECG signal vector. Thus, the PACO algorithm has to solve a subset selection problem. The good subsets (i.e., subsets with positions that have a great potential of containing a large amount of signal information) that have been found by the ants are stored in the population of the PACO. The proposed PACO algorithm uses a so-called time window for fixing certain decisions of the ants during the run of the algorithm. A time window approach for fixing decisions has previously been used for PACO only once in [16]. In that paper the time window for fixing decisions was used to reduce the amount of hardware resources that are occupied by a hardware PACO. Experimentally it was shown in [16] that the time window approach improves the optimization behaviour. Moreover, in the proposed PACO a local search procedure is applied to each iteration best solution. The local search procedure exchanges some randomly selected (potentially) less useful positions with better positions. To prove the usability of the proposed PACO its performance is evaluated with signals from the MIT-BIH Arrhythmia data base [17,18]. The simulation results show that the PACO achieves high reconstruction performance.

Since PACO is related to ant colony optimization (ACO) [13] it should be mentioned that ACO has already been used in the context of ECG signals (see [19,20] for an overview). In particular, ACO was used for data clustering [21] and for finding patterns of heart diseases [22]. To the best of our knowledge PACO and ACO have never been used for the reconstruction of ECG signals.

The article is organized as follows. Section 2 provides a background on CS-based ECG signal compression and on the PACO metaheuristic. The proposed PACO algorithm is introduced in Sect. 3. Section 4 specifies the experimental setup. In Sect. 5, the experimental results are presented. Section 6 concludes the article.

2   Background

In this section we present some background information on the CS-based ECG Compression and on the PACO metaheuristic.

2.1   CS-based ECG Compression

Let $x \in \mathbb{R}^N$ be a real-valued ECG signal with an observing window that contains $N \in \mathbb{N}$ samples and is highly sparse on wavelet domain. Further, let x be expressed with respect to an orthonormal wavelet basis $\Psi = [\psi_1, \ldots, \psi_N]$. In this paper we assume - as it is commonly done for ECG signals - that the orthonormal Daubechies wavelet basis [4,10] is used and that x can be approximated as follows:

$$x = \Psi u \approx \sum_{k=1}^{N} \tilde{u}_k \psi_k,$$

where $u \in \mathbb{R}^N$ is a coefficient vector and vector $\tilde{u} = (\tilde{u}_1, \ldots, \tilde{u}_N)$ is a sparse vector with only K nonzero entries that are selected from u with $K \ll N$. The process of transforming x into Ψu is called Discrete Wavelet Transform (DWT). To compress vector $\tilde{u}$ M samples are collected by using linear combination. Here an $M \times N$ sensing matrix Φ is used ($M < N$). Thus, the CS compressed vector $v \in \mathbb{R}^M$ is obtained by $v = \Phi \tilde{u}$.

According to CS theory, the sensing matrix Φ must satisfy the restricted isometry property (RIP) or the mutual coherence property [1]. A random sensing matrix is a good option. Examples for suitable random matrices are, a matrix with values from a Gaussian random distribution $N(0,1)$ or from a symmetric Bernoulli distribution, or a sparse binary sensing matrix which satisfies another form of RIP (for more explanation see [3]).

In this paper we use an encoding decoding system from [10]. In this system the sparse binary sensing matrix Φ (from [3]) is used. This matrix has λ nonzero elements with value $\mp 1/\sqrt{\lambda}$ in each column. The positions of the λ values in each column are randomly generated with uniform probability. Matrix Ψ which is used for encoding and Ψ^{-1} which is used for decoding are constructed using the well known 4-level Daubechies-4 (db4) wavelets (see, e.g., [23]).

For the decoder the problem to reconstruct the signal x is to find a vector $\hat{u}$ that is as similar as possible to $\tilde{u}$, i.e., to reverse the compression step. When a good vector $\hat{u}$ is found it can be transformed with matrix Ψ^{-1} into the reconstructed signal $\tilde{x}$. This transformation step is called Inverse Discrete Wavelet Transform (IDWT).

Matrix Φ does not have an inverse for $M \neq N$. Therefore, one idea for finding a good vector $\hat{u}$ could be to use the pseudoinverse $\Phi^\dagger$ and to compute $\hat{u} = \Phi^\dagger v$. However, that would typically create many small nonzero values at positions where $\tilde{u}$ has zero values (Recall, that there are $N - K$ zero values in $\tilde{u}$). Therefore, another approach is taken in [10]. In this approach a subset $Y \subset [1:N]$ of K nonzero positions is estimated. Then, matrix $\Phi|_Y$ is generated from Φ by deleting all columns j of Φ with $j \notin Y$. The pseudoinverse $(\Phi|_Y)^\dagger$ of $\Phi|_Y$ is used to compute vector $\hat{u}_Y = (\Phi|_Y)^\dagger v$ (Note, that $|\hat{u}_Y| = K$). Adding $N - K$ zero elements so that a zero is at each of the positions in $[1:M] \setminus Y$ gives vector $\hat{u}$. The quality of vector $\hat{u}$ can be computed by compression with matrix Φ and then comparing the resulting vector $\Phi\hat{u}$ with v. Ideally, $\Phi\hat{u} = v$.

Now, the following optimization problem that we call the reconstruction subset problem (RSSP) occurs. Given are a coefficient vector $u = (u_1, \ldots, u_N)$ of some signal x, the compressed signal v, the sensing matrix Φ, and a sparsity level $K > 0$. The RSSP is to find a subset $Y \subset [1:N]$ of K positions which minimizes

$$f(\hat{u}) = \|v - \Phi\hat{u}\|_0 \text{ where } \hat{u} = (\hat{u}_1, \ldots, \hat{u}_N) \text{ is defined by } \hat{u}_i = \begin{cases} \hat{u}_{Y,i}, & i \in Y \\ 0, & \text{else} \end{cases} \text{ for}$$

$i \in [1:N]$ and $\hat{u}_{Y,i}$ for $i \in [1:N]$ is defined by $\hat{u}_Y = (\Phi|_Y)^\dagger v$.

In [10] a greedy algorithm was proposed to solve the RSSP. In the following subsection we describe the PACO algorithm for solving the RSSP.

2.2   Population Based ACO (PACO)

The population based ant colony optimization (PACO) metaheuristic was proposed in [12] (a recent overview and generalization of PACO can be found in [14]). PACO uses an iterative approach where the main principle is to keep a small population P of promising solutions. Population P is updated at every iteration and then transferred to the next iteration. In each iteration population P is used by m artificial ants to create new (and hopefully better) solutions. Each ant uses a probabilistic rule to create a new solution.

The details of PACO are described in the following for the case of a typical subset selection problem (SSP). An SSP can be defined as follows. Given is a set of integers $X = \{x_1, x_2, \ldots, x_n\} \subset \mathbb{N}$, an objective function $f : \mathbb{P}(X) \to \mathbb{R}$ where $\mathbb{P}(X)$ is the set of all subsets of X, and an integer k. The problem is to find a subset $Y \subset X$ that has size k, i.e., $|Y| = k$, such that $f(Y)$ is minimized. An SSP instance with set X, objective function f, and integer k is denoted by (X, f, k).

To describe the principle of PACO for solving an SSP let $P = \{Y_1, \ldots, Y_l\}$ be the population that contains $l \in \mathbb{N}$ solutions. Each artificial ant builds up a new solution Y by starting with $Y = \emptyset$ and iteratively adding a new element to Y until $|Y| = k$. For the decision which element from X should be added an ant uses the following probabilistic rule. Note, that this probabilistic rule is adopted from the typical probabilistic rule that ACO [13] uses for solving an SSP (e.g., [24, 25]). The probability to select an element $x_i \in X$ that is not included in Y so far is

$$p_i = \frac{(\tau_{init} + n_i\tau_{sol})^\alpha (\eta_i)^\beta}{\sum_{h \in S}(\tau_{init} + n_h\tau_{sol})^\alpha (\eta_h)^\beta}, \tag{1}$$

where $\tau_{init} > 0$, $\tau_{sol} > 0$ are parameters, n_i is the number of occurrences of x_i in the solutions in P, i.e., $n_i = \sum_{j=1}^{l} |Y_j \cap \{x_i\}|$, $\eta_i > 0$ is heuristic value for x_i, and S is the set of all indices h for which x_h is not element of Y so far (i.e., S is the set of indices of all elements of X that are still selectable). The value $\tau_{init} + n_i\tau_{sol}$ is often called pheromone information in the literature. Parameters $\alpha \geq 0$ and $\beta \geq 0$ are used to determine the relative influence of pheromone information and heuristic information, respectively.

After each of m artificial ants has created a new solution, the best of these solutions with respect to the objective function is added to the population P. In addition, the oldest solution is removed from P. This population update strategy has been called age-based strategy [12]. It is the standard strategy for PACO. At the beginning of the PACO algorithm either P is empty or P is filled with randomly constructed solutions. In the former case no solution is removed from P until it has reached its intended size l.

3   PACO for RSSP

In this section the details of the proposed PACO algorithm for the RSSP are described. Assume that a compressed signal v of some signal x, an integer

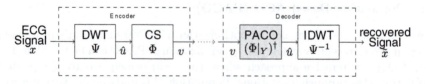

Fig. 1. System block diagram of ECG compression showing the use of PACO for solving the RSSP to detect the K nonzero positions of $\tilde{u}$ as part of the decoder (see gray square).

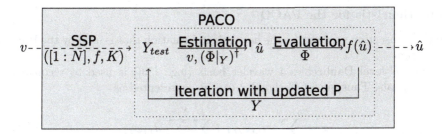

Fig. 2. Diagram of one iteration of the PACO algorithm for solving the RSSP with compressed signal v and reconstructed signal $\hat{u}$ in the wavelet domain. Input and output of PACO are denoted by dashed arrows.

Algorithm 1. Pseudocode of PACO for RSSP

Input : RSSP instance (X, f, k)
Output: $Y^* \subset X$

Initialization of P, $Y_{fix} = \emptyset$;
while *stop condition is not satisfied* **do**
 for $i = 1, \ldots, m$ **do**
  | ant i constructs solution Y_i with $Y_{fix} \subset Y_i$
 end
 set $Y := Y_j$ with $j = \arg\min_{i=[1:m]} f(Y_i)$;
 apply local search to Y to get Y';
 add Y' to P and remove oldest solution from P;
 for *each* $x_i \in Y'$ **do**
  if x_i has to be fixed set $Y_{fix} := Y_{fix} \cup \{x_i\}$;
  if $|Y_{fix}| = k$ return Y_{fix} and STOP;
 end
 if $f(Y') < f(Y^*)$ set $Y^* := Y'$;
end
return Y^*

$K \in [1 : N]$, and a sensing matrix Φ are given. To construct a solution for the RSSP each ant has to select K nonzero positions, i.e., it has to find a subset $Y \subset [1 : N]$. Then, the vector $\hat{u} = (\hat{u}_1, \ldots, \hat{u}_N)$ is computed with $\hat{u}_i = 0$ for $i \notin Y$ and $\hat{u}_i = \hat{u}_{Y,i}$ for $i \in Y$ where $\hat{u}_{Y,i}$, $i \in [1 : N]$ is defined by $\hat{u}_Y = (\Phi|_Y)^\dagger v$, and where $(\Phi|_Y)^\dagger$ is the pseudoinverse of $\Phi|_Y$. The quality of a solution Y is defined by the value $f(\hat{u}) = \|v - \Phi\hat{u}\|_0$ where smaller values correspond to better solutions. A block diagram of the ECG CS-based system with the PACO for a WSBN application is shown in Fig. 1. A scheme of the PACO algorithm for reconstructing a signal in the wavelet domain by solving the RSSP is shown in Fig. 2.

A pseudocode of the PACO algorithm is given in Algorithm 1. More details are described in the following subsections.

3.1   Heuristic for the PACO

In this subsection we describe the heuristic information that is used by the PACO algorithm.

The D-level Daubechies-4 wavelet basis (e.g., [23]) is used to reconstruct ECG signals. Thus, an ECG signal x can be represented as

$$x = \sum_{i=0}^{N_D-1} c_{D,i}\psi_{D,i} + \sum_{j=1}^{D} \sum_{i=0}^{N_j-1} d_{j,i}\varphi_{j,i},$$

where $\varphi_{l,i}$ is a wavelet function and $\psi_{l,i}$ is a scaling function with contraction $j \in [1:D]$ and transition i. At scale D, the number of coefficients is $N_D = N/2^D$. The wavelet representation spreads the ECG information onto the different wavelet levels. A position of a higher level contains more ECG information than a lower level position [4].

The (estimated) probability of a coefficient to be a nonzero term in the wavelet domain is shown in Fig. 3. The probability values in Fig. 3 have been calculated as follows. 10000 signals of MIT-BIH Arrhythmia database have been chosen randomly with uniform distribution. For every coefficient the fraction of signals where the coefficient is a nonzero term in the wavelet domain was computed and is shown in Fig. 3. Note that the probabilities are influenced by the ECG sampling frequency. Here, we have used the MIT-BIH Arrhythmia database with sampling frequency 360 Hz. Figure 3 shows that each scaling coefficient $c_{4,i}$ has a high probability of approximately 1 to be selected as a nonzero term. In contrast, wavelet coefficients at lower levels have a much lower selection probability. Due to the fact that the higher level coefficient contains more ECG information we use the values in the probability vector as heuristic information η_i in Formula (1) of the proposed PACO system. Clearly, a heuristic influence

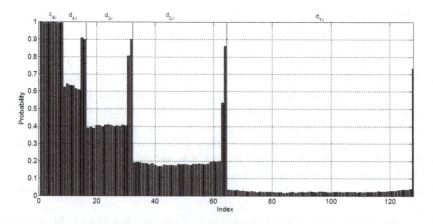

Fig. 3. Probability of a coefficient to be selected as nonzero term in the wavelet domain coefficient vector for $D = 4$. High probabilities of coefficients near the boundaries are caused by discontinuities of the circular convolution.

might also lead the ants in the wrong direction, in particular, in later stages of the algorithm when the ants have already found good solutions that are in population P. Therefore, the influence of the heuristic in the proposed PACO algorithm is decreased to zero after 50 iterations (similarly as done in [26] for ACO), i.e., after 50 iterations we set $\beta = 0$.

3.2 Time Window for PACO

In this subsection we describe a variant of the PACO algorithm that uses a time window for the decisions to fix some parts of the solution (in our case some elements of the subset Y) during the run of the algorithm. In [16] this idea was used for a hardware ACO in order to reduce the amount of hardware that the algorithm uses during a run. It was shown that this does not only reduce the amount of used hardware but can also lead to an improved optimization behaviour. The reason is that the fixed decisions make the remaining optimization problem smaller (and therefore simpler). However, to the best of our knowledge this idea has not been used in any other ACO or PACO algorithm so far.

To decide which nonzero positions $i \in [1 : N]$ should be fixed for inclusion into the solution Y the following criterion is used. It is checked in each iteration j whether a position $i \in [1 : N]$ was included in the last $\omega \in \mathbb{N} \cup \{\infty\}$ iteration best solutions, i.e., whether $i \in Y_{j-t}^*$ for all $t \in [1 : \omega]$ where Y_h^* is the iteration best solution of iteration h. If for position i at iteration j this condition is satisfied then position i is fixed as a nonzero position, i.e., i is included into the solution of every ant in all following iterations. The set of all positions that are fixed at a time is denoted by Y_{fix}. Note, that $\omega = \infty$ means that no time window is used and therefore no position will be fixed. In this $Y_{fix} = \emptyset$ during the whole run of the algorithm.

3.3 Local Search

A local search procedure is applied in each iteration to the best solution Y that the ants have found during this iteration. The local search procedure works such that solution Y' is never worse than Y, i.e., $f(Y') \leq f(Y)$. Solution Y' becomes then the iteration best solution and is included into population P.

The local search procedure works as follows. Let $Y = \{y_1, \ldots, y_K\}$ be the best solution that the ants have found in an iteration. Then, two elements $y_i \in Y$ and $x_j \in X \setminus Y$ are chosen randomly with uniform probability. It is checked if $Y' = (Y \setminus \{y_i\}) \cup \{x_j\}$ is better than Y, i.e., if $f(Y') < f(Y)$. If this is the case Y is exchanged by Y'. The described local search step is repeated s times where $s \in \mathbb{N}$ is a parameter.

4 Experimental Setup

We use records from the MIT-BIH Arrhythmia database to investigate the optimization behaviour of the proposed PACO algorithm. The database contains 48 half-hour excerpts of two channels ambulatory ECG recordings of 47 patients. Files in the database are all digitized with 11-bit resolution over 10 mV range and sampled at 360 Hz. The simulation is extracted from the 11 records number 100, 101, 102, 103, 107, 109, 111, 115, 117, 118 and 119. The signal is single lead and the sampling is done over 10 min. From each record, we selected ECG data of 128 samples ($N = 128$). The sparsity level K is set to 30. The decomposition level $D = 4$ of DWT is selected. For the sparse binary sensing matrix, the number of nonzero elements per column is set to 16, i.e., $\lambda = 16$.

For deciding the PACO parameter setting, in particular for determining how strong the heuristic influence should be (i.e., what value for parameter β should be chosen) it is helpful to study the properties of some wavelet domain ECG signals. Figure 4(a) shows an example of a relatively smooth ECG signal in time domain and Fig. 4(b) is the corresponding signal in the wavelet domain with sparsity level $K = 30$. For this kind of signal the nonzero terms are mostly within the scaling coefficients $c_{4,i}$ and other higher level wavelet coefficients. Thus, the given nonzero positions and the probability of the coefficients to be chosen as a nonzero position (see Fig. 3) are similarly distributed which makes the heuristic helpful. However, this is different for recovering more rough signals as illustrated in Fig. 4(c) and (d). In this case, for the corresponding sparse signal in the wavelet domain some nonzero positions occur also at the lower levels of

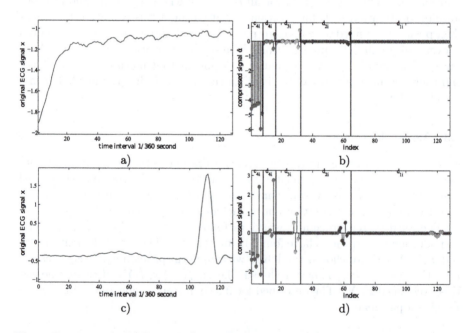

Fig. 4. Two example ECG signals (upper part: smooth signal; lower part: rough signal) of 128 samples in time domain (left) and in wavelet domain (right) with sparsity level $K = 30$. Diamonds mark nonzero values of $\tilde{u}$. Different wavelet level $d_1, \ldots, d_4, c_4$ are bounded by vertical black lines and colored diamonds.

the wavelet domain. Thus, for this type of signal, the distribution of the nonzero positions does not fit so well to the distribution provided in Fig. 3. Hence, if an ant always chooses positions with a strong heuristic influence, then it might fail in finding the good positions in the lower level of the wavelet domain. Therefore, for this type of signals the heuristic might not be so helpful or even misleading.

As a consequence, parameter β which determines the influence of the heuristic was set to $\beta = 1$ for the first 49 iterations and was then reduced to $\beta = 0$ from iteration 50 on. Other parameters of the PACO were set as $\alpha = 1.5$, $\tau_{ini} = 1/N$, and $\tau_{sol} = 2/N$. The population size and the number of ants is 20, i.e., $l = 20$ and $m = 20$. Population P was initialized with solutions $Y_1, \ldots, Y_l$ that were randomly generated based on the probability distribution of the wavelet coefficients (see Sect. 3.1). The number of local search steps for each iteration best solution was set to $s = 200$.

The PACO algorithm was tested for the different time window lengths $\omega \in \{(\infty, \infty), (40, \infty)\}$. For $a, b \in \mathbb{N} \cup \{\infty\}$ the pair (a, b) denotes the case that $\omega = a$ is used if and only if $|Y_{fix}| < 8$ and $\omega = b$ else. Here, value 8 was chosen because the probability to select the 8 coefficients of $c_{4,i}$ as nonzero values is nearly 1 (see Fig. 3). Hence, the PACO might profit when potentially the decision to include these parameters can be fixed soon.

To select the test signals we have classified the database signals into two categories depending on how good the positions of the nonzero values on the different wavelet levels are consistent with the probabilities given in Fig. 3. The reason for this is that the level of consistency might influence how advantageous it is for the PACO to use a time window and therefore to fix some decisions early. For the classification every nonzero position of a wavelet level d_i has been assigned with a weight w_i such that $w_i = 5 - i$ for all $i \in [1 : 4]$. Each of the eight coefficients from c_4 has been assigned the weight 0. A signal is classified as a hard signal if the sum of the weights of all nonzero positions is larger or equal to 60. A signal where this sum is less than 60 is classified as an easy signal. From each of the two categories of signals 100 test signals were selected randomly. With each test signal 10 test runs have been done.

The performance measure that was used for the evaluation of a reconstructed signal is the signal-to-noise ratio (SNR). Where SNR is defined as $SNR(\hat{x}) = -20 \log_{10}(\|x - \hat{x}\|_2 / \|x\|_2)$ where x and $\hat{x}$ denote the original signal and the reconstructed signal, respectively. Note, that a larger SNR value indicates a higher quality of the reconstructed signal $\hat{x}$.

To evaluate the PACO algorithm the following algorithms from the literature have been implemented for a comparison: PKS cOMMP [10], OMP [3], OMMP [9], IHT [4], and CoSaMP [4]. The implementation was done with MAT-LAB on Windows 8. All algorithms have been applied to the same test data sets. Since OMP and CoSaMP are preferred methods for CS signal reconstruction with respect to reconstruction accuracy and computational time (see [2,11] for an overview) we did not compare PACO with the following algorithms: CS reconstruction algorithms [7,8], BP convex optimization [6], and the regularized orthogonal least square methods [5].

We also compare the PACO algorithm with two random search heuristics with local search that are described in the following. For each random search heuristic 20 solutions are randomly created at every iteration (Note, that the number of solutions that are constructed per iteration is the same as for the PACO algorithm) with respect to a uniform random distribution (this heuristic is called random search) or with respect to the probabilities that are given in Fig. 3 (this heuristic is called improved random search). Same as for the PACO algorithm, 200 steps of the local search procedure as defined in Sect. 2.2 are applied to the best solution found at each iteration.

5 Results

The average quality of the solutions found by the proposed PACO algorithm is shown in Fig. 5 for the easy test instances (left part) and the hard test instances (right part). The results of two variants of the PACO - without time window (i.e., $\omega = (\infty, \infty)$) and with time window (i.e., $\omega = (40, \infty)$) - are shown. The average perfect SNR values that are obtained by a comparison of the sparse signal vector $\tilde{x}$ and the original signal vector x can also be seen.

The figure shows that the average perfect SNR for the hard test data is lower than the average perfect SNR for the easy signal data set. The reason is that the wavelet signal vector u of a hard signal has more small nonzero elements that are widely spread within the wavelet domain, whereas the sparse vector $\tilde{u}$ contains only K positions with nonzero elements in vector u. Therefore, for hard

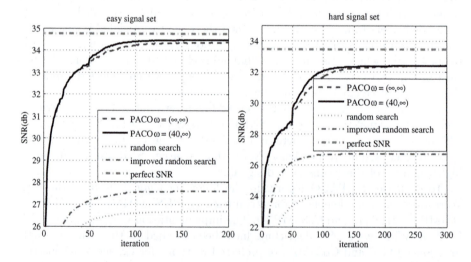

Fig. 5. Average SNR performance of the PACO algorithm with two different window settings ω) and for two random search approaches; perfect SNR values are given for comparison; left: results for the easy signal data set; right: results for the hard data set.

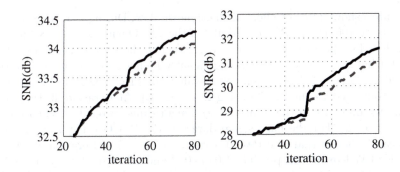

Fig. 6. Detail of Fig. 5 which shows the sudden rise of SNR at iteration 50; easy signals (left), hard signals (right); notation as in Fig. 5.

signals more information is deleted by CS than for easy signals which leads to lower SNR values for the hard signals.

For both data sets Fig. 5 shows a sudden rise of the SNR at iteration 50 (the corresponding section of the curves can be seen in more detail in Fig. 6). This rise of the SNR occurs because the value of parameter β is changed from 1 to 0 after iteration 49 which removes the influence of the heuristic. Thus, the improvements after iteration 49 are based only on the influence of population P. It can also be seen that the heuristic greatly assists the ants in finding high quality solutions with SNR ≥ 33 already within the first 49 iterations for the easy signal data set. In contrast, for the hard signals the heuristic seems to give much less help. This is indicated by the facts that the SNR remains below 29 during the first 49 iterations and that the increase after iteration 49 is much stronger for the hard signals than for the easy signals.

Table 1. Average SNR results for easy signal data set (left) and hard signal data set (right) for the PACO compared with previously presented signal reconstruction greedy algorithms from the literature. The perfect SNR values show the best achievable SNR averages. The best values for each data sets are underlined.

Algorithm	SNR	
	Easy signal set	Hard signal set
Perfect SNR	34.751	33.449
PACO, $\omega = (\infty, \infty)$	34.374	<u>32.414</u>
PACO, $\omega = (40, \infty)$	<u>34.489</u>	32.390
PKS vOMMP [10]	32.758	31.403
OMP [3]	30.541	30.541
OMMP [9]	29.286	28.748
IHT [4]	22.257	20.557
CoSaMP [4]	5.942	5.716

Figure 5 shows also the effect of the time window. During iterations approximately 50 to 100, the SNR of the results of the PACO with $\omega = (40, \infty)$ improve more than for the PACO with $\omega = (\infty, \infty)$. For the easy signal data set the PACO using $\omega = (40, \infty)$ achieves significantly higher SNR results at iterations 100 and 200 (one-sided Wilcoxon test, p-values $9.3 * 10^{-5}$ and $6.4 * 10^{-3}$, respectively) compared to the PACO using $\omega = (\infty, \infty)$. At iteration 49 there is no significant difference in the results of both algorithms (two-sided Wilcoxon test, p-value $= 0.32$). For the hard signal data set, the PACO using $\omega = (40, \infty)$ is significantly better than the PACO using $\omega = (\infty, \infty)$ at iteration 49 and 100 (one-sided Wilcoxon test, p-values $1.03 * 10^{-3}$ and $1.16 * 10^{-6}$, respectively). At iterations 200 and 300 there is no significant difference between both algorithms (two-sided Wilcoxon test, p-values 0.65 and 0.66, respectively). Here the PACO with $\omega = (\infty, \infty)$ shows slightly better results for the hard signal data set (see Table 1 for exact values).

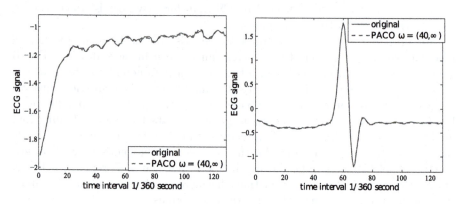

Fig. 7. Reconstruction obtained by the PACO algorithm for two example signals; left: easy signal; right: hard signal

Figure 5 shows also the results of two random search heuristics - random search and improved random search - with local search. Figure 5 shows that the SNR values of the solutions of both PACO versions are much higher than the SNR values of the solutions that have been found by the random search heuristics.

For a comparison with other algorithms from the literature, Table 1 shows SNR values that have been obtained by the PACO algorithm and the algorithms PKS cOMMP [10], OMP [3], OMMP [9], IHT [4], and CoSaMP [4]. It can be seen that the PACO achieves the best SNR results for both data sets.

For a visual impression on the quality of the reconstruction that has been obtained by the PACO algorithm Fig. 7 shows two example reconstructions (one easy signal and one hard signal). The figure shows that both reconstructed signals are very similar to their original signals.

Fogire 8 presents the SNR results of the signal reconstruction for the different algorithms with respect to different oversampling ratios M/K. For every tested

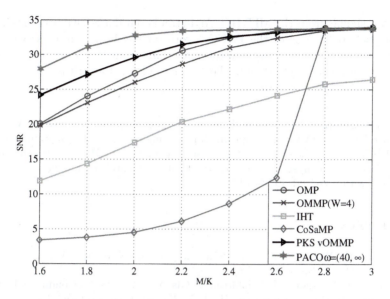

Fig. 8. Average SNR results of signal reconstruction versus oversampling ratio M/K for the proposed PACO, matching pursuit family algorithms, and IHT. Oversampling ratio is sampled for $M \in \{48, 54, \ldots, 90\}$ and $K = 30$.

oversampling ratio exactly 100 signals have been chosen randomly with uniform probability from the MIT-BIH Arrhythmia data base. The depicted SNR results are the average results over all sampled signals per oversampling ratio. Note that for smaller oversampling ratios the compressed ratio N/M is higher which means that the same signal information is transmitted using a smaller amount of measurement. The figure shows that the proposed PACO algorithm achieves the best average SNR results. For algorithms OMP, OMMP, vOMMP, and CoSaMP for the lower oversampling ratios ($M/K \leq 2.6$) the sparse binary sensing matrix has a smaller size which influences the correlation between the sensing matrix and the residual element. This causes a problem for identifying the nonzero entries, see [10] for details. This shows that a metaheuristic as PACO can be advantageous in particular for small oversampling ratios.

6   Conclusion

We have proposed a population based ant optimization (PACO) algorithm for reconstructing electrocardiogram (ECG) signals. The particular problem that the PACO algorithm solves in the reconstruction process is to find a subset of nonzero positions of a sparse wavelet domain ECG signal vector which is used for the reconstruction of a signal. A specific property of the proposed PACO algorithm is that it can use a time window for fixing certain decisions of the ants during the run of the algorithm. It was shown experimentally that this is particularly helpful for the reconstruction of relatively easy signals. For the experiments,

ECG test signals from the MIT-BIT Arrhythmia database have been used. The optimization behaviour of the PACO algorithm has been compared with two random search heuristics and several algorithms from the literature for ECG signal reconstruction. The results show that the PACO algorithm obtained the best results for the test data.

Acknowledgments. YCC received financial support granted by German Academic Exchange Service (DAAD) through the Taiwan Summer Institute Programme within 57190416. TH was funded by the German Israeli Foundation (GIF) through the project "Novel gene order analysis methods based on pattern identification in gene interaction networks" within G-1051-407.4-2013.

References

1. Candes, E., Wakin, M.: An introduction to compressive sampling. Sig. Process. Mag. IEEE **25**(2), 21–30 (2008)
2. Craven, D., McGinley, B., Kilmartin, L., Glavin, M., Jones, E.: Compressed sensing for bioelectric signals: a review. IEEE J. Biomed. Health Inform. **19**(2), 529–540 (2015)
3. Mamaghanian, H., Khaled, N., Atienza, D., Vandergheynst, P.: Compressed sensing for real-time energy-efficient ecg compression on wireless body sensor nodes. IEEE Trans. Biomed. Eng. **58**(9), 2456–2466 (2011)
4. Polania, L., Carrillo, R., Blanco-Velasco, M., Barner, K.: Exploiting prior knowledge in compressed sensing wireless ecg systems. IEEE J. Biomed. Health Inform. **19**(2), 508–519 (2015)
5. Blumensath, T., Davies, M.E.: On the difference between orthogonal matching pursuit and orthogonal least squares. Technical report, University of Edinburgh (2007)
6. Chen, S.S., Donoho, D.L., Saunders, M.A.: Atomic decomposition by basis pursuit. SIAM J. Sci. Comput. **20**(1), 33–61 (1998)
7. Dixon, A.M.R., Allstot, E.G., Gangopadhyay, D., Allstot, D.J.: Compressed sensing system considerations for ECG and EMG wireless biosensors. IEEE Trans. Biomed. Circ. Syst. **6**(2), 156–166 (2012)
8. Mamaghanian, H., Khaled, N., Atienza, D., Vandergheynst, P.: Structured sparsity models for compressively sensed electrocardiogram signals: a comparative study. In: Biomedical Circuits and Systems Conference (BioCAS), 2011, pp. 125–128. IEEE (2011)
9. Wang, J., Kwon, S., Shim, B.: Generalized orthogonal matching pursuit. IEEE Trans. Signal Process. **60**(12), 6202–6216 (2012)
10. Cheng, Y.C., Tsai, P.Y.: Low-complexity compressed sensing with variable orthogonal multi-matching pursuit and partially known support for ECG signals. In: 2015 IEEE International Symposium on Circuits and Systems (ISCAS), pp. 994–997 (2015)
11. Dixon, A.M.R., Allstot, E.G., Chen, A.Y., Gangopadhyay, D., Allstot, D.J.: Compressed sensing reconstruction: comparative study with applications to ECG biosignals. In: 2011 IEEE International Symposium on Circuits and Systems (ISCAS), pp. 805–808 (2011)

12. Guntsch, M., Middendorf, M.: A population based approach for ACO. In: Cagnoni, S., Gottlieb, J., Hart, E., Middendorf, M., Raidl, G.R. (eds.) EvoIASP 2002, EvoWorkshops 2002, EvoSTIM 2002, EvoCOP 2002, and EvoPlan 2002. LNCS, vol. 2279, pp. 72–81. Springer, Heidelberg (2002)

13. Dorigo, M., Gambardella, L.: Ant colony system: a cooperative learning approach to the traveling salesman problem. IEEE Trans. Evol. Comput. 1(1), 53–66 (1997)

14. Lin, Y., Clauss, M., Middendorf, M.: Simple probabilistic population based optimization. In: IEEE Transactions on Evolutionary Computation, no. 99, p. 1 (2015)

15. Weise, T., Chiong, R., Lässig, J.L., Tang, K., Tsutsui, S., Chen, W., Michalewicz, Z., Yao, X.: Benchmarking optimization algorithms: an open source framework for the traveling salesman problem. IEEE Comput. Intell. Mag. 9(3), 40–52 (2014)

16. Janson, S., Middendorf, M.: Flexible particle swarm optimization tasks for reconfigurable processor arrays. In: Proceedings of the 8th International Workshop on Nature Inspired Distributed Computing (NIDISC 2005), p. 8 (2005)

17. Goldberger, A.L., Amaral, L.A., Glass, L., Hausdorff, J.M., Ivanov, P.C., Mark, R.G., Mietus, J.E., Moody, G.B., Peng, C.K., Stanley, H.E.: PhysioBank, PhysioToolkit, and PhysioNet: components of a new research resource for complex physiologic signals. Circulation 101(23), 215–220 (2000)

18. Moody, G.B., Mark, R.G.: The impact of the MIT-BIH arrhythmia database. IEEE Eng. Med. Biol. Mag. 20(3), 45–50 (2001)

19. Bursa, M., Lhotska, L.: The use of ant colony inspired methods in electrocardiogram interpretation, an overview. In: The 2nd European Symposium on Natureinspired Smart Information Systems [CD-ROM], NiSIS (2006)

20. Jafar, O.M., Sivakumar, R.: Ant-based clustering algorithms: a brief survey. Int. J. Comput. Theory Eng. 2(5), 787–796 (2010)

21. Bursa, M., Lhotska, L.: Ant colony cooperative strategy in electrocardiogram and electroencephalogram data clustering. In: Krasnogor, N., Nicosia, G., Pavone, M., Pelta, D. (eds.) Nature Inspired Cooperative Strategies for Optimization (NICSO 2007). Studies in Computational Intelligence, vol. 129, pp. 323–333. Springer, Berlin (2008)

22. Ramo, F.M.: Diagnosis of heart disease based on ant colony algorithm. Int. J. Comput. Sci. Inf. Secur. 11(5), 77 (2013)

23. Walker, J.S.: A Primer on Wavelets and Their Scientific Applications. Chapman and Hall/CRC, Boca Raton (2008)

24. Abd-Alsabour, N.: Binary ant colony optimization for subset problems. In: Dehuri, S., Jagadev, A.K., Panda, M. (eds.) Multi-Objective Swarm Intelligence. Studies in Computational Intelligence, vol. 592, pp. 105–121. Springer, Berlin (2015)

25. Solnon, C., Bridge, D.: An Ant Colony Optimization Meta-Heuristic for Subset Selection Problems. Technical report RR-LIRIS-2005-017, University Lyon (2005)

26. Merkle, D., Middendorf, M., Schmeck, H.: Ant colony optimization for resourceconstrained project scheduling. IEEE Trans. Evol. Comput. 6(4), 333–346 (2002)

EvoINDUSTRY

Can Evolutionary Algorithms Beat Dynamic Programming for Hybrid Car Control?

Tobias Rodemann[1(✉)] and Ken Nishikawa[2]

[1] Honda Research Institute Europe, Carl-Legien-Strasse 30,
63073 Offenbach/Main, Germany
tobias.rodemann@honda-ri.de
[2] Graduate School of Information Science and Engineering,
Tokyo Institute of Technology, O-okayama 2-12-1, Meguro,
Tokyo 152-8552, Japan
nishikawa.k.af@m.titech.ac.jp

Abstract. Finding the best possible sequence of control actions for a
hybrid car in order to minimize fuel consumption is a well-studied problem. A standard method is Dynamic Programming (DP) that is generally
considered to provide solutions close to the global optimum in relatively
short time. To our knowledge Evolutionary Algorithms (EAs) have so
far not been used for this setting, due to the success of DP. In this work
we compare DP and EA for a well-studied example and find that for the
basic scenario EA is indeed clearly outperformed by DP in terms of calculation time and quality of solutions. But, we also find that when going
beyond the standard scenario towards more realistic (and complex) scenarios, EAs can actually deliver a performance en par or in some cases
even exceeding DP, making them useful in a number of relevant application scenarios.

Keywords: Hybrid cars · Dynamic programming · Evolutionary algorithms

1 Introduction

One of the biggest challenges for the global automotive industry is to reduce
fuel consumption and CO_2 emissions. Hybrid cars are potential solutions to this
challenge. They are powered by two separate power sources, a limited battery
supply driving electric motors (EMs) and a conventional internal combustion
engine (ICE). In order to minimize CO_2 emissions, the car's controller has to use
the limited battery power in an optimal way. This optimal control problem can
often be expressed as a simple action for each (discrete) time step, for example
the so-called torque split $u(t)$ – the share of the torque (as requested by the
driver) that is provided by the electric motor (using battery power).

For practical applications, the torque split is normally computed using simple (often fuzzy logic) rule-based systems, that consider the current driving
situation and battery state-of-charge (SOC) to compute the optimal control.

© Springer International Publishing Switzerland 2016
G. Squillero and P. Burelli (Eds.): EvoApplications 2016, Part I, LNCS 9597, pp. 789–802, 2016.
DOI: 10.1007/978-3-319-31204-0_50

Parameters of this rule set can be optimized using a variety of methods, including EAs, see for example [4–6,9,12,13]. At last year's EvoAPPS conference we have presented a study of a many-objective optimization of a controller of this type [14].

A potentially superior, but up-to-now largely academic approach, is to use methods that can consider the complete route (drive cycle) and generate a globally optimal control, for example as a sequence of $u(t)$ with $t \in \{start, end\}$. The most commonly used method is Dynamic Programming (DP) [2,6,10,16], that can quickly provide solutions close to the global optimum. A typical application scenario would use a predefined drive cycle, like the NEDC (New European Drive Cycle), and compute the best sequence of control actions using the DP (see also [11]). The target is a minimization of fuel consumption, under a relatively large number of constraints, like minimum and maximum SOC values, or torque limits. To our knowledge evolutionary algorithms (EAs) have so far not been applied for solving this problem. In this work we will investigate how EAs perform against DP in this basic scenario and a more realistic, more open scenario.

The typical usage of this global, direct optimization of control outputs integrating predictive information is to provide benchmark values for online controllers (i.e. giving an upper limit of fuel efficiency on a certain drive cycle) and as a source of inspiration to derive (in an often manual process) controller architectures for online control.

2   Hybrid Car Modeling

The main application is taken from Sundström and Guzzella [17]. They provided a Matlab model of a simple Hybrid Electric Vehicle (HEV, details see below), a fixed, standardized drive cycle and flexible Matlab code for DP. We will use all three components as the basis for our analysis. The basic application can be described as follows:

- The initial state of charge (SOC) is set to SOC_{ini}.
- The selected drive cycle determines for every point in time, the vehicle speed v. From consecutive speed values the necessary acceleration $a(t)$ is computed. Based on the vehicle model, the required total demanded torque T_{dem} can be derived.
- Next the optimizer (EA or DP) determines the current torque split factor $u(t)$ and the requested motor torque T_{EM} and combustion engine torque T_{ICE} are computed as follows:

$$T_{EM} = u(t) * T_{dem} \tag{1}$$
$$T_{ICE} = (1 - u(t)) * T_{dem} \tag{2}$$

- Through the vehicle model the fuel consumption is computed as a function of v and T_{ICE}. This calculation is based on a look-up-table (normally derived experimentally on a test bench).

- Depending on T_{EM} and $SOC(t)$ the new battery $SOC(t+1)$ is computed. In the model this is handled via measured efficiency maps and a simplified battery model.
- The HEV model also checks if any constraints are violated and, for the DP, will mark the solution $u(t)$ as infeasible if any constraint is not met.
- The process is repeated until the car arrives at its final position (end of drive cycle).
- A final penalty is computed depending on the difference between the final battery level SOC_{fin} and the target value SOC_{tar}. The difference $|SOC_{fin}-SOC_{tar}|$ of all results in this paper are smaller than 0.0001 %.
- The HEV model was modified based on the"torque assisted parallel HEV" model introduced in [18]. The engine is always mechanically connected with the motor in the torque assist hybrid. The modification was done only so that we don't have to consider the decoupling mechanism used in [17].

Due to many physical limitations, a large proportion of actions $u(t)$ will be infeasible. A smart handling of the constraints is therefore highly important for an efficient optimization process. The target is to find the sequence of $u(t)$ actions that leads to a minimal fuel consumption without violating any constraints.

2.1 Drive Cycles

In this work we are using two different drive cycles: the Japanese 10–15 drive cycle (J1015) from [17] and the New European Drive Cycle (NEDC), which is approximately twice as long as the J1015. Each drive cycle provides a specified target speed value for every point in time (or equivalently space). See Fig. 1 for a comparison of the two cycles.

Furthermore, the best gear setting is needed. It might be optimized along with $u(t)$ but would make the setting far more complex and can normally be computed from known engine efficiencies and gear ratios. We have chosen the simple option in this work. Therefore only $u(t)$ is optimized for all timesteps in the drive cycle. An additional optimization of the currently used gear could furthermore improve performance to some degree but would also increase optimization times.

2.2 Car Model

We employ a simple model of a HEV taken from [17]. The car is a parallel hybrid electric vehicle, which means that combustion engine and electric motor can work in parallel to accelerate the car. The modeled car has a total weight of 1800 kg, a battery capacity of 6 Ah (with a battery voltage around 240 V), maximum electric motor torque of 160 Nm and maximum ICE torque 200 Nm.

The simulation model only covers the 1-D (longitudinal) dynamics of the car, i.e. speed $v(t)$ and acceleration $a(t)$. The combustion engine is modeled via a look-up-table of engine efficiencies, typically derived from test-bench measurements. For the torque split u the following conditions are possible:

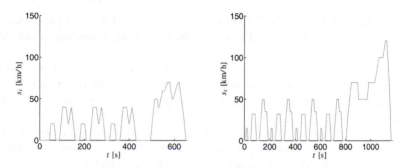

Fig. 1. Pre-defined speed over time for the two used drive cycles. *(left)* Japanese 1015 (J1015) drive cycle and *(right)* New European Drive Cycle (NEDC).

$u = 1$: Using only the electric motor or full brake energy recuperation.

$0 < u < 1$: The torque T_{dem} is provided by both internal combustion engine and electric motor.

$u = 0$: Torque is only provided by the internal combustion engine.

$u < 0$: The internal combustion engine provides more torque than T_{dem} to recharge the battery.

Figure 2 presents a fuel consumption map for the engine and an efficiency map for the electric motor depending on requested torque and crankshaft rotation speed.

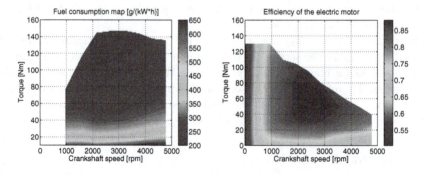

Fig. 2. Efficiency of the two power sources (left: fuel consumption map of engine, right: electric motor efficiency). Horizontal axis is motor/engine rotation speed (rpm) and vertical axis is demanded torque in Nm.

2.3   Constraints and Penalty Functions

For the adaptation of EA to the HEV control, some constraints of the HEV model must be considered in order to find feasible solutions. For the DP it is sufficient to mark any solution that violates one or more constraints as infeasible.

For the CMA-ES, the constraints are handled by adding penalty functions to the fitness values of any solution that violates constraints. The fitness value is given by,

$$J = \Sigma_{k=0}^{T-1}(\Delta m_f(u_k, k) \cdot T_s + P_1(k) + \ldots + P_7(k)) + P_8. \tag{3}$$

The main factor is the fuel mass flow (consumption) $\Delta m_f(u_k, k)$. The constraints and the penalty functions $P_1(k), \ldots, P_8(k)$ are defined as follows;

1, Torque split 1

When crankshaft speed ω_c is smaller than engine idle speed ω_0, the HEV must use only EM, that is, $u = 1$ and $T_{ICE} = 0$.

$$P_1 = \begin{cases} w_1 \cdot |u - 1|, \ \omega_c < \omega_0, \\ 0, \qquad \omega_c \geq \omega_0. \end{cases} \tag{4}$$

2, Torque split 2

When $T_{dem} < 0$, the T_{EM} must also be smaller than 0, that is, $u \geq 0$ and $T_{EM} \leq 0$.

$$P_2 = \begin{cases} w_2 \cdot |u|, \ T_{dem} < 0 \text{ and } u < 0, \\ 0, \qquad \text{otherwise.} \end{cases} \tag{5}$$

3–6, Limitation of the HEV

The engine torque T_{ICE}, electric motor torque T_{EM}, battery current I_{batt} and voltage V_{oc} must be within the feasible region, which is characterized by parameters of the HEV model.

$$P_3 = \begin{cases} w_3 \cdot (T_{ICE} - T_{ICE,max})^2, \ T_{ICE} > T_{ICE,max}, \\ w_3 \cdot (T_{ICE} - T_{ICE,min})^2, \ T_{ICE} < T_{ICE,min}, \\ 0, \qquad\qquad\qquad\qquad \text{otherwise,} \end{cases} \tag{6}$$

$$P_4 = \begin{cases} w_4 \cdot (T_{EM} - T_{EM,max})^2, \ T_{EM} > T_{EM,max}, \\ w_4 \cdot (T_{EM} - T_{EM,min})^2, \ T_{EM} < T_{EM,min}, \\ 0, \qquad\qquad\qquad\qquad \text{otherwise,} \end{cases} \tag{7}$$

$$P_5 = \begin{cases} w_5 \cdot (V_{oc}^2 - 4 \cdot R_{int} \cdot P_{EM})^2, \ V_{oc}^2 < 4 \cdot R_{int} \cdot P_{EM}, \\ 0, \qquad\qquad\qquad\qquad\ \ V_{oc}^2 \geq 4 \cdot R_{int} \cdot P_{EM}, \end{cases} \tag{8}$$

$$P_6 = \begin{cases} w_6 \cdot (|I_{batt}| - I_{max})^2, \ |I_{batt}| > I_{max}, \\ 0, \qquad\qquad\qquad\ \ |I_{batt}| \leq I_{max}, \end{cases} \tag{9}$$

where $T_{ICE,max}, T_{ICE,min}, T_{EM,max}, T_{EM,min}$ and I_{max} are maximum and minimum limits of each value, R_{int} the internal battery resistance, and P_{EM} the electric motor power.

7, Limit of SOC_k

The SOC_k must be in predefined range $[SOC_{min} \ \ SOC_{max}]$.

$$P_7(k) = \begin{cases} w_7 \cdot |SOC_k - SOC_{max}|, \ SOC_k > SOC_{max}, \\ w_7 \cdot |SOC_k - SOC_{min}|, \ SOC_k < SOC_{min}, \\ 0, \qquad\qquad\qquad\qquad \text{otherwise.} \end{cases} \tag{10}$$

8, Final SOC

When $k = T$ (end of drive cycle), the $SOC(T)$ must be bigger than the target SOC_{fin} (charge conserving operation).

$$P_8 = \begin{cases} w_8 \cdot |SOC(T) - SOC_{tar}|, & SOC(T) < SOC_{tar}, \\ 0, & SOC(T) \geq SOC_{tar}. \end{cases} \tag{11}$$

The weights w, shown in Table 1, were selected after some manual tests.

Table 1. Weights of penalty function

w_1	$1.0 \cdot 10^{-3}$	w_2	$1.0 \cdot 10^{-3}$
w_3	$1.0 \cdot 10^{-5}$	w_4	$1.0 \cdot 10^{-5}$
w_5	$1.0 \cdot 10^{-12}$	w_6	$1.0 \cdot 10^{-5}$
w_7	10	w_8	0.5

2.4   Meta Model

A decent part of the total computation time is taken from the actual simulation of the car. We therefore implemented a simple meta-modeling approach in the form of a look-up-table that computed the output of the HEV simulation model (new SOC, fuel consumption) for a specific power split and torque demand setting for a single time step (not the entire fitness evaluation). The look-up table is formed by running the original model for a grid of input values and storing the results. For the optimization table values are interpolated linearly to compute the model output. Table 2 shows the parameters that were used to generate the meta-model for the HEV simulator.

Table 2. Grid number and range of inputs for the meta-HEV model

	Grid number	min	max
SOC_t	21	40 [%]	70 [%]
u_t	21	-1	1
v_t	401	0 [m/s]	40 [m/s]
a_t	41	-2 [m/s^2]	2 [m/s^2]
g_t	7	0	6

The maximum difference in fuel consumption was 1.7×10^{-4} [kg] and the highest fuel deviation 7.1×10^{-2} [%] SOC and no noticeable difference in the final optimization results between original and meta model. As a benefit the mean computation time to evaluate the J1015 drive cycle went down from 1.593 [s] to 0.093 [s], a speed-up factor of 17.13. We see that with a very small deviation from the full model output a substantial speed-up is possible. All consecutive results are therefore based on this surrogate model.

3   Optimization Methods

The optimization problem at hand can be described as finding the sequence of control actions $u(k)$ that minimizes the fuel consumption given a certain drive cycle, with k as a discretized time index. As an example of a typical drive cycle, the J1015 has a total length of 661 timesteps (of length 1 s each). This is a fairly large number of parameters, especially for evolutionary optimization.

3.1   Evolutionary Approach

Evolution strategies (ES) [15], which use vectors of real numbers as representations of solutions, are used in this study. The main benefit of evolutionary approaches is that it is comparatively easy to adapt the algorithm to specific problem domains and find some way to handle constraints, multiple objectives, and simulator inconsistencies. However, finding the best representation of solutions, constraints, and objectives often proves rather challenging. The representation we employ in this work is specifying the control $u(t)$ as a curve parametrized by a number of control points, which are optimized by the ES.

ES have been used for a rule-based control technique in order to find best parameters of the control strategy (e.g. [1]). However, in those approaches the number of parameters to be optimized was rather small. For our test problem, the number of parameters would be rather large. As an example, the J1015 drive cycle requires the specification of $u(k)$ for in total 661 discrete time steps.

Curve Representation. In order to reduce the number of parameters to optimize we represent the sequence of $u(k)$ values as a curve controlled by a number n_p of control points (x_i, y_i), where the x_i are evenly spread along the drive cycle (one point every l_C/n_p seconds, where l_C is the length of the specific drive cycle in seconds). The parameters to optimize are a set of n_p control point y-positions.

We have conducted a number of tests to determine the optimal representation of the control actions u. As alternatives we also tested an optimization of $SOC(t)$ or $\delta SOC(t)$, the absolute value or relative change of the battery state-of-charge in every time-step. From the change in the SOC value it is trivial to calculate the corresponding u value. Our tests have shown that a direct encoding of u provides the best results, mainly because it is easier to handle the large number of constraints given by the car's hardware. Another factor we investigated is how control points are translated into the final control curve. Four different approaches were tested (piecewise linear, nearest neighbor, spline, and pchip (piecewise cubic hermite interpolating polynomial)[1]). In an initial curve fitting test we tried to fit a curve to a target curve, which was actually the solution derived from the DP algorithm. These tests showed that the nearest neighbor is the best of the four types of curve encoding to represent u. We therefore chose the *nearest neighbor* scheme. We also used this test to determine useful settings for other strategy parameters like number of control points and number of generations.

---

[1] from the Matlab *interp1* function.

CMA-ES. For the optimization we employ the Covariance-Matrix-Adaptation Evolution Strategy (CMA-ES) [7]. This method uses information from previous generations to refine the search directions. In our experience it is a very robust method and performs well even when using standard strategy parameter settings. We directly used the Matlab code from the webpage of the original author (https://www.lri.fr/hansen/cmaes_inmatlab.html#matlab).

The following strategy parameters were used for the optimization: number of control points $n_p = 200$ for J1015 and $n_p = 400$ for NEDC, at most 20000 generations each for 10 trials (or when the difference between min and max fitness values over a window is less than 10^{-5}). All other parameters were used at the default value as suggested by the CMA-ES implementation, for example the effective (μ, λ) was computed as $\lambda = 4 + \text{floor}(3 * log(n_p)) = 19$ and $\mu = \text{floor}(\lambda/2) = 9$, as specified in [7].

3.2   Dynamic Programming

Dynamic Programming, proposed first by Bellman [2], is an efficient global exhaustive search method. It relies on a recursive structure of the problem and implements an efficient data storage approach to minimize redundant calculations. In contrast to Evolutionary Algorithms it is not a flexible, almost plug-and-play, approach but a conceptual framework that has to be adapted to the problem at hand manually. We follow the well known approach in hybrid car control [3,11,17] to discretize the problem in state and action space and evaluate all possible action sequences under a chosen discretization. Due to the necessary discretization of real-valued controls, time, and states DP results would be close to, but not exactly on, the global optimium A coarser grid would make the optimization faster but also increase the distance to the global optimum.

In the following we briefly outline the approach taken for the DP, for more details please consult [17]. Due to the problem structure and the simplified simulation model, there is only one state of the system, the battery state-of-charge at every discrete point in time $SOC(k)$. The range of possible SOC values is $SOC_{min} = 40\%$ to $SOC_{max} = 70\%$. Within this range we define a grid of n_{SOC} grid points, evenly spaced between SOC_{min} and SOC_{max}. In time direction we use the full resolution, so that the complete grid is composed of $n_{SOC} \times l_C$ nodes. Please note that other important states of the car, the speed of the vehicle and the speed of the engine, are fully determined by the drive cycle and need not be considered here. We also discretize the range of possible actions $u(k) \in [-1, 1]$ in n_u equally spaced points.

The DP will now investigate for every grid node $SOC(k)$ and every action $u(k)$ the resulting costs (fuel consumption and consumed charge). Since the final SOC value is predefined in the basic scenario ($SOC_{tar} = SOC_{fin} = SOC_{ini}$), it is advantageous to start the computation from the final state and go backwards in time. The DP will remember all intermediate computation results, i.e. the lowest cost to go from SOC_k to the final node. This storage will greatly increase the memory consumption of the algorithm but vastly speed up the computation. Linear interpolation is used to compute costs for actions/states in between the grid nodes.

There are a number of ways to further improve the algorithm, for example by explicitly removing all node/action combinations that would produce infeasible results (before calling the simulation model). In the present work we did not adapt the DP model to the specific application except for ignoring all grid nodes that can't be reached from the initial or (backwards from) the final node. However, we note that the original authors of the DP model [17], developed it specifically for this application.

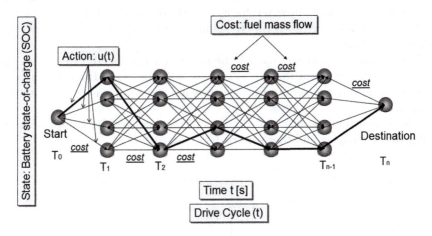

Fig. 3. Basic structure of DP grid approach

4   Results for Basic Scenario

In this section we compare the performance of EA and DP in terms of solution quality (best fuel consumption achieved without violating any constraints), variability of found solutions (restarting with different random values), probability to find any valid solutions and total computation time (including calls to the HEV simulator model).

Table 3. Fuel consumption result of DP and CMA-ES

	DP	$CMA-ES$		
		Best	Mean	Worst
$m_f[kg](J1015)$	0.1831	0.1838	0.1841	0.1847
$m_f[kg](NEDC)$	0.4749	0.4770	0.4774	0.4779

The first item to notice is that the DP is substantially faster than the ES approach, by a large margin: on the J1015 DP finishes after 44 s, while the CMA-ES needs on average 1956 s. For the NEDC cycle the relation is 80 s for DP and

9175 s for CMA-ES. Taking a look at the quality of solutions (Table 3) we see that the DP always provides better solutions, with no variation (randomness) in results. The EA-based approach on the other hand has some variation in the fuel consumption, which means that in order to produce competitive results, the EA would need to be run multiple times with different random seeds. Both methods managed to find solutions that did not violate any constraints, although some runs of the EA failed to satisfy constraints 1 and 2.

Looking at the resulting SOC curves, Fig. 4, the differences between DP and EA outputs are obvious. However, for most of the drive cycle, the actual decisions of the two methods are surprisingly similar. This is also true for the effective fuel consumption.

In summary, we can say that for the basic scenario the DP can reliably find better solutions in much less time than the EA approach. However, in terms of solution quality the results from the EA are not too far away from the DP solutions.

This insight motivated us to go a step further and test both approaches for a little more complex and therefore realistic use case.

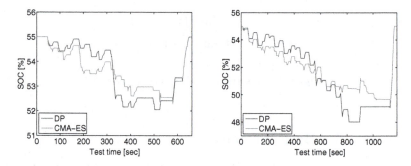

Fig. 4. Resulting $SOC(t)$ of DP and CMA-ES for J1015 (left) and NEDC (right) drive cycles

5   Extended Scenario

In the basic scenario, there is only a single system state, the SOC, and a single action, the torque split. This simple approach is possible since the drive cycle fixes the car's speed and acceleration. In a more realistic situation, car speed would be variable and be determined to some degree by the driver. We now modify our scenario in the following way:

- The target is to drive a certain distance in the most fuel-efficient way.
- As a new constraint, the maximum time to drive the distance is limited.
- We define an initial SOC value SOC_{ini}. The final SOC needs to be $SOC_{tar} \geq$ 55 %

- The controller may flexibly adapt the speed of the car.
- Car speed v_t will be a second state variable, and acceleration a_t a second action.
- The EA approach also encodes the total length of the trip as an optimization parameter.

This simple modification of the scenario greatly increases the size of the computational grid that the DP has to evaluate, see Table 4. It was therefore necessary to set a relatively small distance ($D = 580\,\mathrm{m}$) and time limit ($T_{limit} = 100\,\mathrm{s}$). This allowed us to reduce the number of control points for the EA representation, which in turn reduces the number of generations we need to run and therefore overall computing time. In this experiment we used the *pchip* curve interpolation method from Matlab for optimizing the speed vector. We ran the optimization for three different values of $SOC_{ini} = 51\,\%, 55\,\%, 64\,\%$. The EA was restarted 10 times for getting performance statistics.

Table 4. Grid node number and range of inputs for the DP

	Grid number	min	max
SOC_t	61	50 [%]	65 [%]
u_t	21	-1	1
v_t	37	0 [m/s]	20 [m/s]
a_t	7	$-1.5\,[m/s^2]$	$1.5\,[m/s^2]$
d_t	101	0	600

Furthermore the CMA-ES now optimizes the following three elements:

1. Position of n control points in u_t space, evenly spaced on t ($0 \leq t \leq T$)
2. Position of m control points in v_t space, whose time positions $t_{q,1}, \ldots, t_{q,m}$, are also optimized
3. The total length of drive T

The total number of the controlled parameters is $n + 2 \cdot m + 1$. In our experiments we used $n = 25$ and $m = 7$ (for results see Fig. 5 and Table 5).

First we take a look at the computation times: the CMA-ES on average took 304 s until convergence. This is a strong reduction, due to the much lower number of parameters. The DP on the other hand needed 1220 s, much longer than before (despite a shorter drive range), due to the larger number of states and actions. In the current condition, the CMA-ES was a factor four faster than the DP.

Looking again at the results from DP and EA we see that both approaches used the same strategy - acceleration to an optimal speed value, then driving at this speed until close to the end and braking (including energy recovery) before reaching the target distance. A closer look at the efficiency maps (Fig. 2) shows

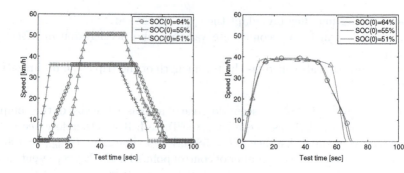

Fig. 5. Speed control calculated by the DP (left) and CMA-ES (right): The red, green and blue lines represent the solutions with $SOC_{ini} = 51\%, 55\%$ and 64%.

Table 5. Fuel consumption results

$m_f[kg]$		DP	$CMA - ES$		
			Best	Mean	Worst
SOC_0	51%	0.0401	**0.0398**	0.0406	0.0419
	55%	0.0246	**0.0239**	0.0249	0.0256
	64%	**0.0030**	0.0033	0.0045	0.0055

that the chosen speed values have a high efficiency rating when considering fuel economy and battery power consumption.

What is surprising is that the DP still has the better performance compared with the average of the EA results, but in some cases the EA finds better solutions than the DP. It also finds these solutions much quicker than the DP, which is effectively already at its computational (especially memory) limit for this degree of problem complexity.

Due to the doubled number of states and actions, the computational grid for the DP had to be made very coarse in order to satisfy computational time and memory limits. Obviously, the DP can no longer find the optimal solution under all conditions despite an enormous increase in the total computation time.

6 Summary and Outlook

In this work we have compared two well known optimization methods for finding the optimal control of a hybrid electric vehicle, an application of high relevance in automotive industry.

Our findings confirmed that for the standard application of minimizing the fuel consumption on a standard drive cycle, DP strongly outperforms our EA-based method in terms of solution quality and computation time. This advantage can be explained through the special structure of the problem and a probably sub-optimal representation of the control curve for our EA approach. EAs might profit from repair functions that turn an invalid solution into a valid one, but the impact on the optimization progress is difficult to assess.

In the second part we made a small modification of the problem in the direction of more realistic driving situations, by removing the constraints on the car's speed. This rather minor change already suffices to bring EA almost on par with DP regarding solution quality and results in lower computation times of EA compared to DP. Our findings suggest that for even less constrained controller optimization tasks, the EA approach might outperform DP in both computation time and solution quality.

It has to be added that the DP is very well adapted to handle the numerous constraints in this application, some of which posed very hard to deal with for the EA. Furthermore, the DP can provide optimal control results for a complete set of conditions, like different initial SOC values, at no additional cost.

In summary, depending on the specific problem, both DP and EA would have their niches. In [14] it was shown that the sole focus on fuel consumption might result in controls that are not optimal for the driver. Rather a many-objective optimization considering other objectives like battery lifetime, noise, or emissions might be required. Here multi-objective EAs promise substantial improvements.

A promising route to follow is a hybrid optimization approach using a combination of evolutionary and dynamic programming methods, for example by using the EA for a coarse control and the DP on a smaller grid for fine tuning. A similar approach for a different application has been presented by [8].

Acknowledgments. The authors want to the thank the reviewers for valuable feedback. Ken Nishikawa acknowledges the financial support from Honda Research Institute Europe.

References

1. Bacher, C., Krenek, T., Raidl, G.R.: Reducing the number of simulationsin operation strategy optimizationfor hybrid electric vehicles. In: Esparcia-Alcázar, A.I., Mora, A.M. (eds.) EvoApplications 2014. LNCS, vol. 8602, pp. 553–564. Springer, Heidelberg (2014)
2. Bellman, R.: Dynamic programming and stochastic control processes. Inf. Control **1**(3), 228–239 (1958)
3. Cassebaum, O., Bäker, B.: Predictive supervisory control strategy for parallel HEVs using former velocity trajectories. In: Vehicle Power and Propulsion Conference (VPPC), pp. 1–6. IEEE (2011)
4. Desai, C., Williamson, S.S.: Comparative study of hybrid electric vehicle control strategies for improved drivetrain efficiency analysis. In: Electrical Power and Energy Conference (EPEC), pp. 1–6. IEEE (2009)
5. Desai, C., Williamson, S.S.: Optimal design of a parallel hybrid electric vehicle using multi-objective genetic algorithms. In: Vehicle Power and Propulsion Conference, VPPC 2009, pp. 871–876. IEEE (2009)

6. Guemri, M., Neffati, A., Caux, S., Ngueveu, S.U.: Management of distributed power in hybrid vehicles based on DP or Fuzzy Logic. Optim. Eng. **15**, 993–1012 (2013)
7. Hansen, N., Kern, S.: Evaluating the CMA evolution strategy on multimodal test functions. In: Yao, X., et al. (eds.) PPSN 2004. LNCS, vol. 3242, pp. 282–291. Springer, Heidelberg (2004)
8. Jacquin, S., Jourdan, L., Talbi, E.-G.: Dynamic programming based metaheuris-ticfor energy planning problems. In: Esparcia-Alcázar, A.I., Mora, A.M. (eds.) EvoApplications 2014. LNCS, vol. 8602, pp. 165–176. Springer, Heidelberg (2014)
9. Krenek, T., Ruthmair, M., Raidl, G.R., Planer, M.: Applying (hybrid) metaheuris-tics to fuel consumption optimization of hybrid electric vehicles. In: Di Chio, C., et al. (eds.) EvoApplications 2012. LNCS, vol. 7248, pp. 376–385. Springer, Heidelberg (2012)
10. Lin, C.C., Peng, H., Grizzle, J.: A stochastic control strategy for hybrid electric vehicles. In: Proceedings of the American Control Conference, vol. 5, pp. 4710–4715, June 2004
11. Millo, F., Rolando, L., Fuso, R., Mallamo, F.: Real CO2 emissions benefits and end users operating costs of a plug-in hybrid electric vehicle. Appl. Energy **114**, 563–571 (2014)
12. Montazeri-Gh, M., Poursamad, A., Ghalichi, B.: Application of genetic algorithm for optimization of control strategy in parallel hybrid electric vehicles. J. Franklin Inst. **343**(4–5), 420–435 (2006)
13. Piccolo, A., Ippolito, L., zo Galdi, V., Vaccaro, A.: Optimisation of energy flow management in hybrid electric vehicles via genetic algorithms. In: IEEE/ASME International Conference on Advanced Intelligent Mechatronics I, vol. 1, pp. 434–439 (2001)
14. Rodemann, T., Narukawa, K., Fischer, M., Awada, M.: Many-objective optimiza-tion of a hybrid car controller. In: Mora, A.M., Squillero, G. (eds.) EvoApplications 2015. LNCS, vol. 9028, pp. 593–603. Springer, Switzerland (2015)
15. Schwefel, H.P.P.: Evolution and Optimum Seeking: The Sixth Generation. Wiley, New York (1993)
16. Sinoquet, D., Rousseau, G., Milhau, Y.: Design optimization and optimal control for hybrid vehicles. Optim. Eng. **12**(1–2), 199–213 (2011)
17. Sundström, O., Guzzella, L.: A generic dynamic programming Matlab function. In: Control Applications, (CCA) Intelligent Control, (ISIC), pp. 1625–1630. IEEE (2009)
18. Sundström, O., Guzzella, L., Soltic, P.: Optimal hybridization in two parallel hybrid electric vehicles using dynamic programming. In: Proceedings of the 17th IFAC World Congress vol. 17, no. 1, pp. 4642–4647, May 2013

NSGA-II Based Auto-Calibration of Automatic Number Plate Recognition Camera for Vehicle Speed Measurement

Patryk Filipiak[1]([⊠]), Bartlomiej Golenko[2], and Cezary Dolega[3]

[1] Institute of Computer Science, University of Wroclaw, Wroclaw, Poland
patryk.filipiak@ii.uni.wroc.pl
[2] Signal Theory Section, Wroclaw University of Technology, Wroclaw, Poland
bartlomiej.golenko@pwr.edu.pl
[3] Neurosoft, Sp. z o.o., Wroclaw, Poland
cezary.dolega@neurosoft.pl

Abstract. This paper introduces an auto-calibration mechanism for an Automatic Number Plate Recognition camera dedicated to a vehicle speed measurement. A calibration task is formulated as a multi-objective optimization problem and solved with Non-dominated Sorting Genetic Algorithm. For simplicity a uniform motion profile of a majority of vehicles is assumed. The proposed speed estimation method is based on tracing licence plates quadrangles recognized on video frames. The results are compared with concurrent measurements performed with piezoelectric sensors.

Keywords: Camera auto-calibration · Multiobjective optimization · Evalutionary approach · Camera model · Vehicle speed measurement

1 Introduction

Automatic Number Plate Recognition (ANPR) is a digital image processing technique that allows for a vehicle identification based on recognizing number (licence) plates on photos or video frames captured with a dedicated camera [1]. Most often some additional basic information about the observed vehicles and/or their motion can also be extracted from such images, including a driving direction, a type of a car body, a vehicle manufacturer, etc. However, one of the most desirable inputs required in many traffic monitoring systems is an accurate vehicle speed measurement.

Vast majority of the methods suggested for a speed approximation using computer vision were proposed for surveillance cameras [2–7] that observe a relatively large fragment of road. ANPR cameras, unlike surveillance ones, are usually mounted in such a manner that only a very small area of interest can be seen through their lenses in order to maximize the performance of the plate recognition task. In practice, monitoring devices of this type are often grouped into systems of cameras placed along the highway so that they could record

© Springer International Publishing Switzerland 2016
G. Squillero and P. Burelli (Eds.): EvoApplications 2016, Part I, LNCS 9597, pp. 803–818, 2016.
DOI: 10.1007/978-3-319-31204-0_51

consecutive passes of the same vehicles and process the data concerning time intervals between each spots, computing average speed, etc.

Generally, processing traffic data from a video camera imposes some prior knowledge about the camera spatial location, orientation, focal length of the optical system, etc. Numerous methods for computing these parameters were proposed [8–10]. However, most of them require some pre-processing and/or a supervision that seem rather inconvenient around a traffic area as it may cause an interference in a traffic flow. Hence, it is tempting to incorporate a fully automatic technique instead.

In this paper an auto-calibration method for a single ANPR camera based on the Non-dominated Sorting Genetic Algorithm (NSGA-II) [11] is proposed. An application of NSGA-II seems a natural choice for that purpose since a multidimensional continuous space of parameters needs to be explored and the solutions to be found are expected to satisfy a number of optimization criteria. However, it has to be pointed out that in the camera calibration task there exists a tiny set of solutions that meet all the specified criteria at once. Finding this set is an ultimate task of the suggested approach.

The proposed auto-calibration procedure estimates both intrinsic and extrinsic camera settings. As a consequence, quadrangles of number plates in the 2D image coordinate system can be projected back to the 3D real world coordinate system. This allows for computing the per-frame distances covered by a given vehicle and thus for estimating its speed.

Prior to the auto-calibration, the camera needs to be *roughly calibrated*[1] and equipped with a software for automatic recognition of licence plates location on a video frame. It has to be pointed out that the above assumptions are fairly easy to satisfy in practice.

The main contribution of the presented approach is twofold: an introduction of the dedicated multi-objective function that allows for estimating the camera settings using NSGA-II and the method for a vehicle speed measurement using 3-dimensional geometric transformations of the observed scene. Within the scope of this paper the speed measurement is used only to asses an accuracy of the auto-calibration mechanism although in a broader perspective it is intended to improve the quality of the Weigh-In-Motion Enforcement (WIM-E) systems [12]. Speed approximations derived from camera image analysis could be used for cross-checking measurements performed simultaneously with units embedded in the WIM-E devices.

Preliminary experiments revealed promising results concerning the high level of an auto-calibration accuracy. Vehicle speed measurements obtained using the suggested method compared to the ones made with piezoelectric sensors were very similar.

---

[1] By the term *roughly calibrated* it is meant that a camera is mounted at its desired location, it points at the road and its focal length is vaguely set with a bare eye such that number plates are readable on the captured images.

2   Camera Projection

From the geometrical point of view, any camera transforms 3D real world objects onto their 2D images. In this section a necessary theoretical background of the above fact will be outlined very briefly.

2.1   Pinhole Camera Model

Pinhole camera model [13] is a commonly used simplified (i.e. assuming no distortions) mathematical model of a camera projection defined as follows

$$
s \begin{bmatrix} u \\ v \\ 1 \end{bmatrix} = \mathbf{A} \cdot [\mathbf{R}|\mathbf{t}] \cdot \begin{bmatrix} x \\ y \\ z \\ 1 \end{bmatrix}, \tag{1}
$$

where $\mathbf{A}$ is the matrix of 3×3 real-valued coefficients representing the five *intrinsic camera settings*

$$
\mathbf{A} = \begin{bmatrix} F_x & \gamma & u_p \\ 0 & F_y & v_p \\ 0 & 0 & 1 \end{bmatrix} \tag{2}
$$

among which a vector (u_p, v_p) defines a principal point, F_x and F_y represent a focal length expressed in horizontal and vertical photosensitive elements distances (respectively), while γ encodes a skewness between the axes of image coordinate system (for perpendicular axes $\gamma = 0$). The matrix $[\mathbf{R}|\mathbf{t}]$ determines the six *extrinsic camera settings* with the sub-matrix $\mathbf{R} = [r_{ij}]_{1 \leq i,j \leq 3}$ encoding the rotation and $\mathbf{t} = (t_x, t_y, t_z)^T$ encoding the translation of a camera with respect to the real world coordinate system. Hence, a pinhole camera model, as defined in Eq. (1), projects each visible[2] point $(x, y, z) \in \mathbb{R}^3$ of the real world space into the point $\left(\frac{u}{s}, \frac{v}{s}\right) \in \mathbb{R}^2$ of the image space with $s \neq 0$ (cf. Fig. 1).

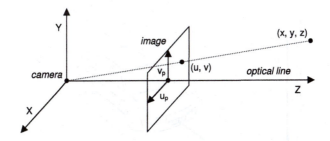

Fig. 1. Illustration of the pinhole camera model with $s = 1$.

---

[2] Infinitely-distant points of the 3D space correspond to $s = 0$ which make them *invisible* in the image space.

2.2   Reversed Projection

As a matter of fact, a reversed projection (or a re-projection for short) of the camera model leads to a certain ambiguity. An inverse of the mapping from Eq. (1) implies that each point $(u, v) \in \mathbb{R}^2$ of the image space maps to a straight line in the 3D real world space. Thus, for practical applications of the camera image re-projection some additional insight is required, e.g. some prior knowledge about the projected space or the use of stereo vision. However, this paper concerns only the re-projection of plate numbers. This fact leads to some key assumptions (described further) that allow for eliminating the ambiguity associated with the inverse mapping.

For the sake of clarity, assume the following notation convention. For the remainder of this paper a camera projection of any point $A = (x, y, z) \in \mathbb{R}^3$ with respect to the real world coordinate system will be noted with overline, i.e. $\overline{A} = (\overline{u}, \overline{v}) \in \mathbb{R}^2$ and vice versa (A will refer to the re-projection of $\overline{A}$). Analogously, a projection of some distance $a > 0$ in the real world coordinates will be noted as $\overline{a}$ and vice versa ($\overline{a}$ will refer to the re-projection of a).

2.3   Calibration

The process of estimating camera settings is referred to as *calibration*. Numerous algorithms dealing with this issue were proposed, e.g. Tsai's method [10] or the analysis of vanishing-points [8,9]. Most of them are based on processing images of some well-known shapes or patterns either presented deliberately in front of a camera or recognized serendipitously. Unfortunately, none of those is usually the case with ANPR cameras, due to their rather small and strictly directed visible area.

Figure 2 presents the typical spatial location of a sample ANPR camera. In this paper, the distances d_x and d_y are assumed unknown whereas the angles α, β and the distance d_z are constrained to fixed yet fairly wide ranges specified by domain experts.

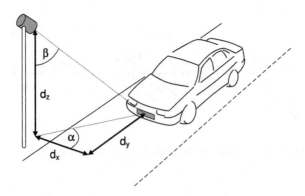

Fig. 2. Spacial location of a sample ANPR camera.

3   Speed Measurement Procedure

A typical ANPR camera covers a road section of ca. 5 m in length. Taking into account the typical road traffic velocities one may expect that no abrupt changes in speed can be observed within such small visible area. As a result, the uniform motion profile is assumed for the vehicle speed measurement in the remainder of this paper.

3.1   Formula

Assume that a given ANPR camera captures $n > 0$ still images of a vehicle passing by a measurement point at the time steps $t_1, t_2, \ldots, t_n$. In practice, the intervals $\tau_i = t_{i+1} - t_i$ between consecutive time steps are usually constant (i.e. $\tau_i = \tau$ for all $i = i, \ldots, n - 1$) due to the camera fixed frames-per-second frequency, although this assumption is not necessary for the presented model.

A dedicated recognition system performs an on-line image processing of all n snapshots then outputs a sequence of plate number locations encoded as quadrangles. For all $i = 1, \ldots, n$ let $\overline{C}_i \in \mathbb{R}^2$ be the center point (with respect to the image coordinate system) of the i-th quadrangle. A sample snapshot is presented in Fig. 3(a) and a sequence of quadrangles is shown in Fig. 3(b).

Let $h \in \mathbb{R}_+$ represent the distance between the center point of a plate number and the ground (with respect to the real world coordinate system). Generally, the exact value of h is unknown, yet frequently it oscillates between 0.3 and 0.5 m. This observation allows for eliminating the ambiguity of a camera re-projection (discussed in the previous section). Note that h is time-invariant for any vehicle. Hence, by assuming some estimation $\hat{h}$ of h, the distances $\overline{s}_i$ between $\overline{C}_i$ and $\overline{C}_{i+1}$ (for $i = 1, \ldots, n - 1$) can be re-projected into their corresponding distances s_i between C_i and C_{i+1} in the real world coordinate system fairly precise. Since $t_1, t_2, \ldots, t_n$ are known, the corresponding speed values $v_1, v_2, \ldots, v_{n-1}$

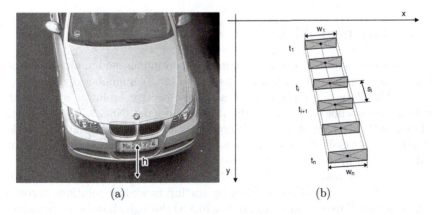

(a)   (b)

Fig. 3. A sample ANPR camera snapshot (a) and a sequence of quadrangles encoding consecutive locations of plate numbers (b).

can be easily computed accordingly using the simple formula $v_i = s_i/\tau_i$ for $i = 1, \ldots, n-1$. Finally, the output value is obtained as the arithmetic mean of $v_1, v_2, \ldots, v_{n-1}$.

3.2   Validation

As stated above, for each $i = 1, \ldots, n-1$, a per-frame vehicle speed v_i depends on the distance $s_i \geq 0$ between consecutive licence plate centers $C_i, C_{i+1} \in \mathbb{R}^3$ and the time interval $\tau_i = t_{i+1} - t_i$. The exact locations of C_i and C_{i+1} are not known thus their approximations are re-projected from the observed image point coordinates $\overline{C}_i, \overline{C}_{i+1} \in \mathbb{R}^2$. This means that the per-frame vehicle speed v_i is not only a function of τ_i and s_i but also depends on the intrinsic camera settings (the focal length F_x and the principal point u_p, v_p), the extrinsic camera settings (d_x, d_y, d_z) and a distance between the centre of licence plate and the ground $h > 0$. The remaining camera settings were neglected, as the focal length F_y is proportional to F_x whereas the skewness γ is typically assumed zero.

Let $\overline{C}_i = (\overline{u}_i, \overline{v}_i)$ and $\overline{C}_{i+1} = (\overline{u}_{i+1}, \overline{v}_{i+1})$ then v_i is indeed a function expressed as follows

$$v_i \equiv v_i\left(F_x, (u_p, v_p), (d_x, d_y, d_z), h, (\overline{u}_i, \overline{v}_i), (\overline{u}_{i+1}, \overline{v}_{i+1}), \tau_i\right). \qquad (3)$$

Assuming that these variables are independent, the speed estimation error can be calculated using the following total differential formula

$$\Delta v = F_x + u_p + v_p + d_x + d_y + d_z + \\ + h + \overline{u}_i + \overline{v}_i + \overline{u}_{i+1} + \overline{v}_{i+1} + \tau. \qquad (4)$$

It is important to discover which of these components have the highest influence on the computed speed value yet the evaluation of the above-mentioned formula in a symbolic form is not feasible in practice. Instead, the empirical analysis of the measurement precision is presented in Subsect. 5.4.

4   Evolutionary Algorithm

The parameter space of camera settings is continuous and multidimensional whereas the solution to be found is expected to satisfy a number of optimization criteria. Therefore, an application of a multi-objective evolutionary algorithm seems a natural choice for solving the problem in hand. NSGA-II is an effective and frequently used algorithm of this type [11]. Additionally, the fact that it does not require a normalization of criteria is considered a strong advantage (e.g. over MOEA/D [14]) in the domain of industrial applications.

Let us briefly remind that NSGA-II aims at estimating a Pareto front for a multi-objective optimization problem by sorting candidate solutions according to their mutual Pareto dominance. It is achieved through clustering a population of individuals into sets of non-dominated solutions. In the first set the solutions that are non-dominated within the scope of a whole population are included.

Then, all the non-dominated solutions are selected among the remaining ones and placed in the second set. The further sets are constructed in the same manner until no individual is left. The numeric labels of sets determine the ordering of solutions while in case of ties the so-called *crowding distance* is used. The latter one is a measure that promotes distant solutions over the closer ones in order to prevent stagnation of a population and enforce a rather evenly distributed estimation of the Pareto front.

All the domain specific details and modifications of the original approach are presented in details in the following subsections.

4.1 Genotype Coding

Candidate solutions are represented as real vectors $(F, \alpha, \beta, \varphi, d_z) \in \mathbb{R}^5$ where:

(a) $F \in [3000, 15000]$ is a focal length (it is assumed that $F = F_x = F_y$),

$\left.\begin{array}{l} \alpha \in \left[-\frac{\pi}{3}, \frac{\pi}{3}\right], \\[6pt] \text{(b)} \quad \beta \in \left[\frac{\pi}{4}, \frac{\pi}{2} - 0.1\right], \\[6pt] \quad d_z \in [200\text{cm}, 800\text{cm}] \end{array}\right\}$ determine a camera pose (cf. Fig. 2),

(c) $\varphi \in \left[-\frac{\pi}{8}, \frac{\pi}{8}\right]$ is an angle of camera rotation over an optical line.

Note that the parameter φ was not included in Fig. 2. In fact, it is added here for practical reasons in order to compensate possible imperfections in horizontal positioning of a camera.

4.2 Multi-objective Evaluation Function

A global goal of the considered optimization problem is to find such a geometric transformation that maps the 2D camera image onto the 3D Euclidean space such that the licence plates quadrangles mapped from 2D to 3D and then projected further onto the planar $z = \hat{h}$ (with $\hat{h} > 0$ being an average altitude of licence plates above the ground) would form a sequence of equidistant parallel segments of an actual length as illustrated in Fig. 4.

The above global goal will be decomposed to the following local goals that can be easily formulated as optimization criteria:

Criterion #1. Let $n \in \mathbb{N}_+$ be the number of frames on which the licence plates were recognized. All n segments are supposed to lie in parallel on the planar $z = \hat{h}$. This implies that the n lines that contain these segments should all have equal slopes. Let $y = a_i x + b_i$ for $i = 1, \ldots, n$ be their respective line equations. The criterion is fully satisfied when $a_1 = a_2 = \ldots = a_n$. Let $l : y = a^l x + b^l$ be the line that best approximates the points $(i, a_i) \in \mathbb{R}^2$ for all $i = 1, \ldots, n$. Thus, an absolute value of the slope of l, i.e. $|a^l|$ shall be the objective to be minimized in order to assure that the criterion #1 is satisfied.

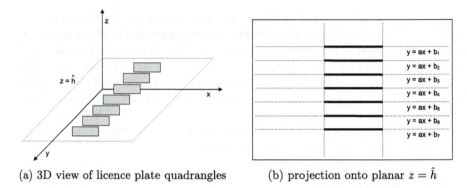

(a) 3D view of licence plate quadrangles   (b) projection onto planar $z = \hat{h}$

Fig. 4. A globally desired state — licence plates recognized on consecutive video frames form a sequence of equidistant parallel rectangles (in 3D) or segments (in 2D).

Criterion #2. The uniform motion profile that is assumed and the fixed frames-per-second rate of a video camera imply that all the n segments should be equidistant. In practice, it is better to slightly reformulate this goal since some of the video frames might be occasionally lost due to technical disturbances. Instead, the $n - 1$ per-frame speed estimations can be compared. Let $s_i > 0$ for all $i = 1, \ldots, n-1$ be the distances between the centers of consecutive segments on the planar $z = \hat{h}$ and let $t_1, \ldots, t_n$ be the timestamps associated with each of the n video frames. The per-frame speeds can be derived as $v_i = s_i/(t_{i+1}-t_i)$. Let $k : y = a^k x + b^k$ be the line that best approximates the points $(t_{i+1} - t_1, v_i) \in \mathbb{R}^2$ for all $i = 1, \ldots, n - 1$. Similarly to the previous case, the absolute value of the slope $|a^k|$ shall be minimized in order to satisfy the criterion #2.

Criterion #3. An orientation of the coordinate system as such is not crucial for a vehicle speed measurement however it is a part of the camera calibration task. In order to achieve this goal let $(x_i, y_i) \in \mathbb{R}^2$ for all $i = 1, \ldots, n$ be the coordinates of the centers of consecutive segments on the planar $z = \hat{h}$. Again, let $m : y = a^m x + b^m$ be the line that best approximates these points in $\mathbb{R}^2$ and let $|a^m|$ be the objective to be minimized. Eventually, a translation over the vector $[0, -b^m]$ can be performed afterwards to assure an alignment of the segments along the y-axis.

Criterion #4. An undesired camera rotation over an optical line can be easily compensated by "straightening" licence plates whose misalignment over y-axis manifests itself in the proportional misalignment of the corresponding segments on the planar $z = \hat{h}$. Once again, let $y = a_i x + b_i$ for $i = 1, \ldots, n$ be the lines that include the respective segments. An imperfection of a camera mounting can thus be reduced by minimizing the value $|\sum_{i=1}^{n} a_i|$.

Criterion #5. From the vehicle speed measurement perspective, images have to be scaled properly so that per-frame distances could be measured in commonly used units, e.g. in meters. Fortunately, the dimensions of licence plates are often regulated by law, hence a majority of observed plates can be assumed to have standardized widths specific to a given area. Let $w_1, \ldots, w_n \in \mathbb{R}_+$ be the widths of segments on the planar $z = \hat{h}$ and let $\overline{w} \in \mathbb{R}_+$ be the median of $\{w_i; i = 1, \ldots, n\}$. A minimization of the criterion $|\overline{w}/w_{target} - 1|$, where $w_{target} \in \mathbb{R}_+$ is a predefined target width, can effectively solve the scaling problem.

All the rotation-related goals (i.e. criteria #1 – #4) are illustrated in Table 1 where each row demonstrates a single criterion (left column) followed by the sample target state that satisfies it (right column). Note that the globally desired state presented in Fig. 4 meets all of the specified criteria albeit the illustrations presented in Table 1 are selected in a manner that emphasizes the respective partial goals.

4.3 Selecting a Single Solution from Pareto Front

It is rather obvious that any particular criterion of a multi-objective optimization problem can be satisfied by a solution that most probably violates some other criteria. Nevertheless, in the camera calibration task there exists a tiny set of solutions that meet all the specified criteria at once. In fact, it could also be found by solving the equivalent single-objective problem obtained through aggregating the optimization criteria at the design level. However, a selection of proper weighting parameters would then be a challenging and highly case dependent supplementary task. Also an application of some multi-objective optimization algorithm based on decomposition (e.g. MOEA/D) could be an option yet a prior normalization of criteria that is often required there seems inconvenient in practice.

In the proposed approach the optimization criteria are normalized and aggregated *ex post*, i.e. out of the Pareto front (POF) approximated by NSGA-II. Let $f_1, \ldots, f_5 : \mathbb{R}^5 \longrightarrow \mathbb{R}_+$ represent these criteria as defined before. For all $x \in \text{POF}$ the normalized optimization criteria $\widetilde{f}_1, \ldots, \widetilde{f}_5$ are obtained through dividing f_i by the 75-th percentile of $\{f_i(x); \; x \in \text{POF}\}$ for all $i = 1, \ldots, 5$. The 75-th threshold level was chosen arbitrarily as a result of preliminary experiments. Obviously, different values can be used instead.

Eventually, an individual with the lowest aggregated value computed as $\sum_{i=1}^{5} \widetilde{f}_i$ is selected as the one that best approximates actual camera settings.

4.4 Handling Uncertainties

Let us remind that the objective function is defined for a single vehicle pass captured at $n > 0$ consecutive video frames represented as a sequence of licence plate quadrangles recognized on images. In real-world applications there is a number of possible uncertainties that can emerge while preparing the above input data. A particular vehicle might be slightly accelerating or slowing down which

Table 1. Visualization of rotation-related optimization criteria (left column) followed by sample target states that satisfy them (right column).

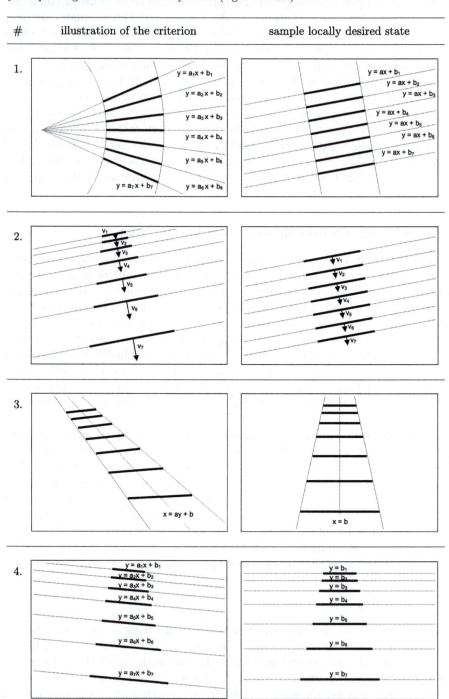

violates the assumption of the uniform motion profile, also its driver might be intending to change a lane or avoid a collision with an obstacle, etc. Eventually, a solution found by the algorithm might be imprecise or erroneous.

In order to alleviate possible uncertainties, an output data smoother based on Simple Moving Average (SMA) is applied. A set of $N_{pass} \gg 0$ vehicle passes is analyzed instead of a single one. Each of the passes is utilized for evaluating candidate solutions within a fixed time period of $N_{sub} > 0$ generations. Thus, the algorithm is run for $N = N_{pass} \times N_{sub}$ iterations producing a sequence of vectors $x^{(t)} \in \mathbb{R}^5$ (for $t = 1, \ldots, N_{pass}$) each of which encodes a best available approximation of camera settings. Let $0 < k < N_{pass}$ be a smoother lag. Define a sequence $\left(y^{(t)}\right)$ such that for all $1 \leq t \leq N_{pass} - k + 1$ holds

$$y^{(t)} = \text{SMA}_k \left(\left\{ x^{(t)}, \ldots, x^{(t+k-1)} \right\} \right) = \frac{1}{k} \sum_{i=t}^{t+k-1} x^{(t)} \in \mathbb{R}^5. \qquad (5)$$

It is a sequence of smoothed solutions out of which $y^{(N_{pass}-k+1)}$ is considered the most exact one since it is produced at the end of an evolutionary process.

5   Experiments

The experiments were performed on video streams from two ANPR cameras mounted at the test site in the city of Wroclaw (Poland). Cameras were assigned to separate lanes, both of which were leading towards the city center. The lanes were numbered from right to left, i.e. the lane $l1$ was located on the right-hand side (lower speeds) whereas the lane $l2$ was on the left-hand side (higher speeds).

A video resolution was set to 1280×720 with 25 frames-per-second rate. The 2 h of video recordings were split into the training and testing sets as presented in Table 2.

The evolutionary algorithm was run in 10 variants of numbers of iterations $N = N_{pass} \times N_{sub}$ where $N_{pass} \in \{443, 556\}$ (depending on the lane) while $N_{sub} \in \{1, 2, 3, 4, 5\}$. The population size was set to 100. Simulated binary crossover (SBX) and polynomial mutation operators were used with default parameters as described in [11].

In order to avoid a possible stagnation of individuals also a slightly modified variant of the algorithm with an introduction of Random Immigrants (RI) at

Table 2. A split of input data into the training and testing sets.

Date	Lane	Training set	Testing set
17th March 2014	l1	12:00:00 – 12:59:59	13:00:00 – 13:59:59
		(556 vehicle passes)	(546 vehicle passes)
17th March 2014	l2	12:00:00 – 12:59:59	13:00:00 – 13:59:59
		(443 vehicle passes)	(409 vehicle passes)

the end of each iteration was considered. Three modes of RI fraction sizes were tested, i.e. $\{0\,\%, 10\,\%, 20\,\%\}$.

Finally, four modes of output vectors smoothing were applied, i.e. SMA with lags $k \in \{50, 100, 150\}$ and *no smoothing* (for comparison).

5.1   Testing Procedure

Each experiment was performed as follows. Firstly, the NSGA-II was run on the training data set utilizing the multi-objective evaluation function and producing one solution per generation as described in Sect. 4. Then, the accuracy of such obtained results was assessed on the testing data set by comparing vehicle speeds approximated with the method described in Sect. 3 and the concurrent measurements performed with piezoelectric sensors mounted in the road.

In order to eliminate "serendipitous" winners and observe a long-term stability of the proposed approach, sequences of latest 100 solutions were analyzed in every run. Particularly, in the cases with SMA smoothers each candidate vector of camera settings among $y^{(N_{pass}-k-98)}, \ldots, y^{(N_{pass}-k+1)}$ was applied for the testing set whereas in the cases with *no smoothing* all the vectors among $x^{(N_{pass}-99)}, \ldots, x^{(N_{pass})}$ were considered[3].

5.2   Quantitative Results

Table 3 summarizes the results averaged over 30 independent runs of each variant of input parameters for an auto-calibration of cameras capturing vehicle passes on the lanes (a) $l1$ and (b) $l2$. The columns of Table 3 represent smoothing modes whereas the various combinations of N_{sub} and *RI fraction* sizes are given in rows. For each variant the corresponding *mean squared error* (mse) of vehicle speed approximation and the associated *standard deviation* (std) are specified.

The above metrics were obtained as follows. Let $\hat{N}_{pass} \in \{409, 546\}$ be the size of the testing set (depending on the lane) and let

$$v_{piezo}^{(\hat{t})} > 0, \quad v_{anpr}^{(\hat{t})} > 0 \quad \text{for} \quad \hat{t} = 1, \ldots, \hat{N}_{pass} \tag{6}$$

represent the speeds of vehicles measured with piezoelectric sensors and speed estimations derived from ANPR camera images (respectively). For all of the latest 100 solutions produced by NSGA-II the errors were computed as

$$\frac{1}{\hat{N}_{pass}} \sum_{\hat{t}=1}^{\hat{N}_{pass}} \left(v_{piezo}^{(\hat{t})} - v_{anpr}^{(\hat{t})} \right)^2. \tag{7}$$

Eventually, the arithmetic mean and standard deviation of these 100 errors were computed and averaged further over 30 runs.

---

[3] The notations of $x^{(\cdot)}$ and $y^{(\cdot)}$ were introduced in Sect. 4.

Table 3. The average results of vehicle speed measurements performed with auto-calibrated ANPR cameras using the NSGA-II algorithm with the specified N_{sub} values, RI fraction sizes, and parameter smoothers.

N_{sub}	RI fraction	no smoothing mse	std	SMA50 mse	std	SMA100 mse	std	SMA200 mse	std
1	0%	5.95	1.95	5.69	0.14	5.70	0.07	5.98	0.16
	10%	6.10	2.75	6.92	0.45	6.94	0.20	6.83	0.07
	20%	16.69	4.19	18.33	0.49	18.75	0.40	19.90	0.43
2	0%	4.13	2.14	3.97	0.23	4.05	0.15	4.04	0.04
	10%	3.57	2.23	3.56	0.39	3.84	0.36	4.08	0.17
	20%	7.79	3.14	8.55	0.89	9.15	0.70	9.84	0.42
3	0%	5.64	2.34	5.31	0.22	5.29	0.05	5.27	0.08
	10%	4.60	2.48	4.44	0.29	4.57	0.20	4.94	0.29
	20%	7.81	3.11	8.36	0.74	8.79	0.52	9.29	0.30
4	0%	8.05	2.35	6.79	1.07	6.19	0.61	5.66	0.41
	10%	5.40	2.63	5.15	0.15	5.08	0.06	5.12	0.05
	20%	9.84	3.73	10.58	0.52	10.93	0.45	11.13	0.13
5	0%	7.88	2.54	6.82	0.56	6.44	0.36	5.99	0.27
	10%	7.69	3.01	7.51	0.25	7.61	0.22	8.10	0.20
	20%	9.33	3.21	9.91	0.74	10.18	0.35	10.12	0.13

(a) lane $l1$

N_{sub}	RI fraction	no smoothing mse	std	SMA50 mse	std	SMA100 mse	std	SMA200 mse	std
1	0%	8.32	1.03	7.22	0.30	7.00	0.07	6.86	0.03
	10%	14.59	2.15	16.34	0.41	16.26	0.17	17.30	0.24
	20%	17.67	3.27	22.83	1.39	23.06	0.74	24.93	0.18
2	0%	8.58	1.30	7.22	0.17	7.12	0.12	7.19	0.10
	10%	7.39	1.13	7.51	0.11	7.52	0.08	7.44	0.08
	20%	14.00	2.12	15.01	1.08	14.53	0.83	14.24	0.45
3	0%	9.08	0.84	8.80	0.67	8.97	0.49	8.66	0.10
	10%	7.11	1.19	6.89	0.44	6.37	0.54	5.48	0.38
	20%	11.87	2.23	12.46	0.38	12.51	0.10	12.65	0.07
4	0%	10.30	1.00	9.18	0.59	8.39	0.67	7.53	0.24
	10%	9.26	1.52	9.29	0.31	8.99	0.30	8.21	0.41
	20%	11.95	2.33	12.43	0.56	11.97	0.47	11.24	0.37
5	0%	10.56	1.14	9.79	0.56	9.38	0.36	8.30	0.29
	10%	11.47	2.02	11.36	0.72	10.82	0.60	9.74	0.56
	20%	10.94	2.17	11.46	0.44	11.07	0.44	10.24	0.42

(b) lane $l2$

5.3   Discussion

It can be clearly seen in Table 3 that the frequency of altering vehicle passes used by the evaluation function (reflected in the N_{sub} parameter) had a large impact on the speed measurement accuracy. The lowest approximation errors were achieved for $N_{sub} \in \{2,3\}$. Indeed, a setting of $N_{sub} = 1$ caused undesired rapid fluctuations of the objective function whereas $N_{sub} > 3$ resulted in a progressing stagnation of individuals. The latter was alleviated through the introduction of random immigrants. Note that the 10 % RI fraction helped reducing mean squared errors in most of the cases. However, the 20 % fraction seemed too large since it often worsened the results.

Even though the *mse* values were rather low for non-smoothed camera settings, the respective standard deviations of errors were unacceptably high. In other words, every single vector of parameters among the set $\{x^{(t)}; \ N_{pass} - 99 \leq t \leq N_{pass}\}$ being estimated by the algorithm without smoothing was likely to be erroneous. Nonetheless, the SMA based smoothers turned out to be extremely effective in reducing the above risk as illustrated in Fig. 5. For instance, a use of SMA with lag $k = 50$ decreased *std* values way below 0.5 in majority of cases although a distinction between the particular lag sizes among 50, 100, and 150 appeared barely significant.

Generally, the best result at the lane $l1$ was achieved with $N_{sub} = 2$, 10 % RI fraction and SMA50 reaching the average error of $\sqrt{3.56} \approx 1.9$ [km/h] among all the vehicles that passed by the measuring point within 1 h. Analogously, the best result at the lane $l2$ was achieved with $N_{sub} = 3$, 10 % RI fraction and SMA200 reaching the mean error of $\sqrt{5.48} \approx 2.3$ [km/h] compared to the piezoelectric sensors. The respective standard deviations were 0.36 at the lane $l1$ and 0.38 at the lane $l2$ which implies that any given vector of camera settings among the last 100 ones produced by the algorithm was very likely to be accurate.

From the vehicle speed measurement perspective, the experimental results suggest that the proposed auto-calibration method is able to reach the satisfying level of precision and thus guarantee fairly accurate speed estimations.

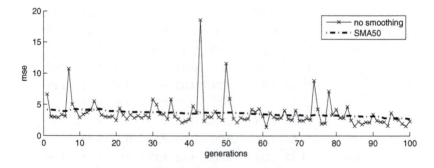

Fig. 5. A comparison of mean squared errors of the vehicle speed measurement performed with smoothed and non-smoothed camera settings.

5.4   Empirical Analysis of the Measurement Precision

Due to the similarity of conditions where ANPR cameras are deployed, it is reasonable to assume that the weights of individual error components of the formula (4) would be similar for every passing vehicle. Applying (4) to the representative sample selected from one of the test sites where vehicle speed v was 83 km/h (according to inductive loop measurement) yields

$$\Delta v[m/s] = 0.0507(\Delta \overline{v}_{i+1} - \Delta \overline{v}_i) - 3.85 \cdot 10^{-5}(\Delta \overline{u}_{i+1} - \Delta \overline{u}_i) +$$
$$+ 0.356\Delta h - 100.19\Delta\tau - 0.005\Delta F_x +$$
$$+ 0.0237\Delta d_x + 2.263\Delta d_y - 3.245\Delta d_z, \tag{8}$$

where the gradients of the coordinates $\Delta \overline{u}_i$, $\Delta \overline{v}_i$ are given in pixels, the time interval $\Delta\tau$ — in seconds and all the distances (Δd_x, Δd_x, Δd_x, Δh) — in meters.

It is clear that the accuracy of a licence plate detection in the horizontal direction is completely irrelevant. In fact, it might be reasonable to focus on the tracking of any other horizontal vehicle feature in the future work. A miscalculation of vertical coordinates of the licence plate by two pixels results in a speed change of about 0.1m/s. Analogously, a variation of the licence plate elevation above the ground level is not significant as well — at least as long as only typical passenger cars are considered.

Since absolute value of F_x is usually relatively high, the accurate measurement of a camera focus is essential despite a very low value of the ΔF_x coefficient. Note that a 1 % error in the camera focus measurement results in a 1 % change of the calculated vehicle speed. It is evident that also the accuracy of a camera altitude d_z and a distance in y-axis d_y are highly significant for the exact speed estimation.

6   Conclusions

In this paper an auto-calibration mechanism for an ANPR camera aimed at the vehicle speed measurement was proposed. Assumptions made about the uniform profile of vehicles motion while passing the visible area of a camera and a constant altitude of number plates above the ground allowed for the re-projection of reference points and thus the estimation of speed.

A calibration task was formulated as the multi-objective optimization problem and solved with NSGA-II. Then the *ex post* method for normalizing and aggregating the optimization criteria was applied to select a globally best solution. Also the smoothing procedure was proposed in order to decrease a level of uncertainty and assure a greater stability of the camera settings found.

A number of experiments were made on the real-world data. The results were compared with the concurrent measurements performed with piezoelectric sensors. A use of the proposed approach allowed to calibrate the tested ANPR cameras achieving such level of precision that the speed approximations performed with the proposed method reached ca. $+/- 2$ km/h accuracy.

It is planned to apply the proposed camera auto-calibration and speed measurement techniques in Weigh-In-Motion Enforcement systems since one of the key factors required for the accurate estimation of a vehicle weight is the precise measurement of its speed while it passes the automatic weighing instruments. It is highly probable that combining the proposed vehicle speed measurement method with the measurements performed simultaneously with the units embedded in the WIM-E devices could improve the quality of the WIM-E system in general without incorporating additional costs.

References

1. Qadri, M.T.: Automatic number plate recognition system for vehicle identification using optical character recognition. In: Proceedings of International Conference on Education Technology and Computer (ICETC), pp. 335–338 (2009)
2. Cathey, F.W., Dailey, D., J.: A novel technique to dynamically measure vehicle speed using uncalibrated roadway cameras. In: Proceedings of IEEE Intelligent Vehicles Symposium, pp. 777–782 (2005)
3. Dailey, D.J.: An algorithm to estimate mean traffic speed using uncalibrated cameras. IEEE Trans. Intell. Transp. Syst. 1(2), 98–107 (2000)
4. Dailey, D.J., Li, L.: Video image processing to create a speed sensor, University of Washington, Seattle, WA, ITS Research Program, final research report (1999)
5. Grammatikopoulos, L., Karras, G., Petsa, E.: Automatic estimation of vehicle speed from uncalibrated video sequences. In: Proceedings of International Symposium on Modern Technologies, Education and Profeesional Practice in Geodesy and Related Fields, pp. 332–338 (2005)
6. Lin, H.-Y., Li, K.-J., Chang, C.-H.: Vehicle speed detection from a single motion blurred image. Image Vis. Comput. 26(10), 1327–1337 (2008)
7. Schoepflinand, T.N., Dailey, D.J.: Dynamic camera calibration of roadside traffic management cameras for vehicle speed estimation. IEEE Trans. Intell. Transp. Syst. 4(2), 90–98 (2003)
8. Caprile, B., Torre, V.: Using vanishing points for camera calibration. Int. J. Comput. Vision 4(2), 127–140 (1990)
9. Kim, Z.W.: Geometry of vanishing points and its application to external calibration and realtime pose estimation. Reseach Report, Institute of Transportation Studies, University of California at Berkeley (2006)
10. Lenz, R.K., Tsai, R.Y.: Techniques for calibration of the scale factor and image center for high accuracy 3D machine vision metrology. IEEE Trans. Pattern Anal. Mach. Intell. 1(5), 713–720 (2000)
11. Deb, K., Pratap, A., Agarwal, S., Meyarivan, T.A.M.T.: A fast and elitist multi-objective genetic algorithm: NSGA-II. IEEE Trans. Evol. Comput. 6(2), 182–197 (2002)
12. Wang, J.: An overview of research on weigh-in-motion system. In: Proceedings of Fifth World Congress on Intelligent Control and Automation (WCICA 2004), pp. 5241–5244 (2004)
13. Hanning, T.: High precision camera calibration. Habilitation thesis. Vieweg+Teubner Verlag, University of Passau (2009)
14. Zhang, Q., Li, H.: MOEA/D: a multiobjective evolutionary algorithm based on decomposition. IEEE Trans. Evol. Comput. 11(6), 712–731 (2007)

Environment-Model Based Testing with Differential Evolution in an Industrial Setting

Annamária Szenkovits[1]([✉]), Noémi Gaskó[1], and Erwan Jahier[2]

[1] Cluj-Napoca Faculty of Mathematics and Computer Science,
Babeş-Bolyai University, Kogalniceanu 1, 400084 Cluj-Napoca, Romania
`szenkovitsa@cs.ubbcluj.ro`
[2] University Grenoble Alpes/CNRS, VERIMAG, 38000 Grenoble, France

Abstract. Reactive systems interact continuously with their environments. In order to test such systems, one needs to design executable environment models. Such models are intrinsically stochastic, because environment may vary a lot, and also because they are not perfectly known. We propose an environment-model based testing framework optimized for reactive systems, where Differential Evolution (DE) is used to fine-tune the environment model and to optimize test input generation. In order to evaluate the proposed method, we present a case study involving a real-world SCADE system from the domain of railway automation. The problem specification was proposed by our industrial partner, Siemens. Our experimental data shows that DE can be used efficiently to increase the structural coverage of the System Under Test.

Keywords: Evolutionary testing · Differential Evolution · Model-based testing · Reactive systems

1 Introduction

Testing is a crucial step in the software development life-cycle. It is common to dedicate at least 50 % of the project resources to this step [1]. Therefore the problem of finding efficient ways to automate different aspects of software testing received a lot of attention in the scientific community [2].

Testing is even more crucial for reactive systems, as they are very often critical (e.g., in embedded systems). Reactive systems present additional challenges in this regard, because of the feedback loop: a reactive system acts on its environment, which in turn acts on the system. Realistic input sequences need to be generated on-the-fly, by executing the system into a simulated environment. Moreover, environments are intrinsically stochastic, because they may vary a lot, and also because they are not perfectly known. Lutin [3] is a language dedicated to the programming of such environment simulators; it allows one to model stochastic reactive systems. Lutin programs perform a guided random exploration of the system under test (SUT) environment state-space. This exploration is

© Springer International Publishing Switzerland 2016
G. Squillero and P. Burelli (Eds.): EvoApplications 2016, Part I, LNCS 9597, pp. 819–830, 2016.
DOI: 10.1007/978-3-319-31204-0_52

parametrised by weights, that define probabilities to various behaviors to occur. Such probabilities are not always easy to define. The central idea of this article is to take advantage of evolutionary algorithms to optimize these parameters automatically.

Indeed, evolutionary algorithms have been successfully used in test automation due to their ability to effectively find solutions in large, poorly defined search spaces. To be able to perform test automation with evolutionary algorithms, the test aim must be transformed into an optimization task. To achieve this, we propose a testing framework which consists of the combination of evolutionary methods and model-based testing. The main goal of this framework is to enable the optimization of the automatic test generation process for reactive systems by maximizing the structural coverage of the SUT.

The system on which we performed model-based testing is a real-world reactive system from the domain of railway automation. The problem was proposed by our industrial partner, Siemens. It was designed and developed in SCADE [4], a synchronous, widely used industrial tool-set, especially for designing avionic and railway systems.

The paper is structured as follows. In Sect. 2 we review some of the existing scientific papers which are relevant for this topic. Section 3 presents the components of the testing framework. Subsect. 3.1 describes some of the fundamental aspects of the language Lutin, focusing on the different properties of the language that will be exploited in our work, while Subsect. 3.2 describes the real-world system on which the testing was performed. Section 4 presents in detail how the Differential Evolution optimization was applied into the environment model-based testing framework. Finally, Sect. 5 summarizes the experimental results, while Sect. 6 closes the paper with our conclusions and ideas for future work.

2 Related Work

The idea of using evolutionary methods to enhance different aspects of software testing has been investigated before, however progress in the use of artificial intelligence in automated test-case generation has been slow. There are a number of significant research papers that study the use of evolutionary methods in software testing. A good overview is given in [5]. We mention a few of the related articles which emphasize the practical applicability of evolutionary methods to real-life problems.

The work in [6] discusses the scalability, applicability, and acceptability of evolutionary functional testing in industry. The problem is investigated through two case studies, drawn from serial production development environments. In [7,8], structural evolutionary testing is applied to some complex real-world software modules frequently used in embedded systems. In [9], evolutionary testing is used for automatic sequence testing of industrial control systems in the automotive domain.

The methods presented in [10,11] use an evolutionary algorithm to automatically generate a test program for pipelined processors by maximizing a given verification metric. Genetic Evolutionary Algorithms (GEA) are also used for test generation by Cheng and Lim [12]. The problem of parameter selection is discussed and a Markov chain based method is used to model the test generation process and to parametrize the process characteristics. The method is used here in particular for generating test cases to verify hardware design for semiconductor industry.

However, if we wish to apply the methods presented in the above mentioned papers to reactive systems, we must face two major difficulties. On one hand, reactive systems have cyclic behavior, meaning that at each cycle they read the inputs coming from their environment, compute the outputs and update the internal state of the system. Considering this, instead of generating a single test input, the tester has to provide test sequences, i.e., sequences of input vectors. Another issue that arises during test input generation is that input sequences cannot be generated offline. Because a reactive system is in continuous interaction with its environment, the input vector at a given reaction may depend on the previous outputs. Thus, input sequences can only be produced on-line, and their elaboration must be intertwined with the execution of the SUT.

In our paper, we present a test framework optimized for reactive systems that deals with both of the above mentioned issues, and uses Differential Evolution (DE) for online test input sequence generation.

3   Environment-Model Based Testing Framework with Differential Evolution

In this paper, we set up an environment-model based test framework for reactive systems in which we use DE to optimize the environment model and generate input sequences. In order to evaluate the proposed method, we focus on a case study involving a real-world, industrial system.

The program on which we performed environment-model based testing with DE belongs to the family of reactive systems. The main characteristics of such systems is that they are in continuous interaction with their environment. During their lifecycle, they continuously read the input data coming from the environment, update their internal state, then write the outputs to their actuators, by which they act upon their environment. The behavior of reactive systems can be best described using synchronous languages [13]. SCADE and Lutin, the languages used to create the SUT and environment model, respectively, are also synchronous languages.

The testing framework used in the current case study consisted of the following components, as illustrated by Fig. 1: the SUT, environment model and MTC analyzer.

The SUT on which the testing was performed was the SCADE model of the train control system 'Transmission Beacon/Locomotive 1 +' (TBL1+). More precisely, we used the C code (with approx. 17000 lines of code) generated from the SCADE model by the SCADE Suite KCG code generator and executed the test cases on it. The TBL1+ system is presented more in detail in 3.2. To model the environment and to generate test inputs, we used the Lutin language.

As illustrated in Fig. 1, the SUT and environment are in continuous interaction. The SUT reads the inputs from the Lutin environment, updates its internal state, and generates some outputs. This cyclic behavior continues for a specified number of iterations. At the end of each series of reactions consisting of a given number of iterations, to evaluate how thoroughly the SUT has been executed, we used the SCADE Suite MTC tool. The MTC tool retrieves structural coverage data of the SUT based on the MC/DC (modified condition/decision coverage) criterion. The MC/DC is a widely-used coverage criterion in the model-based testing of SCADE models and of safety-critical systems in general. It is a stricter form of decision (or branch) coverage which requires that each condition in a decision be shown by execution to independently affect the outcome of the decision [14]. By optimizing the environment model with Differential Evolution, we aimed to generate test inputs that, executed against the SUT, produced a MC/DC value as high as possible. Thus, the fitness function of the DE optimization was the MC/DC value itself.

The SUT and environment were connected by the Lurette tool [15,16] an automatic test generator for reactive systems. Lurette automated the stimulation of the SUT, by executing the Lutin code and feeding in the inputs generated to the SUT. The SCADE MTC was executed in batch mode and communicated with the rest of the framework via Java sockets.

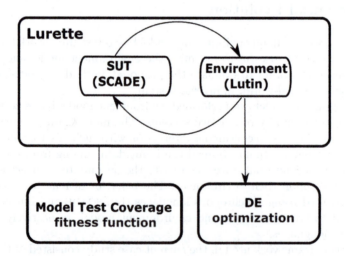

Fig. 1. Components of the test framework: the **SUT** as a C code generated from a SCADE model with SCADE Suite KCG, the **environment** model expressed in Lutin and the **MTC analyzer**. The SUT and environment are in continuous interaction with each other and have a cyclic behavior. The execution of the SUT and environment is done by the Lurette tool.

The DE-driven optimization was added to the testing framework described above. Our goal was to fine-tune the Lutin environment so that the test cases generated and executed against the SUT would cover the structure of the SUT code as much as possible. At the end of each series of reactions, the MC/DC value was retrieved by the SCADE MTC tool and the Lutin environment was updated accordingly. The process of the optimization is described step by step in Sect. 4.

Further on in this section, we present in detail the components of the testing framework.

3.1   Non-deterministic Environment Model in Lutin

By using Lutin for modeling the SUT environment, one can express non-deterministic behaviors, and perform guided random exploration of the environment state-space. Since for most real-life industrial systems, the environment is versatile or unknown (or both), the non-deterministic nature of Lutin makes it suitable to model realistic, or just interesting, test scenarios. Because Lutin models allow to fully automate the testing process, the random exploration can be done nightly, and hopefully trigger a lot of (corner) cases.

Nevertheless, when a completely random exploration fails to increase the SUT coverage, it is necessary to refine the Lutin model to guide the random exploration. This can be done by experts who know how to bring the SUT into some specific state. To some extent, we believe that DE based optimization process could help to refine existing Lutin programs automatically. In the following we briefly review the Lutin language concepts [17], to be able to explain which part of Lutin programs could be computed by DE algorithms. More program examples can be found in [16,18].

A Lutin *node* is a data-flow program that transforms a sequence of input tuples (made of Boolean, integer, or real values) into a sequence of output tuples, exactly as Scade or Simulink block-diagrams do. Lutin node can be activated periodically (time-triggered) or sporadically (event-driven). The main difference with a Scade or Simulink node is that they are made of a set of equations that have (thanks to some syntactic restrictions) exactly one solution, whereas a Lutin node is made of a set of (linear) constraints that can have any number of solutions. Such constraints are relations between inputs, outputs, and memories. The following Lutin program transforms a sequence of 2 reals (min,max) into a sequence of 1 real (x).

```
node between(min, max : real) returns (x : real) = loop
{ min <= x and x <= max }
```

Depending on the values taken by the sequence (min,max), the constraint can have 0, 1, or several solutions. When several solutions exist, one is chosen at random. If there is no solution (max < min), this program stops. For example, if this program is called with the sequence [(1.0,1.0); (0.0, 10); (-10.0,0.); (0.0,-1)], a possible output sequence could be [1.0; 3.14; -7.3].

In order to express different scenarios, Lutin also has control structures based on regular operators: the sequence (`fby`: pronounce followed by), the choice (`|`), and the Kleene star (`loop`). The alphabet of this regular language is made of constraints. The choice operator introduces a second (control) level of non-determinism. The Lutin interpreter therefore works in 3 stages: (1) choose the active constraint in the control structure; (2) solve the constraint; (3) choose a point among the solutions; if no solution exists, backtrack to stage 1. One can attach weights to each alternative of a choice (the default weight is 1) to define the probability of that alternative to be chosen at stage 1. For example, the following (input-free) Lutin node will generate a infinite sequence of integers with 75 % of 42, and 25 % of 1.

```
node choice () returns(x:int) =
loop {
    |3: x = 42
    |1: x = 1
}
```

Weights can be any integer expression made of constants, inputs, or memories. Our proposition is to let a DE algorithm fine-tune the weights of existing Lutin programs. The above mentioned control structures were used to create an environment model for the SCADE SUT. Afterwards, we used the parameters of these non-deterministic operators (for example the weights of the choice operator) in the optimization process with DE, in order to fine-tune the environment.

3.2   System Under Test

Our case study focused on a real-world, industrial problem from the domain of railway automation. The problem specification was proposed by Siemens. The problem is related to the TBL1+ system, a train protection system compatible with the European Train Control System [19] used in Belgium and on Hong Kong's East Rail Line. Its main role is to ensure safe operation in the event of human failure.

More precisely, the TBL1+ system requires the locomotive driver to manually acknowledge a warning when passing a double yellow signal (A double yellow signal means: *Preliminary caution, the next signal is displaying a single yellow aspect*, while a Single yellow aspect signalizes the following: *Caution, be prepared to stop at the next signal*).

The system consists of a beacon on the ground that emits an electromagnetic signal. This signal is received by an antenna underneath the driving cab. As the train driver approaches a red signal, this support system switches on a light in the cab. The train driver must then confirm that they have received the warning by pressing a button. If the driver does not do this, then the emergency brake is automatically activated.

Besides the vigilance checking functionality mentioned above, the TBL1+ system has a speed restriction checking functionality. This feature is activated by a beacon located 300 m up-line from a signal. If the train travels at a speed greater than 40 km/h ahead of a red signal, the TBL1+ system triggers the emergency brake. However, the brake can be deactivated after 20 s by pressing the acknowledgment button, if the danger is no longer present.

Regarding its interface, the SUT has seven boolean inputs indicating the states of the different buttons located on the on-board equipment, and three integers, two of which code the information sent by the beacons from the ground, while the third one represents the speed of the train. Among the outputs of the SUT, four boolean variables represent the state of different on-board lamps, while the fifth one shows whether the emergency brake has been activated or not.

4   Optimizing the Environment Model with Differential Evolution

Evolutionary algorithms (EA) are powerful optimization tools, they can be easily adapted to specific optimization tasks and have good applicability in different fields such as engineering, robotics, biology, economics, etc. EA is inspired by biological evolution, using mechanisms such as reproduction, mutation, recombination and selection. Generally in an EA, a population of candidate solutions (individuals from the population) are evaluated, and a fitness function defines the value ("goodness") of each individual. The evolutionary process of surviving the fittest ("best") paradigm is fulfilled.

Differential Evolution (DE) [20] is a stochastic, population-based optimization algorithm, with good results in many real-world optimization problems, even in non-continuous or dynamic environments (some applications and description can be found in [21]). DE is very popular because of its simplicity, speed, and robustness.

DE enters a loop of generating offspring, evaluates offspring, and selects individuals to create the next generation, until the termination condition is met (for example the maximum number of generation is reached).

4.1   Optimization Steps

Our approach proposed for applying the DE algorithm for the optimization of the parameters of an environment model is composed of the following main steps:

1. Specifying the subject of the optimization and population representation (which parameters are to be optimized);
2. Specifying the fitness function;
3. Specifying the operators.

Population Representation (Parameters Optimized). In order to generate the test inputs and stimulate the SUT, we created an executable environment model in Lutin. We used the information contained in the specification of the TBL1+ system to write the Lutin code. In this environment model, we simulated

```
-- acknowledgement button
node BacButton () returns (bac: bool) =
  loop {
    |p: bac = true
    |100-p: bac = false
  }
```

Fig. 2. Simulating the pushing of the acknowledgment button. The simulation is realized with Lutin's random choice operator. Weight p assigned to the branches is optimized by DE.

the following three aspects: the change of the speed of the train depending on the state of the emergency brake, the random pressing/releasing of the different buttons, and the change in the codes transmitted by the beacons located on the ground.

We created two versions of the Lutin environment. In the first version for the simulation of the random pressing/releasing of the buttons, as well as for generating the beacon information, we used Lutin's choice operator with default weights. This means that the different variables had uniform distribution in each iteration.

In the second version however, we parameterized the weights of the choice operator and let the DE algorithm optimize these weights. We restricted each weight to the range between 1 and 100. For the choice operators with two branches, we assigned weight p to the first branch and $100 - p$ to the second one. In case of the choice operators having more than two branches, we assigned the binomial distribution $B(n, p)$ to the branches, with n denoting the number of branches of the choice operator. In both cases, we let DE optimize p.

As an example, Fig. 2 illustrates the choice operator simulating the pressing and releasing of the acknowledgment button, and the weights assigned to the branches of the operator. As described in Sect. 3.2, the locomotive driver can manually acknowledge a warning by pressing this button, but he can also deactivate the emergency brake, if the danger is no longer present.

Fitness Function. After configuring the Lutin environment with a set of parameters, we let the SUT-environment loop run for a given number of iterations. At the end of each sequence of iterations, we made use of the SCADE MTC tool to measure the structural coverage of the SUT obtained with the given Lutin parameters. The obtained coverage rate was used as the fitness value of the given parameter vector in the DE algorithm.

Operators. The initial population was selected randomly from the admissible parameter ranges. During each iteration, for each individual l from the population, an offspring $O[l]$ was created using the scheme presented in Algorithm 2, where $U(0, x)$ is a uniformly distributed number between 0 and x, CR denotes

---

Algorithm 1. DE

1: Randomly generate initial population P_0 of solutions;
2: **while** (not termination condition) **do**
3:     **for** each $l = \{1, ..., population\ size\}$ **do**
4:         create offspring $O[l]$ from parent l;
5:         **if** $O[l]$ is better then parent j **then**
6:             $O[l]$ replaces parent j;
7:         **end if**
8:     **end for**
9: **end while**

Algorithm 2. DE - the DE/*rand/1/bin* scheme

Create offspring $O[l]$ from parent $P[l]$
1: $O[l] = P[l]$
2: randomly select parents $P[i_1]$, $P[i_2]$, $P[i_3]$, where $i_1 \neq i_2 \neq i_3 \neq i$
3: $n = U(0, dim)$
4: **for** $j = 0$; $j < dim \wedge U(0, 1) < CR$; $j = j + 1$ **do**
5:     $O[l][n] = P[i_1][n] + F * (P[i_2][n] - P[i_3][n])$
6:     $n = (n + 1) \bmod dim$
7: **end for**

---

the crossover ratio, F is the scaling factor, and dim is the number of problem parameters.

The main steps of the algorithm are outlined in Algorithm 1.

5 Experimental Results

Our experimental set-up consisted of two Lutin environments. In the first, plain Lutin environment, no weights were added to the choice operators, so that the different variables were generated with uniform distribution in each iteration. In the second environment, DE was used to optimize the weights of the choice operators, in order to increase the structural coverage of the SUT.

We ran DE with two different parameter settings. These settings are summarized in Table 1. Based on [22], for some problems a smaller value of F is a good setting, therefore we used a small value for F for both Setting 1 and 2. Ten independent runs were conducted for both of the settings. Average results and

Table 1. DE parameter settings.

Parameter	Setting 1	Setting 2
Pop size	50	50
Scaling factor F	0.1	0.25
Crossover rate (CR)	0.9	0.75

standard deviation are described in Table 2, as well as the results obtained with the plain Lutin environment, where no weights were added.

The algorithm converged to the obtained values in 8–10 generations, therefore the maximal number of the generations was 10. The number of parameters optimized, that is, the number of weights added to the choice operators and optimized with DE, varied between 3 and 8.

As shown in Table 2, the coverage rate increased with approximately 10 % for both setting 1 and 2 compared to the coverage rate obtained with the test inputs generated with the plain Lutin environment. Considering that the experiments were run on a real-world, complex system, the rate of improvement obtained with DE can be regarded as significant. Also, a Wilcoxon test indicates that there is a difference between the results of Setting 1 and Setting 2 in the case of 3, 6 and 7 parameters.

Table 2. MTC coverage rates (average results and standard deviation) obtained with the DE-optimized and the plain Lutin environment. The different settings of the DE are described in Table 1. The number of parameters refers to the number of weights optimized in the Lutin code.

Parameters	Setting 1	Setting 2	Plain Lutin environment
	$avg. \pm stdev.$	$avg. \pm stdev.$	
3	70.03 ± 0.45	70.51 ± 0.32	62.70
4	69.86 ± 0.36	70.07 ± 0.45	62.70
5	70.51 ± 0.40	70.72 ± 0.12	62.70
6	71.10 ± 0.24	71.28 ± 0.23	62.70
7	71.52 ± 1.17	72.74 ± 0.22	62.70
8	69.75 ± 0.68	70.17 ± 0.25	62.70

6   Conclusion and Future Work

We presented a model-based testing framework for reactive systems, where an executable, non-deterministic environment model was created in the Lutin language. The structural coverage of the SUT during environment model-based testing is largely dependent on the parameters of the Lutin environment, therefore we implemented a DE algorithm to fine-tune the parameters.

To evaluate the performance of the testing framework, we ran experiments on a real-world, industrial system the specification of which was proposed by Siemens. More precisely, the SUT consisted of the C program generated from the SCADE model of the TBL1+ train control system.

We compared the performance of our model-based testing framework that uses a DE optimized Lutin environment to standard model-based testing where the same Lutin environment was used without any optimization.

The experimental data shows that, as a result of the DE optimization, the code coverage of the SUT as measured after executing a number of generated test sequences, was significantly improved.

The importance of this result is twofold: (1) we did not need any domain specific expert knowledge in order to increase the code coverage during the testing phase of a fairly complex reactive system; (2) the optimized executable environment model can be reused any number of times when testing the same system that has undergone minor modifications.

The proposed testing framework could further be improved by experimenting with different evolutionary algorithms and extending the evaluation criteria to also include oracle coverage rates. As part of our future work we also consider experimenting with different learning techniques in order to improve the Lutin environment parameters automatically.

Acknowledgement. This material is based upon work supported by the Siemens international Railway Automation Graduate School (iRAGS) and the SCADE Academic Program.

References

1. Beizer, B.: Software Testing Techniques, 2nd edn. Van Nostrand Reinhold Co., New York (1990)
2. Ammann, P., Offutt, J.: Introduction to Software Testing, 1st edn. Cambridge University Press, New York (2008)
3. Raymond, P., Roux, Y., Jahier, E.: Lutin: a language for specifying and executing reactive scenarios. EURASIP J. Embed. Syst. **2008**, 1–11 (2008)
4. Dormoy, F.X.: Scade 6 a model based solution for safety critical software development. In: ERTS 2008 (2013)
5. McMinn, P., Holcombe, M.: The state problem for evolutionary testing. In: Cantú-Paz, E., et al. (eds.) GECCO 2003. LNCS, vol. 2724, pp. 2488–2498. Springer, Heidelberg (2003)
6. Vos, T.E., et al.: Evolutionary functional black-box testing in an industrial setting. Softw. Qual. Control **21**, 259–288 (2013)
7. Wegener, J., Buhr, K., Pohlheim, H.: Automatic test data generation for structural testing of embedded software systems by evolutionary testing. In: Proceedings of the Genetic and Evolutionary Computation Conference, GECCO 2002, pp. 1233–1240. Morgan Kaufmann Publishers Inc., San Francisco (2002)
8. Wegener, J., Baresel, A., Sthamer, H.: Evolutionary test environment for automatic structural testing. Inf. Softw. Technol. **43**, 841–854 (2001)
9. Baresel, A., Pohlheim, H., Sadeghipour, S.: Structural and functional sequence test of dynamic and state-based software with evolutionary algorithms. In: Cantú-Paz, E., et al. (eds.) GECCO 2003. LNCS, vol. 2724, pp. 2428–2441. Springer, Heidelberg (2003)
10. Corno, F., Cumani, G., Reorda, M.S., Squillero, G.: Evolutionary test program induction for microprocessor design verification. In: 2012 IEEE 21st Asian Test Symposium, p. 368 (2002)

11. Iwashita, H., Kowatari, S., Nakata, T., Hirose, F.: Automatic test program generation for pipelined processors. In: IEEE/ACM International Conference on Computer-Aided Design, pp. 580–583 (1994)
12. Cheng, A., Lim, C.C.: Markov modelling and parameterisation of genetic evolutionary test generations. J. Glob. Optim. **51**, 743–751 (2011)
13. Halbwachs, N., Caspi, P., Raymond, P., Pilaud, D.: The synchronous data flow programming language lustre. Proc. IEEE **79**, 1305–1320 (1991)
14. Jones, J., Harrold, M.: Test-suite reduction and prioritization for modified condition/decision coverage. IEEE Trans. Softw. Eng. **29**, 195–209 (2003)
15. Jahier, E., Raymond, P., Baufreton, P.: Case studies with lurette v2. Softw. Tools Technol. Transf. **8**, 517–530 (2006). http://www.springerlink.com/content/u02131123x856227/fulltext.pdf
16. Jahier, E., Halbwachs, N., Raymond, P.: Engineering functional requirements of reactive systems using synchronous languages. In: International Symposium on Industrial Embedded Systems, SIES 2013, Porto, Portugal (2013)
17. Raymond, P., Nicollin, X., Halbwachs, N., Weber, D.: Automatic testing of reactive systems. In: Proceedings of the 19th IEEE Real-Time Systems Symposium, pp. 200–209 (1998)
18. Jahier, E., Djoko-Djoko, S., Maiza, C., Lafont, E.: Environment-model based testing of control systems: case studies. In: Ábrahám, E., Havelund, K. (eds.) TACAS 2014. LNCS, vol. 8413, pp. 636–650. Springer, Heidelberg (2014)
19. Official github repository of the open etcs project (2008). https://github.com/openETCS. Accessed November 2015
20. Storn, R., Price, K.: Differential evolution-a simple and efficient heuristic for global optimization over continuous spaces. J. Glob. Optim. **11**, 341–359 (1997)
21. Das, S., Suganthan, P.N.: Differential evolution: a survey of the state-of-the-art. IEEE Trans. Evol. Comput. **15**, 4–31 (2011)
22. Liu, J., Lampinen, J.: On setting the control parameter of the differential evolution method. In: Proceedings of the 8th Internationl Conference Soft Computing, MENDEL 2002, pp. 11–18 (2002)

Workforce Scheduling in Inbound Customer Call Centres with a Case Study

Goran Molnar[1]([✉]), Domagoj Jakobović[1], and Matija Pavelić[2]

[1] Faculty of Electrical Engineering and Computing,
University of Zagreb, Unska 3, 10000 Zagreb, Croatia
{goran.molnar2,domagoj.jakobovic}@fer.hr
[2] Asseco SEE, Ulica Grada Vukovara 269d, 10000 Zagreb, Croatia
matija.pavelic@asseco-see.hr

Abstract. Call centres are an important tool that businesses use to interact with their clients. Their efficiency is especially significant since long queuing times can reduce customer satisfaction. Assembling the call centre work schedule is a complex task that needs to take various and often mutually conflicting goals into account. In this paper, we present a workforce scheduling system suited for small to medium call centres and adjusted to the needs of two real–world client institutions. The scheduling problem is to minimise the difference between allocated and forecasted number of staff members while also caring for numerous legal and organisational constraints as well as staff preferences. A flexible constraint handling framework is devised to enable rapid prototyping methodology used during the development. Based on it, two metaheuristics are devised for schedule construction: GRASP and iterated local search. Performance analysis and comparisons for these two methods are provided, on a real–world problem example. The devised system is successfully implemented in a real world setting of call centres at PBZCard, Croatian largest credit card vendor and PBZ (Intesa Sanpaolo), one of the largest Croatian banks.

Keywords: Staff scheduling · Call centres · GRASP · Iterated local search

1 Introduction

Virtually all types of human organisations occasionally face scheduling problems. Staff scheduling is a classic operations research problem that consists of assigning a set of employees to a set of working times, subject to various constraints and with the goal of finding the schedule of the best quality. Such problems are $\mathcal{NP} - hard$ [1–3] in their very simplest forms, which imposes tractability issues and renders them difficult to solve efficiently. Contact centre staff scheduling is an example of this type of a problem [2,4–6].

Call centres are an instrument many companies use for communication with their clients. They typically consist of a centralised pool of trained staff members

© Springer International Publishing Switzerland 2016
G. Squillero and P. Burelli (Eds.): EvoApplications 2016, Part I, LNCS 9597, pp. 831–846, 2016.
DOI: 10.1007/978-3-319-31204-0_53

called *agents*. When a customer dials the number of a company, if there are available agents in the centre, her call is answered immediately. However, if all staff members are busy at that moment, the customer will have to wait in the queue until an agent becomes available. Long waiting times can cause customer dissatisfaction and it is critical to keep the waiting times as low as possible. On the other hand, to keep the operating cost reasonable, hiring too much staff is also undesired.

In a traditional call centre, staff members are communicating exclusively via telephone. With the increasing popularity of the Internet, many companies added support for other means of contact, such as e-mail or chat, extending call centres into their contemporary generalisation called *contact centres*. Based on the number of supported channels, centres can be *multi–channel* or *single–channel*. Centres that are answering, but never actively initiating communication are called *inbound* contact centres. Conversely, centres that are only initiating communication are called *outbound* contact centres. If both answering and calling are performed, the service is classified as a *mixed* contact centre. With regard to the staff skills, a contact centre can be *single–skilled* or *multi–skilled*. In a single skilled centre, all of the employees have the same training and theoretically, provide a homogeneous service level, independent of the agent. In a multi–skilled centre, there are various profiles of agents, depending on their skill sets.

In this paper, we describe a call centre scheduling algorithm implemented in a commercial software package. It is suited for the needs of a small to moderately sized single–skill inbound call centres. The system is based on a flexible constraint management framework that allows easy addition of new company–specific constraints and two robust, scalable metaheuristics. Related work proposes a diverse range of techniques such as dynamic programming, linear, quadratic and mixed integer programming and relaxations of the linear programs [7–15]. Several heuristic methods are also proposed [16]. However, the aforementioned work is focused on the optimisation part and less on the constraints imposed by the organisation. As an exception, a hybrid heuristic approach with several techniques, including an algorithm inspired by simulated annealing is described in [5] and applied to a real–world problem. Constraint handling appears to be performed through the means of the objective function, therefore defining them as ranked soft constraints that can be violated in final solutions. In [9,10], an integer program is solved using an iterative cutting plane method with evaluations based on simulation while in [17] the authors frame the problem as a mixed integer stochastic program. A detailed literature review on staff scheduling, including specifics of call centres, is available in [7].

We are focused on GRASP and iterated local search (ILS), which were, to the best of our knowledge, never successfully applied to this type of problems. While a direct heuristic to schedule weekend working days is also used, times of arrival are scheduled relying exclusively on metaheuristic methods, instead of using hybrid algorithms. As compared to related work, the problem we are solving is a *highly constrained* example of a *real world* problem. A flexible constraint handling system is a highly prominent feature of our system. Moreover, our approach deals with the constraints differently than any other proposed method we are aware of. Instead of implementing them in the objective function as in [5], we devised a rule based assurance system in each component of the algorithm. In that manner, hard constraints are guaranteed to be satisfied in all of the solutions throughout the execution.

To conclude this analysis, we are focused on improving small to medium contact centres with 20–40 employees, while related work in real world scheduling mentions a higher staff number, up to a thousand. Contrary to the common first notion, small contact centres might prove to be more difficult to schedule than big ones [18]. As a further differentiation, in this stage of our work, we are investigating single–channel, single–skilled centres only, not taking into account complexities of multi–channel and multi-skilled ones. The rationale behind that decision was to provide a quick–to–develop solution for such centres. Solutions suited for larger and more complex centres surely do exist, but our goal was to produce a simple product with low development cost and, therefore, competitive price, affordable for small businesses.

As noted in [19], a noticeable gap was observed between the research output from academia and the needed expertise that can be directly utilised by the industry. Solving real problems is difficult since it necessitates modelling and handling various kinds of constraints which are usually simplified in academic problems. This paper aims to contribute to bringing the worlds of academia and practical implementations closer together.

The remainder of this paper is organised as follows: in Sect. 2, the formal problem description, including the objective function and constraints, are given. Section 3 presents a brief overview of the methodology and Sect. 4 details our development approach. The optimisation algorithms for generating schedules are described in Sect. 5. Section 6 brings the results. Some ideas for future improvements and the conclusion are given in Sect. 7.

2   Problem Descriptipon

The call centre workforce scheduling problem ($CCWSP$) is a scheduling problem whose goal is to maximise the *service level* while satisfying all the imposed constraints. The service level is typically defined as a combined measure consisting of (i) a percentage of answered calls (more is better) and (ii) percentage of dropped calls (less is better) during a time period. In our system, schedules are typically generated one month ahead, with adjustable target service levels. Times are quantized, with the minimum quantum duration given as a parameter and 30 min is used in this case study. If needed, finer granularity could be used. This naturally means that the optimisation problem might be more complex.

The problem is defined as an ordered triplet:

$$CCWSP = (S, s_d(t), C),$$

that consists of a set of staff members S, ideal staff distribution in time based on the demand $s_d(t)$ and a set of constraints. The set of staff members is a fixed set of workers at the contact centre, with their permanent workplace designated for each staff member. Seating and workplace assignment is therefore not a part of the optimisation problem. Each staff member is defined by its identifier in the system and the relevant staff data is stored in a suitable database.

The ideal staff distribution $s_d(t) : \mathbb{N} \rightarrow \mathbb{N}$ is a function which defines a minimum number of employees needed at time t, in order to achieve the desired service level. As the frequency of calls varies throughout the days of the month, staff number that is needed to handle such demand will also vary. In our case study, the demand peaks are usually experienced afternoon, and they are most prominent on Mondays, as seen on Fig. 1. Much fewer calls are placed during weekends, requiring noticeably less workforce.

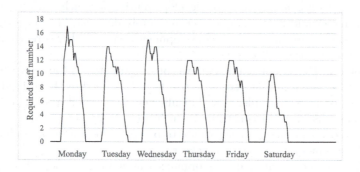

Fig. 1. An example staff distribution curve for one week (Mon–Sun)

The set of constraints is inspired by the labour law in Croatia, company specific policies and regulations. Furthermore, it's easy to notice that many of the constraints were defined out of a desire to keep the employees satisfied, by giving them as much flexibility as currently possible. Most of the constraints are related to a single worker but there are some, that are dealing with an aggregate number of working staff during weekends.

Contact centre constraints include:

- Contact centre open hours, e.g., centre is open 7:00–21:00 during working days, 07:00–17:00 during Saturdays and closed Sundays and holidays.
- Shifts intervals (e.g., morning shift arrivals from 07:00–10:00).
- Maximum allowed daily work time (legally defined to be 8 h).
- Minimum daily break, defined as the time between consecutive presences at work, (currently 12 h as defined in the Croatian labour law).

Basic staff constraints include:

- Suitable arrival times (some workers can only work in the morning while some can work anytime – during the morning, afternoon and night shift).
- Suitable working days (e.g., some workers work Mon–Sat, some Mon–Fri).

- Duration of the work time (can be adjusted for each day of the week and holidays, e.g., worker can work 8 h Mon–Fri and 7 h on Saturdays).
- Prearranged absences (vacations, free days, sick leaves).
- Prearranged presences (a manager might arrange some activities in advance).

Some company specific constraints in our case study include:

- Workers arrive to work at the same hour throughout Mon–Fri.
- The weekend schedule is independent of the work day schedule.
- Simple night shift rotation rules are supported for the night shift staff.
- Employees working on Saturdays need to have at least two working Saturdays, if needed (during 5 Saturday months), an employee can work three Saturdays, but only if he or she didn't work three Saturdays during the previous month.
- Shift work (soft constraint) – employees working in both morning and afternoon shifts need to have at least 5 days in each, to ensure they are entitled to their monthly shift work bonus.
- To ensure equal service levels during weekends, number of staff members during Saturdays needs to be roughly equal.
- Unpopular shifts constraint: an employee cannot be scheduled in unpopular shifts (e.g., 14:00) during two consecutive weeks.

While these constraints are highly specific for the case study call centre, the implementation of the constraint management system is modular and parametrised, in order to enable easy integration into similar call centres and further customisation.

3   Methodology Overview

Similarly to related approaches, the process of scheduling the contact centre workforce is performed in three stages: (1) forecasting, (2) staffing and (3) scheduling. The typical workflow starts by the contact centre manager telling their staff to enter their preferences and absence days in the scheduling system through the staff version of the user interface. Call centre administrators then make further adjustments such as setting up the target service level and constraints, using the administrator version of the user interface. In this stage, the S and C component of the contact centre scheduling problem are defined.

The ideal staff distribution $s_d(t)$, however, isn't known in advance and depends on the number of calls received during the upcoming scheduling period. The call forecasting is based on the inbound call records from the activity history. Currently, we apply a simple forecasting model that assumes that the calls for the current month will match the inbound calls during the same month previous year. To further refine the forecasting and account for the differences among days of the week, the forecasting system automatically aligns working days for the forecasted month with the working days of the reference month. In that way, working days, weekends and fixed non working days are aligned.

The greatest variation in the volume of calls is dependent on the day of the week and the season of the year. Aside from already mentioned weekend–work day load intensity differences, late December peaks and low–intensity summer weeks are common in our case study and this pattern repeats every year. Therefore, even such a simple model provided a good estimate of the future calls.

After forecasting, the staffing is performed. It consists of determining the ideal staff distribution for the upcoming period. Since our call centre has a single queue and is single–skilled, there is an analytical solution to calculating the number of necessary staff. This calculation is computed using the Erlang-C [4] formula. An example of an ideal staff distribution curve calculated by the system for an example real–world week is shown on Fig. 1. The user interface allows further manual staff distribution curve adjustments as specific events such as promotions of new services might change the demand.

After defining an ideal staff distribution, it is needed to find the workforce schedule that matches this distribution as closely as possible. Due to the complex constraints and tractability issues, devising a direct algorithm for the problem is difficult. Therefore, we decided to use a metaheuristic approach.

4 Development Approach

4.1 Bottom–up Methodology

The implementation methodology we decided to use is a *bottom–up* approach in which the constraint assurance system was developed first. This component is essential to allow generating feasible schedules as well as querying which modifications can be done to keep the feasibility preserved in the resulting solution. After the constraints are precisely defined and implemented, the optimisation algorithm that works on top of the constraint manager is developed.

The process of building an optimisation algorithm can be complex and implementing a metaheuristic, from a practitioner's point of view still holds many elements that bear more resemblance to an art rather than to an exact engineering process. For example, when starting a project based on metaheuristics, the question of "Which metaheuristic should I use for my project?" is regularly asked. While some problems can be more naturally represented for certain metaheuristics (e.g., TSP is trivial to represent in the ant colony optimisation framework), when it comes to the efficiency, little can be said about it before the algorithm is implemented and experimentally evaluated [20,21]. Therefore, the metaheuristic choice is typically based on the implementer's level of familiarity with various metaheuristics and a lot of educated guesswork.

We wanted to evade this problem by employing a bottom–up design approach in which we are building reusable components that can be a part of several different metaheuristics. Metaheuristics typically consist of (1) initial solution generator (2) intensifying component, such as local search operator, (3) diversifying component (such as perturbation in the ILS terminology or mutation in the genetic algorithm terminology). We wanted to devise each of these components as a separate software module sharing the common solution representation and the interface with the constraint assurance system.

While it is long known that most metaheuristics fit into the intensification and diversification (I&D) frame which enables component reusability and creating hybrid heuristics [22], we also argue that this fact, combined with the bottom–up approach explained above can be profitably employed to make the metaheuristic implementation faster and thus, cheaper for the practitioners. By gradually increasing the complexity level of our algorithm the software development cycle can end at the earliest point when the product starts producing results within expectations. Starting from the simplest and easiest to implement, such as random restart search and gradually moving to more complex metaheuristics by adding the required operators might save time if simple algorithms work well enough for practical use.

The component requirements are simple to determine for each metaheuristic. To implement random search, only the random solution generator in a loop is needed. Adding the intensification component such as deterministic or random local search is sufficient to implement GRASP and tabu search. If we build the diversification component as well, we can build virtually any metaheuristic and to the very least, those that fit into the I&D frame [22].

When the project was initiated, an initial screening revealed that the schedules were built by hand. The contact centre managers relied mostly on their subjective judgement while determining necessary staff number for each day, as a more objective approach is difficult without the appropriate IT support. Furthermore, since the beginning of the development, the clients insisted that the scheduling system needs to guarantee that most, if not all constraints are satisfied in each schedule. Therefore, enforcing constraints by the objective function was not suitable for this purpose.

We used this methodology and built random restart search based on the initial solution generator. After this, the local search operator was used to produce the GRASP metaheuristic. Surprisingly for us, even this simple metaheuristic was appraised as working well enough by our clients. For research purposes, after a simple modification component was built, some experiments were performed with the iterated local search algorithm.

4.2   Constraint Assurance and Prototyping Approach

While general ideas about the constraints were agreed upon before the project started, they were not detailed enough to allow simple translation into scheduling rules that can be implemented in a software product. The goal was to produce a scheduling system that is able to respond to a plethora of possible events in a highly sophisticated way — in essence, acting as a replacement for the human that was in charge of scheduling. Agreeing on the definition of the necessary constraints for the workforce scheduling system is a difficult management and communication issue. The knowledge of the persons who were manually scheduling the workforce can be characterised as *tacit knowledge*, something that is intuitively clear, but difficult to explain, formalise and automate. Therefore, a prototyping approach was chosen. Prototypes of a scheduling system were presented on regular meetings with the call centre management and further improvements were discussed and agreed upon. Since the definition of the problem varied through time, a flexible constraint assurance system was needed.

We decided to use a system in which all of the constraints except one are satisfied throughout the search process. The only exception is the soft shift–work constraint that was added later and is a combinatorial optimisation problem on its own, in current version partially solved by postprocessing. The devised system is a modular object oriented set of classes and interfaces written in Java and based on the *observer design pattern*. Such approach, that was to the best of our knowledge first used in [23] is a natural way to implement complex interactions of various constraints as scheduling events are occurring. In this approach, each staff member is a subject, to which an arbitrary set of constraint objects (observers) can be added. After scheduling an agent to a certain time slot, all of the observer objects (constraints) assigned to him/her are notified. After receiving the scheduling event notification, each observer updates the list of suitable times for that staff member to reflect the effects of each schedule change.

Using the described architecture, the lists of allowed time intervals are constantly maintained by the observers. In each search step, the algorithm can easily determine which quanta are suitable to schedule a staff member by performing a simple querying of the feasible times list.

For example, the unpopular times constraint removes other unpopular times throughout the previous and next week for that staff member if it receives a notification about scheduling in an unpopular period. Removed time quanta are "invisible" to the search algorithm in the remainder of the run, thus further reducing the size of the search space. If a new type of a constraint is needed, it is easy to implement – it is enough to provide an implementation of the observer interface and simply plug it in the rest of the system.

5   Scheduling Algorithm

In this work, we implement *GRASP* (Greedy Randomised Adaptive Search Procedure) [24] and iterated local search [25] metaheuristics to solve *CCWSP*. The GRASP metaheuristic uses a solution construction component and local search to build different locally optimal solutions while keeping a record of the best so far. *Iterated local search* (ILS) is a simple but effective metaheuristic that systematically explores close proximities of known good solutions using the local search operator while avoiding being stuck in local optima by the means of a *perturbation operator*.

The complete solution algorithm works in three stages: (1) preprocessing, (2) work day scheduling, (3) time scheduling. In the first step, preprocessing is done to find misconfigurations in the input problem and handle prearranged presences and absences. Working days are then split into *statically* and *variably* scheduled days. For statically scheduled days, the schedule is known in advance. Such days are not subject to optimisation and are not modified by the algorithm.

After the preprocessing, weekend working days are chosen using a direct, feasibility–preserving heuristic that determines which Saturdays and Sundays a person is working. When all the working days are known, the solution representation data structures are built. Since staff always comes at the same time Monday – Friday, in most cases Monday can represent the remaining four working days as well. Such preprocessing significantly reduces the dimension of the search space. After building the solution representation, the working days are known, however, the schedule is still empty as variable working times are yet to be determined by the optimisation algorithm.

5.1 Objective Function

An ideal timetable will have exactly $s_d(t)$ staff members available at time t. As ideal staffing curves frequently have short spikes and irregularities (Fig. 1) and the typical working time is around 8 h, it's generally impossible to achieve absolute accordance to the ideal curve. The objective function evaluates how close the real staff distribution is to the ideal one.

In the proposed objective function, we define penalty for each candidate schedule as a sum of penalties for all time quanta in the schedule:

$$p = \sum_{t=0}^{t_{max}} (s_d(t) - s(t))^2,$$

where t_{max} is the last quantum in the scheduling period and $s(t)$ is the number of staff in the generated candidate schedule, as opposed to the goal number of staff members $s_d(t)$ during time t.

Note that the number of missing staff is squared. The rationale for such objective function is the decision to emphasise bigger deviations from the desired curve. For example, let us compare a schedule with one man–hour missing during 5 h, one per each hour and a schedule with 5 man–hours missing during one hour. For that day, they both have a total of five man–hours missing. However, a lack of one person is hardly noticeable in the call centre while lack of five people has a significant impact on the service quality of a small centre with 30 employees. Squaring the number of missing persons helps emphasise this effect. A bigger difference from the ideal staff distribution means bigger penalty and lower solution quality. The best possible solution would perfectly follow the ideal distribution and have a penalty of zero.

5.2 Random Solution Generator

The random solution generator uses the described constraint assurance system to produce an initial solution in which all of the working times are chosen randomly from a set of feasible times. The scheduling is done sequentially, for all of the working days in the schedule, for each staff member that works on that day. As each scheduling decision is made, the constraints handling system is keeping updated information about which working times are suitable for the employee. As the schedule is having more working times determined, less and less time intervals are suitable, since many constraints contradict each–other. In the case of a failed constraint setup for an employee, that employee's schedule might be empty due to a constellation of conflicting constraints which caused the feasibility checker to determine that no working hours are suitable for the staff member. In such cases, users are advised to carefully inspect the setup for that person or ask for help from the support team.

5.3 Local Search and Perturbation Operators

Local search is a simple greedy operator that tests if moving a staff member earlier or later during the day helps make a better schedule. It can operate only on feasible times and does nothing if the current time is the only one that's feasible. Local search is performed for each staff member, for each day in the search space. The local search operator has one parameter: number of passes through the entire schedule. If the number of passes is set to 2, the local search algorithm will be performed twice.

To promote unbiased discovery of good solutions, the order in which local search is applied on staff members is randomised. In that manner, unfair schedules are avoided, since initial solutions tend to have a different distribution than the goal staff distribution. For that reason, some favouritism was discovered as the agents that were first to be optimised had a tendency to be scheduled to unpopular afternoon times when the demand is, in general, higher.

The perturbation operator introduces random changes in the solution, which might both improve or decrease the solution quality. Perturbation operator has one parameter: the percentage of staff members for which the schedules will be modified. For each staff member, and for each time in the dynamic schedule, a random choice of a feasible work time is performed.

5.4 Random Restart Search

After implementing an initial solution generator, a trivial random restart search metaheuristic can be easily implemented by running it in a loop. It is frequently used as a baseline to check initial versions of more advanced algorithms. The more advanced algorithms should, at their very least, be able to significantly outperform random restart search. As the end condition, a timeout is used, currently fixed to 30 s, as described in Sect. 6.1.

5.5 GRASP and Iterated Local Search

An initial solution generator and the local search are sufficient to assemble an implementation of a GRASP metaheuristic [24, 26–28]. It starts by building a random initial feasible solution, on which the local search operator is applied. After that, the procedure is repeated, all the time keeping a record of the best solution found so far. The algorithm pseudocode is given in Algorithm 1. As the end condition, a timeout is used, currently fixed to 30 s, as described in Sect. 6.1.

Adding the perturbation operator to initial solution generator and local search makes it possible to assemble the iterated local search metaheuristic. It uses the perturbation operator instead of generating entirely new solutions to escape local optima. As with the previous two approaches, a timeout of 30 s is used. The algorithm pseudocode is given in Algorithm 2. In our current implementation, perturbation is always applied to the last local optimum, therefore, we employ the *random walk* acceptance criterion for the iterated local search [25, 27, 28].

generate initial solution S_0;
S = local search (S_0) ;
$S_{best} = S$;
while *end condition not met*
do
    generate initial solution S_0;
    S = local search (S_0) ;
    if $eval(S) < eval(S_{best})$
    then
       | $S_{best} = S$;
    end
end

Algorithm 1: GRASP pseudocode

generate initial solution S_0 ;
$S_{best} = S_0$;
while *end condition not met*
do
    S = local search (S_0) ;
    if $eval(S) < eval(S_{best})$
    then
       | $S_{best} = S$;
    end
    S_0 = perturb (S);
end

Algorithm 2: ILS pseudocode

6   Results

The proposed metaheuristics are implemented in Java 1.8 programming language. During the experimental evaluation, a series of experiments was run on a real–world example that we consider to be an appropriate representative of an usual scheduling problem at the client's call centre. The problem is based on the March 2014 schedule, having 32 staff members under a typical workload for that period. In that example, the call centre is open 7:00–21:00 during working days, 7:00–20:00 on Saturdays and closed Sundays and holidays. Most staff members have 7 h work time Mon–Fri and 5 h work time during Saturdays. Total of 13 staff members is using their absence days, 58 days in total, with most of them absent for a week (6 days). There are no public holidays in this month. Time quantum duration is 30 min. This example is representative for most other examples that were encountered during the development.

All experiments were run on a computer equipped with a 2.4 GHz Intel Core i7-4700HQ processor and 16 GB of RAM. First, the initial tuning of the termination criteria was performed. Then, a detailed parameter tuning for the local search intensity and perturbation parameters (ILS only) was done with the best–found termination run time.

6.1   Termination Condition

We decided to use the allowed execution time as the termination condition. This termination condition allows a consistent user experience and near–real–time way of using the system. Before doing the tuning process, we wanted to decide on the allowed time and performed an initial tuning step to investigate the influence of execution time on the performance. Typically, when running a metaheuristic, there is an initial period of fast improvement that soon starts to slow down. After a certain point in time, the improvement rate becomes slow to none and the task of termination time tuning was to determine the time which will allow the algorithm to converge but not waste the effort if improvement rates drop too low. Initial experiments have shown that increasing the GRASP running time from 1 to 10 s gives a quality increase of only 6 % and that 30 s runs give approximately the same quality as the 10 s runs. Based on user requests and support for larger instances, the termination criterion of 30 s is selected as a tradeoff between comfortable user experience and scalability.

6.2 Parameter Tuning

Most metaheuristics have various parameters that need to be set to a certain value. While a quick setup with parameters chosen by the developer's intuition might work well, parameter tuning based on experimental evaluation is an essential part of the metaheuristic implementation. It ensures that the parameters are adapted to the problems representative of those being solved by the final version of the algorithm and, therefore, provide the best possible performance in a production setting [20].

During the GRASP tuning, experimental evaluation consisted of determining the best local search intensity. Running the local search multiple times after each initial solution is constructed might improve the solution. However, focusing on local search too much will cause the search procedure to get stuck in local optima too frequently and not exploring the search space thoroughly enough. A total of 8 experimental setups was evaluated ranging from 1 to 500 iterations, with 30 algorithm runs for each of them, as displayed in Table 1. The best average performance was achieved with only 5 iterations, proving that the local search operator is able to reach the local optimum in a rather low number of passes through the solution.

Iterated local search tuning consists of finding a balance among the local search and the perturbation operator. The intensity of the perturbation should be high enough to ensure that the local search operator doesn't return back to the initial solution, yet not too high, to prevent the algorithm from degrading to random restart local search [25]. For each of the 4 different perturbation configurations, 8 different local search intensities are evaluated, leading to a total of 40 different setups. The average performance for 30 runs of each setup is shown in Table 1. Similarly to the GRASP local search tuning, the results show that the local search converges to the local optimum after a relatively low number of passes. For the perturbation intensity equal to zero, the algorithm degrades to infinite local search around the initial solution, leading to clearly suboptimal solutions. Changing 20 % of the solutions (the percentage of staff members affected with perturbation) with 10 iterations of LS has produced the best results. Setting the perturbation to higher values produces worse solutions,

Table 1. Metahruristic tuning results

GRASP tuning	
LS passes	Average penalty
1	14802
5	**14582**
10	14760
20	14951
50	15188
100	15278
200	15716
500	16263

ILS tuning				
	perturbation intensity			
LS passes	0.0	0.2	0.5	1.0
1	18675	14158	14284	14528
5	18293	13975	14188	14188
10	17944	**13952**	14094	14098
20	18277	13957	14209	14181
50	18916	14108	14180	14149
100	19699	14100	14223	14253
200	19193	14020	14331	14398
500	18972	14349	14580	14530

but only slightly, since the LS operator is able to reach good solutions even with very high perturbation rates.

6.3 Performance Comparison

We compare the best performing configurations of GRASP and ILS with the baseline random restart search as well as with each other. The results obtained in 30 experimental runs are shown in Table 2, including the median penalty (relevant for the statistical test). Additionally, average, minimum, maximum and sample standard deviation of the penalty is provided to give a better illustration of the distributions. From the table, it is clear that both GRASP and ILS are producing better results than the baseline random restart search and that ILS is better than GRASP. This is formally verified using Wilcoxon–Mann–Whitney test to check the H0 hypothesis that distribution functions of the algorithm performances for GRASP and ILS are the same as for random restarts. Both tests resulted in p–values below $1 \cdot 10^{-10}$. Using the same statistical test, it is shown that performance differences of GRASP and ILS are statistically significant ($U = 18$, $p = 1.773 \cdot 10^{-10}$).

In order to test how representative was the example used for tuning, we tested the performance on a larger set of input examples: April, May, July and

Table 2. Random restart, GRASP and ILS performance comparison

	Penalty				
	Avg	Median	Min	Max	Std. dev
Rand. restart	30072	30267	27384	31340	876
GRASP	14563	14558	14180	14902	173
ILS	**13952**	**13937**	**13548**	**14456**	232

November of 2014 and January 2015. Note that these examples weren't available at the time of initial tuning. Overall, while still being better, the median performance of ILS is only 2 % better than in the case of GRASP. On a case–by–case basis, ILS had better median results in 4 out of 5 examined problem instances, while GRASP was more successful in one.

6.4   Customer Adoption

The described system has been successfully implemented in PBZCard, the largest Croatian credit card vendor and in PBZ, one of the largest banks in the country. Previously, the call centre schedules were being created manually, which is highly labour intensive. Managers needed more than one full working day each month to create an initial version of the schedule. While assembling schedules, they were mostly concerned with the organisational issues and didn't have any way to assess the impact on the service quality before the schedule is put to use.

The devised system produces schedules that are ready for use in less than a minute. While the generated schedules usually still need minor manual adjustments, the automation has brought great time savings. The system features a user-friendly schedule editor which graphically represents the effects of any schedule change, highlights understaffed periods and provides simple means to distribute the schedules to the staff members. Performance analysis has shown that the forecasts are accurate and that schedules, in general, achieve the desired performance levels. While comparisons to the performance prior to implementation are difficult due to staffing fluctuations, some performance indicators have increased after implementation. Furthermore, the subsequent implementation at the second institution proven the system to be flexible and generic enough for a relatively quick implementation without major constraint changes.

7   Future Work and Conclusion

In this paper, we present a case study of applying two metaheuristics: GRASP and ILS to solve a call centre scheduling problem as defined with our first two clients. To ensure suitability for small enterprises, minimalism, simplicity and, therefore, reduced development costs were main implementation strategies. The relatively low resulting price makes the suite attractive to institutions that don't require complex and significantly more expensive suites featuring support for multi–channel multi–skilled centres. Despite the simplistic design goals, we believe that the devised problem definition is still quite broad and applicable to a wide range of call centres.

It is also demonstrated that the proposed bottom–up design approach produced surprisingly good results even in the early stages of implementation. While the original plan was to connect the components in some of the more complex metaheuristics such as ant colony optimisation or genetic algorithm, even the initial GRASP version produced results that were satisfactory for our customers. We believe that using a simple algorithm in the aforementioned manner led to great savings in software development costs. An essential component of this approach is an independent constraint assurance system that facilitated rapid prototyping. Since defining the constraints during a scheduling project is a difficult communication issue, prototyping approach that gives users a chance to see the constraints in action after short development cycles significantly simplified the requirement analysis and functional specification for the project.

The implemented system helped the contact centre management to save time on scheduling and focus on other important tasks. The automated scheduling system and related graphical tools have enabled the management to base their schedules on real data from their centre. While the early version of the system has shown promise, a lot more can still be done to make the system more flexible and suited to the needs of modern contact centres, such as implementing the mentioned shift–work rule in the constraint assurance system. Break scheduling could further improve service quality and as a major feature, multi–channel and multi–skilled contact centres could also be supported.

Acknowledgements. We would like to thank our colleague Ivana Horvat Čuka for her invaluable insights and expertise that greatly helped this research and Danko Kozar for building the graphical user interface.

References

1. Pinedo, M., Zacharias, C., Zhu, N.: Scheduling in the service industries: an overview. J. Syst. Sci. Syst. Eng. **24**(1), 1–48 (2015)
2. Slack, N., Chambers, S., Johnston, R.: Operations Management. Pearson Education, Upper Saddle River (2009)
3. Garey, M.R., Johnson, D.S.: Computers and Intractability. Freeman, San Francisco (1979)
4. Ernst, A.T., Jiang, H., Krishnamoorthy, M., Sier, D.: Staff scheduling and rostering: a review of applications, methods and models. Eur. J. Oper. Res. **153**(1), 3–27 (2004)
5. Fukunaga, A., Hamilton, E., Fama, J., Andre, D., Matan, O., Nourbakhsh, I.: Staff scheduling for inbound call and customer contact centers. AI Mag. **23**(4), 30 (2002)
6. Aksin, Z., Armony, M., Mehrotra, V.: The modern call center: a multi-disciplinary perspective on operations management research. Prod. Oper. Manag. **16**(6), 665–688 (2007)
7. Van den Bergh, J., Beliën, J., De Bruecker, P., Demeulemeester, E., De Boeck, L.: Personnel scheduling: a literature review. Eur. J. Oper. Res. **226**(3), 367–385 (2013)
8. Henderson, S., Mason, A., Ziedins, I., Thomson, R.: A heuristic for determining efficient staffing requirements for call centres. Technical report, Citeseer (1999)
9. Atlason, J., Epelman, M.A., Henderson, S.G.: Call center staffing with simulation and cutting plane methods. Ann. Oper. Res. **127**(1–4), 333–358 (2004)
10. Avramidis, A.N., Chan, W., Gendreau, M., L'ecuyer, P., Pisacane, O.: Optimizing daily agent scheduling in a multiskill call center. Eur. J. Oper. Res. **200**(3), 822–832 (2010)

11. Gans, N., Shen, H., Zhou, Y., Korolev, K., McCord, A., Ristock, H.: Parametric stochastic programming models for call-center workforce scheduling. Technical report, Working Paper (2009)
12. Bhulai, S., Koole, G., Pot, A.: Simple methods for shift scheduling in multiskill call centers. Manufact. Serv. Oper. Manag. 10(3), 411–420 (2008)
13. Cezik, M.T., L'Ecuyer, P.: Staffing multiskill call centers via linear programming and simulation. Manag. Sci. 54(2), 310–323 (2008)
14. Alfares, H.K.: Operator staffing and scheduling for an it-help call centre. Eur. J. Ind. Eng. 1(4), 414–430 (2007)
15. Dietz, D.C.: Practical scheduling for call center operations. Omega 39(5), 550–557 (2011)
16. Saltzman, R.M.: A hybrid approach to minimize the cost of staffing a call center. Int. J. Oper. Quant. Manag. 11(1), 1 (2005)
17. Robbins, T.R., Harrison, T.P.: A stochastic programming model for scheduling call centers with global service level agreements. Eur. J. Oper. Res. 207, 1608–1619 (2010)
18. Gans, N., Koole, G., Mandelbaum, A.: Telephone call centers: tutorial, review, and research prospects. Manufact. Serv. Oper. Manag. 5(2), 79–141 (2003)
19. EPSRC/ESRC: Review of research status of operational research in the UK. Eur. J. Oper. Res. 125(2), 359–369 (2004)
20. Birattari, M., Kacprzyk, J.: Tuning Metaheuristics: A Machine Learning Perspective, vol. 197. Springer, Heidelberg (2009)
21. Wolpert, D.H., Macready, W.G.: No free lunch theorems for optimization. IEEE Trans. Evol. Comput. 1(1), 67–82 (1997)
22. Lozano, M., García-Martínez, C.: Hybrid metaheuristics with evolutionary algorithms specializing in intensification and diversification: overview and progress report. Comput. Oper. Res. 37(3), 481–497 (2010)
23. Matijaš, V.D., Molnar, G., Čupić, M., Jakobović, D., Bašić, B.D.: University course timetabling using ACO: a case study on laboratory exercises. In: Setchi, R., Jordanov, I., Howlett, R.J., Jain, L.C. (eds.) KES 2010. LNCS, vol. 6276, pp. 100–110. Springer, Heidelberg (2010)
24. Festa, P., Resende, M.G.: Grasp: an annotated bibliography. Essays and Surveys in Metaheuristics. OR/CSIS, vol. 15, pp. 325–367. Springer, Heidelberg (2002)
25. Lourenço, H.R., Martin, O.C., Stützle, T.: Iterated Local Search. Springer, Heidelberg (2003)
26. Hart, J.P., Shogan, A.W.: Semi-greedy heuristics: an empirical study. Oper. Res. Lett. 6(3), 107–114 (1987)
27. Glover, F., Kochenberger, G.A.: Handbook of Metaheuristics. Springer, Heidelberg (2003)
28. Luke, S.: Essentials of Metaheuristics. Lulu, Raleigh (2013)

Author Index

Printed in the United States
By Bookmasters